# SITING OF NUCLEAR FACILITIES

PROCEEDINGS SERIES

# SITING OF NUCLEAR FACILITIES

PROCEEDINGS OF A SYMPOSIUM
JOINTLY ORGANIZED BY
THE INTERNATIONAL ATOMIC ENERGY AGENCY
AND THE
OECD NUCLEAR ENERGY AGENCY
AND HELD IN
VIENNA, 9-13 DECEMBER 1974

INTERNATIONAL ATOMIC ENERGY AGENCY
VIENNA, 1975

SITING OF NUCLEAR FACILITIES
IAEA, VIENNA, 1975
STI/PUB/384
ISBN 92-0-020175-X

Printed by the IAEA in Austria
April 1975

# FOREWORD

The continually increasing demand for power and the resulting increase in the number of nuclear facilities foreseen make the role played by siting in the nuclear industry more important than ever before.

This volume contains the record of the Symposium on the Siting of Nuclear Facilities held in Vienna on 9 - 13 December 1974, organized jointly by the International Atomic Energy Agency and the Nuclear Energy Agency of the Organization for Economic Co-operation and Development.

Former symposia on this subject sponsored by the IAEA had identified various important siting parameters. The present symposium provided the opportunity to discuss technical aspects of site selection of nuclear facilities and to exchange information on recent experience. The first six sessions dealt with the subject under the headings of (a) national criteria and practices, (b) siting and environment, and (c) reprocessing plants. The remaining three sessions were devoted to new approaches to siting policy, such as quantifying the social criteria, and to a number of special problems.

During the week 38 papers were presented. The symposium was attended by 241 participants from 39 countries and 10 international organizations. This large attendance demonstrates the importance and timeliness of the subject. The IAEA and the NEA are grateful to those who initiated and contributed to the discussions both during the sessions and in the corridors afterwards.

# CONTENTS

## SITING AND ENVIRONMENT (Sessions IV and V)



## SITING AND REPROCESSING PLANTS (Session VI)

NEW APPROACHES TO A SITING POLICY AND SPECIAL PROBLEMS
(Sessions VII, VIII and IX)

# SITING AND NATIONAL CRITERIA AND PRACTICES

(Sessions I, II, III)

Chairmen:

R. GAUSDEN  (United Kingdom)
M. NASIM     (Pakistan)
H. SCHNURER (Federal Republic of Germany)

# REGIONAL APPROACHES TO POWER PLANT SITING IN THE UNITED STATES OF AMERICA

J.J. DiNUNNO
NUS Corporation,
Rockville, Md.,
United States of America

Abstract

REGIONAL APPROACHES TO POWER PLANT SITING IN THE UNITED STATES OF AMERICA.

The selection and evaluation of sites for power plants in the United States of America have become increasingly difficult in recent years as pressures from various societal segments have resulted in governmental restraints on selection and burning of fossil fuels, methods of heat dissipation, acquisition of transmission rights of way, and on environmental impact of industrialization in general. New legislation at both Federal and state levels has been enacted that influences power plant siting. In addition to environmental requirements that must be satisfied, implementing procedures require documented justification for sites chosen and public disclosure of the basis for selection. Some states have consolidated their regulatory activities in the power plant siting area to provide for a more unified approach to these problems. Although nuclear plants have by far the most rigorous requirements for documentation of site selection and plant design, the application of the same general philosophies to fossil plants has been made in several states and can be anticipated elsewhere. Individual site-related investigations have not so much changed in basics as they have been enlarged in scope. Whereas in the past the search for siting alternatives was frequently confined to a utility's service area, the additional siting constraints represented in environmental laws, the economies of size of nuclear power plants, and the sharing of plant capacities among utilities have contributed to a widening of the search area. Several states have assumed the responsibility for site search and investigation and their efforts extend state-wide. This paper discusses applications of regional approaches to power plant siting in the United States of America using case studies made by NUS Corporation, an engineering/environmental consulting firm. The universality of these approaches is indicated, leaving to national policies and goals the importance of values assigned to the basic siting factors.

## 1. INTRODUCTION

An appreciation of what is involved now and what seems to be evolving further on regional approaches to power plant siting in the United States requires some understanding of both the nature of the electric power industry in the United States and recent environmental legislation which has resulted in new ground rules for the siting of power plants.

## 2. LEGISLATIVE BACKGROUND

The site selection process for power plants in the United States has become increasingly more difficult in recent years as pressures from various societal segments have resulted in new governmental restraints on selection and burning of fossil fuels, methods of heat dissipation, acquisition of transmission rights-of-way, and on environmental impact in general. New legislation at both the Federal and state levels has been enacted over the past several years that influences power plant siting [1].

A.    <u>Federal</u>

At the Federal level, two Acts are of particular significance - the National
Environmental Policy Act of 1969 (NEPA) and the Water Quality Improvement Act
of 1970. A national dedication to the promotion of efforts to prevent or eliminat
damage to the environment was articulated in the NEPA. In addition to setting
forth broad goals toward which the nation's efforts should be directed, the Act
also includes a requirement (Sec. 102(c)) that all agencies of the Federal
Government shall:

"Include in every recommendation or report on proposals for legislation
and other major Federal activities affecting the quality of the human environmer
a detailed statement by the responsible official on -

    (i)      the environmental impact of the proposed action,

    (ii)     any adverse environmental effects which cannot be avoided should
        the proposal be implemented,

    (iii)    alternatives to the proposed action,

    (iv)     the relationship between local short-term use of man's environment
        and the maintenance and enhancement of long-term productivity, ar

    (v)      any irreversible and irretrievable commitments of resources which
        would be involved in the proposed action should it be implemented

Prior to completing a final environmental statement, the responsible Federal
agency is required to obtain comments of other Federal agencies which have
jurisdiction by law of special expertise with respect to any environmental impa
involved. Copies of the environmental statement together with the comments a
views of the appropriate Federal, state and local agencies which are authori
to develop and enforce environmental standards must be submitted to a Council
on Environmental Quality (an advisory body reporting to the Office of the Presic
and must be made public. The licensing of nuclear power plants and fuel repro
cessing facilities is being treated by regulatory authorities within the US Atom
Energy Commission[1] as a 'major Federal action', and thus requires the prepar
tion of environmental statements.

The Water Quality Improvement Act of 1970 is less encompassing than NEPA bu
does have one feature particularly noteworthy because of its impact upon the
nuclear power plant licensing process. Under prior statutes, the AEC lacked
authority to regulate nuclear power facilities with respect to their thermal
effluents. Under the new Act, the AEC will not set thermal water standards bu
will require certification from those who do before a facility is licensed. The
procedures involved are briefly as follows. Under the recently enacted Water
Quality Improvement Act, general water quality criteria have been established
at the Federal level by the United States Environmental Protection Agency with
the individual states authorized and encouraged to develop implementing them

---

[1]  To be set up as a separate agency effective 1 Jan. 1975 as the Nuclear Regulatory Commission (NRC).

standards for water bodies within their boundaries.  Federal approval is required
of state standards.  Where states fail to take the initiative, Federal standards
can be imposed.  What is perhaps more pertinent to the nuclear plant licensing
process are two features:

1.      An applicant for a construction permit must obtain a water discharge
permit from the Environmental Protection Agency (EPA), in effect certifying that
approved standards will be met.

2.      The AEC must independently weigh the environmental implications
of the thermal discharges, with the obligation to impose additional constraints
on either plant design and/or operation if necessary to protect the aquatic
environment.


B.     State

Heightened interest in the problems of siting and operating power plants is
reflected not only at the Federal  but at the state level.  A survey in 1972 showed
five states had legislation involving the siting of power plants and/or transmission
lines.  As of September 1974, eighteen states have passed such legislation.  A
significant number of additional states are likely to act during 1975.  These
actions are part of a growing awareness at other than the Federal level that the
availability and location of energy facilities may well determine where people
live and work, the site of recreational and industrial complexes, as well as the
trade-offs involved when land and water resources are committed to a specific
purpose for many years.

State activities are far too numerous to catalogue and vary considerably from
state to state [2], but the following are a few examples of actions taken by
states:

(a)     In New York a law enacted in 1972 requires utilities wishing to
        construct a power plant in the state to obtain a state certificate
        of environmental compatibility and public need.  A public hearing
        is mandatory, conducted under the auspices of a Siting Board.  The
        Siting Board provides a state-consolidated review by the various
        offices of the state having more specialized and narrowed responsi-
        bilities.

(b)     In Oregon, a power plant siting and environmental protection bill,
        passed in 1971, created the Oregon Nuclear and Thermal Energy
        Council.  The council has authority to approve sites for thermal
        power plants with associated transmission lines, based on independent
        studies and a ten-year plan for the region.

(c)     Washington State in 1973 set up a Thermal Power Plant Site Evaluation
        Council.  The council is charged with making all regulatory reviews
        prior to granting an approval to a siting application.  The council
        conducts hearings and submits a recommendation to the governor
        concerning the site application.  The governor is then the final
        authority for the state to approve or reject site applications.

(d)     California passed legislation in 1974 dealing with both conserva-
        tion and the development of energy in its various forms.  The bill
        becomes effective in January 1975.  Under the terms of this act,
        there is established a State Energy Resources Conservation and
        Development Commission (ERCDC).  The ERCDC is authorized to
        conduct an ongoing assessment of opportunities and constraints
        presented by all energy forms.  This includes studies, research
        projects, data collection and other activities necessary to develop
        proper energy conservation practices.  The ERCDC also has exclusive
        power to certify all site-related facilities in-state, except for
        facilities proposed to be located in the coastal zone of the state.

(e)     Maryland siting legislation was passed in 1969.  Maryland pioneered
        the concept of having the state assume the responsibility for the
        selection and qualification of sites for power plants within the state.
        Under this arrangement the state applied a surtax to electric utility
        bills, income which goes for the purchase of land for sites and the
        necessary impact studies that are required to determine their suit-
        ability.  The sites are then purchased by utilities when they have
        need.

(f)     Ohio legislation was passed in 1972 that established a Power Siting
        Commission to control the location of major utility facilities.  In
        order to obtain a certificate of environmental compatibility and
        public need, the proposed facility must meet all air pollution,
        water pollution, and solid waste disposal laws, regulations and
        standards, in addition to other siting criteria prescribed by the
        power siting law itself.  Application for certification must be
        filed 2-5 years in advance of construction.

(g)     Arizona legislation was passed in 1971 that created the Arizona
        Power Plant Siting Committee.  This group enters into procedural
        hearings only upon receipt of an application.  The committee has
        broad overseer authority on power plant sites and the installation of
        transmission lines.

Common to both Federal and state requirements for information that must be su
mitted as part of the application for permit to construct power plants is the
documented basis for sites chosen and public disclosure of the information.
nuclear power plants, AEC guidelines [3] encourage a regional approach to s
selection and require that the basis for the choice of a preferred site from amo
viable alternatives be documented.


## 3. STRUCTURE OF THE POWER INDUSTRY IN THE UNITED STATES [4]

The electric power industry in the US is made up of many utility systems, sor
owned by private companies (investor-owned utilities), some owned by the Ur
States government or other public bodies such as municipalities, and some owi
by cooperatives (users).  In all, there are approximately 3,500 individual ent

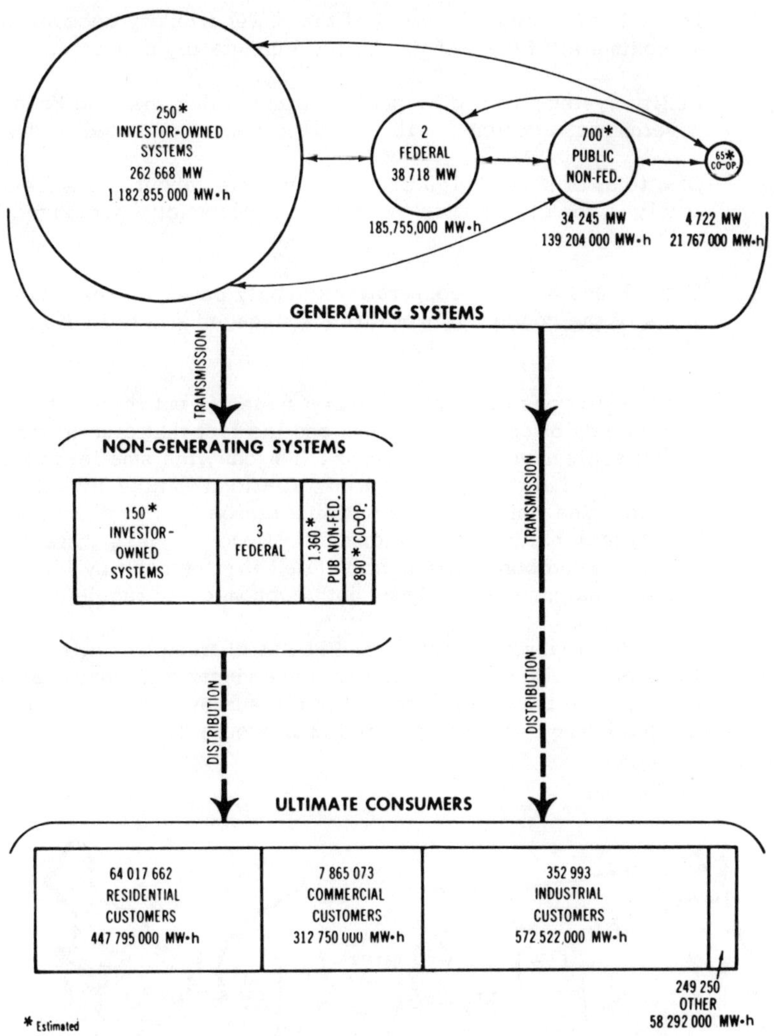


FIG.1. The electric power industry in the United States of America, 1970. Note: power generated at other Federal facilities is marketed by the five major Federals shown.

prises. The general makeup of the industry is shown in Figure 1. As may be noted from Figure 1:

(a)    Investor-owned systems (250) accounted for 77% of the total generating capacity in 1970 and served 78% of the ultimate customers.

(b)    A substantial number (150) of investor-owned utilities are engaged solely in distribution of electricity.

(c)     The Federally-owned segment of the power industry accounts for
        approximately 11.5% of the US total generating capacity.

(d)     Utility systems owned by public bodies, other than the Federal
        Government, account for 10.5% of the power generated in the US.

(e)     Of a total of some 2,100 separate systems, about 2/3 are engaged
        solely in the distribution and resale of electricity purchased from
        bulk suppliers.

(f)     About 1,000 electric cooperatives supply power in many rural
        areas of the country.  Over 90% are engaged solely in distribution
        of power.

There is a trend toward consolidation of utility systems, but to date this has b
gradual.  The existence of so many systems and the fact that a great majority
operate on a small scale have led to problems.  For one, the smaller systems a
not able individually to build large enough installations to take advantage of
economics of scale.  The many systems within a region tend to complicate the
job of coordinating regional power development.  None-the-less there exists a
great deal of pooling and coordination, stimulated first perhaps by interest in
reliability of supply and more recently by both economic and regulatory pressu

In the USA, the Federal Power Commission (FPC) is charged by law to survey
periodically the state of the power industry.  In conducting the national power
survey, FPC has divided the continental US into 6 survey regions and 48 powe
supply areas.  This survey scheme is depicted in Figure 2.

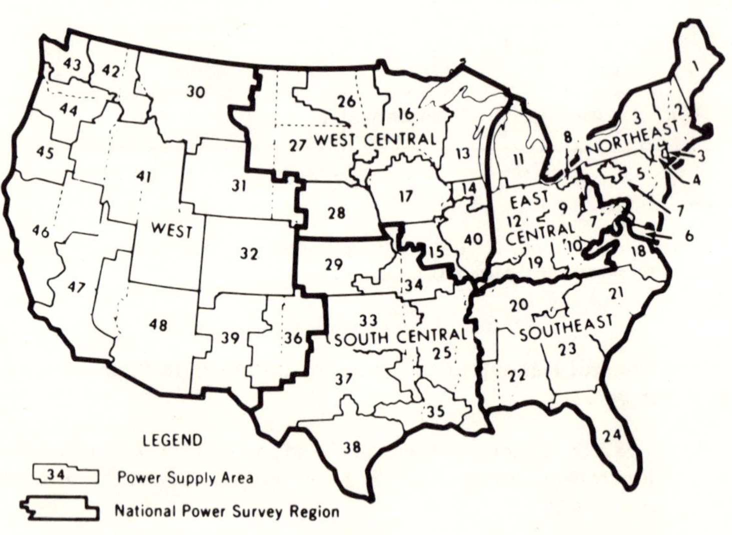


FIG.2.  National power survey regions and power supply areas.

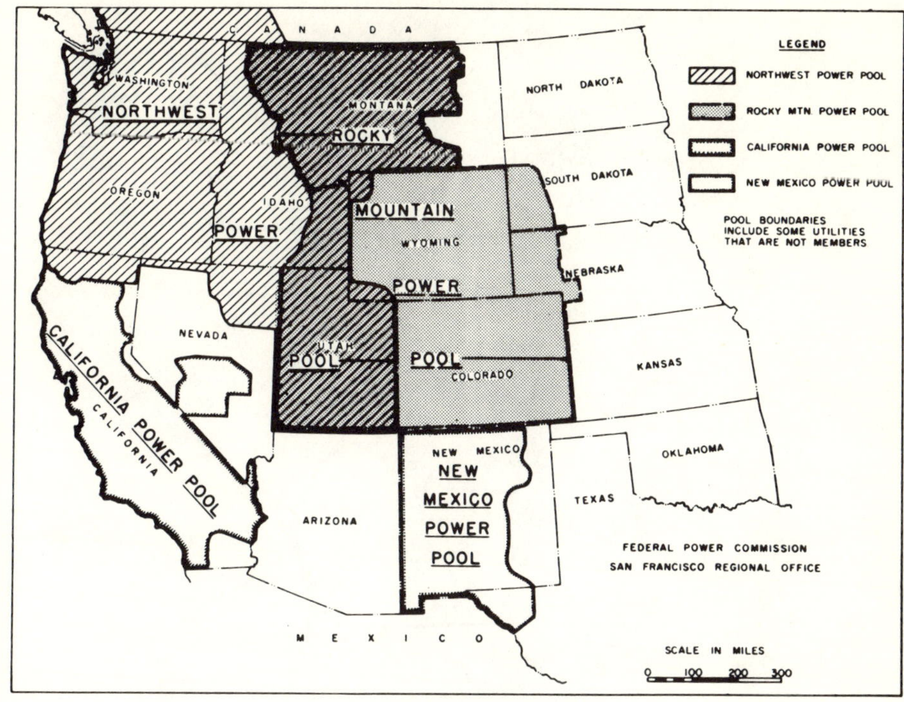


FIG.3.  Western loop power supply pools.

Each power supply area in turn consists of a considerable number of utilities,
each providing electricity to a well defined service area.  The utilities within a
power supply region commonly are connected by interties and frequently pool
supplies.  Such pooling arrangements are voluntary.  Figures 3 and 4 show
several sections of the USA by power pool groupings [4]. Within the pools,
service areas of individual utilities vary widely in size as depicted in Figures
5 and 6.

## 4. REGIONAL SITING

The interdependence of site-related characteristics and nuclear power plant design
was recognized from the earliest days of application of nuclear power for generation of electricity.  This is evidenced in the records of the ad hoc reviews of the
first nuclear plants granted construction permits and the establishment shortly
after, in 1962, of AEC's general siting criteria [5]. Under these criteria, an applicant
for a construction permit (and subsequently an operating license) was required to
examine meteorology, hydrology, geology, demography and land use for the site
and its environs and to relate such site factors with the intended design and
operation of the facility.  The emphasis in such analysis was the protection of
the health and safety of the public.

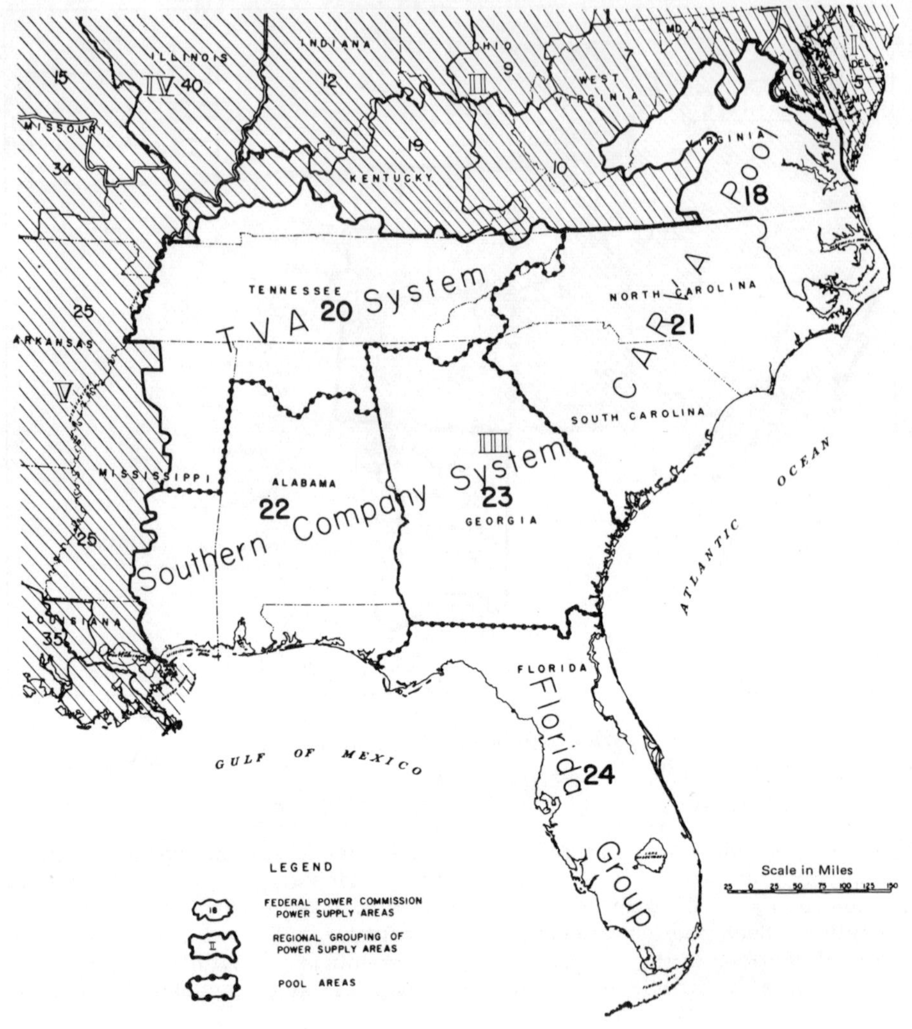



FIG.4. Power supply areas and power pools.

With the advent in the United States of recent national and state environmenta
legislation, individual site-related investigations have not so much changed
in basics as they have been enlarged in scope.  Whereas in the past the searc
for siting alternatives was frequently confined to a utility's service area, the
additional siting constraints represented in environmental laws are leading mo
and more frequently to examination of siting possibilities over wider areas.  T
economics of size of nuclear power plants, the sharing of plant capacities am
utilities, and the assumption of site selection responsibilities by some states
also contributed to a widening of the search area for suitable sites.  Efforts to

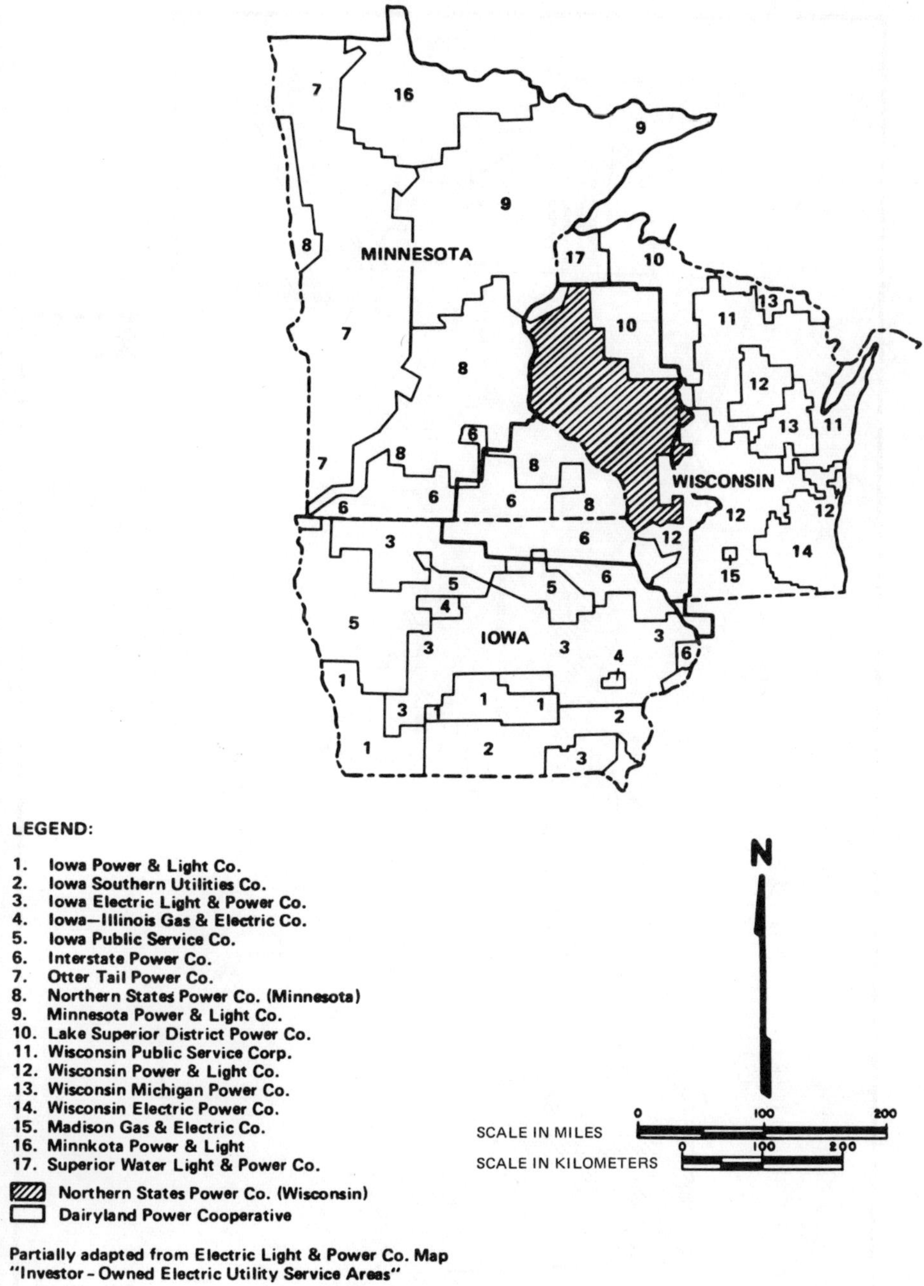

**LEGEND:**

1. Iowa Power & Light Co.
2. Iowa Southern Utilities Co.
3. Iowa Electric Light & Power Co.
4. Iowa—Illinois Gas & Electric Co.
5. Iowa Public Service Co.
6. Interstate Power Co.
7. Otter Tail Power Co.
8. Northern States Power Co. (Minnesota)
9. Minnesota Power & Light Co.
10. Lake Superior District Power Co.
11. Wisconsin Public Service Corp.
12. Wisconsin Power & Light Co.
13. Wisconsin Michigan Power Co.
14. Wisconsin Electric Power Co.
15. Madison Gas & Electric Co.
16. Minnkota Power & Light
17. Superior Water Light & Power Co.

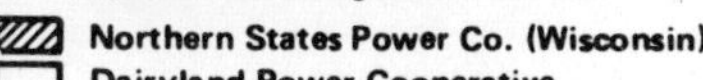

Northern States Power Co. (Wisconsin)
Dairyland Power Cooperative

**Partially adapted from Electric Light & Power Co. Map
"Investor-Owned Electric Utility Service Areas"**



FIG.5. Northern states power company service area.

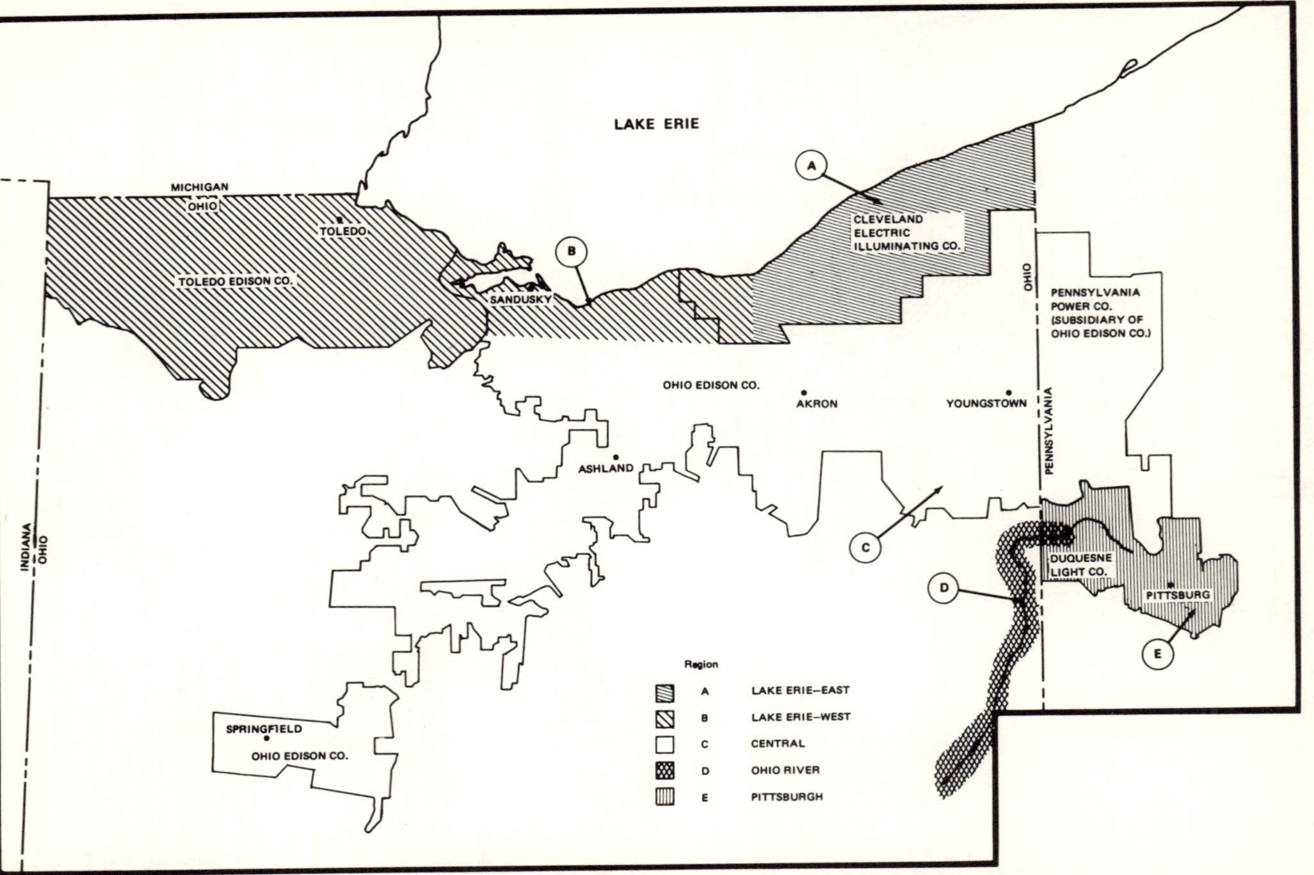


FIG. 6. Map of Cogen regions for siting analysis for 1979–1980 generating capacity additions.

shorten the licensing schedules are creating additional pressures to decouple site and specific plant reviews to enable site qualification studies based upon standardized plant parameters to proceed well in advance of plans to commence construction.

## A.    Generic Methods

Three generic methods for selecting candidate sites as practised in the US have been identified [6] as follows:

(a)    Successive Screening

Several sets of siting factors are selected against which in succession each potential site is evaluated. Sites which meet the first set are then candidates for further evaluation against the second set, etc. Each set of considerations in effect acts as a go, no-go screen.

(b)    Comparative Evaluation

A group of siting factors is treated collectively. Each site is rated for each factor. The ranked aggregate for each site reflects relative preferability. This method requires the assignment of relative weighting to each siting factor.

(c)    Classification and Rating

These are generally more formalized methods than comparative evaluations. The basic siting factors are often broken down into a finer sub-factor structure. Siting factors and sub-factors pertinent to the siting situation are identified, criteria for judging each factor established and a weighting value assigned to the factor. Each site is then evaluated for each sub-factor, and a numerical value is assigned to reflect its relative degree of favorability, the finding is multiplied by the weighting factor, and the aggregate of weighted numerical values for all factors summed to provide a 'composite' figure of merit.

NUS has provided consulting services to a number of utilities in the survey of sizeable regions in the search for licensable plant sites. We have found it useful to use a combination of the successive screening technique and the classification and rating scheme.

## B.    Basic Siting Considerations

All our studies have had in common a balance among certain basic siting considerations [5-7]. These involved institutional, safety, engineering/economics, environmental and socio-economic factors.

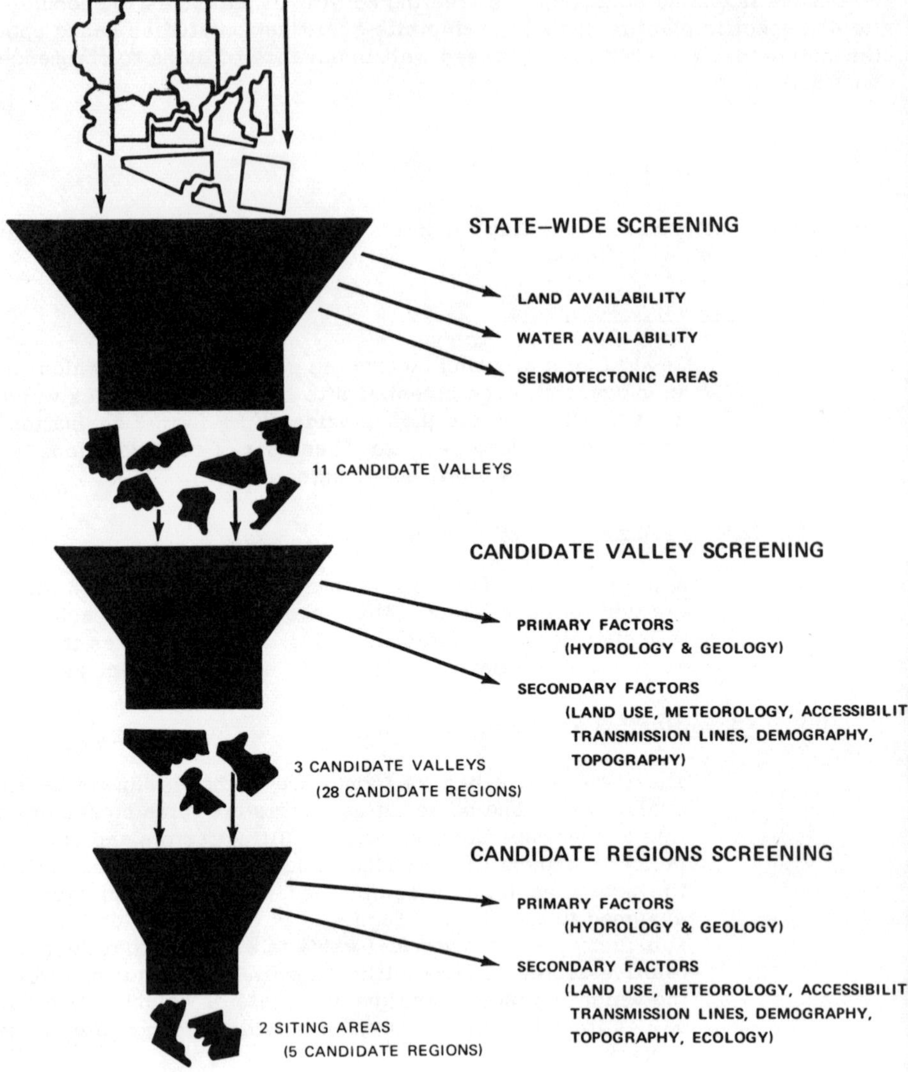



FIG.7. Selective screening procedure.

Broadly speaking, key factors can be grouped as follows:

(a)   <u>Institutional</u>

    (i)    Required service date or on-line availability
   (ii)    System reliability requirements
  (iii)    Size and number of units/sites
  (iv)    Search area boundaries (e.g.   utility franchise area or
        power pool bounds)

(b)  <u>Engineering</u>

   (i)    Safety

          Geology (Seismic)
          Hydrology (Flooding and Effluent Dispersion)
          Demography
          Meteorology (Dilution and Dispersion)

   (ii)   Functional

          Cooling-Water Availability
          Geology (Foundation, Soil Characteristics)
          Accessibility (People, Materials and Components,
          Transmission Grid)

(c)  <u>Environmental</u>

   (i)    Ecological Sensitivity (Site, Transmission Corridors,
        Site Environs)

          Terrestrial
          Aquatic (Thermal and Chemical)

   (ii)   Land Uses (Site and Environs)

          Compatibility
          Dedicated Lands (Parks, Natural and Scenic Area,
          Forests, Reservations)
          Areas of Historic and Archaeological Significance
          Water Quantities and Qualities
          Climatology
          Demography
          Aesthetics

(d)  <u>Economic</u>

   (i)    Land Costs
   (ii)   Cooling-System Alternatives
   (iii)  Site Preparation Costs (Geology & Topography)
   (iv)  Transmission Line Corridors
   (v)   Site-Dictated Special Engineered Safeguards

(e)  <u>Socio-Economic</u>

   (i)    Land-Owner Dislocations
   (ii)   Competitive Use of Resources (Water and Land)
   (iii)  Community Attitudes and Public Acceptance
   (iv)  Economic Influence on Existing Life Styles (e.g.  Homes,
        Schools, Traffic)

## C.    Case Studies

A regional approach was used in a site study we conducted for the Arizona Nucle
Power Project[2][8]. In this study the entire State of Arizona constituted the searc
area. The immediate need was for a site that could be advanced for licensing
considerations for a plant for a 1981 start-up. A multiple-plant site was preferr
looking to siting needs in the early 1980's for several plants. The initial task
was one of searching roughly 73 million acres of land for 5 or 6 of the most prom
ising prospects for a power plant site.

The screening process used in Arizona is graphically illustrated by Figure 7.
It consisted of several basic steps:

1.    Statewide screening to narrow the search area to the most
promising portions of the state. A gross screening mesh consisting of
hydrology, geology and land availability was used. Eleven areas of roughly
7,500 square miles each were identified.

2.    The candidate areas were ranked on the basis of information
available on potential water availability (hydrology), seismotectonic
characteristics, meteorology, land use, demography, accessibility, transmissi
line access and routings and topography. Three top-ranked areas were selected

3.    Within the three broad areas, twenty-eight siting possibilities
encompassing some 5,000 acres each were identified. A ranking of the twenty-
eight candidate sites was done. Ecology was added to the site-ranking factors

Having narrowed the search area to the order of square miles, aerial flights and
ground surveys for study of relative ecological sensitivities could be more
meaningfully done. Through this effort, siting areas were ranked in order of
promise and the top-ranking 5 siting possibilities were selected for closer
examination. At this point, the study emphasis shifted from surface examinat
to sub-surface exploration. More detailed on-site geological examinations of
the top ranked possibilities shifted the apparent ranking order developed from
surface investigations. The end result was the selection of one site as havin
characteristics most likely of best satisfying the functional requirements neces
to a generating station and the regulatory restraints deriving from concerns abo
the potential impact on people and their surroundings. Another backup candida
site was similarly identified, but not subjected to the same detail of sub-surfa
exploration. These two siting areas were equipped with met.towers and on-site
data collection [8] begun in parallel with detailed site characterization studie
of the principal site.

This site selection study combined selective screening to narrow down the
search area and a classification and rating scheme to rank siting possibilities
in order of preference for detailed consideration. This combination for large-
sized regional studies appears to be most cost effective in that the more detail
and costly field investigations are directed to a relatively small number of
specific siting areas. Detail of examination in effect is inversely proportione
to the size of the area under study.

----

[2] The Arizona Nuclear Power Project (ANPP) is a combined utility project with present participating
members consisting of Arizona Public Service Co., Salt River Project and Tucson Gas & Electric Co., Public
Service Co. of New Mexico and El Paso Electric Co., and The Arizona Electric Power Cooperative.

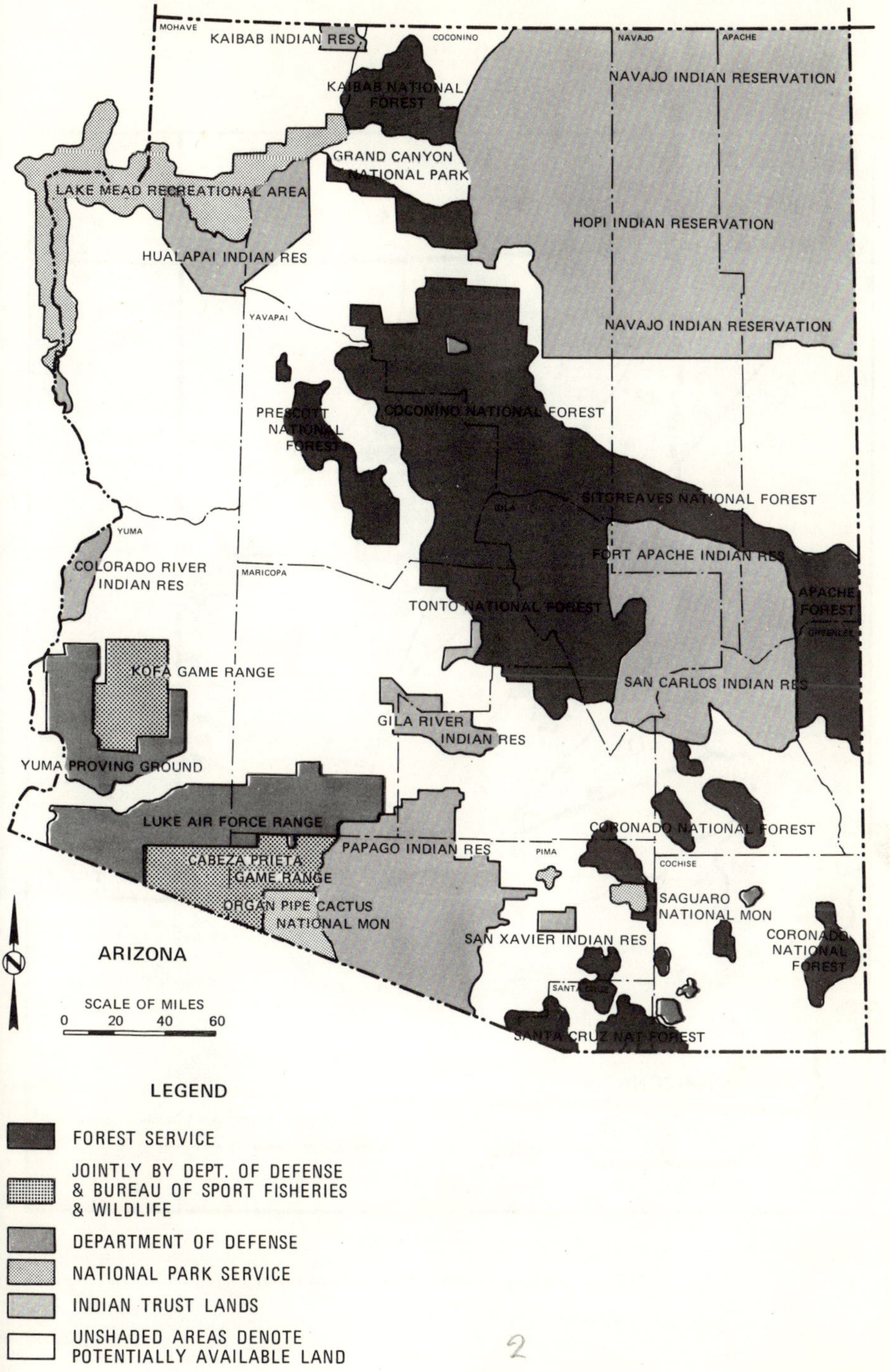

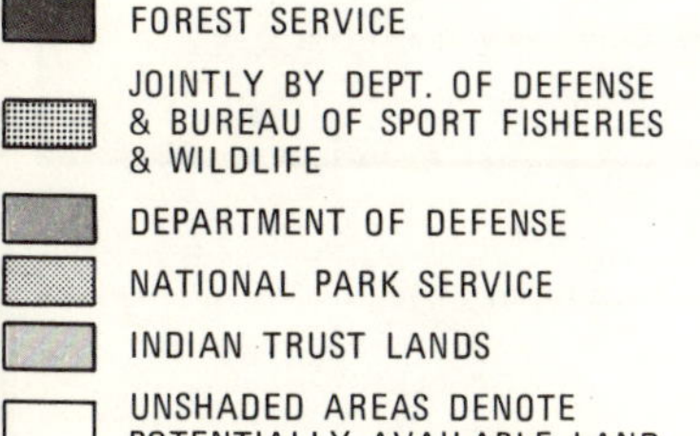


FIG.8. Land availability in the state of Arizona.

     DiNUNNO

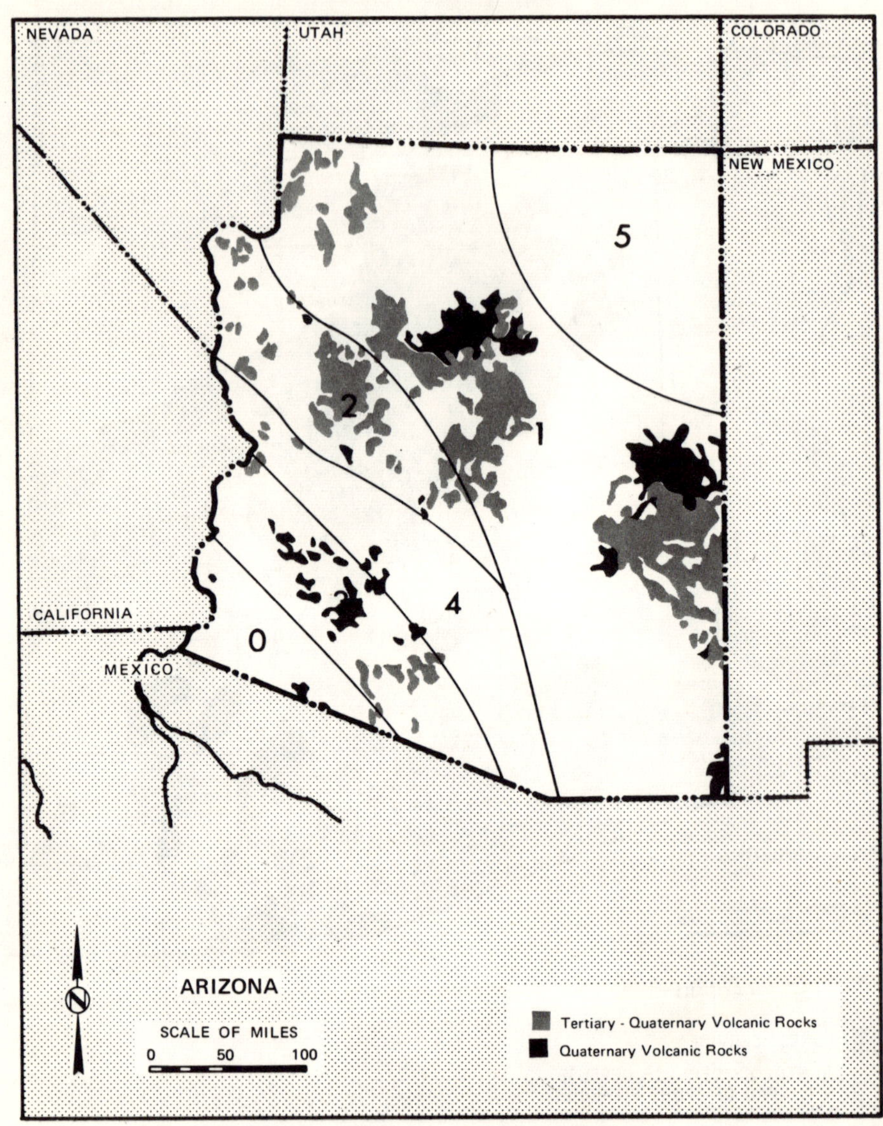


FIG. 9.  Tertiary and quaternary volcanic rocks.

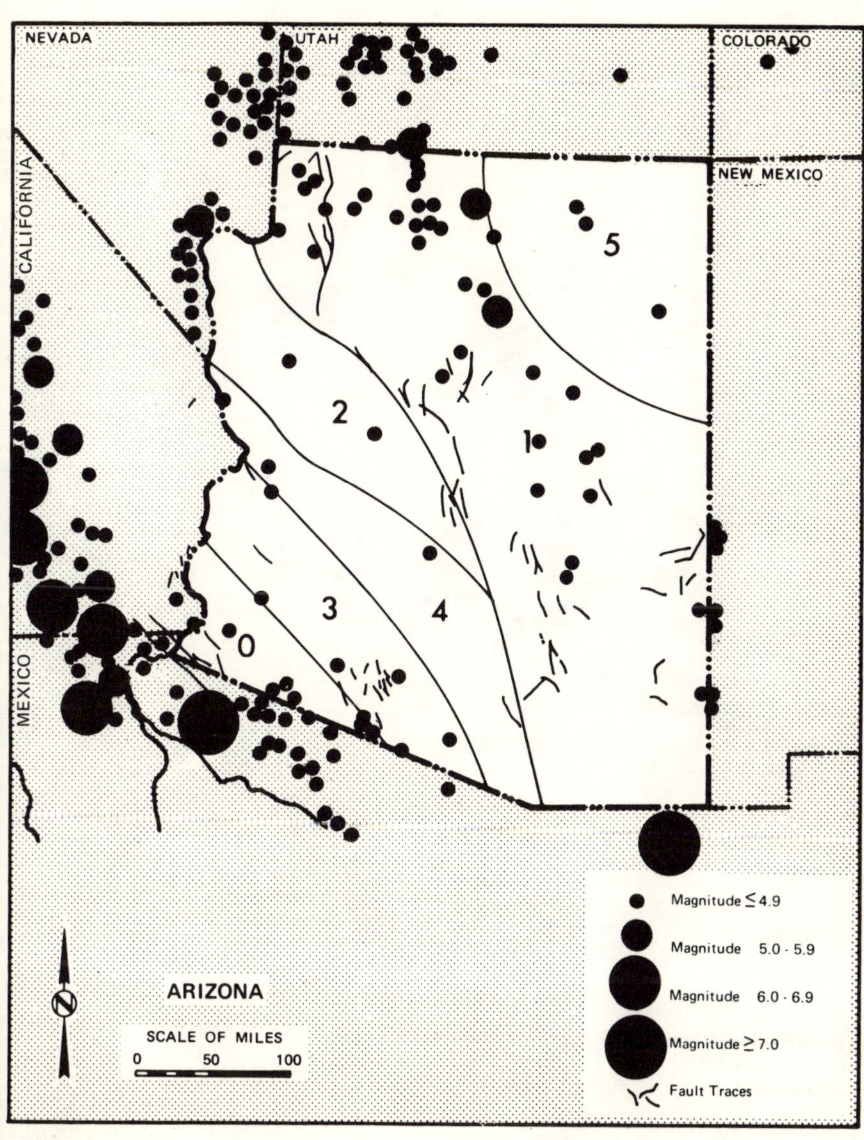


FIG.10. Earthquake epicenters and quaternary fault traces.

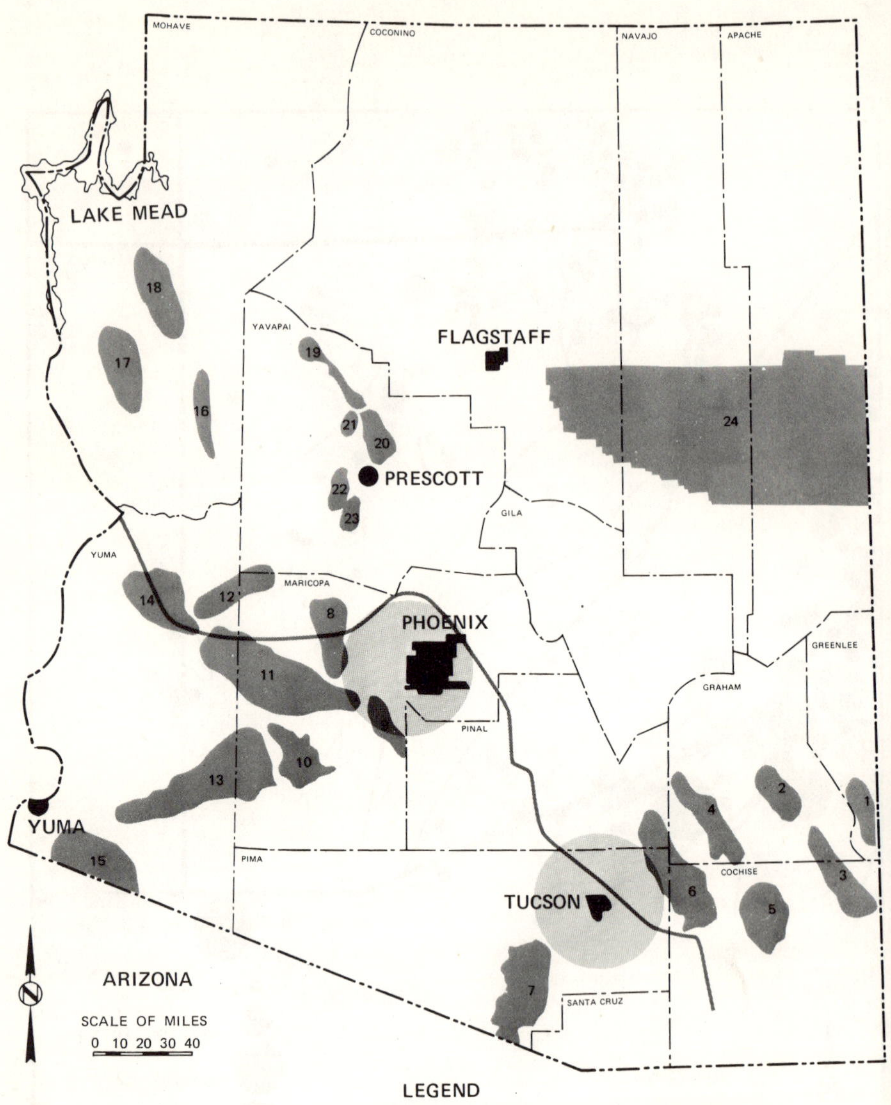


FIG. 11. Statewide water availability screening.

Figures 8 through 15 have been selected to graphically illustrate some of the key
aspects of the study.  Briefly summarized, these figures illustrate the following:

Figure 8 shows lands already dedicated to specific public and
reservation uses.  The unshaded areas in turn are the lands that
were identified as potentially available.

Figures 9 and 10 show some of the safety-related geological
characterizations of the site.  The bands numerically identified
from 0-5 indicate in ascending order the relative favorability of
the areas for focusing the site search.

Figure 11 identifies potential sources of water for condenser cooling
purposes.  Treated sewage water from the city of Phoenix figured
importantly in the potential supply picture.

Figure 12 shows existing and proposed transmission lines as well
as major load centers.

Figure 13 shows major highways and railroads offer varied possibilities
for transport of major components, materials for construction and for
commuter travel.

Figure 14 shows the principal candidate areas that fell out of the
state-wide screening process.

Figure 15 shows typical site identification in a candidate area.

The basic methodology used in the Arizona site selection study has been applied
also to a similar search for sites in the eastern United States [9]. In this case,
the search was for sites for thermal power plants (fossil and nuclear).  While
the basic siting factors were the same and the methodology quite similar, the
application was very different.  The character of the Ohio area in terms of the
key siting factors such as geology, hydrology, demography, meteorology and
land use was so different that the factors of primary significance were not the
same; i.e.  the sub-set of siting factors most appropriate for initial screening
was not the same.


Further, the diversity of cooling-water resources in the East offered the possibility
of sites with different cooling options.  One of the reasons for comparing these
two studies is to emphasize that siting factors of greatest importance in one case
may not be the greatest in others.


Figure 16 summarizes the general study methodology.  The basic factors used in
the preliminary screening included hydrology, geology, demography/land use,
meteorology, ecological sensitivity, geography/topography and general acceptability.  Accessibility was not evaluated in detail in the preliminary screening.
This was because railroads and major (paved four-lane) highways were found
to be highly developed in the state and no obvious areas existed which would be
ruled out by accessibility problems.  To perform the preliminary screening,
general siting criteria were established for each factor.  These criteria were

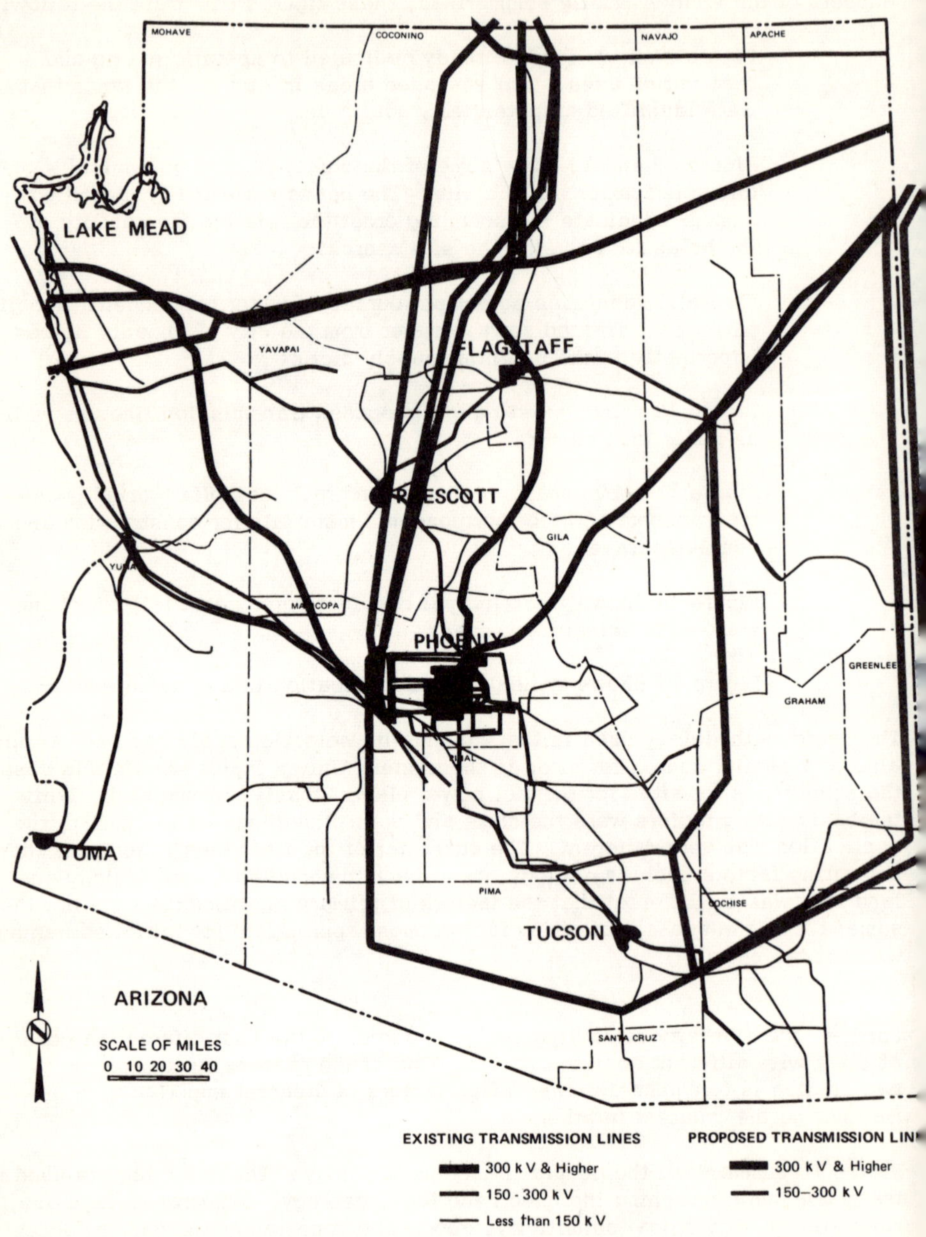


FIG. 12. Existing and proposed transmission lines.

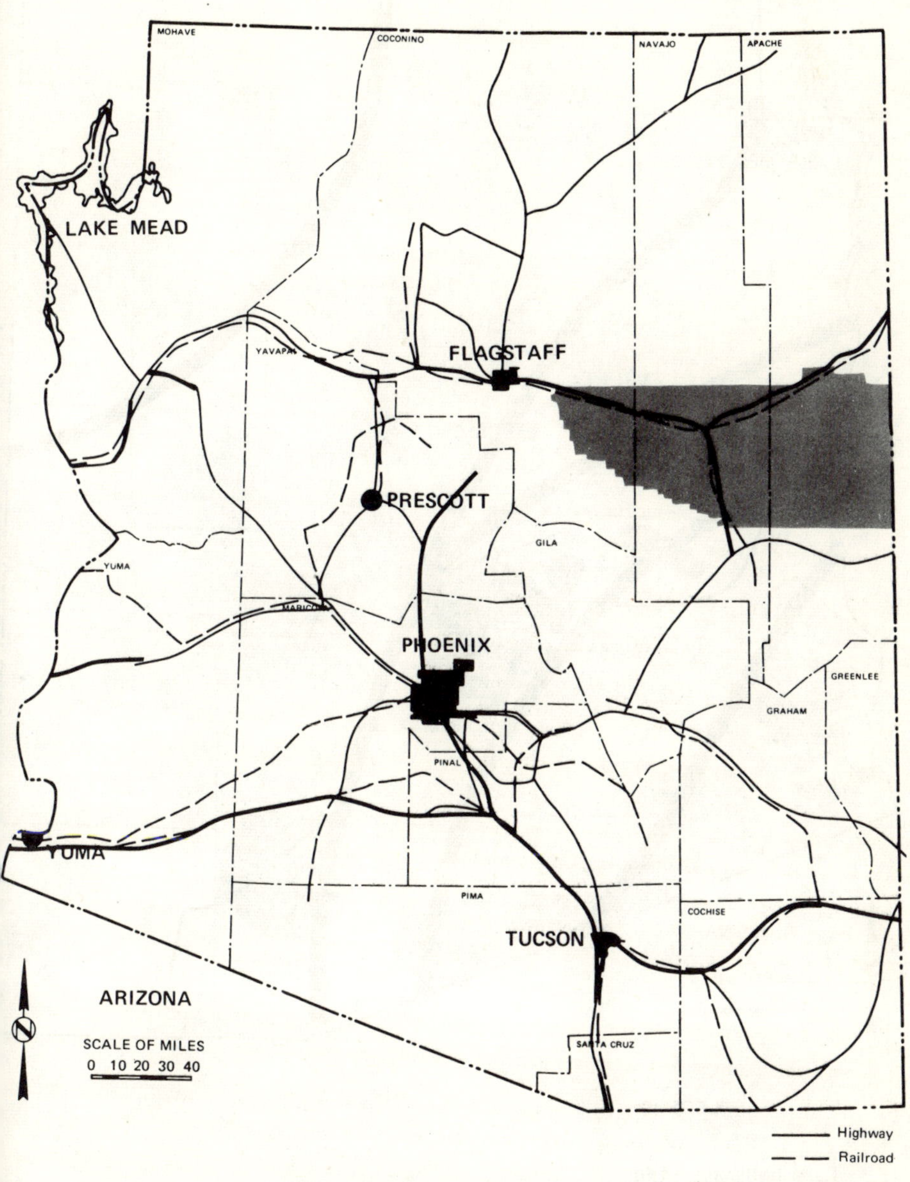



FIG.13. Major highways and railroads.

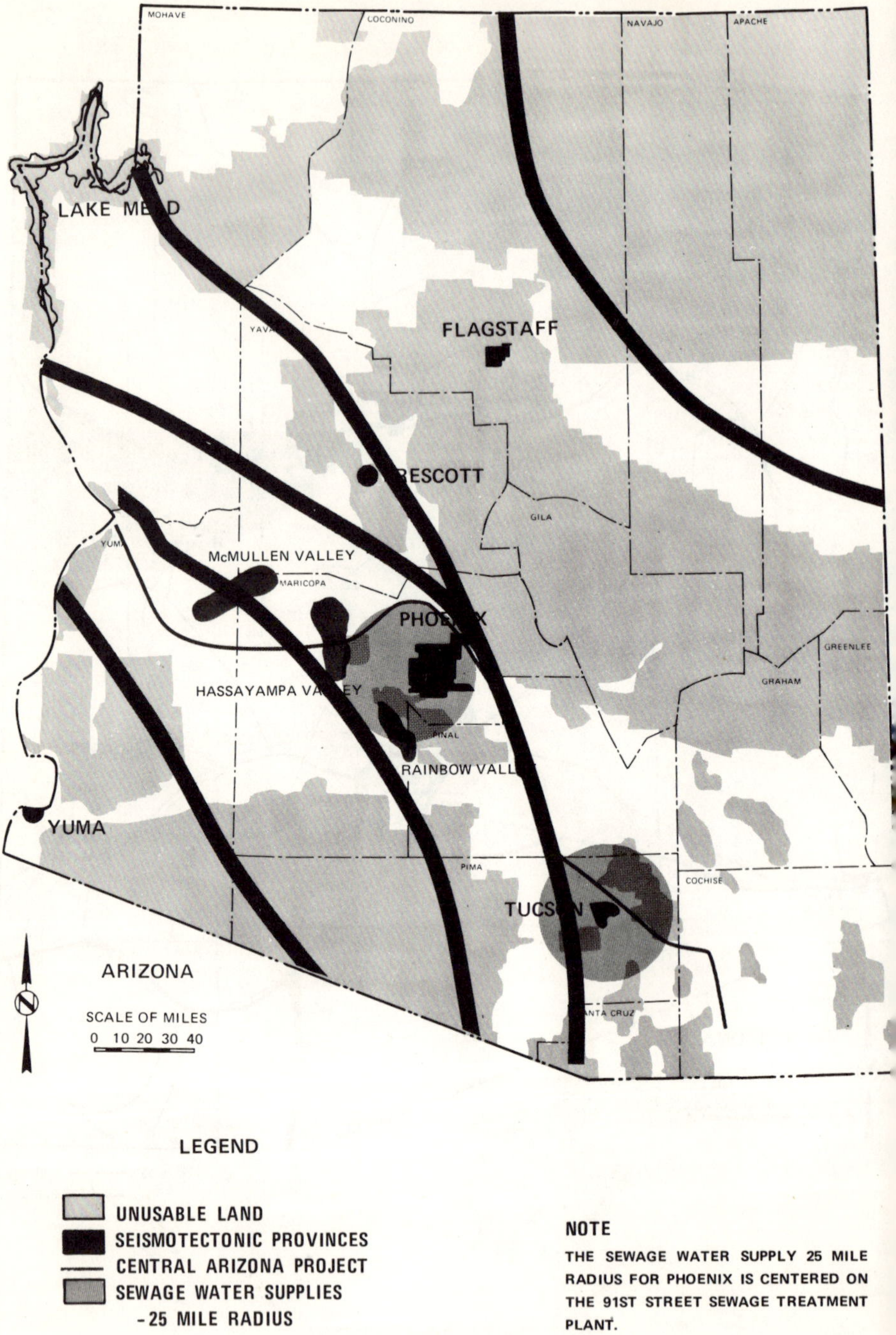

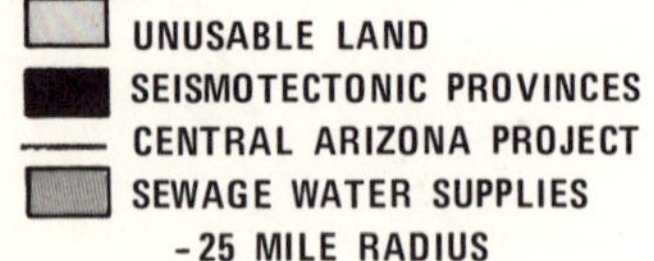



FIG. 14. Relationship of prime candidate valleys — statewide screening factors.

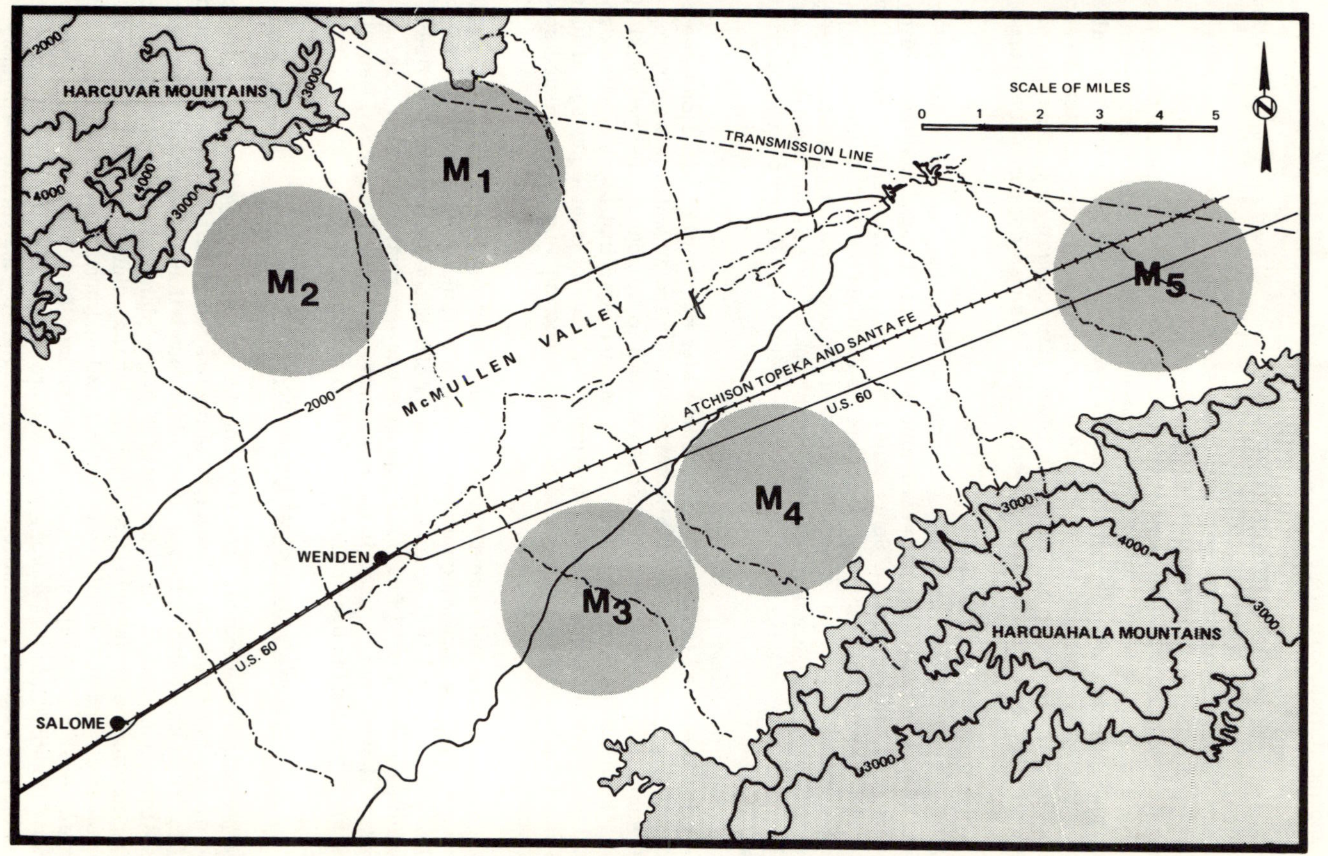


FIG.15. Candidate regions in McMullen Valley.

then used to examine areas within the state. The seven basic screening factors
as applied in this case are as follows:

### 1.    Statewide Hydrological Screening

Surface water, sewage effluent and groundwater initially were considered as
sources of cooling water in the preliminary screening process. A literature
survey showed that, of the water resources potentially available, surface water
offered the most promise for satisfying power plant needs. In contrast, the
maximum expected groundwater withdrawal from bedrock and buried river valleys
which has  previously been investigated was not found to be sufficient to suppo
an additional 1100 MWe to 4400 MWe power plant capacity. Under these circu
stances, the search for new groundwater resources was not pursued as part of t
study.

The three types of cooling systems considered for either nuclear or fossil-fuele
power plants in the preliminary screening were once-through cooling, cooling
towers  and cooling ponds. Spray canals and spray ponds were not explicitly
considered because their water requirements were similar to the other options.
Consideration of cooling towers **was** further broken down into cooling towers
with and without reservoirs. The reservoir concept entails the creation of an
impounded water volume sized to provide makeup water during periods of critica
low flows.

The criteria applied to rivers as a cooling-water source were based on three cor
siderations:  potential biological entrainment damage as water is withdrawn,
allowable temperature rise as water is returned to its source, and potential effe
of consumption on downstream users. It was assumed that lakes act like cooli
ponds in dissipating waste heat. The criterion used for treated sewage effluent
was that the plant must be capable of being supported only by the treated sewa
effluent.

For cooling towers and once-through cooling, the lowest recorded daily average
flow was used for screening purposes. Using cooling towers with reservoirs
allowed consideration of areas on the basis of average rather than low flow.
Because a cooling pond functions both as a cooling system and as a reservoir,
average flows could be used in the preliminary screening. Figures 17 and 18
graphically show results of such a hydrological screening.

### 2.    Statewide Geological Screening

The preliminary geological screening considered depth of overburden, seismici
and geological substrata. The depth of overburden affects the relative ease or
difficulty of designing suitable power plant foundations. Seismic danger incre
the engineering and construction costs of power plant development. Geologica
considerations also included possibilities for the solution of bedrock, the distr
bution of oil and gas withdrawal, and vertical and horizontal distribution of su
surface coal mines. These aspects could present foundation problems for powe
plants if such mineral resources were withdrawn. In Ohio, these parameters w
sufficiently localized or sufficiently low in intensity that in the preliminary
screening no area of the state was excluded from further consideration on the k
of geological factors for either nuclear or fossil-fueled power plants.

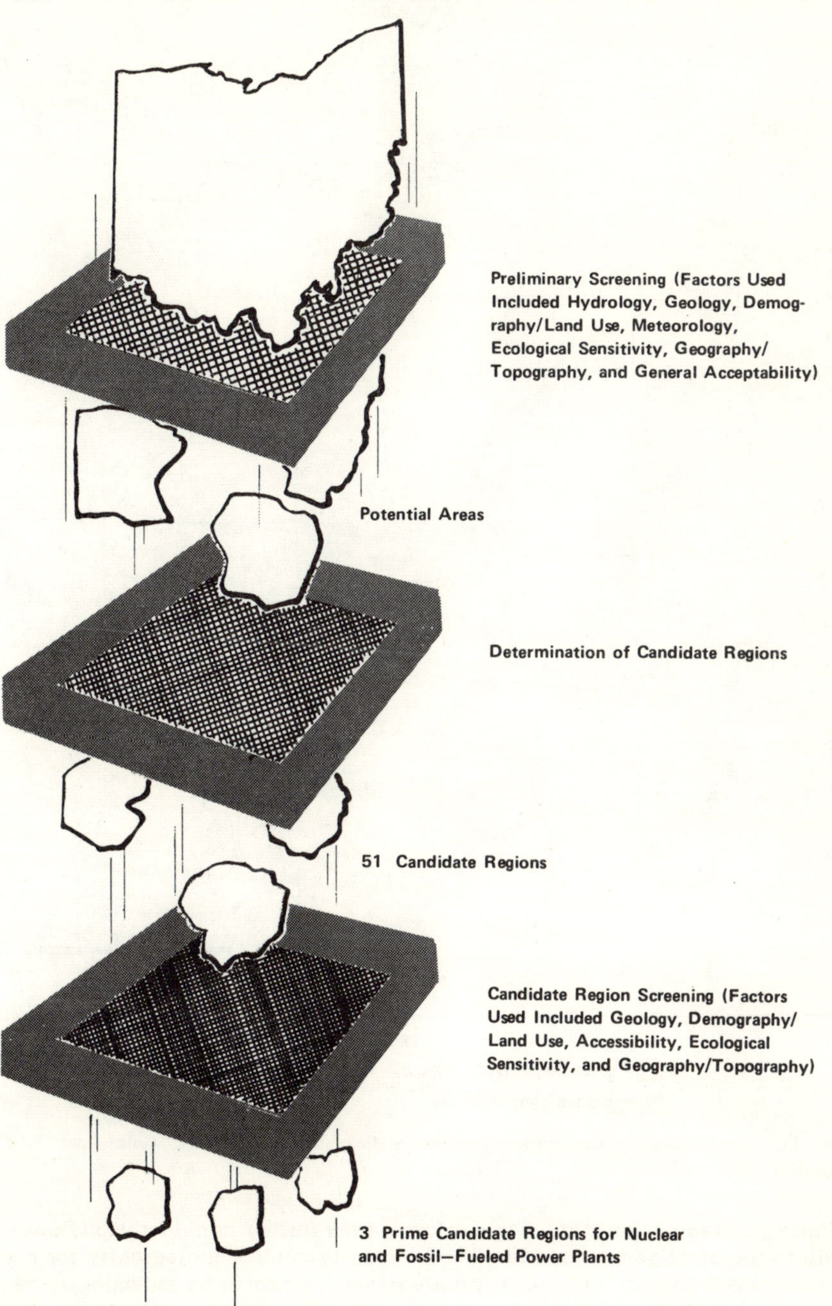


FIG.16. General study methodology.

  DiNUNNO

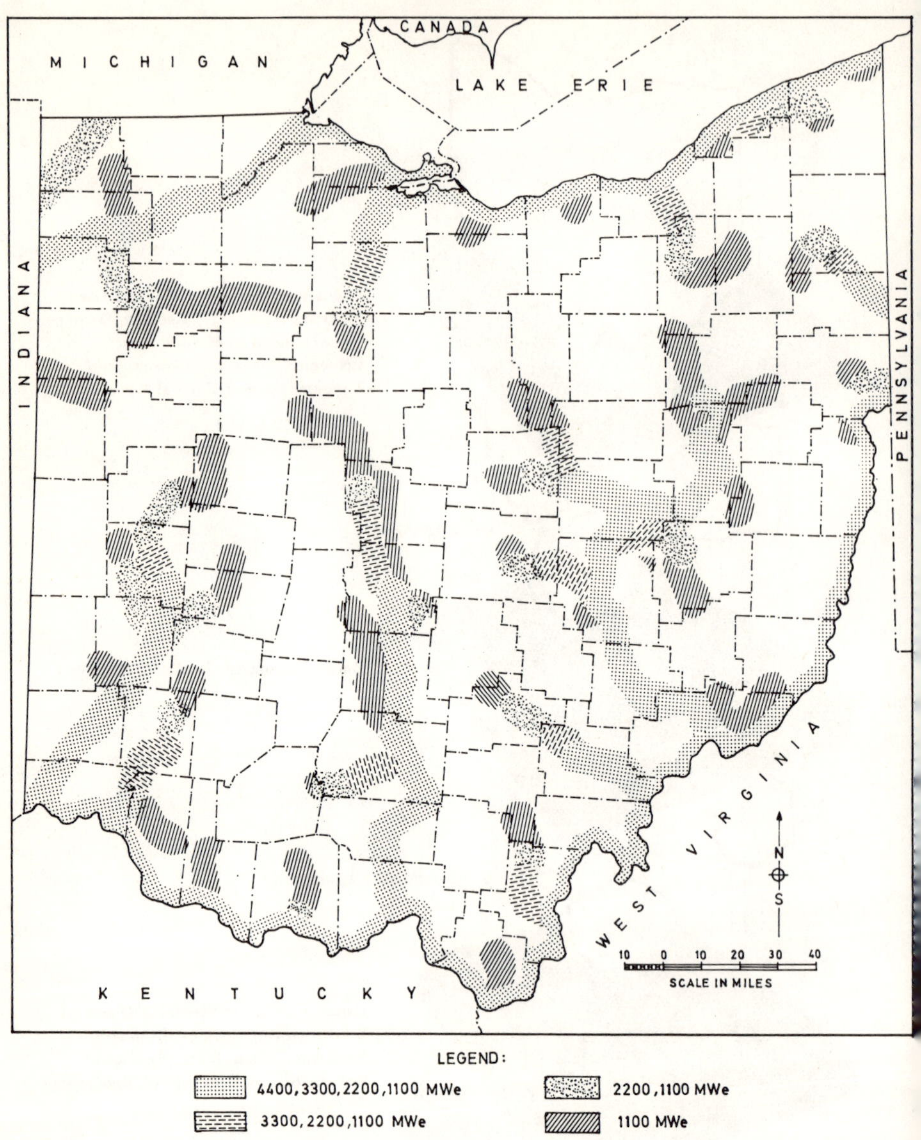


FIG. 17.   Areas suitable for nuclear power generating stations using cooling ponds.   Broken lines represent county borders.

While no area of the state was excluded from further consideration, some area within the state were considered to be less favorable geologically for power pl development than others.   Certain areas were found to be seismically 'active' (recorded history) and other small isolated areas were identified as known epicenters.   Therefore, although not eliminated, these areas were taken to be less promising for nuclear power plants at this stage of the study, and principa search efforts were diverted elsewhere.

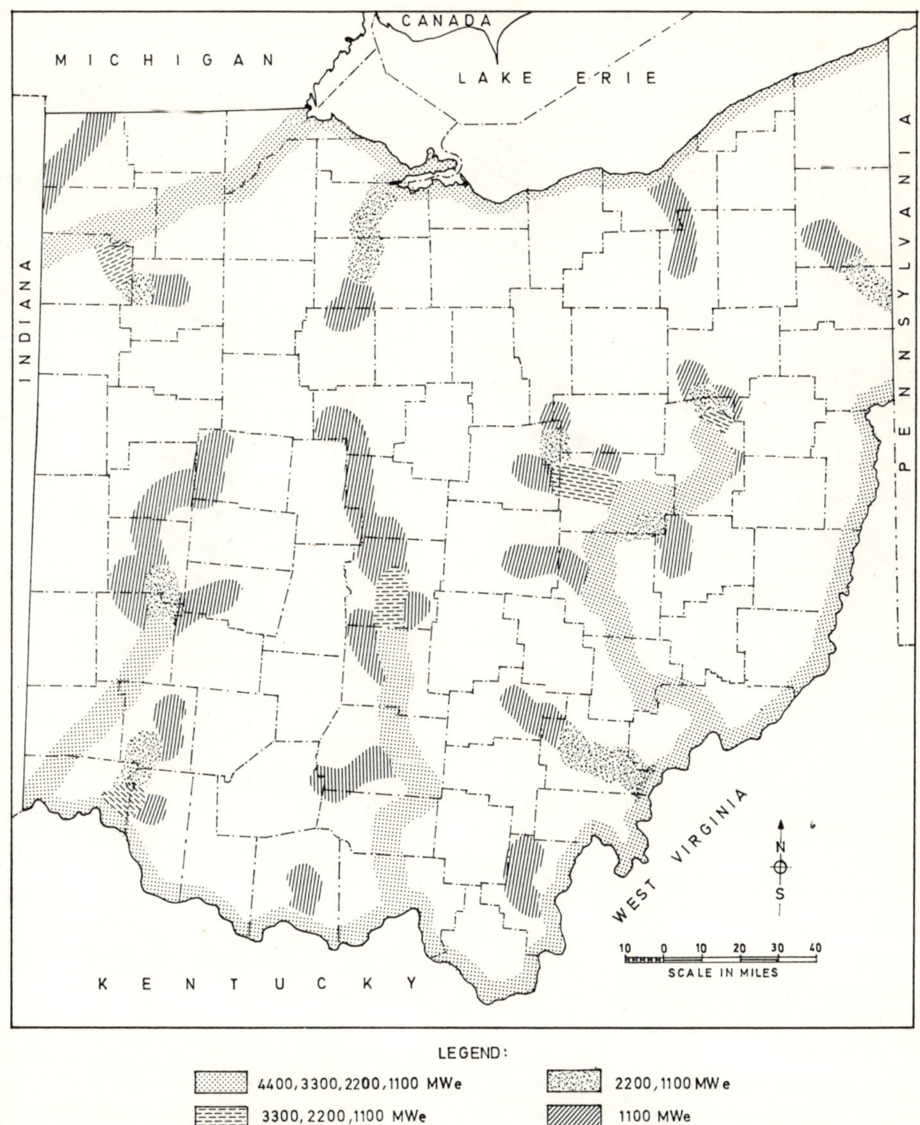



FIG. 18. Areas suitable for nuclear power generating stations using cooling towers with reservoirs. Broken lines represent county borders.

## 3.  Statewide Demography/Land Use Screening

The preliminary demography/land use screening considered 1970 population density, estimated 2020 population density, growth patterns of metropolitan areas, and special-use lands (See Figures 19 and 20).  The population density was derived for each county within the state to identify areas of high concen-

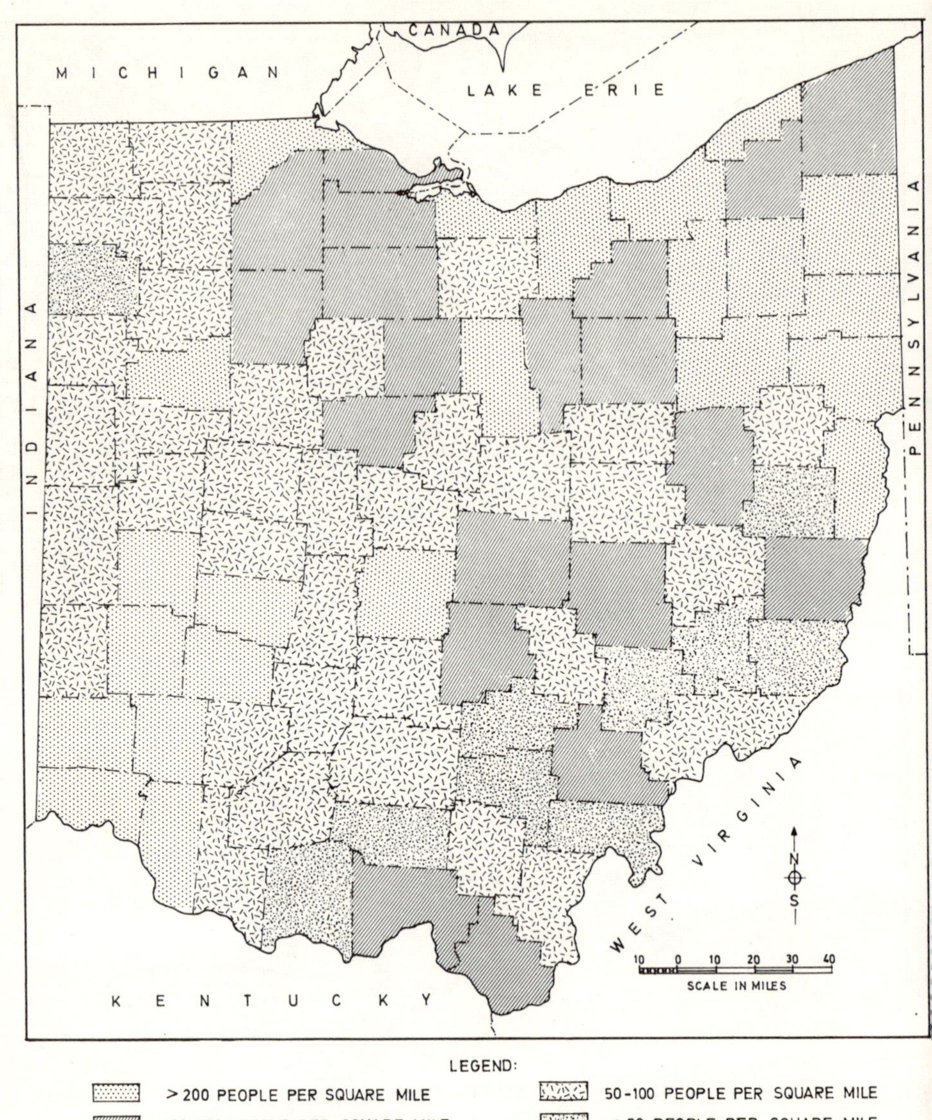



FIG.19. Population density by county, 1970.

trations of people. Since the prospective plant would not be operational for
about ten years after the starting date and is expected to remain in operation
for approximately 40 years, projected population densities to the year 2020 were
taken into consideration. Growth patterns of metropolitan areas were also exami
to identify areas likely to face strong competitive use problems. Additionally,
special land uses that could be affected by the plant were identified. These incl
mineral lands, parks and recreation areas, historical sites, Indian lands, US
Federal lands and US military reservations.

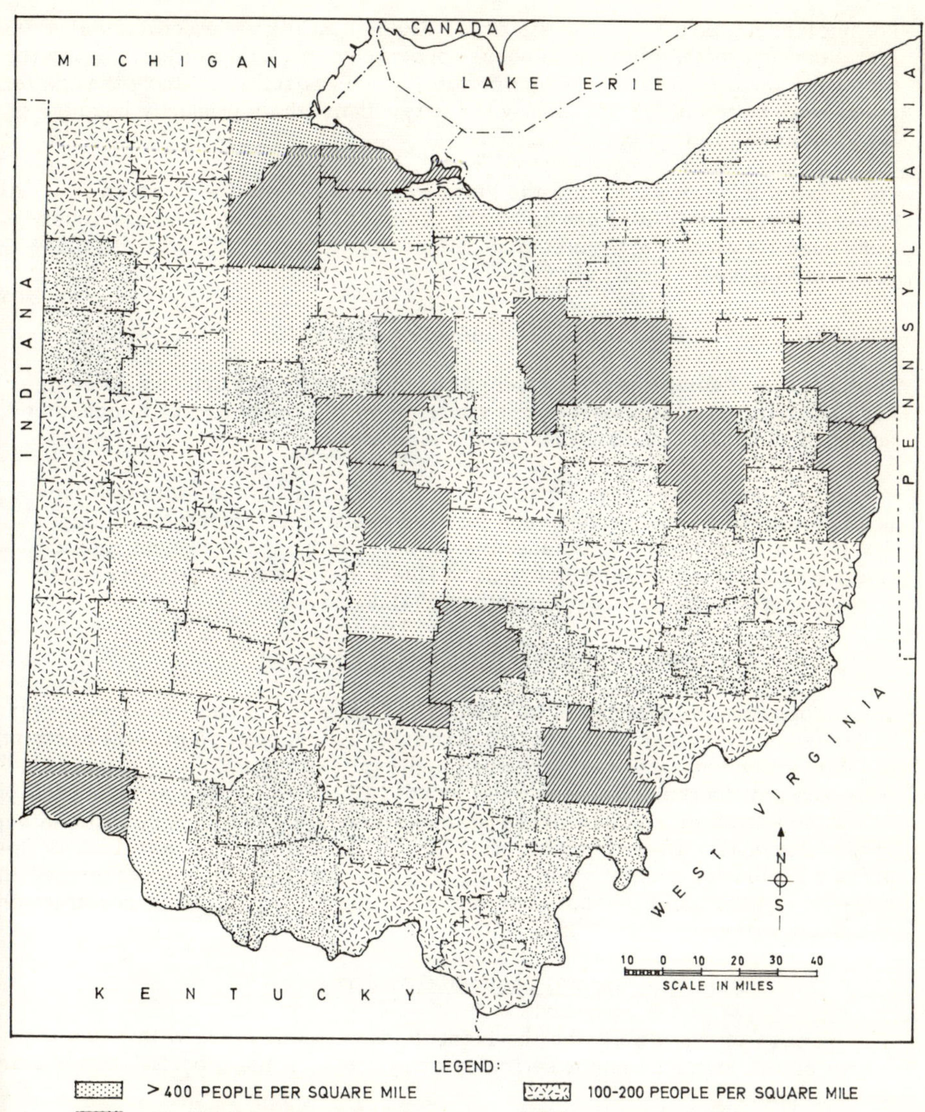


FIG.20.  Population density by county, for the year 2020.


## 4.    Statewide Meteorological Screening

The preliminary meteorological screening used six basic parameters. These were
topography as it affects meteorology, dilution potential, air quality priority level
(degree of compliance necessary to conform to the various state and Federal
ambient air quality standards), frequency of air pollution episodes, and potential
for increased fog and for decreased visibility due to evaporative cooling systems.

For nuclear power plants, the state was divided into five meteorological areas s
that each had relatively homogeneous topography and meteorology. Using the
aforementioned parameters as appropriate, it was possible to show that certain
areas were meteorologically somewhat better than others primarily because of
the increased dilution potential.

For fossil-fueled power plants the state was divided into the fourteen Air Qualit
Control Regions (AQCR's) set up by the Federal Government. The previously
discussed meteorological screening parameters were used to evaluate these
fourteen regions.

####     5.     Statewide Ecological Sensitivity Screening

Terrestrial and aquatic ecology were considered in the preliminary screening.
For the purposes of this part of the study, it was not meaningful to consider
nuclear and fossil-fueled power plants separately.

For terrestrial ecology, the locations of natural areas in the state that have
been preserved or are likely to see future pressures for preservation were
determined in order to avoid sensitive areas. For a similar reason scenic
rivers which have unique or scientific value and threatened species of fish
were considered in the aquatic ecology evaluation.

####     6.     Statewide Geography/Topography Screening

The purpose of this effort was the identification of any areas where power plant
siting would be impractical from a topographic standpoint. A general descriptic
of Ohio's physiography was developed and implications for siting examined. F
principal physiographic regions were found. However, there was no topograph:
basis for excluding any part of the state from consideration. Although one exp
offered fewer flat sites than the remainder of the state, there were locally flat
areas suitable for generating-station siting. In addition, it was recognized tha
non-flat terrain could present good opportunities for impoundment construction
the small valleys.

####     7.     Statewide General Acceptability Screening

The purpose of this effort was the identification of any areas within the state
where public attitudes might seriously complicate, if not prohibit, power plant
development.

The selective screening technique was used to make the first cut at determinir
candidate regions. For nuclear plants, first selective screening was done prim
on the basis of hydrology and demography. However, for fossil-fueled plants,
the meteorological considerations discussed were also of primary importance.
In effect, regions which exhibited favorable hydrological characteristics and
were not located within metropolitan areas were identified as candidate region
for nuclear power plants.

For fossil-fueled power plants, regions that were located in an area whose
atmosphere is classified as having essentially zero capacity for absorbing
additional effluents from a fossil plant were considered unacceptable.

The remaining factors indicated in Figure 16 did not eliminate large enough areas to be a prime factor in the selection of the candidate regions. They were, however, considered.

Because of the large number of potential candidate regions that survived the first screening, the following additional selective criteria were established to narrow down the candidate regions to a more manageable number:

> All candidate regions capable of being developed only for a
> 1100 MWe power plant would not be considered further.
>
> The search area would be narrowed to the candidate regions in
> or adjoining the client's service area.

As a result of the second selective screening, 51 candidate regions were identified for further evaluation. The screening of these 51 candidate regions used geology, demography/land use, accessibility, ecological sensitivity and geography/topography in a rating matrix. Hydrology was not included in the numerical ratings because the 51 candidate regions were selected primarily on the basis of hydrological potential and, as a result, the candidate regions did not differ enough hydrologically to provide a means for discrimination at this phase of the study. Meteorology was not included in the numerical ratings for basically the same reasons. General acceptability was not included in the ratings because not enough data were available to offer a good means of discrimination at this point in the study, and inquiry might have prematurely revealed potential interest in real estate in the regions of study.

Rating criteria were established for these five factors and ratings were determined for each candidate region. Each rating factor was broken down, as appropriate, into parameters and sub-parameters, which were then evaluated and rated. Table I lists the site factors and the parametric breakdown used for the third screening of candidate regions.

A classification and rating scheme was applied. The numerical scale used in this study ranged from a low rating of 0.00 to a high rating of 1.00. Table II describes what constitutes a particular numerical rating in relation to favorability for plant development.

Using criteria established for each of the parameters as a definition of favorability, numerical classification indicated in Table II was applied, i.e. what constituted highly favorable geology, hydrology, etc.,were defined. Recognizing that some parameters and subparameters were more important than others, a weighting was assigned to each parameter and subparameter to arrive at the overall rating for each siting factor for both nuclear and fossil-fueled plants. The weighting system for each type plant was designed so that the sum of all the weightings for the parameters and subparameters of any one factor would equal 1.00. Table III contains an example of this procedure.

Each candidate region was evaluated for each of the five factors included in the rating base. Our experience has shown that all siting factors are not of equal importance. To clearly show this in the decision process, it was necessary to assign a weighting value to each. Such weighting factors, in turn, can be region-dependent, influenced not only by the judgment of relative technical importance of a factor, but reflecting also the likely public acceptance aspects of the region affected.  <u>Weighting values are very likely to be country dependent as well, for the relative emphasis placed upon one set of factors versus another will undoubtedly reflect national policies and goals</u>.


## TABLE I.   REGIONAL SCREENING PARAMETERS

<u>Hydrology</u> – not used (since all regions – promising)

<u>Geology</u>:

Foundations
Mine hazard
Oil hazard
Seismology
Groundwater (dewatering)
Tectonic

<u>Demography</u>:

1970 population density
2020 population density
Proximity to nearest population center

<u>Land use</u>:

Agriculture
Special land use (parks, wildlife refuge, etc.)
Residential character
Public facilities
Sensitive industries (oil and gas pipelines)
Weapons testing
Land values

<u>Accessibility</u>:

Transmission line network:
    Land characteristics along route
    Distance to service area
    Distance to transmission line interties and substations

Transportation network:
    highway
    railroad
    waterways

Fossil-fuel resources (fuel supplies)

<u>Meteorology</u> – Not used (existing regional data not detailed enough to allow discrimination)

<u>Ecology</u>:

Aquatic:
    River classifications
    Endangered species
    Biological conditions

<u>Terrestrial</u>:

Quality of natural areas
Proximity to candidate regions

<u>Geography/Topography</u>: Number of potential sites/regions
Terrain for power plant
Terrain for impoundments

## TABLE II.   NUMERICAL RATING SCALE FOR CANDIDATE REGIONS

| Rating | Explanation |
|---|---|
| 1.00 | Development conditions are unusually favorable |
| 0.75 to < 1.00 | Development conditions are favorable |
| 0.50 to < 0.75 | Development conditions are acceptable |
| 0.25 to < 0.50 | Development must overcome deficiencies which require special engineering and/or analysis |
| 0.00 to < 0.25 | Development is costly but practical |
| 0.00 | Development is not impossible but is not encouraged in light of available alternatives.  Acceptability, if achievable, is likely to entail extensive and time-consuming field investigations, engineering feasibility studies and economic penalties to establish site qualification. |

## TABLE III.   EXAMPLE OF PROCEDURE USED TO ARRIVE AT AN OVERALL RATING FOR A FACTOR

| Geology parameters | Importance factor | Rating |
|---|---|---|
| Foundation | 0.200 | 0.30 |
| Mine hazard | 0.125 | 0.90 |
| Oil hazard | 0.125 | 1.00 |
| Seismology | 0.200 | 0.60 |
| Groundwater | 0.100 | 0.10 |
| Tectonics | 0.250 | 0.80 |
| Total geological rating |  | 0.63 |

## TABLE IV.   RATING AND RANKING OF CANDIDATE REGIONS: OHIO STUDY

| | Site rating factors | | | | | |
|---|---|---|---|---|---|---|
| Region No. | Geology | Demography | Accessibility | Ecology | Topography | Composite ratings[a] |
| 1 | 0.58 | 0.54 | 0.94 | 0.80 | 1.0 | 0.77 |
| 2 | 0.68 | 0.64 | 0.66 | 0.80 | 0.85 | 0.74 |
| 3 | 0.48 | 0.63 | 0.76 | 0.30 | 0.85 | 0.63 |
| 4 | 0.54 | 0.51 | 0.80 | 0.60 | 0.85 | 0.66 |
| 5 | 0.58 | 0.56 | 0.86 | 0.40 | 0.48 | 0.54 |
| 6 | 0.58 | 0.59 | 0.73 | 0.40 | 0.40 | 0.51 |
| Factor weighting values | 0.1 | 0.3 | 0.1 | 0.2 | 0.3 | |

[a] For the composite rating, each factor rating is multiplied by the factor weighting value and all weighted factor values summed.

TABLE V.  IDENTIFICATION OF THE TOP TEN CANDIDATE REGIONS
FOR NUCLEAR AND FOSSIL-FUELED POWER PLANTS

| Candidate region/cooling option[a] | | Rating | Ranking |
|---|---|---|---|
| Region 200 | Lowland/cooling towers | 0.78 | 1 |
| Region 204 | Lowland/cooling towers | 0.77 | 2 |
| Region 110 | Lowland/cooling towers with reservoir | 0.77 | 2 |
| Region 110 | Upland/cooling towers with reservoir | 0.75 | 4 |
| Region 111 | Upland/cooling towers | 0.74 | 5 |
| Region 5 | Lowland/cooling towers with reservoir | 0.74 | 5 |
| Region 201 | Lowland/cooling towers | 0.74 | 5 |
| Region 202 | Lowland/cooling towers | 0.74 | 5 |
| Region 5 | Upland/cooling towers with reservoir | 0.73 | 9 |
| Region 21 | Lowland/cooling towers | 0.72 | 10 |
| Region 204 | Lowland/cooling towers | 0.82 | 1 |
| Region 200 | Lowland/cooling towers | 0.79 | 2 |
| Region 202 | Lowland/cooling towers | 0.75 | 3 |
| Region 5 | Lowland/cooling towers with reservoir | 0.74 | 4 |
| Region 25 | Lowland/cooling ponds | 0.73 | 5 |
| Region 113 | Lowland/cooling towers | 0.72 | 6 |
| Region 201 | Lowland/cooling towers | 0.72 | 6 |
| Region 109 | Lowland/cooling towers with reservoirs or cooling ponds | 0.71 | 8 |
| Region 111 | Upland/cooling towers | 0.71 | 8 |
| Region 111 | Lowland/cooling towers | 0.70 | 10 |
| Region 112 | Lowland/cooling towers | 0.70 | 10 |
| Region 4 | Lowland/cooling towers with reservoirs | 0.70 | 10 |

[a]  The highest ranking for a candidate with cooling ponds as a cooling option in this study is for
candidate region 109 rated 0.70 and ranked 12th.

The ratings for each of the siting factors were multiplied by their weighting
values and summed to arrive at a composite rating for each candidate region for
both nuclear and fossil-fueled power plants.  These candidate regions were
then ranked on the basis of their composite numerical rating.  Table IV illustrate
the compositing of factor ratings.  Table V illustrates the overall ratings and ran
of the top candidate regions that emerged from the application in the Ohio study
the methodology described.  It may be noted that no region had a perfect score,
indicating that some trade-off of site characteristics would be necessary in all
cases.

One of the frequent complaints aimed at US utilities has been that engineering/
economic considerations have dominated site selection practices rather than
concern for the environment.  For the Ohio study, an evaluation was made of th

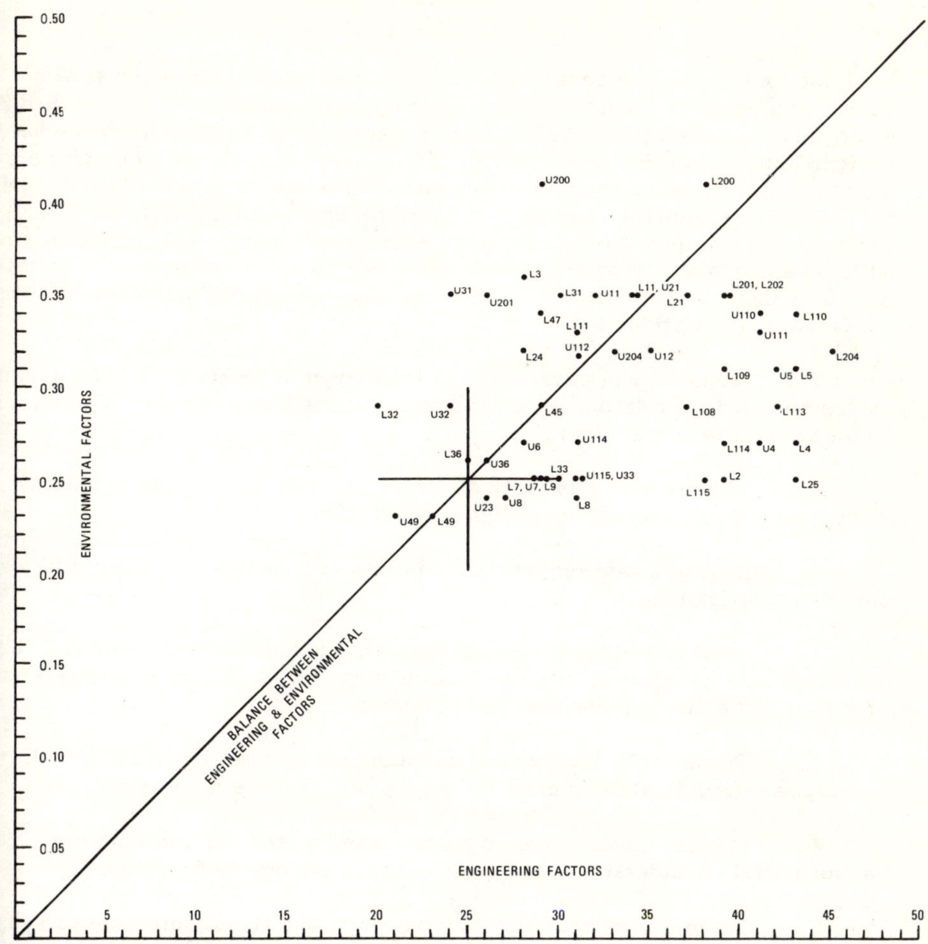


FIG.21. Engineering versus environmental rating for candidate regions for nuclear power plant.
U: upland; L: lowland.

engineering versus the environmental rating of the candidate regions for both
types of power plants. Geology, accessibility and geography/topography were
grouped as engineering factors. Demography/land use and ecological sensitivity
were grouped as environmental factors. Figure 21 summarizes the engineering
versus the environmental rating of the candidate regions examined in this study
for nuclear power plants. This kind of analysis allows the identification of
candidate regions with a good balance of siting considerations. Similar calculations can be done for sites rated in the same way.

## 5. SUMMARY

As illustrated by the two case studies summarized herein, the regional site
selection process is a balance of engineering, environmental, economic and
socio-economic factors.  This balance can be reflected in the importance value
assigned siting factors.  Our experience has shown that all siting factors are
not of equal importance in every siting case.  Importance values are very likely
not only to be regionally dependent but country dependent as well, for the
emphasis placed upon one factor versus another will undoubtedly reflect nation
policies and goals.  What kind of alternatives is one forced to examine in making this balance?  The following situations encountered in studies we have ma
are perhaps illustrative:

1.      Economic and environmental incentives to locate near to load cent
versus the need for relative isolation from urban centers and areas of relativel
intense use of land resources.

2.      Regions with relatively clean air and able to disperse pollutants
versus constraints against 'significant degradation'.

3.      Desire to rate regions on hydrology and meteorology versus lack o
significant difference.

4.      Good power plant site characteristics, but because of special lan
use (forest, etc.) between site and load centers, transmission line access is
very poor and a site may have to be eliminated.

5.      Competitive use for land or water in the same region for a similar
development (region already planned by other utility for a power plant).

6.      Cooling ponds versus cooling towers - land requirements and
environmental considerations can dictate locale preference for towers.

7.      Hydrological cost of supplying water and flood protection versus
geological cost of establishing a good foundation.

8.      Ease of getting equipment and work force into a site versus isolat
for environmental and safety reasons.

9.      Land availability - a site which is average on engineering ground
can be very good because the site has very few owners.

10.      Groundwater withdrawal for plant cooling versus competitive dem
for suburban growth or agriculture.

11.      Land use for nuclear plant restricted zone - particularly shorelin
marsh lands - versus public recreational use or ecological preserves.

Looking to the future, it becomes quite apparent that the siting problem is go
to become more difficult as the competition for available resources becomes
more intense.  Whereas the most pressing problem today stems from the fuel
picture and the safety and environmental laws influencing fuel options, the

availability of water could well be another principal determinant of the future.
Geological considerations in some regions may severely limit the choice of sites
for nuclear plants. As siting options narrow and siting possibilities diminish,
as they will eventually because of land and cooling-water limitations, long-range
forecasting and early site acquistions of future sites will take on ever-increasing
importance. The needs of the electric power industry for land, water and fuel
resources must be integrated into the overall growth patterns planned for the
regions they serve.

## ACKNOWLEDGEMENT

The author is grateful to the utilities sponsoring the Arizona and Ohio
studies for permission to use graphic material illustrating the regional
studies for their projects.

The NUS effort was that of an interdisciplinary team. Principal technical
support was provided by Paul Morgan and M. Bland (Ecology), P. Altomare
(Meteorology), L. Perez (Hydrology), A. DeAgazio (Demography & Land Use),
P. Goldstein (Water Quality) and their technical staffs. M. I. Goldman,
M. Elkins, and L. Klein provided project coordination and overall guidance.

Principal geological consultants were J. Smith and J. Scott of FUGRO, Inc.
on the Arizona project, and P. Rizzo and A. J. Eggenberger of E. D'Appolonia
Consulting Engineers on the Ohio project. A. J. Harshbarger of the University
of Arizona contributed to the water resources portion of the Arizona study.

## REFERENCES


[1]   DiNUNNO, J.J., LEVINE, S., Environmental matters concerning nuclear electrical power production,
      Nucl. Eng. 15 (1970) 607.
[2]   SOUTHERN INTERSTATE NUCLEAR BOARD, Power Plant Siting in the United States, topical monograph,
      Sept. 1974.
[3]   USAEC, Preparation of Environmental Reports for Nuclear Power Plants, Regulatory Guide 4.2, March 1973.
[4]   FEDERAL POWER COMMISSION (HSA), The 1970 National Power Survey, Parts I-IV.
[5]   USAEC, Code of Federal Regulations, 10 CFR 100, Reactor Site Criteria.
[6]   ATOMIC INDUSTRIAL FORUM, INC., Nuclear Power Plant Siting – A Generalized Process, August 1974.
[7]   USAEC, Regulatory Guide 4.7, General Site Suitability Criteria for Nuclear Power Stations.
[8]   Environmental Report, Palo Verde Nuclear Generating Station, AEC Docket File Nos. STN 50-528 to 530.
[9]   ELKINS, M., DiNUNNO, J., "A regional siting study for thermal power plants in the state of Ohio",
      Conf. on Nuclear Power Plant Siting, Portland, 1974, Trans. Am. Nucl. Soc., Suppl. 1 (1974).


## DISCUSSION

Y. NISHIWAKI: Some time ago I was given to understand by a person
associated with the United States Environmental Protection Agency that an
' environmental impact statement ' must be submitted in order to obtain
Government permission in the United States of America. At what stage
does the statement have to be submitted to the Government – during the
selection of a site from a number of promising choices, or after the choice
has been made?

J.J. DiNUNNO:  Environmental impact assessment reports for nuclear power plants must be supplied to the Atomic Energy Commission in support of the application for a construction permit.  The impact studies and base line studies are performed on the specific site upon which the utility plans to build.  The environmental report filed by the utility must also include information on the site studies which led to the selection of the specific site.

J. EDWARDS:  How much effort, in terms of time and cost, is involved in such a site selection process?  Are all the basic data available for detailed assessments?  And who bears the cost of selection — the Federal Government, the States or the utilities concerned?

J.J. DiNUNNO:  A regional screening study such as the two case studies I have discussed takes something like five to six months, the cost ranging from US $ 125 000 to US $ 175 000.  Such a study is preliminary, and the sites ranked highest must be confirmed by a geological rating carried out largely on the basis of research in the literature, aerial observations and limited field studies of surface indications.  Final ranking of the most promising sites cannot be performed reliably without on-site explorations beneath the surface.  Since such subsurface work can be time-consuming an difficult, particularly where evidence of faulting in the area requires the establishment of datable horizons, geological confirmation of preliminary site selection and ranking can vary widely in cost, running into figures of the order of a million dollars or more in the most difficult cases.

C.J. van DAATSELAAR:  Weighting values are assigned to different factors in the process of comparing sites.  Do they tend to be the same for different regions, and could you quote some numerical values for us by way of example?

J.J. DiNUNNO:  Weighting values for the various siting factors are likely to be different for different regions, a point which is discussed in more detail in the paper.  Examples of numerical values of weighting factors are shown in Tables III and IV.  It may be noted that in Table IV demography was given an importance factor of 0.3 because the search area in question had a relatively dense population, in contrast to case 1, which involved a sparsely populated search area.

E.H. HUBERT:  In referring to your screening factors, you did not mention public acceptance.  Is it included in demography?  What sort of weighting is it likely to receive in different areas?

J.J. DiNUNNO:  We did recognize that public acceptance could be an important factor, particularly in the classification and ranking of sites. However, we were not able to quantify it in any meaningful way when search over large areas.  Public acceptance problems are generally most intense the local level, i.e. after a specific site has been identified.  One cannot enquire into local  sentiment during early surveys and screening without revealing site areas of potential interest and running the risk of an escalati of land values.

Our experience has shown that the opposition frequently springs from concern over competition for land and water resources, intrusion into sensitive ecological areas (e.g. spawning grounds) or disruption of local 'life-styles' in a specific site area.  To the extent that these matters are considered in the demography/land uses, ecological and hydrological factor they give a relative measure of the likelihood of public acceptance or opposition.

P. CANDES: In describing your methods of screening, in the first stage, you made no mention of a factor associated with the proximity of load centres. Do you take this factor for granted or do you disregard it?

J.J. DiNUNNO: The location of the site in relation to transmission lines and load centres is an important factor, and is regularly taken into account. You may note that Table 1 lists accessibility as a factor.

H.A.G. KOHLER: You mentioned the San Andreas fault. May I ask to what extent the design and construction of nuclear power stations in that area is influenced by it? In the vicinity of the fault, I believe, the Oconee Nuclear Power Station is now under construction.

J.J. DiNUNNO: Several prospective reactor sites in relative proximity (by USAEC standards) to the fault never successfully completed the construction permit process — e.g. Bodega Head. For the San Onofre site, the regulatory authorities have laid down requirements which significantly increase the seismic design loadings for any additional plants that may be located there (0.67g).

M. NELKEN: Do you consider cooling towers that use treated sewage water as a proven technology accepted by the public? And do you regard dry cooling towers as representing an unproven technology?

J.J. DiNUNNO: There is not yet enough experience with cooling towers using treated sewage water to allow any safe generalization about public acceptance. However, it is expected that the water will, before use in the tower, be subjected to treatment comparable to what it would have to receive if it were to be discharged directly to the public domain. The question of potential dispersal of viruses in the drift has been raised. Impact assessments indicate that no unacceptable environmental effects are likely to be encountered. As a precautionary measure, a monitoring requirement might be imposed until more experience is obtained.

Dry cooling towers have not yet become proven technology for the size of power plants generally being built. One dry tower rated for roughly 300 MW is being built and should provide useful experience for the design of larger towers.

# SITING – MEANS BY WHICH NUCLEAR FACILITIES ARE INTEGRATED INTO A CANADIAN COMMUNITY

J.W. BEARE, R.M. DUNCAN
Atomic Energy Control Board of Canada,
Ottawa, Ontario,
Canada

## Abstract

SITING – MEANS BY WHICH NUCLEAR FACILITIES ARE INTEGRATED INTO A CANADIAN COMMUNITY.
   In Canada, the Atomic Energy Control Board considers the siting implications of a number of different kinds of nuclear facilities. Its overall objective is to ensure that nuclear facilities are sited, designed, constructed and operated so that the level of risk to the health and safety of people and any adverse impact on the environment are acceptable. This requires in part the establishment of siting criteria, which involves the balancing of safety, environmental, sociological and economic factors. The application of these criteria is performed in the light of the need to integrate the plant with its surroundings in a manner acceptable to all those involved, particularly the public. This paper discusses the implications of siting two of the major types of facilities in Canada: nuclear power plants and heavy-water production plants, the former presenting radiological hazards, the latter being of concern due to the severe chemical toxicity of one of the process fluids. The paper indicates the similarities and differences in safety standards and regulatory approach for the two types of plant, the site approval procedure, the intergovernmental relationships required for efficient regulation and the current status of public participation.

## 1. INTRODUCTION

In Canada, the Atomic Energy Control Board (the Board) considers the siting implications of a number of different kinds of nuclear facilities. Some of these facilities, however, such as large gamma irradiation machines, accelerators and fuel fabrication plants, present very few problems in siting because of the limited hazards associated with them. Of major concern, from the radiological safety point of view, are research and power reactors. In addition, there are other facilities, some related to the fuel cycle and others to the production of heavy water, which are also considered because of the significant public hazards posed by the chemical toxicity of the working fluids, the most important of these at this time being the heavy-water production plants. To date, we have not licensed spent-fuel reprocessing plants, but the experience gained in assessing radiological and chemical hazards associated with these other facilities should provide sufficient background to enable us to deal with reprocessing plants should the occasion arise.

Our overall objective is to ensure that nuclear facilities are sited, designed, constructed and operated so that the level of risk to the health and safety of people and any adverse impact on the environment are acceptable. This requires the establishment of siting criteria for these facilities, which involves the balancing of safety, environmental, sociological

and economic factors.    The application of these criteria is performed in the
light of the need to integrate the plant with its surroundings in a manner
acceptable to all those involved, particularly the public.

This paper deals with the siting of two types of facilities
which are currently of primary significance to the Canadian nuclear
program   -   nuclear power plants and heavy-water production plants.

## 2.    LEGISLATION

In any nuclear licensing process there are legislation and
regulations which give a regulatory agency the legal basis on which to set
requirements governing the safe design, construction and operation of
nuclear facilities and to establish guidelines which outline such requirements.

The degree to which these safety requirements are documented depends on the obligations imposed by relevant legislation, the
legal system, the size and complexity of the industry, the working
relationship between the regulatory agency and the licensees, public
interest and, to an increasing degree, international interest, particularly
on the part of international customers.

In Canada, the Atomic Energy Control Act allows, among
other things, the Atomic Energy Control Board to make regulations governing the production and use of prescribed substances.    Uranium used in
nuclear reactors and heavy water produced in heavy-water plants are two of
the prescribed substances of interest.    The regulations made under the Act
stipulate the requirement for an operating licence for a nuclear facility[1],
which refers to a plant which uses, produces or processes prescribed substances, and further stipulate that no operating licence will be issued unless
prior permission is obtained from the Board for construction of the facility,
part of which involves approval of the site.

## 3.    APPROACH

Although the primary objective, as stated earlier, of
reducing risk to the public and workers to an acceptably low level applies to
both nuclear power plants and heavy-water production plants, the basic
safety and siting considerations differ because of the characteristics of the
processes, the nature of the hazards and the way in which safety criteria
evolved for the nuclear industry compared to other similar industries.
Typically, a nuclear power plant is energy producing, a heavy-water plant is
energy consuming; a nuclear power plant presents mainly a radiological
hazard, the hazard associated with a heavy-water plant is one of chemical
toxicity; the nuclear industry has been strictly regulated and externally
overseen since its inception, the chemical industry has been mainly self-
controlled with some basic regulations evolving over the years.

---

[1] The Atomic Energy Control Regulations define 'nuclear facility' as follows: 'nuclear facility'
means a nuclear reactor, a subcritical nuclear reactor, a particle accelerator, a plant for the separation,
processing, re-processing or fabrication of fissionable substances, a plant for the production of deuterium
or deuterium compounds, a facility for the disposal of prescribed substances and includes all land, buildings
and equipment that are connected or associated with such reactor, accelerator, plant or facility.

Since heavy-water plants are essentially chemical plants, their regulation has been a compromise between that normally applied to the chemical industry and the highly regulated nature of the nuclear industry. The difference becomes evident when comparing the siting criteria.

The siting criteria for heavy-water plants are deterministic, that is, the characteristics of an acceptable site are defined by a number of guidelines, and the suitability of proposed sites is judged by the degree of conformity   of their characteristics to these guidelines.   The criteria were selected on the basis of judgement to ensure that the risk to the public would be acceptably low.   The siting criteria for nuclear power plants, on the other hand, are in part probabilistic and in part deterministic.   The basic safety criteria for the plant are expressed in terms of probability vs. acceptable consequences (i.e. acceptable risk);   however, there are some guidelines concerning the site characteristics and the design of the plant which indicate how the basic safety criteria are to be met.

The regulatory approach taken by the Board is to stipulate the basic safety criteria and to require the applicant to show how his proposed plant will meet the basic safety criteria.   Figure 1 shows a simplified 'hierarchy' of the safety considerations for a nuclear facility with the more general considerations at the top and with an increasing degree of detail going from top to bottom.   The word 'considerations' is meant to imply safety-related considerations for design, construction, commissioning and operation of the facility.

The documentation associated with licensing of a nuclear facility does not divide along such clear-cut lines as indicated by Figure 1 but, at some risk of oversimplification, may be grouped as follows:

Level 1:    Legislation   -   applies to all uses of
            atomic energy.

Level 2:    Basic safety and siting criteria   -
            apply to all nuclear facilities, but may
            be different basic criteria for different
            types of nuclear facilities.

Level 3:    Licensing documents   -   for a
            particular nuclear facility:

            Construction Licence /Operating Licence[2]
            Safety Report[3]   (design description and
                                       accident analysis)

            Operating Policies and Principles[3]
            Radiation Protection Regulations
            (where applicable)[3]
            Emergency Procedures[3]   (on-site)[4]
            'Other submissions'[3]   -   (detailed
            reports and correspondence bearing on
            the design, construction, commissioning
            and operation of the facility).

---

[2]   Issued by the Board.
[3]   Prepared by the applicant; reviewed and approved by the Board.
[4]   Off-site emergency procedures are prepared by provincial government authorities.

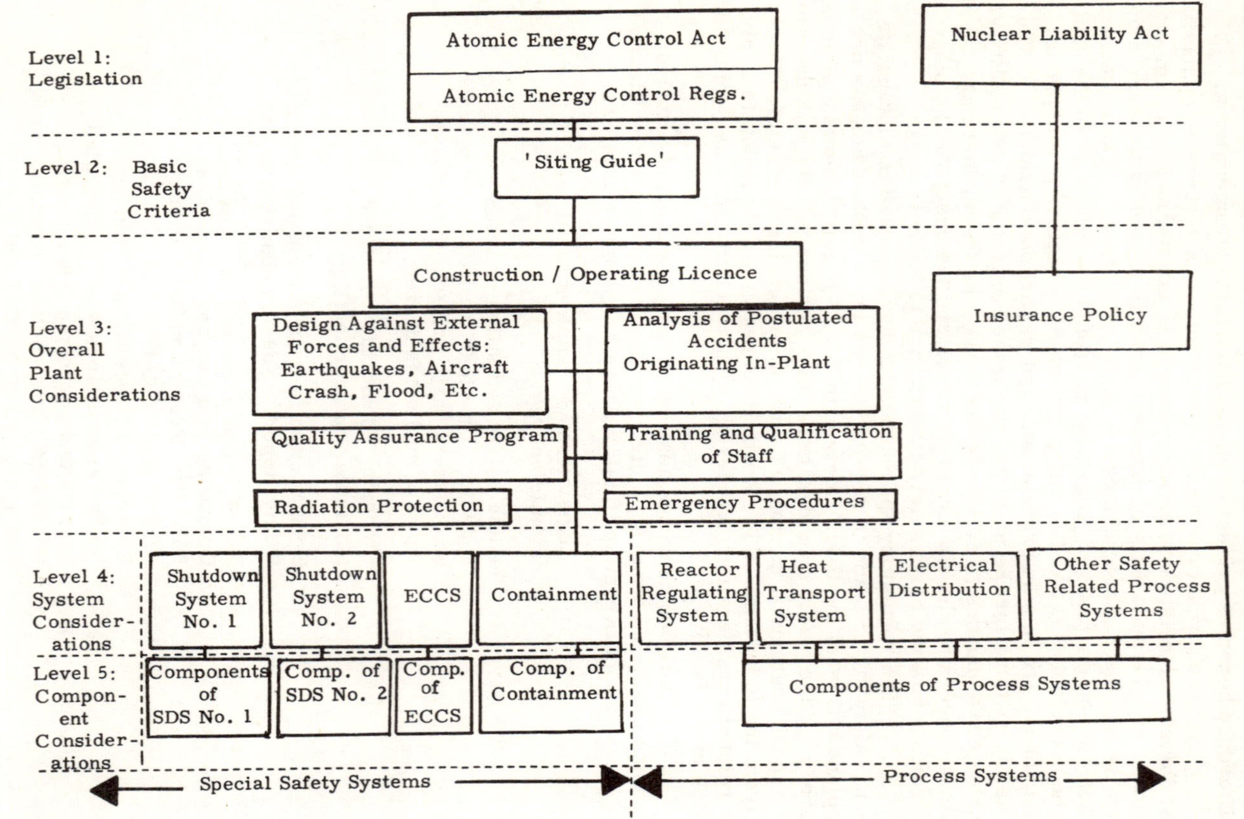


FIG.1.  Simplified 'hierarchy' of nuclear power plant safety considerations.

Levels 4 & 5:     Reference documents:

> Design Manuals
> Construction Documents
> Commissioning Manuals
> Operating Manuals

The foregoing approach works well when there are only a few plants in existence and only a few organizations involved in the design, construction, commissioning and operation of those plants.   Such is the case in Canada but the situation is changing because of the rapidly expanding national nuclear power program and because of international interest in the CANDU-type nuclear power plant.

## 4.    SITE APPROVAL PROCEDURE

### 4.1.  Governmental Relationships

Although Board approval of a site is not a legal requirement before an application for a construction licence is made, it has become customary for applicants to seek approval of a site at an early stage.   Indeed, when federal funding is involved, it is now federal government policy that the site must be approved.

The Board is concerned largely with health and safety aspects of nuclear facilities, although the Atomic Energy Control Act permits broad consideration of the national interest.   Some aspects of siting are also reviewed by other federal government agencies which also become involved in site approval in the event of federal funding.

Provincial governments also exercise authority in the site approval, the extent of provincial government involvement varying from province to province.

Generally speaking, environmental and other societal effects are evaluated by both levels of government and there is a developing trend to holding public meetings on such matters.

Good communication between levels of government and co-ordination of federal and provincial activities minimize duplication of effort. A single environmental impact report prepared by or on behalf of the applicant serves the purposes of both federal and provincial environmental agencies, and arrangements are often made for joint evaluation with one of the agencies taking the lead.

### 4.2.  Procedure

A site evaluation report dealing with health and safety matters is submitted to the Board by the applicant.   This report is reviewed by the appropriate one of a number of safety advisory committees and the Board's staff, and recommendations are made to the Board on the suitability of the site for the nuclear plant or heavy-water production plant as the case may be.   Such committees are composed of independent experts and appropriate representatives from federal, provincial and local government agencies. This arrangement has been very satisfactory in providing a wide range of expertise for the committees and in facilitating inter-governmental communication and cooperation.

The applicant is then informed by the Board whether or not there are any major obstacles to eventual site approval. At this time, if he has not already done so, the applicant must announce publicly his intention to build the facility at the particular site, giving the public an opportunity to comment on the proposal. If no substantive safety issues are raised, the Board issues final site approval after allowing a suitable period of time for public comment.

### 4.3. Public Participation

Our limited experience with public participation in Canada indicates that most public comment can be expected at the site approval stage. Public participation programs are a relatively recent development and are conducted by the applicant. Public meetings have been held regarding nuclear power plants and a heavy-water plant in the Province of Ontario, a heavy-water plant in Quebec and a nuclear power plant in New Brunswick. It is of interest to note that in some provinces, public meeting are now held regarding site approval for any type of proposed power plant. Meetings are relatively informal and their purpose is to inform the public and to give members of the public a chance to ask questions and present their views. Even so, planners must allot several months in their project schedules for the public participation program. In the case of nuclear facilities, much of this process occurs in parallel with the Board's licensing and safety review process.

Public attitudes to nuclear facilities have been generally favorable but there are signs that opposition is growing in certain parts of Canada. For example, eight environmental groups in New Brunswick and adjoining Nova Scotia have combined to oppose the proposed nuclear power plant in New Brunswick.

### 5.   NUCLEAR POWER PLANTS

### 5.1. General

From the health and safety point of view, the suitability of a site for a nuclear power plant depends on the design of the plant and the way it is to be operated and maintained. The Reactor Safety Advisory Commit (RSAC), which advises the Board on the safety of nuclear power plants, has established basic safety criteria which define in a quantitative way what is regarded as 'safe enough' or, put another way, 'low risk to the public'. These criteria were first published in 1964[1] and have been applied to the Pickering Generating Station and later plants. Earlier plants were design to similar safety criteria proposed by the designers. The 'Reactor Siting Guide', as the basic safety criteria, collectively, have come to be called, has been modified since 1964 but the basic concepts remain and are tabulated in Appendix A. Supplementary safety criteria and principles are listed in Appendix B[2].

## 5.2.    Basic Safety Criteria

### 5.2.1.  Background

A nuclear power plant is envisaged as consisting of process systems and special safety systems.    The process systems are those
systems required for normal operation of the plant and would be provided,
though not necessarily to the same standards, even if the result of nuclear
fission did not include radiation and radioactive fission products.    However, nuclear fission does produce radiation and radioactive fission products which are a potential hazard to the public and the plant staff.    The
special safety systems are designed to mitigate the consequences of failures
and malfunctions in the process systems which, in the absence of special
safety system action, could lead to fuel failures and a release of radioactivity to the environment.    Such failures are called 'serious process
failures'.

The 'defence-in-depth' approach inherent in the Reactor
Siting Guide is developed as follows (refer to Appendix A):

(1)    Process systems are designed and operated
to high standards so as to minimize the frequency of serious process failures.

(2)    The total frequency of all serious process
failures must not exceed once per three plant-
years and the consequences as mitigated by the
action of the special safety systems must not
exceed the values listed in the second situation
referred to in Appendix A.    Because of the
relatively high assumed frequency of serious
process failures, the risk is combined with
that associated with radioactive releases from
normal operation of the plant (first situation
listed in Appendix A).

(3)    The total frequency of serious process
failures coincident with failure of any one
special safety system must not exceed once
in three thousand plant-years and the consequences must not exceed the values listed
in the third situation of Appendix A.    Such
postulated failures are often described as
'dual failures'.

### 5.2.2.   Application of Basic Safety Criteria

The criterion regarding the frequency of serious process
failures is too general to be used in the detailed design of individual process
systems but it is useful in assessing the adequacy of the safety of plants
which are operating.    The design of individual process systems is generally
done by reference to proven design techniques which have given satisfactory

results in the past;  or else a reliability target for a given process system
is assigned which is a small fraction of once in three years.

Consideration of process failures and dual failures leads
to several requirements on the design of special safety systems:

(1)    The unavailability of each special system
       must be less than $10^{-3}$.

(2)    Special safety systems must be physically
       and functionally separate from process
       systems and from each other to the maxi-
       mum extent practicable in order to mini-
       mize the possibility of cross-linked or
       common mode failures under normal and
       accident conditions.

(3)    The operation of the special safety systems
       must be analysed for a spectrum of serious
       process failures and dual failures with the
       required effectiveness of each system being
       determined by the most exacting case analysed.
       Analysis of a broad spectrum of postulated
       accidents ensures that each special safety
       system will cope with the 'worst case ',
       which is not always an obvious case.

## 5.3.    Limits and Targets for Radioactive Releases from Normal Operati

The Board subscribes to the internationally held view that
releases of radioactive materials from normal operation of nuclear faciliti
should be minimized to the extent considered practicable.   Experience ha
shown it to be practicable to limit releases from a four-unit station such as
the Pickering G. S[5] 'A' to 1% or less of the Derived Release Limits, whic
are the licensed limits derived from the dose limits for normal operation
in gaseous and in liquid effluents.   Accordingly the Board has specified
that   new   nuclear power stations be designed and operated with a target
of 1% of the Derived Release Limits (DRL's) for each effluent pathway.
Should the target value be exceeded, it is envisaged that the action taken to
correct the situation would be graded depending on the circumstances.

On most sites in Canada there are two or more nuclear
facilities on one site (Appendix D).   Since the design and operating target
for radioactive releases is based on practicability, each type of facility
or each group of similar facilities (e.g. a four-unit nuclear power
station) can have  an individual target which is a small percentage of
the DRL's for the site.   The licensed limits, i.e. the DRL's themselves, are applied to the site as a whole.

There may be some argument that the part of the popula-
tion which lives around a certain site should not be exposed to more than a
certain part of the whole risk associated with the nuclear power industry.

_______

[5]  G.S. = generating station.

In practice other considerations, such as a utility's desire to avoid excessively large blocks of power on one site, availability of cooling water and other practical problems, would probably limit the concentration of nuclear facilities at a particular site before it could be argued that the risk to a particular part of the population is excessive.


## 5.4.   Codes and Guides

The preparation of codes and guides for nuclear power plants at Levels 3, 4 and 5 in Figure 1 was begun largely through the initiative of the nuclear industry itself in cooperation with, and sometimes provoked by, the Board.   The Canadian Nuclear Association (CNA), a body representing the Canadian nuclear industry, formed the Committee on Codes, Standards and Practices which was subsequently accredited as a Sectional Committee of the Canadian Standards Association (CSA), a nationally accepted standards-issuing body.   The Board is also preparing Licensing Guides which will cover topics not covered by industry codes and guides or will promote the use of codes and guides prepared by the industry.

It is expected that such activities will help to streamline the licensing process by consolidating accumulated experience with nuclear power plants and by avoiding repetition of effort where the safety considerations from plant to plant are similar.


## 5.5.   Licensing and Site Considerations

The licensing steps consist of site approval, as described earlier in the paper, construction licence and operating licence, and are described fully in reference 2.

At the time of issuance of a construction licence, many aspects of the design may not be firm but the general characteristics of the plant are known.   This situation has led to the practice of having a standard-sized exclusion zone with a radius on the landward side of 900 to 1000 metres. In other words, the applicant has some confidence that his site will be accepted by the Board (considering the preliminary nature of the plant design) if he proposes a site with that size of exclusion zone (see Appendix C) and provided the other characteristics of the site are acceptable.

Appendix A lists design-basis dose limits for the most exposed individual and the surrounding population for normal operation and for accident conditions.   Either the population dose or the individual dose is limiting in determining the design-basis releases under normal and accident situations, depending on the density of the surrounding population. For accident conditions, the Pickering site near Toronto is 'population dose limited'.   All other present sites in Canada are 'individual dose limited'.   In the absence of peculiar geographical features, such as deep valleys, the concept of a standard-sized site is of some assistance in a utility's long-term planning for nuclear power plants.   Of course, size is only one of the many characteristics of a site that must be considered.

## 6.    HEAVY-WATER PLANTS

### 6.1.    Background

Heavy-water plants are essentially conventional chemical plants, but are under the Board's jurisdiction because they are defined as nuclear facilities in the Atomic Energy Control Act and Regulations    (see Section 3).

The process currently used is one of dual-temperature isotope exchange between ordinary water and hydrogen sulphide ($H_2S$), taking place in large contacting towers, some of which are approaching 300 feet high and 30 feet in diameter, containing over 100 tons of $H_2S$.

The potential for hazard from heavy-water production plants was recognized in Canada at the time the first plant at Glace Bay, Nova Scotia,was proposed, but it was not until the second plant at Point Tupper, Nova Scotia,was under way and the Glace Bay plant had experienced some difficulties that the Board stepped in to take a leading role in the safety assessment and licensing of these plants.

The hazard is not radiological, but is associated with the large quantities of highly toxic and corrosive hydrogen sulphide held up in the process under elevated temperature and pressure.    Concentrations of $H_2S$ in air in excess of approximately 500 ppm   are considered to be lethal if breathed for a very few minutes.

### 6.2.    Basic Safety Considerations

Minimization of exposure of the plant workers and surrounding public to excessive quantities of $H_2S$ and its combustion product, sulphur dioxide ($SO_2$), is the basic safety consideration relative to the current design of heavy-water plants.    The occurrence of excessive releases of $H_2S$ from the process system must therefore be minimized and provision must be made to mitigate the consequences of such a release should it occur. This is the same basic safety philosophy which is applied to nuclear power plants.

The process itself does not have the potential for fast pressure or temperature transients, its control being relatively sluggish and its energy sources being limited to that imparted to it by the external steam source.    Hence, a heavy-water plant does not require the complexity  or depth of protective measures as does a nuclear power plant.   This same consideration applies to the safety assessment of a heavy-water plant, wherein the analyses performed are concerned with single events only, major emphasis being placed on an accidental release due to failure of the process envelope and on other events which could cause an abnormally high gas release from normal process effluent release points, e.g. by lifting of safety valves.   However, the risk to the public and workers is substantial, which necessitates application of a similar approach to safety as for nuclear power plants.

The integrity of the process system is the first line of defence against inadvertent releases of $H_2S$.   Proper process and equipment design and material selection are essential, as is an effective quality control program which begins at the preliminary design stage and continues throughout the plant lifetime.

Provision is made to isolate sections of the plant quickly in order to minimize the quantity of $H_2S$ released to atmosphere should a leak occur.    Other means used to minimize the release from a leak are to reduce quickly the quantity of $H_2S$ in the affected part of the plant by pumping the $H_2S$ back into a storage tank or to divert the $H_2S$ to a flare system via blowdown valves where the $H_2S$ is burned and released mainly as $SO_2$ at high elevation for better atmospheric dispersion.    The dispersion, of course, reduces the concentration as the gas moves downwind from the plant.

Steam strippers, aeration lagoons  and chemical treatment are used to reduce the $H_2S$ concentration in the liquid effluent from the plant.

To aid in reduction of $H_2S$ concentration in the atmosphere in the remote event of a large gas release, some plant owners are proposing installation of a unique gas dispersion system, which consists of a ring of propane burners around the plant to provide a large thermal input into the discharging gas to cause it to rise, and thus obtain better atmospheric dispersion.    This system could be triggered automatically by plant sensors or manually by the plant operator.

Personal protective equipment for plant workers, emergency plans for both workers and public and land use control for a significant distance outside the plant fence are also employed to provide protection against the possible effects of a gas release.

## 6.3.    Codes and Guides

Heavy-water plants are designed, built and inspected to a similar combination of codes and standards as other chemical plants, with special provisions being made for $H_2S$ - $H_2O$ service conditions where appropriate.    The pressure vessels are built basically in accordance with ASME Boiler and Pressure Vessels Code Section VIII and the piping to USAS B31.3 for Petroleum and Refinery Piping as a minimum standard.

Material specifications and selection for $H_2S$ - $H_2O$ service have always been contentious items in heavy-water plant design. In order to correlate the current information, Atomic Energy of Canada Limited (AECL), the crown company  mainly responsible for research, development, design and promotion of nuclear energy in Canada, has recently compiled a document which is currently being assessed by the Board with the possibility of its being adopted as a Guide for future plants.

Guides have been, or are in the process of being, issued to the industry by the Board to cover siting, emergency plans, safety report content, content of other licensing documents, as well as an outline of the licensing procedure.    These guides should aid the applicant in  proceeding through the licensing procedure with a minimum of delay. The latter is further facilitated by coordination at the preliminary design stage of the Board's licensing schedule and the applicant's project schedule.

## 6.4.    Siting Considerations

The siting of heavy-water plants in Canada has, in some cases, caused as much concern to regulatory bodies and the public as the siting of nuclear power plants.    As mentioned previously, the severe toxi-

city of $H_2S$, its corrosiveness, and the large quantities which are held up
under elevated temperature and pressure in the process and in storage at a
heavy-water plant cause these plants to be regarded as potentially very
hazardous.

In order to provide those interested in establishing a
heavy-water plant with some guidance as to the Board's concerns regarding
siting of these facilities, the Board issued, on 29 August, 1973, 'Siting
Guidelines for Heavy-Water Plants Utilizing the Girdler-Sulphide Process',
which have just been revised and reissued (see Appendix E).

This document reflects the safety considerations men-
tioned previously and outlines possible protective provisions which should
be employed to reduce the risk to the surrounding population to an acceptable
level.

The guidelines are concerned with the effects within the
'zone of influence' of the plant, this zone extending out to the point where
the risk posed by the plant is judged  to be reduced to an acceptable level.
The risk, of course, involves exposure to high concentrations of $H_2S$ and/or
$SO_2$ in the atmosphere, the concentrations being reduced by dispersion as
the gas moves away from the plant.

The size of the zone of influence depends upon the provisions made to mitigate the consequences of a major gas release, and
could extend out as far as five miles from the plant.

The zone of influence is divided into three areas, A, B
and C.   Areas A and B are restricted to controlled population only, i. e.
persons under the direct supervision of the heavy-water plant management
or a management which has committed to cooperate with the heavy-water
plant management in the event of an emergency.   This combined area
extends out to approximately one mile from the plant.   The restriction to
controlled population within this area relates to the belief that most individuals would be unable to respond and protect themselves from the
consequences of a gas release with normally available means within the sho
time it would take for the gas to travel from the point of release out to abou
one mile.   Beyond this distance, in area C, it is felt that  individuals wou
be able to cope, utilizing the means at hand, provided they were aware
the hazard and were provided with ample warning. The most appropriate for
of protection in any event is felt to be a stay-in procedure wherein persons
enter a building or other shelter which is relatively leak-tight and remain
there during the period of gas passage.

It is of prime importance that the number of persons
which could be subjected to additional risk be minimized.   To facilitate
this, a low population guideline has been set down, limiting population
density in the zone of influence to that found in a typically rural area.

As mentioned before, reduction in the size of the zone
of influence would be considered, depending on the effectiveness and reliability of protective measures provided to reduce the risk to the surroundings.   Currently, consideration is being given to the extent of reduction
warranted if a highly reliable and effective gas dispersion system were to
be supplied.   This reduction could even be to the extent of eliminating
area C altogether.

It is evident, from the siting considerations, that
locating a heavy-water plant in a particular area can have a widespread

impact, possibly even more so than a nuclear power plant, particularly
when considering the local sociological implications of restricted land usage
and establishment of extensive emergency plans.

## 7.    STANDARDIZATION AND EVOLUTION OF SAFETY REQUIREMENTS

For reasons of economy there is a trend toward repeating
proven facility designs with relatively few changes.    The trend towards
standardization of designs should facilitate licensing;  however,  safety
requirements are evolving.    Experience with earlier plants is reflected in
the requirements for later plants.    The spectrum of postulated failures
and accidents is being widened and made more detailed.     Postulated
events are being examined in greater depth.    Increased attention is being
given to events originating outside the plant, such as earthquakes and the
risk posed by the possibility of aircraft crashes onto the site.    The latter
example is gaining increased attention because some nuclear facilities are
being located closer to centres of population where one can expect a higher
density of aircraft movements.

The evolution of safety requirements presents two
dilemmas.    One is to the applicant who, for example, faces new requirements for a repeat of a design which was earlier licensed.    The second is
to the regulatory agency itself which must justify the continued operation of
earlier plants when (usually) more stringent requirements are applied to
later plants.    There is no complete answer to the two dilemmas but much,
if not most, of the answer includes the following:

1)    The nuclear power industry is regulated on
      the basis of risk posed by postulated events,
      most of which have never happened at any
      nuclear facility.    (For most other industries
      the definition of 'credible events' is 'those
      events which have actually happened '.)

2)    Consideration is given to the cumulative risk
      posed by all nuclear facilities, particularly
      nuclear power plants.    The incrementally
      higher risk posed by the relatively few early
      plants designed to less stringent requirements
      than later ones may be negligible.

3)    The larger size of later plants and the tendency
      to locate them closer to centres of population
      warrant the imposition of more stringent
      requirements in order to maintain the same low
      level of risk to the public.

4)    When deemed necessary, new requirements can
      be backfitted to existing plants.

## 8.    PUBLIC EDUCATION

In Canada, as in other countries, public opinion  or, more
exactly, certain elements of public opinion, are influencing the licensing of

nuclear power plants.    One of the comments frequently expressed is,
'Tell us what we should know (about nuclear power plant safety)'.    One
group of high school students went so far as to suggest that the teaching of
nuclear technology should be mandatory in schools.

One of the problems of dealing with the public is to determine what is 'the public'.    It is clear, however, that graphic displays,
models and brochures, while necessary, are no longer sufficient means of
communicating with the public.    It is necessary to establish some sort of
dialogue with the public which is based on reasoned argument.    Such a
dialogue can be established only if the public is given an understanding of
nuclear power technology.    The establishment of this basic knowledge
should begin in the schools, as suggested earlier.

## 9.    CONCLUSION

The ever-increasing complexity of the systems of nuclear
facilities and of the process for licensing these facilities could, if not given
careful consideration, cause many delays in the ordered development of the
nuclear industry.

It is  evident that a great amount of communication, co-
ordination and cooperation is required among the owners of a nuclear
facility, the regulatory bodies and the public in order to ensure that no
unnecessary delays occur, while still ensuring that the facilities are properly sited.

The establishment of guides, the involvement of all
pertinent regulatory bodies in the Board's advisory committees, the continued working-level contact between owners and regulatory bodies throughout the facility's schedule and provision of opportunities for public comment
are only a few examples of the means employed in Canada to ensure that the
facilities will become an accepted part of a community.

## APPENDIX A

Design-Basis Dose Limits For Normal Operation and Accident Situations

| Situation | Assumed Maximum Frequency | Meteorology to be Used in Canada | Maximum Individual Dose Limits | Maximum Total Population Dose Limits |
|---|---|---|---|---|
| Normal Operation | | Local long-term average weather | 0.5 rem/yr whole body 3 rem/yr to thyroid * | $10^4$ man·rem/yr $10^4$ thyroi d· rem/yr |
| Serious Process Equipment Failure | 1 per 3 years | Either worst weather existing at most 10% of time or Pasquill F condition if local data incomplete | | |
| Process Equipment Failure Plus Failure of any Safety System | 1 per $3{\times}10^3$ years | Either worst weather existing at most 10% of time or Pasquill F condition if local data incomplete | 25 rem whole body 250 rem thyroid ** | $10^6$ man·rem $10^6$ thyroid· rem |

* For other organs use 1/10 ICRP occupational values.

** For other organs use 5 times ICRP annual occupational dose (tentative).

## APPENDIX  B


Power Reactor Safety Criteria and Principles


1.  Design and construction of all components,  systems and structures
essential to or associated with the reactor shall follow the best
applicable code, standard or practice and be confirmed by a system
of independent audit.


2.  The quality and nature of the process systems essential to the
reactor shall be such that the total of all serious failures shall not
exceed 1 per 3 years.   A serious failure is one that in the absence
of protective action would lead to serious fuel failure.


3.  Safety systems shall be physically and functionally separate from the
process systems and from each other.


4.  Each safety system shall be readily testable, as a system, and
shall be tested at a frequency to demonstrate that its (time)
unreliability is less than $10^{-3}$.


5.  Radioactive effluents due to normal operation, including process
failures other than serious failures (see  2 above), shall be such
that the dose to any individual member of the public affected by
the effluents, from all sources, shall not exceed 1/10 of the allowable dose to Atomic  Radiation  Workers and the total dose to the
population shall not exceed $10^4$ man·rem/year.


6.  The effectiveness of the safety systems shall be such that for any
serious process failure the exposure of any individual of the population
shall not exceed 500 mrem and of the population at risk, $10^4$ man rem.


7.  For any postulated combination of a (single) process failure and
failure of a safety system, the predicted dose to any individual
shall not exceed (i) 25 rem, whole body, (ii) 250 rem, thyroid,
and to the population, $10^6$ man·rem.


8.  In computing doses in 6 and 7 the following assumptions shall be
made unless otherwise agreed to:

(i)      meteorological dispersion that is equivalent to
Pasquill category F as modified by Bryant [3]

(ii)      conversion factors as given by Beattie [4].

## APPENDIX C

### Exclusion Zone

**Definition**

An Exclusion Zone is an area, specified by the Atomic Energy
Control Board, immediately surrounding a nuclear facility and under the
control of the licensee or the operator.

**Conditions**

1.  There shall be no permanent habitation within the Exclusion Zone.

2.  Use of the land for purposes other than the licensed activities
    shall require separate AECB approval.

3.  Exclusion Zones shall be posted in a manner acceptable to the
    Board.

4.  Radiation safety within the Exclusion Zone is the responsibility of
    the licensee, or, subject to AECB approval, his designate.  Methods
    and measurement for ensuring radiation safety are subject to review
    as required by the Board.

NOTE: For all power reactors licensed to date the Exclusion Zones ex-
      tend from the reactor core to a radius of 3000 feet with the
      exception of navigable waters and minor other exceptions.

### APPENDIX D

### Major Multi-Nuclear-Facility Sites in Canada

1.  Bruce Nuclear Power Development Site

    1.1.  Douglas Point G. S.* - operating - 200 MWe, CANDU - PHWR

    1.2.  Bruce G. S. 'A' - under construction - 4 x 750 MWe, CANDU -
                                                              PHWR

    1.3.  Bruce G. S. 'B' - construction started - 4 x 750 MWe, CANDU -
                                                              PHWR

    1.4.  Intermediate term waste management area

---

* G. S. = generating station.

1.5.   Bruce heavy-water production plants:

- 2 units of 400 t/a capacity each, operating
- 4 units of 400 t/a capacity each, under construction
- 2 units of 400 t/a capacity each, planned

( no release of radioactivity associated with heavy-water
production plants )

2.   <u>Pickering Site</u>

2.1.   Pickering G. S. 'A' - operating - 4 x 500 MWe, CANDU - PHWR

2.2.   Pickering G. S. 'B' - starting construction - 4 x 500 MWe, CANDU
PHWR

3.   <u>Gentilly Site (and environs)</u>

3.1.   Gentilly - 1 NPS** - operating - 250 MWe, CANDU - BLW

3.2.   Gentilly - 2 NPS - starting construction - 600 MWe, CANDU PHW

(more CANDU - PHW units being considered )

3.3.   Intermediate term waste management area

3.4.   La Prade heavy-water production plant - 2 units of 400 t/a
capacity  each,  starting  construction


APPENDIX E

Siting Guidelines For Heavy-Water Plants Utilizing The

Girdler-Sulphide Process


1.   INTRODUCTION

Heavy-water production plants which employ the Girdler-
Sulphide (G. S.) process contain large inventories of hydrogen sulphide
($H_2S$).   $H_2S$ and its combustion product, sulphur dioxide ($SO_2$), are highly
toxic at relatively low concentrations.   Both also have an offensive odour
at extremely low concentrations, and $SO_2$ can be injurious to vegetation.
Radiological safety problems might be significant if these plants are used
to process irradiated heavy water.   However, a complete assessment of
the situation, particularly with regard to tritium, would be required before
such use.

It is necessary to ensure that minor releases of $H_2S$ and
$SO_2$ from the plant into the surrounding environment are controlled and
larger releases prevented, particularly those releases which could have a

** NPS = nuclear power station.

harmful effect on the workers at the plant site and the public in the vicinity of the plant. This is achieved by strict attention to safety considerations in design, commissioning and operation, but there always remains a remote possibility that a dangerous release could occur.

To mitigate the consequences of such a release, the siting of the plant and the provision of appropriate protective measures for the public and workers, including contingency plans, are of primary importance.

## 2. RELEASES INTO WATER

Releases into water occur mainly when the liquid effluent treatment system malfunctions. Provincial and federal regulations strictly limit the allowable concentrations of $H_2S$ in the effluent stream and also the total quantity of $H_2S$ allowed in the particular receiving waters during a specific time period. Also, the heat input to the receiving waters is closely regulated.

## 3. RELEASES INTO ATMOSPHERE

Releases into the atmosphere can result from gas coming out of solution during liquid effluent treatment, from the flare stack in the event of equipment blowdown or as a result of failure of the system pressure envelope.

The two former situations would likely result in only nuisance-type releases, whereas the pressure envelope failure could have serious effects on the surroundings. Released gas would disperse over the countryside, decreasing in concentration as it moved downwind from the plant. For large releases, the effects could be serious for a significant distance from the plant. For this reason, care in site selection and provision of protective measures are necessary.

## 4. SITING GUIDELINES

The following guidelines make use of the experience gained from the safety assessments of the existing heavy-water plants:

4.1. The site should be chosen with a very low population density within the zone of influence of the heavy water plant (e.g. typically rural, about 20 to 30 persons per square mile). Population concentrations such as villages, major service and recreational facilities, particularly overnight camping facilities, schools and hospitals, should be avoided. Municipal plans for future land use should be reviewed by the appropriate governmental agencies and provision should be made to control population growth in this area during the plant lifetime.

4.2. The topography, hydrology and meteorology of the area should be such that there are no abnormal conditions which could result in a substantially increased risk in one sector as opposed to the surroundings general, (e.g. a plant should not be located in a deep valley which would channel a gas release with little dispersion occurring and possibly cause a hazard to a populated area which would otherwise have been sufficiently distant as not to be at risk).

4.3. The possibility of the plant being subjected to external forces which could cause major damage, such as earthquakes, tornadoes, floods or aircraft crashes should be assessed for the sites being considered. Plant design should account for these external forces where appropriate.

4.4. All use of water by the plant for feedwater or for receipt of effluents must be in accordance with the applicable provincial and federal regulations. Both chemical and thermal effects must be considered.

4.5. Effluents released into the atmosphere must be controlled in accordance with applicable provincial and federal regulations.

4.6. The 'Provisions for protection of the population' listed in Table I should be applied (see pages 62 and 63).

4.7. The plant should be surrounded by a Restricted Zone which is an area, specified by the Board, immediately surrounding a heavy-water production facility and under the direct control of the licensee or the operator. The conditions applied to the Restricted Zone are:

4.7.1. Access to the Restricted Zone shall be limited to persons forming part of a Controlled Population.

4.7.2. There shall be no permanent habitation within the Restricted Zone.

4.7.3. Use of the land for purposes other than the licensed activities shall require the prior approval of the Board.

4.7.4. The Restricted Zone shall be posted in a manner acceptable to the Board, including traffic warning signs on all roads entering the Restricted Zone.

4.7.5. Protection from toxic substances within the Restricted Zone is the responsibility of the licensee, or, subject to the prior approval of the Board, his designate. Methods and measurement for ensuring toxicological protection are subject to review as required by the Board.

TABLE I. PROVISIONS FOR PROTECTION OF POPULATION WITHIN ZONE OF INFLUENCE OF A HEAVY-WATER PLANT IN THE EVENT OF GAS RELEASE  (Definitions are given below)

| | Area | Uncontrolled population | Controlled population |
|---|---|---|---|
| A | 0 to 1/2 mile from $H_2S$-containing part of plant | No uncontrolled population is acceptable in this area | (1) Direct warning<br>(2) Training for emergencies<br>(3) Respiratory protection<br>(4) (i) Stay-in procedure with controlled ventilation area  <u>or</u><br>    (ii) Evacuation procedure |
| B | 1/2 to 1 mile from $H_2S$-containing part of plant | No uncontrolled population is acceptable in this area | (1) Direct warning<br>(2) Training for emergencies<br>(3) (i) Respiratory protection  <u>or</u><br>    (ii) Stay-in procedure with controlled ventilation area  <u>or</u><br>    (iii) Evacuation procedure |
| C | 1 mile to x[a] miles from $H_2S$-containing part of plant | (1) Warning mechanism<br>(2) Written information on emergency procedures<br>(3) Stay-in procedure for enclosure with leak-tightness equivalent to a normal residence | (1) Warning mechanism<br>(2) Written information on emergency procedures<br>(3) (i) Stay-in procedure for enclosure with leak-tightness equivalent to a normal residence  <u>or</u><br>    (ii) Evacuation procedure |

[a] Distance x could extend to 5 miles for plants of current design depending on the extent of provisions to mitigate the consequences of a gas release.

DEFINITIONS

(i) 'Controlled population' refers to persons who are under direct supervision of the heavy-water plant management or are under the direct supervision of a management which has made a commitment to co-operate with the heavy-water plant management in the event of an emergency.

(ii) 'Controlled ventilation area' refers to a building which has special controls on the ventilation system and all other openings to atmosphere which would allow immediate closure to provide a relatively leak-tight enclosure in which persons could withstand the passage of a plume of gas from a major release from the heavy-water plant.

(iii)  'Direct warning' means that the controlled population receives a known warning signal from a device which is actuated directly from the control room of the heavy-water plant without any further decisions having to be made by any other group regarding the need for emergency action.

(iv)  'Evacuation procedure' refers to a procedure wherein people move in an orderly fashion to a safe position away from the source of hazard.

(v)  'Respiratory protection' refers to provision of some means of adequately purifying the air breathed by each individual or supplying him with fresh air from uncontaminated sources during the period of exposure to excessive $H_2S$ in the atmosphere. Sufficient resuscitation equipment should also be available.

(vi)  'Stay-in procedure' refers to a procedure wherein people remain in, or enter into, an enclosure close at hand which can be isolated from the source of hazard.

(vii)  'Traffic warning signs' are signs erected on all roads which enter the restricted zone, for the purpose of indicating the action to be taken if a warning is announced.

(viii)  'Training for emergencies' involves instruction in emergency procedures and periodic practices. For plant operating staff and supervisory personnel of subcontractors, this should include first aid and resuscitation training.

(ix)  'Uncontrolled population' includes all persons not covered in the definition of 'controlled population'.

(x)  'Warning mechanism' refers to a means of communicating a prompt and effective warning to the population either directly from the heavy-water plant or by a responsible public authority upon receipt of information from the heavy-water plant.

(xi)  'Zone of influence' is an area within which emergency procedures or other special precautions are necessary due to the presence of the heavy-water production plant. The extent of the zone of influence (out to distance x mentioned in Table) depends on the extent of the provisions taken to mitigate the consequences of a gas release from the plant.

## REFERENCES

[1]   LAURENCE, G.C., BOYD, F.C., JENNEKENS, J.H., SUTHERLAND, J.B., HAMEL, P.E., "Reactor
      safety practice and experience in Canada", 3rd Int. Conf. Peaceful Uses of Atomic Energy (Proc. Conf.
      Geneva, 1964) 13, UN, New York (1965) 317.
[2]   HURST, D.G., BOYD, F.C., "Reactor licensing and safety requirements", Atomic Energy Control Board
      Paper presented at 12th Annual Conf. Canadian Nuclear Association, June 1972, Ottawa.
[3]   BRYANT, P.M., UKAEA Rep. AHSB (RP) R42 (1964).
[4]   BEATTIE, J.R., UKAEA Rep. AHSB(S) R64 (1963).

## DISCUSSION

A. MERTON: You have mentioned that public opposition to the
acceptance of nuclear power is growing in Canada. Can you give any
reasons for this? Are the arguments peculiar to the situation in Canada,
or are they the same as those advanced in other countries encountering
such opposition?

R.M. DUNCAN: The arguments are generally the same as those put
forward in other countries, although the opposition is possibly not as
organized or as vocal; but the concerns expressed are the same.

E.F.F.W. USHER: In view of the vast quantity of feedwater required
for heavy-water production plants, water supply is a major siting consideration. What are the chemical and biological requirements placed on the water
supply? If the water is drinkable, is it still of potable quality when it has
passed through the plant and returned to source?

R.M. DUNCAN: The feedwater must have an acceptably high deuterium
content to make the process economic, and a very low level of normal
contaminants, particularly organic material; this latter requirement is
aimed at minimizing any problems which might upset the process.

The effluent from the plant is purified to the extent that the hydrogen
sulphide content is reduced below the level where it could have any effect on
the most sensitive aquatic life; this is the limiting factor.

Pamela M. BRYANT: I should like to ask Mr. Duncan what the basis is
for the population dose limit of $10^4$ man·rem/a, whether it applies to any
plant regardless of size, and whether there is a threshold below which the
dose rate need not be regarded as a contributor to population dose.

R.M. DUNCAN: This value was chosen in the light of a consideration of
somatic as well as genetic effects and on the assumption of a linear dose-
effect relationship, as suggested in the 1964 report of the United Nations
Scientific Committee on the Effects of Atomic Radiation (particular attention
being given to the incidence of leukaemia). But the man·rem values are
currently under review and further information should be available in 1975.
The limit does apply to all plants regardless of size.

As to the last question, the integration should extend over all areas
outside the exclusion area in which the individual external dose exceeds
$5 \times 10^{-3}$ rem in one year.

C.F.G. DELANEY: I should like to ask what problem has been particularly stressed in public questioning in Canada — small continuing releases of
activity, major releases or reprocessing?

R.M. DUNCAN: There has been no concern about reprocessing plants
as yet. Concern has been expressed about the environmental consequences
of continuous minor releases, and the more 'philosophical' question of the

need for nuclear power in general has been aired, attention being given alike
to the possible genetic effects of chronic exposure to low levels of radiation
and the extreme somatic effects of acute exposure to large doses from highly
improbable accidents.

D. STOIAN: What extra precautions are taken for the construction of
power plants in seismic zones, and how do these measures affect the cost
of the plants?

R. M. DUNCAN: The National Building Code of Canada specifies the
seismic considerations to be taken into account in various parts of the country,
and these are currently applied to the structures of nuclear plants.

One of the Canadian bodies concerned with standards is drafting a set of
guidelines which will prescribe a more detailed approach to seismic problems
for the nuclear plants of the future. The first draft of these guidelines is
nearing completion.

A. DE ACHA ARACAMA: What is the public information period in
Canada?

R. M. DUNCAN: The period between preliminary site approval and final
site approval, during which the applicant for a licence must make a public
announcement regarding his intention to build a plant at a particular site and
also allow for public response, is about four months on average.

H. A. G. KOHLER: How long do the site approval and construction licence
procedures take in Canada, from submission of the application to the issue of
the approval or licence?

R. M. DUNCAN: From the time when the applicant submits his letter of
intent to build a facility, which is accompanied by a site-evaluation report,
about three months normally elapse before preliminary site approval is
granted. Thus about six to seven months would pass between the submission
of the letter of intent and the granting of final site approval. At about this
latter time the Preliminary Safety Analysis Report would be presented, and
some fifteen months after the letter of intent, construction authorization would
be issued.

A. P. HULL: Your paper contains a reference to a 'design target' of 1%
of the derived release limits for normal operation. To what extent does this
correspond to the United States of America 'Appendix I' limits, and has it
made additional design features necessary which would otherwise not have
been required?

R. M. DUNCAN: In Canada, the Atomic Energy Control Regulations take
into account the recommendations of the ICRP, and the recommendation that
doses should be 'as low as is readily achievable' serves as a guide in the
licensing of nuclear plants. In keeping with this recommendation, design
targets have been set to keep doses to the public from nuclear plant effluents
below 1% of the statutory dose limits. This target value is consistent with
the measured releases from the Pickering Generating Station. The value does
compare favourably with the 1% value published in the United States of America,
but was not derived from it.

D. DAGAN: Is co-operation between federal, provincial and local bodies
based simply on mutual agreement or is their relationship regulated by law?

R. M. DUNCAN: The Atomic Energy Control Act, a federal law, takes
precedence over other federal and provincial legislation. However, the
Atomic Energy Control Board has declared that a licensee must comply with
all other pertinent legislation which, were it not for the Atomic Energy Control
Act, would apply, as long as it is not in conflict with that Act. A practical

way of co-ordinating all these regulatory schemes is through a committee
such as the Reactor Safety Advisory Committee.  Good co-operation between
the different bodies does depend, of course, on mutual understanding.

# SAFETY POLICY FOR SITING NUCLEAR POWER STATIONS IN THE FEDERAL REPUBLIC OF GERMANY

K.P. BACHUS, H. SCHNURER
Federal Ministry of the Interior,
Bonn, Federal Republic of Germany

## Abstract

SAFETY POLICY FOR SITING NUCLEAR POWER STATIONS IN THE FEDERAL REPUBLIC OF GERMANY.
Sites for nuclear power stations must meet various requirements  related, e.g., to economy, environmental protection, nuclear safety or regional planning.  So far, new sites in the Federal Republic of Germany have been selected by the utilities and have to be licensed by different responsible authorities at federal, Länder or lower level.  There are serious problems in selecting suitable sites in such a densely populated country.  The Federal Government's energy programme, however, requires up to thirty new sites for large nuclear power plants within the next five or six years.  To avoid delays in licensing procedure and at the same time  to select optimal sites from limited possibilities, a long-term policy of making provisions for sites is under development.  This will  include (a) an optimal site preselection process which takes relevant aspects of sites into account, and (b) a process to guarantee a later license as far as possible.  It is intended to develop an abundant stock of sites from which sites can be chosen as necessary, combined with a standardized nuclear power station concept.  This will allow a construction permit to be issued much sooner than at present. As a tool for facilitating site planning, safety-orientated site evaluation data are under development.  It is not intended to establish any stringent licensing criteria but to offer site evaluation data  which will indicate for each safety-related site property the trend which site planning should follow.  This will be achieved by three categories of data for judging individual site properties.  Due to the far-reaching actual or potential influence of a nuclear power station, certain information and co-ordination mechanisms are needed in order to harmonize site planning in frontier regions.

## 1. INTRODUCTION

The nuclear age began with the establishment of small research and experimental reactors most of which were built in remote areas.  Although these installations were at first looked at in wonder as new technological miracles, the wide public showed little interest in them.  Today numerous large nuclear power stations are being built in many countries, and they are of greater visual significance within the landscape and in their effects upon the environment.  They also arouse great attention from the public.

## 2. PREVAILING SITING PRACTICE

The planning of sites for nuclear power stations in the Federal Republic of Germany (FRG) has hitherto been carried out by energy supply companies organized on the principles of private industry.  In selecting locations for their facilities, these companies take into account the possible problems that may arise from subsequent licensing procedures, but primarily they make their choice in accordance with criteria of energy resource management, which include in particular:

## TABLE I.   POPULATION ROUND NUCLEAR POWER PLANTS IN THE FEDERAL REPUBLIC OF GERMANY; NUMBER OF INDIVIDUALS WITHIN CIRCULAR BELTS

| Nr. | | 0 - 1 km | 1 - 2 km | 2 - 3 km | 3 - 4 km | 4 - 6 km | 6 - 8 km | 8 - 10 km | 10 - 15 km | 15 - 20 km |
|---|---|---|---|---|---|---|---|---|---|---|
| 1 | BASF-Mitte | | 7 000 | 53 200 | 79 500 | 170 400 | 152 200 | 99 300 | 254 600 | 297 700 |
| 2 | Biblis | | 500 | 1 760 | 5 355 | 16 225 | 24 350 | 72 000 | 102 730 | 203 200 |
| 3 | Breisach | | | 926 | 178 | 7 512 | 7 730 | 3 478 | 29 034 | 51 733 |
| 4 | Brokdorf | 160 | 403 | 316 | 748 | 4 005 | 4 802 | 30 554 | 57 978 | 46 532 |
| 5 | Brunsbüttel | 54 | 1 089 | 2 421 | 3 684 | 6 123 | 4 758 | 5 461 | 33 019 | 54 547 |
| 6 | CWH | 30 | 2 050 | 11 250 | 32 290 | 45 480 | 20 290 | 69 680 | 388 600 | 492 500 |
| 7 | Grafenrheinfeld | 20 | 461 | 8 735 | 1 739 | 16 885 | 34 217 | 47 507 | 46 573 | 38 617 |
| 8 | Grohnde | | 2 320 | 3 562 | 1 322 | 5 680 | 25 125 | 41 416 | 69 476 | 59 87 |
| 9 | Gundremmingen | 24 | 960 | 1 000 | 1 918 | 14 340 | 15 228 | 58 860 | 53 190 | 71 55 |
| 10 | Jülich | 132 | 1 975 | 2 046 | 5 711 | 25 322 | 19 249 | 37 910 | 196 960 | 283 99 |
| 11 | Kahl | 724 | 11 615 | 26 275 | 7 295 | 24 730 | 43 734 | 75 834 | 256 360 | 540 16 |
| 12 | Karlsruhe | | 51 | 3 189 | 13 467 | 15 402 | 38 521 | 129 286 | 281 000 | 202 53 |
| 13 | Krümmel | 373 | 1 714 | 1 030 | 12 684 | 13 615 | 4 873 | 13 907 | 72 262 | 182 19 |
| 14 | Lingen | 30 | 572 | 2 730 | 6 228 | 27 713 | 8 578 | 6 515 | 27 328 | 84 00 |
| 15 | Mülheim-Kärlich | 27 | 5 963 | 24 050 | 11 181 | 37 278 | 74 860 | 71 516 | 127 184 | 107 1 |
| 16 | Neckarwestheim | | 5 438 | 3 935 | 8 866 | 29 282 | 17 707 | 55 592 | 232 051 | 238 7 |
| 17 | Niederaichbach | 587 | 1 298 | 826 | 1 515 | 4 800 | 8 650 | 8 600 | 106 370 | 46 9 |
| 18 | Obrigheim | 150 | 3 993 | 3 095 | 7 188 | 18 235 | 9 554 | 15 649 | 67 145 | 132 7 |
| 19 | Philippsburg | 4 | 3 152 | 5 334 | 10 810 | 8 803 | 67 338 | 27 085 | 83 028 | 275 4 |
| 20 | Schmehausen | 282 | 903 | 1 090 | 1 676 | 6 250 | 13 819 | 53 613 | 183 343 | 126.6 |
| 21 | Stade | 100 | 400 | 100 | 2 650 | 37 285 | 3 995 | 11 295 | 86 355 | 235 3 |
| 22 | Unterweser | 50 | 150 | 1 231 | 4 426 | 8 984 | 16 269 | 24 127 | 77 854 | 60 0 |
| 23 | Würgassen | 18 | 2 240 | 5 765 | 5 460 | 7 280 | 7 280 | 12 443 | 40 508 | 76 2 |
| 24 | Wyhl | | 10 | 2 420 | 1 800 | 3 065 | 9 966 | 12 147 | 46 045 | 61 3 |

TABLE II.  POPULATION ROUND NUCLEAR POWER PLANTS IN
THE FEDERAL REPUBLIC OF GERMANY;  NUMBER OF INDIVIDUALS
WITHIN RADIUS

| Nr. | | 0 - 1 km | 0 - 2 km | 0 - 3 km | 0 - 4 km | 0 - 6 km | 0 - 8 km | 0 - 10 km | 0 - 15 km | 0 - 20 km |
|---|---|---|---|---|---|---|---|---|---|---|
| 1 | BASF-Mitte | | 7 000 | 60 200 | 139 700 | 310 100 | 462 300 | 561 600 | 816 200 | 1 113 900 |
| 2 | Biblis | | 500 | 2 260 | 7 615 | 23 840 | 48 190 | 120 190 | 222 920 | 426 120 |
| 3 | Breisach | | | 926 | 1 104 | 8 616 | 13 073 | 16 551 | 45 585 | 91 297 |
| 4 | Brokdorf | 160 | 563 | 879 | 1 627 | 5 632 | 10 434 | 40 988 | 98 966 | 145 498 |
| 5 | Brunsbüttel | 54 | 1 143 | 3 564 | 7 248 | 13 371 | 18 129 | 23 590 | 56 606 | 111 156 |
| 6 | CWH | 30 | 2 080 | 13 330 | 45 620 | 91 100 | 111 390 | 181 070 | 569 670 | 1 062 170 |
| 7 | Grafenrheinfeld | 20 | 481 | 9 216 | 10 955 | 27 840 | 62 057 | 109 564 | 156 137 | 194 754 |
| 8 | Grohnde | | 2 320 | 5 882 | 7 204 | 12 884 | 38 009 | 79. 425 | 148 901 | 208 772 |
| 9 | Gundremmingen | 24 | 984 | 1 984 | 3 902 | 18 242 | 33 470 | 42 330 | 145 520 | 217 070 |
| 10 | Jülich | 132 | 2 107 | 4 153 | 9 864 | 35 186 | 54 435 | 92 345 | 289 305 | 573 295 |
| 11 | Kahl | 724 | 12 339 | 38 614 | 45 909 | 70 639 | 114 373 | 190 207 | 446 567 | 986 727 |
| 12 | Karlsruhe | | 51 | 3 240 | 16 707 | 32 109 | 70 630 | 200 116 | 481 116 | 683 646 |
| 13 | Krümmel | 373 | 2 087 | 3 117 | 15 801 | 29 416 | 34 289 | 48 196 | 120 458 | 302 653 |
| 14 | Lingen | 30 | 602 | 3 332 | 9 560 | 37 273 | 45 851 | 52 366 | 79 694 | 163 694 |
| 15 | Mülheim-Kärlich | 27 | 5 990 | 30 040 | 41 221 | 78 499 | 153 359 | 224 875 | 352 059 | 459 243 |
| 16 | Neckarwestheim | | 5 438 | 9 373 | 18 239 | 47 521 | 65 228 | 120 820 | 352 871 | 591 651 |
| 17 | Niederaichbach | 587 | 1 885 | 2 711 | 4 226 | 9 026 | 17 676 | 26 276 | 132 646 | 179 566 |
| 18 | Obrigheim | 150 | 5 031 | 7 238 | 14 426 | 32 661 | 42 215 | 57 864 | 125 009 | 257 754 |
| 19 | Philippsburg | 4 | 3 187 | 8 490 | 19 300 | 28 103 | 95 441 | 122 526 | 205 554 | 480 982 |
| 20 | Schmehausen | 504 | 1 771 | 2 286 | 3 962 | 10 212 | 24 031 | 77 644 | 260 987 | 387 593 |
| 21 | Stade | 100 | 750 | 1 100 | 3 250 | 40 535 | 44 530 | 55 825 | 142 180 | 377 560 |
| 22 | Unterweser | 50 | 300 | 1 431 | 5 907 | 14 891 | 31 170 | 55 297 | 133 151 | 157 749 |
| 23 | Würgassen | 18 | 3 114 | 8 023 | 13 483 | 21 469 | 28 749 | 41 192 | 81 700 | 157 930 |
| 24 | Wyhl | | 10 | 2 430 | 4 230 | 7 295 | 17 261 | 29 508 | 72 553 | 136 803 |

Availability of cooling water;
Favourable location of feed-in points into the electric power grid;
Nearness to the focal points of electric energy consumption; and
Availability of suitable land.

In the FRG, it has not yet been possible to grant approval for the
location of a nuclear power station without detailed specification of the
facility to be established. In practice, therefore, the energy supply
companies undertake detailed planning of the site and power plant to be erected
before they file an application for a licence under atomic legislation for the
site and construction of the facility with the licensing authority of the federal
Land in which the power station is to be built. In addition, there is the
possibility of issuing a preliminary siting decision which, in the event of
approval, gives the energy supply company an assurance that it may at a later
time build a nuclear power station specified as to type, size and safety
concept, but which does not imply any right to take up building operations.
Since a preliminary siting decision of this kind has a maximum validity of
only four years, little use is made of it.

The licensing authority responsible under atomic law examines a site
applied for and the installation to be built on it with regard to nuclear safety
and radiation protection, i.e. it assesses the radiological impact of an
installation upon the surrounding area and the possible effects of the
immediate environment upon the installation. In the field of nuclear
technology, the licensing authority of the Land is subject to supervision by
the Federal Government, represented by the Federal Minister of the Interior.
The Federation thus exercises a function of co-determination on the approval
or rejection of a site both legally and in practice.

Since a nuclear power station must comply with numerous other
regulations apart from the nuclear aspects, it is obvious that a number of
further approvals are needed for the final acceptance of a site, in particular
with regard to the protection of the environment, landscape and nature, as
well as the integration of the facility in regional (Länder) planning schemes.
Thus the approval of a site depends on the consent of a large number of
authorities.

## 3. SITING PROBLEMS

Siting problems are encountered much more frequently in the Federal
Republic of Germany than in other countries because the conditions there are
less favourable than, for example, in the United States of America, France
or the United Kingdom. In the first instance, the FRG lacks the necessary
coastal areas and rivers suitable for cooling purposes. Along the few large
rivers available, the built-up areas, densely populated as they are, have
developed into vast urban agglomerations (see Tables I and II). Moreover,
riverside locations near focal points of consumption, though much coveted
under the aspects of energy resources management, are in most cases
unfavourable for reasons of environmental protection, landscape conservation
and safety.

The decision taken by a licensing authority on an application for site
approval may only be either 'yes' or 'no'. As a rule, relocation of a planned
installation to a potentially more favourable site is not possible within
the licensing procedure. It requires a rejection of the originally
delimited site and a new procedure. This entails special difficulties.

The rejection of a site means considerable economic disadvantage to the applicant, since at the time of the application great effort and expenditure will have been made (premises purchased, concept for the installation worked out, subcontracts awarded). On the other hand, expert opinions which take a long time to obtain (e. g. of a meteorological nature), as well as appeals and actions in rescission before the courts at such a late stage, cause delays and thus lead to an economically disadvantageous increase of costs, possibly also to risks in energy supply. It has become a difficult matter to accept a given location against the resistance of a very critical public. Mass protests by as many as 100 000 objectors from the area surrounding a site naturally also involve political problems.

However, a large number of sites for nuclear power stations will be needed to implement the FRG's energy programme. Nuclear power stations with a total electric output of 20 000 MW are to be in operation in the FRG by the year 1980. The availability of an installed nuclear power station output of 50 000 MW is scheduled for 1985. This means that in the next few years until 1980 the sites for some thirty new nuclear power station units with up to 1300 MW electric output will have to be planned and licensed. About fifty nuclear power stations are to be in operation or under construction in the year 1980.


## 4.  PLANNING OBJECTIVES

In the further development of the technical design of nuclear power stations as such, it appears possible only to a limited extent to take into account all the important aspects of a given site. In addition, this would be contrary to the desired standardization of installations, unless it were intended to standardize all nuclear power stations under the assumption of the most unfavourable conditions at the site. However, this would be contrary to the principles of economy and energy resources management. For this reason it would be inexpedient to develop a type of power station completely independent of the respective site, which would meet all requirements regarding safety, environmental protection etc. from the very beginning and could be installed in greater numbers at any location, even the most unfavourable. This does not affect the possibility of developing specific standardized types of power stations for different appropriate categories of sites.

Instead of such approaches, long-range provision should be made for the selection of sites. For this purpose, all bodies concerned, in particular the energy supply companies and the responsible authorities, should co-operate to determine optimal locations within larger regions by way of a suitable procedure taking all important aspects into account. Thus it will be essential to select the most favourable location from a larger number of potential sites under consideration. Such policy should be planned in advance over a longer period according to the demand for sites and would at the same time provide an opportunity to arrange for the necessary expert opinions to be made and to carry on the dialogue with the population affected without any time pressure.

The ideal approach would be to determine a number of locations agreed upon in advance and legally secured against other uses, and to hold them 'in stock'. Should there be a demand for energy at the regional level, it would then be sufficient to take such a location 'out of the drawer' and combine it

with a standardized type of nuclear power station similarly agreed upon in
advance.   Apart from permitting an optimal selection of sites within the
practical limits existing in a Land, this policy would be likely to shorten the
licensing procedures.


## 5.   SOLUTION BY MAKING PROVISION FOR SITES (Standortvorsorge)

Making early provision for sites may already be an effective means of
mitigating the conflict of objectives between energy supply and environmental
protection by an optimal selection of locations for energy supply facilities,
due account being taken right from the beginning,and to an equal extent,of the
aspects of environmental protection and energy resources management.
Making provision for sites must consist of the planning of locations and of
legally securing them against other uses.

### 5.1.   Selection of sites

The selection of sites should cover a longer period, e. g. up to the
year 2000, and should take into consideration not only those locations needed
for energy supply but also other potential sites.   This is to provide
alternatives for decision which would allow the establishment of priorities in
terms of time.   For the time being, linear planning along the large rivers
will be necessary on account of cooling-water requirements.   In the long run
however, when it will be possible to use dry-cooling towers, planning is to
extend to areas away from surface waters.   The individual sites would have
to be selected and precisely delimited within location zones.   It appears
expedient to extend the planning to other power stations and to consider not
only nuclear facilities.   As planning criteria, basic values in the form of
rough estimates should be made available, so that it will be possible, even
during the preselection of sites, to take largely into account those aspects
which will play an essential part in the ultimate licensing procedure.   In the
light of the subject of this paper, I shall later refer to such basic planning
values relating to reactor safety and radiation protection.


### 5.2.   Securing sites against other uses

The main purpose of securing a site in this context should be to ensure
that sites, once they have been selected, will not be put to any other use
and to influence the development of the surrounding area in such a way that
the properties of a site will not subsequently deteriorate to any considerable
extent.   In the framework of such a policy the necessary investigations shou
be made and co-ordination with all bodies and agencies concerned should be
sought.   After clarification of all essential and basic problems, the remaini
sites should be reserved for later use, e. g. by means of regional planning.


## 6.   PLANNING AIDS

In selecting locations for nuclear power stations many different objecti
have to be reconciled;  they concern the following fields:

(a) <u>Reactor safety and radiation protection:</u>
    Effects of the nuclear power station on the environment under normal
    operation and in case of incidents, and notable risks to the
    installation caused by existing or future effects emanating from the
    surrounding area;
(b) <u>Energy resources management and supply:</u>
    Location close to consumer, cooling possibilities, linkage to power
    grid, availability of land, accessibility;
(c) <u>Environmental protection:</u>
    Water pollution control, air quality management, noise abatement,
    landscape and nature conservation;
(d) <u>Regional planning:</u>
    Integration of the installation in the surrounding area and its
    influence on the development of this area.

In the important field of the technical precautions taken for reactor safety,
site evaluation data are now being compiled which are to provide rough
estimates in the form of basic planning values for the selection of sites while
minimizing nuclear risks.  It is not at present intended to define any firm
criteria for the licensing of sites.  Owing to the variety of site properties to
be considered, there would be reason to fear that stringent and effective
criteria could hardly be met in every respect at a given site.  On the other
hand, less stringent criteria would not be a means of ensuring that only the
most suitable sites would be chosen.

The overall safety of a nuclear power station is based on the properties
of the installation and those of the site.  In this connection account should be
taken of the normal operation of the installation, of the possibility of incidents
which may even take the form of design-basis accidents, and of a residual
risk beyond that stage, a risk which, though very small, is contingent upon
extremely improbable but not completely excludable accidents by which
radioactive material might be released into the environment.  Primarily on
account of this residual risk, it has become an international practice to apply
special standards in assessing the suitability of sites for nuclear power
stations.  So long as no comprehensive experience has been gained in the
operation of large-scale nuclear power stations, it would not appear advisable
to waive specific safety requirements for sites without any compelling reason.
However, since hypothetical accidents of this kind are most unlikely, it does
not seem necessary to exclude specific sites from the very beginning; it
should rather be sufficient, as a first approach, to make use of the siting
resources of a country by selecting those sites which are more favourable
from the point of view of safety.

Consequently, evaluation data in respect of the following properties of
a site are now being compiled:

Meteorology;
Distribution of the population and emergency action plans;
Hydrology;
Geology;
External impact;  and
Transport problems

It is not proposed to lay down any strict limits for these properties but to
indicate a certain preferential trend for action.  This is achieved by dividing
each property into three categories:

Category 1:     favourable property
Category 2:     conditionally favourable property
Category 3:     unfavourable property

The following is to be expected in the course of a licensing procedure for sites having properties in the categories 'conditionally favourable' and 'unfavourable':

(a)  Special verifications and expert opinions will be required;
(b)  It must be possible to compensate the unsatisfactory property by other favourable properties or by additional technical means;
(c)  In the case of properties in category 3, the expenditure to be anticipated (costs and/or delays) is considerably greater than in category 2.

Site properties found to be unfavourable may lead to the rejection of the entire site if no compensation is possible.

It is not proposed to undertake an overall evaluation of a site but only to offer an evaluation pattern.  This is to create an incentive for the planning authorities and bodies to look for sites possessing, where possible, the properties referred to under category 1.  Linked with this is a warning against sites with category 3 properties.

Bearing in mind the procedure of providing rough estimates, the selection of a site possessing exclusively favourable or conditionally favourable properties does not imply any entitlement to a licence.  Properties under category 1, however, do not give reason to expect any special licensing problems, unless particular features of the site have remained unnoticed during the broad assessment, features which, e. g. , might not be discovered until a thorough analysis of the soil has been made.

In the first stage, such site evaluation data are being compiled for light-water reactors, which are the most eligible nuclear facilities in the FRG. After a certain experimental phase, it will be possible to extend the survey to other nuclear power stations or installations.  Since evaluation data of this kind take account of the present safety standard, a periodical review would be advisable.

In the light of medium-term planning, it appears reasonable to orientate such site evaluation data along the lines of a nuclear risk concept.  Finally, it should be noted that similar evaluation data would also be useful for the remaining non-nuclear areas.  In this connection, attention should be drawn to the thermal load plans for rivers which are already available or are being prepared as well as to the 'immission cadastres'.


## 7.   HARMONIZATION OF ACTION TO MAKE PROVISION FOR SITES

In accordance with the federative structure of the FRG, responsibility for implementing measures designed to make provision for sites rests with the Länder.  Most of the Länder have initiated appropriate action and have begun to select sites on a long-term basis in advance, especially for nuclear power stations, and secure them against any other uses.  Along the borders between these Länder there are problems of co-ordination of the various planning schemes, in particular where a river forms the common border. In the field of nuclear energy, the establishment of site evaluation data

applicable within the whole federal territory is of great significance in view of
the necessary harmonization.

In principle, there is a similar need for co-ordination of the various
siting schemes in relation to neighbouring countries.  This is of particular
importance at the international level, since conditions in neighbouring
countries often differ, thus leaving more scope for the selection of sites,
e. g.  with regard to the admissibility of specific population densities.  The
problem is rendered more difficult by the fact that, in the sector of nuclear
energy, the pressure of public criticism results in a certain compulsion to
implement also in our own national sphere the most stringent safety
requirements existing in any other country.  Intensive international contacts
are therefore especially necessary in view of the principles and practices of
siting.  The FRG is prepared to make available its practices and experience
in the selection of sites for nuclear power stations in discussions with
neighbouring and other countries.


## 8.   FUTURE OUTLOOK

In the Federal Republic of Germany a number of specific research and
development projects are being carried out which appear to be a promising
approach to the improvement of the critical siting situation in the densely
populated federal territory.  Today the lack of suitable surface waters
already makes it necessary, with but few exceptions, to use cooling towers.
Wet-cooling towers are now in operation  but, owing to shortage of water,
the installations to be built in the early 1980s will be equipped with dry-
cooling towers.  This will in principle open up a possibility of selecting
sites at a distance from rivers.  On-shore or off-shore concepts for the
siting of nuclear power stations (e. g. on floating platforms or artificial
islands) are being studied in order to examine the possibilities of developing
new siting regions.

Finally we should not leave unmentioned the fact that additional safety
precautions for nuclear power stations are being discussed in the FRG (e. g.
reactor concept with burst-protected primary circuit, underground siting),
which might possibly make sites in areas of agglomeration acceptable from
the point of view of safety.  Such considerations should also be seen in
connection with multipurpose facilities for producing electricity as well as
process steam or thermal energy for district heating.  In principle, systems
of this kind appear to be a suitable means of curbing the rapid growth of the
primary energy demand, of mitigating the enormous problem of waste heat
affecting large power stations with several blocks, and thus of helping to
facilitate the selection of sites for such installations.


## DISCUSSION

D.C. COOPER:  One of the difficulties I foresee in your proposal that
nuclear sites be 'earmarked' several decades in advance is that the sites
might very well be used for different purposes.  This would be particularly
true if there were changes in land use in the vicinity of the site once it had
been earmarked.  Comprehensive land-use planning measures seem to me to

be the only way to ensure with some degree of certainty that sites designated
as being suitable would remain suitable in the future.

H. SCHNURER:  I fully agree with your comment.  My proposal that
selected sites be secured in advance by means of land-use planning will rule
out other uses of suitable sites and prevent deterioration of the surrounding
area in a manner which would make it impossible to obtain a licence at a
later date.

S.O.W. BERGSTRÖM:  Tables I and II do not give the census years for
the population figures.  It seems that such figures should apply to the
maximum extrapolated population density during the lifetime of the plant.
I should like to hear your comments about that.

H. SCHNURER:  Tables I and II give examples of population density at
sites where a nuclear power station is in operation, under construction, or,
in a few cases, at the stage of preliminary discussion (but not agreed upon).
The tables are designed to show the types of sites we have to consider in our
densely populated country.  (The site with the highest population density so
far approved by the Federal Government is No. 15 at Mülheim-Kärlich.)
The figures relate to the time of application for a licence.  No allowance has
been made for future growth of population in the surrounding area.  Such
extrapolations are evaluated in the licensing process.

J. PELSER:  Is it intended to explore the entire territory of the Federal
Republic of Germany for possible sites or only areas along the main rivers?
Who makes the preliminary selection, and on what criteria is this selection
based?

H. SCHNURER:  In the long run the entire territory will be examined
for possible sites.  During the next five to ten years, however, sites along
the rivers will be needed, since dry-cooling towers have still to be developed.
Hence, priority is now being given to sites on or close to the larger rivers and
on or near the coast;  sites located some distance away from surface water
are at the moment accorded only secondary importance.

T.F. SÖRENSEN:  Do the figures in your tables include the populations
in neighbouring countries or do you take into account only the population in
the Federal Republic of Germany?

H. SCHNURER:  The populations across the borders have been included,
at least as estimates, in the evaluation of population density.  Co-operation
with the competent authorities of the neighbouring countries will naturally be
valuable as a means of obtaining more exact information about population
density — figures which can be used in deciding on the acceptability of a site
in a border region.

M. NELKEN:  Could I ask you to comment on sites 1, 6 and 11 (Tables I
and II), which are situated in high-population areas?  What limitations are
placed on installed capacity at the sites considered?

H.  SCHNURER:  No. 11 refers to the Kahl nuclear power station, which
was the first such station to be built in the Federal Republic of Germany.
It went into operation in 1963, and has a rather low capacity:  25 MW(e).
It is not intended to take Kahl as a reference point for the acceptability of
population density around a large nuclear power station.  As a matter of fact
an application for a new large nuclear power station close to Kahl has not yet
been approved, mainly because of thermal pollution problems.

No. 1 refers to the BASF project involving a nuclear power plant within
a large chemical plant for process steam supply and power generation.  It is
very close to the densely populated areas of Ludwigshafen and Mannheim.

The applicant has offered additional safety systems to compensate for the high population density (which limits the scope for emergency actions) and for hazards to the reactor from the chemical plant. However, the Federal Government has not yet decided whether the scheduled safety precautions can be accepted as sufficient for that site. For several years a detailed research and development programme has been under way with a view to evaluating these operations. The project appearing as No. 6 is being planned by the Chemische Werke Hüls (CWH) at Marl. The open decision about the BASF project has a bearing on other similar projects like CWH, licences for which will be sought if and when a positive decision is taken on the BASF site.

In reply to your second question: in view of the scarcity of suitable sites in the Federal Republic of Germany, we assume that licensed sites will be used for more than one unit; let us say for up to four units, each with a capacity of about 1300 MW(e). The limiting factor will be cooling capacity, even when wet-cooling towers are used. From the radiological standpoint, we follow the concept of limiting the dose to the individual at the worst point in the surrounding area to a maximum value of 30 mrem/a (whole-body dose). This includes the dose burdens due to other sources in the surrounding area and to effluents from several units at one site.

E. H. HUBERT: If a utility makes its preliminary assessment and earmarks sites for possible construction in ten or twenty years' time, how can it be sure that the sites will not be used for something else? Would the utility buy the sites or would the Federal Ministry of the Interior give some kind of guarantee?

H. SCHNURER: In my opinion, land-use planning is adequate as an instrument for securing selected sites against other uses. The inclusion of nuclear power station sites in regional development plans has to be carried out by the Länder (States) in the Federal Republic, and will involve at least a preliminary agreement of all responsible authorities concerned.

Purchase of the designated land by the utilities should not be mandatory, though that would, of course, be the strongest means of securing a site against other uses.

Gloria CAMPOS VENUTI: What are the maximum total population dose limits in the Federal Republic of Germany?

H. SCHNURER: The licensing authorities limit the maximum dose to the individual to 30 mrem/a (whole body) and 90 mrem/a (thyroid). This has to be seen as an emission value at the worst place in the area surrounding a site, and takes into account all possible sources of radiation (e. g. several units at one site or in one region). The Radiation Protection Decree now under review would extend this concept to other nuclear facilities like reprocessing plants and nuclear research centres incorporating different facilities.

D. E. ANDERSON: Are there any statutory regulations regarding the ownership and operation of nuclear power stations in the Federal Republic of Germany? Is ownership limited to electrical utilities with the responsibility for power supply to a particular region of the country, or can any private company be licensed to construct and operate a nuclear power station for its own use?

H. SCHNURER: Licences are not restricted to utilities producing electricity but can also be issued to other industries which intend to operate a nuclear power station, e. g. the chemical industry for process steam supply. Under our Atomic Energy Act, however, a licence may be issued

only if the owner and operator of the plant can prove that he has the necessary
technical know-how (as well as the financial ability) to operate the plant in a
safe manner.

Y. NISHIWAKI: In Europe it is often the case that a country is bounded
by several neighbours and that a large river flows through a number of
countries. The site most favourable to one country may not always be
favourable to another. Are efforts being made in the Federal Republic of
Germany to co-ordinate with other countries the selection of sites for nuclear
power stations or other nuclear facilities such as reprocessing plants? When
nuclear sites close to a frontier are being considered, is it necessary under
the present regulations to obtain official approval from the neighbouring
country? I think it would be important to organize a regional committee to
consider long-range siting problems which could affect several countries in
order to minimize any possible friction and maintain smooth relations.

H. SCHNURER: You have raised a very important point. In our country
problems may arise when a river forms the frontier with a neighbouring
country, as does the Rhine with Switzerland and France, and the Elbe with
the German Democratic Republic, or when a river flows to another country,
as the Danube to Austria and the Rhine to the Netherlands. We maintain
bilateral contacts with some of our neighbours, especially Switzerland,
France and the Netherlands, exchanging information on safety and radiation
protection. It would, of course, be useful for all parties to intensify such
contacts and include information on the siting of nuclear facilities. However
official approval of a neighbouring country for the licensing of a site in one's
own country is not necessary under the present regulations.

A positive development worth mentioning is that there now exists an
international commission to keep the Rhine clean, on which all countries
along the Rhine are represented and which has prepared recommendations
with a view to preventing unacceptable thermal pollution of the river by
further power stations.

# КРИТЕРИИ ВЫБОРА ПЛОЩАДОК ДЛЯ ИССЛЕДОВАТЕЛЬСКИХ ЦЕНТРОВ, ЭНЕРГЕТИЧЕСКИХ УСТАНОВОК И УСТАНОВОК ТОПЛИВНОГО ЦИКЛА В СССР

Н.П.ДЕРГАЧЕВ
Государственный Комитет по использованию
атомной энергии СССР,
Москва,
Союз Советских Социалистических Республик

**Abstract—Аннотация**

CRITERIA FOR THE SITING OF RESEARCH CENTRES, POWER PLANTS AND FUEL-REPROCESSING FACILITIES IN THE USSR.
Given the unavoidable necessity of releasing some radioactive isotopes into the atmosphere and open water-bodies, nuclear facilities represent a potential hazard to the environment and the population. This paper considers problems associated with keeping the gaseous and liquid radioactive wastes released to the external environment at permissible levels by means of a comprehensive set of measures which rely not only on the recovery and purification of wastes but also on correct siting of facilities, so as to ensure the safety of the population both in normal operation and in emergency situations. The establishment of a health protection zone around a nuclear facility reduces to an insignificant level the radiation hazard to the population living in the vicinity.

КРИТЕРИИ ВЫБОРА ПЛОЩАДОК ДЛЯ ИССЛЕДОВАТЕЛЬСКИХ ЦЕНТРОВ, ЭНЕРГЕТИ-ЧЕСКИХ УСТАНОВОК И УСТАНОВОК ТОПЛИВНОГО ЦИКЛА В СССР.
Предприятия атомной промышленности при наличии неизбежных выбросов радиоактив-ных изотопов в атмосферу и открытые водоемы могут представлять потенциальную опас-ность для окружающей среды и населения. В докладе рассматриваются вопросы, связанные со снижением до допустимых уровней газообразных и жидких радиоактивных отходов, сбра-сываемых во внешнюю среду, за счет комплекса мероприятий по улавливанию и очистке от-ходов, а также по правильному выбору места расположения предприятия, при котором обес-печивается безопасность населения как при нормальной работе, так и в случае непредви-денных аварийных ситуаций. Организация санитарно-защитной зоны вокруг атомного пред-приятия снижает опасность облучения населения, проживающего в непосредственной близос-ти, до незначительной величины.

## 1. ВВЕДЕНИЕ

Развитие атомной энергетики и предприятий топливного цикла выд-вигают серьезные задачи по выбору площадок для размещения АЭС и предприятий топливного цикла.

Особые требования к площадкам обуславливаются наличием потенци-альной опасности таких установок в связи с наличием неизбежных выб-росов радиоактивных изотопов во внешнюю среду как при нормальных условиях эксплуатации, так и в условиях возможных аварийных ситуа-ций.

## II.  ОСНОВНЫЕ ПРИНЦИПЫ

Выбор площадок для размещения в заданном районе предприятий атомной промышленности, атомных энергетических установок, заводов по изготовлению тепловыделяющих элементов, а также исследовательских атомных реакторов подчинен двум основным положениям [1,4], которым должна удовлетворять такая площадка:

1. Удовлетворение техническим требованиям таким как:

1.1. Наличие удобных транспортных связей.

1.2. Обеспеченность технической и питьевой водой. Для АЭС — наличие охлаждающей воды для конденсаторов турбин.

1.3. Удобные энергетические связи как для транспортирования вырабатываемой электроэнергии, так и для обеспечения надежной работы установки.

1.4. Удобство площадки для размещения АЭС (завода).

1.5. Геология площадки.  Сейсмичность.

2. Удовлетворять условиям безопасности размещения площадки для окружающей территории и окружающего населения:

2.1. Возможность организации санитарно-защитной зоны, где исключено проживание населения.

2.2. Метеорологические условия площадки, исключающие возможность образования устойчивых застойных зон в атмосфере.

2.3. Удаление от зон водоснабжения больших городов.

2.4. Удаление от городов с большим числом жителей.

2.5. Возможность организации подземного захоронения жидких отходов, особенно для площадок при размещении заводов по переработке тепловыделяющих элементов.

## III.  АТОМНОЕ ПРЕДПРИЯТИЕ, КАК ИСТОЧНИК ПОТЕНЦИАЛЬНОЙ ОПАСНОСТИ

Предприятия атомной промышленности представляют потенциальную опасность облучения населения и загрязнения прилегающей территории за счет наличия источников радиоактивного излучения радиоактивных изотопов, которые в процессе работы неизбежно сбрасываются во внешнюю среду и приводят к радиационному воздействию на окружающую территорию и на человека.

Накопленный к настоящему времени опыт эксплуатации атомных предприятий, обеспеченных сложными и надежными системами, предотвращающими сброс радиоактивности во внешнюю среду выше допустимых пределов, позволяет утверждать, что при нормальных условиях работы такие объекты не представляют опасности для населения.  И мы говорим о потенциальной опасности таких предприятий для населения в связи со случайными аварийными ситуациями, при которых могут иметь место повышенные выбросы радиоактивности.

В соответствии с этим положением потенциальная радиоактивная опасность объекта будет различаться в зависимости от следующих факторов:  общего количества радиоактивных изотопов, находящихся на объекте, агрегатного состояния, в котором находится радиоактивность, возможных аварийных ситуаций, оценки выделения радиоактивности во внешнюю среду при случайных маловероятных авариях.

В соответствии со степенью опасности предприятия атомной промышленности можно распределить следующим образом:

1) Радиохимические заводы по переработке облученного урана и получения наиболее токсичных $\alpha$-излучателей.

2) Атомные электростанции с реакторами на тепловых и быстрых нейтронах.

3) Научно-исследовательские центры, оснащенные ядерными реакторами и радиохимическими лабораториями.

Для радиохимических заводов мощностью порядка 1000 т/год в цикле переработки находится несколько десятков тонн облученного урана, содержащего $\sim 10^8$ Ки долгоживущих осколочных продуктов деления даже после длительной выдержки [2].

В соответствии с технологическим процессом переработки облученного урана осколочные изотопы содержатся в технологических растворах, так что имеется определенная опасность возникновения утечек радиоактивных растворов в рабочие помещения завода и затем во внешнюю среду. Осколочные продукты деления после переработки урана направляются в долговременные хранилища высокой, средней и низкой активности. Высокоактивные растворы после концентрирования (упарки) остекловываются или битумируются для длительного хранения, низкоактивные отходы могут быть подвергнуты подземному захоронению.

Радиохимический завод оборудуется надежной системой сбора и переработки протечек оборудования. Специальная система вентиляции завода обеспечивает сбор воздуха из помещений, где возможно его загрязнение радиоактивными изотопами, и последующую тщательную очистку перед выбросом в атмосферу. Выброс воздуха через высокие (100-150 м) вентиляционные трубы обеспечивает достаточное разбавление радиоактивности в приземном слое атмосферы.

В топливе реактора атомной электростанции мощностью 100 МВт(эл) радиоактивность продуктов деления оставляет около $10^{10}$ Ки в том числе $\sim 7 \cdot 10^7$ Ки иода-131.

Защитные покрытия тепловыделяющих элементов снижают ыход радиоактивных изотопов в контур теплоносителя до малых величин и следовательно в значительной мере снижают опасность попадания радиоактивности во внешнюю среду. Возможные небольшие протечки радиоактивного теплоносителя системой специальной вентиляции выводятся из помещений установки, подвергаются выдержке для обеспечения распада короткоживущих радиоактивных изотопов, очистке на специальных фильтрах и выбрасываются в атмосферу через вентиляционную трубу высотой 100-120 м.

Перечисленные мероприятия позволяют снизить радиоактивность выбрасываемого воздуха до незначительных величин, обеспечивая приземные концентрации радиоактивности значительно более низкие, чем это допускается нормами.

Меньшую опасность для окружающей среды и населения представляют исследовательские реакторы в связи с тем, что их мощность обычно в 10-100 раз ниже, чем мощность реакторов атомных электростанций. Поэтому, очевидно, требования к выбору места размещения оказываются менее жесткими.

Как отмечалось выше, практика эксплуатации атомных предприятий позволяет утверждать, что лишь при маловероятных случайных аварийных ситуациях они могут представлять определенную опасность для окружающей территории и населения.

## IV. ПРИНЦИПЫ ОРГАНИЗАЦИИ ЗАЩИТЫ ВНЕШНЕЙ СРЕДЫ И НАСЕЛЕНИЯ

В СССР "Нормами радиационной безопасности" [3] регламентируются предельно допустимые дозы (ПДД) для различных групп населения, которое с этой точки зрения делится на ряд категорий.

К категории "А" относят персонал, занятый непосредственно на атомном предприятии.

К категории "Б" относят отдельных лиц из населения, ограниченное количество лиц, проживающих в непосредственной близости к предприятию атомной промышленности.

К категории "В" относят население в целом для оценки генетически значимой дозы облучения.

Для категории "А" установлена предельно допустимая доза – ПДД, равная 5 бэр/год.

Для отдельных лиц из населения (категория "Б") установлена предельно допустимая норма ПДД', равная 0,1 ПДД, установленной для категории "А".

Этот норматив определяет комплекс технических мероприятий, обеспечивающих допустимые уровни излучений и концентраций радиоактивных изотопов в атмосфере и на поверхности земли в непосредственной близости к предприятию при его нормальной эксплуатации.

Аварийный выброс радиоактивности во внешнюю среду может привести к облучению населения за счет внешнего воздействия от проходящего облака, а также за счет поступления радиоактивности с выдыхаемым воздухом. Более длительное воздействие радиации связано с радиоактивными выпаданиями на грунт и возможностью поступления изотопов с водой и пищей.

Наиболее радикальным способом защиты населения является организация санитарно-защитной зоны вокруг предприятия, где проживание населения исключено.

Эффект от организации санитарно-защитной зоны связан с величиной снижения степени опасности для населения, проживающего за ее границей в зависимости от ее радиуса.

Естественно, что затраты на организацию такой защитной зоны возрастают с увеличением ее радиуса.

Снижение степени опасности, обусловленной движением радиоактивного облака в зависимости от расстояния от предприятия при случайной аварии зависит от ряда факторов, таких как метеорологические условия, высота подъема и движения радиоактивного облака, скорость выпадания радиоактивных частиц, состав радиоактивности и т.д.

В общем случае снижение степени опасности с расстоянием при выборе радиоактивного облака может быть описано зависимостью

$$P = P_0 X^{\alpha}$$

Где

$P$ – кратность снижения величины опасности,

$X$ – расстояние от места выброса (м),

$\alpha$ – некоторая величина больше единицы,

$P_0$ – для нормировки, принимается равное 1.

Для наиболее опасных погодных условий, категория устойчивости F по Пасквиллу, величина $\alpha$ может быть принята равной 1,6.

(Степень опасности определяется, как обратная величина концентрации радиоактивности на расстоянии X от точки выброса). Если величину затрат на организацию санитарно-защитной зоны принять пропорциональной площади отчуждаемой зоны $R = AX^2$, то величина относительных затрат на единицу снижения степени опасности составит:

$$\frac{\dfrac{dR}{dX}}{\dfrac{dP}{dX}} = \frac{2AX}{\alpha P_0 \, X^{\alpha-1}} \simeq X^{2-\alpha} \sim X^{\beta} \qquad \beta > 0$$

Из данного соотношения следует, что относительные затраты на единицу степени опасности растут с увеличением радиуса санитарно-защитной зоны. Поэтому экономически целесообразно стремиться к снижению степени опасности лишь за счет увеличения радиуса защитной зоны.

В соответствии с этим положением устанавливаются некоторые предельные значения величины санитарно-защитной зоны, при которых обеспечивается существенное снижение степени опасности.

Так для радиохимических заводов, наиболее радиационно опасных предприятий, предельный радиус санитарно-защитной зоны можно установить равным $X_1 = 5$ км со снижением степени опасности на ее границе, по сравнению с условной границей зоны промышленной застройки $X_0 = 0,5$ км, примерно в 35 раз.

Для атомных электростанций радиус санитарно-защитной зоны может быть сокращен до $X_2 = 3$ км со снижением степени опасности примерно в 15 раз. Для исследовательских центров, оснащенных реакторными установками, $X_3 = 1 \div 3$ км в зависимости от мощности исследовательского реактора [5], снижение степени опасности в $3 \div 15$ раз. При этом должны быть приняты все необходимые технические меры, ограничивающие величину возможного случайного выброса при аварийной ситуации так, чтобы при максимально возможной аварии на границе санитарно-защитной зоны доза внешнего облучения не превысила величину 25 бэр.


ВЫВОДЫ

Безопасность атомного предприятия в значительной степени связана с выбором места его размещения. Естественно, предпочтение отдается районам с невысокой плотностью населения.

Организация санитарно-защитной зоны существенно повышает безопасность атомного предприятия для населения. Однако эффективность затрат на снижение степени опасности при организации санитарно-защитной зоны падает с ростом ее радиуса.

Экономически целесообразно ограничить величину санитарно-защитной зоны некоторым предельным значением ее радиуса, направляя основные усилия и соответствующие средства на повышение надежности и исключение аварий на атомном предприятии.

## ЛИТЕРАТУРА

[1] Основные санитарные правила работы с радиоактивными веществами и другими источниками ионизирующих излучений, ОСП-72, Атомиздат, М., 1973г.
[2] STEHN, CLEANCY, 2nd Int. Conf. Peaceful Uses At. Energy (Proc.Conf.Geneva,
1958) 13, UN, New York (1958) 57.
[3] Нормы радиационной безопасности (НРБ-69), Атомиздат, М., 1972 г.
[4] Санитарные нормы проектирования промышленных предприятий (СН-245-71), Изд-во
литературы по строительству, М., 1972 г.
[5] Санитарные правила работы с радиоактивными веществами и источниками ионизирующих излучений, Гос.издат. в области атомной науки и техники, М., 1960 г.

## DISCUSSION

A.P. HULL: You seemed to indicate a maximum permissible dose (MPD) of about 0.1 rem/a for the dose to the population nearby due to routine effluents. Is this for external whole-body exposure, and do you make a distinction for $^{131}$I and the dose to the thyroid?

N.P. DERGACHEV: For individual members of the population living in the immediate vicinity of a nuclear facility, we have established MPD' $= \frac{1}{30}$ MPD $= \frac{1}{30} \times 5 = 170$ mrem/a for external whole-body exposure. The dose to the thyroid is set separately at a value not exceeding 5 rem/a from inhalation and intake of $^{131}$I through water and food.

J.B. BURNHAM: You mentioned the storage of high-level wastes. These can be dangerous for up to one million years. Have you considered alternative methods for the ultimate disposal of these wastes and criteria for the selection of the optimum method?

N.P. DERGACHEV: After temporary storage the high-level wastes from radiochemical plants are concentrated and then vitrified or bituminized. The blocks so obtained are sent to special areas for long-term storage. Both methods of solidification are now being developed. Owing to radioactive decay, the activity of the solidified wastes decreases in much less than a million years to levels representing a less serious hazard.

H. SCHNURER: You cited a maximum dose limit of 25 rem to the public in accident conditions. Could you please say a little more about the type of accident (e.g. the design-basis accident) to which such a dose limit applies?

N.P. DERGACHEV: According to our estimates, a dose of 25 rem at the boundary of the health protection zone could ensue from partial melt-down of the reactor core, leakage from the building and release into the atmosphere of more than 0.1% of the fission products accumulated in the reactor of a 1000-MW(e) nuclear power plant. It is assumed that from the molten fuel the yield of noble gases would be 100%, that of volatile elements (I, Cs) 10%, and of other fission products 1%. A burst pipe in the coolant circuit with impairment of heat removal from the core fuel is taken as the model for such a low-probability accident.

R. GAUSDEN (Chairman): Do you have the necessary powers to maintain a 3-km health protection or exclusion zone for the lifetime of the station?

N.P. DERGACHEV: Yes, the 3-km health protection (exclusion) zone is maintained during the entire operating life of nuclear power plants. The State supervisory authority has the right to change (increase) the radius of this zone if it is demonstrated during operation that the dose at the boundary of the zone will exceed the maximum permissible dose.

# SITING OF NUCLEAR POWER STATIONS IN INDIA

L. VENKATESH
Site Selection Committee,
Department of Atomic Energy,
Bombay

T. P. SARMA
Health Physics Division,
Bhabha Atomic Research Centre,
Bombay, India

Abstract

SITING OF NUCLEAR POWER STATIONS IN INDIA.

In India, a National Committee appointed by the Government has evolved certain guidelines on the principal criteria for evaluation of sites in relation to Indian conditions, taking into account the available experience in the country on the construction and operation (as appropriate) of the initial stations (at Tarapur, Ranapratapsagar and Kalpakkam) as well as experience abroad. The philosophy pursued with regard to the safety of the environment from radioactive pollution arising from the release of wastes from the station is one of containment of wastes rather than of dilution and disposal, which would enable utilization of the environment without undue restriction. Applying the criteria thus evolved, the National Committee has examined a number of sites both coastal and inland and generally situated in areas far away from coalfields, keeping in mind the design requirements of Candu-type stations, and has identified some of them as suitable on techno-economic grounds. These sites, spread over a large part of the country, will remain suitable so long as the facilities and the environmental conditions now obtaining there are preserved. It may be some years before all the selected sites are utilized. If the site conditions change in the meantime, fresh evaluation would be necessary. The paper describes some of the principal criteria and guidelines evolved and the present effort for selection of sites for future stations, together with the problems faced and the experience gained.

## 1. INTRODUCTION

The growth of generation and utilization of electricity is an index of a country's economic growth, and the annual per capita rate of consumption of electricity has come to be recognized as a yardstick in this regard. Although tremendous strides have been made in the field of power development in India since Independence, the country is still far behind the minimum required level. The per capita consumption according to statistics for 1971-72 was 90 kW·h, which, both in itself and in comparison with the levels of consumption in developed countries, is very low. Concerted efforts are therefore required to increase the per capita consumption, which is considered to be at least 500 kW·h for an acceptable minimum standard of living [1]. This calls for large-scale increase in generation; to achieve this objective, long-term perspective planning has to be made and implemented in stages.

A good part of the hydro resources in India is concentrated in the north-east and in the foot-hills of the Himalayas, much of which is inaccessible and

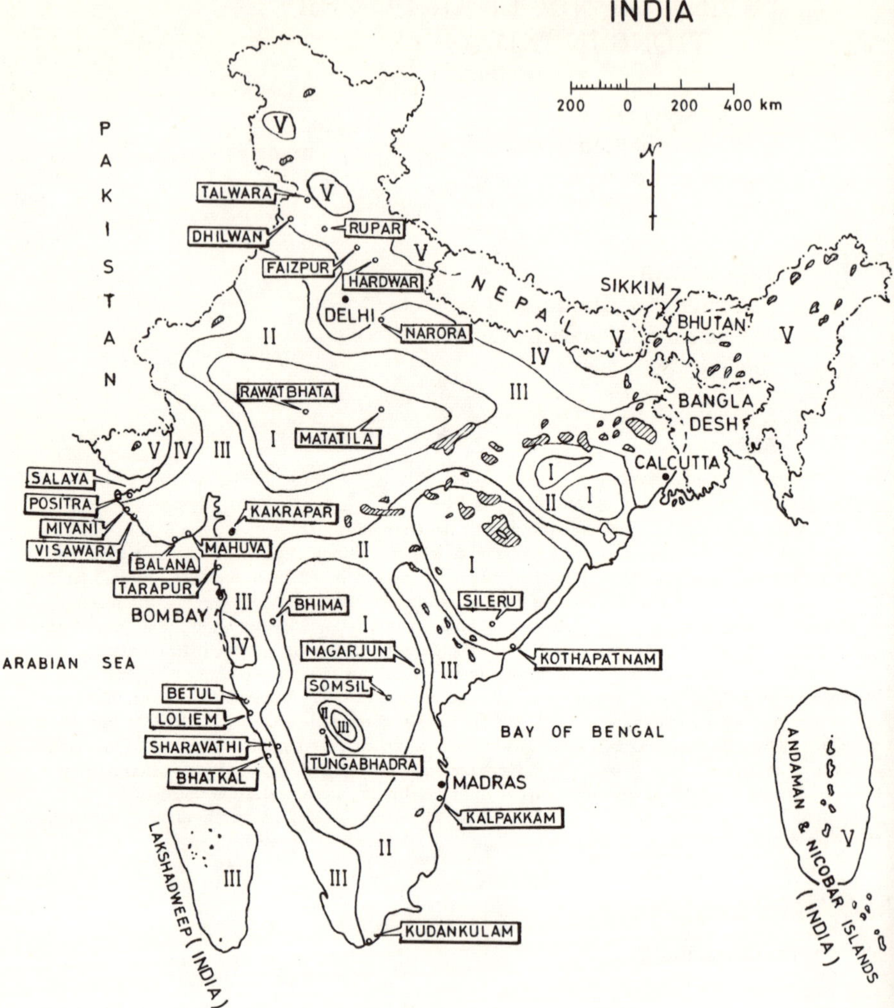


Fig. 1. Main coal deposits and seismic zones in India.

not economically exploitable.  Main coal deposits are concentrated in the east
(Fig. 1) and long haulage of large quantities of coal poses strain on transport
system, besides freight costs.  The known oil and gas resources are limited.
In these circumstances, a large part of the country which is far removed
from coalfields and also has limited hydro resources would have to depend on
nuclear energy for development.

Nuclear power stations are normally far away from coalfields for reasons
of overall economy.  The break-even distance from coalfields beyond which
nuclear power would be cheaper would depend upon a number of factors, such
as unit size adopted for nuclear stations, escalation in construction costs,
pit-head prices and transport costs of coal, available technology for nuclear
power plants, etc., and would naturally vary with time.  Present calculations
indicate that this break-even distance may be about 500 km [2].  Apart from
economics, diversity in generation would be necessary to provide stability
in the power systems.  Thus, need for nuclear power will start to become
more urgent in different regions at different times;  perhaps, in time to come,
nuclear power stations may be located even closer to coalfields.  This,
incidentally, is also expected to reduce strain on transport and enable the
available transport infrastructure to be utilized in other ways.  Keeping these
factors in mind, the Government of India constituted a National Committee
in September 1970, associating experts in the various aspects involved, to
study in depth and recommend a panel of suitable sites for setting up nuclear
power stations in areas remote from coalfields.


## 2.  SITING CRITERIA

Even though the basic philosophy in siting nuclear power stations shows
similarity throughout the world, the approach may vary from one country
to the other, depending upon national resources and requirements as well as
industrial and technological capabilities.  While the basic considerations in
siting conventional thermal stations also apply to siting nuclear stations, the
latter involves special evaluation of the site in relation to aspects concerning
the design and safe operation of nuclear reactors.  Besides, the foundation
and seismic conditions obtaining at a site are more critical for a nuclear
station than for a conventional thermal station.  It is heartening that guidelines
for siting nuclear power stations have in fact preceded the setting up of the
stations themselves.  This is not the case for thermal stations, where even
now there is scope for evolving some of the necessary guidelines.  The
Committee took into consideration experience gained in India [3] and abroad
and has formulated the criteria for siting inland and coastal nuclear power
stations with thermalized reactors [4].

The siting criteria are to some extent conditioned by the type of reactors
proposed to be set up, particularly in regard to their impact on the environmental factors.  At present, there is one BWR station (400 MW(e)) in
operation at Tarapur and two Candu-type stations each of 400 MW(e) in
various stages of construction at Ranapratapsagar and Kalpakkam.  A fourth
station, also of the Candu type (400 MW(e)), has just been taken up at Narora.
The development of technology is at present oriented towards the construction
of these natural-uranium-fuelled, heavy-water-moderated and cooled
Candu-type stations.  This position is expected to be sustained for some
years to come until the breeder-reactor technology, which is expected to

take over the next stage of nuclear power development has been developed
adequately. In accordance with this policy, the evaluation of the site-reacto:
combination has been based on Candu-type stations for the present and the
sites examined have been assessed for suitability for setting up stations of
1000 MW(e) capacity in stages. Very briefly, the main siting considerations
are the following:

(a)  Water requirements and availability

It is economical to have condenser cooling on the once-through system
for nuclear power stations. In a tropical country like India, where the
ambient temperatures are sometimes high, it will be necessary to limit the
temperature rise in the condenser cooling water to around 8°C. A station of
about 1000 MW(e) capacity requires a steady and continuous supply of roughl
2500 ft$^3$/s of water for keeping the temperature rise in the condensers at abc
8°C [4]. This is the reason for siting large stations on the sea coast or on
large reservoirs or on perennial canals of adequate capacity. While at coas
sites an unlimited supply of cooling water would be available from the sea fo
secondary cooling purposes, it would be difficult to obtain such large requir
ments inland, where water is needed for irrigation and other community use
Where a sufficient supply of cooling water is not likely to be available for once-
through cooling, it should be feasible to use cooling towers or cooling ponds
provided the climatic conditions and the topography permit, but this would
involve an economic penalty. At some places it may be worth paying this
economic penalty and deriving other benefits such as containment of radioactivity in the station itself. Experience has shown that for satisfactory results
the area of the cooling pond would have to be about 1.2-1.5 acres/MW(e) [5] anc
the make-up water requirement would be of the order of 2% [6].

In the case of coastal sites, the problems relating to littoral drift,
accretion or erosion phenomena, sea-bed characteristics, tidal and non-tid:
currents, storms, waves,etc., have to be investigated in detail and the
feasibility and economics of satisfactory cooling-water arrangements have
to be worked out. At Tarapur, one of the existing coastal stations, the
condenser cooling water is drawn from the sea through an open channel
about 1 km long, and the outfall is led away through a double discharge
system to correspond with the tidal oscillations. At Kalpakkam, another
coastal station under construction, an off-shore intake and a submarine
tunnel connecting the intake with the shore-based pump house had to be
provided in view of very heavy littoral drift and of the cyclonic conditions
obtaining there.

Until now, the condenser cooling water is being utilized for disposal of
low-level radioactive liquid wastes during normal operation, in view of the
extensive dilution it is capable of providing. This practice is tending to
become difficult in view of the possibility of our going in for higher-power
stations in the future and hence complete containment of radioactive liquid
wastes may have to be considered. Short spells of abnormal releases
(~ 10-100 Ci) completely offset the safety of the practice of dilution as
sufficient water is not always available for dilution to meet the permissible
levels of activity [7]. Where disposal along with condenser outfall is
adopted, utilization of the outfall (water) for some distance downstream is
controlled to ensure safe and adequate dispersal of the activity in the
environment. Where such disposal with the condenser outfall is not feasibl

special arrangements for disposal by evaporation, concentration,
immobilization and burial, or other suitable means, may have to be provided,
which would involve substantial additional costs. At one of the inland
stations, a large part of the low-level liquid wastes is chemically treated
and concentrated to avoid risks of contamination of the lake water which
supports fisheries and some irrigation downstream. For off-standard
conditions, however, it is necessary to provide for adequate storage, treatment and subsequent disposal, consistent with the acceptable capacity of the
environment. Such storage capacities are provided at all the stations now
on the ground. Coastal locations provide flexibility of operation for disposal
of liquid wastes compared with inland sites.

Experience has shown that, in a tropical country like India, solar heat
could be utilized advantageously to provide evaporative ponds to deal with low-
level radioactive liquid wastes. The solar evaporation averaged over one
year can be as high as 1 cm/d in dry inland areas. In areas where rainfall
is not very heavy (less than 100 cm/a) an average evaporation rate of 5 mm/d
is available [8, 9]. With these evaporation rates, a pond 60 m $\times$ 60 m wide
and 1 m deep is adequate to retain the normal liquid releases from a
station of about 470 MW(e) capacity and the solar evaporation rate will
compensate for the average normal discharges into the pond. This concept
is now under consideration for adoption at one of the inland stations. In
certain areas, however, the yearly evaporation rate and the rainfall are very
close, but in such regions the rainfall is generally confined to a part of the
year (as in Bombay) and it may still be possible to adopt the open-pond
approach for containing liquid effluents with adequate provision to deal with
the precipitation.

The chemical properties as well as the silt content of the cooling water
would have to be of acceptable standard for the required degree and type of
water treatment to be judged prior to its use in the various circuits of the
station. The silt content assumes significance not only for condenser and
heat-exchanger tube erosion, but also for activity pick-up and contamination
build-up along the outfall channels. The silt charge in some of the rivers
traversing alluvial country has been found to be as high as 5000 ppm during
the flood season. At certain coastal locations also, it could reach 2000-
3000 ppm. Desilting the cooling water would entail construction of large
ponds and would also involve disposal of substantial quantities of the deposited
silt which for a large station may be of the order of $10^6$ t/a. However, it has
been found that, generally, the fraction higher than 60 $\mu$m size should be
limited, and finer silt could be passed through without much difficulty provided
the outfall flows away without allowing the suspended silt to drop nearby.
Where the silt charge in the cooling water is higher than the permissible
limit, adequate desilting arrangements would have to be provided, involving
additional initial as well as running costs.

Inland fisheries are being developed in the country at a large number of
reservoirs and lakes which are formed by damming rivers and streams for
irrigation and/or power generation. If stations are to be located on such
lakes/reservoirs, thermal pollution and ecological problems arising from
dumping large quantities of heat into a water body would have to be studied to
ensure that by the rise in temperature of the water body there is no danger
to aquatic life in the environment. The outfall water when let into a lake is
likely to form warm-water zones affecting the dissolved oxygen content of
the water. Questions relating to the possible recirculation of the (warm)

condenser outfall into the (cold) intake are being studied and measures are being devised, where necessary, to overcome the recirculation problem. Experience has shown that hot-water plumes may be traced up to 1 km from the outfall point [10]. The existence of any natural physical barriers would be an advantage in this regard.

Since the cooling-water requirements are large, the pumping head would have to be reasonably low for reasons of economy. Sites on the sea coast with low tidal ranges and those on balancing reservoirs or canals have an advantage in this regard.

Where sea-water cooling is adopted for condensers in a Candu-type station, the process cooling requirements have to be met from fresh-water supply sources and are of the order of 150 $ft^3/s$ for a 500 MW(e) station. Where this supply is not available, it is necessary to adopt recirculation, involving considerable additional equipment as well as initial and recurring costs. Apart from the above station uses, adequate fresh water supply is necessary to meet the station needs including boiler make-up.

## (b)   Electrical system

The primary object of any power programme is to ensure, as far as possible, uninterrupted supply to the consumer at the lowest cost. To evaluate the various power dispersal aspects, the system characteristics which may obtain at the time the nuclear station may be expected to go on line would have to be studied in depth. Nuclear power stations are basically capital intensive with comparatively low fuel costs and hence it is best to utilize them for base load and at a high annual plant factor. As such, they could be located with advantage in a system with adequate hydro and conventional thermal stations which could act as a back-up. Since fuel cost in the nuclear stations is lower, it would be preferable, for overall system economics, to provide high loading on the nuclear stations, even by reducing the load on the thermal stations to some extent and utilizing storage-based hydro stations as peaking stations.

For the satisfactory operation of a nuclear power station, an assured supply of adequate start-up power at an appropriate voltage from the grid is essential. It has been found that to achieve this objective it would be desirable to interconnect the nuclear power station with two independent (and adequate) sources of generation. If this is not feasible, perhaps the provision would have to be considered of a captive generating unit within the station to meet the requirements. It is also necessary that a good part of the nuclear power should be absorbed quickly after start-up to avoid problems of poison in the reactor. This could be achieved satisfactorily if there are adequate industrial loads in the grid in the vicinity of the nuclear station which could come on line swiftly. Sometimes this aspect poses problems in a predominantly agricultural country where industries are neither concentrated nor intensive in many areas otherwise suitable for nuclear power development.

## (c)   Foundation conditions

The geological stability, the seismic susceptibility and the tectonic features of the area are extremely important in locating a nuclear power station. A good part of the northern and western regions of the country is underlain by recent deposits of alluvium and coastal sands. The central pa

and the Deccan plateau are formed of igneous, sedimentary or metamorphic
rock structures.   While the northern parts near the foothills of the Himalayas
and some areas south thereof are normally known to be prone to seismic
activity, the peninsular shield was until recently considered to be one of the
stablest parts of the earth and free from seismic disturbances.   However,
after the Koyna earthquake (Dec. 1967) in the Deccan plateau, there has been
some rethinking about the seismotectonics of the area.   New interpretation
of the potential seismicity has become available.   It how appears that no area
in India could be considered free from seismic activity.   The continued
integrity of the containment structure without any chances of development of
cracks and the safe operability of the station would, to a large extent, depend
upon the foundation conditions.   The foundation loading of the reactor building
is as much as 160 $t/m^2$ in certain locations and the loading is also uneven.
This poses special problems in alluvial areas where special precautions have
to be taken.   The working conditions require construction of suitable rigid
foundations capable of resisting differential settlement.   The geological
conditions at the site would have to be investigated to ascertain the depth to
bedrock, lithology, thickness of formations, presence of shears/faults,
engineering properties of the substrate, presence of dykes, any structural
deformities or mineralogical alterations, etc.   The nature of the subsoil,
whether rocky or alluvial, and the groundwater levels, have considerable
influence on the design of foundations, particularly in seismic zones. Adverse
foundation conditions do add to costs and time, both in design and construction
of the station.
      Sites in seismic zones must be specially studied to ensure safety during
and after an emergency.   After a thorough examination of the factors involved,
it is felt that a nuclear power station can be designed within reasonable
limits of safety provided the (tectonic) fault is not very close to it and its
activity is not too high.   A thorough geotechnical examination of a site would
be necessary in view of the potential hazard the failure of a containment can
bring about to the environment.   Investigations to determine the seismogenic
potential might become necessary at certain sites owing to the postulated but
unconfirmed seismic faults in the vicinity of the sites.   At one of the coastal
sites in western India, which was otherwise attractive, the Committee had
to promote extensive investigations to establish the existence or otherwise
of a postulated fault and evaluate its seismogenic potential.   Such investigation
involves surface geological as well as stratigraphic survey together with
geophysical traversing to generate adequate corroborative data to establish
the presence or otherwise of the fault/fracture.   Microseismic observations
at appropriate locations for a prolonged period would also be useful to monitor
energy release from any faults or fractures that may be present in the vicinity
of the site area.   The nature and the magnitude of any possible future
earthquake in the vicinity would have to be estimated with the help of
parameters evolved from geophysical field experiments.   This is particularly
necessary in cases where the available local seismic history is rather short.
Evaluation of the strongest earthquake motion that can occur in the site area
is important in developing appropriate design parameters for the station
structures, equipment, and safety systems.

(d)   Environmental conditions

      The health and safety aspects of nuclear power reactor siting in India
have been discussed in detail by several workers in various journals and at

a number of IAEA symposia [11-14].  The environmental factors such as
population density in the surroundings, agriculture, dairy farming,
pisciculture, and other human utilization of the environment in the neighbourhood, are significant in assessing the problems involved in safe dispersal of
radioactivity into the environment.  As these are matters of health and safety
extra care is given to these aspects in evaluating the site.

Experience has shown that gaseous radioactive releases under normal
operation are to a large extent predictable and can be controlled well within
safe limits.  The acceptability of such releases at a given site depends upon
the dispersal capacity of the environment, the concentration and transport
processes of radioactive products in the neighbourhood of the site, utilization
of the environment and radiation safety standards.  Under accident conditions
plant behaviour is not entirely under control and may lead to significant
release of radioactivity.  It is mainly the effect of possible release of fission
products in a serious accident (however remote it may be) that determines th
siting criteria.  The evaluation of the site reactor combination over accident
conditions involves (a) stipulation of acceptable emergency dose (AED) to
the public based on somatic exposure;  (b)  study of site parameters such as
population distribution, meteorological conditions (worst), hydrological
conditions,  environmental utilization; and (c) knowledge of reactor
parameters such as type of accidents and consequent release of fission
products, engineered safeguards and overall containment capacity of the
system.

According to present standards, no residential population is permitted
within 1.6 km of the plant [4].  With progressive development of containment
technology and operational experience of the engineered safety features,
this distance is subject to review.  It is also important that the area between
1.6 km and 4.8 km from the station is only thinly populated.  The current
practice is to control the growth of utilities in this area by suitable
administrative measures.  The water table level and its fluctuation, as well
as the groundwater movement, would influence arrangements for disposal of
the solid wastes from the station.  Extensive irrigation is practised in semi
arid areas by pumping subsoil water, and this aspect causes additional
concern and calls for thorough investigation.  Meteorological conditions at a
site indicate the consequence of release and govern the means for adequate
dispersal of gaseous wastes into the atmosphere.  Micrometeorological
and climatological data have to be carefully studied to see if the predominan
wind directions are towards any concentration of population, which may be
subject to radiation exposure.  These measures are taken to control radiati
exposure of the population in the neighbourhood.  At all the power stations
now on the ground, environmental survey laboratories have been established
which undertake micrometeorological and microterrestrial studies in great
detail commencing from the preconstruction stage and carried on during the
operation of the station.  The data and information collected by these
laboratories provide a feedback to the Committee's deliberations.

(e)  Access

It is always an advantage to have a good and all-weather approach by
road and/or rail to a project site for movement of materials and personnel
Road transport of certain equipment and components from the places of
manufacture and/or ports of entry to the site becomes particularly importa

to nuclear power stations in the event that their size exceeds the permissible
dimensions for transport by railway.  These problems are likely to be greater
in the developing countries.  During construction of one of the inland stations,
a number of heavy and large consignments had to be transported by road
from the port of entry over distances of the order of 1000 km.  This problem
is particularly significant in a large country like India with equipment-
manufacturing centres and power  station sites spread all over and not
necessarily close to the coast.  Site fabrication and assembly of such equipment and components to meet the exacting standards also present problems.
Under such conditions, it appears that long road haulage is inescapable even
for future stations.  It is of interest to cite here that the largest (not the
heaviest) single piece involved for a 200/235 MW(e) unit was about 7.5 m long,
about 7 m dia., weighing about 75 t.  The matter is even more severe for a
larger unit of 500 MW(e) capacity  for which, from present indications, a
single piece may weigh anything up to 280 t  and may need to be transported
by road.

Movement of such heavy equipment to the site by tractor-trailer
combination over long distances presents many problems.  This calls for a
thorough study of the road routes involved for evaluating the carrying capacity
of the roads, particularly the cross-drainage works, the negotiability of
curves and gradients, the passability through towns and villages or the need
for alternative diversions, clearance requirements, etc., and remedying the
weaknesses identified.  It would be of interest to mention here that nearly
4750 km of road has so far been surveyed by the Department of Atomic Energy,
and improvements to bridges and culverts en route as well as to road alignments and curves have been carried out.  Moreover, similar surveys and
improvements are also being taken up for the proposed power stations.

The feasibility of safe transport of irradiated fuel from the station to
the possible sites of reprocessing plants have also to be examined.  In view
of the transport problems, with economic penalty, as well as the radiological
hazards involved, it might become expedient to think in terms of locating
reprocessing plants close to every station in the future.  The questions
relating to the satisfactory containment and controlled disposal of radioactive
wastes would have to be resolved before the reprocessing plants are
located inland.


3.    PROBLEMS IN SELECTION OF SITES

Applying the criteria briefly described above, the Committee carried out
investigations in areas which are prima facie economically favourable for
nuclear power development.  It examined a large number of promising sites
(Fig. 1) located on inland reservoirs, lakes, large canals and on the coast,
where the primary criteria relating to foundation conditions, environmental
conditions and cooling-water availability, among others, were generally
satisfactory.  The modus operandi adopted by the Committee was to have the
site investigations carried out through the State agencies and collect adequate
data for examination.  It has been the experience of the Committee that public
enthusiasm and acceptance came forth in good measure.  In order to make a
techno-economic study, it became necessary to undertake a fairly detailed
study of the various site factors including the dispersal of power from the
station at the alternative sites.  In addition, the aspects of construction

methods compatible with each site and their effect on construction time and
cost had also to be taken into account.   Certain special studies and
investigations had to be carried out in relation to the conditions peculiar to
some of the sites.

No site could be considered ideal from all points of view, and, as could
only be expected, each had some weakness or other;  it was necessary to
identify the weaknesses and see if it were possible to remedy them
economically, consistent with the required standards of safety.   In the course
of the investigations and relative assessment of the alternative sites, the
Committee came across a number of problems which had to be investigated
and resolved to the extent feasible [15].   Some of them are briefly discussed
as follows:

(a)   It was realised that the Candu design being adopted for the stations
under construction at Ranapratapsagar and Kalpakkam could not be adopted
in a major part of the country which is subject to moderate seismic conditions
(Fig. 1).   It would not be practicable to exclude such a large part of the
country from the benefits of nuclear power, particularly because nuclear
power would be the only answer in those areas owing to the non-availability
(or if available, the inadequacy) of alternative sources of generation.
Recognizing this need, it was decided that the design should be developed to
suit such site conditions also.   The redesign effort is now in progress initial
for adoption in India's fourth atomic power station at Narora and it is
ultimately expected to become a standard design for all future stations with
Candu-type reactors for some time to come.   This calls for considerable
development work including prototype tests before framing up detailed design
Some of the important changes contemplated are:  (i)  a double containment
to improve the leakage rate and thus adhere to the stringent regulatory
requirements;  (ii)  integral calandria and end shields which would make it
possible to lower the elevation of these assemblies as well as ensuring freed
from relative movement between lattice positions in the end shields and thos
in calandria;  and (iii) certain other changes in the reactivity mechanisms a
associated reactivity control systems.

(b)   At one of the sites it was found that a tear fault was passing very
close and geological reports indicated that this tear fault had down-thrown t
sediment by a vertical displacement of nearly 1000 m and a horizontal
dislocation of about 6.4 km.   Though available data were scanty, the area w
considered geologically unstable and slow secular movement along the fault
in the area could not be ruled out.   Significant movements in a sinistral (lef
lateral) direction had in fact been computed by the Geological Survey of Indi
from the precision levelling and geodetic triangulation carried out by the
Survey of India in 1964.   It would be very difficult to ascertain whether this
phenomenon was still continuing or had stopped, as a close and constant
watch was necessary for a number of years before a conclusion could be
reached.   Such geological instability would be harmful for maintaining
containment integrity and would present difficulties in the design of the stat
From this point of view the site is not considered acceptable at present.

(c)   Yet another site in this region, which was favourable on major
site factors, was found to be far away from the load centres as now existi
and as could be anticipated in the near future.   This involved considerable
additional transmission network to transmit power to the load centres,
entailing large investments and heavy power losses in transit.   Hence the s
cannot be taken up at present.

(d)   One of the sites was situated in an area well suited to nuclear power development.  At the instance of the Committee, investigations were carried out which indicated that the substrata consisted of miliolite limestone with frequent occurrences of solution channels.  The limestone was found to be friable and non-homogeneous and interspersed with clayey and gritty material to a considerable depth.  Investigations revealed that extensive and expensive treatment of the substrata would be necessary.  Further, a major fault had been postulated in this area, but no conclusive proof of existence or activity was available.  If the fault were to be found to exist close to the site and to be active, it could present a potential hazard to a nuclear power station. Extensive investigations had to be undertaken to determine the seismogenic potential of the area.  No conclusive results were available up to the completion of this paper.

(e)   The Committee considered the extensions to the existing stations among the alternative sites.  Such extensions of existing stations always have economic advantages, as the sites will have been developed and will afford economy in expenditure besides saving considerable time.  However, at such sites other nuclear installations such as fuel-reprocessing plants, waste treatment facilities, etc., would usually spring up, which would add to the site's radiological burden.  The optimization of the site is achieved by allocation of the acceptable site radiological burden to the various installations through the different routes.  It may be necessary to postpone an extension in such cases until the actual radiological burden levels are known after all the planned facilities go into operation.

(f)   As already mentioned, transport of heavy and large equipment to the station involves the survey of a very great length of roads concerned and improvements to bridges and culverts as well as road alignments and curves. It has been found from experience that the time taken to complete the road improvements necessary to meet the required movement would be substantial; hence survey and improvement work has to be initiated well in advance.  Since it is likely that larger unit sizes of say 500 MW(e) may be adopted for future stations, which involve larger consignments of equipment, a special study team from the Department of Atomic Energy is at present going into the relevant aspects, in co-operation with the Ministry of Transport and the State Public Works Departments concerned, to devise ways and means of transporting such consignments by the available roads with minimum improvements.

(g)   On the east coast of India, frequent storms and resulting hurricanes and tidal waves, heavy littoral drift, possible existence of off-shore tectonic features, etc., restricted the availability of sites for a large part of the coast.

(h)   The suitability of a site is dependent on the existing environmental conditions and the available facilities.  It was recently found that, at a site which was in a generally favourable area, the State authorities had of late prepared plans for development of the salt industry on a big scale and had acquired large areas for the purpose.  The site in question had to be dropped. At another site, it had been assumed during the preliminary investigations that water for once-through cooling would be available.  It later became known that the State requires the same waters to flow along another route for hydro generation.  The Committee has to be cognizant of the development plans in the area.

(i)   For comparison of the acceptable alternative sites, the Committee considered at length the various methods that may be adopted and came to

the conclusion that the relative cost basis was the best, provided the additiona
costs for the feasible remedial measures to meet the site weaknesses could
be determined.  Unfortunately, costs depend so much on the designs, as also
on a number of other factors for which sufficient data are not always availabl
at the investigation stage.  It was therefore difficult to frame cost estimates
accurately.  Further, the costs vary with time.  Even so, the Committee
attempted to evaluate the site weaknesses in monetary terms so long as the
weaknesses could be mitigated technologically.  The sites were graded from
a comparative analysis of these costs as well as the other factors for which
costs could not be estimated and only a good judgement had to be made.

## 4.    FUTURE NUCLEAR POWER PROGRAMME

In the context of our nuclear power programme in the future, the site
selection effort is of significant value.  The Committee has carried out
extensive and thorough investigations in the various regions with the active
help and co-operation of the States concerned, bringing together expert
technical opinion on the different aspects involved.  With the reports of the
Committee in their hands, the Government will have a ready list of a number
of suitable sites to choose from in the various regions whenever they decide
to set up nuclear power stations in these regions, depending upon the nationa
power policy.  With the sites identified, it should be possible to develop
roughly 10 000-15 000 MW(e) of additional nuclear power, thereby meeting to
large extent the requirements of a nuclear power programme for some year
to come.

## ACKNOWLEDGEMENTS

The authors are indebted to Shri V.R. Vengurlekar, Chàirman, Site
Selection Committee, and Dr. A.K. Ganguly, Director, Chemical Group,
Bhabha Atomic Research Centre, for their encouragement and helpful
criticism in the preparation of this paper.

## REFERENCES


[1]    MINISTRY OF IRRIGATION AND POWER, INDIA, Power Development Programme for 1972-77 (1972).
[2]    DEPARTMENT OF ATOMIC ENERGY, INDIA, Feasibility of a Nuclear Power Station in West Bengal,
       a Review, Internal Rep., Programme Analysis Group (1973).
[3]    CHAKRAVARTI, M.N., SRINIVASAN, M.R., "Selection of site for the Tarapur Atomic-Power Station"
       Siting of Reactors and Nuclear Research Centres (Proc. Symp. Bombay, 1963), IAEA, Vienna (1963) 457
[4]    DEPARTMENT of ATOMIC ENERGY, INDIA, Interim Report on Selection of Sites for Future Atomic
       Power Stations, Part I (1972).
[5]    KAVANAGH, G.M., "Environmental effects of producing electric power", Selected Materials on
       Environmental Effects of Producing Electric Power 1, Joint Committee on Atomic Energy Congress of the
       United States, 91st Congr., 1st Session, Washington (1969).
[6]    "Problems in disposal of waste heat from steam electric plants" (Federal Power Commission Staff Study -
       1969), Selected Materials on Environmental Effects of Producing Electric Power 8, Joint Committee on
       Atomic Energy Congress of the United States, 91st Congr., 1st Session, Washington (1969) 285.
[7]    SARMA, T.P., SHIRVAIKAR, V.V., GANGULY, A.K., "Site selection and optimization of its use
       in India", Principles and Standards of Reactor Safety (Proc. Symp. Jülich, 1973), IAEA, Vienna (1973) 2

[8]   RAMAN, P.K., SATAKOPAN, V., "Evaporation in India calculated from other meteorological factors",
       Scientific Notes 6 61, India Meteorological Department, Delhi (1935) 51.
[9]   BHABA ATOMIC RESEARCH CENTRE, Six-Yearly Climatological Summary for the Rajasthan Atomic
       Power Project Site for the Period 1964-69, Rep. BARC-567 (1971).
[10]  KAMATH, P.R., BHAT, I.S., GURG, R.P., ADIGA, B.B., CHANDRAMOULI, S., "Seasonal features
       of thermal abatement of shoreline discharges at nuclear sites", Environmental Effects of Cooling Systems
       at Nuclear Power Plants (Proc. Symp. Oslo, 1974), IAEA, Vienna (1975) 217.
[11]  RAO, A.S., KAMATH, P.R., GANGULY, A.K., "Health and safety criteria for siting power reactors
       in India", Siting of Reactors and Nuclear Research Centres (Proc. Symp. Bombay, 1963), IAEA,
       Vienna (1963) 299.
[12]  SHIRVAIKAR, V.V., GANGULY, A.K., "Containment and power reactor siting", ibid., 211.
[13]  SOMAN, S.D., KRISHNAN, L.V., SUBBARATNAM, T., GOPINATH, D.V., "Power-reactor design
       evaluation from the standpoint of radiation safety for a given site", ibid., 107.
[14]  SHIRVAIKAR, V.V., GANGULY, A.K., "Systematics of reactor siting", Health Physics, Progress in
       Nuclear Energy Series XII, 1, Pergamon Press, Oxford (1969) 393.
[15]  VENGURLEKAR, V.R., VENKATESH, L., "Problems in siting future nuclear power stations", National
       Seminar on Power: Present Problems and Planning for Future, India, No.3.02 (1974).

# DISCUSSION

R.R. BOUSSARD: You mention problems of safe transport as a justification for establishing reprocessing facilities near nuclear power stations. I think the hazard inherent in reprocessing units is greater than that associated with transport, because, by definition, reprocessing involves a discharge of effluents and storage of wastes, some of which contain very long-lived isotopes ($\alpha$-emitting wastes). Such dispersion of waste centres is, in my opinion, justifiable only where there is a very large power-generating capacity at a single site.

L. VENKATESH: This matter is still under study. Some thoughts on the subject are contained in paper SM-188/26.

D. DAGAN: Who initiates the erection of a nuclear power plant in your country, and what sort of relationship is there between the Central Government and the State in this connection?

L. VENKATESH: The Indian Atomic Energy Act confers on the Central Government exclusive responsibility for the development of nuclear energy, so it is the Central Government which builds and operates nuclear power stations. The power from such stations is supplied to the electrical grids controlled by the States, and thereafter the distribution of power to consumers is the statutory responsibility of the States concerned. The association of the States in the site investigation effort provided a much-needed opportunity for public participation and has in fact been responsible to a large extent for the enthusiasm that emerged. During the construction of a station, the active help and co-operation of the State in which the site is situated will be necessary for certain activities such as acquisition of land, resettlement of any population displaced from the exclusion area and so on.

G. HAKE: Referring to the Koyna earthquake, you said that, had you chosen a site in this area, you would probably have decided it was the most stable one you could have selected. I believe the Koyna earthquake was related to increased crustal loading resulting from a dam in the neighbourhood. How do you intend to cope with such situations when you are selecting sites in the future?

L. VENKATESH: The Deccan Plateau (where the Koyna dam is situated) had long been considered by geologists to be one of the stablest areas on

earth.  The recent (1967) Koyna earthquake has thrown new light on this
subject.  The causes of the 1967 earthquake are still under discussion.  It is
however, always possible to take into consideration the latest knowledge
available on the subject when selecting future sites.  In fact, data derived
from the seismotectonic studies carried out in India after the Koyna earthquake were taken into account in the site selection work discussed in
the paper.

# CRITERES DE SURETE NUCLEAIRE ORIENTANT LE CHOIX DES SITES
## Pratiques françaises

P. CANDES, Ph. AUSSOURD
Service central de sûreté des installations nucléaires,
Ministère de l'industrie et de la recherche,
Paris, France

## Abstract—Résumé

NUCLEAR SAFETY CRITERIA APPLIED IN SITE SELECTION — THE PRACTICE IN FRANCE.
In France, the safety of nuclear facilities is the responsibility of the Ministry for Industry and Research (Central Department for the Safety of Nuclear Facilities). The first part of the paper deals with the conception and contents of the site studies which are included in a safety report with the object of obtaining authorization to go ahead with work on the establishment of a facility. The conception is governed by the following two considerations: (a) the site is a place where the natural elements and living organisms occur and which is characterized by the permanent presence of the human factor, while the proposed nuclear facility will — like any industrial facility — present risks and have an impact on the site, particularly through the discharge of radioactive effluent and potentially in consequence of a nuclear accident; (b) the site exercises an influence — in fact, it even imposes constraints — on the nuclear facility. The site study as submitted by the operators to the authorities responsible for the safety evaluation traditionally consists of six sections, covering: (I) description and history of the site; (II) meteorological conditions; (III) hydrology of the area; (IV) geological and seismological conditions; (V) ecological factors; (VI) natural and/or previous radioactivity at the site. These six sections contain the data which serve as a basis for applying the two considerations spelled out above. However, the two corresponding directions of study and analysis do not settle the fundamental problem of the distribution of the population around the site. Methods for dealing with this problem are suggested in the second part of the paper; they take into account the efforts made so far at the international level. The authors consider that limiting criteria should not be based solely on the radioactive effluent discharges associated with normal operation but on the radioactivity releases associated with accidents. The methods proposed by them constitute a gradual approach to the solution of this problem.


CRITERES DE SURETE NUCLEAIRE ORIENTANT LE CHOIX DES SITES — PRATIQUES FRANÇAISES.
La sûreté des installations nucléaires est placée en France sous la responsabilité du Ministère de l'industrie et de la recherche (Service central de sûreté des installations nucléaires). La première partie du mémoire traite de la conception et du contenu des études de site, telles qu'elles sont présentées dans le rapport de sûreté de l'installation en vue d'obtenir l'autorisation de création. Cette conception repose sur les deux constatations suivantes: a) le site est considéré comme un milieu d'éléments naturels et vivants marqué par la présence permanente du facteur humain; l'installation nucléaire qu'on se propose d'y implanter présentera, comme toute installation industrielle, des risques et aura un impact sur le site, notamment par le rejet d'effluents radioactifs, ou potentiellement en cas d'accident nucléaire; b) le site exerce une influence, voire des contraintes sur l'installation nucléaire. L'étude du site telle que la soumettent les exploitants aux autorités chargées de l'évaluation de la sûreté comprend traditionnellement six paragraphes descriptifs: I — description et évolution du site, II — météorologie, III — hydrologie, IV — géologie et sismologie, V — écologie, VI — radioactivité naturelle ou préalable du site. Ces six paragraphes fournissent des données de base pour l'application des principes exposés ci-dessus. Toutefois le double faisceau d'études et d'analyses ainsi décrit ne règle pas le problème jugé primordial de la répartition des populations autour d'un site. La deuxième partie du mémoire propose une méthodologie pour aborder ce problème. Cette méthodologie prend en compte les efforts fournis jusqu'ici au plan international. Les auteurs pensent que des critères limitatifs dans ce domaine ne peuvent être fondés sur la seule prise en compte des rejets en fonctionnement normal, mais doivent l'être principalement sur celle des rejets liés aux situations accidentelles. La méthode fournit une approche progressive pour régler ce problème.

## 1. GENERALITES

Le programme électronucléaire français va entraîner la mise en
service de centrales de production d'électricité représentant une puissance
électrique de 18,5 GW entre 1975 et 1980. Ces centrales seront réparties
sur huit sites: Fessenheim (Haut-Rhin), Bugey (Ain), Saint-Laurent-des-Eaux
(Loir et Cher), Gravelines (Nord), Dampierre en Burly (Loiret), Tricastin
(Drôme), Braud et Saint-Louis (Gironde), Creys-Malville (Isère).

Entre 1980 et 1985, huit ou neuf autres sites devront être ouverts.
Enfin, vers la fin du siècle, notre pays comptera une puissance électrique
installée d'environ 170 GW d'origine nucléaire, répartie en trente ou
quarante centrales.

L'importance de ce programme explique que le choix des sites nucléaires
soit devenu une préoccupation majeure et que le temps est révolu où, devant
implanter des centrales peu nombreuses, les techniciens pouvaient, sans
analyse préalable trop poussée, choisir quelques sites peu nombreux dans
un relatif «désert» français.

L'implantation d'une centrale nucléaire sur un site donné crée inévitable
ment des risques pour les personnes vivant sur ce site ou dans son voisinage.
Elle ne diffère pas, en cela, de toute autre activité industrielle importante
et, d'ailleurs, les risques sont pour une part de même nature: accumulation
et mise en œuvre d'énergie, de fluides dangereux, etc. Mais l'installation
nucléaire induit également des risques particuliers, liés à la radioactivité.
En dehors des rejets d'effluents radioactifs résultant de l'exploitation
normale ou de légers incidents, il n'est pas douteux que l'accumulation de
substances radioactives au sein d'une centrale représente un danger potentiel
important et il n'est pas possible d'exclure totalement a priori l'éventualité
d'accidents graves conduisant à la dispersion d'une partie de ces substances.

Dans un passé récent, certains accidents, et notamment celui de
Windscale, ont attiré l'attention sur les conséquences graves que pourrait
avoir la dispersion d'une fraction notable des produits de fission contenus
dans le cœur d'un réacteur, et mis en évidence la complexité des voies
par lesquelles la radioactivité ainsi dispersée dans le milieu naturel peut
atteindre l'homme.

Dès lors, un regain d'attention et de rigueur a été apporté à la définition
et à l'application des mesures à prendre pour éviter toute dispersion de
produits de fission ou à défaut pour en limiter les conséquences. Les
principales mesures consistent à:
- apporter une grande rigueur à la conception et la construction des
  installations comme aux tâches d'exploitation ou d'entretien, et à en
  contrôler l'exécution,
- s'assurer notamment de la qualité des composants et de la fiabilité des
  systèmes,
- perfectionner, tant sur le plan des performances que sur celui de
  l'entretien, l'ensemble des dispositifs assurant le confinement des
  substances radioactives et la protection contre les rayonnements,
- étudier dans le détail les différents types d'accidents concevables pour
  déterminer les dispositions techniques permettant de les prévenir,
- choisir les sites de préférence dans des zones de faible densité de
  population en prenant en considération les problèmes soulevés par les
  influences réciproques entre l'installation et son voisinage,

— prévoir enfin les mesures à prendre en cas d'accident grave, tant à
   l'intérieur des installations (plans d'urgence) qu'à l'extérieur de celles-
   ci (plans ORSECRAD), en vue de minimiser les conséquences de l'accident.


Les résultats des efforts de prévention ont été, semble-t-il, très
satisfaisants, puisque depuis l'avertissement donné par l'accident de
Windscale, on n'a eu à déplorer, sur le plan mondial, aucun accident dont
les conséquences sur le public puissent être considérées comme importantes.
   Cependant, il faut bien reconnaître que, lorsqu'il s'agit de décider de
certaines dispositions de sûreté ou des exigences que l'on doit avoir dans
le choix des sites nucléaires, les exploitants d'une part et les autorités
chargées de la sûreté d'autre part se trouvent devant des choix qu'il est
difficile d'effectuer de façon totalement rationnelle.
   Il est donc logique qu'un site nucléaire, s'il est convenablement choisi,
soit en premier lieu tel que les conséquences des accidents graves
susceptibles de se produire dans l'installation, si faible que soit leur
probabilité, restent aussi réduites que possible pour les populations résidant
alentour. La zone d'influence de tels accidents, telle qu'on l'évalue à partir
d'hypothèses prudentes, peut s'étendre sur des distances notables, de
l'ordre de la centaine de kilomètres pour les cas extrêmes. La sagesse
commande donc d'éviter, dans toute la mesure du possible, de construire
les grandes installations nucléaires, et notamment les réacteurs de
production d'électricité, trop près de grandes agglomérations ou à l'intérieur
de zones à forte densité de population. Au-delà de cette indication de bon
sens, qui pourrait d'ailleurs être nuancée en fonction du type de l'installation
et des précautions prises pour en assurer la sûreté, le souci de rationaliser
les décisions relatives au choix de sites relativement nombreux dans un
territoire limité conduit à la recherche de critères quantitatifs, ou du
moins de critères d'appréciation et de comparaison aussi objectifs que
possible. Ce problème ne se limite d'ailleurs pas aux choix initiaux: il
convient en effet que la situation acceptée au moment de la création de
l'installation ne se «dégrade» pas trop sensiblement dans le temps compte
tenu de l'évolution démographique dans l'environnement.
   Mais il convient, en second lieu, d'être attentif aux autres facteurs de
l'environnement. En effet, parmi les causes qui peuvent provoquer un
accident grave conduisant à une importante dispersion de radioactivité,
si les unes ont leur origine au sein même de l'installation et ne sont pas
traitées dans la présente note, les autres proviennent précisément du site
lui-même. Le site peut, par ses manifestations naturelles, par celles qui
sont liées à la nature des modifications que l'homme lui a fait subir, par
celles enfin qui sont caractéristiques des activités (industrielles ou autres)
situées au voisinage de l'installation, exercer sur elle diverses influences,
voire conduire à de véritables agressions capables de déclencher l'enchaîne-
ment des phénomènes entraînant une avarie majeure.
   Les considérations qui précèdent ont conduit l'Autorité chargée de la
sûreté nucléaire en France à tracer un cadre général pour l'étude des
sites nucléaires. Nous allons en premier lieu exposer cette conception
des études de sites.
   Nous tenterons ensuite de proposer une méthodologie permettant
d'aboutir à des critères de répartition de densité de la population autour
des sites nucléaires.

## 2.   CONTENU DES ETUDES DE SITES NUCLEAIRES

L'étude d'un site nucléaire doit être effectuée essentiellement en deux
stades:

a) Au moment du choix préalable, des éléments éventuellement
sommaires, mais suffisants, doivent permettre de juger si le site
est acceptable ou non.

b) Au moment de la demande d'autorisation de construction, l'étude
approfondie doit permettre d'apprécier avec suffisamment de détail
et de précision:
- l'influence et les contraintes que l'environnement peut exercer sur
  l'installation nucléaire,
- l'impact de l'installation nucléaire sur son environnement.

L'étude du site doit être orientée et ordonnée en vue d'atteindre les
deux objectifs ci-dessus, jugés essentiels.

L'impact de l'installation nucléaire sur l'environnement et notamment
sur les populations résidant dans la zone d'influence possible de l'installation
se manifestera essentiellement par les rejets radioactifs possibles en cas
d'accident et à un moindre degré par les rejets radioactifs de routine.

Il est bien évident également que les nuisances non radioactives dont
on ne traite pas ici (bruits, rejets thermiques, problèmes esthétiques, etc.)
doivent être prises en compte et que leur impact peut être important dans
certains cas.

Les voies de transfert des rejets vers les populations doivent être
décrites.  Il s'agit principalement:
- de l'atmosphère,
- des eaux de surface,
- des eaux souterraines,
- du milieu vivant.

Les influences et les contraintes que l'installation peut subir sont
d'origine:
- atmosphérique (orages, tornades, etc.),
- hydrologique (inondations, ruptures de barrages),
- géologique (séismes),
- industrielle (incendies, explosions, émissions de produits toxiques ou
  déflagrants).

Il est clair que certains dommages que peut subir l'installation nucléaire
peuvent entraîner un accident nucléaire et c'est bien entendu ce qui motive
ce deuxième aspect de l'étude de site.

C'est donc à partir d'une telle conception d'ensemble que doit être
défini le contenu des études ci-après.


### 2.1.  Description du site et évolution

Emplacement.  Cette section doit comporter une explication d'ensemble
des raisons qui ont motivé le choix du site:  demande d'énergie, proximité
de centres de consommation, facilité de trouver de l'eau de refroidissement, etc.  La description de l'emplacement doit être assortie de cartes
à différentes échelles selon les détails à mettre en évidence.

Zone d'influence de l'installation.  Le choix de son étendue et de ses
contours est justifié principalement par la considération des conséquences
d'accidents et à un moindre degré par l'évaluation des doses dues aux
rejets de routine.  Compte tenu de ce qui est actuellement pratiqué au
niveau international, on peut admettre que cette zone s'étend jusqu'à
50 kilomètres au moins autour du site.  Il faut se garder d'être trop
«géométrique» en définissant par exemple un cercle de 50 km de rayon.
Si une ou des agglomérations importantes sont situées à une distance légèrement supérieure, il faut bien entendu les inclure, de même qu'on peut se
tenir en deçà pour d'autres secteurs peu peuplés ou physiquement protégés
par le relief.

Ce qui vient d'être dit ci-dessus est surtout valable, on le devine,
pour ce qui concerne les doses dues aux transferts atmosphériques.  Pour
celles dues à la voie hydrologique, les secteurs à prendre en compte
pourront être totalement différents.  On pourra par exemple se préoccuper
des villes successives en aval du site sur un cours d'eau, ou, si le site est
en bordure de mer, d'une frange côtière dont la définition tiendra compte
des caractéristiques physiques et humaines de l'environnement marin.

En outre, comme nous l'avons souligné dans l'introduction, il y a lieu
de se préoccuper des effets extérieurs sur l'installation et dans ce cas
certains d'entre eux peuvent avoir leur origine loin de l'installation (par
exemple, rupture de barrage).  Il y a donc lieu d'examiner certains aspects
dans un cadre «régional».


Population.  La distribution des habitants permanents de la zone
d'influence selon la direction et la distance doit être indiquée au moyen
de graphiques et de tableaux faciles à exploiter.  L'étude d'un critère de
limitation de la densité de population nécessite une description soigneuse
de cette répartition de population.  Une autre communication [1] expose
ce point avec les justifications correspondantes.

Des renseignements analogues doivent être fournis pour les résidents
saisonniers lorsque cela revêt une importance significative.  Une prévision
du développement futur de la zone est donnée, en tenant compte éventuellement de l'influence que l'installation elle-même pourra exercer sur ce
développement dans le proche avenir.


Utilisation des terrains.  Cette section doit contenir des informations
sur les utilisations actuelles et futures des terres situées sur la zone
d'influence.

La végétation naturelle et la production agricole doivent être décrites
en essayant de dégager les facteurs caractéristiques de transfert vers
l'homme, notamment par la chaîne alimentaire.

L'utilisation industrielle doit être décrite en mettant particulièrement
en lumière les influences réciproques de l'installation nucléaire et de
l'environnement industriel et les dangers que ce dernier peut recéler.

L'utilisation touristique peut, dans certains cas, poser des problèmes
importants.

Les voies de communication et d'accès peuvent créer des risques
(transports dangereux, ruptures d'ouvrages d'art) ou au contraire favoriser
l'intervention en cas d'accident.  Ces aspects doivent être décrits.

## 2.2. Influence de l'environnement naturel et humain sur l'installation

Comme on l'a signalé, le site peut être pour l'installation une source d'influences, voire d'agressions risquant de conduire dans certains cas à des accidents graves. Non seulement le réacteur lui-même peut être menacé, mais également des dispositifs annexes de la centrale dont l'importance pour la sûreté est considérable (salle de commande, piscine de stockage des combustibles irradiés, etc.).

Les manifestations provenant du site peuvent avoir des origines diverses qui peuvent être naturelles:
— phénomènes atmosphériques tels que tornades, orages, chutes de neige inhabituelles, brouillards givrants, etc.,
— phénomènes hydrologiques tels qu'inondations,
— phénomènes géologiques tels que séismes, glissements de terrain,
ou humaines:
— modifications importantes du cadre naturel, telles que création de grands ouvrages (barrages, voies navigables, ports) pouvant être eux-mêmes sujets à des accidents parfois catastrophiques (par exemple, rupture de barrage),
— création d'un environnement industriel (usines chimiques, ports pétroliers, raffineries, poudreries, voies de communications terrestres, navigables ou aériennes) risquant de constituer une source importante d'accidents, dont certains peuvent affecter la centrale si certaines précautions ne sont pas prises.

Il ne peut être question, dans le cadre du présent exposé, d'énumérer le détail des études qu'impose la prise en compte de ces différents facteurs. De telles études ne sauraient d'ailleurs être menées de façon systématique et générale, pour tous les facteurs simultanément. Leur utilité et leur définition précise se dégagent des problèmes concrets que pose — ou qu'a déjà posés — l'implantation de centrales sur des sites déterminés: elles sont conduites au premier chef par l'exploitant, puis elles sont jugées et éventuellement complétées à l'occasion des procédures d'examen de sûreté qui précèdent la délivrance des autorisations.

L'une des difficultés du problème réside dans la réalisation — puis la vérification — d'une protection suffisante et relativement homogène à l'égard des divers risques externes, en fonction de leur probabilité et de leurs conséquences.

Comme on le voit, les éléments d'une telle étude apparaîtront à partir de la description du site (paragraphes précédents) ou des études sur la météorologie, l'hydrologie ou la sismologie dont il sera question plus loin. La sélection des points ou aspects importants dans tous ces domaines est donc essentielle.


## 2.3. Météorologie

Comme cela découle des principes précédemment exposés les élément à prendre en compte dans cette section doivent viser un double but:
— établir l'influence des paramètres météorologiques sur la dispersion des effluents radioactifs durant l'exploitation normale et au cours de situations accidentelles,

— connaître les manifestations climatiques capables d'influencer le
comportement de la centrale sur le plan de la sûreté (vents très violents,
orages, neige, verglas, etc.),

Les éléments d'investigation vont donc se situer au plan de la météorologie
régionale et au plan local.

Le rayon d'investigation autour de la centrale sera fonction des zones
importantes de population à une distance pas trop lointaine (par exemple,
on ne doit pas se désintéresser de la présence d'une ville de un million
d'habitants si celle-ci est située à moins de 100 km de distance).

En conséquence, il y a lieu de prévoir la collecte de données
météorologiques:
— avant la construction et en permanence sur le site,
— éventuellement en quelques points éloignés de façon temporaire (un an)
si cela peut éclairer la connaissance des cheminements d'effluents.

Dans certains cas, des mesures par traceurs (gazeux notamment)
pourront apporter des informations complémentaires sur ces trajectoires,
tout en précisant les valeurs des coefficients de diffusion en certains points.

L'ensemble de ces études vise à prendre en compte l'influence,
souvent très importante, de la topographie; dans les cas où, pour des
raisons valables, une étude moins détaillée pourrait se justifier, il y aura
lieu malgré tout, pour l'évaluation des transferts, de prévoir certaines
marges de sécurité dans les calculs.

## 2.4. Hydrologie

Cette section doit donner un résumé des renseignements relatifs aux
eaux de surface et souterraines du site et des alentours qui peuvent avoir
une influence sur la sécurité, ou concerner le refroidissement.  On déterminera
les sources de l'eau et leur volume total, et les utilisations — tant présentes
que prévues — de l'eau ayant son origine dans la zone ou la traversant.
Une grande partie des renseignements nécessaires proviendront de relevés
existants, mais des études sur le terrain peuvent être nécessaires sur
certains sites.  Les facteurs à prendre en considération sont les suivants:

Eléments descriptifs.  Il s'agit des eaux superficielles:  eaux côtières,
lacs, fleuves, rivières, ruisseaux et marais;  et des eaux souterraines:
la ou les nappes aquifères présentes sous le site et les points d'eau s'y
alimentant.

Evaluation des transferts.  Pour les eaux superficielles, les paramètres
de transfert sont parfois difficiles à évaluer.  On s'efforcera néanmoins de
les préciser en fonction des données disponibles.  Des études expérimentales
pourront s'avérer nécessaires si le niveau de risque le justifie.

Pour les eaux souterraines, il y aura lieu de se préoccuper des
caractéristiques de transfert, notamment en cas d'accident.  La connaissance
de ces transferts peut être acquise:
— grâce à l'étude hydrogéologique détaillée du site permettant de déterminer
les directions et vitesses d'écoulement des eaux souterraines et leurs
exutoires,
— par la mesure du pouvoir d'échange d'ions des terrains — des échantil-
lons en provenance des sondages peuvent être testés en laboratoire en
vue de connaître leur pouvoir de rétention à l'égard des ions radioactifs,
— éventuellement par des mesures directes in situ au moyen de traceurs.

## 2.5.  Géologie et sismologie

Ce chapitre doit contenir un résumé des données géologiques existantes ou d'études faites spécialement à cette fin sur le terrain.  On insistera sur les aspects géologiques particuliers à prendre en compte pour la conception des structures et de la sécurité de l'installation.

Lorsque le site se trouvera dans une zone d'activité sismique notable, l'étude sismique devra être orientée, outre la recherche précise des épicentres importants, vers l'investigation des caractéristiques de transmission des ondes sismiques dans la zone du site.  On tentera aussi de préciser le séisme caractéristique à prendre en compte pour le dimensionnement de l'installation.

Actuellement l'intensité de chaque site est repérée d'après les effets produits, dans l'échelle internationale MSK.  Mais l'expérience, notamment celle des derniers séismes destructeurs observés, tend à montrer que la connaissance de telles intensités fondée sur des observations historiques ne suffit pas à prévoir l'intensité des séismes risquant de se produire dans l'avenir avec une probabilité non négligeable.  Des études se poursuivent en vue d'améliorer cette prévision.  Elles sont fondées davantage sur les notions de magnitude des séismes (énergie libérée au foyer) et de provinces tectoniques que sur la notion, très empirique, d'intensité ressentie sur le site.

Jusqu'ici, et pour tenir compte des incertitudes de la méthode actuelle, les centrales ont été conçues de façon que le maintien des fonctions de sûreté du réacteur, le confinement des substances radioactives en cas d'accident, la protection sanitaire et le contrôle des rayonnements ionisants soient assurés pour un séisme d'intensité supérieure d'une unité (dans l'échelle MSK) à l'«intensité maximale probable» sur le site, déduite des intensités observées en tenant compte des données géologiques locales.

Une autre difficulté réside dans la nécessité de passer de l'intensité prise comme référence, qui est de nature qualitative, aux mouvements réels du sol engendrés par le séisme correspondant à cette intensité et aux effets qu'ils engendrent sur les structures.  Les études en cours ont également pour objet de mieux résoudre ce problème, à partir de la connaissance de la magnitude du séisme de référence et de la distance de son foyer par rapport au site.  Elles distinguent les fréquences des mouvements[2] (notion de spectre de fréquences) et tiennent compte de leur amortissement au sein des structures selon ces fréquences.

A partir des accélérations, des vitesses et des déplacements imprimés aux structures, ainsi déterminés en fonction de la fréquence, les calculs de génie parasismique permettront mieux de déterminer les dispositions à prendre au stade de la conception et de la construction pour assurer la sûreté dans les conditions imposées.

La section pourra comprendre, suivant les besoins, les renseignements suivants:
— géologie régionale,
— stratigraphie,
— existence de failles sur le site et au voisinage,
— histoire sismologique de la région,
— études géologiques et sismologiques sur le terrain,
— sources des données géologiques et sismologiques.

## 2.6. Ecologie

Ce chapitre doit traiter des processus de transfert de la contamination
par le milieu biologique (flore et faune et plus particulièrement agriculture
et élevage) vers les groupes de population qui peuvent être exposés. Il
est donc fondé sur:
— la connaissance de la flore et de la faune,
— la description de l'agriculture et de l'élevage de la région,
— les résultats des transferts directs depuis l'installation par la voie
  atmosphérique ou la voie hydrologique.

On sélectionnera les voies critiques de transfert de la contamination
afin d'évaluer les conséquences radiologiques correspondantes.

## 2.7. Etat initial du site

Il est utile d'établir une carte de l'irradiation naturelle autour du site.
On évaluera la dose d'irradiation que le public reçoit en dehors de toute
activité nucléaire. Par la suite on tentera de comparer les doses dues au
fonctionnement des installations avec celles dues à la radioactivité naturelle.

## 3. RECHERCHE D'UN CRITERE DE JUGEMENT SUR LA DENSITE DE POPULATION AUTOUR D'UN SITE

La caractéristique commune aux grandes installations nucléaires réside,
on l'a dit plus haut, dans la présence au sein de l'installation de quantités
très importantes de produits radioactifs. Dans les réacteurs nucléaires,
cette radioactivité se trouve, pour une grande part, intimement mêlée au
milieu générateur d'énergie: le cœur. Dans l'hypothèse où à la suite d'une
avarie grave, une fraction notable de cette radioactivité serait dispersée
dans l'environnement de manière à atteindre les populations résidant dans
les zones proches de l'installation, de nombreuses personnes pourraient
être irradiées, dont certaines gravement. L'estimation des conséquences
de tels accidents a été effectuée dès 1957 par la Commission américaine
de l'énergie atomique (AEC) (Rapport WASH-740, mars 1957), reprise plus
récemment par Forbes et al. (Nucl. News, septembre 1971) et a été tout
récemment remise en cause par les conclusions du rapport WASH-1400
(projet, août 1974), dit «Rapport Rasmussen».

La probabilité d'apparition de tels accidents est difficile à estimer
avec précision mais elle est très faible: on avance des limites supérieures
de l'ordre de un événement par installation et par dix millions d'années
de fonctionnement. Cependant rares sont les cas où il est possible d'affirmer
qu'une telle probabilité est nulle, même en tenant compte de la conception
de l'installation et des dispositions de sûreté qui sont prises. C'est pourquoi
il est sage de tenir compte d'une telle situation à l'égard des populations
vivant dans l'environnement immédiat ou proche de l'installation.

Actuellement, d'après ce que l'on sait des pratiques dans les pays
industrialisés où l'énergie nucléaire connaît un développement important,
aucun gouvernement n'a jusqu'ici voulu prendre la responsabilité d'autoriser
l'implantation de grandes centrales nucléaires en bordure des villes ou
dans des zones très peuplées. Des problèmes précis ont été examinés aux
Etats-Unis d'Amérique (Ravenswood), en République fédérale d'Allemagne

(Ludwigshafen notamment) et en Suède (site de Vartan dans la zone urbanisée de Stockholm). A notre connaissance ces problèmes ont été résolus par la négative.

Les Etats-Unis possèdent une réglementation prévoyant autour de chaque installation une zone d'exclusion entourée elle-même d'une zone à faible densité de population: les dimensions de ces zones sont évaluées de façon à éviter toute irradiation importante de la population, dans l'hypothèse d'un accident grave, du type de ceux qui sont retenus pour le calcul du dimensionnement (Design Basis Accident). Dans les autres pays, les décisions ont été prises par cas d'espèces mais toujours dans le sens d'une assez grande prudence.

Il est donc difficile de se référer, en cette matière, à une pratique bien établie. La réglementation américaine n'est pas directement adaptable à la France. Elle est typique d'un pays où les régions peu ou très peu peuplées sont nombreuses. C'est pourquoi on essaie de mettre au point une méthode qui permettrait de classer les sites au stade précoce de la «présélection» et d'éliminer ceux qui pourraient se présenter de façon défavorable.

Les propositions de Farmer au Colloque de Vienne (AIEA), en 1967 [3], sont bien connues. Ces propositions ont fait depuis l'objet de développements, de perfectionnements et, il faut le dire, de critiques. L'idée de départ était qu'il ne fallait pas refuser de considérer des accidents très graves (par exemple un rejet de $10^6$ Ci d'iode pour un réacteur de puissance mais que leur probabilité d'apparition devait leur être associée. Il était proposé (comme intuitivement admissible) que la probabilité soit supposée d'autant plus faible que le rejet radioactif était plus grand.

On peut par exemple noter ici que la probabilité de rupture de grands ouvrages (barrages par exemple) est en général plus importante que celle d'un accident majeur de réacteur nucléaire et que la parade aux conséquenc est généralement incertaine.

L'étude actuellement conduite par Rasmussen pour le compte de l'AEC utilise de façon plus élaborée le couple probabilité d'apparition-conséquence [4].

Récemment, Le Quinio, au Colloque de Portorož (AIEA), a présenté une approche intéressante proposant un «facteur de pondération des densité de population sur un site» [5].

La plupart des études que l'on peut rencontrer actuellement proposent des approches de plus en plus élaborées. Cependant, Le Quinio et la plupar des auteurs le reconnaissent, la recherche d'une théorie très élaborée se heurtera encore longtemps à des difficultés. En effet, il est bien clair que d'une part la saisie de certaines données sera pendant longtemps encor très imprécise (par exemple les probabilités d'apparition des accidents graves), d'autre part, la nature des conséquences, c'est-à-dire des effets sur le milieu naturel et humain, est très complexe, et par conséquent difficile à cerner par une formulation physique.

Il faudra donc se contenter de rester empirique dans ce domaine.

### 3.1. Position du problème

Rasmussen a donné récemment aux Journées AIF de Lucerne une présentation du problème de l'impact des accidents dont les données de départ peuvent être adoptées, nous semble-t-il, comme schéma.

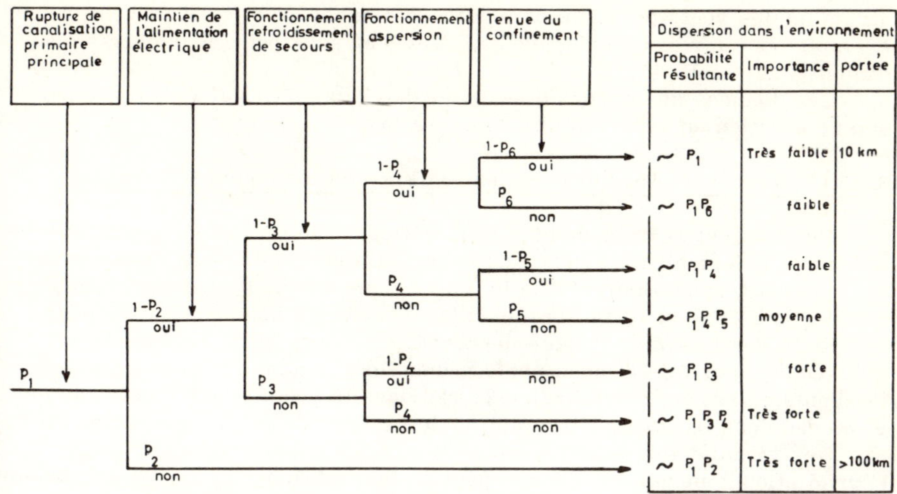



FIG. 1. Arbre d'événements conduisant à un rejet donné. On a supposé que la probabilité de non-tenue du confinement était différente suivant que l'aspersion fonctionnait ou non. Pour la composition des probabilités on suppose les termes en 1-P très proches de l'unité.

Pour donner une idée du problème, il est difficile de raisonner d'emblée sur le plan général de toutes les installations nucléaires. Aussi choisirons-nous comme exemple l'un des plus communs actuellement: celui de la perte de réfrigérant dans un réacteur à eau ordinaire. On peut proposer le schéma de la figure 1.

Ce schéma montre que, dans la recherche logique qui conduit à établir l'arbre d'événements, il y a lieu de ne pas négliger des situations dont on ne tient pas compte généralement, parce qu'elles surviendraient après l'accident principal (ou de dimensionnement), mais dont la probabilité partielle d'apparition n'est finalement pas très faible. Bien entendu, la méthode utilisant le schéma des arbres d'événements permet également de prendre en compte tous événements en «amont» de l'accident de dimensionnement et notamment toutes les agressions en provenance du site (agressions atmosphériques, hydrologiques, sismiques ou d'environnement industriel). On peut ainsi classer à divers niveaux de l'échelle de probabilités divers groupes d'événements ayant des conséquences déterminées.

## 3.2. Météorologie et répartition géographique des populations

Un accident nucléaire grave peut avoir des conséquences sur les populations environnantes à travers deux vecteurs initiaux qui sont l'eau et l'atmosphère. On peut bien entendu considérer l'ensemble des deux vecteurs. Mais il est aisé de remarquer que dans la quasi-totalité des cas, le vecteur critique est l'atmosphère. Le maintien du confinement atmosphérique est plus difficile et le transfert par l'air beaucoup plus rapide.

Le spectre des situations possibles pourrait être détaillé; néanmoins il nous paraît raisonnable d'utiliser la distribution des fréquences de direction des vents pour une ou plusieurs classes de vitesse de vent, par exemple pour les sites français 1 m/s et 5 m/s, et pour deux classes de diffusion verticale [1,6,7].

Les données démographiques sont individualisées autour du site par une répartition en secteurs de 20 degrés [6].

### 3.3.  Critères radiologiques — Seuils d'irradiation pour les personnes

Ce point est certainement le plus délicat et le plus sujet à controverse si l'on se réfère à l'expérience du passé.  La notion de seuil de dommage radiologique, dès qu'on prétend lui donner une assise biologique, échappe rapidement à une analyse simple.  L'une des voies souvent choisie pour définir le seuil au-delà duquel une irradiation sera prise en compte est de prendre deux fois la dose professionnelle d'irradiation concertée pour l'organisme entier, c'est-à-dire 25 rad dans le cas de l'irradiation externe et 30 rad dans le cas de l'irradiation de la thyroïde par l'iode radioactif [8].

Des irradiations supérieures à 10 fois ces valeurs (250 rad pour l'irradiation générale ou 300 rad pour la thyroïde) seront prises en compte pour définir un deuxième niveau.

Si N est la norme professionnelle précédemment citée, l'irradiation du premier type (type I)  serait définie par:

$$2 \, N < \text{irradiation de type I} < 20 \, N$$

L'irradiation de type II serait définie par:

$$20 \, N < \text{irradiation de type II}$$

On peut se demander s'il est souhaitable d'adopter un double niveau d'irradiation.

Les irradiations de niveau II pourront se produire de façon significativ pensons-nous, dans le cas où des agglomérations proches ou très proches du site existeraient.  La prise en compte d'un deuxième niveau devrait permettre de mieux identifier et souligner de tels cas.  Mais il est trop tôt pour dire si ce concept s'avérera nécessaire et il faut attendre l'étude de cas réels pour être mieux fixés sur son utilité.

### 3.4.  Présentation des graphes de conséquences

Nous aurons donc à examiner sur un site les conséquences d'événemen de gravité croissante se situant sur une échelle de probabilités décroissant

Comme nous l'avons vu, il sera possible de calculer les conséquences radiologiques de chaque accident ainsi défini, soit dans un secteur angulair soit à l'intérieur d'un secteur, dans un «élément», c'est-à-dire un trapèze compris entre deux rayons déterminés.  Le calcul de ces conséquences radiologiques fait bien entendu intervenir la fréquence d'apparition des situations météorologiques, qui vient se combiner avec la probabilité d'apparition de l'accident.

Si dans une première série d'analyses nous retenons la présentation au seul niveau des secteurs angulaires nous obtiendrons une double série de graphiques (fig.2).

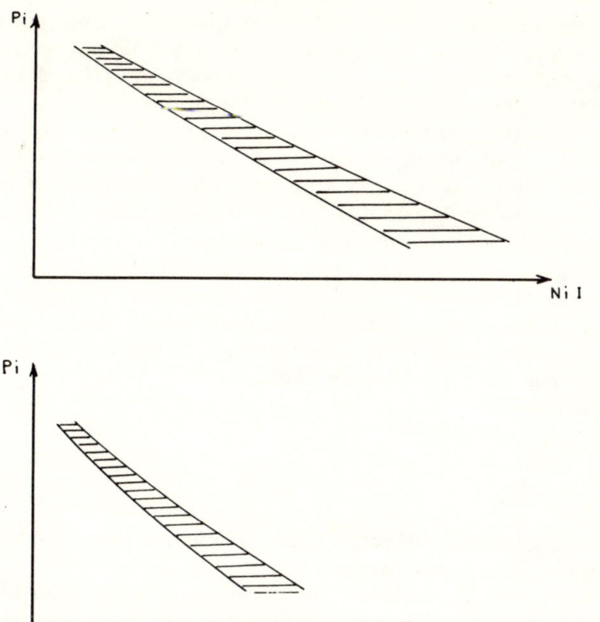

FIG. 2.  Graphes des conséquences radiologiques.

$P_i$     : Probabilité d'apparition d'un accident donné dans le secteur angulaire de rang i;  $N_i$ I:  nombre d'irradiations de type I pour le secteur i;  $N_i$ II:  nombre d'irradiations de type II pour le secteur de type i.

En fait, si l'on superpose pour un secteur les diverses situations météorologiques possibles, les points seront situés à l'intérieur d'une certaine zone hachurée représentée sur le diagramme.

Une autre présentation, suggérée par Le Quinio, consisterait à prendre la moyenne stochastique des nombres d'irradiés dans un secteur pour les différents cas possibles de situations météorologiques dans ce secteur. Si $P_i$ est la probabilité d'apparition d'une des situations dans le secteur i et $N_i$ (I ou II) le nombre d'irradiations de type I ou II pour le secteur i, on écrit:

$$\left( \sum P_i \right) \overline{N}_i \text{ (I ou II)} = \sum P_i \, N_i \text{ (I ou II)}$$

On obtient ainsi une valeur $\overline{N}_i$ (I ou II) unique pour toutes les situations météorologiques du secteur.  Le graphe de représentation deviendrait une courbe au lieu d'une zone comprise entre deux courbes.

## 3.5.  Approche d'une définition de critères

La première démarche vers la définition d'un critère est de comparer une situation projetée (ici le tableau probabilités-accidentés) à une situation définie comme acceptable.

La définition d'un site acceptable comportera de toutes façons une décision basée sur un jugement et il ne peut en être autrement. Néanmoins, il faut proposer une approche raisonnable pour une telle décision. Aussi peut-on à notre avis proposer la démarche suivante.

Pour une nation européenne comme la France, compte tenu de son contexte démographique, quelques sites pour grandes centrales électro-nucléaires ont été choisis. Ces sites ont été acceptés et la construction des centrales nucléaires décidée. Il est raisonnable, pensons-nous, de considérer de tels sites comme bases de comparaison (ils se présentent favorablement du point de vue des conséquences d'accidents). Pour les installations de même type sur d'autres sites, les diagrammes probabilités-irradiations peuvent être tracés. La comparaison dès lors établie entre les sites «acceptés» et les sites «proposés» devrait faciliter l'établissement de critères de comparaison permettant de guider les responsables au moment du choix.


4.   CONCLUSION

Nous avons tenté de décrire dans cet exposé la méthode pratiquée en France pour l'étude des sites nucléaires.

Cependant le choix de nouveaux sites incite l'Autorité administrative à s'assurer l'appui de critères de choix de caractère rationnel, sans toute-fois perdre de vue les réalités géographiques, économiques et humaines.

La méthode proposée ici s'efforce de contribuer à atteindre ce but. Elle nous semble devoir être fructueuse si l'on veut bien noter les remarque suivantes:
- les niveaux d'irradiation proposés comme seuils sont basés sur des valeurs communément admises,
- les probabilités d'accident sont encore mal établies en valeur absolue mais sont peut-être mieux cernées au niveau relatif, plus utile dans notre présentation,
- la méthode incite naturellement à une évaluation précise des conséquences d'accidents,
- la méthode tient compte des situations météorologiques, géographiques et démographiques de sites réels dans un pays déterminé, elle est donc basée sur les résultats d'observations objectives concernant de tels sites,
- la méthode tente d'aboutir à la définition de critères par comparaison à des choix dûment acceptés dans un pays déterminé.

Bien entendu l'objet de cette méthodologie est de permettre notamment aux responsables de proposer une présélection des sites à un stade d'investigation très préliminaire. L'étude pourrait, à ce stade, être effectué avec des éléments d'information (notamment météorologiques) éventuelle-ment sommaires. En tout état de cause, elle ne saurait dispenser de l'étude approfondie cas par cas, qui prend en compte la totalité des éléments du site susceptibles d'intervenir dans la définition des dispositions de sûreté à mettre en œuvre sur l'installation.

Il faut également souligner que l'intérêt d'une telle méthode, à notre avis, réside dans la souplesse qu'elle peut apporter dans la prise de décision. En effet, dans les cas «peu défavorables» le site pourra être encore retenu en exigeant un renforcement de la sûreté de l'installation.

Le graphique montre en effet qu'on peut atteindre une «zone acceptable»
en se déplaçant sur l'échelle des probabilités ou sur celle des conséquences,
et ceci ouvre la voie à une large gamme de solutions envisageables.

Enfin pour les sites jugés acceptables mais dont on pourrait craindre
une évolution démographique défavorable au sens des critères présentés,
la méthode devrait permettre de proposer des mesures appropriées pour
limiter l'expansion démographique soit sur l'ensemble du site, soit simplement sur certains secteurs.


## REFERENCES


[1] DOURY, A., GERARD, R., « Recherche d'un critère d'évaluation de site par la pondération des données
    radiologiques, météorologiques et démographiques», ces comptes rendus, IAEA-SM-188/18.
[2] BARBREAU, A., FERRIEUX, H., MOHAMMADIOUN, B., «Les études sismologiques effectuées au CEA
    dans le domaine de la sûreté des sites nucléaires», ces comptes rendus, IAEA-SM-188/17.
[3] FARMER, F.R., «Siting criteria — A new approach», Containment and Siting of Nuclear Power Plants
    (C.R. Coll. Vienne, 1967), AIEA, Vienne (1967) 303.
[4] RASMUSSEN, N.C., «The approach of the USAEC study to the public risks of power reactors», Journées
    AIF (At. Int. Forum), Lucerne, 5-8 mai 1974.
[5] LE QUINIO, R., «Influence de la puissance et de la distance sur les risques présentés par un réacteur
    nucléaire — Facteur atmosphérique de site», Population Dose Evaluation and Standards for Man and
    his Environment (C.R. Journées d'études Portorož, 1974), AIEA, Vienne (1974) 83.
[6] Annexe du Rapport DSN n° 42, Département de sûreté nucléaire, CEN de Saclay (1974). (Cette note,
    rédigée par A. Doury, propose des valeurs de base pour les transferts atmosphériques.)
[7] DOURY, A., Une méthode de calcul pratique et générale pour la prévision numérique des pollutions
    véhiculées par l'atmosphère, Rapport CEA-R-4280 (1972).
[8] Décret n° 67-228 du 15 mars 1967, Annexe III, Equivalents de dose maximaux, J. Off. Répub. Fr.



## DISCUSSION

M. NASIM (Chairman): Are the IAEA guidelines relating to the siting
of nuclear power plants followed in France? In what way do you depart
from them, if at all?

P. CANDES: The IAEA guidelines serve as the basis for defining the
content of the site safety report. To that extent there is little difference
between the IAEA recommendations and French practice. However, at
each specific site greater importance will be attached to one aspect or
another, depending on the more or less serious problems that come up.
An illustration of what I mean is to be found in the seismic problems which
affect certain sites, and in the problems of industrial environment presented
by other sites investigated recently.

R. COULON: Could you clarify the term 'exposure' in relation to the
threshold of 250 rem?

P. CANDES: The purpose of the suggested double threshold (25 rad
and 250 rad) is essentially to obtain a classification of irradiated persons
as a means of comparison of the sites studied.

# TERRITORIAL AND ENGINEERING REQUIREMENTS IN NUCLEAR STATION SITING

C. DELLARCIPRETE, V. MORELLI
Ente Nazionale per l'Energia Elettrica,
Rome, Italy

**Abstract**

TERRITORIAL AND ENGINEERING REQUIREMENTS IN NUCLEAR STATION SITING.

In view of the very large number of factors to be considered in siting a nuclear station, it is necessary to proceed through a succession of selections carried out by introducing discriminating factors to simplify and rationalize the work. The first selection is made on the basis of a number of factors capable of characterizing the territory in general. Then, among the prospective sites, a further selection is made, taking engineering requirements into account. A survey is made of the primary characteristics of the territory and engineering requirements affecting the choice. This paper proposes a scheme for collection and examination of primary territorial data; a functional diagram is then presented to illustrate the activities to be carried out to ascertain that the site is suitable also from the engineering standpoint.

## 1. PLANT SITING PROCESS

The factors which combine to determine the choice of a nuclear plant site are so numerous that it would not be convenient to collect them and consider them together for a vast territory such as a whole nation or a whole region. A rational choice can only be reached by steps, i.e. by proceeding through a series of intermediate preselections based on the gradual introduction of more and more discriminating factors that narrow down the area of investigation and simplify the selection process. It is thus advisable to organize the investigation in a succession of phases, and at each phase consideration should be given only to the sites singled out in the preceding phases.

Some of the elements that combine in the choice are of general interest to the life of the country in which a plant is to be sited; these are the elements that define the territory in its broad outlines and can be indicated as 'primary characteristics'. On the other hand, there are other, more specific, elements whose interest is more circumscribed: these are the elements that are closely linked to the peculiar requirements of the activities to be carried out and can be called 'secondary sectorial characteristics'.

Because of their general scope, primary characteristics are normally recorded systematically for the entire national territory by the various pertinent bodies; they are thus known and can be obtained more or less easily, even though they are not always in a form suitable for a territorial investigation. However, sectorial characteristics very often concern punctiform areas and can only be recorded case by case.

For these reasons, it is advisable, in making a site selection, to adopt the criterion whereby the territory is first analysed through its primary characteristics and a preliminary set of choices made; then a localized study is carried out on the areas singled out, in the course of which all the

data are collected and in-depth investigations performed in relation to the
sectorial characteristics affecting the plant.  In brief, the sites are first
selected from the standpoint of the relationship between territory and plant,
and then the selected sites are further investigated to ascertain that the
characteristics of the territory are compatible with the requirements of
the plant.  This means fulfilling first the general needs of the country and
then those of the sector of activity to be developed,  subordinating a particular interest to the general interest and fitting industrial plans into the
broader national plans.

This being so, the work relating to the qualifications of a nuclear plant
site can be organized into three phases [1]:

(a)  The first is aimed at a preliminary screening based on general
     studies of vast areas of territory;
(b)  The second examines the preselected sites on the basis of sectorial
     and engineering studies;  and
(c)  The third is implemental and includes all the activities preceding
     plant construction.


2.   SITE SELECTION BASED ON PRIMARY TERRITORIAL CHARACTER-
     ISTICS (FIRST PHASE)

The first phase is subdivided into three groups of activities developed
in series and concerning:

(a)  Investigation to identify the primary characteristics of the
     territory;
(b)  Synoptic representation of these characteristics;  and
(c)  Selection based on the collected data.

## 2.1.  Primary territorial characteristics

The primary characteristics of the Italian territory have recently been
identified,  co-ordinated and described by the Ente Nazionale per l'Energia
Elettrica (ENEL) [2] in a document covering the whole coastal belt, which
has been indicated [1] as the part of Italian territory of greatest interest
for constructing conventional thermal and nuclear power plants[1].  For
completeness and order, we shall briefly describe the main lines of this
work and refer whoever may be interested in the details to the document
itself.

The primary territorial characteristics identified by ENEL are:

(a)  Present territorial configuration:
     (i)  Topographical,  altimetric and bathymetric characteristics;
     (ii)  Land use:  distribution and extent of inhabited centres,  types of
          cultivation,  archaeological areas,  national parks and natural
          preserves,  shooting preserves,  oases for fauna protection, etc. ;

---

[1]  The need for cooling water calls for a first basic selection that, here again, will take into account
the availability and utilization of the resources (sea, rivers, lakes, etc.) on one hand, and the essential
engineering requirements on the other.

(iii)  Infrastructures: ports, roads, railways, airports, power systems,
       etc.;
(iv)   Seismicity: maximum recorded intensity, epicentre locations
       and intensities, legal seismicity classification;
(v)    Hydrography;
(vi)   Administration division: region, provinces, municipalities.
(b)  Land planning:
   (i)    National plans: natural assets to be protected, biotope maps,
          proposed parks and natural preserves;
   (ii)   Regional plans: territorial co-ordination plans;
   (iii)  Municipality plans: town planning and construction programmes;
   (iv)   Limits: landscape, archaeology and monuments, military,
          airports, hydrogeology.
(c)  Demographic characteristics:
   (i)    Resident population;
   (ii)   Population variation;
   (iii)  Seasonal population variations, equivalent inhabitants.
(d)  Development indexes:
   (i)    Agricultural development indexes;
   (ii)   Industrial development indexes;
   (iii)  Touristic development indexes.


These are very general characteristics and can profitably be used in
the site selection for any type of activity, be it industrial, touristic or
other.  At any rate, it is our experience that a study restricted to the
primary characteristics alone, even though they form a rather limited
basis, is an efficient means for a selective investigation of the possibilities
offered by Italian territory for nuclear plant sites.

## 2.2.  Criteria for representation of the territory

A national territory is an extremely complex 'system' and can be
regarded schematically as an aggregate of subsystems — its primary
characteristics — stratified over it and linked one to another in complex
relations of interdependence.  A set of superimposed topical maps (one for
each characteristic) provides an accurate representation of this type of
territorial structure, but does not allow all the characteristics to be viewed
simultaneously, so that a synthesis must be made for every point of the
territory.

Because of its typically linear structure, the coastal belt lends itself to
synthesized representation when it is considered to unfold, within certain
limits, in one dimension.  It is thus possible to make out a 'vertical cross-
section ' of the territory across this ideal 'linear' coast, so that all the
primary characteristics can be transferred to a synoptic chart consisting
of a set of successive diagrams placed one on top of the others against a
common abscissa representing the development of the coast (Fig. 1).

This graphic representation makes it possible to:


(a)  Gather and co-ordinate all the main information on the territory
     in a schematic format;
(b)  Promptly provide the stored information on each point of the coast;

(c) Carry out a rapid comparison of all the data pertinent to one coastal area with others; and

(d) Determine the necessary correlations for correct 'reading' of the territory[2].

## 2.3.  Site selection procedure

Once the geographical subdivision (e. g. the region) in which the plant is to be sited has been singled out on the basis of the network development plans, the charts produced by ENEL can be used for a tentative choice based on the primary territorial characteristics.

In the charts the territorial characteristics are denoted chromatically, so that they are more evident and easily distinguishable. (For technical reasons, the chart is not reproduced in colour here. ) Colours can be used to denote selective tasks or limits (e.g. grade elevation, special uses of the land, nature protection, etc.) which generally indicate unfavourable conditions for plant siting. In this manner, by successive eliminations, identification of the most suitable areas for plant siting is immediate.

It should be pointed out that the presence of a colour does not constitute an absolute preclusion, but merely an indication that a certain problem requires further attention, as it may be more difficult to find a solution in that area than elsewhere. This criterion was further developed in the representation of seismicity, so that a deeper colour means a high seismic intensity. The projection of all the colours on the abscissa emphasizes the uncoloured areas, which indicate absence of adverse factors, and are thus best suited for plant siting. In view of the numerous requirements, restraints and limitations conditioning site eligibility, it may be that an area in which no unfavourable factors exist is not available. In such a case it is necessary to compare the areas having the least number of adverse elements (i. e. the least coloured areas). The comparison is then completed by comparing the factors that can be expressed numerically (population classes, equivalent inhabitants, development indexes, etc.). This scrutiny makes it possible to identify the areas that, from a merely territorial standpoint, are eligible for plant siting. The aggregate of areas thus identified constitutes the 'site bank'.

## 2.4.  Territorial data bank

When a territory is subject to rapid evolution, as is the case for the coasts of the Italian peninsula where the population density is about 470 inhabitants/km[2] and grows at a rate of approximately 12% every ten years[3], the conclusions drawn at a certain date are not valid in an absolute sense, but require periodical verification. The ultimate purpose of the first phase of investigation is the formation, not so much of a site bank, but rather of a territorial data bank to be kept up to date at all times so that the evolution of the territory can be followed closely and a real-time analysis can be performed.

---

[2]  The graphic representation referred to can easily be applied to any linear territorial structure and has thus been adopted by ENEL also to study lakesides and riverbanks.

[3]  Growth rate recorded between the 1961 and 1971 censuses.

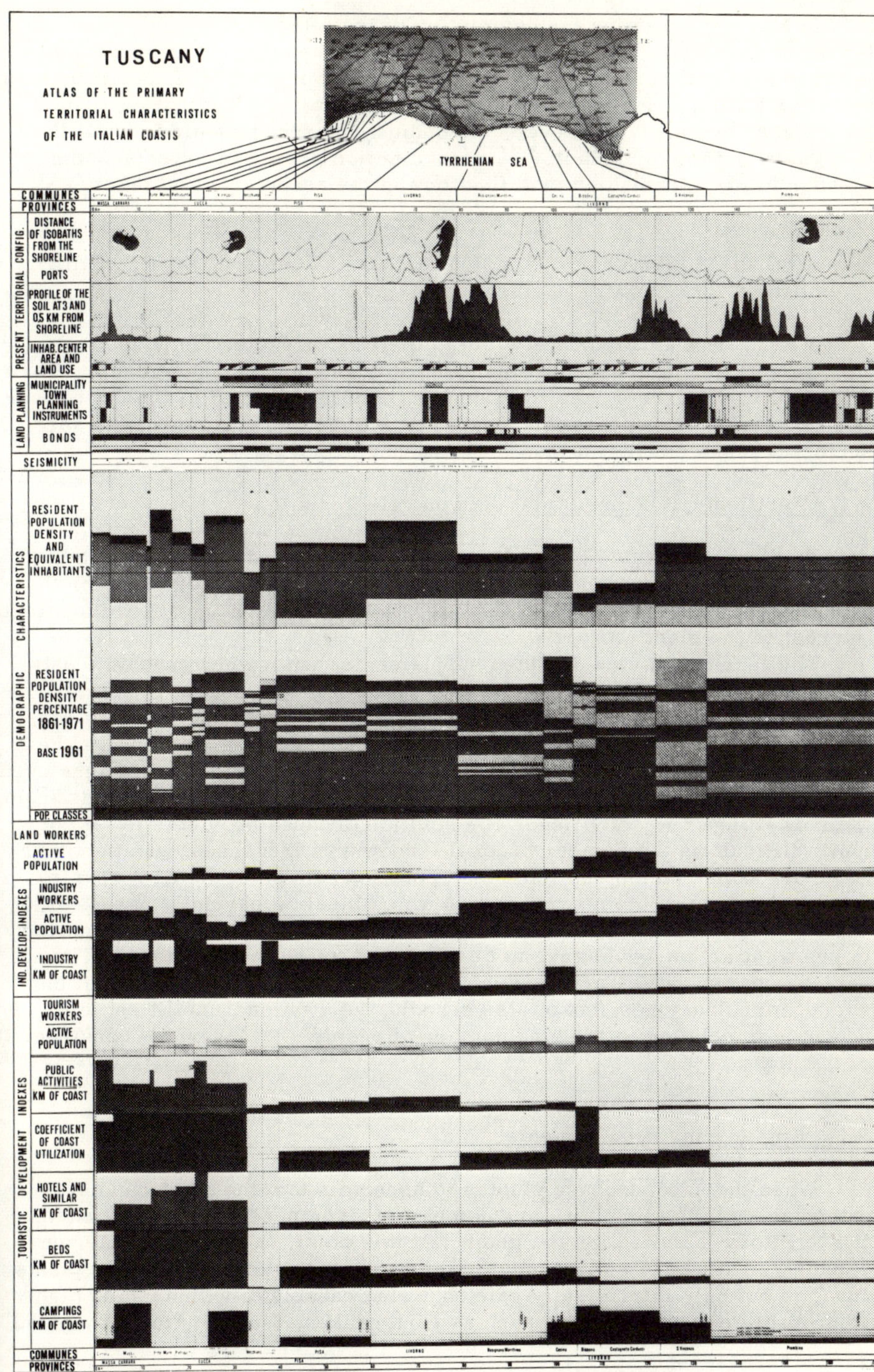


FIG. 1.    Example of schematic representation of a linear territorial structure.

In this way, pending the completion of the allocation of the whole territory also from the energy point of view[4], the authorities responsible for land planning are provided with the means to make well founded choices, based on a complete and up-to-date understanding of the situation. Once the choices are made, however, the bodies concerned should take all the necessary action (by territorial co-ordination plans and town-planning means) to prevent further modifications in the ascertained situation.

## 3. SITE QUALIFICATION IN RELATION TO ENGINEERING REQUIRE-MENTS (SECOND PHASE)

The areas identified as eligible in the first preselection phase must then be scrutinized from the standpoint of engineering requirements. Before this scrutiny is carried out, the territorial data collected in the first phase must be verified over a wider area; then a full understanding of the specific sectorial requirements in that area must be acquired.

### 3.1. Verification of territorial data on selected sites

The land surrounding the eligible site is surveyed over a vast radius, and the same information collected on the coastal belt is recorded for the inland area, with the addition of any territorial data that may be of particula interest to the plant project.

The territorial data are then supplemented with socio-economic data, which will bring to the fore any impact of the plant on the surrounding areas. In building a nuclear plant, it is necessary to provide a rather large constru tion area and to impose some limitations on the use of the peripheral belt. The physical insertion of a plant in the territory thus entails problems such as the possible alternative uses of the land and removal of inhabitants from their dwellings and activities. The solution of these problems is all the more difficult as the number of local landowners increases. On the other hand, the construction and later the operation of a plant act as a spur to the development of the nearby areas, which must be carefully assessed to ensure harmonious integration into the social context. It is thus necessary to see what action would have to be promoted towards town planning (roads, parking spaces, green areas, receptivity, etc.) and to ascertain that the social infrastructures (hospitals, schools, recreational facilities, etc.) are adequate in relation to the high number of people working on the construction of the plant.

### 3.2. Engineering requirements

When the second-phase studies to ascertain that the plant can be integrated into the territory are completed, it is necessary to see whether the territory is suited for the plant. At this point, the guiding criterion is safety, both for the plant and for the nearby population. Nuclear plants must be not only intrinsically safe, but also capable of withstanding highly improbable external events, such as earthquakes, tsunami, tornadoes and

---

[4] This data bank may also be useful for land planning.

floods, without creating a hazard to the population[5]. From this point of view knowledge of the seismicity and tectonics in Italy is a selective factor of paramount importance. In other countries, of course, the main selective factor may be something else. Because of the complexity of the studies required to determine the reference acceleration for plant design and owing to the shortage of available data in Italy today on the geology of the quaternary, studies have so far been made on well defined areas of the coast that are being considered for possible plant siting. However, since the number of sites under examination is rather high, we feel we shall be able to complete the general assessment of all the Italian coast within the coming years. Knowledge of the seismic and tectonic characteristics may then become an important sectorial selective tool on a national scale. On such a scale, seismicity is for the moment merely a means for a first broad selection, as the exclusions are based only on the maximum intensities recorded to date and on probabilistic considerations.

Investigations on a national scale may also be carried out for other events such as tsunami and tornadoes, of which no orderly historical record exists today and which will be of great assistance to this type of survey.

Other sectorial elements change from one point to another and must therefore be considered only after actual selection of the site. The most important of these elements, because of their high selectivity, are, as will be shown, the local geology and soil mechanics.

Local geology aims essentially at checking for the presence of active faults near prospective sites. An aid to this investigation is provided by the results of geophysical and stratigraphic campaigns performed over wide areas for other purposes, e.g. oil prospecting. Unfortunately, there are no national publications yet that provide a co-ordinated synthesis of the data, and some of the data available were obtained without the advantage of modern methods, so that they are of limited use. For this reason it is necessary to supplement any data available for a particular area with field tests and soil dating (micropaleontology, paleomagnetism, radioisotopes) to establish when any observed tectonic disturbance took place.

In any case, the critical path goes through the activities aimed at determining the geotechnical properties of the soil. These generally vary from one point to another in even a small area; therefore, they are first determined on broad lines, by means of a widely spaced mesh of boreholes, to find which area is most suited for plant construction, and then by means of a thicker mesh in that area, to set the preliminary design data. This is one of the most delicate stages of the survey because, after months of work, a negative result may come up and the site have to be abandoned.

For coastal sites, hydrological and hydrographical data combine to define the plant gradient in relation to the possibility of extraordinary floods. Determination of the plant gradient is also useful to establish the additional load on the soil when earthfills are required.

Meteorological and thalassographic studies serve to establish possible interactions with the surrounding air, land and sea [3] and to provide the information required for the design of the plant structures.

_________

[5] Once the plant site has been chosen correctly on the basis of the land features, and when plant safety in respect of external events has been ensured, with the engineered safeguards currently prescribed for nuclear plants there should be no problem of population safety.

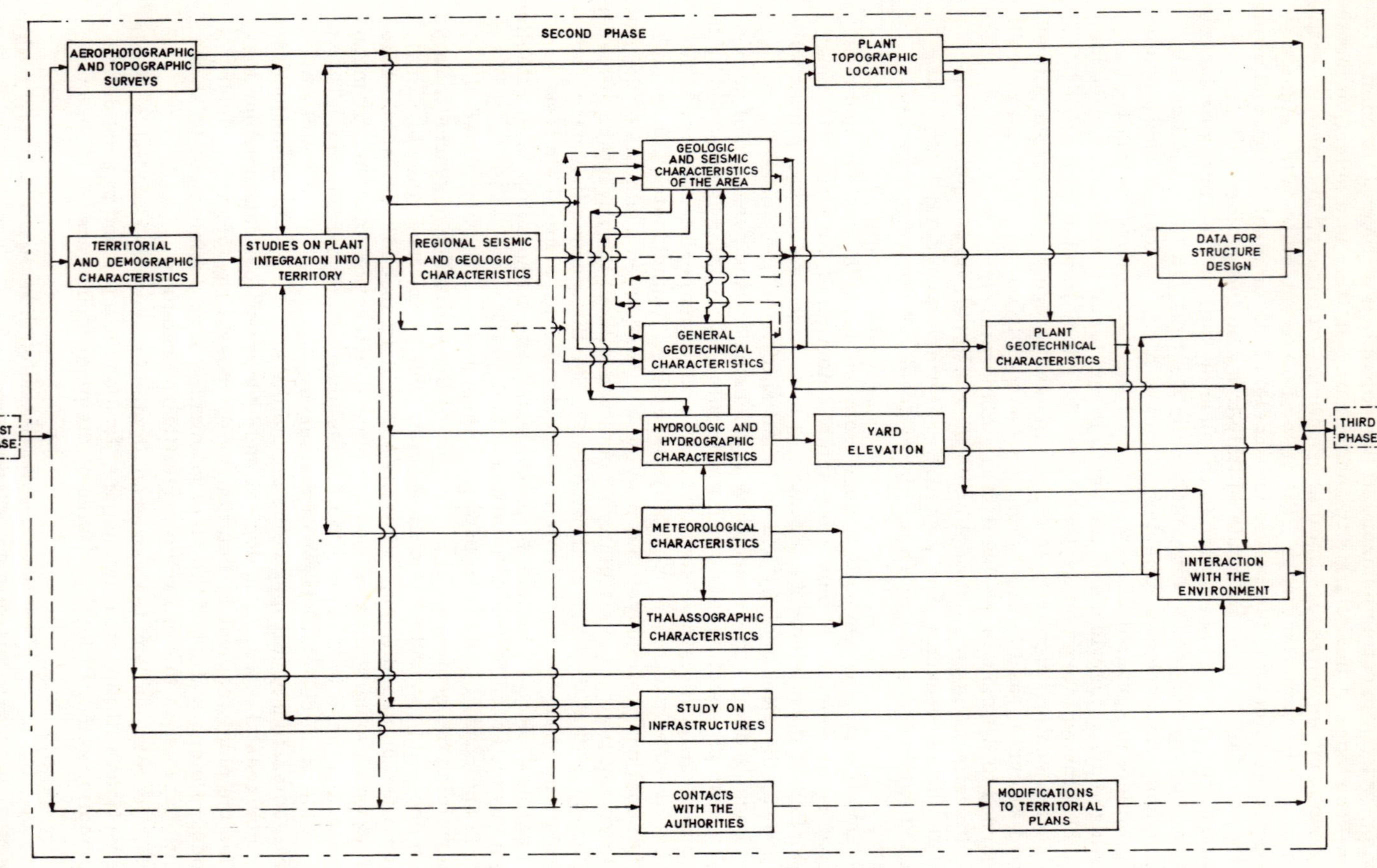


FIG. 2.   Simplified scheme of second-phase activity groups.

The infrastructures referred to in socio-economic investigations constitute a factor of preference for the choice, but not necessarily a condition
for exclusion.  One important infrastructure is a high-voltage transmission
line, as the line tying into it should be as short as possible to reduce transmission losses and to limit such side effects as land occupation, landscape
defacement, etc.

## 3.3.  Scheme of second-phase elementary activities

To better qualify the foregoing illustration, the elementary activities
required to reach a full understanding of the site characteristics have been
gathered and organically plotted in the Appendix, as a schematic diagram
explained by a list of the activities, under seven main headings:  demography
and territory, geology and seismicity, geotechnical characteristics,
meteorology, thalassography, hydrology and hydrography, topography and
infrastructures.

In addition to the functional links connecting the activities of each group,
the main diagram shows the links between activities of different groups so
that the 'subordinate activities' can be identified immediately.  These are
the activities in one group that cannot be carried out unless the information
to be developed from the 'instrumental activities' of another group is available.  A typical case of instrumental activity is represented by aerophotographic surveys, which are required not only to update the maps, but also
to allow the performance of activities belonging to other groups, such as
studies on land use, identification of archaeological areas, collection of data
on coast utilization, geological photo-interpretation, etc.

## 3.4.  Chronological sequence of second-phase activities

The only time connotation in the scheme shown in the Appendix stems
from the logical sequence of activities in one group and from their
instrumental dependence on the activities in other groups.  The performance
of the various groups of activities — in parallel or in series — depends on
the particular circumstances encountered by the surveyor.  Of course,
parallel performance wherever possible shortens investigation times but
increases the cost of the site qualification process.  In fact, in this case,
work is carried out on activities that would not even be started if, in a
series performance, the preceding activities had given negative results.
The choice of chronological sequence is thus merely dependent on considerations of convenience and company strategy in relation to the time available
for site qualification.

Clearly, it is always advisable to give priority to the activities aimed
at ascertaining the environmental features which cannot normally be handled
without recourse to non-traditional and extremely expensive methods — and
sometimes not even then — these are mainly the territorial characteristics,
the general and local seismotectonic characteristics and the soil geotechnical
properties.  The activities aimed at defining what we shall conventionally
term non-controllable characteristics (in the technical-economic sense
mentioned above) can be appropriately marked in the scheme and list shown
in the Appendix, together with any antecedent activities.  Thus, in the
second phase we can reach a subdivision of activities that leads to a further,

intermediate preselection process in which priority is given to the activities
that define the non-controllable characteristics.

On the basis of this criterion and in consideration of the restraints of
functional and instrumental dependence, it is possible to plot a general
diagram showing the chronological sequence of elementary activities reflecting the company's decision to proceed by parallel or series performance.
Since the Appendix is so complex, Fig. 2 shows a simplified diagram plotted
by groups of activities according to the above criteria, in which the links
that can be followed optionally are indicated by dashed lines.


## 4.   ACTIVITIES PRECEDING PLANT CONSTRUCTION  (THIRD PHASE)

When the investigation to determine the suitability of the site for plant
construction is completed, the third phase begins, i. e. the activities that
lead directly to the construction of the plant.  It is pointless here to
examine in detail the activities of the third phase, as they are aimed at
peculiarities.  We shall merely mention the main groups to which these
activities belong.

The first group includes the detailed studies and investigations (e.g. the
final geognostic campaign) required to complete the information gathered
in the preceding investigations and to prepare the working drawings.  The
second group comprises the activities associated with acquisition of land.
The third concerns the procurement of the construction permit and operating
license from central and local authorities (this group of activities is
governed in Italy by Law No. 1860 of 31 Dec. 1962, President of the
Republic's Decree No. 185 of 13 Feb. 1964 and Law No. 880 of 18 Dec. 1973).
The short time allowed by law for site selection makes it necessary, as we
have seen, to carry out most of the suitability investigations during the
second phase.  However, in view of the time and money effort required to
complete the second phase, contacts are established with the authorities
as early as possible so that a preliminary approval of the site, and possibly
an allocation in the land plans, can be obtained pending subsequent technical verifications.  This means introducing this activity at the end of the
first phase, or at the end of the territorial studies in the second phase by
means of a series link.  Once knowledge of the geological and seismic
characteristics has been extended to the whole national territory, it will
be possible to start dealing with the authorities as soon as a site has been
selected as suitable from this standpoint (Fig. 2).

To limit the total duration of the third phase, an attempt is made to
perform in parallel the groups of activities forming it; thus when the
permits have been formally issued and the land purchased, the technical
activities for preparation of the working drawings are at such a stage of
progress that it is possible to start breaking ground immediately.

APPENDIX

## LIST AND SCHEME OF ELEMENTARY ACTIVITIES FOR SECOND-PHASE SITE QUALIFICATION

The scheme is shown in three parts on pages 129-131

### I.  TERRITORIAL DEMOGRAPHIC CHARACTERISTICS

#### A.  Demography

A.1   Population classes
A.2   Total population (distribution density, etc.)
A.3   Comparison with regional, national and other plant data
A.4   Resident population
A.5   Annual variations in resident population
A.6   Equivalent inhabitants
A.7   Average population density
A.8   Scattered population density
A.9   Average population age
A.10  Territorial population distribution
A.11  Inhabited centre distribution
A.12  Population censuses (1861-1971)
A.13  Seasonal population variations
A.14  Population census
A.15  Municipality data on population variations
A.16  Hotel and extra-hotel tourism
A.17  Residential tourism
A.18  Excursionists
A.19  Tourism statistics
A.20  Dwelling census and visitors' tax
A.21  Direct observation by specimens

#### B.  Typical development indexes

B.1   Active population
B.2   Agricultural development indexes
B.3   Industrial development indexes
B.4   Touristic development indexes
B.5   Ratio between employed and active population
B.6   Enterprises per km of coast
B.7   Ratio between employed and active population
B.8   Coast occupation coefficient
B.9   Public activities and shops per km of coast
B.10  Hotels and beds per km of coast
B.11  Camping sites and touristic villages per km of coast
B.12  Population census
B.13  Industry and commerce census
B.14  Specialized publications
B.15  Direct observation of the coast touristic facilities
B.16  Industry and commerce census
B.17  Statistical data on internal commerce
B.18  Data drawn from specialized publications

#### C.  Social and economic enquiries

C.1.  Food chain
C.2   Receptivity
C.3   Recreational facilities
C.4   Sanitary infrastructures
C.5   Educational infrastructures
C.6   Transport
C.7   Manpower
C.8   Present facilities
C.9   Facility utilization coefficient
C.10  Forecasts on facility inventory
C.11  Present availability
C.12  Forecasts

#### D.  Present territorial configuration

D.1   National parks
D.2   Regional natural and recreation parks
D.3   Nature preserves
D.4   State forests and other similar territories
D.5   Areas for repopulation and hunting, shooting preserves
D.6   Oases for fauna protection
D.7   Fauna refuges (World Wildlife Fund)
D.8   Land use
D.9   Archeological area detection
D.10  Legislation
D.11  Park limits
D.12  Boundaries of natural preservation areas
D.13  Boundaries of the protection oases
D.14  Specialized publications
D.15  Biotopes
D.16  Existing cultures
D.17  Inhabited centres, small communities and scattered dwellings
D.18  Geophysical survey and borings
D.19  Water supply
D.20  Discussions by regional councils
D.21  Decrees on natural preservation areas
D.22  Decrees
D.23  Specialized publications
D.24  Direct detection

D.25   Ancient cartography
D.26   Specialized publications
D.27   List of regional parks
D.28   List of areas subject to nature preservation
D.29   List of protection oases
D.30   Institute of Ancient Cartography
D.31   State Archives
D.32   Italian Geographical Society
D.33   Region
D.34   Specialized publications
D.35   Central authority
D.36   Region
D.37   Specialized publications

**E.  Land planning**

E.1   Land allocation
E.2   Territorial limitations
E.3   Proposed parks and natural reserves
E.4   Natural areas to be protected
E.5   Regional co-ordination plans
E.6   Territorial plans for sectorial co-ordination
E.7   Plans for touristic development
E.8   Municipality town-planning instruments (town-planning schemes and construction programmes)
E.9   Landscape planning
E.10   Landscape limitations
E.11   Archeological limitations
E.12   Monument limitations
E.13   Hydrogeological limitations
E.14   Airport limitations
E.15   Military limitations
E.16   List of parks and natural preserves of major national importance
E.17   Regions
E.18   Other organizations
E.19   Map of Italy biotopes
E.20   Regional biotope maps
E.21   Studies by central offices for regional public works
E.22   Studies by regional offices
E.23   Specialized publications
E.24   Fund for the Improvement of Southern Italy
E.25   Other organizations
E.26   Fund for the Improvement of Southern Italy
E.27   Municipality
E.28   Official Regional Bulletin
E.29   Specialized publications
E.30   Superintendence of Monuments
E.31   Region
E.32   Studies by the Fund for the Improvement of Southern Italy
E.33   Specialized publications
E.34   Bounds of areas subject to limitations
E.35   List and bounds of areas proposed for limitations
E.36   Control at the Real Estate Registry Office of the limitation area

E.37   List and bounds of areas proposed for limitations
E.38   Superintendence of Monuments and Galleries
E.39   List and bounds of limited areas
E.40   Regulatory ordinances on air space
E.41   Maps of height limits
E.42   List and bounds of limited areas
E.43   Project 80 (Preliminary Report on National Economic Programme 1971-1975)
E.44   National Council for Research, Ministry of Public Works
E.45   Decrees on limitations
E.46   List and bounds of limited areas
E.47   Central authority
E.48   Regional Forest Bureau
E.49   Forest Division Inspectorate
E.50   List of airports
E.51   Laws
E.52   Territorial Military Headquarters
E.53   List of limited areas
E.54   Region
E.55   Superintendence of Antiquities
E.56   Archeological map of Italy
E.57   Specialized publications
E.58   Ministry of Transport and Civil Aviation
E.59   Aerial region
E.60   Official Gazette
E.61   Central authority
E.62   Superintendence of Monuments and Galleries
E.63   Specialized publications

## II.  SEISMIC AND GEOLOGICAL CHARACTERISTICS

**F.  Regional geological characteristics**

F.1   Tectonic scheme, geological and fault maps on regional scale
F.2   Geological photo-interpretation on regional scale
F.3   Existing data on regional geology
F.4   Aerial photographs on regional scale

**G.  Seismic characteristics**

G.1   Ground acceleration and design earthquake spectrum
G.2   Seismotectonic correlation on regional scale
G.3   Safe shutdown earthquake
G.4   Local seismotectonic correlations
G.5   Seismic recordings
G.6   Epicentre map
G.7   Earthquake statistics
G.8   Correlations of seismic quantities
G.9   Seismic equipment location
G.10   Data on regional seismic history
G.11   Seismic instruments

H. Local geological and hydrogeological
   characteristics

H.1    Tectonic scheme, local geologic and
       fault maps
H.2    Slope stability (avalanches, landslides)
H.3    Hydrology
H.4    Stratigraphy
H.5    Soil dating (paleontologic, paleomagnetic
       and radiocarbon examinations)
H.6    Analysis of surface faulting
H.7    Geologic formation reconstruction
H.8    Geoseismic investigations
H.9    Geo-electric investigations
H.10   Gravimetric investigations
H.11   Deep stratigraphic boring
H.12   Geological site surveys
H.13   Geological and geomorphological photo-
       interpretation
H.14   Existing wells
H.15   Planning of site surveys
H.16   Authorizations
H.17   ENEL mining service
H.18   Existing data on local geology


III.  GEOTECHNICAL CHARACTERISTICS

L.   Data on structure design

L.1    Yard elevations
L.2    Foundation planes
L.3    Load-settlement curves
L.4    Soil instability (sand liquefaction)
L.5    Studies on scarp stability and excavation
       dewatering
L.6    Tests on earthfill materials
L.7    Stratigraphic and geotechnical data
L.8    Quarry investigation
L.9    Data on water tables and soil permeability
L.10   Identification and preliminary classification
       of soils
L.11   Laboratory tests
L.12   In-situ examination
L.13   Inspection with TV probes
L.14   Static tests
L.15   Dynamic tests
L.16   Standard penetration tests
L.17   Dynamic soil characteristics
L.18   Sampling
L.19   Penetrometric tests
L.20   Vane borer tests
L.21   Pressiometric tests
L.22   Loading tests in boreholes
L.23   Measurement of shear and pressure-wave
       velocity
L.24   Density measurements with radioisotopic
       probes
L.25   Piezometer measurements

L.26   Microflowmeter measurements
L.27   Discharge assessment
L.28   Reflux in the borehole
L.29   Boring
L.30   Field tests
L.31   Planning of geotechnical campaign


IV.  TOPOGRAPHICAL, CADASTRAL AND
     INFRASTRUCTURAL DATA

M.  Topographical data

M.1    Aerophotographic return
M.2    Up-dating of existing maps
M.3    Site surveys
M.4    Mosaic
M.5    Aerophotographic survey (colour, low sun
       angle)
M.6    Existing maps


N. Data on property distribution

N.1    Cadastral data processing
N.2    Cadastral maps
N.3    Cadastral research
N.4    Cadastre


O. Infrastructures

O.1    Normal and exceptional transport during
       construction
O.2    Fuel supply and transport
O.3    Plant connection to the national grid
O.4    Possibility of transport by road
O.5    Possibility of transport by railway
O.6    Possibility of transport by rivers or lakes
O.7    Landing-place availability
O.8    Possibility of air transport
O.9    Project planning
O.10   Project planning
O.11   Project planning
O.12   Present plans
O.13   Future plans and schedules
O.14   Project planning
O.15   Existing road and motorway network
O.16   Roads and motorways designed or under
       construction
O.17   Possibility of site-to-roads connections
O.18   Existing railway network
O.19   Railways designed or under construction
O.20   Possibility of site-to-railway connections
O.21   Existing waterways
O.22   Inland navigation designs
O.23   Existing ports, jetties and other facilities
O.24   Ports and jetties designed or under
       construction
O.25   Airports near the site
O.26   Refineries near the site
O.27   Oilducts

O.28   Substations
O.29   Power lines
O.30   State Road Authority
O.31   Other organizations
O.32   State Railways
O.33   Associations for inland navigation
O.34   Maritime authorities
O.35   Other organizations
O.36   Ministry of Transport and Civil Aviation
O.37   Oil companies
O.38   ENEL

## V. HYDROLOGICAL AND HYDROGRAPHICAL CHARACTERISTICS

### P. Maximum foreseeable levels

P.1    Curves of section flows
P.2    Riverbed surveys

### Q. Maximum foreseeable flows

Q.1    Hydrological data statistics
Q.2    Possibility of barrage failures
Q.3    Possibility of lake formation by valley
       barrage
Q.4    Geomorphological characteristics of
       catchment areas
Q.5    Maximum precipitation statistics
Q.6    Diversions for irrigation and aqueducts
Q.7    Hydrological data
Q.8    Hydroelectric reservoir operating data
Q.9    Irrigation basins and aqueduct operating data
Q.10   Barrage structural data
Q.11   Ministry of Public Works
Q.12   ENEL
Q.13   Other organizations

### R. Cooling-water availability for riverbank sites

## VI. THALASSOGRAPHIC CHARACTERISTICS

### S. Interaction with marine environment

S.1    Sea bottom stratigraphy
S.2    Detailed bathymetry
S.3    Chemical and biological data
S.4    Type and quantity of suspended materials
S.5    Sedimentary characteristics
S.6    Water density
S.7    Water salinity
S.8    Water temperature
S.9    Sea current velocities and directions
S.10   Geophysical surveys at sea
S.11   Oceanographic campaigns
S.12   Site survey planning

### T. Data for structure design

T.1    Statistics
T.2    Exceptional events (tsunami, etc.)
T.3    Wave planes
T.4    Coastal morphological evolution
T.5    Wave motion recording
T.6    Seawater temperature
T.7    Wave motion
T.8    Present and past bathymetric surveys
T.9    Tide-gauge measurements
T.10   Wavemeters
T.11   ENEL — thermal station operation
T.12   Other organizations
T.13   Air Force
T.14   Hydrographic Institute of the Navy
T.15   University Thalassographic Institutes
T.16   Existing data

## VII. METEOROLOGICAL CHARACTERISTICS

### U. Interactions with atmospheric and terrestrial environment

U.1    Statistics
U.2    Wind velocity and direction
U.3    Particular phenomena (tornadoes, etc.)
U.4    Temperature
U.5    Humidity
U.6    Cloudiness
U.7    Visibility
U.8    Rainfall and snowfall
U.9    Wind velocity and direction
U.10   Temperature
U.11   Humidity
U.12   Background pollution, environment
       radioactivity, etc.
U.13   Air Force
U.14   Hydrographic Institute of the Navy
U.15   Ministry of Agriculture and Forests
U.16   Ministry of Public Works
U.17   Weather stations at site
U.18   Existing weather stations
U.19   Site survey planning

### V. Data for structure design

V.1    Exceptional atmospheric events

## X. SITE QUALIFICATION

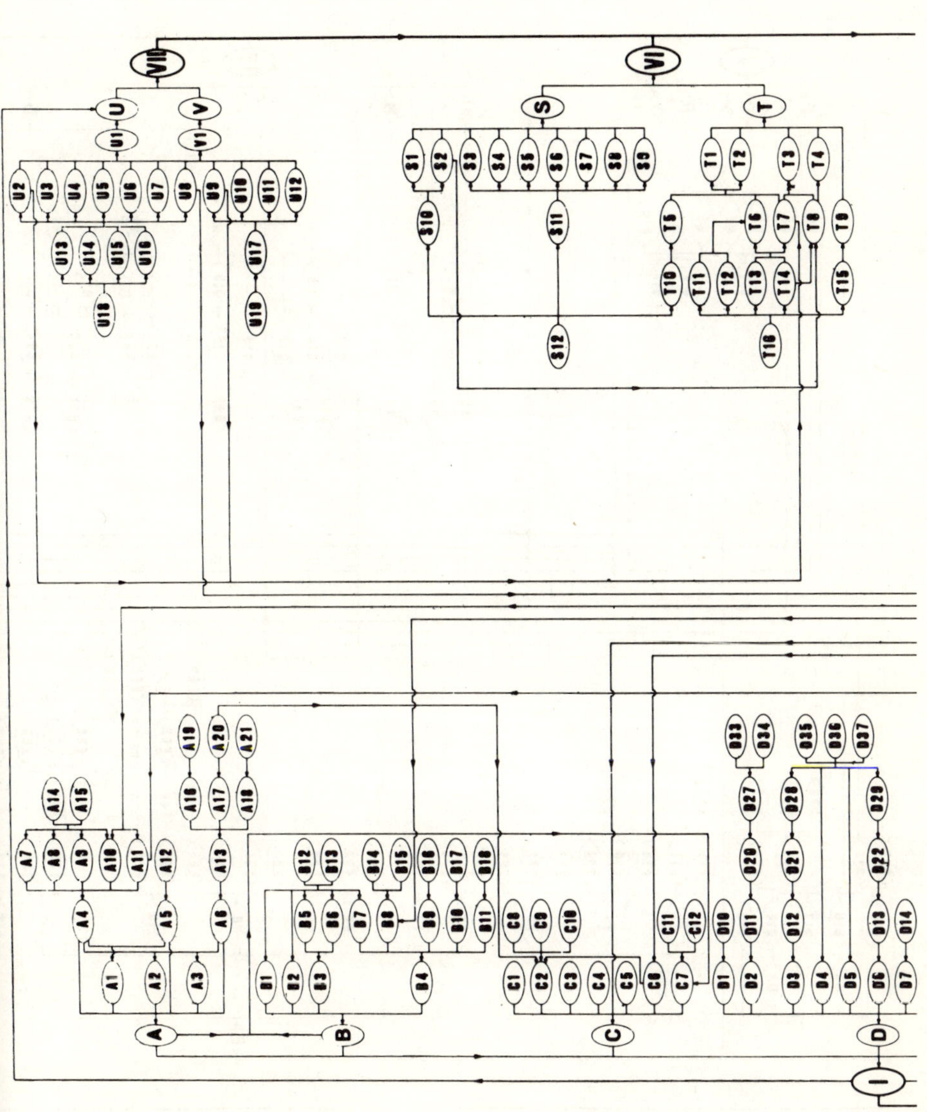

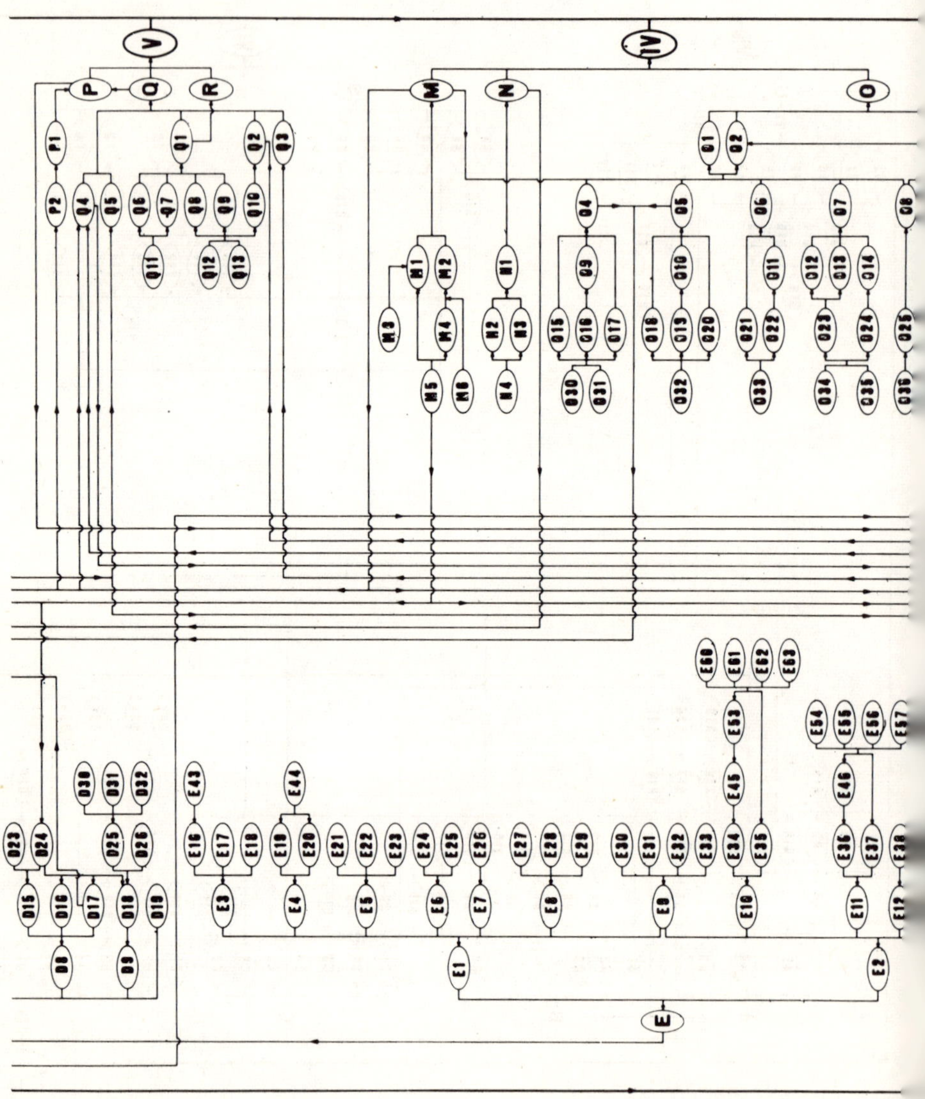

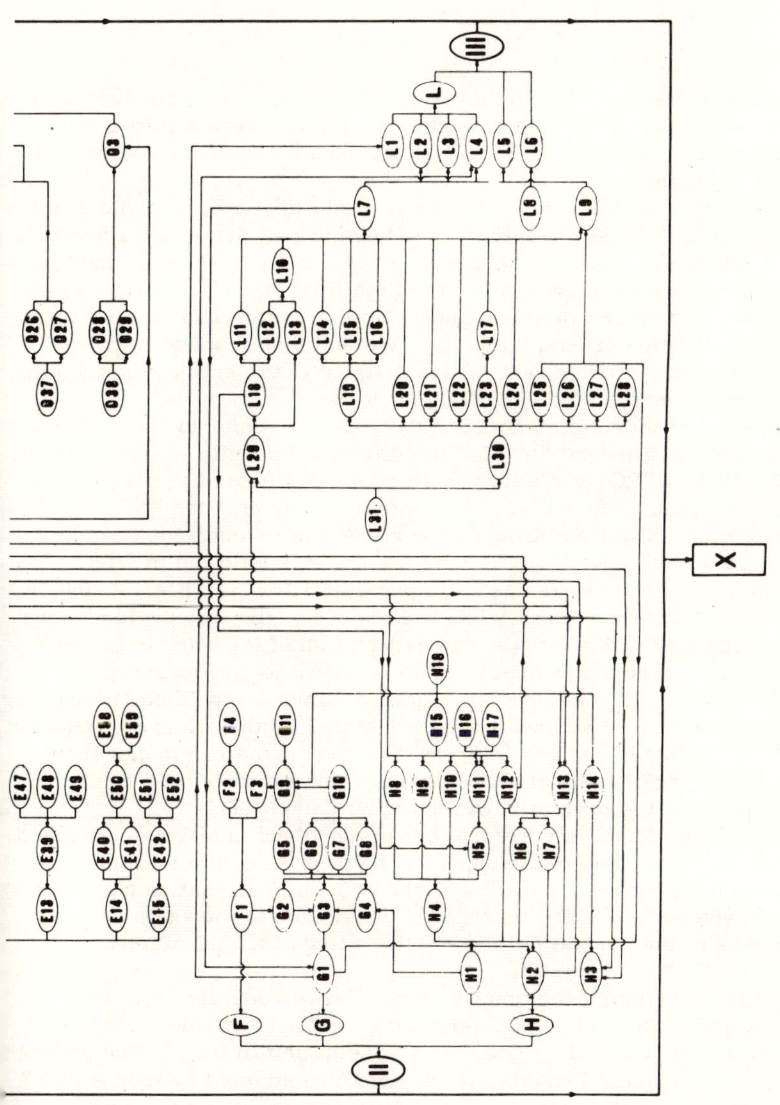

REFERENCES

[1]  ANGELINI, A.M., "L'Energia Elettrica e l'Ambiente", ENEL Rep. (March 1972).
[2]  ENEL Rep., "Atlante delle Caratteristiche Territoriali Primarie delle Coste Italiane" (1973).
[3]  ENEL Rep., "L'ENEL e l'Ambiente" (June 1973).

DISCUSSION

A. DE ACHA ARACAMA:  Could you please indicate what seismotectonic areas have been established in Italy and what criteria were followed in determining them?  Also, what intensity-acceleration correlations do you use in seismic studies?

V. MORELLI:  In Italy, seismotectonic studies follow a scheme similar to that set forth in the USAEC guides.  Studies such as those described in the paper are being carried out at present over a large area all round the sites which had been selected by the end of the first phase.  Therefore, the criteria followed in determining seismotectonic areas depend on the particular tectonic situation around the sites.  We feel that in a few years we shall be able to complete our general assessment of the entire coastal belt of Italy by connecting the studies now under way.

For intensity-acceleration correlations, too, we refer to the USAEC guides, suitably modified to allow for the different intensity scales adopted.

S.O.W. BERGSTRÖM:  Could you give us some idea of the amount of work that goes into the survey and description of a given site?

V. MORELLI:  The time needed to perform the second-phase studies on each selected site depends strictly on the criteria followed in site qualification, i.e. either in series or in parallel paths (see Fig. 2), and on the specific site characteristics.  The time accordingly varies widely; our estimates suggest 12 - 24 months.  The description of the work to be performed for each selected site qualification is shown in the Appendix.

H. SCHNURER:  In your paper you have outlined a very detailed assessment of the properties of a coastal area in relation to the siting of a nuclear power station.  I should be very interested to hear some comments about the mutual balancing of individual properties — for example, the number of tourist beds against the proximity of industrial complexes.

V. MORELLI:  There are several phases involved in site selection and qualification, as described in the paper.  The purpose of the first phase is to carry out a preliminary screening on the basis of information about the primary characteristics of the territory.  This leads to the exclusion of large stretches of the coastal belt where the siting of a nuclear plant is evidently in conflict with existing features.

As regards tourism, development indices were considered in the territory analysis with a view to determining, in the first phase of the study, what activities take place along the entire coastal belt of Italy.  The principal activity in any case being tourism, it has been investigated by means of six indices (in addition to the data on population increase during the summer: equivalent inhabitants).  At locations where the indices are especially high in comparison with regional and national values it can be assumed that the siting of a nuclear station will indeed meet with very considerable difficulties. It would therefore be advisable to exclude such locations.

# EXTERNAL HAZARDS AS THEY AFFECT NUCLEAR POWER PLANT SITING

B.K. GRIMES
Accident Analysis Branch,
Directorate of Licensing,
United States Atomic Energy Commission,
Washington, D.C.,
United States of America

Abstract

EXTERNAL HAZARDS AS THEY AFFECT NUCLEAR POWER PLANT SITING.

The USAEC Regulatory Staff (soon to be known as the Nuclear Regulatory Commission) has recently issued guidance to utilities in the areas of man-made hazards and natural phenomena which may influence availability and development costs of candidate nuclear power plants. They are in the form of regulatory guides and revisions to the standard format and content of safety-analysis reports. It has also issued guides to its own staff in the form of review plans, some of which have been recently completed and made available to the public. This paper summarizes the USAEC guides now available on the subject of external hazards and discusses typical hazards which are given attention by the regulatory staff in the course of the review of an application to construct a nuclear power plant.

External hazards are receiving increasing attention as factors which limit availability of nuclear power plant sites or which can require special design measures to make sites acceptable. Man-made hazards such as aircraft activity and transport and storage of gaseous material, as well as natural phenomena such as earthquakes and floods, may influence availability of candidate nuclear power plant sites and the relative cost of development of such sites. The United States Atomic Energy Commission Regulatory Staff (soon to be known as the Nuclear Regulatory Commission) has recently issued guidance in these areas to utilities in the form of regulatory guides and revisions to the standard format and content of safety-analysis reports. It has also issued guides to its own staff in the form of review plans, some of which have been recently completed and made available to the public. This paper summarizes the USAEC guides now available on the subject of external hazards and discusses typical hazards which are given attention by the regulatory staff in the course of the review of an application to construct a nuclear power plant.

There are two ways in which external hazards influence the selection of sites for nuclear power plants. First, the magnitude of the potential hazard may be so great that no foreseeable design measures could make the site acceptable and the site would be rejected as unsuitable. Examples of this would be siting on an active fault where ground displacement is expected, siting at the base of a large dam whose failure could destroy the facility, or siting adjacent to a very large munitions storage area. Any such hazards which could make sites absolutely unsuitable are usually easy to identify and are usually eliminated very early in the site selection. It is, however, becoming increasingly difficult to predict the future location and nature of hazardous industrial and transport activities.

The second way in which external hazards influence site selection is through the relative cost of development of various alternative sites. Examples of additional costs incurred because of external hazards are the substantial cost of additional concrete which might be required to harden the plant against an aircraft impact and the cost of relocating a large natural-gas pipeline away from the site area.

These additional costs can be substantial and may be the deciding factor in site selection when two or more sites are found equally acceptable with respect to environmental impact and other development costs.

The information required to assess hazards relating to natural and man-made phenomena are listed in Section 2 of the USAEC document entitled "Standard Format and Content of Safety Analysis Reports for Nuclear Power Plants − Revision I". This document provides guidance to utilities who wish to submit safety-analysis reports to the USAEC Regulatory Staff for review. A recent further revision to Section 2.2 of this document was issued as Regulatory Guide 1.70.8 [1]. The revision provides more detailed guidance on the information needed about man-made hazards from nearby industrial, transport and military facilities.

In addition to a description of the location of various potential hazards and specification of the nature of these hazards, a determination of which events should be considered as design-basis events for the plant is required. A detailed analysis of the effects of those accidents chosen as design-basis accidents and the measures taken in the plant design to mitigate their consequences is also required. To quote from the guide:

"Design basis events external to the nuclear plant are defined as those accidents which have a probability of occurrence on the order of about $10^{-7}$ per year or greater and have potential consequences serious enough to affect the safety of the plant to the extent that Part 100 guidelines could be exceeded."

This definition has two implications which bear emphasis. First, small events need not be considered. For example, most truck accidents do not have the potential for affecting the safety of adjacent nuclear facilities. Second, in determining design-basis events, both the probability of the event and the probability of the event's leading to serious consequences should be considered. An example of this would be a release of a large quantity of toxic gas at some distance from the nuclear plant. Not only the probability of release but also the probability of reaching the plant in a hazardous form may be considered in comparing against the $10^{-7}$ per year guidelines. In this case factors such as wind direction and stability conditions might be involved. However, in the case of a large munitions explosion, these same factors would not reduce the probability of a hazard to the nuclear plant.

The information received from a utility in a safety-analysis report is reviewed by the Regulatory Staff in accordance with Standard Review Plans, which are now being issued. These review plans were written for the guidance of Regulatory Staff members but they are also being released to the public as they are completed. The review plans are numbered to correspond to sub-sections of the Standard Format and Content document and each plan consists of: (1) a description of the subject to be reviewed; (2) a list of the acceptance criteria by which the subject matter is to be judged; (3) a list of procedures for the staff reviewer; and (4) a set of evaluation findings. We hope these Review Plans will be useful to others in understanding the depth of review and the important issues involved in each subject area.

## TABLE I.  NATURAL PHENOMENA

1.  Earthquakes

2.  Tornadoes

3.  High water from:

    Floods
    Dam failure
    Hurricanes
    Tsunamis
    Seiches

4.  Low water

5.  Snow, ice, high winds

6.  Meteorology

## TABLE II.  MAN-MADE HAZARDS

1.  Aircraft impacts

2.  Ship collisions

3.  Explosions

4.  Flammable gases and vapour clouds

5.  Toxic chemicals

6.  Fires

## TABLE III.  AIRCRAFT

HAZARDS

1.  Impact

2.  Fire

PROBABILITY CONSIDERATIONS

1.  Proximity to airports

2.  Bombing ranges

3.  Low-level training routes

The external hazards which are considered in the USAEC Regulatory Staff Review are given here in the form of tables.  They are briefly as follows:

Table I lists the natural phenomena considered in our review of sites. These areas will not be treated further in this paper except to note that several Regulatory Guides have been issued on these subjects.  In particular, Regulatory Guide 1.76 [2], which deals with the design-basis tornado wind speeds should be noted.  These wind speeds are derived on a probabilistic basis for various regions of the United States of America.  Table II summarizes the man-made hazards which will be discussed in more detail in this paper.

The hazards characteristic of aircraft considerations are indicated in Table III.  The potential injection of flammable gases into facility air intakes gives rise to a fire hazard as well as a hazard due to the physical impact of the aircraft on the facility structures.  The type of aircraft activity strongly influences the type of probabilistic analysis that is made for aircraft hazards. Table III lists several types of aircraft activity that may present special hazards.  Of course there is also some low 'background' level of probability resulting from general overflights in air corridors.  Unless traffic in these corridors is extremely heavy and at relatively low altitude it has been our

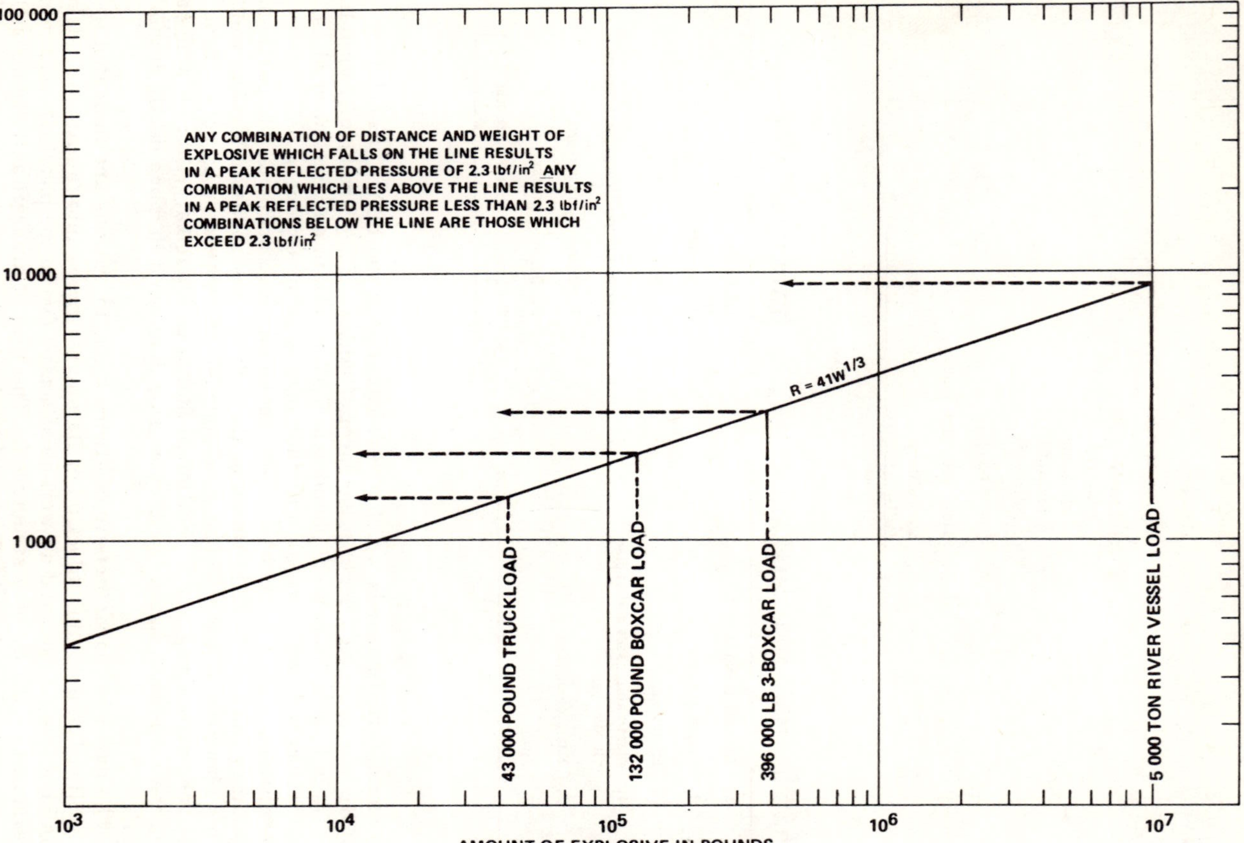

DISTANCE FROM EXPLOSION IN FEET
100 000
10 000
1 000
ANY COMBINATION OF DISTANCE AND WEIGHT OF
EXPLOSIVE WHICH FALLS ON THE LINE RESULTS
IN A PEAK REFLECTED PRESSURE OF 2.3 lbf/in² ANY
COMBINATION WHICH LIES ABOVE THE LINE RESULTS
IN A PEAK REFLECTED PRESSURE LESS THAN 2.3 lbf/in²
COMBINATIONS BELOW THE LINE ARE THOSE WHICH
EXCEED 2.3 lbf/in²
R = 41W^1/3
43 000 POUND TRUCKLOAD
132 000 POUND BOXCAR LOAD
396 000 LB 3-BOXCAR LOAD
5 000 TON RIVER VESSEL LOAD
10³
10⁴
10⁵
10⁶
10⁷
AMOUNT OF EXPLOSIVE IN POUNDS

experience that such activity need not affect the plant design.  The basic
elements of the equation used to compute aircraft impact probability are:

$$P = \begin{pmatrix} \text{Probability} \\ \text{of an} \\ \text{accident} \\ \text{per movement} \\ \text{per area} \end{pmatrix} \times \begin{pmatrix} \text{movements} \\ \text{per} \\ \text{year} \end{pmatrix} \times \begin{pmatrix} \text{effective} \\ \text{area of} \\ \text{plant} \end{pmatrix}$$

The type of potential consequences which might result from ship collision
accidents and some factors affecting the probability analysis of ship collisions
are given in Table IV.

TABLE IV.   SHIP COLLISION ACCIDENTS

POTENTIAL CONSEQUENCES OF SHIP ACCIDENTS[a]

1.  Explosion

2.  Missiles

3.  Fire

4.  Flammable vapour cloud

5.  Toxic chemicals

6.  Corrosive or cryogenic chemicals

7.  Spilled liquids entering water intake

8.  Collision with intake structure or breakwater

FACTORS AFFECTING PROBABILITY ANALYSIS

1.  Number of vessels carrying a specific cargo

2.  Present and projected traffic estimates

3.  Accident statistics for particular body of water adjacent
    to site

4.  Future changes in waterway

5.  Given a ship is in an accident, what is likelihood of
    cargo becoming involved?

[a]  Shipping accidents are unique because of the large quantities
    involved.  Projections are for even larger vessels.

TABLE V.   EXPLOSIONS (OVERPRESSURE AND MISSILES)

1.  Transport of high explosives, munitions, liquid and gaseous fuels

    (a)  Highway:    Trucks
    (b)  Railroad:   Munition trains, propane tank cars
    (c)  Waterway:  Munition ships, barges

2.  Tank farms storing liquid and gaseous fuels

3.  Industrial/chemical complexes

4.  Pipelines

5.  Mine and quarry operations

6.  Rocketing rail-car phenomenon

## TABLE VI.  FLAMMABLE GASES AND VAPOUR CLOUDS

---

       1.  Transport of volatile liquids.

       2.  Storage of volatile liquids.

       3.  Pipelines

Extent of vapour clouds must be determined.

Vapour clouds may deflagrate or detonate.

Delayed ignition may occur.

---

Several sources of off-site explosion hazards (overpressure and missiles) are listed in Table V.  Item 6 on Table V refers to the projection of large parts of propane tank cars up to 1200 feet from the rail line which carried the cars after suffering a tank failure in one car which allowed compressed gases to escape.  This has happened in the United States of America at least three times in the last several years.

With respect to explosive overpressures which might result from off-site sources, Regulatory Guide 1.91, recently issued [4], gives guidance on screening criteria which can be used to eliminate explosive overpressures as a consideration in plant design.  The distance-explosive size relationship presented allows determination of the distance beyond which particular activities need not be considered in the design of the plant.  The relationship is derived on the basis of the inherent capability of the plant to withstand a certain static external overpressure based on the design of the plant against tornadic winds.  This technique allows elimination of many off-site hazards from consideration which might otherwise require a detailed probabilistic analysis (see Fig. 1).  It should be noted that not all explosive sources which are at a closer distance than those indicated in Regulatory Guide 1.91 [3] would necessarily present a hazard to the plant.  The plant design may be found to have the capability to withstand greater explosions based on a detailed structural analysis of the particular load transient and the orientation of the facility structures with respect to the explosion sources.

Table VI presents several considerations on flammable gases and vapour clouds; more detailed considerations on pipeline accidents are given in Table VII.  This subject is now under review by the USAEC Regulatory Staff with regard both to pipeline accidents and the transport and storage of large quantities of liquefied natural gas (see Table VIII).

Regulatory Guide 1.78, recently issued [4], presents a quantity-distance table for the release of toxic chemicals which, when adjusted for the toxicity limit of a particular chemical, may be used to rule out hazardous material as a consideration in plant design.  It should be noted that most plants in the United States of America have self-contained breathing apparatus available to plant operators so that the only design conditions imposed on the plant by nearby toxic chemical sources may be that sensors for the particular chemical be provided in air intakes combined with automatic closure of these intakes when the chemical is detected.

Chlorine gas is often stored on-site to provide chemical treatment of condenser cooling water.  Because of the common nature of this hazard, a

## TABLE VII.  PIPELINE ACCIDENTS

POTENTIAL CONSEQUENCES OF PIPELINE ACCIDENTS

1. Explosion

   (a) Blast overpressure
   (b) Missiles
   (c) Ground shock

2. Fire

   (a) Thermal radiation
   (b) Smoke – may affect control room or diesel operation

3. Flammable vapour cloud – delayed ignition

   (a) Cloud may explode (detonate) or burn (deflagrate)
   (b) Gas concentrations at site may affect diesel operation

TYPICAL INFORMATION REQUIRED FOR ANALYSIS OF PIPELINE ACCIDENTS

1. Type of gas or liquid carried

2. Size and age of pipeline

3. Operating pressure

4. Location and mode of operation of isolation valves

5. Are pipelines used for storage at higher than normal operating pressure?

6. Future use of pipeline

SOME PROBLEMS IN ANALYSING PIPELINE ACCIDENTS

1. Determining maximum amount of gas which could be released

2. Determining gas release rate, direction of release

3. Buoyancy effects, plume rise, for different types of gas at varying temperatures

4. Dispersion of gas vapour cloud

## TABLE VIII.  TOXIC CHEMICALS

1. Release from onsite storage facilities

2. Release from major depots or storage areas

3. Release due to transport accidents

## TABLE IX.  FIRES – SOURCES

1. Onsite oil storage facilities

2. Nearby transport, industrial and storage facilities

3. Forest and brush fires

4. Floating fires due to ship and barge accidents

special regulatory guide, dealing specifically with design provisions against chlorine release, will be issued in the near future.

The potential off-site sources of fires are indicated in Table IX. The potential hazards resulting from fires include thermal flux, heavy smoke and hazardous fumes.

In summary, all these sources of external hazards may impose design requirements which are reflected in additional facility costs. In some cases these costs may be substantial enough to influence site selection.

## REFERENCES

[1]   USAEC, Regulatory Guide 1.70.8,   Additional Information Nearby Industrial, Transportation, and
      Military Facilities.
[2]   USAEC, Regulatory Guide 1.76,   Design Basis Tornado for Nuclear Power Plants.
[3]   USAEC, Regulatory Guide 1.91,   Evaluation of Explosions Postulated to Occur on Transportation Routes
      Near Nuclear Power Plant Sites.
[4]   USAEC, Regulatory Guide 1.78,   Assumptions for Evaluating the Habitability of a Nuclear Power Plant
      Control Room During a Postulated Hazardous Chemical Release.

## DISCUSSION

R. M. DUNCAN:  You have discussed, in connection with your siting
considerations, the external hazards that could affect a nuclear power plant.
Have you, in estimating the risk to the local population around any site,
considered the possibility of hazardous releases from nearby sources of
toxic chemicals or flammable vapours combining with the hazards posed by
the nuclear plant itself?

B. K. GRIMES:  We consider only the effect which releases of hazardous
chemicals might be expected to have on the plant, not the effect they might
have on the general public.  The plant is assumed to be so designed that no
major emission of radioactivity will occur when hazardous chemicals are
released.

P. CANDES:  In your accident probability calculation for aircraft traffic
in the vicinity of a site you use the relation $P \leq 10^{-7}$.  If the calculations
showed a probability of, say, $P = 10^{-6}$ or $10^{-5}$, what steps would you envisage
– to improve the level of protection or reject the site?

My second question relates to the case of an accident due to the explosion
of a cloud of natural gas issuing from a leakage of liquefied gas.  Can the
USAEC recommend a minimum safe distance?  Are studies being carried out
with a view to making such a recommendation?

B. K. GRIMES:  If the probability of an aircraft strike were determined
to be greater than $10^{-7}$ per year, the first step would be to ask the utility to
examine the inherent capability of the nuclear facility structures to withstand
a strike of the type of aircraft under consideration.  This alone might be
enough to reduce the probability of catastrophic consequences to the degree
required (e. g. if very small aircraft were involved).  If additional protection
were required, it would take the form of concrete and protection of air intake
against ingestion of combustible fuels.  An example of this is the Three Mile
Island nuclear plant, where up to six feet of reinforced concrete was provided
on flat-walled structures to protect against the impact of a 200 000 lb aircraft
striking at a velocity of 200 knots.

Now to your second question.  We have not yet developed a quantity-
distance relationship for flammable gases which would allow the determination
of safe distances for various quantities of hazardous gases.  We are currently
reviewing several utility studies.  On the basis of these reviews, our own
calculations and the work of others (in other countries), we shall attempt to
arrive at such a relationship within about a year.

K. CEHAK:  In Table I you have listed snow, wind and so on as natural
hazards, the last item being 'meteorology'.  What phenomena do you have in
mind when you say meteorology?

B.K. GRIMES:  By meteorology I meant the dispersion conditions which would affect the doses computed for postulated accidents. If the doses resulting from design-basis accidents were calculated to exceed our allowable siting guideline doses, additional engineered safety features would be required. This would affect the cost of the facility and could in turn affect the utility's choice of site.

A. FERNANDES FORTE:  I should like to ask if you have any regulatory guide on tsunamis.

B.K. GRIMES:  I do not believe we have any regulatory guide dealing only with tsunamis. However, we have Standard Review Plans, issued for the guidance of regulatory staff reviewers, which deal with tsunamis (Section 2.4.6 of Standard Review Plan, which refers to Regulatory Guide 1.59: "Design-Basis Floods for Nuclear Power Plants").

C.J. van DAATSELAAR:  With regard to the escape of flammable explosive gases from tankers and similar vessels, is there any special meteorological dispersion formula in use to determine deflagration or detonation risks?

B.K. GRIMES:  There are several meteorological dispersion models which have been proposed for this purpose; we are at present evaluating these to determine which should be adopted as a standard technique. Under review are several proposals submitted by utilities, relating to the dispersion of flammable gases.

D. DAGAN:  How do you control subsequent external activities in the vicinity of nuclear facilities and in the air?

B.K. GRIMES:  The USAEC has no control over off-site private activities. If something comes to our attention which might affect plant safety, we do have the authority to require changes in the nuclear facility to cope with the hazard. In addition, we may work with other government agencies on certain problems. For example, an environmental statement must be issued by the Federal Aviation Agency (FAA) prior to the construction of a new airport. If the airport presented a hazard to a nuclear plant, we would comment on the environmental report and indicate this to the FAA. (We do try to predict, ahead of time, new activities over the lifetime of the plant, as far as we are able).

D.E. ANDERSON:  Do you in the Regulatory Guides require that consideration be given to the possibility of simultaneous occurrence of several of the 'natural' or 'man-made' phenomena? One might consider, for example, that during a hurricane the probability of an aircraft crashing into the plant would be greater than otherwise.

B.K. GRIMES:  Generally, we do not. There are a few cases where this might be needed. For example, a $10^{-3}$ hurricane coincident with a $10^{-3}$ earthquake might be important because the increased water level resulting from the hurricane might result in an increased possibility of liquefaction during an earthquake. In general, however, we find that it is adequate to design individually for each low-probability phenomenon; this seems to give a design conservative enough to cope with combinations of less severe but more probable events. However, in individual cases where a very high frequency of occurrence is identified for a given severe event (say $10^{-1}$ per year), it might well be appropriate to consider combinations of events.

H. SCHNURER:  Bearing in mind the large area of the United States of America, I wonder how you are able to calculate the probability of an aircraft crashing into the small area of any designated site. We have tried

to make similar evaluations but have not been able to find a correlation
between a potential crash point and its distance from an airport.  Therefore,
a standard requirement has been laid down to protect nuclear power stations
against crashes of fast-cruising military aircraft, for which the probability
of hitting a nuclear power station site away from the neighbourhood of an
airport is about $10^{-6}/a$.  Did you take military aircraft into account in your
calculations?

B.K. GRIMES:  Our calculations of aircraft crashes away from airports
indicate probabilities in the range $10^{-7} - 10^{-8}$ per year.  We have an arrangement with our military authorities in accordance with which they plot their
low-level training routes some distance away from nuclear power stations.
Of course, this does not guarantee that all such flights will stay away from
the plant area, but it does ensure that most military air activity will be
distant from the site.  The fact that we have a relatively large country
naturally helps to reduce the probability of military aircraft hitting plants
which are distant from airports.  I may add that if there is special military
activity nearby, such as in the Boardman site-analysis case, a special
analytical study is made to determine the probability of the aircraft hazard.

# NUCLEAR POWER STATION SITING
# EXPERIENCE IN THE UNITED KINGDOM
## Past and present and
## proposals for the future

T.P. HAIRE, E.F.F.W. USHER
Nuclear Health and Safety Department
 and Planning Department,
Central Electricity Generating Board,
London, United Kingdom

**Abstract**

NUCLEAR POWER STATION SITING EXPERIENCE IN THE UNITED KINGDOM:  PAST AND PRESENT AND
PROPOSALS FOR THE FUTURE.
    Foremost of the many factors in site selection considerations are population distribution, cooling-
water availability and amenity.  Others are safety of potable water sources, geological stability and the
risk of external hazards.  Where cooling-water supplies are a limiting factor, the choice of reactor system
is of major importance.  To determine as early as possible the effect a station might have on its environment,
desk studies, visual surveys and wind-tunnel tests are carried out.  The Central Electricity Generating Board
places great importance on obtaining the fullest degree of acceptance by the public for its nuclear stations
and ensures that full consultation is provided with the relevant authorities at all stages of power-station
development.  It also provides public exhibitions, public meetings and liaison with the local inhabitants.
Recruitment of station staff where possible from the immediate area of the station  and formation of sports
and social clubs are two of the practical steps which help to integrate the station into the local community.
Whilst the current energy crisis has reinforced the need for a substantial nuclear programme, possible ways
of further reducing the impact of nuclear stations on the environment are being considered.  The paper
concludes that sufficient nuclear sites can be provided for future needs but that continuing effort will be
required to ensure public acceptance.

## 1.    INTRODUCTION

The British Government recently announced a third programme of nuclear
power station construction of 4000 MW based on the Steam Generating Heavy
Water Reactor.   It is nearly 20 years since the first British nuclear programme was announced.   During this period the Central Electricity Generating
Board (CEGB) has complied with the relevant legislation governing design, manufacture and construction of nuclear power plants and has ensured that the
prevailing nuclear siting requirements were met.   With the exception of
Hartlepool and Heysham, all CEGB nuclear licensed sites have been selected
on the basis of siting criteria proposed in 1955.   These criteria limited
the areas available for consideration and resulted in the use of rural land
often of high amenity value.   As a consequence, opposition to nuclear stations
was more on the amenity and environmental intrusion aspects than on nuclear
safety grounds.   The siting requirements were re-examined following advances
in reactor design, and the policy which emerged permitted the use of sites,
such as Hartlepool and Heysham, nearer to centres of population.   Implicit
in public acceptance of nuclear power is the maintenance of the highest
standards of nuclear safety.   The Board also has a duty to have regard to
the effect their proposals may have on the environment and amenity.

The paper reviews the measures taken to meet these obligations and obtain public acceptability by pursuing a 'good-neighbour' policy and providing good communications with local interests, and suggest means by which the effect of nuclear power stations may be limited.

## 2.    DEVELOPMENT OF SITING CRITERIA

The Government's White Paper presented to Parliament in February 1955 allowed for between 1500 and 2000 MW of nuclear plant.   Nuclear sites were subject to the approval of the Reactor Location Committee, which was made up of members from the United Kingdom Atomic Energy Authority and the Central Electricity Generating Board, until the Nuclear Installations Act 1959 came into force in 1960.    This Act provides for the licensing of nuclear installations by the Minister of Power (now Secretary of State for Energy). For the purposes of advising on the implementation of the Act, the Minister established the Nuclear Installations Inspectorate (NII).    The Minister also set up the Nuclear Safety Advisory Committee (NSAC) comprised of independent experts to give advice on matters of nuclear safety.

Although formal statutory requirements only applied late in the Board's magnox programme, all the magnox reactors and the first two AGR's were sited in accordance with the principles provided by Marley and Fry in 1955  [1] . These criteria meant that it was necessary to site reactors in remote rural areas, often of high amenity value.    Changes in the approach to siting were in progress, and in 1959 [2]    Farmer and Fletcher gave a more realistic view of the risks involved with total fission-product release from gas-cooled graphite-moderated reactors.    A further and more radical development was provided by Farmer in 1962  [3]    when he proposed a method of deriving four classes of site, based on population distribution, the dispersal characteristics of a radioactive plume and the related risk to health.    The early magnox sites were rated as Class I on this scale.    In 1963 the NSAC recommended that AGR and magnox reactors in concrete pressure vessels would be acceptable for Class II sites.

The Government's Second Nuclear Programme issued in 1964 provided for 5000 MW of plant, and the Generating Board proceeded to order four AGR stations. The reactors were encased in pre-stressed concrete pressure vessels as were the last two magnox reactors.    The first AGR stations were built at the existing magnox sites at Dungeness and Hinkley Point.

In 1968 the Minister of Power stated that 'gas-cooled nuclear reactors in pre-stressed concrete pressure vessels could be constructed and operated much nearer to built-up areas than so far permitted'.    No specific definitio of site requirements was given at this time, but the interpretation used is that emergency measures comprising iodine distribution and population evacuat required by the Nuclear Site Licence must be achievable within two hours for any 30° Sector, up to a distance of 1 km from the reactor and that no special difficulty is presented by topography or population characteristics to the extension of these arrangements up to about 3 km from the reactor.

Proposals for siting AGR's at Hartlepool and Heysham were made by the Board well before the Government announcement in 1968, and full discussions wi both the NII and the NSAC were held during this time.    The emergence of rela siting criteria confirmed the acceptability of these sites.

The method of deriving site and sector characteristics was further devel by Charlesworth and Gronow in 1967   [4]    with refinements added by Gronow 1969 [5]    on the criteria defining the acceptability of sites.    Limiting curves for various reactor types have been prepared by the NII and the

characteristic risk curves for the proposed sites prepared by this method are
compared against these standards.   If the characteristic curves are higher
than the limiting curve the site is unacceptable, but if they are lower then
the site merits further investigation.   It is of interest to note that the
criteria have been recently assessed by Shaw and Palabrica  [6]  using later
data to obtain new values for Site and Sector weighting factors.   This suggests
that increased limits of population around the site would be acceptable.

Although it is a CEGB requirement that all new reactors must be licensable
on relaxed sites it is a continuing requirement of the NII that reactor systems
new to commercial operation in the United Kingdom are sited on Class I sites.
It is planned to build the Board's first commercial SGHWR at Sizewell, one of
the original Class I sites.

Siting criteria apply to a station throughout its operational life, and
population changes must not be allowed to adversely affect the implementation
of adequate emergency measures.   For this purpose the local planning authority
for the area around nuclear sites consults the NII before giving planning permission for development which increases permanent or temporary population.

Industrial development is also subject to continuing consideration to
ensure that it does not present a hazard to the nuclear plant and that the
work force can be readily evacuated.

3.    IMPLEMENTATION OF SITING POLICY

Major factors in selecting nuclear sites are population distribution [7],
cooling-water availability, suitable routes for transmission lines, the amenity
of the area and economics.   The ten sites licensed for nuclear generation in
England and Wales are shown in Fig. 1 related to the main centres of population,
the rivers and areas of land which are afforded statutory protection because of
amenity or scientific value.

To determine a site's population characteristics and its licensability
requires judgement and experience.   The population numbers are plotted on
large-scale maps from the most recent census supplemented by the Government's
annual report of population changes and local voters register.  Usually aerial
photographs are taken of the area and detailed checks made.   Information is
obtained on the overnight tourist and caravan population, the size and type of
hospitals and schools.   Details of future plans affecting population growth,
recreational activity and industrial development are obtained from the local
authority together with any special arrangements necessary for road access in
an emergency.

Before a licence is granted, the Chief Constable of the area must confirm
that the police can evacuate all inhabitants within any 30° sector out to 1 km
within two hours and provide supervision of the public out to greater distances.
This is an essential requirement and potential sites have been abandoned because
suitable assurance was not received.   Examples of conditions which have made
sites unacceptable are:

(a)  A college at 1 km from a site drawing large numbers of teenage pupils
     in from a wide area;

(b)  a site alongside the causeway to an island from which the population
     would have to be evacuated; and

(c)  the inability of the police to effectively clear an extensive area of
     remote beach and sand dunes frequented by sunbathers.

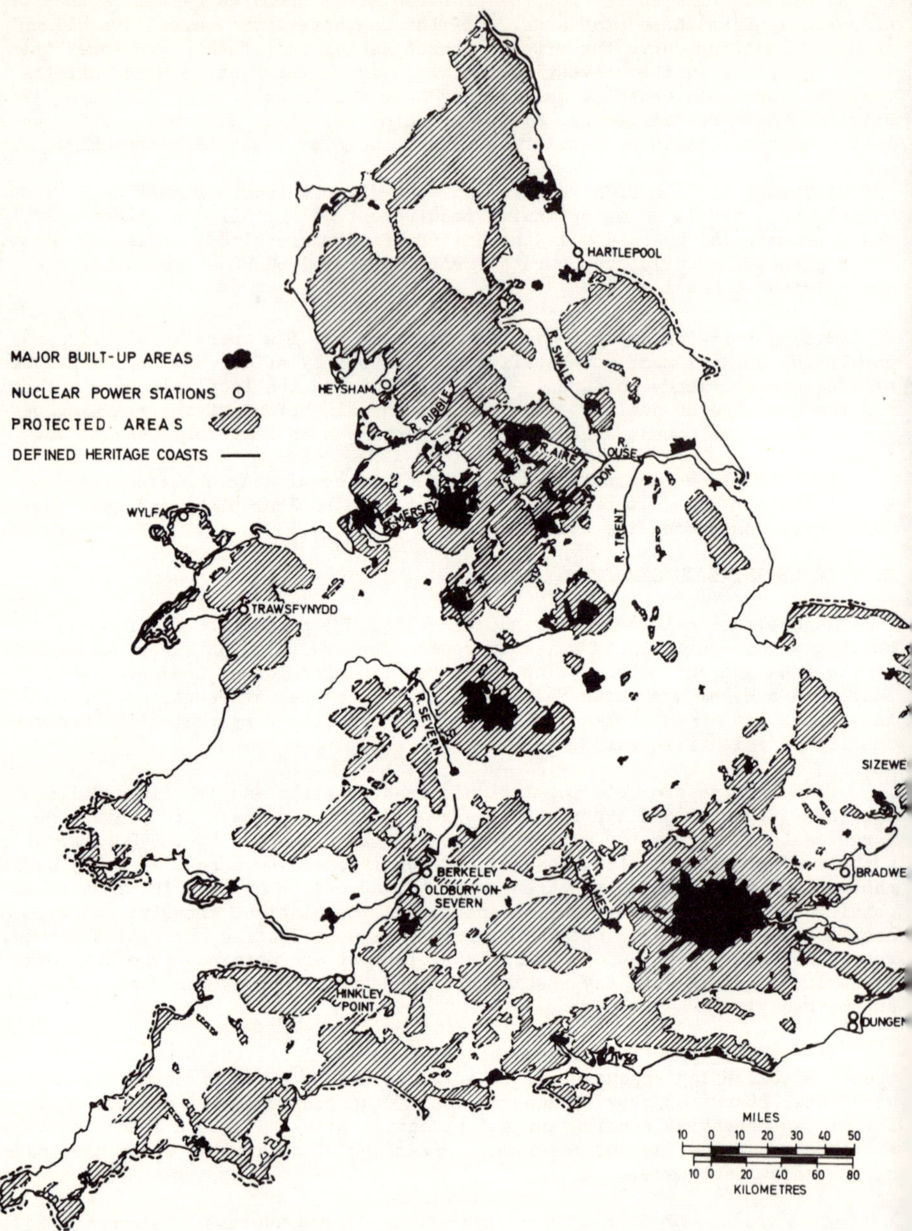


FIG. 1.   Nuclear power stations and protected land in England and Wales as at June 1974.  The percentage of England and Wales made up of protected land is:  green belt 10%; national parks 9%; areas of outstanding natural beauty 10%; forest parks 1%; areas of great landscape, historic or scientific value 15%.

Other matters which have to be considered before submitting an application
for a nuclear site licence include risk to potable-water sources, geological
stability, the frequency and magnitude of earth tremors and the proximity of
activity likely to produce explosions, explosive gas clouds or potential hazards
such as aircraft which could affect the integrity of the reactors.

In England and Wales, rivers and their catchment areas are small and the
amount of water available for power station requirements in inland areas is,
after meeting the essential needs of population and industry, limited. It was
therefore logical to use inland supplies of cooling water to produce the maximum
amount of electricity from coal-fired stations using cooling towers.   Where
water economy is a major consideration the reactor type selected can significantly
influence the output.   Magnox reactors produce steam at low pressure and
temperatures compared with contemporary fossil-fuelled power stations and require
about twice as much water per unit of electricity generated but because of
their relatively small size this was not a serious limitation.   Uranium-
dioxide fuels in stainless steel cans in the AGR permit higher fuel temperatures
which produces steam conditions and cooling-water requirements comparable with
those of conventional practice.   The SGHWR produces low steam conditions and
has a cooling-water requirement nearly 50% greater than a fossil-fuelled plant
and could prove difficult to site inland.

The enormous weight of gas-cooled graphite-moderated reactors, and the
need to prevent differential settlement for operational and safety reasons,
imposes severe limitations on the types of subsoil which are economically
acceptable.   Hard rock at a suitable depth is to be preferred but marl, stiff
London clay, and densely packed sands have all been used successfully.   To
avoid extensive pumping costs, site levels should not be much above high water
but for reasons of nuclear safety the risk of flooding must be extremely low.
On the East Coast, storm surges of 2.7 metres are not unknown and the coastline
is slowly sinking relative to the sea level, making long-term flood level
protection essential.   Flood protection is usually provided against the 'one
in a 1000-year' risk with the protective works capable of extension.

## 3.1. Site Selection

Desk studies of the area of interest and early confidential discussions with
planning and other statutory authorities to identify sites worthy of detailed investigation are first undertaken. When this interest is announced through the press,
the reactions and interests of the public will normally be made known by their
representatives on local authorities.   The local authority views do not
necessarily coincide with those of the local or national amenity societies who
often provide the core of opposition to power stations.   Consultations and
joint studies are also carried out with statutory authorities on such matters
as water supplies, road and rail access etc., in order to identify the problems
of local services and to ensure that both the Board's and local needs are
reconciled.

Subsoil investigations are used to check the foundation conditions; marine
surveys of tidal movements, depth of water and littoral drift provide information to
determine the effect of cooling-water extraction and discharge on the coastal
regime, and the dispersion of heated discharge is often predicted with the use
of mathematical models.   The zones of visual influence are prepared and, with
topographical models, help the subjective assessment of alternative station
building and transmission arrangements.

Where the formation of the land or meteorological conditions are likely to
affect the normal dispersion of ventilated gases, tests are carried out in wind
tunnels to determine the optimum height of the stack to ensure acceptable

ground-level concentrations are not exceeded.   Representatives of the authorit
concerned are invited to see the work and inspect the findings.

The recommendation of a site for development is based on a carefully considered appraisal of the capital and operating costs for each alternative layou
with an objective assessment of the effects each proposal will have on the envi
ment and amenity.   In addition to drawing on its long experience of assessing
all the environmental aspects of such plants, the views of Government Department
local authority and other bodies who also have responsibility for protection of
the environment, are carefully weighed.

The recommendation is considered by the Generating Board,
which includes an eminent practising architect to provide independent advice
on environmental matters.   When a formal application for statutory consent is
submitted, the Board must be prepared to justify all aspects at a public inquir
where it may be challenged by the general public, local authorities or statuto
bodies.

The following examples illustrate some of the problems encountered in
seeking consent for nuclear stations:

## Bradwell

This station was sited on the remote marshland area of the Blackwater
estuary some 77 km from London.   Oysters have existed in the estuary
for thousands of years and opposition from the oyster fishermen, ornithologist
and other users of the Blackwater produced many newspaper articles and emotive
headlines such as 'Dead Sea Fears of Nuclear Station'.   Numerous objections
made because of the intrusion into the countryside, the possible effect on oys
and on the Brent Geese, caused a public inquiry at which a leading biologist
categorically stated that '...the power station at Bradwell would soon extingu
oyster culture...'.   Since 1959 the conditions in the estuary have been under
constant review by the Ministry of Agriculture, Fisheries and Food.   No adver
affect on the hydro-biology has been found although the station has been opera
at high load factor since 1962.   The numbers of geese have increased to the
of becoming a nuisance to farmers, whilst the station  is generally accepted a
part of the locality.

The road built for construction has improved access and has brought new
demands for yacht harbours and weekend homes, producing further intrusion into
the countryside and necessitating some control of development.

## Trawsfynydd

The one exception to coastal siting is located in the Snowdonia National
Park some 200 m above sea level and provides an interesting illustration of
conflicting sectional interests.   The local populace strongly supported the
station for the prosperity it would bring to an area of high unemployment, but
national amenity interests opposed its intrusion into a National Park.

Trawsfynydd lake is used for surface cooling and a society formed to opp
hydro-electric development in the area paradoxically opposed the station on t
grounds it would destroy the beauty of the lake which was originally created
hydro-electric generation!   In their campaign the society published their id
of the appearance of the station which is shown in Fig.2(A). To provide an
impression of the effect of the station on its surroundings and the architect
and landscape treatment proposed, the Board provided the artist's impression,
Fig.2(B).The resemblance between prediction and reality is shown in the photo
Fig.2(C).

(A)  AS VISUALISED BY OBJECTORS

(B)  AS VISUALISED BY CEGB

(C)  AS BUILT

FIG.2.  Trawsfynydd nuclear power station.

### Connah's Quay

This site in Flintshire was partly developed by a coal-fired station, but
with space for nuclear generation, and in 1969 an application was made to build
a 2500-MW AGR station with cooling towers using water from the shallow Dee
Estuary.   The existing population distribution was within the limits of nuclea
siting and the Board understood that any development proposal would take a
possible nuclear station into account.

One year later a report on development studies was published and the loca
authority announced their intention to build a new town in a way not compatibl
with the station.   A public inquiry was held into various planning schemes an
the effect of control of development.

In 1972, three years after the application, the Minister refused his cons
because the restraints that would be imposed on future urban development could
not be accepted.   The new town proposal has since been indefinitely postponed
but because of the limited nuclear programme now authorised the site is being
considered for an oil-fired station.

### Sizewell

This site, on a low sand plateau on the Suffolk coast, is partly occupied
a  magnox  station with foundations on deep dense crag and drawing cooling wat
from the North Sea.   The station was located back from the cliff face to pres
the long-distance views along the beach and carefully located within existing
pine woods which form part of an extensive landscape scheme.

Since the station has been commissioned, the area has been designated one
of outstanding natural beauty and lately as part of the Heritage Coast.

To complete the site development, consent was obtained for a 2500-MW AGR
station in 1969.   In 1973 alternative applications were submitted for using
either HTR, LWR or SGHWR.   At the time the problems of LWR's were debated in
press, on radio and television, with critics from America participating and th
was some criticism against nuclear development at national level.   The local
planning authority engaged a Nuclear Consultant to advise them, but after full
discussions they decided to give their planning clearance.   Five objections
public and private organisations were lodged, with a number of letters seeking
information.   Subsequently the Government announced the choice of the SGHWR
for the United Kingdom, and the Secretary of State's consent to build the Boar
first SGHWR station is confidently awaited.

## 4.    PUBLIC ACCEPTANCE

An important aspect of the Board's activities not covered by regulatory
requirements, which is difficult to quantify in specific terms, is that of
public acceptance, and every effort is made to secure the co-operation of bot
the local authorities and general public by keeping them fully informed at al
stages.   The Board applies this policy to all its operations, both conventio
and nuclear, but nuclear power stations provide a special problem both from t
siting and amenity aspects and the public fear of the unknown radiological ha
It is therefore essential to ensure that such concern is fully discussed with
those involved as early as possible.

During the final stages of site selection a public exhibition and confer
may be held.   Physicists, engineers and planners describe the Board's propos
and answer questions.   Leading landscape consultants and architects engaged
by the Board to advise on the layout, orientation, texture and colour of buil

and the landscape often present, in model form, their proposals for the site to the
local authority and general public.   Local representatives are invited to visit
other nuclear stations to meet the local authority and see the surrounding area.
If the Secretary of State eventually decides to hold a public inquiry, the Board
continues to make every effort to maintain rapport with the local inhabitants.

The Board's first nuclear reactors commenced operation in 1962 and over
140 reactor years of operation have been accumulated.   During that time there
has never been an accidental release of radioactivity giving significant effects
beyond the site boundary.   Based on this record, Board staff can provide firm
reassurance to people with genuine worries about the nuclear power station
being built in their area.

When construction starts, the Board takes every practical step to ensure
its activities provide the minimum disturbance and the Site Manager makes
arrangements for liaison with the local inhabitants to ensure that complaints
are quickly dealt with.   Interest in construction is encouraged by providing
viewing towers equipped with diagrams, photographs etc., and a recorded
commentary explaining the work.

The dialogue between the Board and the local authorities is started during
the early planning stages and continues throughout the operational phase, when
the Local Liaison Committee is established.   The Committee, which includes
representatives of the local statutory bodies, farming interests, police, fire
brigade, and any other local bodies with special interest, is used to provide
information and assurance to the local inhabitants on the operation of the
power station and any possible effects it may have on the local environment,
the significance of district survey measurements, the results of emergency plan
demonstrations and any other matters which may be of local concern.   When the
stations are operational, guides are recruited from the locality, including the
wives of station staff, to conduct the general public around the station.   This
is becoming increasingly popular and many thousands of visitors are received
annually at nuclear stations.

The Board does not provide houses for its staff, who may number 500 for each
nuclear station, but encourages them to live in the local community.   A large
number of staff are recruited locally.   The station has its own sports and
social club and by joining in all local activities the station staff quickly
become integrated into the community and with the station form an important part
of local life.

Additionally other social, educational and environmental benefits are
promoted and together with the measures outlined above are part of the overall
care and concern provided by the Board to obtain public acceptance of its
nuclear power stations and to become a 'good neighbour' in every sense of the
word.

## 5.   THE WAY AHEAD

The energy crisis highlighted by the concerted action of OPEC and the
difficulties of increasing coal production has cast grave doubts on the
availability of economic fossil fuels in future.   Fortunately, in the longer term
a greater proportion of the overall energy requirements can be met by electricity
based on substantial nuclear programmes which by gradually reducing the use of
fossil fuels will also lessen the products of combustion and so improve the
atmospheric environment.   Whilst there are proved or prospective nuclear sites
in England and Wales with sufficient potential to meet short and middle term
needs, the use of all these sites will bring increasing conflict with the growing
conservation interests who are opposed both to the further encroachment on the

countryside and to nuclear energy on the grounds of safety and the long-term
storage of active waste.

The Board is fully conscious of its responsibilities for amenity and is
currently considering means by which large increases of generating capacity ca
be provided, with the minimum effect on the environment.   One method by which
this objective may be achieved is to concentrate large blocks of plant on one
site in a 'Large Concentration of Generation'.   The problems of 10 GW or more
on one site are obvious and studies are in hand to quantify the effects of
massive heat discharges and other effluents  and of large-scale irradiated fue
movements together with the restructuring of the electrical transmission syste
needed to accommodate such large transfers of power.   When operating at full
load 10,000 MW of nuclear capacity could require over 400 $m^3$/s of cooling
water heated through 12$^o$C when discharged.   Only the sea can provide such a se
and dissipate the heat.   The risks of such levels producing local ecological
changes cannot be ignored and must be thoroughly explored.   The advantages o
such nuclear complexes are, however, compelling reasons for their study.   Some
60% of the 4,500 km of England and Wales is protected coastline (Fig.1) and, as
much of the remainder is developed, there is little coast available on which to
find sites with the depth of water and massive flows necessary.

Another means of reducing environmental effects and providing abundant
cooling water with good thermal dispersion is to site stations offshore.   Th
offers advantages such as reduced visual impact, avoids control of developmen
and the need to provide for the evacuation of the population within one kilom
but could present formidable problems in construction, communication and emer
evacuation of staff.   Consideration is being given to the construction of st
on artifical islands formed in shallow waters where suitable tidal conditions
exist.   Most of the problems of constructing a station in the offshore envir
ment could be overcome by the use of stations built on floating platforms in
dockyard where the cost and production control benefits of centralised constr
could be enjoyed, although these would require standard design and repetitive
duction.   Such stations when completed might be floated into position and bu
into an island where the seas are turbulent or moored within suitable protect
in sheltered waters.   Although offshore islands are being investigated in En
and elsewhere, they appear to be extremely expensive and no substantial relief
the need for on-shore sites can be expected in the near future.

In a densely populated country the main safeguards for the public must b
very high standard of design, construction and operation rather than the remo
location of the sites.

Two developments could give relief from some siting constraints.   The u
of double containment to limit radioactive discharge to the atmosphere in the
unlikely event of an accidental release has been thoroughly explored and appl
to light-water reactors.   Its adoption for some other types of reactor in th
United Kingdom could be advantageous.   Past studies of undergrounding reacto
have proved unrewarding but application of recently developed civil engineeri
techniques such as placing a concrete membrane to form a caisson and excavati
down from the surface might make undergrounding viable in some circumstances.
Such reactors may have further advantages when they are decommissioned as the
could be sealed underground and the surface works removed.


6.    CONCLUSION

In finding sites for its nuclear power stations the Board is required t
comply with established siting criteria and to take into account the effect
its stations may have on the environment.   These requirements  may prove co

flicting and it is the Board's responsibility to provide the right balance
consistent with its statutory duties.    Every endeavour is made to obtain
public understanding of its proposals and to become a 'good neighbour'.

In view of the general public acceptance which now prevails for the
existing stations, it is believed that this policy has been successful and
will establish a basis of mutual trust for the future.    Some modifications
to the present siting strategy matched to improved design should provide the
nuclear sites required without promoting excessive public opposition.    There
is, however, an ever growing concern for the environment and it is increasingly
necessary to provide continuing effort to ensure that nuclear power remains
acceptable by publicity, education and the encouragement of public participation.

## REFERENCES


[1]  MARLEY, W.G., FRY, T.M, "Radiological hazards from an escape of fission products and the impli-
     cations in power reactor location", Int. Conf. Peaceful Uses Atomic Energy (Proc. Conf. Geneva,
     1955) 13, UN, New York (1956) 102.
[2]  FARMER, F.R., FLETCHER, P.T.,   Siting in Relation to Normal Reactor Operation and Accident
     Conditions,  UKAEA Rep. NP-7753 (1959).
[3]  FARMER, F.R.,  The Evaluation of Power Reactor Sites, UKAEA Rep. DPR/INF/266 (1962).
[4]  CHARLESWORTH, F.R., GRONOW, W.S.,  "A summary of experience in the practical application
     of siting policy in the United Kingdom",  Containment and Siting of Nuclear Power Plants (Proc. Symp.
     Vienna, 1967), IAEA, Vienna (1967) 143.
[5]  GRONOW, W.S.,  "Application of safety and siting policy to nuclear plants in the United Kingdom",
     Environmental Contamination by Radioactive Materials (Proc. Seminar Vienna, 1969), IAEA,
     Vienna (1969) 549.
[6]  SHAW, J., PALABRICA, R.J.,  "A critical review and comparison of the nuclear power plant siting
     policies in the United Kingdom and the United States of America", Ann. Nucl. Sci. Eng. 1 (1974) 241.
[7]  WILLIAMSON, P., "The environment of nuclear stations in England and Wales", XVII Nuclear Congress
     of Rome, Nuclear Energy and the Environment, 1972, CNEN, Rome.


## DISCUSSION

W.N. LABLANS:  I find from your paper that evacuation is planned for
a 30° sector within two hours.  Would it not be wise to expect contamination
of a wider sector owing to changes in wind direction in two hours?

T.P. HAIRE:  The plume sector is constantly monitored to ensure
that there is no undetected spread of contamination outside the 30° sector,
and in the event of a wind change (wind direction is also constantly monitored) the evacuation area would be enlarged as required.

A. MERTON:  Are evacuation plans kept secret from the public?
If not, what is the public response to the idea of evacuation?

T.P. HAIRE:  The public is fully informed about the emergency
arrangements  including the public evacuation plans  during the initial
public information period, at any public enquiry which may be held and
at the periodic meetings of the local liaison committee, where details of
emergency plan rehearsals held since the last meeting may be discussed.

There has been no strong reaction from the public to the evacuation
plans.  Great care is taken by the CEGB to explain the reason for such
plans and to emphasize that the stations are built and operated in such a
way as to ensure that no such evacuation will ever be required.  If, however,
for some inconceivable reason there is an accident,  then the plan is

available to minimize the effect of any radioactive release on the local
population.

B. OBERBACHER:  What do you estimate to be the operational life of
a nuclear plant?  In your safety considerations, have you taken into account
the fact that the risk of failure increases with the age of the plant?

T.P. HAIRE:  The early Magnox plants were designed for a safety life
of 20 years; the later plants were designed for 25 years.  These were of
course prudent estimates, subject to review in the light of operational
experience.

All components of the CEGB reactors which have a bearing on safety
are subject to regular inspection and maintenance, and all aspects of their
operation relevant to safety are regularly reviewed; plants are therefore
maintained at the required level of effectiveness throughout their life.  At
present there are no indications that nuclear safety levels are likely to
be reduced by a plant aging process, but this matter, as I say, is kept
under constant review.

H. AAMLID:  I should like to ask you three questions.  First, what
would be the typical number of people to be evacuated from a nuclear power
station zone within two hours in an emergency situation?  Second, what
in general is the British view on evacuation – are you pessimistic or
optimistic about the possibility of moving the people involved in time?
Lastly, supposing that a site has been approved with certain restrictions on
population density close to the site, how do you ensure that the population
density does not increase at a later stage?

T.P. HAIRE:  A typical number of people in an affected zone would
be 2000-3000.

The CEGB is quite confident that the necessary evacuation can be
carried out by the local police in the required time.

As to the last question, increases of population within the evacuation
area are controlled by monitoring the applications for planning permission
which are required for all permanent buildings, whether domestic or
industrial.  These are seen by the Nuclear Installations Inspectorate, and
if they consider that a development is likely to lead to difficulties with
evacuation, there is little chance of its being approved.  This form of
control is exercised throughout the lifetime of the nuclear site.

W. HEINZ:  Have you ever carried out an evacuation exercise, and,
if so, what was the experience?

T.P. HAIRE:  No exercise in the evacuation of the local population
has been carried out.  However, the police have wide experience of moving
large numbers of people in a short time and their judgement on the practicability of evacuation is acceptable to the CEGB.

It is important, however, to demonstrate that the emergency arrangements for the station itself function satisfactorily and these are rehearsed
regularly, culminating in an annual demonstration based on a mock maximum
credible accident.  Every effort is made to ensure that such rehearsals
are as realistic as possible.

The provision of an emergency plan is a regulatory requirement for
each station and the annual rehearsals are witnessed by inspectors from the
Nuclear Installations Inspectorate.

G. ARABIAN:  It seems that among the general public there is now a
greater willingness to accept the installation of nuclear power plants than
there used to be; either that, or their objections and questions are of a

kind that can be answered more easily.  At the same time one notes a
growing concern among technologists and people from our own circle
(waste management experts and so on), who are more difficult to satisfy.
Do you find that this is true in the United Kingdom?  If so, what is being
done about it?

T.P. HAIRE:  There is a growing awareness of environmental problems
in the United Kingdom, but this has to be seen against a long history of
environmental protection provided by statutory legislation.  The British
record in this respect is good, and the CEGB has had, since its inception
in 1957, a specific statutory duty to show due regard for the environment.

With the growing need for nuclear sites there will be a corresponding
need to retain public acceptance and confidence by education, assurance
and effective public relations.

E.H. HUBERT:  One problem we shall all have to face in the fairly
near future – especially you in the United Kingdom – is plant decommissioning.
What would normally be your policy:  to dismantle the old station and build
a new one at the same site or to choose a new site?  The latter solution
would probably be difficult owing to the limited availability of sites.

T.P. HAIRE:  This is certainly a problem which will have to be
tackled sooner or later in the United Kingdom.  The current approach
would be to dismantle the reactor superstructure, which would reduce
the height by about one half, and then, if necessary, to earth the remainder
over.  It is of course a question of economics.  It would be possible to
dismantle and store the radioactive components; this would be costly,
but it might be worth while as a means of recovering the site for future
nuclear development.  To the best of my knowledge there is no experience
of dismantling or decommissioning large power reactors as such, although
there is wide knowledge of the remote-handling techniques which would be
so necessary for such an operation.

E.F.F.W. USHER:  I should like to emphasize, in connection with
Mr. Hubert's question, that sites with the necessary attributes for nuclear
power generation are rare and constitute valuable assets in England and
Wales.  We would not, therefore, consider abandoning a site when the
reactors are decommissioned.  Studies for redevelopment of earlier nuclear
station sites are in hand and would entail the steps briefly outlined by
Mr. Haire: complete dismantling of the turbine house and auxiliary
buildings, strip-down of the reactor to the biological shield, earthing
over and landscaping.  The site, with a new reactor, would continue to
be licensed and fully controlled.

# SITING CONSIDERATIONS FOR NUCLEAR FACILITIES IN PAKISTAN

M. NASIM, S. A. HASNAIN
Karachi Nuclear Power Plant,
Karachi, Pakistan

## Abstract

SITING CONSIDERATIONS FOR NUCLEAR FACILITIES IN PAKISTAN.
With a growing nuclear power programme in Pakistan, the siting of reactors and other facilities has become important from safety and economic viewpoints.  A review is given of the considerations in site selection of the 137-MW(e) Karachi Nuclear Power Plant and other reactors.  The location of a complex of nuclear power plants to be built in the next decade has been decided on grounds of availability of cooling water, distance from load centres, and acceptable meteorological and geophysical factors.

## 1. INTRODUCTION

The problem of finding suitable sites for nuclear facilities in Pakistan has been faced since the inception of the Pakistan Atomic Energy Commission (PAEC) in 1956.  At first it was a matter of siting nuclear research centres. Later, when a programme for constructing nuclear power plants was drawn up, the question of finding appropriate sites for them became important.  The first sites were relatively easy to determine, but criteria had to be established for acceptable meteorological and other environmental factors affecting safety.  An expanded nuclear power programme is now in the offing, and a systematic approach is being defined not only for siting the power plants and supporting nuclear facilities but also in determining standards to limit their environmental impact.

The present paper reviews the experience in Pakistan in the siting of nuclear power plants and allied facilities.  This experience would be of interest to developing countries embarking on a programme of nuclear power.

## 2. NEED FOR NUCLEAR POWER

The population of Pakistan, according to the preliminary results of the 1972 census, is 64.89 million, which is expected to grow at approximately 3% per year for the next several decades, reaching a population of about 150 million in the year 2000.  This rapid growth of population is accompanied by continued industrialization.  Consequently, studies made in Pakistan by the National Water and Power Development Authority (WAPDA) and PAEC show that the load growth of 6% is expected to continue for the foreseeable future.  Long-term forecasts show an ultimate demand of 28 000 MW in the country by the year 2000.  These figures are based on a nominal per-capita consumption of about 730 kW·h.  By 1981 the forecast shows an expected gross maximum demand of about 3800 MW (at present 1900 MW) for which a generating capacity exceeding 4700 MW would be required [1].

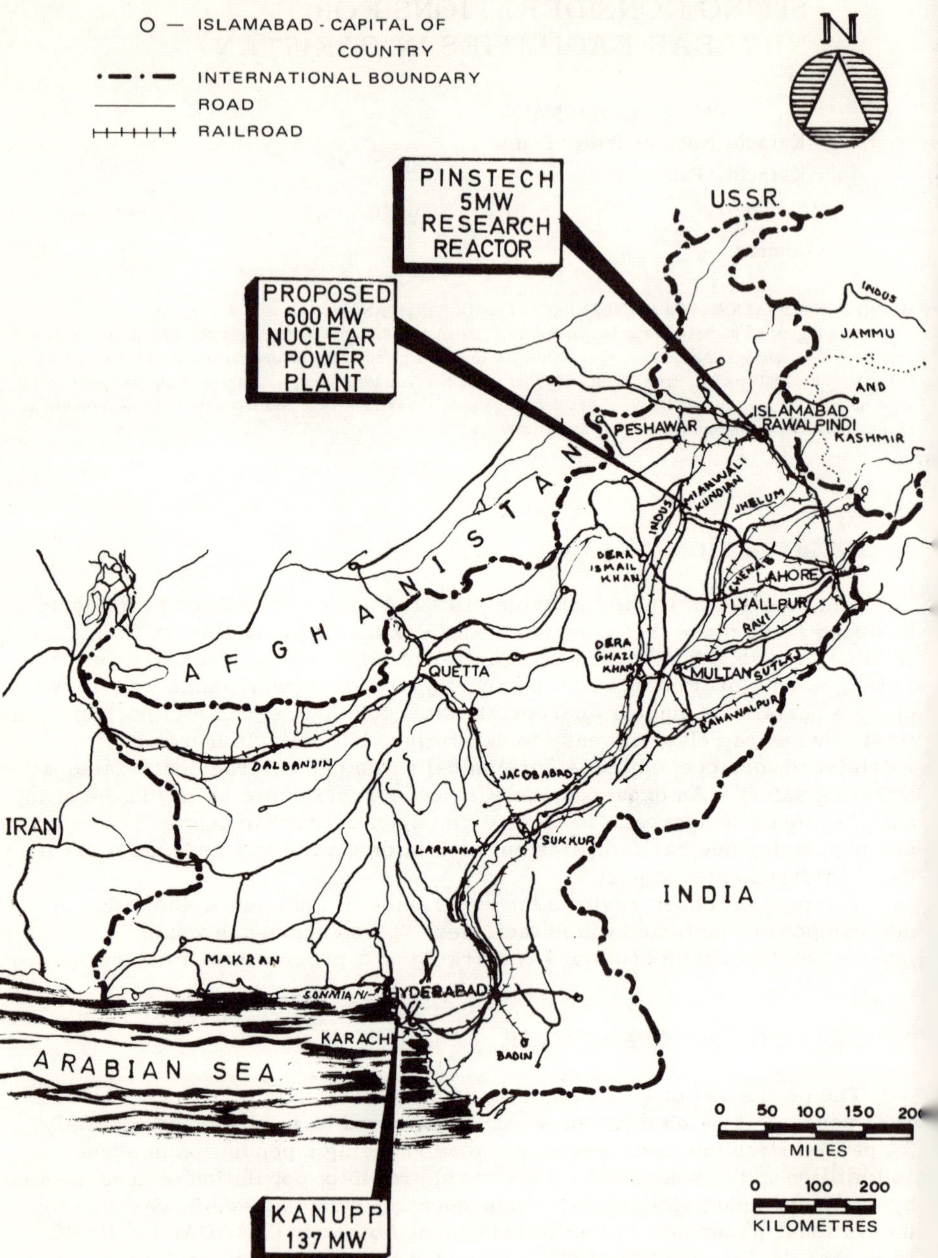



FIG. 1.   Location of research and power reactor sites in Pakistan.

Studies also show that the available and future indigenous energy resources most suitable for power generation in Pakistan are hydro-power and natural gas, though it may be possible to develop some coal. For reasons of distance and variability, hydro-power is largely confined to the northern regions of the country. Efforts are being made to develop all possible water resources, but it is likely that not more than 1100 MW may be economically developed by 1981 by this means. The natural gas resources are limited to about $5 \times 10^{11}$ m$^3$ ($18 \times 10^6$ ft$^3$) and unless additional reserves of suitable quality are found the demand for gas for industrial and commercial uses will soon reach the level where further allocation of gas for power generation would not be warranted. The coal is remotely located and of low quality.

In view of the limited indigenous fuel supply, the alternatives facing Pakistan are imported oil-fired plants and nuclear power. One nuclear generating station, Karachi Nuclear Power Plant (KANUPP) with a capacity of 137 MW, is currently in operation and supplying power to the Karachi grid. PAEC foresees that because of the high cost of imported oil, a long-term commitment to nuclear power will have to be made.

Based on these studies, PAEC had drawn up an expansion plan which calls for the first of a series of large nuclear power stations to be operational by 1981 in the northern part of the country. Network analysis done for the feasibility of this plant shows that unit size of 600 MW can be accommodated by 1981 [1]. Studies done for the Karachi area in the southern part of the country show a requirement of a 500-MW dual-purpose water-power nuclear station by 1981, although it is recognized that system-stability considerations and other factors may delay this unit for a few years [2].

In view of this requirement of nuclear power stations in Pakistan, several prospective sites, as shown in Fig. 1, in the northern areas and the Karachi area were investigated.


3. SITING

A number of specific site locations for nuclear plants have been investigated in various regions of Pakistan. The criteria which have normally been applied to site selection are presented as follows [3]:

3.1. General requirements

Most of the general criteria which influence site selection for a conventional thermal power plant apply equally to the choice of site for a nuclear plant. The major common factors are:

> Location relative to load centre
>    and transmission network;
> Availability of cooling water;
> Accessibility and transport;
> Soil conditions;
> Elevation, terrain and flood levels; and
> Availability of labour and material

A nuclear plant in the northern zone of Pakistan would probably have to be equipped with cooling towers because of the variable flow in the rivers and the annual maintenance of the irrigation canals. Although suitable sites do exist upstream of one or two barrages, where sufficient cooling water is available all the year round, it may be economical to rely on cooling towers and to locate the plant near the major load centres.

For plants close to the coast, sea water would be used for condenser cooling. These plants should be located as near to the load centres as will be permitted by safety considerations.

Transport of nuclear plant components would present a problem in the northern zone of Pakistan, as most of the possible sites would not be accessible by river. Railway clearances would not be adequate for the more voluminous and heavy equipment (pressure vessel, etc.). Transport by road was therefore considered to be the most suitable method, at least for the heavier equipment, although restrictive bridge loadings and other obstacles requiring engineering solution would be encountered. Karachi is directly accessible by sea, and good roads are available. Present loading facilities at Karachi port are inadequate for a typical pressure vessel. Heavy lift facilities would have to be made available for such items.

## 3.2. Radiation

Site selection for a nuclear power plant is also affected by the hazards of radioactivity released to the environs. Two types of release are considered: the routine liquid and airborne effluents which are released during normal plant operation and the gross release of fission products which might occur in the event of a serious accident.

## 3.3. Plant effluent

The radioactive wastes accompanying normal plant operation would be concentrated and stored, or treated and allowed to decay, before disposal to the environment. Their release would be monitored and regulated so that the resulting activity concentrations do not exceed specified permissible values.

Airborne effluent would be discharged through a tall stack so that permissible concentrations are not exceeded at ground level.

Liquid wastes would be treated by filtration, ion-exchange and evaporation and by storage for sufficient time to allow decay. Their discharge would be monitored and they would be mixed with the condenser cooling water so that station effluents do not exceed the drinking-water tolerances before the discharge water is returned to the river or sea.

In the northern zone of Pakistan, the downstream irrigation uses of water and the absence of a condenser coolant stream could lead to unusually restrictive design of a liquid-waste disposal facility. For this reason, at certain sites it may be necessary to purify and recycle all liquids to the demineralized water-storage facility.

Solid wastes, which would include activity removed from liquid, would normally be collected and stored in underground tanks on the site.

Design of waste-disposal facilities would preclude leakage of contamination to the soil. Nevertheless, consideration must be given to the drainage patterns of the site with regard to the ultimate disposal of any leakage.

## 3.4.  Exclusion

Certain improbable but credible accidents could occur in a nuclear
power plant, releasing large quantities of fission products which would
represent much greater public hazard than is normally encountered in
industrial accidents.  Until extensive safe operating experience is accumulated
for large nuclear plants, it will be necessary to locate them at safe distances
from large centres of population.  The site selection criteria adopted for
KANUPP, as well as future nuclear power stations, involve two distances
from the reactor that must meet specified conditions.  The exclusion distance
is defined such that a child located at any point on its boundary during the
entire time of passage of activity from the maximum design accident would
not receive more than 250 rem to the thyroid or 25 rem to the whole body.
A population distance is stipulated between the reactor site and an area of
dense population such that the total population dose will not exceed a reference
value of $10^6$ man·rem.

The philosophy of the maximum design accident for KANUPP has been
based on the risk-probability approach, and it is expected that for future
plants similar criteria would be used.  The anticipated results of the accident
are evaluated on the basis of engineered safety features and other safeguards,
as well as meteorological and other conditions, considered on a conservative
basis.  This analysis permits determination of the requirements of a
monitoring area, notably the distance between the site and the public [4].

## 3.5.  Environmental conditions

The suitability of a nuclear plant site may also be influenced by prevailing
wind conditions and by the history of naturally occurring phenomena such as
floods, severe windstorms and earthquakes.

In general, the plant should not be located where prevailing wind
conditions would prevent adequate dispersal and dilution of gaseous effluents
in the atmosphere or where prevailing wind directions might threaten large
population centres.

Consideration must eventually be given to the proximity of actual sites
to earthquake faults.  The seismic history of the areas considered indicates
that none of the possible site locations for nuclear plants is in areas of
intense seismic activity.

## 4.  REVIEW OF SITES FOR POWER PLANTS

For KANUPP, early studies established that two alternative sites located
east of the city of Karachi (over 3 million population) were not appropriate,
on account of the growth of population in that direction and restricted cooling-
water supplies from creeks.

Preference was given to the present site lying 24 km (15 miles) west of
the city, as the area round the site has very low population out to about 8 km
(5 miles).  The load-bearing capacity at the site was $9$-$11 \times 10^4 \, kg/m^2$
($8$-$10 \, tons/ft^2$) and was found to be suitable.  The recorded seismic history
of the area indicates that while earthquakes occur rather frequently they have
been consistently of shallow and distant origins.  The maximum horizontal
acceleration experienced in the area from recorded earthquakes was less
than 0.05g [5].

For the proposed nuclear dual-purpose water-power plant for Karachi, it is appropriate to give careful consideration to the siting of the plant, especially as the choice of site would have a large bearing on the costs of the plant intake and outfall and on the costs of transmission of power and conveyance of water from the site to the point of juncture with the existing system. Five possible sites covering the area of Karachi were investigated [2].

For sites close to Karachi the most appropriate was the one located near KANUPP. Cape Monze (western part) and Cape Monze (Buleji) are in a remote and unpopulated area and were considered favourable from the point of view of nuclear safety. The geological and foundation conditions were found to be suitable. Enough cooling water would be available as the sea is about 9 m (30 ft) deep about 600-700 m (2000-2500 ft) from the coast. Another site, at Sonmiani, located about 48 km (30 miles) west of Karachi, was considered mainly because of the possible development of an industrial and urban centre associated with the construction of a major port, a preliminary feasibility study of which is being carried out. The area is thinly populated. Agricultural activities are of minor importance and there are no natural resources requiring protection. There is enough sea water as a depth of more than 9 m (30 ft) is available at the proposed site. As in the case of Karachi, this area is located in the active seismic zone. Some faults have been recorded at about 160 km (100 miles) in the northern part of this region. Two other sites east of Karachi close to the industrial area were also investigated. The bearing capacity on the Bundal Island site was found to be limited as no bedrock was encountered up to 36 m (120 ft) below ground level. Other characteristics of the site are normal. The Gharo area site was found to be better in load-bearing capacity. Ample water supply is available. Some problems in connection with the liquid-waste disposal may be encountered as the development of heavy industry (including the Karachi Steel Mill) may eventually contaminate the sea water.

In 1972, a co-operative study was launched by PAEC and WAPDA for selecting sites for future nuclear power plants in Pakistan. The region of interest was the northern zone of the country on account of the burgeoning power demand. Because of the variable flow in the river Chenab, the possibility of cooling towers was investigated for a site near Chiniot and Lyallpur. Other factors considered included geology, seismology, depth of water table, load-bearing capacity of soil and availability of bedrock, as they greatly influence the cost of construction. Good foundation conditions were found in the area near Chiniot, Kalabagh and Chashma [3].

Nearness to load centres, as they are developing in the northern grid, was also an important consideration, as this contributes to the economy, reliability and stability of the power system. Four prospective sites along river Indus in the Kot Adu and Kalabagh areas were rejected due either to lack of sufficient cooling water or the presence of geological faults. The site near Chiniot was found adequate as it has good load-bearing capacity of $9\text{-}13 \times 10^4$ kg/m$^2$ (8-12 tons/ft$^2$), assured water supply, good rail and road communication, and is near the load centre of the city of Lyallpur. Near Chashma, the right bank of the river Indus was found unsuitable because of non-availability of suitable land and closeness to the barrage and lake. But the left bank an ideal location is available with low water table, no risk of flooding or erosion, and low seismic activity [3]. The Chashma site has accordingly been selected for the establishment initially of a 600-MW nuclear power plant and a fuel fabrication plant.

## 5. SITING OF RESEARCH FACILITIES

For the Pakistan Research Reactor-I at the Pakistan Institute of Nuclear Science and Technology (PINSTECH), safety studies were performed by the Commission's own scientists, and their report was approved in 1962 by an Advisory Panel on Reactor Safety assembled by the IAEA. It was confirmed that reactor operation would pose no hazard to the public, including the population of Islamabad located about 19 km (12 miles) to the west of PINSTECH. Some public concern was engendered at the time by minor earthquakes in the region, but it was demonstrated that the seismic design of the reactor structures was adequate.

## 6. MANAGEMENT OF NUCLEAR SAFETY

In 1965 the Federal Government passed the Pakistan Atomic Energy Ordinance empowering the PAEC to make safety regulations and take all necessary measures to ensure that they are properly implemented. To carry out its dual role as the owner/operator and the licensing authority for the nuclear installations, an independent group called the Pakistan Nuclear Safety Committee, consisting of specialists in nuclear safety, reactor engineering and physics, metallurgy, health physics, reactor operation, etc., has been set up within the PAEC. This Committee carries out a thorough safety evaluation of the plant, from the siting and conceptual design stage to approval of the final design and operation. In addition, continued surveillance of plant operations is made to assess the reliability of safety systems, unusual occurrences, radiation exposure and radioactive releases to the atmosphere.

The standards and guides adopted by the Committee are in conformity with recommendations of the International Commission on Radiological Protection and the IAEA.

A proposal to establish an independent Nuclear Safety and Radiation Protection Control Board is under consideration by the Government to further enhance the effectiveness of the present system of safety evaluation and licensing of nuclear installations.

## 7. CONCLUSION

The safety criteria used in Pakistan for selecting sites for nuclear installations do not differ in principle from those adopted in other countries. The basic requirement is the careful analysis of the environment with regard to population, use of water, use of land, meteorological and seismic characteristics, groundwater movement, etc. This is essential so that the routine as well as the maximum design accident conditions, particularly with regard to releases of radioactive materials from the plant and disposal of radiation wastes, do not present any radiological hazards to the population and the environment. The Pakistan Nuclear Safety Committee has reviewed the design, construction and operation of the 137-MW KANUPP, near Karachi. It reviewed the installation and operation of the 5-MW pool-type research reactor at PINSTECH and examined siting considerations for the conceptual 500-MW dual-purpose water-power nuclear power plant near Karachi. The

Committee has also studied the characteristics of the site near Chashma for
a 600-MW nuclear power plant in the northern grid of Pakistan and considers
it acceptable for setting up a plant of this size.


## ACKNOWLEDGEMENTS

The authors wish to thank Mr. Munir A. Khan, Chairman, PAEC, for his
interest and encouragement in the work discussed in this paper. The authors
are also grateful to Mr. M. Shafique and Dr. Ahsan Mubarak for valuable
suggestions during the preparation of the paper.


## REFERENCES


[1]     PAKISTAN ATOMIC ENERGY COMMISSION, Preliminary Feasibility Study for a 600 MW Nuclear Power
        Plant in the Northern Grid of Pakistan (Feb. 1974).
[2]     INTERNATIONAL ATOMIC ENERGY AGENCY, Review of Dual-Purpose Nuclear-Power Water Desalination
        Plant for Metropolitan Karachi, Report on an IAEA Mission to Pakistan, June 1972.
[3]     PAKISTAN ATOMIC ENERGY COMMISSION, A Preliminary Study on the Location of a Nuclear Power
        Plant in the Northern Zone of West Pakistan (March 1973).
[4]     NASIM, M., "Nuclear safety principles and practices at KANUPP", Principles and Standards of Reactor
        Safety (Proc. Symp. Jülich, 1973), IAEA, Vienna (1973) 169.
[5]     HARRISON, W.C., A Dynamic Analysis for the KANUPP Reactor, CGE Rep. No. R68CAP3 (1968).



## DISCUSSION

H. SCHNURER (Chairman): I should imagine that other industrial
facilities would normally be located in the vicinity of a nuclear power station.
Considering the population increases in your country, the acceptability of an
approved site may change significantly in the future as industrialization
progresses and the population becomes denser. Do you have any
administrative means of controlling such developments?

M. NASIM: In Pakistan the development of industrial and recreational
facilities is usually co-ordinated by different governmental organizations.
These organizations consult the Nuclear Safety Committee, established by the
PAEC, if any such facilities are to be developed in areas close to the power
reactor sites.

P. THOMAS: Do you perform extended meteorological measurements at
the site of a future power plant for several years before the approval in order
to obtain statistical data on weather conditions? Do you also accurately
measure the natural radioactivity before the plant starts operation so that it
can be compared later on with the total radioactive burden in the neighbourhood and any increment due to the plant can be determined?

M. NASIM: At present we do not perform extended meteorological
measurements before approval of a particular site if meteorological
information is available for areas close to the site; the data already available
are then used to judge the meteorological characteristics of the site. For
certain sites, if meteorological information is not available, we assume
the worst weather conditions (usually Pasquill F category) in evaluating
accidental releases.

We do establish the base-line radioactivity before the plant starts
operation. For this purpose we usually follow ICRP recommendations.

# CRITERIOS ADOPTADOS EN LA EVALUACION DEL EMPLAZAMIENTO DE TRES CENTRALES NUCLEARES ESPAÑOLAS

Consuelo PEREZ DEL MORAL
Junta de Energía Nuclear,
Madrid, España

Abstract—Resumen

CRITERIA ADOPTED IN SELECTING THE SITES FOR THREE NUCLEAR POWER STATIONS IN SPAIN.
The main object of this report is to outline one of the criteria applied in Spain in evaluating a site from the demographic point of view. It shows how these criteria were applied to the sites chosen for three nuclear power stations under construction in Spain which use pressurized-water reactors with a unit power of 900-1000 MW(e). Each of these sites has its own particular problems: a knot of population close by, high population density beginning only 2 km away, and a busy railway line crossing the site. These site characteristics naturally affect the minimum requirements for containment and associated technical safeguards that must be fulfilled to meet the criteria referred to above.

CRITERIOS ADOPTADOS EN LA EVALUACION DEL EMPLAZAMIENTO DE TRES CENTRALES NUCLEARES ESPAÑOLAS.
El principal objeto de este informe es presentar uno de los criterios adoptados en España para evaluar un emplazamiento en su aspecto demográfico. Se presenta la aplicación que se ha hecho de estos criterios a los emplazamientos elegidos para tres centrales nucleares españolas en periodo de construcción, que incorporan reactores de agua a presión de una potencia unitaria del orden de los 900-1000 MW(e). Cada uno de estos emplazamientos tiene sus problemas particulares, como son un núcleo cercano de población, una alta densidad de población a partir de los dos kilómetros y una vía férrea de importancia que atraviesa el emplazamiento, respectivamente. Estas características propias de cada emplazamiento afectan a los requisitos mínimos relativos a la contención y salvaguardias técnicas asociadas, que deben exigirse para que se cumplan los criterios mencionados anteriormente.

## INTRODUCCION

En España no se ha hecho hasta ahora, por parte de la Administración, un estudio sistemático de todo el territorio nacional a fin de seleccionar las áreas más favorables para el establecimiento de centrales nucleares, sino que cada futuro explotador hace una propuesta concreta del emplazamiento que considera más conveniente.

Una vez recibida la propuesta sobre un emplazamiento dado, la Administración estudia las características del mismo para determinar si no existe ningún aspecto que lo excluya como emplazamiento de una central nuclear, y en caso contrario, establece los límites y condiciones que deberá cumplir la futura central, para que de su operación no se derive un riesgo indebido al medio ambiente y a la población de los alrededores, tanto en funcionamiento normal como en caso de accidente.

Para prevenir los efectos de un accidente se delimitan alrededor de la central dos zonas: la «zona de exclusión» y la de «baja densidad de población», definidas del mismo modo que lo hace la USAEC en el Federal Code Regulation 10CFR 100 [1], y la distancia mínima a un centro mayor de 25 000 habitantes.

Es necesario que el proyecto del reactor sea tal que, en caso de ocurrir
el máximo accidente previsible (MAP), las dosis potenciales de exposición
a un individuo situado en la frontera de las zonas antes mencionadas no
excedan los valores límites definidos o bien que se disponga de un tiempo
razonable, después del accidente, para evacuar la población antes de que
ningún individuo reciba dosis de exposición radiactiva excesivas.  Se aplica
el criterio más limitativo de los dos anteriormente citados.

Los valores límites de dosis de exposición admitidas después del MAP
no coinciden con las definidas en 10CFR 100, sino que se admiten 25 rems
a todo el cuerpo y 150 rems al tiroides, recibidas por la exposición durante
dos horas después del accidente por un individuo situado en los límites de
la zona de exclusión o durante todo el tiempo de paso de la nube radiactiva
en el límite de la zona de baja población.

La 10CFR 100 no define explícitamente las dosis límites de exposición
a toda la población $-$ rem·hombre $-$ para caso de accidente, como por
ejemplo especifican los criterios canadienses.  No obstante implícitamente
está incluido este concepto en la definición de distancia mínima a centros
de población mayores de 25 000 habitantes.  Aceptando este criterio
implícito, se adopta un valor límite de la dosis al tiroides a toda la población
que normalmente coincide con $2,5 \times 10^6$ rem·hombre.


## EL CODIGO DE CALCULO EVATO

El código de cálculo EVATO [2] desarrollado en la Junta de Energía
Nuclear española (JEN), calcula las dosis potenciales de exposición
radiactiva a todo el cuerpo y al tiroides, para diferentes distancias al
reactor y diferentes intervalos de tiempo después de ocurrido el accidente,
con las hipótesis sobre tipo de accidente, características del reactor y
condiciones meteorológicas que aparecen en las Regulatory Guide de la
USAEC, Nº 1.3, 1.4, y 1.5 [3].  El código admite como opciones los
diferentes tipos de salvaguardias tecnológicas asociadas a la contención
que puedan incorporarse a cada reactor como son:  el rociado, los filtros
de recirculación, las salas de penetraciones con sistema de ventilación y
filtrado, etc.  También incluye la posibilidad de emplear coeficientes de
dilución atmosféricos propios del emplazamiento en estudio, en caso de
que sean conocidos.


## CENTRALES NUCLEARES ESPAÑOLAS

Las tres últimas centrales nucleares, evaluadas recientemente por la
JEN y que han obtenido por parte del Ministerio de Industria el permiso de
construcción correspondiente, son las de Almaraz, Lemoniz y Ascó I, en
todos los casos del tipo de agua a presión y con una potencia nominal por
reactor de 2686 MW(t).


## CENTRAL NUCLEAR DE ALMARAZ

La central nuclear de Almaraz, situada al oeste de la península
Ibérica, en la provincia de Cáceres, constará de dos unidades refrigeradas

en circuito abierto por agua de un estanque artificial construido en un afluente del río Tajo. La distancia de los reactores al pueblo de Almaraz, con 950 habitantes en el momento actual, es de 1700 m aproximadamente.

Por su situación en la meseta castellana, el emplazamiento posee unas condiciones de dilución atmosférica bastante malas y tendencia a situaciones de calma. Existe por tanto una probabilidad lo suficientemente elevada como para considerarla en el análisis de accidentes, de que durante las dos horas siguientes al accidente se presente una F de Pasquill como categoría de estabilidad atmosférica, coincidente con un viento de velocidad inferior a 1 m/s. Esta hipótesis no ha podido ser de momento confirmada con datos de la micrometeorología del emplazamiento, ya que la estación meteorológica instalada, que funciona desde primeros de 1972, es incompleta y sólo desde principios de este año se ha puesto en servicio una estación meteorológica suplementaria con los requisitos que aparecen en la Regulatory Guide 1.23 de la USAEC.

Por los motivos expuestos se tomaron como condiciones meteorológicas del emplazamiento, después del accidente, las contenidas en la Regulatory Guide 1.4 de la USAEC, no descartándose la posibilidad de que durante las dos primeras horas haya unas condiciones de dilución atmosférica peores que las supuestas, lo que daría lugar a valores de dosis más altos que los calculados.

El Informe Preliminar de Seguridad (PSAR) presentado por el solicitante propone como salvaguardias para limitar la emisión de radiactividad a la atmósfera exterior, después de ocurrido el accidente, un sistema de rociado con adición de $N_aOH$ que consta de dos unidades con capacidad de refrigeración del 100% cada una, y un caudal unitario de 3600 gal./min. Suponiendo que después del accidente sólo actúa una unidad, se calcula la efectividad del rociado para la eliminación del yodo en forma elemental y para el yodo en partículas, no dando crédito a la eliminación del yodo en forma orgánica por este medio. Aunque el sistema de rociado esté actuando durante más tiempo, sólo se supone efectivo para la eliminación del yodo hasta que la concentración de éste en la contención disminuye hasta un cierto valor.

En la tabla I aparecen los valores que se han empleado para el cálculo, respecto a constantes de eliminación del yodo por rociado y factor de descontaminación que se alcanza por este medio, sección recta mínima de la contención, potencia a la que ha estado actuando el reactor desde la puesta en marcha hasta que la actividad de los productos de fisión ha alcanzado un equilibrio, y los dos valores que se han supuesto para la velocidad de fugas de la contención bajo las condiciones de presión y temperatura después del accidente, con vistas a determinar los requisitos mínimos necesarios para que se cumplan los criterios que sobre dosis máximas se han descrito en la introducción a este informe.

Los radios de las zonas de exclusión y de baja población propuestos por el solicitante son de 1 y 5 km respectivamente. A estas distancias y con las hipótesis mencionadas se alcanzan los valores de dosis al tiroides que aparecen también en la tabla I. Puede apreciarse que para el caso de que la fuga máxima de la contención sea 0,2%/día, la dosis al tiroides que recibirá un individuo situado en el límite de la zona de exclusión durante las dos horas siguientes al accidente superaría los 150 rems, aún en el caso de que la velocidad del viento durante esas dos horas fuera de 1 m/s.

En el pueblo de Almaraz, si la velocidad del viento se considera también de 1 m/s y la fuga máxima de la contención de 0,2%/día, se tardaría

TABLA I.   CARACTERISTICAS DE LA CONTENCION Y
SALVAGUARDIAS ASOCIADAS DE LOS REACTORES DE LA
CENTRAL NUCLEAR DE ALMARAZ

| Potencia | Efectividad del rociado [a] | Sección recta mínima de la contención | Fuga máxima de la contención | |
|---|---|---|---|---|
| | | | Caso 1 | Caso 2 |
| 2820 MW(t) (105% potencia nominal) | $\lambda r^e = 11{,}6\ h^{-1}$ <br> $\lambda r^p = 0{,}45\ h^{-1}$ <br> $D = 100$ | 2000 m$^2$ | 0,1% vol./d | 0,2% vol./d |
| Dosis tiroides a 1 km recibidas después de 2 h | | | 85,5 rems | 171 rems |
| Dosis tiroides a 5 km recibidas después de 30 d | | | 28,6 rems | 57,2 rems |
| Tiempo que se tarda en recibir 150 rems al tiroides a los 1700 m | | | | 4,3 h |

[a]
$\lambda r^e$ = constante de eliminación del yodo en forma elemental.
$\lambda r^p$ = constante de eliminación del yodo en forma de partículas.
  D   = factor de descontaminación.

4,3 h para que sus habitantes recibieran 150 rems después del accidente,
que es un tiempo excesivamente corto para poder evacuar la población con
los medios de que se dispone; tanto más si,por ejemplo, se diera el caso
de un viento con velocidad de 0,5 m/s, lo que alargaría el tiempo que tarda
la nube radiactiva en alcanzar el pueblo, pero una vez que llegase se alcan-
zaría la dosis límite en 2 h.   En conjunto se dispondría sólo de un tiempo
de 3 h para evacuar la población después de producirse el accidente, lo
cual es insuficiente.

Por todo lo expuesto y si no se aumentan las salvaguardias disponibles
para eliminar la radiactividad del interior del recinto de contención, se
hace preciso que la velocidad de fugas máxima de dicho recinto se reduzca
a 0,1%/día.

## CENTRAL NUCLEAR DE LEMONIZ

La central nuclear de Lemoniz situada al Norte de España en la
provincia de Vizcaya y refrigerada por el mar Cantábrico, constará
tambien de dos unidades.   Según puede apreciarse en la figura 1, la
densidad de población de la zona que rodea el emplazamiento a partir de
los 2 km es elevada, excediendo con mucho a la del resto de los emplazamientos españoles.   A 18 km del emplazamiento se encuentra el centro
de la ciudad de Bilbao con una población de un millón de habitantes prevista

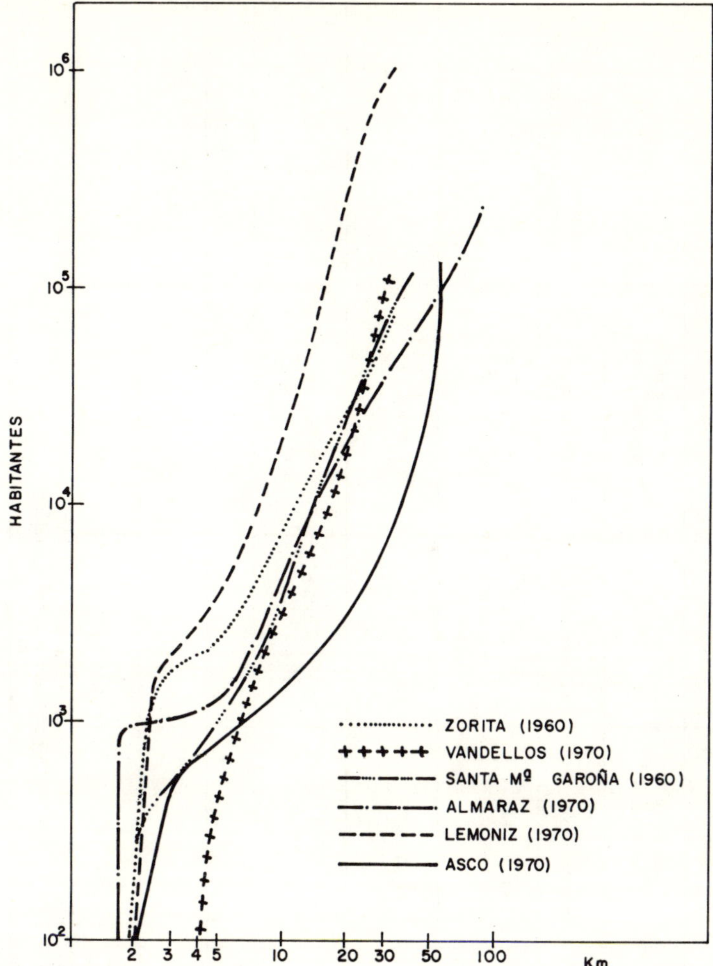



FIG.1.  Población en función de la distancia a los emplazamientos de varias centrales nucleares.

para el año 2000.  Sus suburbios, incluyendo el pueblo de Baracaldo y el
aeropuerto, se encuentran a unos 10 km del emplazamiento, siendo toda
esta zona muy poblada.

Se ha tomado como zona de baja población un entorno de 2 km de radio
con centro en el emplazamiento.  La población en esta zona es baja, aunque
muy distribuida, lo que hace difícil su evacuación.  El radio de la zona
de exclusión se toma de 750 m.

Se realizó un estudio paramétrico incorporando al reactor diferentes
salvaguardias tecnológicas y tipos de estructura de la contención, calculando
para cada caso las dosis individuales fuera de las zonas de exclusión y de
baja población, así como las dosis individuales fuera de las zonas de
exclusión y de baja población, así como las dosis totales a la población,

TABLA II.   DOSIS AL TIROIDES A TODA LA POBLACION AFECTADA, DESPUES DEL MAP, TENIENDO EN CUENTA DIFERENTES SALVAGUARDIAS TECNOLOGICAS ASOCIADAS A LA CONTENCION

| Tipo de contención | Caso 1 | Caso 2 | Caso 3 | Caso 4 | Caso 5 |
|---|---|---|---|---|---|
| | Simple | Cámara de penetraciones | Simple | Cámara de penetraciones | Doble |
| Fuga máxima de la contención | 0,1% vol./d | 0,1% vol./d | 0,1% vol./d | 0,1% vol./d | primario 0,5% vol./d secundario 100% vol./d |
| Efectividad del rociado | $\lambda r^e = 5,3\ h^{-1}$ [a] $\lambda r^p = 0,45\ h^{-1}$ $\lambda r^0 = 0$ | $\lambda r^e = 5,3\ h^{-1}$ [a] $\lambda r^p = 0,45\ h^{-1}$ $\lambda r^0 = 0$ | $\lambda r^e = 5,3\ h^{-1}$ [a] $\lambda r^p = 0,45\ h^{-1}$ $\lambda r^0 = 0$ | $\lambda r^e = 5,3\ h^{-1}$ [a] $\lambda r^p = 0,45\ h^{-1}$ $\lambda r^0 = 0$ | - |
| Filtros del sistema de ventilación de la sala de penetraciones | - | $R_{e,p} - 90\%$ [b] $R_0 - 70\%$ | - | $R_{e,p} - 90\%$ [b] $R_0 - 70\%$ | - |
| Filtros de recirculación interiores a la contención | - | - | $\lambda f^{e,p} = 2,35\ h^{-1}$ [c] $\lambda f^0 = 0$ | $\lambda f^{e,p} = 2,35\ h^{-1}$ [c] $\lambda f^0 = 0$ | - |
| Altura de la chimenea | - | - | - | - | 100 m |
| Filtros de la chimenea | - | - | - | - | $R_{e,p} - 90\%$ [b] $R_0 - 70\%$ |
| Dosis a toda la población (rem·hombre) al tiroides | $5,87 \cdot 10^6$ | $3,54 \cdot 10^6$ | $4,12 \cdot 10^6$ | $2,62 \cdot 10^6$ | $2,04 \cdot 10^5$ |
| Dosis límite a toda la población | $2,5 \cdot 10^6$ rem · hombre al tiroides | | | | |

[a] $\lambda r$ = Constante de precipitación por rociado.    e = Yodo elemental.    p = Yodo en forma de partículas.    0 = Yodo en forma orgánica.
[b] R = Eficiencia de los filtros.
[c] $\lambda f$ = Constante de eliminación por filtrado.

TABLA III.  CARACTERISTICAS DE LA CONTENCION Y
SALVAGUARDIAS ASOCIADAS DEL REACTOR DE LA
CENTRAL NUCLEAR DE ASCO I

| Potencia | Efectividad del rociado [a] | Sección recta mínima de la contención | Fuga máxima de la contención |
|---|---|---|---|
| 2820 MW(t) (105% potencia nominal) | $\lambda r^e = 5,3\ h^{-1}$ $\lambda r^p = 0,45\ h^{-1}$ $D = 100$ | 1500 m$^2$ | 0,1% vol. |
| Dosis al tiroides a los 500 m después de 2 h | | | 207 rems |
| Dosis al tiroides a los 750 m después de 2 h | | | 150 rems |
| Dosis al tiroides a los 2500 m después de 30 h | | | 80 rems |

[a] $\lambda r^e$ = constante de precipitación por rociado del yodo en forma elemental.
$\lambda r^p$ = constante de precipitación por rociado del yodo en forma de partículas.
D = factor de descontaminación.

para poder definir los requisitos mínimos que sobre salvaguardias y
contención debe disponer cada unidad a instalar, para adaptarse a los
criterios descritos.

En la tabla II aparecen los valores que se toman para los diversos
parámetros y sistemas en cada uno de los cinco casos considerados, así
como las dosis al tiroides que se alcanzan en cada caso.

Como puede observarse, se ha tomado un valor de la efectividad del
rociado menor que en el caso de la central nuclear de Almaraz, ya que
tanto la altura libre de caída de la gota como el caudal de rociado son
menores en el caso de la central nuclear de Lemoniz.

La existencia de cámaras de penetraciones adyacentes a la parte
exterior del recinto de contención, incluyendo todas las penetraciones salvo
las de mayor tamaño, permite reducir el escape de radioisótopos a la atmósfera exterior, ya que suponiendo que un 50% del escape de la contención se
produce a través de las penetraciones, esta fracción de radioisótopos pasa
a la atmósfera a través del sistema de ventilación y filtrado de las cámaras
que mantiene a éstas a una presión ligeramente negativa.

En el caso de que se provea de sistemas de filtrado interiores a la
contención, éstos deben estar adaptados para ser efectivos en las condiciones
ambientales posteriores al accidente.  En este caso se supone que poseen
una eficacia del 90% para la eliminación de yodo en forma elemental o
partículas y que el yodo orgánico no puede eliminarse por este medio.

Según se deduce de la tabla II se hace necesario adoptar unos requisitos
mínimos de salvaguardias como las del caso 4, mejorando incluso la eficacia
de alguno de los sistemas, por ejemplo con un aumento del caudal del
rociado.

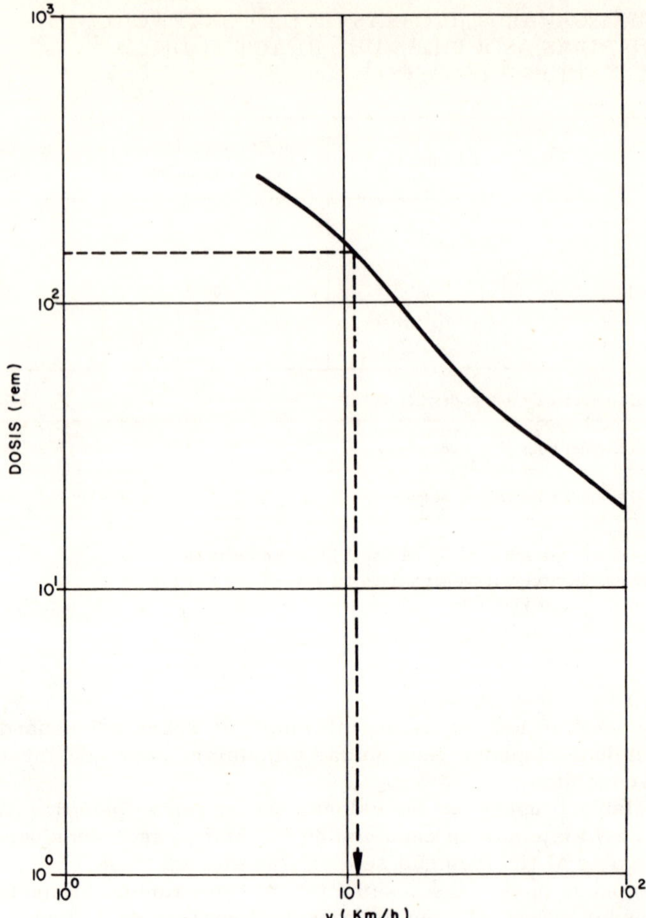


FIG.2.  Dosis al tiroides recibida por los viajeros a causa del MAP en el reactor, para diferentes velocidades del tren.

## CENTRAL NUCLEAR DE ASCO I

La central nuclear de Ascó I está situada al NE de España, en la provincia de Tarragona, y se refrigerará en circuito abierto por aguas del río Ebro.

La línea de ferrocarril Madrid-Barcelona, con un tráfico anual estimado de 1 350 000 personas, atraviesa el emplazamiento, siendo la distancia mínima del ferrocarril al edificio del reactor de 225 m.

Las características de la contención del reactor y salvaguardias asociadas así como el nivel de potencia a que se supone ha operado el reactor aparecen en la tabla III.

El radio mínimo de la zona de exclusión se ha estimado en 750 m, con
lo cual los solicitantes han tenido que ampliar la zona de 500 m de radio
propuesto inicialmente. La zona de baja población propuesta, con 2500 m
de radio, se considera aceptable.

Se ha evaluado el efecto que sobre los viajeros del tren tendría el
máximo accidente previsible ocurrido en la central, suponiendo que actuaran
las salvaguardias descritas en la tabla III. Como se aprecia en la figura 2,
los viajeros recibirían dosis al tiroides superiores a los 150 rems para
velocidades del tren menores a los 14 km/h. Las dosis al tiroides recibidas
por los viajeros, aún para velocidades del tren elevadas, son superiores a
los 20 rems, por lo cual se ha exigido el establecimiento de medidas para
restringir el tráfico ferroviario en caso de accidente nuclear en la central.

## REFERENCIAS

[1]  USAEC, Reactor Site Criteria, Rules and Regulations, 10CFR 100 (1973).
[2]  DIAZ DE LA CRUZ, F.,  Programa de cálculo EVATO para la evaluación del emplazamiento de
     centrales nucleares tipo BWR o PWR, JEN Rep. SS/06/72 (1972).
[3]  USAEC, Regulatory Guides nº 1.3, 1.4 (Rev.1, 1973), 1.5 (1971).

## DISCUSSION

H. M. DRESSE: Your calculation takes into account inhalation doses
of iodine. The thyroid doses resulting from milk ingestion in children,
even outside the low population-density areas, may require protection
measures. Are such measures considered in site selection?

Consuelo PEREZ DEL MORAL: We have taken into account the effect
of iodine ingestion on thyroid doses during normal power plant operation
but not in the case of accidents, because it is assumed that contamination
of milk would then only result in the prohibition of milk consumption until
the iodine concentration in the milk had fallen sufficiently.

D. DAGAN: Are the delineation of an exclusion area and a low population-
density zone essential for every site approved? What body is responsible
for evacuation in the case of an accident?

Consuelo PEREZ DEL MORAL: Such zones are delineated for each
approved site, and the minimum distances in fact vary not only for the
different types of power plant but also as a function of demographic,
topological and meteorological site characteristics.

The Civil Guard (Guardia Civil) units stationed in the area are
responsible for evacuation, although the National Traffic Police and
municipal authorities would also collaborate.

P. CANDES: In the case of the Ascó site you have considered the
hazard to train passengers. Have you, conversely, analysed the hazard
to the power plant from rail traffic?

Consuelo PEREZ DEL MORAL: The hazard to the power station from
rail traffic was evaluated by the applicant in his preliminary safety report
to the Spanish Nuclear Energy Board. The hazard was considered too slight
to be taken into account.

W. VINCK: Could you please confirm whether in the dose assessment
for accident conditions you have considered the most recent criteria (e. g.
regulatory guides)?

How did you, in practice, assess the doses that would be received by
the passengers of a passing train at the time of the postulated accident?
Did you, for instance, take into account the shielding and ventilation effects
which the train itself might be expected to provide?

Consuelo PEREZ DEL MORAL: We have taken into consideration the
latest revisions of Regulatory Guides 1.3 and 1.4 of the USAEC.

We did not allow for the shielding and ventilation effects of the train,
but calculated the doses as though the passengers were in the open. We
adopted other conservative assumptions, too: for example, a variation in
wind direction resulting in the plume centre remaining consistently above
the train. In any case, the only requirement deriving from this study was
to establish rapid communications between the power plant and the two
nearest railway stations.

# EXPERIENCIA ARGENTINA EN LA SELECCION DE EMPLAZAMIENTOS DE CENTRALES NUCLEARES

B.J. CSIK
Comisión Nacional de Energía Atómica,
Buenos Aires,
Argentina

*Presentada por Ambretta Migliori de Beninson*

## Abstract—Resumen

ARGENTINIAN EXPERIENCE IN SELECTING SITES FOR NUCLEAR POWER STATIONS.

One nuclear power station is in operation in the Republic of Argentina, a second is under construction, and the decision to build a third has been taken. According to existing plans, about ten nuclear power stations should go into operation during the next decade. The present paper analyses the experience acquired in selecting sites for the first units, commenting on the criteria and methods applied, the studies that were carried out, the specific problems encountered and the solutions adopted, as well as on the question of acceptance of the chosen sites by the public. It goes on to describe the current programme of selection and study of sites for future nuclear power stations.


EXPERIENCIA ARGENTINA EN LA SELECCION DE EMPLAZAMIENTOS DE CENTRALES NUCLEARES.

En la República Argentina se encuentra una central nuclear en operación, otra en construcción y se ha decidido la instalación de una tercera. De acuerdo con los planes existentes, se prevé la puesta en operación del orden de diez centrales nucleares durante la próxima década. En el presente trabajo se analiza la experiencia obtenida en la selección de las ubicaciones de las primeras unidades. Se comentan los criterios y las metodologías aplicadas, los estudios realizados, los problemas específicos encontrados y las soluciones respectivas adoptadas, como así también la aceptación por parte del público de las ubicaciones elegidas. Se describe a continuación el programa de selección y estudio de las ubicaciones de las futuras centrales nucleares, que actualmente se encuentra en marcha.

En la República Argentina, una central nuclear (Atucha I - 319 MW) se encuentra en operación comercial. La segunda (Córdoba - 600 MW) se encuentra en construcción y la instalación de una tercera (Atucha II - 600 MW) ha sido decidida. De acuerdo con los planes elaborados, se estima la puesta en operación del orden de diez centrales nucleares durante la próxima década.

El estudio y selección de las ubicaciones de las centrales nucleares ha sido y sigue siendo responsabilidad de la Comisión Nacional de Energía Atómica (CNEA), que a lo largo de los últimos diez años adquirió cierta experiencia en la materia. Las circunstancias y los criterios que definieron la selección de los emplazamientos han ido variando en función del tiempo transcurrido. Asimismo, no eran idénticas las condiciones de contorno correspondientes a cada caso particular. Al enfrentar el problema de la selección de emplazamientos de las futuras centrales nucleares se recurre a la experiencia pasada, tanto del país como mundial, sin tratar de establecer metodologías o reglas rígidas e inflexibles, pero determinando ciertos criterios generales que se consideran adecuados para el caso específico de la Argentina.

A continuación se presenta un breve análisis crítico de la experiencia argentina en el tema y se comenta el programa de estudio y selección de emplazamientos nucleares en marcha.


## 1.  EMPLAZAMIENTO DE ATUCHA I

El estudio de factibilidad de la primera central nuclear argentina, que incluye el estudio y selección de su ubicación, se realiza en el año 1965.  Es de notar que la zona en que esta central se instalará quedó definida antes de la iniciación de dicho estudio de factibilidad, como resultado de un análisis previo.

La selección de la zona se basó exclusivamente en la consideración del beneficio económico derivado del incremento de tamaño de las centrales nucleares, que facilita su competitividad económica comparada con las unidades convencionales.  En aquella época, entre los diferentes mercados eléctricos del país sólo uno, el del Gran Buenos Aires-Litoral (GBA-L), admitía la consideración de la instalación de una unidad de tamaño relativamente grande a corto plazo, del orden de 300-500 MW, cuya potencia, en aquel entonces, se consideraba comercialmente atractiva para las centrales nucleares.  Los demás mercados eléctricos del país eran sustancialmente menores.  Por ende, se preseleccionó la zona mencionada.

Definida la zona, se procede a la determinación de ubicaciones posibles empleando como primer criterio de selección la disponibilidad de agua de refrigeración.  El río Paraná y el Río de la Plata ofrecen la única posibilida de refrigeración por circuito directo en la zona, sistema que se considera preferible por razones económicas.

Los análisis tecnicoeconómicos de las fundaciones, sistemas de refrige ción, acceso e infraestructura local, determinan cinco ubicaciones posibles de las cuales finalmente se eligen dos por razones de cercanía a la principa concentración de demanda eléctrica en la zona.  La ubicación de la central nuclear en áreas densamente pobladas se descarta a priori, por razones de riesgo nuclear y política de descentralización industrial.  Es de notar que e país dispone de grandes extensiones despobladas o escasamente pobladas, p lo que no parece razonable tratar de ubicar dicha unidad en zonas urbanas o suburbanas.

Las dos ubicaciones seleccionadas (Atucha y Magdalena) se analizan del punto de vista del riesgo radiológico, resultando ambas satisfactorias. Una de ellas, Atucha, se recomienda como óptima por razones económicas pero ambas se mantienen como alternativas a considerar, hasta la adopció de la decisión final sobre la instalación de la central.

Tal política tiene dos consecuencias, que son dignas de mencionar. Primero, se discute dónde se pondrá la central y pierde intensidad la polémica sobre si conviene instalar una central nuclear o bien una convencional.  No hay que olvidar que se trata de la primera central nuclear del país, un proyecto promovido por la CNEA y visto con ciertas reservas por parte de las empresas eléctricas y autoridades del sector energético-eléctrico.

Segundo, se produce una rivalidad entre las poblaciones más cercanas las alternativas de emplazamiento elegidas, cuyo efecto es netamente bene cioso para el proyecto.  Una central nuclear es sinónimo de progreso, tier «glamour» y coloca al pueblo más cercano sobre el mapa del país y del

mundo. Además, la realización de una obra de esa envergadura aporta
beneficios económicos directos al comercio e industria del área, creando
nuevas fuentes de trabajo y en general promoviendo el desarrollo local.

La reacción de las autoridades, prensa, fuerzas vivas y del público en
general es altamente favorable a la instalación de la central nuclear. Se
presentan petitorios y se ofrecen los terrenos necesarios para la central en
donación. Riesgo radiológico, contaminación, conservación del medio
ambiente, peligro atómico, son temas que si aparecen, no toman estado
público.

Quizás, si se hubiera seleccionado directamente una sola ubicación y
no se hubiera producido en consecuencia la rivalidad mencionada, la reacción
positiva del público habría sido la misma, pero naturalmente esto ahora sólo
es especulación.

La aceptación por parte del público de la central nuclear durante su
construcción, puesta en marcha y operación, siguió siendo positiva, sin que
la psicosis antinuclear haya ganado adeptos en la zona. Por otra parte,
también es cierto que las esperanzas iniciales en cuanto al desarrollo local
inducido por la central nuclear no han sido colmadas en la medida esperada.
Hubo cierta evolución de las localidades más próximas, pero nada espectacu-
lar que permitiese considerar una central nuclear como factor determinante
de desarrollo de sus alrededores inmediatos.


## 2. EMPLAZAMIENTO DE LA CENTRAL EN CORDOBA

El nacimiento y posterior evolución de este proyecto tuvo profunda
influencia en la determinación de la ubicación correspondiente.

En 1967, visto el éxito obtenido con el estudio de factibilidad de Atucha,
la Empresa Provincial de Energía de Córdoba, que cubre un mercado
eléctrico segundo en importancia en el país, se dirige a la CNEA, solicitando
la realización de un estudio de factibilidad para una central nuclear en la
provincia de Córdoba. De esta manera queda definida a priori la zona en que
dicha central posteriormente se instalará.

Parte del estudio de factibilidad, que se realiza en 1968, es el estudio
y la selección del emplazamiento de la central. Para la preselección de
ubicaciones posibles, se da preferencia a sistemas de refrigeración directa
en lugar de torres de enfriamiento. Como en la zona no existen ríos con
caudales adecuados, se recurre a los embalses existentes.

Una de las condiciones de contorno del estudio de factibilidad es la
consideración exclusiva del mercado y sistema eléctrico de la Provincia,
que en aquel entonces no se encuentra interconectado con otros sistemas
eléctricos del país. De ahí surge una limitación del orden de 150 MW en la
potencia máxima instalable en una unidad y, consecuentemente, la selección
del embalse «Los Molinos» como emplazamiento recomendado. Sobre dicho
embalse se determina un sitio apto y conveniente, en función de estudios
económicos comparativos, considerando fundamentalmente el sistema de
refrigeración, fundaciones, acceso-transporte e integración al sistema
eléctrico.

El proyecto de la central nuclear de 150 MW en Los Molinos nunca se
llegó a aprobar, pero la intención de instalar una central nuclear en la
Provincia de Córdoba mantuvo su vigencia. En efecto, entre 1969 y 1971

se realizan estudios complementarios de factibilidad, esta vez analizando
la conveniencia de integrar el sistema eléctrico de la provincia al sistema
nacional interconectado.  Como resultado, surge el incremento de la potencia
de la unidad nuclear a 600 MW.

Las diferencias respecto a una central de 150 MW, en particular en
cuanto a requerimientos de agua de refrigeración, como así también peso y
tamaño de componentes que incrementan las dificultades de transporte, hacen
que se reexamine la selección del emplazamiento.  Manteniendo siempre la
decisión respecto a la zona (Provincia de Córdoba) y el criterio de asignar
prioridad al sistema de refrigeración por circuito directo, se elige como
ubicación recomendada el embalse Río III.  Esta ubicación se encuentra algo
más alejada del centro de consumo que Los Molinos, pero es el único lago
de la zona cuyo tamaño y volumen resultan adecuados para permitir la
refrigeración directa de una central de 600 MW.  También es más favorable
desde el punto de vista del acceso.

Se realizan distintos estudios sobre el emplazamiento, principalmente
en cuanto al sistema de refrigeración, incremento admisible de temperatura
del agua, efectos biológicos, análisis de suelos, sismicidad, acceso y
transporte, seguridad nuclear, infraestructura del área adyacente y régimen
hidrológico.  La aptitud del lugar se confirma, pero es de observar que,
comparado con Atucha, las características del sitio resultan en general
menos favorables.

A lo largo de su evolución, el proyecto ha recibido siempre el apoyo
entusiasta de las autoridades, de la empresa eléctrica y de las fuerzas vivas
locales, con quienes la CNEA mantuvo contacto continuo y cordiales rela-
ciones de cooperación.  Asimismo, se tuvo el apoyo de los medios de difusión
cordobeses.  Es de notar que la Provincia de Córdoba se caracteriza por
intensos sentimientos localistas; la central nuclear cordobesa adquirió el
carácter de un símbolo del progreso y proyecto prioritario.

La aceptación por parte del público, tanto en la zona adyacente al
emplazamiento como en la provincia en general, fue buena.  Sin embargo,
en algunos círculos, en especial entre profesionales y estudiantes universi-
tarios, hubo manifestaciones de temor, objeciones o dudas vinculadas con el
peligro de contaminación radiactiva y térmica.  En particular, varios físicos
que habían estado estudiando en los EE.UU. de Norteamérica volvieron
«contaminados» por la polémica relativa a la seguridad nuclear y manifestar
signos de preocupación por la conservación del medio ambiente.

La CNEA realizó numerosas conferencias, reuniones y trató en general
de aclarar conceptos, dando explicaciones en un nivel técnico.  Es poco
probable que esta acción haya resultado en la «conversión» de los «contamina
dos», pero, por otra parte, los temores de unos pocos no han sido transmitidos
al público en general, ni se ha creado un ambiente adverso a la instalación
de la central nuclear.  Muy por el contrario, la reacción del público siguió
y sigue siendo favorable.

Hubo otras objeciones al emplazamiento de la central en la zona de
Córdoba, pero no por razones de seguridad nuclear, sino por otras derivadas
de diferencias de opinión respecto a la demanda del mercado eléctrico
nacional y el equipamiento óptimo correspondiente.  Tales objeciones
surgieron más bien de intereses vinculados a otras empresas eléctricas.
Teniendo en cuenta la existencia de cierta rivalidad y desencuentros inter-
empresarios y el hecho de que la ubicación de esta central ha sido definida
para la zona de Córdoba a priori, sin un análisis detallado a nivel nacional,

las objeciones son explicables.  Es de notar que no se objetaba el requeri-
miento de instalar una unidad nuclear en el país, sino la zona elegida como
ubicación de dicha central.


## 3.   DECISION DE ATUCHA II

Siguiendo el ejemplo de las centrales Atucha I y Córdoba, autoridades
o empresas eléctricas de varias provincias se aproximan en diversas oportuni-
dades a la CNEA, solicitando el estudio u opinión sobre la eventual
conveniencia de instalar sendas centrales nucleares en sus respectivas zonas
de influencia.  Se realizan así varios análisis preliminares, sin que de los
mismos surja un proyecto en firme.

Paralelamente, avanzan los programas respecto a la interconexión de
los mercados y sistemas eléctricos del país, de manera que actualmente se
prevé que antes del año 1980 más del 90% del mercado eléctrico argentino se
integrará en un sistema nacional interconectado.

En estas condiciones, en 1972/73 se replantea el análisis electro-
energético del país, planeando el equipamiento óptimo del sistema nacional
interconectado.  De ahí surge la necesidad de instalar una unidad nuclear
adicional de 600 MW para el año 1981.  La decisión de instalar esta tercera
central nuclear se adopta a principios de 1974, pero esta vez sin determinar
la zona donde se ubicará.

En la CNEA se encara el análisis del mercado eléctrico del país, cuyo
resultado indica dos zonas de interés inmediato: el Gran Buenos Aires-
Litoral y la región de Bahía Blanca (Sur de la Provincia de Buenos Aires),
en este orden de preferencia.  Se elige una ubicación en cada zona (en el
GBA-L se preselecciona Atucha) y se realiza un estudio económico compara-
tivo entre ambas.

Se recomienda instalar la tercera central nuclear en el emplazamiento
Atucha, adyacente a la primera unidad.  La existencia de la infraestructura,
acceso, capacidad de refrigeración y demás condiciones favorables de la
ubicación señalan netas ventajas económicas.  Tampoco resulta necesario
realizar estudios específicos sobre el emplazamiento ni obtener terreno
adicional, ya que al encarar Atucha I se había previsto la ampliación de esta
central nuclear instalando una segunda unidad.  Teniendo en cuenta, además,
que la instalación de una central nuclear, por sí sola, no crea un polo de
desarrollo, todo lo anterior compensa el interés de llevar adelante una
política de descentralización, fomentando el desarrollo del interior del país.
En consecuencia, en Agosto de 1974 se decide construir Atucha II.

La reacción del público en la región y pueblo adyacentes, puede cali-
ficarse de «indiferente».  Allí, una central nuclear ya no es novedad, no hay
temores por lo «atómico», ni se crean esperanzas exageradas o infundadas
de beneficios económicos derivados de la construcción y ulterior operación
de la central.

Incluso a nivel del país, las repercusiones de la decisión de la tercera
central nuclear no son grandes.  Sólo durante la etapa de la selección de la
zona de ubicación se manifiestan ciertas presiones en el nivel político
desde provincias o regiones que aspiran a obtener la obra.  En todos los
casos se trata de una reacción positiva frente a la central nuclear, no
habiéndose registrado oposición ni psicosis antinuclear.

## 4.  EMPLAZAMIENTO DE FUTURAS CENTRALES NUCLEARES

Atucha I nace como la primera central nuclear de un país que se inicia en el campo nucleoeléctrico.  Forma parte de una política nuclear vislumbrada a largo plazo, pero no de un programa orgánico inmediato de centrales nucleares que haya sido formulado, aceptado y decidido.  La Central de Córdoba nace del deseo e insistencia de una provincia, que se conjuga con las condiciones del mercado eléctrico y el interés de la CNEA en seguir instalando centrales nucleares en el país, fundado en el convencimiento de la conveniencia de seguir este camino en beneficio de la nación.

Atucha II, por otra parte, se decide como la primera de las unidades nucleares propuestas en un programa de equipamiento eléctrico oportunamente formulado.  Tal programa prevé la puesta en operación de una decena de centrales nucleares de módulo 600 MW, durante la década del ochenta.

El análisis regional del mercado eléctrico del país indica, en principio, cuatro zonas (Gran Buenos Aires-Litoral,  Córdoba, Sur de la Provincia de Buenos Aires, Cuyo), que requerirán unidades nucleares durante la próxima década, produciéndose las demandas más inmediatas en las zonas Sur de la Provincia de Buenos Aires (Bahía Blanca) y Cuyo, en este orden de priorida.

Una de las consecuencias inmediatas de un programa nuclear de esta envergadura es tener que encarar el estudio y la selección de emplazamient en forma rutinaria y de acuerdo con previsiones a corto y mediano plazo. E 1974 se inician los trabajos en este sentido.

Los lineamientos y criterios generales en cuyo marco se desarrollan los estudios son los siguientes:
- Ubicadas las zonas de interés potencial a corto y medio plazo sobre la base de la demanda prevista de los mercados y equipamientos eléctricos programados, se preseleccionan áreas para emplazamientos posibles.
- En la preselección de emplazamientos posibles se tienen especialmente en cuenta, además de otros factores, la disponibilidad de agua de refrigeración, acceso, sismicidad, distancia a centros de carga, infraestructura de apoyo e integración en el sistema nacional interconectado, todos ellos factores «convencionales».
- Considerando el riesgo nuclear, los emplazamientos posibles se ubican a priori, a distancias razonables de centros poblados, empleando un criterio generoso para determinar lo «razonable».
- Para aprovechar en el futuro la ventaja económica de la posibilidad de ampliación de centrales, los emplazamientos se dimensionan para acomodar hasta cuatro unidades del orden de 600 MW cada una.
- En cada zona se preseleccionan por lo menos dos alternativas posibles.
- En la primera etapa de los estudios se realiza la recopilación de datos disponibles, con ayuda de los organismos, entes y empresas nacionale y provinciales correspondientes.
- Se establece un sistema para seguir recibiendo y catalogando datos estadísticos, en particular respecto a meteorología, sismicidad y registro de temperaturas y caudales de agua.
- Los estudios específicos de los emplazamientos, tales como investigaciones geológicas y mecánica de suelos, levantamientos topográficos, determinación de parámetros y diseño conceptual de sistemas de refrigeración, y análisis de seguridad nuclear, se realiza en una segunda etapa.

—   Aún cuando en principio se trata de estudiar y seleccionar emplaza-
    mientos en todas las zonas de interés en forma simultánea, se asigna
    tratamiento prioritario a las zonas que así resultan del estudio del
    mercado.
—   Un sector especial de la CNEA organiza, coordina y realiza las diversas
    tareas correspondientes, recurriendo a otros sectores y especialistas
    de la Institución, entes oficiales, o bien empresas privadas, en caso de
    requerir asesoramiento en temas específicos.

En cuanto a la aceptación por parte del público de los emplazamientos y
la posterior instalación de las centrales nucleares respectivas, por ahora —
sobre la base de la experiencia obtenida — no se prevén problemas.  Se
descuenta el apoyo de autoridades políticas, fuerzas vivas y la prensa oral y
escrita.  La CNEA seguirá realizando su tarea de difusión técnica consistente
en explicaciones, conferencias, reuniones, atención de visitantes, exposiciones, etc.  En lo posible se evitará la selección de emplazamientos en
lugares que potencialmente tiendan a infundir temores en el público, aún
cuando esto involucre alguna penalidad económica.  Se apoya con entusiasmo
todo intento de crear desarrollo e introducir mejoras en el área aledaña al
emplazamiento nuclear, fomentando la participación de la mano de obra,
comercio e industria locales, en las obras que se realizan.

Se espera que estas medidas preventivas sean adecuadas para evitar la
«contaminación» del público por el temor desmedido a la «contaminación
atómica».


# DISCUSSION


D. STOIAN:  You mentioned a lake which is used for cooling purposes
at the Córdoba nuclear power plant.  Could you please give us an idea of
the cooling capacity of the lake and indicate the increase in its temperature,
as well as the basis on which that increase is calculated (model study, if
any)?

Ambretta MIGLIORI DE BENINSON: The cooling capacity of the lake can
be expressed in many ways, one being the area needed to cool the discharges,
which is about $10^7$ m$^2$ (about a third of the lake area).  The rise in condenser
temperature was set at 7°C, and surface temperatures decrease to normal
from the injection point as the water runs into the lake.

Two studies form the basis of the decisions on thermal effects: (a) a
heat dispersion study (including both a computer model and electric analogue
simulation experiments); and (b) a biological study of the heat sensitivity of
the most important species in the lake.

H. SCHNURER (Chairman):  With regard to this cooling lake, I should
like to ask whether special safety measures have been taken to avoid a dam
failure or to limit the consequences of such an occurrence, which would
lead to a loss of cooling capacity.

Ambretta MIGLIORI DE BENINSON: The dam failure hypothesis was of
course analysed.  But enough water would remain available as a final heat
sink — by a very considerable margin in fact — to take care of the decay
heat of the reactor.

A. DE ACHA ARACAMA: It appears from your presentation that geological
studies are limited to specific evaluation of a site already chosen.  What
importance is attached to them at earlier stages of the work?

Ambretta MIGLIORI DE BENINSON: Geological factors are considered at two stages: (a) during the preliminary selection studies (here it is mainly a matter of collecting and analysing existing data); and (b) during specific assessment of a site and at the time of the final siting decision (here the studies are much more detailed and include borehole drilling, measurement of seismic oscillation spectra and so on).

# SITES OF LIGHT-WATER REACTOR POWER PLANTS IN THE FEDERAL REPUBLIC OF GERMANY

H. A. G. KOHLER
Kraftwerk Union AG,
Frankfurt, Federal Republic of Germany

**Abstract**

SITES OF LIGHT-WATER REACTOR POWER PLANTS IN THE FEDERAL REPUBLIC OF GERMANY.

The nuclear power plant site requirements formulated for environmental protection in the Federal Republic of Germany allow nuclear power plants to be built at any site provided these requirements are duly taken into account in preparing and monitoring the site and in the design of the proposed power plant. After a brief discussion of site planning, site selection and licensing procedures, this paper reports in some detail on the interpretations placed by expert advisers and by the licensing authorities, in the past and at present, on the criteria, rules and guidelines used in the licensing procedures relating to proposed sites for power plants with light-water reactors.

## 1. SITES WITH LIGHT-WATER REACTORS

In the Federal Republic of Germany (FRG) there are nineteen power plant sites, each with one or more power units with light-water reactors. Of these, seven units with an aggregate gross electric power rating of 3410 MW are in operation and eight units with an aggregate gross electric power rating of 8239 MW are under construction. Orders have been placed and/or applications for construction permits pursuant to the Atomic Energy Act have been filed for eight further light-water reactor power units with an aggregate gross electric power rating of about 10100 MW. Two of these nineteen sites (Biblis, Philippsburg) will each have two power units, at present under construction, and one site (Gundremmingen) will have three power units (see Fig. 1 and Table I).

In addition, planning work is in progress on new sites (Emden, Bad Breisig, Schwörstadt) for light-water reactor power plants and on the extension of sites (Lingen, Neckarwestheim, Ohu, Wyhl), which have already been approved for the first power units (see Fig. 1).

## 2. SITE PLANNING

In the Federal Republic of Germany the planning of specific sites for nuclear power plants is at present largely in the hands of the power supply utilities. So far, the licensing authorities of the federated States (Länder) and of the Federal Government have usually only acted after a power supply utility had filed an application for a construction permit pursuant to Section 7 or for approval of a proposed site for the construction of a nuclear power plant pursuant to Section 7A of the Atomic Energy Act.

FIG. 1.   Sites of light-water reactor power plants in the Federal Republic of Germany.
I–IV: demonstration plants listed in Table I(a).
1–19: commercial plants listed in Table I(b).

## Table I. SITES OF LIGHT-WATER REACTOR POWER PLANTS IN THE FEDERAL REPUBLIC OF GERMANY

### ( a ) Demonstration Nuclear Power Plants

| NO. | SITE | PLANT | REACTOR TYPE/ GROSS ELECTRICAL OUTPUT | ORDER/ COMMIS- SIONING |
|---|---|---|---|---|
| I | KAHL | NPP Kahl (VAK) | BWR/  16 MW | 1958/1961 |
| II | GUNDREMMINGEN | NPP Gundremmingen I (KRB I) | BWR/ 252 MW | 1962/1966 |
| III | LINGEN[a] | NPP Lingen (KWL) | BWR/ 252 MW | 1964/1968 |
| IV | OBRIGHEIM | NPP Obrigheim (KWO) | PWR/ 345 MW | 1965/1969 |

### ( b ) Commercial Nuclear Power Plants

| NO. | SITE | PLANT | REACTOR TYPE/ GROSS ELECTRICAL OUTPUT | ORDER/ COMMIS- SIONING |
|---|---|---|---|---|
| 1 | STADE | NPP Stade (KKS) | PWR/  662 MW | 1967/1972 |
| 2 | WÜRGASSEN | NPP Würgassen (KWW) | BWR/  670 MW | 1967/1972 |
| 3 | BIBLIS | NPP Biblis A | PWR/1213 MW | 1969/1974 |
| 4 | LUDWIGSHAFEN | NPP BASF | PWR/2330 MW[e] | 1969/ — |
| 5 | PHILIPPSBURG | NPP Philippsburg I (KKP) | BWR/  900 MW | 1970/1976 |
| 6 | BRUNSBÜTTEL | NPP Brunsbüttel (KKB) | BWR/  805 MW | 1970/1975 |
| 7 | PHILIPPSBURG | NPP Philippsburg II (KKP II) | BWR/  900 MW | 1971/1978 |
| 8 | NECKARWESTHEIM[b] | NPP Neckar (GKN) | PWR/  813 MW | 1971/1976 |
| 9 | ESENSHAMM | NPP Unterweser (KKU) | PWR/1299 MW | 1971/1976 |
| 10 | BIBLIS | NPP Biblis B | PWR/1299 MW | 1971/1976 |
| 11 | OHU[c] | NPP Isar (KKI) | BWR/  907 MW | 1971/1976 |
| 12 | KRÜMMEL | NPP Krümmel (KKK) | BWR/1316 MW | 1972/1978 |
| 13 | MÜLHEIM-KÄRLICH | NPP Mülheim-Kärlich | PWR/≈1350 MW | 1973/1978 |
| 14 | WYHL[d] | NPP Süd I (KWS I) | PWR/1362 MW | 1973/1979 |
| 15 | GRAFENRHEINFELD | NPP Grafenrheinfeld (KKG) | PWR/1296 MW | 1973/1978 |
| 16 | GROHNDE | NPP Grohnde (KWG) | PWR/1361 MW | 1974/1980 |
| 17 | BROKDORF | NPP Brokdorf | PWR/1363 MW | 1974/1981 |
| 18, 19 | GUNDREMMINGEN | NPP Gundremmingen II (KRB II) | BWR/2x1310 MW | 1974/1979-80 |

[a] Extension by two 1300 MW LWR Units planned
[b] Extension by one 976 MW PWR Unit planned
[c] Extension by one 1300 MW LWR Unit planned
[d] Extension by one 1362 MW PWR Unit planned
[e] Thermal Reactor Output

In view of the rapid increase in the demand for electric power and of the
relative paucity of suitable power plant sites in the FRG many of the parties
concerned have stressed the need for a long-term availability planning for
prospective sites for nuclear power plants.

It is suggested that the Federal Government, acting through the permanent
Federal and State Co-ordination Agency, should encourage all the State
Governments to carry out systematic work on the planning and securing of
suitable sites for nuclear power plants. The federated States should
co-ordinate their planning work with each other and with the Federal
Government. This could be done, for the various planning aspects involved,
through existing administrative machinery, e.g. for regional planning and
land use through the Regional Planning Committee of the Interstate Ministerial
Conference, for the power supply planning through the competent State
commissions at the Federal Ministry for Industry, and for matters connected
with licensing procedures under the Atomic Energy Act through the Interstate
Committee for Atomic Energy at the Federal Ministry for Internal Affairs.

It is further suggested that the Federal Government, the State Governments and the power-supply utilities should work together to prepare a map
of prospective sites for nuclear power plants in the FRG. The Federal
Government, jointly with the competent State authorities, expert advisers
and advisory agencies, should evaluate these sites in accordance with the
licensing procedures under the Atomic Energy Act, and assign appropriate
merit scores. In this manner all the aspects of such prospective sites,
relevant in connection with the licensing procedures under the Atomic Energy
Act, will be on record in advance and can be taken into account in planning.

According to the energy policy planning of the Federal Government, it
is necessary to find by 1978 about 35-45 sites for nuclear power plants in
the FRG. This includes the fifteen sites on which large-capacity nuclear
power stations are already in operation or under construction. Some of
these fifteen sites are suitable for several power units. This means that
within the next 5-6 years decisions must be reached on some 20-30 additional
power plant sites [1].

A map showing possible sites for large power plants in Baden-
Württemberg has recently been published [2]. As shown in Fig.2, this map
defines seventeen locations in four areas. Of these seventeen possible sites
at least eight must be reliably available by 1990, while seven are alternative
sites. The power supply utilities had originally put forward a list of
41 prospective sites. These sites were assessed by an interministry working
group, which excluded some sites as unsuitable. Further sites were excluded
in advance negotiations between the power supply utilities and the communes
concerned. This two-step elimination procedure left the seventeen sites
shown in Fig.2.


3.   SITE SELECTION CRITERIA USED BY THE POWER SUPPLY UTILITIES

The selection by a power supply utility of a site for a nuclear power
station is a very complex procedure entailing the assessment and evaluation
of a large number of factors the overall effects of which must be optimized.
According to Ref.[3] the factors governing site selection may be classified
as first-order and second-order factors. The first-order factors are those
which govern the general area where the plant site should be located. These

FIG.2.   Map of possible power station sites in Baden-Württemberg:  status - May 1974 [2].

factors are closely linked to the intended function of the plant.  Thus, the
site for a plant serving for the generation of process heat must of necessity
be very close to the consumption area because heat cannot be transported
economically over distances of more than a few km.  Site selection for a
plant for the generation of electric power, on the other hand, can be much
more flexible because the decisive factors are the location of the site in
relation to the power grid and the availability of cooling water.  Factors of
the second order pinpoint the location of the site more closely within a chosen
general area.  These factors include, e.g., geological conditions, transport
links, nature conservation and landscape protection, drinking-water procurement areas, etc.  Further considerations on site selection criteria are
presented in Refs. [4-12].

Amongst the suggested possibilities for a methodical site selection
procedure are a qualifying and a quantifying method described in Ref. [13].
The result of a study according to the qualifying method is a list of the
studied locations, catalogued according to a set of characteristic 'site merit'
assessment scores.  The drawback of this method lies in the difficulty of
establishing an evaluation scale.  The quantifying method is in essence a
cost optimization analysis also taking into account the costs of environmental
protection.  The drawback of this method is that it is difficult to estimate
the environmental protection costs accurately.

## 4.   LICENSING PROCEDURES UNDER THE ATOMIC ENERGY ACT

The very brief description of the licensing procedures under the Atomic Energy Act, presented below, is intended only as a background for the better understanding of the licensing procedures as applied in the Federal Republic of Germany in practice.

In the FRG the construction and operation of a nuclear power station is subject to an elaborate licensing procedure under the Atomic Energy Act. The entire licensing procedure is carried out by the competent State authorities on behalf of the Federal Government.  This includes not only the practical implementation of the licensing procedure pursuant to the Atomic Energy Act but also the co-ordinated processing of the numerous individual licences and permits, bearing in mind that all the agencies of the Federal Government, the States (Länder), the communes and all other regional authorities whose competence is in any way touched upon by the project are involved and take part in the licensing procedure.  The Federal Government is assisted in an advisory capacity by the Reactor Safety Advisory Commission (RSK) on which are represented experts in the various relevant scientific and technical disciplines.  The State authorities have available for inspection and acceptance functions, and for rendering expert opinions, the Technical Safety Control Services (Technical Inspectorates, TÜV) and, in addition, all the above parties may call in the Institute for Reactor Safety (IRS) as another organization rendering expert advice.  A schematic chart of the parties and authorities involved in the licensing procedure under the Atomic Energy Act is shown in Fig.3.  Finally, the licensing procedure also involves the various non-nuclear aspects of the project insofar as matters of public law are concerned in the areas of construction, accident prevention and industrial safety, water protection and riparian rights, immission control, nature conservation and landscape protection and land-use planning.

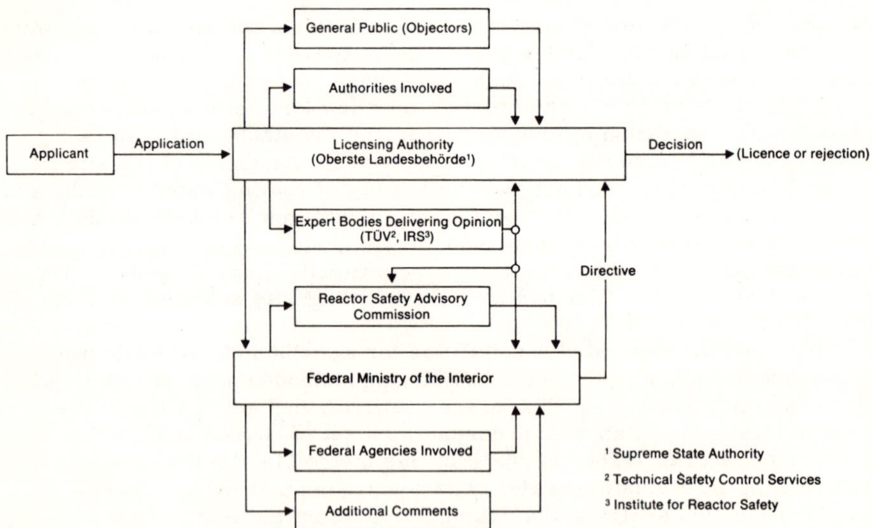


FIG. 3.   Schematic representation of parties involved in the licensing procedure under the Atomic Energy Act.

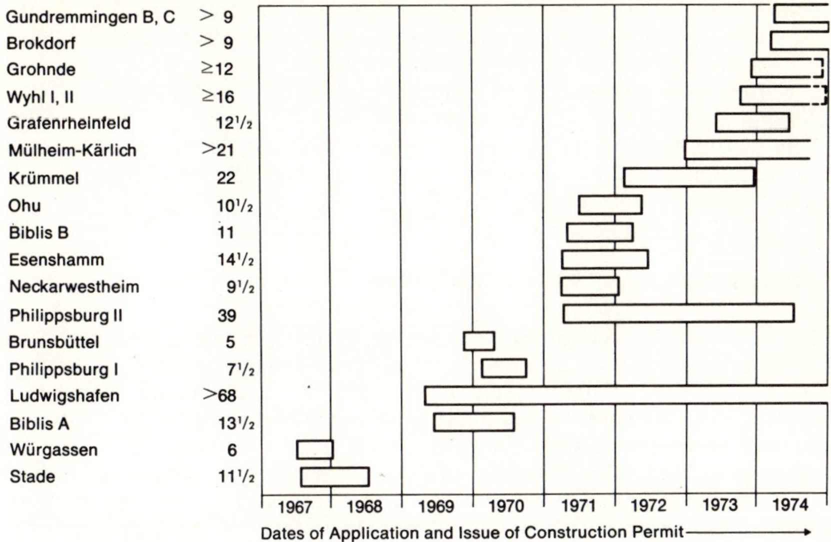


FIG.4.    Time from application to first construction permit for light-water reactor power plants in the
Federal Republic of Germany.

As a rule, a construction permit and an operating licence for a nuclear
power station are granted separately, and each of these in turn is usually
granted in several partial stages.  Since 1969 it has been possible to obtain
a preliminary site ruling and/or a concept approval in advance of the
construction permit, whereas previously these preliminary documents
formed part of the construction permit.  Figure 4 shows the time elapsed
between filing the application and receipt of the first partial construction
permit (which included, in most cases shown, the site approval and the
concept approval) for nuclear power plants with light-water reactors.  A
trend towards longer delays can clearly be seen.

To avoid any further increases in the delays involved, efforts are now
being made to standardize the practical implementation of licensing procedures in all the federated States.  One of the groups particularly active
in this field is the Working Group for the Standardization of Licensing
Procedures set up by the electric power utilities.  Its object is to persuade
the licensing authorities to formulate uniform requirements, to undertake
to keep these requirements immutable for fixed periods of time and to make
joint use of the available limited expert opinion capacities instead of each
State relying only on its own experts.

An important point of the licensing procedure is participation of the
general public.  It is one of the statutory obligations of the licensing
authorities to give public notice of each pending nuclear power plant project.
This notice must include the following points:

(a)  It must state where the application and the supporting documents
     for the application are available for scrutiny;
(b)  It must inform the public that any objections to the project should
     be lodged within 30 days of the date of the notice;

(c)  It must fix the date, time and place of a public hearing at which
     the objections will be discussed, and it must make it clear that
     these objections will be discussed at that hearing regardless of the
     presence or absence of the applicant or of the objectors.

The public hearing is intended to ensure that the licensing authorities
are cognisant of all the arguments relating to the project and that no rights
of third parties are violated.


## 5.  CRITERIA, RULES AND GUIDELINES

In the Federal Republic of Germany no binding rules on reactor sites
have yet been formulated by any of the authorities concerned with the
licensing procedure.  Instead, each particular case has been assessed on
its merits, i.e. on whether the particular projected plant at the particular
chosen site presented a sufficiently low risk for the population.  According
to Kellermann [14], the steady improvement in safety engineering measures
makes it unnecessary to formulate any site acceptability criteria as such
for nuclear power plants, because adverse site factors can be counterbalanced by appropriate engineering means.
The licensing procedures in the FRG under the Atomic Energy Act
nevertheless require compliance with the following criteria, rules and guidelines, amongst others, all of which lay down more or less general site-
specific requisites:

Safety criteria for nuclear power plants, laid down by the IRS [15],
recently superseded by:

   Safety criteria for nuclear power plants, laid down by the Federal
   Ministry for Internal Affairs [16];

   Guidelines for pressurized-water reactors, laid down by the
   RSK [17];

   Guidelines for boiling-water reactors, laid down by the RSK [18];

   Emission source strength of nuclear power plants, recommendation
   of the Working Group on Radiation Protection Engineering of the
   Deutsche Atomkommission, for the determination of permissible
   emission levels of radioactive substances from nuclear power
   plants into the atmosphere [19];

   Guidelines for the immission of radioactive substances into waters,
   laid down by the Interstate Working Group 'Water' (LAWA) [20];

   Rules regarding protection against earthquakes, laid down by the
   IRS [21] and by the Nuclear Engineering Committee on the anti-
   seismic design of nuclear power plants [22];

   Ordinance on the protection against injury by ionizing radiation
   (Radiation Protection Ordinance) [23].

In the FRG, the requirements for environmental protection, formulated
in the criteria, rules and guidelines [15-23] listed above, allow nuclear

power plants to be built at any site provided these requirements are duly
taken into account in preparing and monitoring the site and in the design of
the proposed power plant.  This basic principle is also supported by
Kellermann and Franzen [7].

In the safety criteria [15] the site-specific conditions in Section 2.1
are formulated as follows:  "The design of the reactor plant must take into
account the meteorological, hydrological, geological, seismic and ecological
conditions at the plant site, the population density distribution in the vicinity
of the plant and any further relevant site-specific conditions.  The site-
specific hazards owing to, e.g., earthquakes, subsidences of old mine
workings, flooding or storms, should be taken into account to an extent
ensuring that the actual occurrence of these hazards in the most severe form
to be expected at the site cannot cause an accident likely to endanger the
population in the vicinity of the plant or the operating staff of the plant and
cannot unduly impair the functioning of the facilities ensuring the safety of
the plant."  Sections 5.34 and 11.1, dealing with the monitoring of the
escape and release of radioactive substances, stipulate that:  "... In addition,
appropriate measures must be taken for monitoring and measuring radioactivity in the soil, the water and the air in the vicinity of the power plant..."
and "... facilities must be provided ensuring that the release of radioactive
substances can be kept within the permissible limits in respect of quantity
and concentration stipulated on the basis of local conditions".  Section 8.1
stipulates that "... the consequences of any accidents which may possibly
occur in the reactor plant, up to and including the design-basis accident,
must be kept within the lowest possible limits in respect of the radiation
exposure of the population ...."

In the safety criteria of Ref. [16], criterion 2.3, relating to radiation
levels in the vicinity of the plant, is formulated as follows:  "... that the
radiation load on the environment due to direct radiation from the plant and
to the release and possible escape of radioactive products, taking into
account the rules of science and technology, must be not only within the
permissible limits but also as far below these limits as possible."  A foot-
note relating to the currently permissible limits for a site states:  "The
radiation load due to external and internal radiation exposure, referred to
the whole body and calculated for the most unfavourable point along each
exposure route, and taking into account all the emitters of radioactive
substances relevant for the area considered, must not exceed 300 $\mu$J/kg
(= 30 mrem) per year owing to the release of radioactive substances with the
liquid wastes, and must not exceed 300 $\mu$J/kg (30 mrem) per year owing to
the release of radioactive substances with the exhaust air.  The radiation
load on the thyroid gland of small children  owing to radio-iodine ingested
via the food chain  must not exceed 900 $\mu$J/kg (90 mrem) per year;  this
limit value applies to the most unfavourable point in the vicinity of the
nuclear power plant, taking into account all the emitters of radioactive
substances relevant for the area, regardless of whether or not the point
considered allows permanent human habitation or the grazing of cattle."
Criterion 2.6, relating to effects of external occurrences, lays down that
the design of the plant must take into account the most severe possible
consequences of natural occurrences such as earthquakes, landslides, storms,
floods and flood waves, of sundry other natural hazards, e.g. flocks of
birds or blocking of cooling-water lines by mussels or algae, and of man-
made hazards such as spillage of noxious substances or explosions in the

vicinity of the plant, or subsidences of old mine workings.  Criterion 10.3
lays down the requirements for radiation protection monitoring in normal
operation, in exigencies and in the event of unforeseeable occurrences and
accidents.

In the guidelines for pressurized water reactors [17] and for boiling-
water reactors [18], Section 2.3.1 (of both references) lays down that,
taking into account the local meteorological and orographic conditions and
the effects of built-up areas in the vicinity of the plant, the release of
radioactive substances with the exhaust air must be kept as low as possible.
It must also be shown by means of a radio-ecological study that the individual
exposure originating in the release of radioactive liquid wastes from the plant
by the exposure pathways via the drinking water, the food chain and external
radiation, remains as low as possible.  These guidelines also stipulate for
the whole-body exposure owing to internal and to external radiation the limit
of 300 $\mu$J/kg (30 mrem) per year and for the thyroid gland of small children
the limit of 900 $\mu$J/kg (90 mrem) per year as maximum limits which must
not be exceeded.  Section 2.7 of both Refs [17] and [18], relating to effects
due to human agencies, lays down requirements for protection against the
effects of aircraft crashes, chemical explosions, poisonous and explosive
gases and actions by third parties, which are discussed in greater detail
later, in Section 6.  Sections 10.2 and 10.3 of both Refs [17] and [18] lay down
the requirements for the monitoring of radioactivity in the exhaust air, in
the released liquid wastes and in the surroundings of the power plant.

The recommendations in Ref. [19] suggest methods for calculating
long-term and short-term spread factors based on meteorological observations at the site considered.

The guidelines in Ref. [20] lay down, amongst other stipulations, the
guideline values of 0.5-1.0 Ci for the yearly release of radioactive liquid
wastes (not counting tritium) and of 50 Ci for BWR plants and 200 Ci for PWR
plants for the yearly release of tritium while stressing that the actual
release rates should be kept as low as possible.  These values apply per
100 MW electric power rating of the plant.

The rules in Refs [21] and [22] lay down specifications for the seismic
design of nuclear power plants in various earthquake zones in the FRG,
which are discussed in greater detail in Section 6 of this paper.

The Radiation Protection Ordinance [23] lays down in Section 3.2
(paragraphs 43-47) specifications for the protection of the population and of
the environment from the hazards of ionizing radiation, and in Appendix IV,
amongst other data, the maximum permissible activity incorporation rates
of radioactive substances and the maximum permissible concentrations of
radioactive substance in the air.

At present, the starting point for evaluation of a proposed site for a
nuclear power plant by the expert advisers and the licensing authorities is
the documentation on the characteristics of the site, consisting of the documents listed in the IRS lists, Refs [24] and [25].


6.    CONSIDERATIONS TAKEN INTO ACCOUNT BY EXPERT ADVISERS
      AND LICENSING AUTHORITIES IN ASSESSING SITES

A proposed site for a nuclear power station is assessed in essence
according to the extent to which the safety of the population in the vicinity

can be safeguarded both in normal operation of the plant and in the event of
exigencies or accidents. The assessment criteria are governed by the
possible radiation exposure of the population owing to release of radioactive
substances by the plant [6, 7, 26-30].

The individual considerations relevant to site assessment [24, 25, 28,
31-34] applied so far to sites for power plants with light-water reactors,
are discussed as follows:

## 6.1. Topographical conditions

An important topographical consideration for the assessment of a site
is whether the terrain in the vicinity is flat or hilly. If the site is in a
valley it is important to know the width of the valley floor and the height of
the enclosing hill ranges. This information can be obtained from topographical maps, e.g. scale 1:100 000 or scale 1:250 000. In addition, expert
advisers request a site plan showing the outline of the plant perimeter and
data on the local construction height restrictions, if any. For urban site
locations, detailed information is required on the location and height of
nearby buildings.

## 6.2. Geological conditions

Information on the location and properties of soil strata is obtained by
exploratory borings. These data, and especially any unusual geological
conditions, have a direct bearing on the civil and structural engineering
design of the plant. The expert advisers request copies of bore-hole profiles
as well as data on the mineralogical stratigraphy of the site and its vicinity.
Further requested data are the results of soil mechanics investigations
which provide a basis for assessing the structural safety and the anticipated
settlements of the planned buildings. Geological and/or soil mechanics
reports were prepared, e.g., for the power plant sites at Obrigheim [35],
Philippsburg [36], Grafenrheinfeld [37], Ohu [38] and Wyhl. Today, such
reports are generally required for all sites.

## 6.3. Seismological conditions

In view of the relatively slight seismic activity in the Federal Republic
of Germany, earthquakes do not present major safety problems. The expert
advisers request in this connection a seismological report on the proposed
site, which should also contain, if possible, data on any old mine workings
at and near the site as well as an assessment of the effects of possible
blasting in the vicinity of the site. The question of protection of light-water
reactor nuclear power plants against the effects of earthquakes is discussed
in Section 6.12 of this paper.

## 6.4. Hydrological conditions

In view of the particular importance of guarding against pollution of
waters, accurate information on the hydrological conditions at the proposed
site is indispensable. This includes, where applicable, data on the depth
and thickness of groundwater-bearing strata together with data on the
direction and velocity of the groundwater flow, data on the water flow rate

and water temperature in the outflow channel, data on the effects of tides
on the water flow rate, data on the water levels in storage reservoirs, data
on flood levels (100-year or 1000-year levels), data on the possibility of the
site being flooded by flood waves, floods or cloudbursts, data on canals,
retention dams, retention and storage reservoirs, and data on the waste-
heat dumping schedule for the outflow channel. In addition, data are required
on drinking-water procurement facilities in the environment of the proposed
site.

## 6.5.  Meteorological conditions

To be able to arrive at sufficiently conservative guideline values for the
release of radioactive substances through the exhaust air stack, the expert
advisers request data on the following: weather situations and their kind and
frequency (Pasquill, Klug); wind direction frequency (12-point windrose)
and corresponding wind velocity distribution with height; frequency and
duration of weather categories correlated with the wind directions and wind
velocities; frequency of inversion layers and their height above ground
level; dependence of wind conditions on the topographical conditions; effect
of cooling towers on the spread conditions (mixing of plumes); and air
temperatures, temperature profiles, air humidity and precipitation. For
complex topographical conditions it is usually considered necessary to
compile meteorological records by means of on-site measurements covering
a period of about two years. For the first presentation of documents it is
permissible to refer to the records of one or more meteorological stations
considered representative for the site. It is only if a plant is designed for
the most unfavourable weather conditions that a full analysis may be of
secondary importance. Meteorological reports were prepared, e.g. for
the nuclear power plant sites at Obrigheim [35], Neckarwestheim [39]
and Wyhl. Today, such reports are generally required for all sites.

## 6.6.  Ecological conditions

An ecological report may be requested for a site located in or near a
nature-conservation or landscape-protection area, to provide a basis for
assessing the possible effects of the proposed plant on rare animal or plant
species found in that area. For example, the relevant stipulations for the
site at Grafenrheinfeld are as follows:

"Stipulation 1.8: ... Microbiological investigations must be carried
out at several suitable points in the near proximity and in the further
vicinity of the site in order to obtain information on the existing soil
microflora and thus to make it possible to devise appropriate measures
for ensuring the continuing stability of the existing inventory and social
structure of the vegetation."
"Stipulation 1.9: ... Extensive ecological inventory protection
measures must be carried out including in particular a floral and
botanical inventory, a hydrological and soil inventory and a faunal
inventory, with mapping."
"In addition, to ensure early detection of possible long-term detrimental effects, observations must be continued over a sufficiently long
period of time including, if necessary, repeated inventorying and
mapping."

## 6.7.  Pre-existing radiological load

Under this heading, expert advisers request data on the approved and
actual emission rates of radioactive substances by other plants located within
a radius of 50 km of the proposed plant.

## 6.8.  Population density and distribution

One of the most important assessment criteria for a proposed site for
a nuclear power plant is the population density and distribution in the vicinity.
This consideration has played an important role from the earliest days of
nuclear engineering.  The first reactor plants were almost all built in
sparsely populated areas on sites remote from population centres.  Then,
after extensive positive operating experience had become available, sites for
nuclear power stations were also chosen in more densely populated areas.

The current practice in the FRG is that a proposed site for a nuclear
power station is compared, for population density and distribution in its
vicinity, with sites which have already been approved.  In the absence of any
unfavourable conditions, and if there are no major centres of population in
the immediate vicinity of the proposed site, the site is usually accepted
[34,38].

Figure 5 is a schematic population density chart showing population
numbers plotted against the distance from a point on the map, e.g. from a
prospective plant site.  The solid line is an envelope line of the maximum
population densities in the vicinity of sites for nuclear power stations with
light-water reactors approved in the FRG so far, i.e. all the sites approved
so far have surrounding population density patterns located to the right of
this line.  The two dashed lines show the population density patterns in the
vicinity of the sites of the Biblis and of the Krümmel Nuclear Power Stations,
taken as two typical examples.  The dotted-and-dashed line represents the
conditions in the vicinity of the Ludwigshafen site, discussed in greater
detail below.  The three dotted lines represent typical population density
patterns in a very densely populated area (central zone of a conurbation), a
densely populated area (edge zone of a conurbation) and a rural area, as
presented in 1974 by Ritter [1] at a meeting of the FGR Nuclear Engineering
Association.  According to this classification all the nuclear power plant
sites in the FRG, except the Ludwigshafen site, are located outside the
densely populated areas.  The city of Ludwigshafen, where the Badische
Anilin und Sodafabrik (BASF) want to build a nuclear power plant, has a
population density pattern corresponding partly to that of a central zone and
partly to an edge zone of a conurbation.  The city of Berlin has entirely the
character of a central zone of a conurbation.  Table II shows numerical data
for the rating of prospective nuclear power plant sites according to the
population density in their vicinity;  the table shows the classification
according to Ritter [1], discussed above, and a similar classification
according to Wirtz [27].

An application for a permit to build a nuclear power station in a densely
populated area in the FRG was filed by BASF in May 1969.  Before filing the
application, BASF had obtained from the IRS an expert opinion on the
suitability of three prospective sites.  This expert opinion [40] stated that
the plant could be built at any of the three sites, and also stressed that, in
addition to measures and facilities for protection against plant-internal

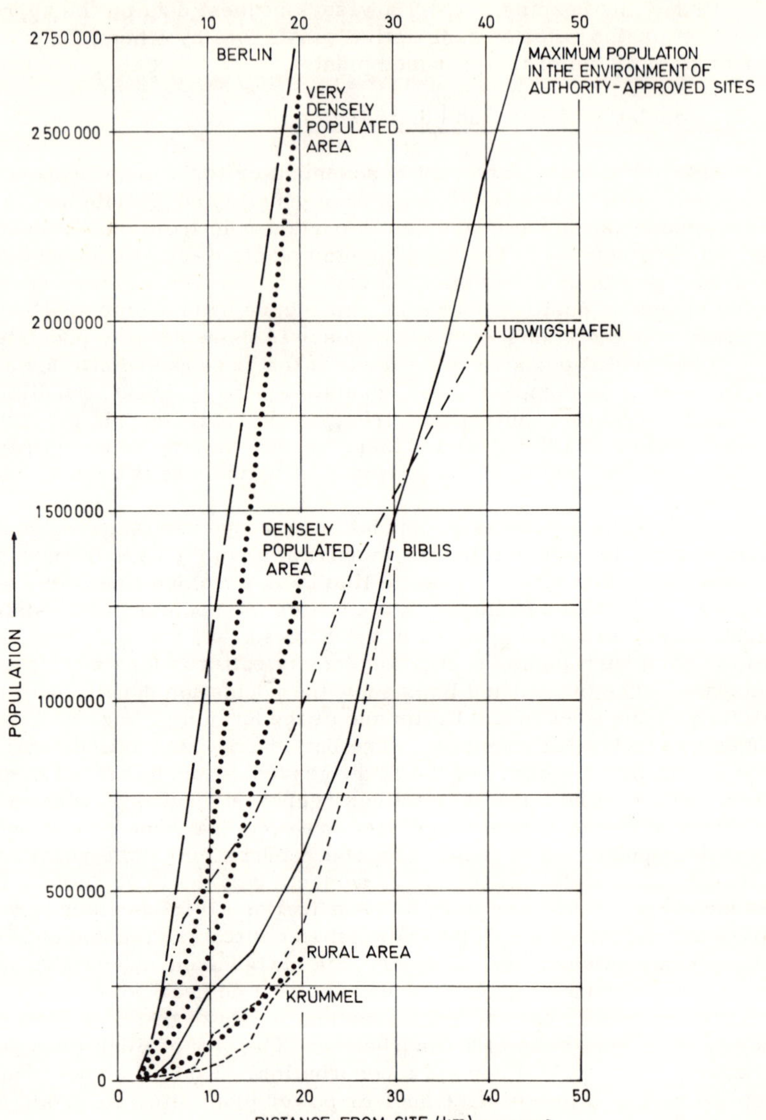


FIG.5.   Population in the environment of power plant sites with light-water reactors in the Federal Republic of Germany.

exigencies and accidents, it would be indispensable to take appropriate
safety precautions against the effects of external occurrences.[1]  At a press
conference in August 1970, the Minister responsible for these matters at
that time stated that the final decision on granting a permit under the Atomic
Energy Act for the nuclear power plant planned by BASF in Ludwigshafen
would be postponed by about three years.  The reason given for this post-
ponement was that, despite elaborate safety engineering precautions, there
still remained a residual risk owing to factors such as human error,
insufficient experience and statistical error sources.  This residual risk
must be reduced either by providing a sufficiently wide safety zone between
the nuclear power plant and the nearest city area or by providing suitable
safety facilities and installations.  The Federal Ministry concerned therefore
stipulated that before the BASF project is implemented an extensive research
programme must be carried out.  This programme included a thorough
review of the safety engineering requirements imposed on nuclear power
plants and a study of the possibilities of further improvement of the safety
engineering facilities and installations and of structural safety measures.

According to the recommendations of the RSK, formulated at its 90th
sitting in January 1974 [ 34 ], the present position in connection with the
BASF project is as follows:

The site can be classified as an industrial site near a city.  At such
a site it is possible to count on the availability of emergency help with
little delay.

This site can only be recommended for a nuclear power plant on
condition that appropriate facilities are installed and measures taken
to reduce to a negligible extent the risk to persons employed on the
plant premises and to the general public outside these premises.  The
supplementary safety facilities required include in particular an integral
bursting safeguard for the primary system.

Subject to favourable findings of the investigations still required,
the RSK considers that promising prospects exist for the fulfilment of
the requirements for the safety concept for a nuclear power plant at the
proposed site.

Before the RSK can issue an opinion on the safety concept and on
the granting of the first partial construction permit, the applicant must
submit the results of extensive analytic and experimental investigations.

For assessing a proposed site in the context of the surrounding
population density, the expert advisers usually request population density
breakdowns by sectors of 30° of arc (12-point compass rose) and by concentric
zones 0-1 km, 1-2 km, 2-3 km, 3-4 km, 4-5 km, 5-10 km, 10-20 km and
20-50 km, as well as data on residential areas and development trends.  The

---

[1]  On the question of sites for nuclear power plants, a Working Group of the Technical Safety Control
Services (TÜV) issued as early as 1962 a favourable opinion on a site in West Berlin, with the proviso that
several special design conditions must be fulfilled [41].  Another expert opinion [27] states that there are
no fundamental reasons against building nuclear power stations near or in large cities, but that the practical
implementation of such projects should be postponed until the safety analysis methods and the safety engineering
facilities and installations have reached a certain development level which is laid down in detail in the
expert opinion.

Table II.   CHARACTERIZATION OF NUCLEAR POWER PLANT SITES ACCORDING
POPULATION IN THE FEDERAL REPUBLIC OF GERMANY

( a ) according to WIRTZ [27]

| Type | Population | |
|------|-----------|---|
| | 0—2 km | 0—10 km |
| Urban Site | 10 000—200 000 | 200 000—1 000 000 |
| Conurbation | 0    — 10 000 | 100 000—  300 000 |
| Federal Average | 0   —  3 000 | 75 000 |
| Suitable Site | 0   —  3 000 | 30 000 |

( b ) according to RITTER [1]

| Type | Population | | | |
|------|-----------|---|---|---|
| | 0—2 km | 0—5 km | 0—10 km | 0—20 km |
| Very Densely Populated Area | 24 000 | 160 000 | 640 000 | 2 600 000 |
| Densely Populated Area | 12 000 | 80 000 | 320 000 | 1 300 000 |
| Rural Area | 3 000 | 20 000 | 80 000 | 320 000 |

population density calculations must include, under the heading 'commuters
persons working in medium-large and large enterprises and living in
residential areas in the vicinity of the site.

6.9.  Land use in the vicinity of the site

Required under this heading are data on the areas under silviculture,
fields, pastures and fallow land within a radius of 10 km of the site, as wel
as data on development trends.  Also required, where applicable, are data
on livestock, fisheries, nature conservation and landscape protection,
recreation areas and archaeological sites.
A typical landscape protection stipulation, in Ref. [37], is formulated
as follows:

"Stipulation 1.7:  A landscaping plan must be prepared by a landscape
architect and submitted to the competent nature conservation authorities.
This plan must show the intended landscaping, replanting and other
measures aimed at preventing or at remedying the landscape damage
associated with the proposed project."

A similar stipulation in Ref. [42] is formulated as follows:

"Stipulation 5:  Alterations of parts of the landscape subject to
landscape protection must be limited to the unavoidable minimum
associated with the kind, size and design of the buildings and structures
necessitated by operating and safety engineering requirements.  In
particular, the shapes and outer finishes of the buildings should fit
harmoniously into the landscape."

The first partial construction permit for unit B of the Biblis nuclear
power station [43] lays down that the entire construction area must be
planted with trees and shrubs appropriate for the locality and that a plan
showing the proposed landscaping must be submitted to the competent nature
conservation authorities before the operating licence is issued.

## 6.10.  Neighbouring industrial and military installations

Required under this heading are data on neighbouring industrial installations, especially on such premises where explosive, toxic, stupefying,
corrosive and/or inflammable substances are handled or stored, and data
on neighbouring military exercise areas, if any.  Data should also be
presented on any anticipated future developments, especially of chemical
and petrochemical facilities in the vicinity of the site.

## 6.11.  Traffic and transport conditions

Required under this heading are data on traffic and transport arteries
in the vicinity of the proposed site (railway lines, waterways, motorways
and roads) on which dangerous goods (explosive, toxic, stupefying, corrosive
and/or inflammable) are or may be transported.  The traffic densities for
trucks and for passenger cars should be shown separately;  this also applies
to the frequencies of goods and passenger trains.  The locations of the nearest
civilian and military airports and airfields, and air traffic lanes, should
also be shown.  If the proposed site is located near an airport, data may
also be required on the types and speeds of the aircraft using the airport [25].

## 6.12.  Effects of external occurrence on the nuclear power plant

In the FRG the licensing authorities impose stipulations on the design of
nuclear power plants aimed at ensuring the safety of the plant surroundings
in the event of external occurrences (especially of earthquakes, aircraft
crashes and gas explosions) affecting the power plant, it being noted that the
relevant assessments were originally based on the probability of such
occurrences whereas at present they are increasingly based on the possible
consequences [44-49].  Adequate protection against the effects of external
occurrences is specified in the form of stipulations forming part of pre-

liminary site approvals and of partial construction permits issued by the
competent licensing authorities [36-38, 50-53] or, in the particular case
of the BASF project, as part of the expert opinion [40]. This protection
against external occurrences is achieved by a combination of measures aimed,
first, at preventing damage to plant components important from the safety
engineering point of view and, second, at controlling the consequences of
any damage which may nevertheless occur.

The various types of possible hazards are discussed individually as
. follows:

## (a)  Floods

It is stipulated under this heading, where applicable, that to avoid flood
damage the terrain level of the power plant site should be raised by means
of fill to a level above that of 100-year floods, and that accesses to parts of
the plant important from the safety engineering point of view should be
located at a sufficient elevation above the terrain level. The supply and
return flow of cooling water required for residual heat removal and rundown
cooling of the reactor must be safeguarded even in a 1000-year flood with
concurrent storm and heavy rain. If the site is located next to a storage
reservoir or to a retention weir, appropriate measures must be taken to
protect parts of the plant important from the safety engineering point of
view against flood wave damage in the event of a possible failure of the dam
or weir. In addition, the buildings forming part of the power unit complex
must be provided with appropriate  durable external insulation, proof against
water under pressure, to protect them from ingress of water. Access to the
controlled-access area of the plant must remain ensured under all possible
flood conditions. The return flow of water from the plant must be so laid
out that, in the event of a flood, liquid wastes carrying radioactive substance
do not flow over flooded agricultural areas. Flood-related stipulations are
formulated, amongst other stipulations, in provisional site approvals and in
partial construction permits, e.g. in Refs [36-39, 43, 52].

## (b)  Low water

If the nuclear power plant site is located at a retention dam or weir, and
if the cooling-water intake is planned from the upstream level, an emergency
cooling-water supply must be ensured, in the event of failure of the dam or
weir, by means of a secured intake from the downstream level or from
wells [51].

## (c)  Groundwater

At the Neckarwestheim site it was stipulated that the reactor building,
the auxiliary reactor building and the switchgear building must be designed
taking into account the groundwater pressure [50].

## (d)  Earthquakes

Draft rules have been formulated regarding the safety-engineering
design of nuclear power plants against seismic effects [21, 22]. In planning
the dimensions of buildings, facilities and component parts of the plant which

are important from the safety engineering point of view, a distinction is
drawn between the design-basis earthquake and the safe-shutdown earthquake:

### (i)   Design-basis earthquake

The design-basis earthquake is defined as the maximum-intensity earthquake definitely known to have occurred in the past at the site or in its near
vicinity (in the same seismotectonic unit up to a distance of about 50 km).
Component parts of the plant important from the safety engineering point
of view must be of such dimensions that continued operation remains
possible even after repeated occurrence of earthquakes of this intensity.

### (ii)  Safe-shutdown earthquake

The safe-shutdown earthquake is defined as the maximum-intensity earthquake which will not be exceeded at any time, so far as can be predicted
on the basis of available scientific knowledge, at the site or in its remote
vicinity (up to a distance of about 200 km).  If these predictions are
based on statistical calculations, the temporal and local probability of
occurrence must be stated (together with the confidence limits) and must
not be greater than the probability of the design-basis accident.  Component parts of the plant important from the safety engineering point of
view must be of such dimensions that they can withstand an earthquake
of this intensity without impairment of function.

For the purpose of assessing the seismic hazards for nuclear power
stations, the FRG has been subdivided into four earthquake zones: 0, 1, 2
and 3.  Plants located in zones 1, 2 and 3 must have seismographs installed
at appropriate points in the reactor building.
Considering the safety aspects of the nuclear power plant as a whole,
the component parts of the plant are classified as follows:

Class I:   Component parts of the plant relevant from the safety engineering
           point of view  which serve to shut down the reactor and to hold it
           in the shutdown condition, to remove residual heat and to prevent
           the escape of radioactive substances.

Class II:  All other component parts of the plant.

The seismic calculations required for an earthquake-safe design must
be carried out by appropriate dynamic calculation methods.  For sites at
which the maximum acceleration of the safe-shutdown earthquake is less than
100 cm/s$^2$, simplified calculation methods are permissible.
Examples of anti-seismic design specifications at power plant sites with
light-water reactors in the FRG are shown in Table III.  Details are often
listed in the partial construction permits.

### (e)  Aircraft crashes

Discussions on the protection of nuclear power stations against aircraft
crashes started in the FRG in about 1969 and were at first concerned with
sites with a particularly high air-traffic density.  Probability calculations

Table III    EXAMPLES OF ANTI-SEISMIC DESIGN SPECIFICATIONS
             AT POWER PLANT SITES WITH LIGHT WATER REACTORS
             IN THE FEDERAL REPUBLIC OF GERMANY

SSE = safe shutdown earthquake                PCP = date of 1st partial construction permit
DBE = design basis earthquake                 HA   = horizontal acceleration

| Site (Plant code name) PCP | Plant components designed against SSE/DBE with HA as shown | | |
| --- | --- | --- | --- |
| **BIBLIS A & B**<br><br>31. 7. 70 / 6. 4. 72 | 150 cm/s² / 100 cm/s²<br>• Reactor building<br>• Switchgear building<br>• Service cooling water supply | • Emergency power supply<br>• Reactor auxiliary building<br>• Exhaust air stack | |
| **NECKARWESTHEIM** (GKN)<br><br>24. 1. 72 | 170 cm/s² / 90 cm/s²<br>• Reactor building<br>• Switchgear building<br>• Service cooling water supply<br>• Emergency power diesel building | • Emergency operation building<br>• Reactor auxiliary building<br>• Exhaust air stack | |
| **OHU** (KKI)<br><br>16. 5. 72 | 75 cm/s² / 50 cm/s²<br>• Reactor building<br>• Switchgear building<br>• Service cooling water supply | • Auxiliary cooling water pump house<br>• Emergency power diesel building | |
| **KRÜMMEL** (KKK)<br><br>18. 12. 73 | 50 cm/s² / —<br>• Reactor building<br>• Service cooling water supply for 2 RHR strands | • Emergency power diesel building with 2 diesels | |
| **WYHL** (KWS)<br><br>— | 250 cm/s² / 125 cm/s² *)<br>• Reactor building<br>• Switchgear building<br>• Service cooling water supply | • Emergency power diesel building<br>• Emergency feed building<br>• Exhaust air stack | |

*not yet approved

RHR = residual heat removal

for aircraft crashes on nuclear power plant sites were carried out in order
to assess whether protective measures are required [54-60].  The calculated
crash probabilities are shown in Table IV.  Comparing the data for the sites
at Biblis, Ludwigshafen, Philippsburg, Esenshamm, Ohu, Krümmel and
Grafenrheinfeld with the overall averages for the FRG as a whole, it can be
seen that the probabilities of the most dangerous crashes, i.e. of large
military machines and of civil airliners, considered together, are of the
same order or smaller at these sites than for the FRG as a whole.

Despite these low probabilities, structural safety measures have been
imposed to safeguard nuclear power plants against the effects of a crash of
a fast-flying military machine, with definite specifications regarding the
weight of the machine, the impact velocity, the impact surface area and the
peak dynamic load.  Examples of design specifications for safety against
aircraft crashes at power plant sites with light-water reactors in the FRG are
shown in Table V.

The recent guidelines ([17] and [18]) of the Reactor Safety Advisory
Commission lay down that all nuclear power plants with light-water reactors
must be protected against aircraft crashes.

**Table IV. PROBABILITY OF AIRCRAFT CRASHES ON NUCLEAR POWER PLANT SITES IN THE FEDERAL REPUBLIC OF GERMANY**

| PLACE | PROBABILITY OF AIRCRAFT CRASHES PER YEAR AND $10^4$ $m^2$ AREA | | | | | | | |
| --- | --- | --- | --- | --- | --- | --- | --- | --- |
| | FRG [56] − [60] | BIBLIS [54] | LUDWIGS-HAFEN [55] | PHILIPPS-BURG [56] | ESENS-HAMM [57] | OHU [58] | KRÜMMEL [59] | GRAFEN-RHEIN-FELD [60] |
| 1. Military Airplanes | $52 \cdot 10^{-8}$ | $10 \cdot 10^{-8}$ | $64 \cdot 10^{-8}$ | $52 \cdot 10^{-8}$ | $29.4 \cdot 10^{-8}$ | $15.8 \cdot 10^{-8}$ | $0.7 \cdot 10^{-8}$ | $3.3 \cdot 10^{-8}$ |
| 2. Large Civil Airplanes | $0.6 \cdot 10^{-8}$ | $1 \cdot 10^{-8}$ | $5.7 \cdot 10^{-12}$ | $0.6 \cdot 10^{-8}$ | $0.12 \cdot 10^{-8}$ | $0.32 \cdot 10^{-8}$ | $2 \cdot 10^{-8}$ | $2 \cdot 10^{-8}$ |
| TOTAL 1 + 2 | $52.6 \cdot 10^{-8}$ | $11 \cdot 10^{-8}$ | $64 \cdot 10^{-8}$ | $52.6 \cdot 10^{-8}$ | $29.5 \cdot 10^{-8}$ | $16.1 \cdot 10^{-8}$ | $2.7 \cdot 10^{-8}$ | $5.3 \cdot 10^{-8}$ |
| 3. Military Heli-copters | $20 \cdot 10^{-8}$ | − | − | $10 \cdot 10^{-8}$ | $3.9 \cdot 10^{-8}$ | $1 \cdot 10^{-8}$ | $9.8 \cdot 10^{-8}$ | $15.3 \cdot 10^{-8}$ |
| 4. Small Civil Airplanes | $13 \cdot 10^{-8}$ | − | $200 \cdot 10^{-8}$ | $50 \cdot 10^{-8}$ | $3.8 \cdot 10^{-8}$ | $3.8 \cdot 10^{-8}$ | $3 \cdot 10^{-8}$ | $2.4 \cdot 10^{-8}$ |

**Table V. EXAMPLES OF DESIGN SPECIFICATIONS FOR SAFETY AGAINST AIRCRAFT CRASHES AT POWER PLANT SITES WITH LIGHT WATER REACTORS IN THE FEDERAL REPUBLIC OF GERMANY**

| Site (Plant code name) PCP | Plant components designed against an aircraft crash; impact loads and impact areas as shown |
| --- | --- |
| BIBLIS B 6. 4. 72 | 1700 Mp (static); $2.14$ $m^2$<br>● Reactor building |
| OHU (KKI) 16. 5. 72 | 1700 Mp (static); $2.14$ $m^2$<br>● Reactor building     ● Satellite control room in the switchgear building |
| ESENSHAMM (KKU) 28. 6. 72 | 1700 Mp (static); $2.14$ $m^2$<br>● Reactor building |
| KRÜMMEL (KKK) 18. 12. 73 | 11 000 Mp (dynamic); 7 $m^2$<br>● Reactor building |
| GRAFENRHEINFELD (KKG) 21. 6. 74 | 11 000 Mp (dynamic); 7 $m^2$<br>● Reactor building     ● Emergency feed building |

PCP = date of first partial construction permit

Mp = Mgf

The discussions on the precise safety measures required still continue. For example, the IRS is considering nuclear power plant design specifications taking into account not only an impact load of 11 000 Mgf distributed over an impact surface area of 7 $m^2$ for the main impact, but also an impact load of 8500 Mgf distributed over an area of only 1.5 $m^2$ for the impact of wreckage with, however, a very short action time in the millisecond range [61].

(f)   Gas pressure shock waves

Buildings and component parts of the plant which are important from the safety-engineering point of view must also be protected against external explosions. These provisions envisage a detonation of explosives or of easily inflammable gases as a result of traffic accidents on road, rail or river, or an explosion in an industrial plant in close proximity to the site of the nuclear power plant. This question was raised for the first time in connection with the Ludwigshafen site  because of the hazard of explosion in the adjacent chemical plant of BASF.  It was raised again in connection with the site of the Brunsbüttel Nuclear Power Station  because of the danger of explosion in the event of collision of liquefied-gas tankers.

Table VI  EXAMPLES OF DESIGN SPECIFICATIONS FOR SAFETY AGAINST GAS PRESSURE SHOCK WAVES AT POWER PLANT SITES WITH LIGHT WATER REACTORS IN THE FEDERAL REPUBLIC OF GERMAN

| Site (Plant code name) PCP | Plant components designed against gas pressure shock waves with reflection pressure/quasi-static pressure as shown | |
|---|---|---|
| BRUNSBÜTTEL (KKB) 2. 4. 70 | 1.45 bar / 1.30 bar<br>● Reactor building<br>● Switchgear building | ● Service cooling water supply<br>● Emergency power diesel building |
| NECKARWESTHEIM (GKN) 24. 1. 72 | 1.20 bar / 1.10 bar<br>● Reactor building<br>● Emergency power diesel building | ● Emergency feed building<br>● Reactor auxiliary building |
| BIBLIS B 6. 4. 72 | 1.45 bar / 1.30 bar<br>● Reactor building<br>● Switchgear building<br>● Service cooling water supply | ● Emergency power diesel building<br>● Reactor auxiliary building |
| OHU (KKI) 16. 5. 72 | 1.20 bar / 1.10 bar<br>● Reactor building<br>● Service cooling water supply | ● Emergency power diesel building with 2 dieseln |
| ESENSHAMM (KKU) 26. 8. 72 | 1.45 bar / 1.30 bar<br>● Reactor building<br>● Emergency operation building | |
| KRÜMMEL (KKK) 18. 12. 73 | 1.45 bar / 1.30 bar<br>● Reactor building<br>● Service cooling water supply for 2 RHR strands | ● Emergency power diesel building with 2 diesels |
| GRAFENRHEINFELD (KKG) 21. 6. 74 | 1.45 bar / 1.30 bar<br>● Reactor building<br>● Emergency feed building | |

PCP = date of first partial construction permit

RHR = residual heat removal

Following these discussions, expert advisers have specified that nuclear power plants must be protected against gas pressure shock waves by designing the buildings for an assumed reflection pressure of 1.45 bar and a quasi-static pressure of 1.30 bar on sites presenting a risk of liquefied-gas explosions, and for 1.20 bar and 1.10 bar, respectively, for other sites. Stipulations on pressure shock wave design are contained, e.g., in the partial construction permits [37, 43, 51]. Examples of design specifications for safety against gas pressure shock waves at power plant sites with light-water reactors in the FRG are shown in Table VI.

The recent guidelines ([17] and [18]) of the Reactor Safety Advisory Commission lay down uniformly that all nuclear power plants with light-water reactors must be designed against a reflection pressure of 1.45 bar and a quasistatic pressure of 1.30 bar.

(g)  Poisonous, stupefying and potentially explosive gases
_______________________________________________________

Stipulations under this heading require proof that the intake of poisonous, stupefying and potentially explosive gases through the air intake ports can be prevented [17, 18, 24, 37, 52, 53].

(h)  Surface fires
__________________

The effects of surface fires, particularly of fires caused by a spillage of fuel, e.g. in the event of an aircraft crash, must be taken into account in the design of the power plant [17, 18, 52, 53].

(i)  Actions by third parties
_____________________________

Malicious or unintentional damage to the plant by third parties must be prevented or limited by the consistent application of the principle of redundant duplication and spatial separation. According to these stipulations the redundantly duplicated reactor shutdown and residual heat removal systems must be accommodated in different compartments of the plant in such a manner that destruction of more than one system by the same causal agency can be excluded. Supplementary engineering installations and supporting administrative measures serving the same purpose are also required [17, 18].

6.13.  Radiation load on the environment
_________________________________________

The safety criteria of the Federal Ministry for Internal Affairs [16] and the guidelines issued by the Reactor Safety Advisory Commission [17, 18] lay down stipulations regarding the monitoring of radioactivity in the exhaust air, in the released liquid wastes and in the vicinity of the power plant. According to these stipulations the radiation load on the environment due to the release of radioactive substances with the liquid wastes and with the exhaust air, in normal operation of the power plant, must not exceed 30 mrem/a by either of these two routes, and the radiation load on the thyroid gland of small children must not exceed 90 mrem/a. Other relevant stipulations are contained in the LAWA guidelines on the immission of radioactive substances into waters [20], which set limits of 0.5-1.0 Ci/a for the release of radioactive liquid wastes, not counting tritium activity, and of 50 Ci/a for the release of tritium by boiling-water reactor plants, and of 200 Ci/a for pressurized-water reactor plants (all data referred to 100-MW electric power rating) and the recommendations [19] regarding the determination of permissible radioactivity emission rates from nuclear power plants (see also Section 5 of this paper).

In addition, the release of radioactive substances with air and with water must be limited to the lowest rates achievable by up-to-date engineering means, and measurements and investigations of the environmental radio-activity in the vicinity of the power plant site must be carried out to prove compliance with the stipulations [36-39, 43, 50, 53].

## 6.14.  Warming-up of waters by waste heat from nuclear power plants

The Interstate Working Group on the Conservation of Rivers has published thermal load schedules for the Danube (partial reaches), the Elbe (partial reaches), the Main, the Neckar, the Rhine and the Weser [62, 63] which lay down the following limits for the ecologically permissible thermal loads:

(a)  The temperature increase of the river water, caused by effluent cooling water discharged by one or more power plants into a river system, after complete mixing, must not exceed 3°C or, in excep-tional cases, 5°C at any point and at any season.

(b)  The maximum temperature of the mixed water must not exceed the following limits:

> For water with mean summer temperatures between 17°C and 20°C but which may occasionally reach natural maximum temperatures up to 23°C, the limit is 25°C;

> For waters with natural summer temperatures up to 25°C, the limit (in summer) is 28°C.

For example, the stipulations for the rivers Main, Rhine and Weser are shown in Table VII.

There is a strong trend in favour of cooling by means of cooling towers. The Rhine-Neckar Regional Planning Commission, e.g., no longer allows the construction of power plants anywhere on the Rhine except with cooling towers [66].

TABLE VII.  MAXIMUM PERMITTED WATER TEMPERATURE

| River | Max. warming-up range | Max. temp. of mixed water | Remarks |
|---|---|---|---|
| Main [37, 64] | 3°C | 28°C | |
| Rhine [63] | 3°C | 25°C | |
| | 5°C | 28°C | For specified reaches |
| Weser [52, 65] | 3°C | 28°C | Freshwater reaches |
| | 2°C | 26°C | Brackish reaches |

### 6.15.  Effects of cooling towers

According to Hübschmann and Nester [67], the waste heat dissipated
by cooling towers on flat terrain causes only slight changes in the mean
climatic conditions.  These changes are limited to the immediate vicinity
of a cooling tower and are usually small in comparison with the natural
fluctuations of the important parameters.  Noticeable effects such as precipitation, heavy misting, raised moisture and temperature, occur only
occasionally under unfavourable meteorological conditions, with the cooling-
tower exhaust often acting only as a precipitating factor.  The effects of
cooling towers on the climatic conditions in locations with elevation differences
of a few hundred metres are less well known, so that meteorological studies
appear to be necessary.  Expert opinions on the effects of cooling towers on
the surroundings of nuclear power plant sites are already available [37].

### 6.16.  Effects of nuclear power stations on groundwater

As a safety precaution against penetration of radioactive substances
into the subsoil and groundwater, all the buildings forming part of the power
unit complex must be provided with an appropriate permanent waterproofing
insulation.  Compartments in which are located tanks containing radioactive
liquids must be designed as waterproof catchment troughs with a capacity
sufficient to accept the quantity of liquid contained in the largest tank in each
compartment [35-39, 43, 50, 68, 69].

### 6.17.  Noise

It must be ensured by the most up-to-date engineering means that the
power plant, including all its ancillary facilities, emits as little noise as
possible into the neighbourhood.  Noise abatement measures must also be
taken during the construction of the plant.  General data on this subject are
found in the paper by Thomassen [70] and in the Technical Ordinance on
Protection against Noise [71].  Examples of specific stipulations are found
in the partial construction permits [35-37, 42, 50].

### 6.18.  Rescue and emergency facilities

Required under this heading are data on accessibility for the nuclear
rescue brigade, on the distance to the nearest fire station with heavy
equipment, and on the distances to other rescue and emergency help organizations and facilities such as police, Red Cross and military authorities.
A direct telephone line to the nearest police station is desirable.

### 6.19.  Public relations

It is desirable to establish good relations with the population in the
vicinity of nuclear power station sites as well as with the public in general.
As can be seen from Fig.6, the number of objections against sites with
commercial nuclear power stations has steadily increased since the first
such plant was built until the present.  In considering these data it should
be borne in mind that by far the largest majority of objectors were signatories
of petitions sponsored by organizations of nuclear power opponents;  thus,

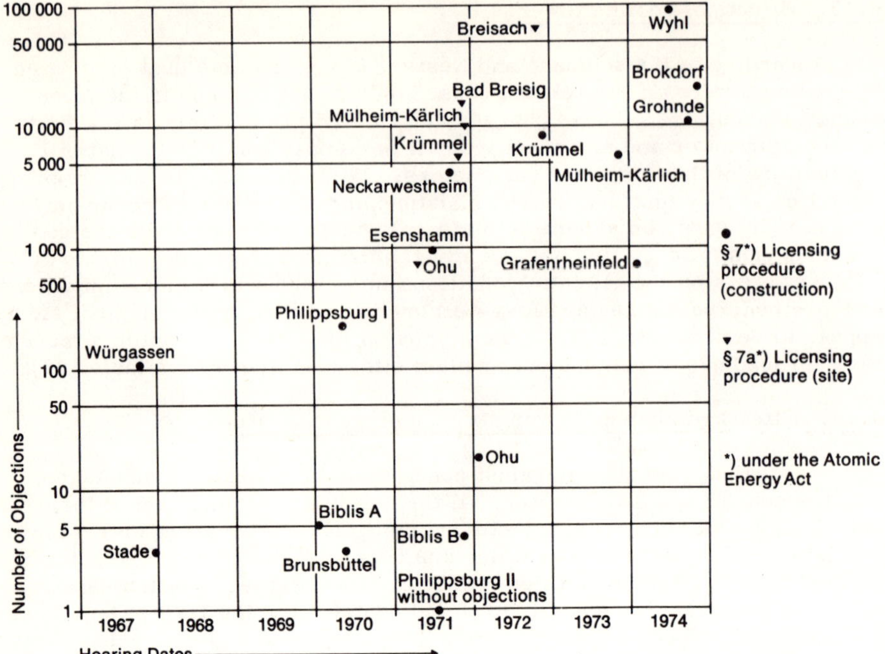



FIG. 6.   Number of objections against sites with commercial light-water reactor plants in the Federal Republic of Germany.

in round figures, of the 4430 objections to the Neckarwestheim site, lodged within the stipulated time, about 4360 were by signatories of stereotyped petitions and the rest by individual objectors for individual reasons, and of the 8535 objections to the Krümmel site only about 90 were by individual objectors for individual reasons. Despite this preponderance of organized opposition, it should be possible, by means of explanations and factual information, to reduce the number of objections based on ignorance or misinformation. This should help to improve public relations in the long term.

Following is a list of site-specific objections lodged against nuclear power stations with light-water reactors in the course of licensing procedure under the Atomic Energy Act:

(1) Damage to the health of the public owing to release of radioactive substances [36-39, 42, 43, 51, 72].

(2) Fear of uncontrolled release of radioactivity in the event of an accident [39, 43, 72].

(3) No radioactivity load schedule for the river [37, 72].

(4) Fear of contamination of water-procurement facilities, groundwater and health springs by radioactivity [35, 37-39, 42, 43].

(5) Hazard for the vicinity owing to transport and storage of nuclear wastes [38, 42].

(6) Nuclear power plants endanger densely populated areas; a more suitable site should have been chosen [37, 39, 42, 53].

(7)  Nuclear power plants no longer in use continue to endanger the
     vicinity [37, 53].
(8)  The disaster-protection planning shows that there is no effective
     protection [38].
(9)  Adverse effects on the climate owing to immission of warm water or
     to cooling-tower operation (fog, reduced insolation, increased air
     humidity, more frequent precipitation, air temperature changes,
     increased incidence of bad weather, icing) and thus damage to the
     health of the public [35, 37, 50].
(10) Apprehension of excessive noise emission into the vicinity [35, 37].
(11) Adverse effects on the development of new residential and industrial
     areas; nuclear power plants should not stand in the way of improving
     rivers for navigation [35, 37, 51].
(12) Nature-conservation, landscape-protection and recreation areas in
     the vicinity will be adversely affected [37, 39, 42, 53].
(13) Health resorts and tourism in the vicinity will be adversely
     affected [42].
(14) Financial losses will be caused by the anticipated drop in land prices
     in the vicinity of the site [42].
(15) The quality and sales of wine, fruit, vegetables and other agricultural
     produce from the vicinity of the site will be adversely affected
     [35, 37, 42, 50, 72].
(16) Immission of warm water is detrimental to most species of fish and
     results in the death of aquatic organisms, especially of micro-
     organisms [36, 37, 39, 42, 50, 51].
(17) The self-cleaning capacity of rivers is impaired; this might impose
     an additional burden on sewage treatment plants at locations down-
     stream from the site [42].
(18) The intake and return of water will create transverse currents
     endangering navigation [42].
(19) Traffic hazards owing to icing and fog [39, 42].
(20) Nuclear power stations are insufficiently protected against the
     effects of external occurrences such as earthquakes, floods, air-
     craft crashes, enemy action in the event of war, and sabotage
     [37-39, 42, 43, 53].
(21) Alternative sites have not been investigated; no justification has
     been presented for choosing this particular site [37].
(22) Floating nuclear power plants should be built [53].

Some of these objections were rejected as unfounded, whereas others
were taken into account in the form of stipulations, restrictions and
recommendations.


## 7.  CONCLUSION

More sites for nuclear power stations will be needed in the Federal
Republic of Germany in the coming years. One of the possibilities being
studied at present is the construction of nuclear power stations on artificial
islands off the coast of the FRG. The stipulations formulated by German
expert advisers and licensing authorities in respect of environment protection
can be satisfied by engineering means, so that a nuclear power plant can

be built on almost any site. It appears necessary, however, that the
licensing authorities should be induced to formulate uniform and realistic
requirements and to maintain these requirements unchanged over fixed
periods of time.

This condition must be satisfied for planning in advance a number of
sites for nuclear power stations sufficient for the implementation of the
power supply policies of the FRG in the coming years.

## REFERENCES


[1]   RITTER, H.A., Vortragsveranstaltung der Kerntechnischen Gesellschaft über Standortwahl von Kernkraft-
      werken in Verdichtungsräumen (Site selection for nuclear power plants in densely populated areas),
      Stuttgart (20 Feb., 1974).
[2]   DAtF (Dtsch. Atomforum Atominf. 6/74 (1974) 4.
[3]   MANDEL, H., Standortfragen bei Kernkraftwerken (Problems of site selection for nuclear power plants),
      Atomwirtsch. 16 (1971) 22.
[4]   MANDEL, H., Gesichtspunkte bei der Standortwahl für Kernkraftwerke (Factors relevant to site selection
      for nuclear power plants), Brennst.-Wärme-Kraft 18 (1966) 493.
[5]   GUCK, R., Standorte für Kernkraftwerke (Sites for nuclear power plants), Atomwirtsch. 18 (1973) 386.
[6]   SCHIKARSKI, W., Zur Frage des Standortes bei Reaktoranlagen (On the question of sites for reactor
      plants), Kerntechnik 3 (1961) 458.
[7]   KELLERMANN, O., FRANZEN, L.F., Auswahl und sicherheitstechnische Beurteilung des Reaktorstandorte
      (Selection and safety engineering assessment of sites for reactor plants), Atomwirtsch. 11 (1966) 380.
[8]   KARWAT, H., Sicherheitsbehälter und Standortwahl für Kernreaktoren (Containment and site selection
      for nuclear reactors), Atomwirtsch. 12 (1967) 448 (report of IAEA Symp. Vienna, 1967).
[9]   COURVOISIER, P., Standortfaktoren (Site factors), SVA-Informationstagung über die Sicherheit von
      Kernkraftwerken und die Probleme der Radioaktivität, Bern, Nov. 1970.
[10]  AILLERET, P., Standortprobleme für Kernkraftwerke (Problems of site selection for nuclear power plants)
      At. Strom 16 (1970) 149.
[11]  VINCK, W.F., MAURER, H.A., LEONARDINI, L., "Engineering safety factors and their influence on
      siting practices for nuclear power plants in the European Communities", Environmental Aspects of
      Nuclear Power Stations (Proc. Symp. New York, 1970), IAEA, Vienna (1971), 661.
[12]  GERSDORFF, B. von, SOMMER, W., Kraftwerke und Umweltbeeinflussung in Ballungsgebieten (Power
      plants and their effects on the environment in high population density areas), Brennst.-Wärme-Kraft
      26 (1974) 393.
[13]  STÄBLER, K., Vortragsveranstaltung der Kerntechnischen Gesellschaft über Standortwahl von Kernkraft-
      werken in Verdichtungsräumen (Site selection for nuclear power plants in high population density areas),
      Stuttgart (20 Feb. 1974).
[14]  KELLERMANN, O., "Sind Kernkraftwerke in Verbrauchszentren notwendig?" (Is it necessary to locate
      nuclear power plants in the centres of consumption?), 4. IRS-Fachgespräch Düsseldorf, 1968,
      IRS-T 16 (1969).
[15]  INSTITUT FÜR REAKTORSICHERHEIT, Sicherheitskriterien für Kernkraftwerke (Safety criteria for nucle.
      power plants), IRS-R-2 (1969).
[16]  BUNDESMINISTER DES INNERN, Sicherheitskriterien für Kernkraftwerke (Safety criteria for nuclear
      power plants), (25 June 1974).
[17]  Reaktor-Sicherheitskommission, Leitlinien für Druckwasserreaktoren (Guidelines for pressurized-water
      reactors) 1. Ausgabe (24 April 1974).
[18]  Reaktor-Sicherheitskommission, Leitlinien für Siedewasserreaktoren (Guidelines for boiling-water
      reactors), Entwurf E 7.74 (31 July 1974).
[19]  BUNDESMINISTER FÜR BILDUNG UND WISSENSCHAFT, Emissionsquellstärke von Kernkraftwerken
      (Emission source intensity of nuclear power plants), Schriftenreihe Kernforschung 6 (1972), (Empfehlung
      zur Bestimmung der zulässigen Emission radioaktiver Stoffe aus Kernkraftwerken in die Atmosphäre).
[20]  LÄNDERARBEITSGEMEINSCHAFT WASSER (LAWA), Richtlinien für das Einleiten radioaktiver Stoffe aus
      kerntechnischen Anlagen in die Gewässer (Guidelines on the immission of radioactive substances from
      nuclear engineering facilities into waters), (Dec. 1973).

[21]  (a)  INSTITUT FÜR REAKTORSICHERHEIT,  Voruntersuchung zu Regeln für die Auslegung von Kernkraft-
          werken gegen seismische Einwirkungen (Rules on the design of nuclear power plants against seismic
          effects), (Oct. 1973).
      (b)  Auslegung von Kernkraftwerken gegen seismische Einwirkungen (Design of nuclear power plants
          against seismic effects), Entwurf einer sicherheitstechnischen Regel des Kerntechnischen Ausschusses,
          KTA 2201.1, Ausgabe 9/1974.
[22]  BORK, M., SCHWARZER, W., "Earthquake safety of nuclear power plants based on a balanced risk
      concept", Principles and Standards of Reactor Safety  (Proc. Symp. Jülich, 1973), IAEA, Vienna
      (1973), 181.
[23]  Verordnung über den Schutz vor Schäden durch ionisierende Strahlen (Strahlenschutzverordnung),
      (Radiation Protection Ordinance), (17 July 1974), (to be published).
[24]  INSTITUT FÜR REAKTORSICHERHEIT, Unterlagen über die Eigenschaften des Standortes (Documents
      on site characteristics), 3. Entwurf (20 Feb. 1974), (to be published).
[25]  INSTITUT FÜR REAKTORSICHERHEIT, Unterlagenswünsche zur Erstellung eines Standortgutachtens für
      ein Kernkraftwerk (Desired documentation for preparing an expert opinion of a site for nuclear power
      plant), (12 Sept. 1974), (not published but will be included in Ref. [24]).
[26]  STAUBER, E., Standortwahl und grösster anzunehmender Unfall bei Reaktoranlagen (Site selection and
      maximum credible accident), Atomkernenergie $\underline{6}$ (1961) 165.
[27]  WIRTZ, K., Gutachten über die Errichtung grosser Kernkraftwerke in oder in der Nähe von Bevölkerungs-
      zentren (Construction of large nuclear power plants in or near population centres), Auftrag des Hessischen
      Ministeriums für Wirtschaft und Verkehr (1968).
[28]  SCHWARZER, W., Beziehungen zwischen Standort und Sicherheit von Kernkraftwerken (Site and safety
      of a nuclear power station), 4. IRS-Fachgespräch Düsseldorf, 1968, IRS-T 16 (1969).
[29]  WACHSMANN, F., SCHWIBACH, J., "Considerations for siting nuclear power plants in areas with high
      population density in the Federal Republic of Germany", Environmental Aspects of Nuclear Power Stations
      (Proc. Symp. New York, 1970), IAEA, Vienna (1971) 791.
[30]  WIRTZ, K., Genehmigungsverfahren, Standortbeurteilung und Sicherheitskriterien (Licensing procedure,
      site assessment and safety criteria), Atomwirtsch. $\underline{16}$ (1971) 70.
[31]  STAUBER, E., "Standortbedingungen" (Site conditions), Abschnitt 4.1, Grundlagen und Massnahmen
      der Sicherheitstechnik bei Kernkraftwerken, VWEW Frankfurt (1970).
[32]  BUNDESMINISTERIUM FÜR BILDUNG UND WISSENSCHAFT, Reaktorsicherheit im Vordergrund (Reactor
      safety is foremost), BMBW 11/72 (16 Nov. 1972).
[33]  FRANZEN, L.F., Vortragsveranstaltung der Kerntechnischen Gesellschaft über Standortwahl von
      Kernkraftwerken in Verdichtungsräumen (Site selection for nuclear power plants in high population
      density areas), Stuttgart (20 Feb. 1974).
[34]  IRS-Kurzinformationen 1974/C/9; Bekanntmachung von Empfehlungen der Reaktor-Sicherheitskommission
      vom 28.5.1974 (Recommendations of the Reactor Safety Advisory Commission), Bundesanzeiger Nr. 116
      (28 June 1974).
[35]  WIRTSCHAFTSMINISTERIUM BADEN-WÜRTTEMBERG, Erste Teilerrichtungsgenehmigung für das
      Kernkraftwerk Obrigheim (First partial construction permit for the Obrigheim Nuclear Power Station),
      Stuttgart (16 March 1965).
[36]  WIRTSCHAFTSMINISTERIUM BADEN-WÜRTTEMBERG, Erste Teilgenehmigung für die Errichtung des
      Kernkraftwerkes Philippsburg I (First partial construction permit for the Philippsburg I Nuclear Power
      Station), Stuttgart (9 Oct. 1970).
[37]  BAYERISCHES STAATSMINISTERIUM FÜR LANDESENTWICKLUNG UND UMWELTFRAGEN, Erster
      Teilgenehmigungsbescheid/Kernkraftwerk Grafenrheinfeld (First partial construction permit for the
      Grafenrheinfeld Nuclear Power Station), München (21 June 1974).
[38]  BAYERISCHES STAATSMINISTERIUM FÜR LANDESENTWICKLUNG UND UMWELTFRAGEN, Vorbescheid
      nach § 7a Atomgesetz zur Wahl des Standortes für die Anlage/Kernkraftwerk Isar (Preliminary ruling
      under Section 7A of the Atomic Energy Act on the selection of the site for the Isar Nuclear Power Station),
      München (25 Nov. 1971).
[39]  HESSISCHER MINISTER FÜR WIRTSCHAFT UND TECHNIK, Erste atomrechtliche Teilgenehmigung für
      das Kernkraftwerk Biblis A (First partial construction permit for the Biblis A Nuclear Power Station),
      Wiesbaden (31 July 1970).
[40]  INSTITUT FÜR REAKTORSICHERHEIT, Gutachterliche Stellungnahme zur Sicherheit eines Industrie-
      kernkraftwerkes im Raume Mannheim-Ludwigshafen (Safety of an industrial nuclear power plant in the
      Mannheim-Ludwigshafen area), (Feb. 1969).
[41]  KUHLMANN, A., LINDACKERS, K.-H., Zehn Jahre Reaktorsicherheit und Strahlenschutz (Ten years of
      reactor safety and radiation protection) TÜV Rheinland, Köln (1968) 29.

[42] ARBEITS- UND SOZIALMINISTER DES LANDES NORDRHEIN-WESTFALEN, Bescheid Nr. 7/1, Erste
     Teilgenehmigung Kernkraftwerk Würgassen (First partial construction permit for the Würgassen Nuclear
     Power Station), Düsseldorf (19 Jan. 1968).

[43] HESSISCHER MINISTER FÜR WIRTSCHAFT UND TECHNIK, Erste atomrechtliche Teilgenehmigung für
     das Kernkraftwerk Biblis, Block B (First partial construction permit for Unit B of the Biblis Nuclear
     Power Station), Wiesbaden (6 April 1972).

[44] KELLERMANN, O., Technische Sicherheitsmassnahmen und -vorkehrungen (Engineering safety measures
     and precautions), SVA-Informationstagung über die Sicherheit von Kernkraftwerken und die Probleme
     der Radioaktivität, Bern, Nov. 1970.

[45] SCHNURER, H., Schutzmassnahmen für äussere Einwirkungen bei Kernkraftwerken (Measures for the
     protection of nuclear power plants against effects of external occurrences), VGB-Konferenz Bautechnik
     in Wärmekraftwerken, 1973.

[46] MAYINGER, F., Sicherheitsforschung in der Bundesrepublik Deutschland (Safety research in the Federal
     Republic of Germany), Atomwirtsch. 19 (1974) 288.

[47] TSCHERNER, M., DECKERS, J., Schutz gegen äussere Einwirkungen, Abschnitt 5, (Protection against
     external occurrences), Atomwirtsch.-Broschüre Nr. 3, S. 11.

[48] STÜGER, R., Sicherheit gegen Grossschäden in Kernkraftwerken, Teil 2: Massnahmen gegen Schäden
     durch äussere Einwirkungen (Measures for protection against damage by external occurrences),
     7. Technisches Allianz-Kolloquium, München, Dec. 1973, Allianz-Bericht Nr. 20 (1974).

[49] BUNDESMINISTERIUM FÜR FORSCHUNG UND TECHNOLOGIE, Studie über die wirtschaftlichen Aus-
     wirkungen des Schutzes von Kernkraftwerken gegen Einwirkungen von aussen (Study on the economic
     consequences of protection of nuclear power plants against the effects of external occurrences), durch-
     geführt von F. Krupp GmbH, Essen, Forschungsvorhaben RS 65 (1973).

[50] WIRTSCHAFTSMINISTERIUM BADEN-WÜRTTEMBERG, Erste Teilgenehmigung für die Errichtung des
     Gemeinschaftskernkraftwerkes Neckar (First partial construction permit for the Neckar Communal Nuclear
     Power Station), Stuttgart (24 Jan. 1972).

[51] BAYERISCHES STAATSMINISTERIUM FÜR LANDESENTWICKLUNG UND UMWELTFRAGEN, Erster
     Teilgenehmigungsbescheid/Kernkraftwerk Isar (First partial construction permit for the Isar Nuclear
     Power Station), München (16 May 1972).

[52] DER NIEDERSÄCHSISCHE SOZIALMINISTER, Erste Teilgenehmigung zur Errichtung des Kernkraftwerkes
     Unterweser (First partial construction permit for the Unterweser Nuclear Power Station), Hannover
     (28 June 1972).

[53] SOZIALMINISTER DES LANDES SCHLESWIG-HOLSTEIN/MINISTER FÜR WIRTSCHAFT UND VERKEHR
     DES LANDES SCHLESWIG-HOLSTEIN, Vorbescheid nach §7a Atomgesetz zur Wahl des Standortes/Kern-
     kraftwerk Krümmel (Preliminary ruling under Section 7A of the Atomic Energy Act on the selection of
     the site for the Krümmel Nuclear Power Station), Kiel (7 Sept. 1972).

[54] BRAUN, W., Wahrscheinlichkeit und Auswirkung eines Flugzeugabsturzes auf die Reaktoranlage
     RWE-Biblis (Probability and consequences of an aircraft crash on the Biblis Nuclear Power Station of
     the RWE), Siemens-RT 5 - Bericht. (1969).

[55] TÜV Rheinland, Bericht IfU 2/70, Studie zur Beurteilung des Kernkraftwerk-Standortes, BASF-Mitte,
     Abschnitt Flugzeugabsturzwahrscheinlichkeit (Study on the assessment of the site for the proposed BASF
     Nuclear Power Plant, section on the probability of an aircraft crash), (1970).

[56] FRANKE, T., Wahrscheinlichkeit eines Flugzeugabsturzes auf das Kernkraftwerk Philippsburg
     (Probability of an aircraft crash on the Philippsburg Nuclear Power Station), AEG-Telefunken-E 313 -
     Bericht 1625/II, (1970).

[57] KOHLER, H.A.G., Wahrscheinlichkeit eines Flugzeugabsturzes auf das Kernkraftwerk Unterweser
     (Probability of an aircraft crash on the Unterweser Nuclear Power Station), Kraftwerk Union Bericht
     KWU/V 915 (1972).

[58] KOHLER, H.A.G., Wahrscheinlichkeit eines Flugzeugabsturzes auf das Kernkraftwerk Isar (Probability
     of an aircraft crash on the Isar Nuclear Power Station), Kraftwerk Union Bericht KWU/V 915 (1972).

[59] KOHLER, H.A.G., Wahrscheinlichkeit eines Flugzeugabsturzes auf das Kernkraftwerk Krümmel
     (Probability of an aircraft crash on the Krümmel Nuclear Power Station), Kraftwerk Union Bericht
     KWU/V 915 (1972).

[60] KOHLER, H.A.G., Wahrscheinlichkeit eines Flugzeugabsturzes auf ein Kernkraftwerk am Standort
     Grafenrheinfeld (Probability of an aircraft crash on a nuclear power station on the Grafenrheinfeld site),
     Kraftwerk Union Bericht KWU/V 915 (1972).

[61] DRITTLER, K., "Technisch-physikalische Modelle für äussere Einwirkungen und Ableitung der Last-
     annahmen", Bild 15 (Models for determination of the effects of external occurrences and of the load
     assumptions to be made in the design), 10. IRS-Fachgespräch, Köln, Oct. 1974.

[62] DAtF (Dtsch. Atomforum) Atominf. 6/74 (1974) 4.

[63] ARBEITSGEMEINSCHAFT DER LÄNDER ZUR REINHALTUNG DES RHEINS, Wärmelastplan Rhein (Thermal load schedule for the Rhine), bearbeitet von Landesstelle für Gewässerkunde und wasserwirtschaftliche Planung, Karlsruhe, (1971).

[64] DAtF (Dtsch. Atomforum) Atominf. 7/8 (1974) 6.

[65] Wärmelastplan Weser (Thermal load schedule for the Weser), Atomwirtsch. 19 (1974) 326.

[66] HOSSNER, R., Kernenergie und Umwelt (Nuclear power and the environment), Beilage zur Atomwirtsch. 17 (1972) II.

[67] HÜBSCHMANN, W., NESTER, K., Meteorologische Auswirkungen der Abwärme aus Kühltürmen (Meteorological effects of waste heat from cooling towers), Dtsch. Atomforum S-11 (Nov. 1973).

[68] NIEDERSÄCHSISCHER SOZIALMINISTER/NIEDERSÄCHSISCHER MINISTER FÜR WIRTSCHAFT UND VERKEHR, Erste Teilgenehmigung zur Errichtung des Kernkraftwerkes Stade (First partial construction permit for the Stade Nuclear Power Station), Hannover (5 July 1968).

[69] SOZIALMINISTER DES LANDES SCHLESWIG-HOLSTEIN/MINISTER FÜR WIRTSCHAFT UND VERKEHR DES LANDES SCHLESWIG-HOLSTEIN, Erste Teilgenehmigung/Kernkraftwerk Krümmel (First partial construction permit for the Krümmel Nuclear Power Station), Kiel (18 Dec. 1973).

[70] THOMASSEN, H.G., "Berücksichtigung der Schallemissionen bei der Planung und beim Bau von Kernkraftwerken" (Taking into account noise emission in the planning and construction of nuclear power plants), VGB-Konferenz Bautechnik in Wärmekraftwerken, 1973.

[71] BUNDESMINISTER DES INNERN, Technische Anleitung zum Schutz gegen Lärm (Engineering guidelines on protection against noise), TA Lärm (1968).

[72] MINISTER FÜR ARBEIT, SOZIALES UND VERTRIEBENE DES LANDES SCHLESWIG-HOLSTEIN/MINISTER FÜR WIRTSCHAFT UND VERKEHR DES LANDES SCHLESWIG-HOLSTEIN, Genehmigungsbescheid Nr. 1 für das Kernkraftwerk Brunsbüttel (First partial construction permit for the Brunsbüttel Nuclear Power Station), Kiel (2 April 1970).


## LITERATURE TO WHICH NO REFERENCE IS MADE IN THE TEXT:


[73] BLÄSSER, G., WIRTZ, K., Nukleare Grundlagen für Standort- und Gebäudewahl von Kernreaktoren (Nuclear plant building and siting criteria), Nukleonik 3 (1961) 164.

[74] ERGEN, W.K., German practices with respect to reactor siting, Nucl. Saf. 10 (1969) 377.

[75] RITTER, H.A., Einholung meteorologischer Gutachten über die Auswirkungen von Kühltürmen bei Kernkraftwerken (Meteorological expert opinions on the effects of cooling towers), Kernenergie und Umwelt, Atomwirtsch. 18 (1973) III.

[76] SAHL, W., Über die Bedeutung der Bevölkerungsdichte für die Standortwahl von Kernkraftwerken (Importance of the population density for the selection of sites for nuclear power plants), Kernenergie und Umwelt, Atomwirtsch. 17 (1972) III.

[77] BÖRNKE, F., Eingliederung des Kraftwerkes in seine Umgebung (Harmonization of power plants with their surroundings), VGB Kraftwerkstech. 53 (1973) 360.

[78] BÖRNKE, F., Gestaltung energiewirtschaftlicher Anlagen im Hinblick auf das Landschaftsbild (Architectural aspects of power plants in relation to the landscape), Brennst.-Wärme-Kraft 26 (1974) 381.

[79] SCHWIBACH, J., Strahlenschutzrichtwerte für die Genehmigung der Ableitung radioaktiver Stoffe (Radiation protection guidelines), Atomwirtsch. 17 (1972) 153, 196, 280.

[80] BRESSER, H., DICK, C., LINDACKERS, K.-H., TSCHNERNER, M., "Site selection for nuclear power plants", Containment and Siting of Nuclear Power Plants (Proc. Symp. Vienna, 1967), IAEA, Vienna (1967) 753.

[81] HANDGE, P., SCHWARZER, W., Probleme der Konzentrierung von Kernkraftwerken auf engem Raum (Problems posed by the concentration of nuclear power plants in a small area), Grundlagen für Belastungspläne, 7. IRS-Fachgespräch, Köln, 1971, IRS - T 23 (1972).

[82] KLOTTER, H.E., "Abwärme und Gewässerschutz" (Waste heat and protection of waters), Die Kernenergie und die Umwelt, (Proc. Int. Symp.) Lüttich, Belgium, Jan. 1973.

[83] SCHIKARSKI, W., Kernenergie und Umwelt (Nuclear power and the environment), At. Strom 18 (1972) 37.

[84] BÜKER, H., JANSEN, P., SASSIN, W., SCHIKARSKI, W., Kernenergie und Umwelt (Nuclear power and the environment), Studie im Auftrag des BMFT, JÜL-929-HT/WT; KFK-1366 (1973).

[85] ZÜND, H., Äussere Einwirkungen (External effects), SVA-Informationstagung über die Sicherheit von Kernkraftwerken, Zürich, Nov. 1974.

[86] FRANZEN, L.F., Zur Bedeutung hypothetischer Störfälle in Kernkraftwerken (Importance of hypothetical accidents in nuclear power plants), ibid.
[87] SÜTTERLIN, L., "Über die Gefährdung von Kernkraftwerken durch äussere Einwirkungen" (Hazards to nuclear power plants owing to external occurrences), 10. IRS-Fachgespräch, Köln, 1974.
[88] BLÜMEL, W., Standortwahl und -sicherung (Selection and securing of suitable sites), 3. Deutsches Atomrechts-Symposium, Göttingen, Oct. 1974.

## DISCUSSION

B.K. GRIMES: Could you please explain the difference between the value for the probability of military aircraft crashes quoted in Table IV and that given by Dr. Schnurer (paper SM-188/56) of $10^{-6}$ per year?

H.A.G. KOHLER: My highest value, $6.4 \times 10^{-7}$ per year, differs only by a factor of 2 from that of Dr. Schnurer. This difference is actually very small.

J.B. BURNHAM: Figure 6 indicates an exponential growth of the number of objections to nuclear plants with time. Have you analysed the reason for this growth?

H.A.G. KOHLER: The increase in the number of objections against nuclear power plant sites shown in Fig.6 reflects enhanced activity on the part of objector groups and improved organizational ability, above all in collecting signatures for stereotyped objections. The number of individual objectors has not increased in the same measure.

A. BELLIN: You referred to nuclear power plants on artificial islands. Could you comment on the economic, operational, safety and environmental characteristics of such plants?

H.A.G. KOHLER: The siting of nuclear power plants on artificial islands is being studied at present, but no data are available yet on the characteristics you mention.

W. VINCK: What is to be gained in your opinion, from comparing newly proposed sites with ones that have been accepted previously? I should think the procedure unlikely to yield practical results, because no site evaluation is meaningful without reference to the project which is to be developed on it. In other words, new technical developments and new engineering concepts would have to be taken into account. A mere comparison with older sites which left the specific project out of account would be of no particular use.

H.A.G. KOHLER: Any comparison of newly proposed sites with those accepted previously by the authorities (after evaluation in terms of the density and distribution of population in the vicinity of the site) must, as you say, be made with an eye to current practice, in engineering and technology as in other things.

H. SCHNURER (Chairman): Mr. Kohler mentioned that the authorities have so far been guided, in their acceptance of sites, by population densities at sites licensed previously. This may have been true in the past; but for the future, I have outlined a new concept (see paper SM-188/56) which is designed to indicate an optimum selection of new sites.

Another point: when considering the probability of a military aircraft crash, we should bear in mind that the pilots of such aircraft apparently use nuclear power stations to take their bearings, since the stations are visible over long distances above the landscape. This may increase the probability of a crash into such a station.

Lastly, I should add that the Technical Inspection Associations (Technische Überwachungsvereine (TÜV)) are non-profit-making private bodies which not only give expert advice to the licensing authority (e.g. on sites and safety concepts as well as on the construction and operation of plants) but also inspect the design and manufacture of components in the workshop and supervise the construction, start-up and operation of the plant on behalf of the licensing authority, to which they communicate their results in the form of advice.

R. GAUSDEN: Could you please say whether objections — apart from those made on amenity grounds — are directed against nuclear power in general or against specific types of plant at particular sites?

H.A.G. KOHLER: The overwhelming majority of the objections have been against nuclear power in general. For instance, there were 8535 objections to the Krümmel site, of which about 8450 were from the organized opponents of nuclear power and the rest from individual objectors.

J. EDWARDS: In the protection guide relating to external explosive events, I imagine that protection is foreseen mainly in the form of structural hardening of the building. But have you considered the possibility of failure of the emergency diesel power supply due to the diesels taking in combustion gases rather than air? If so, what protective measures could be taken to counter this event?

H.A.G. KOHLER: Our authorities do in fact prepare to require the protection of plant buildings against the intake of potentially explosive gases. Studies have been performed on the use of appropriate monitors with a view to shutting off the air-intake ports and thereby satisfying this requirement. The intake ports are, moreover, installed in safe positions to rule out failures of the emergency diesel power supply due to intake of combustion gases instead of air.

# SITING AND ENVIRONMENT

## (Sessions IV, V)

Chairmen:

A.J. GAUVENET (France)
G. HAKE        (Canada)

# EXPERIENCE IN ASSESSING ACCEPTABILITY OF NUCLEAR SITES SELECTED BY THE UTILITY INDUSTRY IN THE USA: 1971-1974*

P.F. GUSTAFSON
Argonne National Laboratory,
Argonne, Ill., United States of America

## Abstract

EXPERIENCE IN ASSESSING ACCEPTABILITY OF NUCLEAR SITES SELECTED BY THE UTILITY INDUSTRY IN THE USA: 1971-1974.

Considerable first-hand experience has been accumulated during the past decade in the USA on selection of suitable sites for nuclear facilities. Initially, sole consideration was given to safety aspects in site selection and approval. Since 1970, the National Environmental Policy Act has required that consideration also be given to potential adverse environmental impacts. Over 100 applications for construction permits had been or were being processed by the USAEC up to autumn 1974. Safety remains the foremost factor in site acceptability, followed by environmental considerations, and, although not a formal requirement, socio-economic issues are also considered in site evaluation. Nuclear safety guides and criteria are covered by existing regulations; hence, decisions on safety aspects of site selection are fairly straightforward. The prime safety issues are public health and safety under routine operating conditions and under all conceivable accident conditions. The constraint against siting nuclear facilities in or near heavily populated areas minimizes the problem of public safety. Siting in regions where the probability of natural events such as earthquakes, tidal waves or floods is high must be avoided. The methodology for site acceptance from an environmental standpoint is still in a developmental stage. The objective is now to reduce the probable environmental impact due to construction and operation of nuclear facilities. Reduction may be accomplished in a number of ways, the most extreme being the use of another site. Less severe changes may involve one or more of the following: primary cooling system, water intake and discharge design, radwaste and chemical waste systems, and transmission-line routing. Early site selection on a regional basis is also being considered. Such an approach will lead to a balance between the need for land and the environmental costs associated with nuclear power production and other land and environmental uses which also benefit mankind.

## Introduction

The methodology for determining the safety-related acceptability of a given site/nuclear facility combination is fairly well established, and has been formalized by appropriate federal regulations. In addition, guidelines have been prepared by the U. S. Atomic Energy Commission (AEC) regarding the specific details to be considered by the applicant in preparing his Preliminary and Final Safety Analysis Reports (PSAR's and FSAR's, respectively). These reports form the basis for an independent Safety Analysis Report (SAR) which is prepared by the AEC staff. As experience accrues with new power reactor designs and from their actual operating histories, new requirements are added to the safety analysis, or former requirements are modified. However, the basic environmentally related safety considerations have remained essentially unaltered, and new requirements relate almost exclusively to plant hardware, emergency systems, and the like.

Starting in 1970, the National Environmental Policy Act[1] required the AEC to prepare an independent assessment of the probable environmental

---

* Work sponsored by the USAEC.

[1] NEPA.

impacts arising from the construction and operation of nuclear facilities
(power plants and fuel cycle facilities).  The starting point for the AEC
assessment is a detailed Environmental Report (ER) prepared by the applicant.
In addition to assessing the probable environmental impact of the facility
as designed by the applicant and his contractors, it is incumbent upon the
AEC to examine practical alternatives to this design (including the use of
another site) which will produce less impact, and to balance the benefits
to be derived from the facility against the possible risks (costs) to the
environment.  A final judgement as to the best way to proceed is then made
upon this cost-benefit analysis.

The initial implementation of NEPA by AEC was done largely on an ad hoc
basis, and the AEC staff sought and has obtained the assistance of staff
from several of the national laboratories (Argonne, Oak Ridge  and Battelle
Northwest) in making the detailed independent assessment of environmental
impact.  A guide has been prepared to aid the applicant in developing the
Environmental Report, and guidelines have also been established for the
preparation and content of Draft Environmental Statements (DES's) and Final
Environmental Statements (FES's).  Due to the comments made on the DES's
and the issues raised at public hearings, the scope and treatment of specific
environmental parameters has varied as a function of time.  There is no
reason to believe that such evolution will not continue in a constructive
vein as operating experience grows and the results of environmental monitoring programs become available.


## Past Practice (Experience)

Prior to 1970 there was no statutory requirement that the AEC (or
anyone else for that matter) consider the environmental consequences of
nuclear facility construction and operation other than those arising from
the release of radioactivity.  Even in this regard, public health, rather
than environmental effects, was the primary concern, albeit that the two are
closely related.  When NEPA went into effect, there were 23 power reactors
in an operational state, not all of which had operating licenses, however,
and some of those that did had only a temporary license.  All told, only 12
plants were exempt from any NEPA review of environmental effects, demonstrate
or predicted.  Of the 23, none were in the 1000 MWe class, and several were
200 MWe or less.  Hence, those environmental impacts that did exist or might
arise were less than those likely to be associated with plants currently
under construction in the U S A.  This is particularly true of the potential
effects due to the discharge of waste heat which varies directly as the
generating capacity of the plant.  An exception existed in the case of
radioactive emissions due to significant improvements in radwaste management
technology over the past few years.  Some of the small, older plants actually
released appreciably larger quantities of radioactivity to both air and
water than do present plants of 10 to 20 times greater generating capacity.
In line with the 'as low as practicable' concept, these older plants are
being backfitted so that their radioactive releases are greatly reduced.
In the area of nonradiological environmental impacts from pre-NEPA nuclear
plants, Indian Point-1 in New York State bears mention [1].  Indian Point-1
utilizes once-through cooling, drawing water from  and discharging it back
to the Hudson River.  Impingement and entrainment (of fry) have been a
recurring problem at this plant, particularly as regards striped bass, which
are an important sport and commercial fish.  Remedial measures in the form
of fish screens and other alterations in the water intake structure have
reduced the impact.  However, Indian Point probably comes closest of all
the pre-NEPA sites as being one which might be unacceptable on the basis
of adverse environmental impact.

It should be stated that in addition to backfitting to meet the radiological requirements of the 'as low as practicable' concept, pre-NEPA plants
may also have to employ improved technology or otherwise alter their
facilities and operating procedures to comply with the requirements imposed
by recent federal air and water quality legislation.  Hence, any disparity
between the environmental effects of pre- and post-NEPA nuclear plants should
largely disappear in the course of time.

<u>Current Practice (Experience)</u>

Assessment of environmental impact was applied to a broad spectrum of
nuclear power plant actions in 1970, ranging all the way from those where
only a site had been chosen by the applicant to situations where a plant was
ready to load fuel and only awaited an operating license to do so.  Obviously,
the same breadth and depth of assessment could not be given to such a range
of cases.  Consideration of another site is not truly a feasible alternative
for a completed nuclear plant whose construction costs were in the neighbor-
hood of a quarter of a billion U. S. dollars.[2]  On the other hand, such
considerations were a viable option in the case of an application for a
construction permit.  This is not to say that substantial additions or
changes in proposed operating procedures to minimize environmental effects
were not imposed in the case of all plants if required, regardless of the
state of construction.

In some cases, backfitting or changes in operating procedures were
required as part of the Environmental Technical Specifications (Tech Specs)
for the operating license, and their accomplishment is mandatory on the
part of the operator.  In other instances, specific operational environmental monitoring requirements were written into the Tech Specs.  The results
of these monitoring programs then serve as a basis for requiring modifications in design or operating procedure if the environmental impact is judged
to be sufficient as to warrant such action.  In all cases at the present
time, detailed field monitoring programs are required under the construction
permit and operating license.  In addition, an applicant for a construction
permit must now have a full year of field data described and analyzed in the
Environmental Report before the ER will be accepted by the AEC.  These field
data under preconstruction conditions provide a base against which subsequent
environmental measurements may be compared to assess the degree of impact.
Implicit in the monitoring program requirement is the fact that field data
may indicate unacceptable environmental effects, the correction (minimization)
of which within a reasonable time frame is the obligation of the facility
operator.  The underlying rationale of environmental monitoring programs is to
validate the environmental impact assessment made prior to plant operation,
and to provide a basis for determining if corrective actions are needed in
order to protect the environment.  Those persons doing environmental assessment use the field results to improve the validity of future impact assessments.  Ideally this reiterative process will lead to a focus on the substantive environmental issues and the means to alleviate them, all done with
a more efficient use of time, money, and human and technical resources.

Since NEPA became effective, no sites have been rejected on the basis
of potential environmental impact being unacceptable.  There have been two
instances of sites being questioned on safety issues.  In the case of
Newbold Island, the proximity of the site to the Philadelphia metropolitan
area led to the denial of a construction permit [2].  In the case of the
Mendocino site (on the California coast), the issue of proximity to an

---

[2]  A billion US dollars = $10^9$ dollars.

active fault zone has led to suspension of the site review process, pending
further investigation of seismic conditions [3].  The Mendocino site also
has the potential for a severe local socioeconomic impact, but the seismic
question overshadows this at present.

Some rather profound alterations have been required for plants already
operating or nearing the operating stage.  The Indian Point-1 (265 MWe)
situation was discussed above; Indian Point-2 (an 873 MWe unit) was also
designed for once-through cooling.  This unit is required to backfit with a
closed-cycle cooling system by 1978, principally to protect the striped
bass which use the Hudson River as a passage to and from their spawning
areas [1].  Quad Cities 1 and 2 (809 MWe each) in Illinois were designed to
use once-through cooling, discharging back into the Mississippi River, via a
cooling canal [4].  Hydrodynamic modeling of the thermal plume indicated the
potential for thermal blocking of much of the river and a tendency for heated
water to spread into shallow spawning areas during certain river flow conditions.  To minimize this effect, the operator installed a high-velocity
diffuser system on the river bottom.  If the problem persists with this
system as shown by field studies, the operator will be required to install
spray modules in the original discharge canal.  Turkey Point Station,
located on Biscayne Bay in Florida, illustrates a situation involving a
multi-unit facility [5].  Units 1 and 2 are fossil-fueled and have operated
for several years discharging directly to Biscayne Bay.  The Bay is shallow
and subject to appreciable solar heating.  The additional thermal input from
Units 1 and 2 caused destruction of eel-grass growing in the vicinity of the
discharge.  This aquatic vegetation comprises a support system for a number
of important aquatic species in the area.  Units 3 and 4 are 693 MWe nuclear
units, the direct discharge from which would further denude the Bay of eel-
grass.  A complex cooling canal system is being built to alleviate the
problem, and, depending upon the indications from environmental studies at
the canal outfall, it may be necessary to install a helper system in the
canal itself.  The three examples cited above all involve modification of
the condenser cooling-water system to protect portions of the aquatic
ecosystem from impacts either at the intake or the discharge end of the cycle

There have also been instances where operating or near-operational
facilities have been required to install additional hold-up or clean-up
systems to reduce radioactive emissions to levels in keeping with the 'as
low as practicable' concept.  Similar reductions in the levels of biocides
discharged to the environment have been effected through the required use
of present state-of-the-art technology.

Obviously, the impact of NEPA is greatest, and the intent of NEPA is
most readily carried out, at the construction permit or site selection
stage.  Most of the present applications under review are in this stage.
The constraint against building nuclear plants in or near population
centers is factored into all siting plans.  There are also certain public
attitudes and perceptions which strongly influence the industry in site
selection.  Even before NEPA, such areas as Yellowstone Park, Yosemite  and
Big Sur were clearly out of bounds because of public opinion.  Granting that
there are lots of sparsely populated areas well removed from unique natural
preserves and historic landmarks, there is only a finite number of locations
which are simultaneously properly situated as to load centers where land is
available for purchase and where the supply of water is adequate for cooling
large steam-generating plants.  Further constraints are introduced in going
to closed-cycle cooling in that more land is needed for cooling towers or
cooling ponds, and the use of large cooling ponds requires that proper soil
and hydrological characteristics exist at the site.  The use of natural draft
cooling towers may influence how close a plant can be sited to some scenic

or historic spot because of the intrusion of 500-foot-high towers into the
field of view.  In the AEC-NEPA review, the environmental (and aesthetic)
impact of transmission lines is also considered.  This puts a whole new
dimension on both site selection and transmission-line routing from such
sites.  Many excellent sites exist in areas remote from any load center, but
there is a fairly sizeable body of public opinion that feels strongly that
one region should not bear the environmental cost for producing power to
be used in another region.  In other words, the region benefiting from the
power should also bear the environmental burden.  Since region is not
really defined, this concept can reduce to rural areas on the one hand and
urban areas on the other.  Another concept involves the continued development of a suitable site; as more power is needed in the service area in
question, more units are added at the site, forming in essence a nuclear
park  or power park as the case may be.  There are undoubtedly limitations as
to size imposed by the heat dissipation capability of the local area.

Many of the problems and partially contradictory constraints discussed
above will be resolved by detailed impact assessment, including a cost-
benefit analysis of the various trade-offs involved.  Such assessments must
then be reviewed by hearing boards, the courts, and in some cases the final
verdict rendered by public opinion.

Two examples will indicate some typical problems in site selection and
their resolution by the present system.  First is the La Salle County Nuclear
Site in Illinois, which was selected to accommodate two 1078-MWe units, with
possible future expansion to include two additional units of this general
size.  Cooling was to be done using a 4500-acre lake which is somewhat larger
than would normally be required for Units 1 and 2, but would have the beneficial feature that it would be a managed recreational resource for the area.
The recreational aspect represented an important trade-off because part of the
land involved was prime agricultural land in active production.  Farmer
opposition, coupled with the staff judgement that effective management of
a sport fishery in the pond was questionable, led to a compromise solution,
namely that the size of the pond be reduced to 2200 acres, that marginal
farm land be used for this purpose, and that a managed fishery not be
implemented -- a fairly rational solution, all things considered.  The other
example is the Bailly Generating Station, a 660-MWe unit, to be sited on
Lake Michigan in northern Indiana.  Two coal plants totaling about this
capacity are already in operation at the site using once-through cooling into
Lake Michigan.  The applicant intends to use a natural draft cooling tower
for the nuclear unit.  The site adjoins the Indiana Dunes National Lakeshore,
and there was opposition to building a nuclear plant this close to a national
recreational area.  There was also opposition to a natural draft tower on
aesthetic grounds, and on the basis that acid mist from the combination of
stack gas and plume vapor would damage vegetation in the park.  The applicant
could build a fossil unit using a cooling tower so the latter two impacts
would pertain to any steam-generating facility built on the site.  The
case was resolved in favor of granting a construction permit, the judgement
being that the benefits outweighed the environmental costs.

Future Trends

It is apparent that more specific guidelines are required for selecting
environmentally acceptable nuclear sites.  It is also apparent that the site
selection and site approval system varies from one state to another over the
U S A.  Within states the system is often cumbersome and time-consuming.  One-
stop siting in which a new state agency or an existing one is given a lead
role and the authority to decide once and for all for or against a particular
site is a useful approach.  The preselection or designation of sites as

being suited for nuclear facilities before they are actually needed is also
a useful concept.  Such an approach really amounts to long-range land use
planning even though applied only to the area of nuclear power production.

Designation of sites in a regional context would seem an appropriate
further step, and might lead to a rational way in which to develop regional
power parks.

## REFERENCES

[1]  USAEC 1973 Annual Report to Congress, Vol. 2, Regulatory Activities, p. 13.
[2]  Ibid., p. 15.
[3]  Ibid., p. 14.
[4]  USAEC 1972 Annual Report to Congress, Regulatory Activities, p. 11.
[5]  Ibid., p. 9.

## DISCUSSION

E.H. HUBERT:  Could you give us a rough estimate of the cost of an
Environmental Report, including the time its preparation requires (in
man·years) and the delays it entails in the issue of construction permits?

P.F. GUSTAFSON:  At present the applicant must spend between one
and two million dollars for the preparation of the Environmental Report
(including collection of the environmental field data).  The USAEC then
spends about US $250 000 in preparing an Environmental Impact Statement.
This sum normally includes the cost of the time spent on a public hearing.

Delays are more difficult to predict.  Strictly speaking the applicant
has to include one year's environmental data in his Environmental Report.
However, under the present system, six months' data together with their
interpretation may initially be included in the report;  USAEC then
takes seven months to prepare its final statement and hold a hearing.
Thus, for the final statement, twelve months' field data will be available
from the applicant.  Safety considerations, however, rather than environmental matters, are what usually cause the delay.

D. DAGAN:  How will the outgoing heavy transmission lines from a
power park influence the environment?

P.F. GUSTAFSON:  Transmission lines may of course create an
unwelcome aesthetic problem.  Very high tension lines (765 kV or 1000 kV)
may also produce some biological effects.  This needs further research.

D. DAGAN:  What body is responsible for planning long-range land
use?

P.F. GUSTAFSON:  Land-use planning is a complex matter in the
United States of America.  It is, consitutionally speaking, a matter for
the individual States.  But practical land-use laws (zoning laws) are a
local prerogative, and local communities do not want to give up their
authority to the states, primarily for political reasons.  Regional land-use
planning groups exist but do not have much legal or financial strength.
Education will be necessary in this area.

D. DAGAN:  At what stage should a negative decision be taken?

P.F. GUSTAFSON:  A negative decision should be taken early;  otherwise the time and money spent on a project may tend to affect the cost
benefit analysis in the direction of permitting the use of an undesirable site

W. HEINZ: If I understand you correctly, you propose to install a
power plant park with a capacity of 10 000 - 15 000 MW. This is equal to
some twelve power stations. Is this feasible, and what impact is it likely
to have on the environment?

P.F. GUSTAFSON: The power park concept involves acceptance of
the idea that the environmental impact may be relatively severe in the
region immediately surrounding the installation. However, the number of
such parks will be limited, and the land will be dedicated, in a sense, to
continued power production — old units will be replaced by new ones as
time goes on. The concept thus threatens environmental degradation in
relatively few areas, while the rest of the region remains unaffected by
power production (except for transmission lines and switch-yards). The
10 000 - 15 000 MW(e) limit on park size is based on estimates of the ability
of the atmosphere to dispose of large blocks of heat without undesirable
changes in the weather. In the United States of America, the supposition
is that parks will involve some form of closed-cycle cooling (towers or
ponds), so that waste heat will be released to the air over a limited area.
However, it is also possible to consider using large lakes and ocean (coastal)
sites with once-through cooling, thus minimizing the water-air heat transfer
problem (i.e. spreading it out over a larger area). This procedure would
probably lead to undesirable biological conditions in the vicinity of the
outfall from such large facilities.

The park concept also involves fuel fabrication and reprocessing, as
well as storage of waste on the site, at least temporarily.

L. VENKATESH: You have referred to suspension of the site review
process in respect of the Mendocino site pending further investigations of
seismic conditions. Could you please say what investigations are being
carried out and whether any results have become available yet?

P.F. GUSTAFSON: The seismic conditions at the Mendocino site are
being examined by both the U.S. Geological Survey and the applicant. The
main effort is directed to obtaining more precise data on recent activity,
and the intensity of that activity, at the site itself. The applicant is
formulating structural modifications to cope with seismic conditions. The
reports from the Geological Survey and the applicant should be issued
within the next six months or so.

J. EDWARDS: Have you any experience of a direct choice having to
be made between a fossil-fired installation and a nuclear plant at a given
site? If so, which plant won the battle and why? Did the public hearings
give any clue to which plant the public favoured, or which plant it disliked
least on grounds of environmental impact and general risk?

P.F. GUSTAFSON: A fossil-fired plant is always considered as an
alternative to a nuclear one. Analysing the choice strictly from the standpoint of environmental impact, we have never, as yet, encountered an
instance where one was clearly more desirable than the other. The influence
of waste heat, chlorine and transmission lines, and the effect on land use,
are common to both. Provided stack clean-up systems are incorporated
in fossil-fired plants for particulate material and gases, their public health
impact is not significantly greater during routine operation (as far as we
can judge now) than that of a nuclear plant. The present cost of fossil fuel,
particularly clean oil, and the availability of low-sulphur coal do, however,
favour nuclear plants, strictly on grounds of economics and fuel resource
conservation.

Since public hearings are concerned with nuclear plants only from the
legal standpoint, the fossil-fired plant is an option that tends to be favoured
by interventionists. However, in those few cases where a fossil-fired plant
has been substituted for a nuclear one, some opposition has also been
expressed to the fossil-fired plant; in other words, the dyed-in-the-wool
opposition is to increased power production as such, not just to nuclear
power, and comes from the advocates of energy conservation. There is
another group, however, which is opposed to nuclear plants only and finds
fossil-fired plants an acceptable alternative (so long as they are not too
close geographically).

# AN INTEGRATED APPROACH
# TO SITE SELECTION
# FOR NUCLEAR POWER PLANTS

E.M.A. HASSAN
Nuclear Power Division,
Atomic Energy Establishment,
Cairo,
Egypt

**Abstract**

AN INTEGRATED APPROACH TO SITE SELECTION FOR NUCLEAR POWER PLANTS.

A method of analysing and evaluating the large number of factors influencing site selection is proposed,
which can interrelate these factors and associated problems in an integrated way and at the same time establish
a technique for site evaluation. The objective is to develop an integrated programme that illustrates
the complexity and dynamic interrelationships of the various factors to develop an improved understanding of
the functions and objectives of siting nuclear power plants and would aim finally at the development of an
effective procedure and technique for site evaluation and/or comparative evaluation for making rational
site-selection decisions.

## 1.  INTRODUCTION

The problems of siting nuclear power stations and the provision of
ancillary facilities are closely related to the population pattern as well as
to the general development pattern, national economic planning, policies
concerning the location of industry, regional, metropolitan and urban
physical planning, and community and social development. This necessitates
the consideration of nuclear power programmes and siting of nuclear power
reactors in co-ordination with regional requirements and planning programmes in an integrated way so that the population demands and the quality
of the environment might both be maintained.

Site selection surveys and studies have shown that the choice of a
suitable site for a nuclear power station is determined by numerous and
varied considerations that involve compromise between conflicting requirements. In fact, a great deal of emphasis has been placed on safety and
economic factors, which are related to some extent to social factors in
making site-selection decisions. However, it is now widely recognized that
all types of nuclear power reactors raise environmental problems caused
by the interaction between the plant location and its local and wider environment. Indeed, studies and observations of the impact of nuclear plants on
the environment have revealed that it has a greater effect on the environment
than had previously been recognized. Therefore environmental protection
and control aspects should be taken into account, turning nuclear plant site
selection into a complex affair. This complexity is further increased when
considering the development patterns and long-term plans of power,
industries, housing and various urbanization schemes that will be required
for the proposed site area. These developments will affect in varying

ways the economy, social aspects and the environment, which in turn will affect the location of power plants.

In fact, site selection is so complex that the major factors should not be considered in isolation. Although each of these factors (e. g. safety, economy, etc.) can be judged independently, the final decision on site suitability involves the interrelationships and interdependence of all the factors to obtain an optimum balance between elements of the overall network. In this paper the various factors influencing site-selection decisions are analysed and an integrated programme approach is developed.

## 2.  IMPACT OF NUCLEAR ENERGY ON THE ENVIRONMENT

The choice of nuclear energy for power generation (including plant location and methods of power distribution) has major implications in terms of physical, social, economic and visual impact on people.

The impacts of nuclear reactors on the internal and external environment may arise from:

(a)  Nuclear hazards and their radiological effects; the potential sources of contamination which may lead to environmental problems may be from the gaseous, liquid and solid radioactive releases.

(b)  Thermal pollution and its possible adverse environmental effects on the ecology and aquatic community.

(c)  The interaction of plant buildings and the environment in terms of visual impact.

### 2.1.  Nuclear hazards and their effects

One of the main problems which a nuclear power station may bring to the area to be built on is the possible radiological hazards and their effects. Contamination of the environment by radioactive releases may have harmful effects on people. Chemically toxic radioactive wastes may be produced by nuclear reactors, laboratories handling radioactive material and chemical plants preparing or reprocessing nuclear fuel. However, the routine release of these radioactive materials into the atmosphere through vents or stacks may comprise gas, vapour and very small solid particles which could react with vegetation and water and endanger health if absorbed into the human system. The conditions leading to maximum fallout at distances of a few miles are quite likely to occur in practice and radioactivity may be found well beyond the boundaries of the nuclear plant site. The public may consequently be affected by both radiation from the effluent cloud and by fallout that leads to contamination of activities required for human survival. Figure 1 shows the principal means of transport and dispersion of radioactive wastes and how contamination reaches man. This poses a delicate problem concerning public relations, health and safety, even when considering the low level of radioactive emissions.

### 2.2.  Thermal pollution

The adverse environmental effects of thermal pollution, which is a result of waste heat discharged into water sources, poses a problem of gre

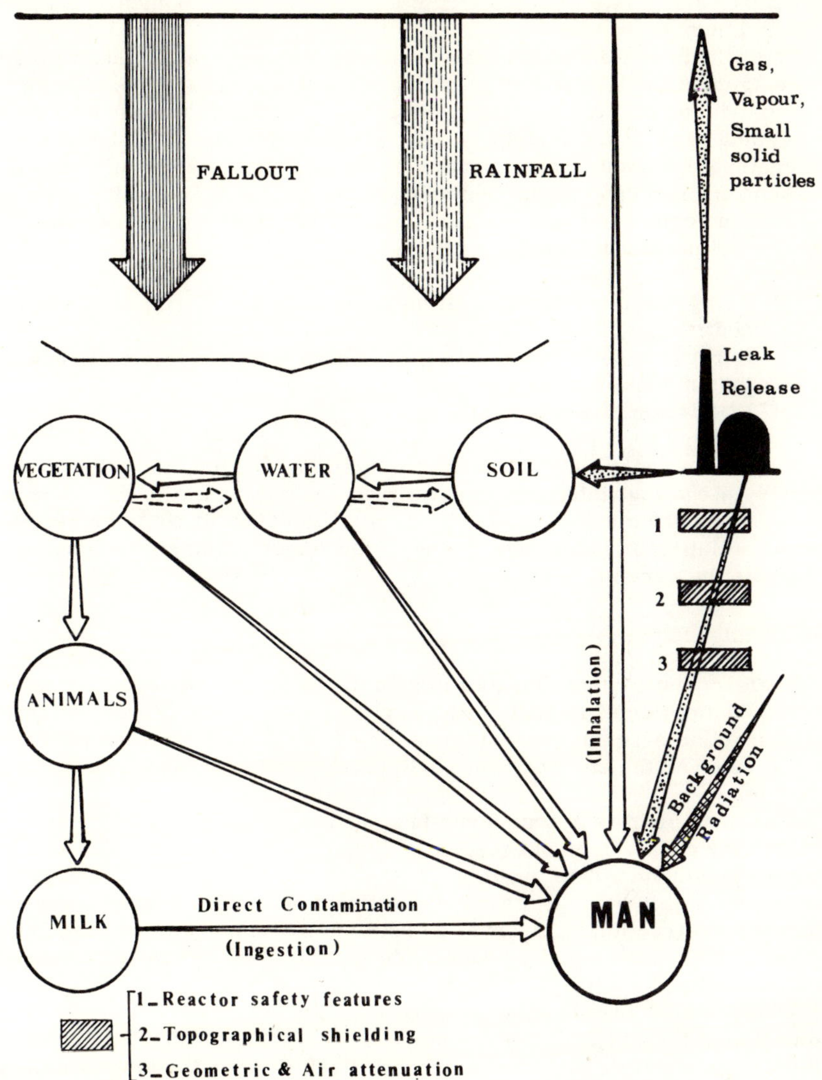


FIG.1.   The principal means of transport and dispersion of radioactive materials into man's environment; how contamination reaches man.

concern due to its possible impacts on the ecology and aquatic community. While the problem exists for both nuclear and fossil-fuelled plants, the rejected waste heat is some 30 - 40% greater in nuclear reactors because of their lower efficiency and because more of the heat is discharged to the atmosphere through the stack in fossil-fuelled plants.

In future, as the number of larger nuclear power plants increases, enormous amounts of waste heat will be produced, added to amounts dis-

charged from fossil-fuelled plants.  This would result in much greater
concentration of waste heat to be dumped into the water bodies at a greatly
increased number of sites.

The effects of waste heat may be harmful, beneficial or insignificant,
depending upon the specific site and the measures and methods used in the
plant design and operation.  The main impact of harmful waste heat on the
ecology and aquatic community is the biological damage involved when the
temperature change is large.  However, in most cases it is quite small
but when imposed upon an environment which can be near a critical point
of a certain segment of aquatic life it may affect it in various degrees.
Thermal pollution can be dealt with by:

    (a)  Using cooling towers, cooling ponds etc. for disposing of excess
        heat.
    (b)  Improving power-generating efficiency with the object of reducing
        the amount of waste heat.
    (c)  Utilization of waste heat.

It should be noted that the different systems for supplying the power
station with cooling water can affect the environment, while the environment
itself can affect the cooling water system.  Therefore it could be seen that
thermal pollution of waterways is one of the major plant-siting consideration
which must be carefully evaluated.

## 2.3.  Visual impact

Nuclear power plants located in remote areas or near urban centres
would have a powerful impact on the skyline both physically and aesthetically.
The interaction between the giant nuclear power plant, its associated transmission line and the environment could produce an adverse visual impact
on people.

A nuclear plant is a huge structure, and in terms of both size and
function may result in an unacceptable visual conflict in both local and
wider environment.  This problem is no less important than the obvious
factors of possible hazards of nuclear effluent and waste heat if unnecessary
damage or destruction to external natural environment is aimed for.  This
problem can be dealt with by proper site location, architectural treatment
and landscaping.

It can be seen that to allow uncontrolled growth of nuclear power would
lead to major threats to the environment.  The impact of nuclear energy
on the environment is such that a re-appraisal of basic thinking is called
for and the environmental protection aspects should be fully integrated with
other factors influencing site-selection decisions.

## 3.  AIMS OF AN INTEGRATED PROGRAMME

The main research activity of the last ten years in the field of siting
has been directed to the location of nuclear power plants nearer to load
centres or off-shore locations from which the power generated can be
transmitted inland.  These two approaches have important implications
in terms of planning and must be considered along the following lines:

(a) Locating nuclear power plants near big cities facilitates their integration into the public utility system of an urban development.

(b) Off-shore locations, where power can be transmitted anywhere, may affect the distribution and shape of cities.

(c) As the demands for electricity rise in proportion to its increasing potentialities and the growing number of applications, large nuclear power centres could be planned. This may affect the distribution of industry, recreational and other activities, as well as land-use plans.

(d) The environmental problems that might affect the quality of air and water, as well as aesthetic values in the region, need control through proper planning which may affect land use. Land-use control, however, requires a study of such measures in the interest of public health and safety.

The concept of an integrated approach presented in this paper, based on the above considerations, also aims to provide for:

(i) The changing pattern of world population and the need to have a decent life in a decent environment.

(ii) Regional planning for industrial, residential, recreational and educational facilities. To pursue one of these activities at the expense of the other is unconstructive and unrealistic.

(iii) Smooth and dynamic interchange between short- and long-term planning programmes.

(iv) Environmental protection and control measures to be an integral part of the site-selection process.

(v) A high benefit/cost ratio.

(vi) Development of an effective procedure and technique for making rational siting decisions.


## 4.    INTEGRATED APPROACH

### 4.1.  General

Because of the inherent complexity of the location of sites for nuclear power plants and associated planning problems, it is not an easy task to find a site which can meet and satisfy the criteria of safety, economy, constructional and operational requirements, as well as environmental protection and control. This means that sometimes one main factor such as safety or economy, or even social aspects, may be the priority factor for site-selection decision, which may depend on the criteria adopted by the authorities. However, in selecting the best location, a proper evaluation of all the factors is required, through an integrated approach, which will enable the decision-makers to study the degree of change effected by one complex factor on the others and to evaluate and select the most appropriate site.

To deal with this problem, the first step would be to examine the factors and criteria adopted for site selection of nuclear power plants in different countries. On the basis of this survey, it would be possible to extend such data to cover other aspects required for planning in terms of allocation of resources related to the site and its region, in terms of sociological, environmental and ecological predictions concerning human demands and development.  As the main aim is to obtain a radical solution to this problem, and in particular to find the relationships between such a large number of factors, the logical step is to categorize each factor with its subfactors under a main heading, i. e. under a major complex factor.  These factors are considered as follows:

  (i)  Site physical characteristics complex;
  (ii)  Engineering complex;
  (iii)  Economic complex;
  (iv)  Social complex;
  (v)  Environmental protection complex;
  (vi)  Nuclear hazards complex.

The most common theoretical approach to the location of nuclear power plants, apart from population density and distribution considerations, has been the economic approach, so as to minimize costs or maximize profits. Such an approach produces the wrong answer to long-range siting policy. The six major complexes mentioned above have therefore been discussed, since they are the main activities which are considered important to siting studies.

This approach can be accurately described in terms of the values or conditions of these six complexes and their relationships, i. e. by changing the values or conditions of one or more of them, or their relationship to each other, it may be possible to investigate the effect of these changes on the rest of the factors or the system.

As the form of the interrelationships between these factors is very complex, it was necessary to develop a framework and a conceptual model that shows the interactions of these complex major factors, which can be readily expanded to include additional factors.

## 4.2.  The conceptual model

The first representation of the conceptual model (Fig.2) shows the relationship between two factors and the other four.  For instance, diagram (a) shows the relationship between the site physical characteristics, engineering and economics.  The boxes represent the major factors and the links represent the relationship between them.  The feedback relation means that the engineering factors can be controlled or influenced by economics, while the site physical characteristics can be modified and controlled by engineering.  The process of the influence and control will continue to obtain the optimum performance output.  However, each of these major factors consists of a set of secondary factors which are limited in some way due to the objective which has to be achieved.  To establish the relationships between such a large number of factors, a graph was developed as shown in Fig. 3.  This is a simple undirected graph and is complete [1] since every two distinct nodes are joined by a link.  Assuming that each node (vertex) represents a major complex factor and that a link is a similarity of relations joining pairs of complexes, then the interrelationships between the six complex factors under consideration are shown in Fig.3.  However, in this study the approach to siting nuclear power plants

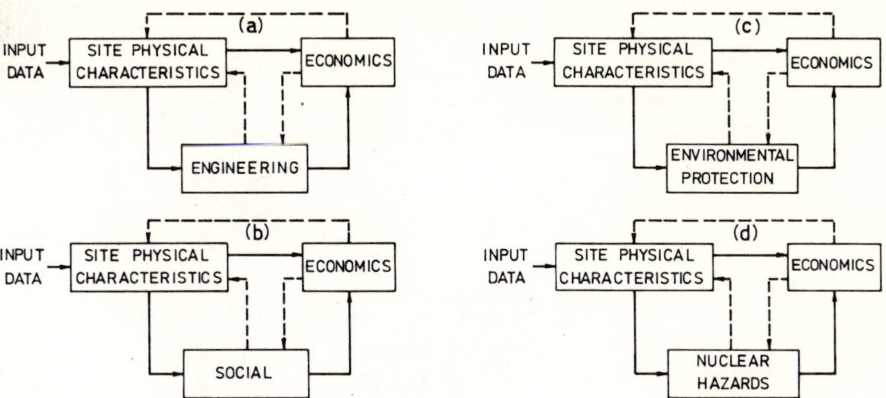



FIG.2.   Relationship between two factors and the rest of the major factors.  The boxes represent the major factors and the links represent the relationships between them.  (a) Relationship between 'site physical characteristics', 'economics' and 'engineering'.  (b) Relationship between 'site physical characteristics', 'economics' and 'social'.  (c) Relationship between 'site physical characteristics', 'economics' and 'environmental protection'.  (d) Relationship between 'site physical characteristics', 'economics' and 'nuclear hazards'.

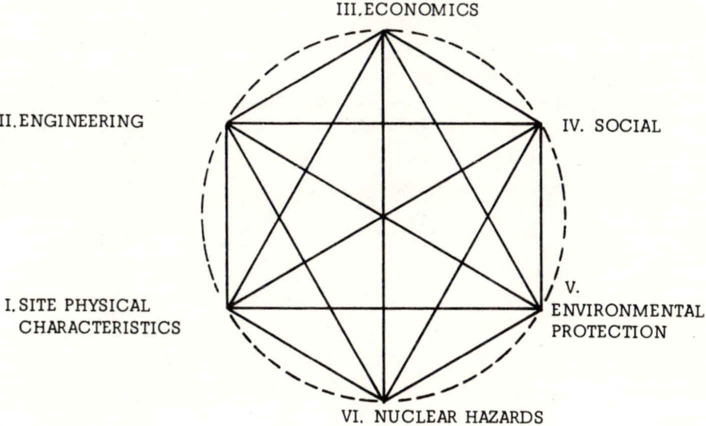



FIG.3.   Conceptual model showing relationship between the major complex factors influencing site selection for nuclear power stations.

is based on considering all the basic factors and subfactors and the inter-relationships between them.  For programming, a survey and collection of all the factors and requirements which influence site selection for nuclear power plants was carried out.   A list of these factors was compiled to establish a common check list which meets a set of criteria for safety, economic, constructional and operational requirements, social and environmental aspects.   Figure 4 shows the basic structure of the model representing the six major complexes, their subfactors and secondary factors and that there are no short cuts to site-selection decisions.

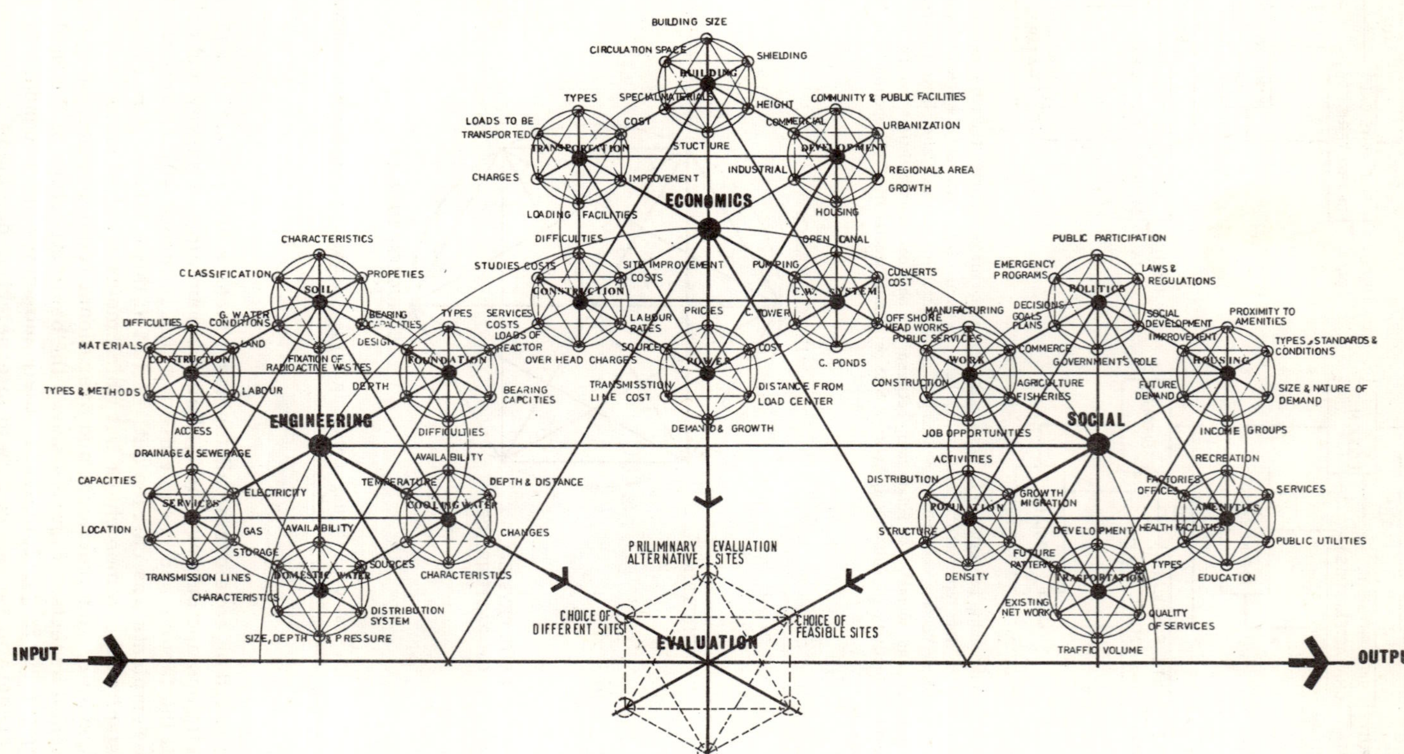
INPUT
OUTPUT
EVALUATION
PRILIMINARY ALTERNATIVE EVALUATION SITES
CHOICE OF DIFFERENT SITES
CHOICE OF FEASIBLE SITES
ECONOMICS
ENGINEERING
SOCIAL
BUILDING SIZE
CIRCULATION SPACE
SHIELDING
BUILDING
TYPES
SPECIAL MATERIALS
HEIGHT
COMMUNITY & PUBLIC FACILITIES
LOADS TO BE TRANSPORTED
COST
COMMERCIAL
URBANIZATION
TRANSPORTATION
STUCTURE
DEVELOPMENT
CHARGES
IMPROVEMENT
INDUSTRIAL
REGIONAL & AREA GROWTH
LOADING FACILITIES
HOUSING
DIFFICULTIES
OPEN CANAL
STUDIES COSTS
SITE IMPROVEMENT COSTS
PUMPING
CONVERTS COST
CONSTRUCTION
C.W. SYSTEM
SERVICES COSTS
LABOUR RATES
PRICES
C. TOWER
OFF SHORE HEAD WORKS
MANUFACTURING
LOADS OF REACTOR
SOURCE
COST
PUBLIC SERVICES
OVER HEAD CHARGES
POWER
C. PONDS
TRANSMISSTION LINE COST
DISTANCE FROM LOAD CENTER
DEMAND & GROWTH
CHARACTERISTICS
CLASSIFICATION
PROPETIES
SOIL
DIFFICULTIES
G. WATER CONDITONS
BEARING CAPACITIES
TYPES
MATERIALS
LAND
DESIGN
SERVICES COSTS
CONSTRUCTION
FIXATION OF RADIOACTIVE WASTES
FOUNDATION
TYPES & METHODS
LABOUR
DEPTH
BEARING CAPCITIES
ACCESS
ENGINEERING
DIFFICULTIES
DRAINAGE & SEWERAGE
AVAILABILITY
CAPACITIES
ELECTRICITY
TEMPERATURE
DEPTH & DISTANCE
SERVICES
COOLING WATER
LOCATION
GAS
AVAILABILITY
CHANGES
STORAGE
TRANSMISSION LINES
DOMESTIC WATER
SOURCES
CHARACTERISTICS
CHARACTERISTICS
DISTRIBUTION SYSTEM
SIZE, DEPTH & PRESSURE
PUBLIC PARTICIPATION
EMERGENCY PROGRAMS
POLITICS
LAWS & REGULATIONS
DECISIONS GOALS PLANS
SOCIAL DEVELOPMENT
PROXIMITY TO AMENITIES
IMPROVEMENT
TYPES, STANDARDS & CONDITIONS
COMMERCE
GOVERNMENT'S ROLE
HOUSING
WORK
AGRICULTURE FISHERIES
FUTURE DEMAND
SIZE & NATURE OF DEMAND
CONSTRUCTION
SOCIAL
INCOME GROUPS
JOB OPPORTUNITIES
ACTIVITIES
RECREATION
DISTRIBUTION
GROWTH MIGRATION
FACTORIES OFFICES
SERVICES
POPULATION
DEVELOPMENT
AMENITIES
HEALTH FACILITIES
PUBLIC UTILITIES
STRUCTURE
FUTURE PATTERN
TYPES
EDUCATION
DENSITY
TRANSPORTATION
EXISTING NET WORK
QUALITY OF SERVICES
TRAFFIC VOLUME

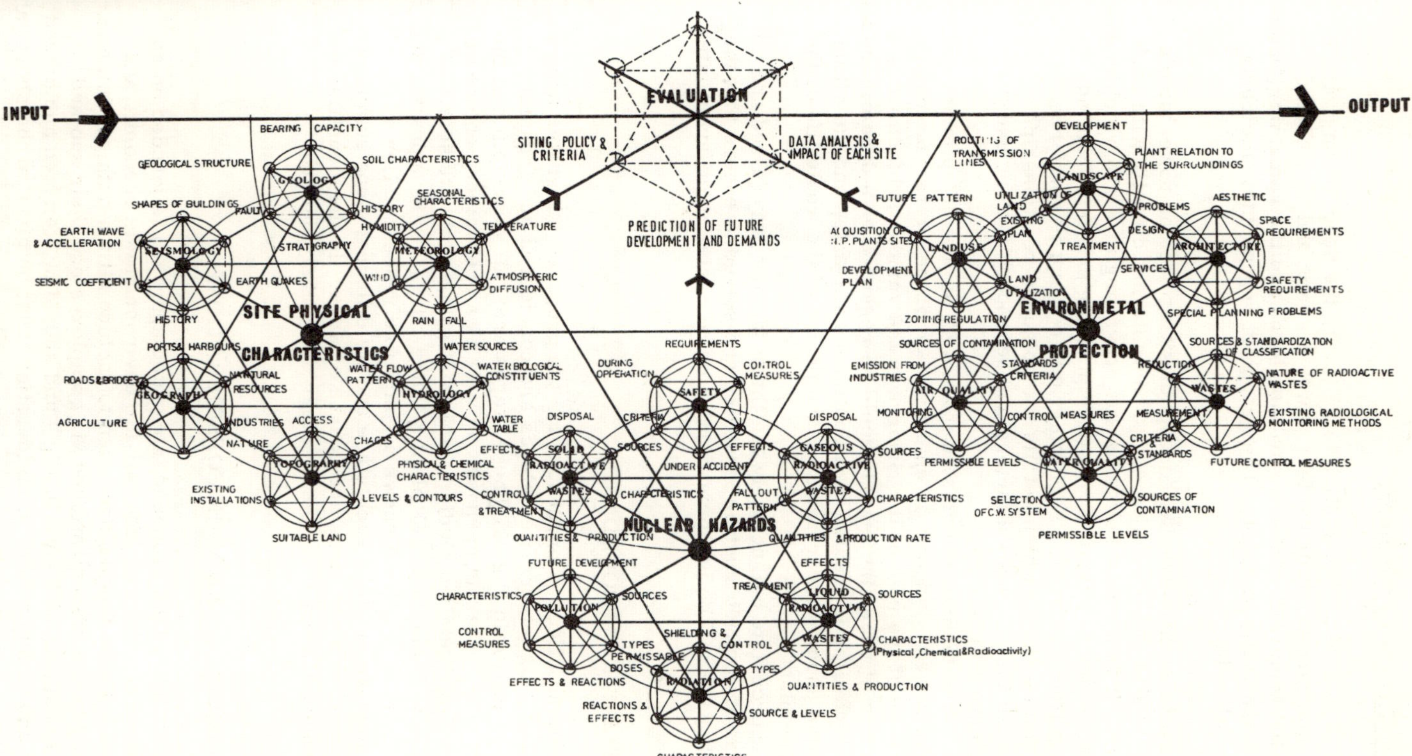


FIG.4. Factors which influence site selection for nuclear power plants, showing the six major complexes and interrelationships of factors.

### 4.3.  The significance of the model

It is the objective of this section to reiterate the need for an overall programme approach to siting nuclear power stations and to show the interactions between each of the major complexes and the remaining factors as shown schematically in Fig.4.  The graphical representation clearly indicates the following:

(a) An oversimplified series concept of siting nuclear power plants is not valid and the major complexes interact directly with each other.  It is the interrelationships prevailing in Fig. 3 which make it necessary to examine the siting planning problem from an integrated viewpoint.  This integrated concept is reinforced by examining Fig.4, which details the principal relationships between the various subfactors and secondary factors and their connecting links.

(b) The siting factors are related intimately so that an optimum result can only be obtained from an evaluation of the overall factors.

(c) Because of the role of the changes in the various subfactors as a function of time, the model should be regarded as a dynamic model in which time is treated as a continuous variable, and the siting studies should be carried over the period of the whole life of the power plant.

(d) An appropriate balance between these factors is required and the areas of conflict must be identified and solved to ensure that the evaluation will be performed on an equal basis for different sites.

## 5.    SITE SELECTION PROCEDURE

To develop an effective procedure for the proposed integrated approach to site selection, six interrelated stages are considered and organized in a framework, as shown in Fig.5.  These stages are:  (a) description of the nuclear plant type, size and purpose, as well as a preliminary choice of different sites; (b) detailed studies and investigation; (c) site selection process; (d) site evaluation; (e) plant site lay-out; (f) site appraisal and final site selection.  Figure 5 shows a simplified representation of the site selection procedure in which each stage and its different steps summarize a whole series of smaller steps and ranges of work.  In the meantime the interdependences of each stage on the other are clearly defined.

## 6.    DATA REQUIRED AND MAIN CONSIDERATIONS

The application of the proposed integrated approach is founded on an analysis of area(s), which includes all the environmental, economic and social factors.  These factors are taken to include all the data and information pertaining to the aforementioned six complexes and their sub-and secondary components as shown in Fig.4 and should be itemized and stored in a detailed and ordered way.  Consideration must be given to the short- and long-term planning programmes of the proposed area and region.

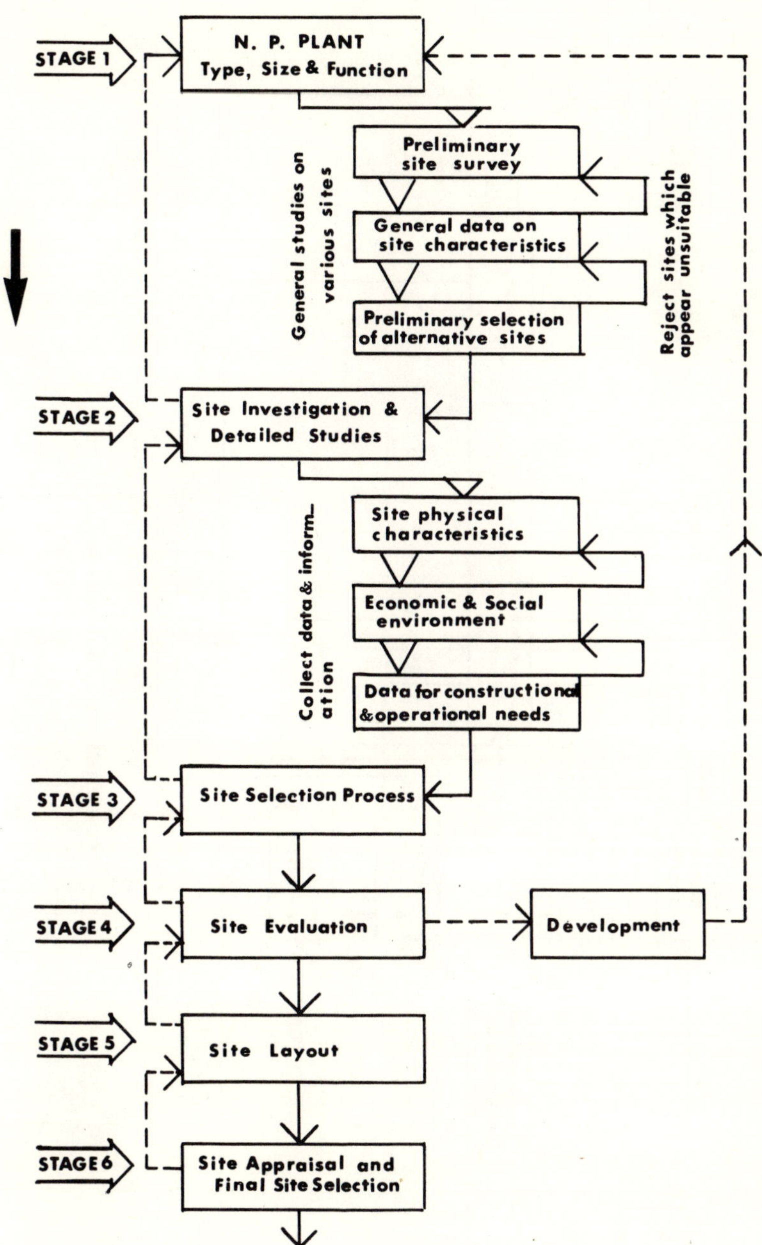


FIG.5.   Site selection procedure chart.

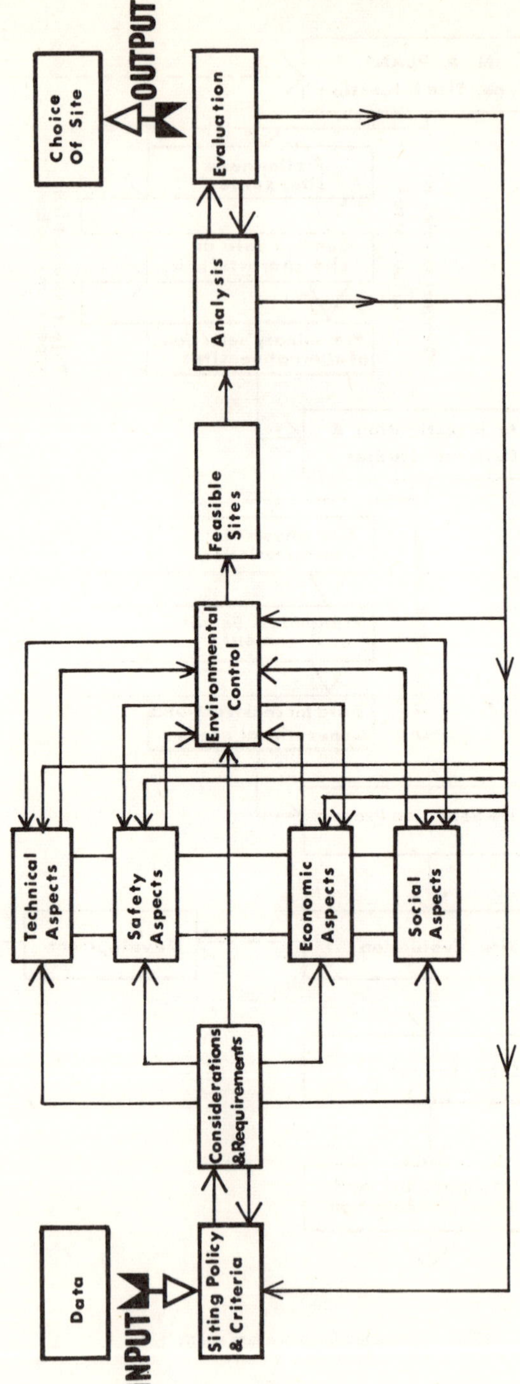


FIG. 6.   Information flow in site selection.

The flow of information and the steps involved to provide for a consistent set of decisions to be made at the various stages presented in the process diagram are shown in Fig.6, which illustrates how the technical, safety, economic, social and environmental protection aspects interact. The flow diagram starts with the input data, including all information on the six complexes, the siting policy and all criteria relating to the requirements and considerations which in turn influence the various aspects of site selection. In the processing stage the interface study team will be able to hypothesize, predict, decide, and carry out a feasibility study of alternative sites. The evaluation stage involves the combined considerations of the results of site analysis and feasibility study and weighing this against site criteria, considerations and requirements. The 'output' would be the choice of the best site. It also produces a set of design criteria for plant lay-out.


## 7.   SITE EVALUATION

An attempt was made to find a method for evaluating the great number of factors and variables as well as their interactions. This method, however, will not deal with a specific site or reactor type and size, but it can be applied to any case study on a regional basis. The factors and subfactors presented in the form of complexes and subcomplexes, considered at the time as optimum links, are taken as the basis for evaluation.

The process starts by evaluating the subfactors and their components. Each component presents a specific problem and can be evaluated separately. The next step is to evaluate the factors, taking into account the results of subfactors. Having evaluated the factors, one can then evaluate the complex. The final step is to evaluate each site being considered in relation to the evaluated 'six-complexes'. This procedure helps to consider, remember and complete every step of the process and to make the overall evaluation supported by the optimum combinations of the successive evaluation stages shown in Fig.7.

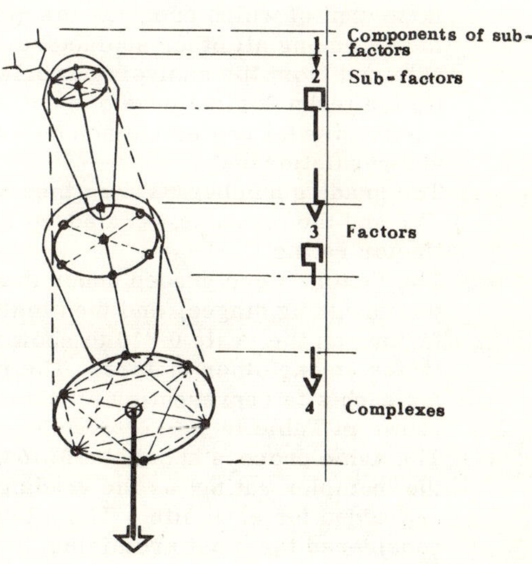


FIG.7.   Stages in evaluation.

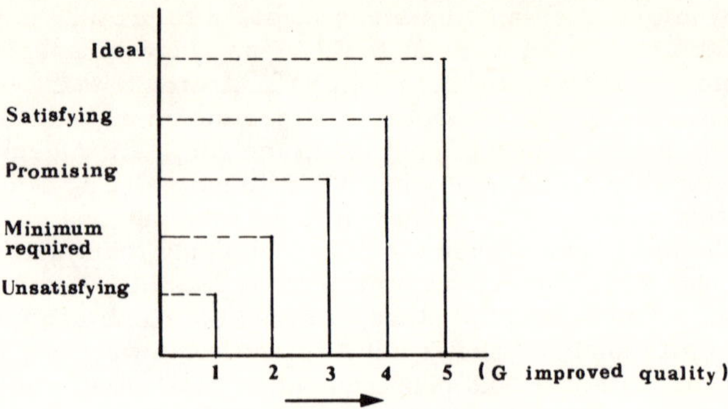



FIG.8.   Grade scale.

The technique for evaluating the possible solution given by Eder [2] and others can be adopted and used for site evaluation as follows:

(i) The subfactors were graded under five grades of quality and compliance. However, the grades should be assigned according to criteria and facts of the situation of every subfactor as a basis for comparison. The grades were considered as the value of the specific subfactor to satisfy the criteria, standards and quality. Number 1 is the lowest and can be considered as unsatisfying. Number 5 is the ideal (Fig.8). The criteria for grading can be partly objective (e.g. a good soil-bearing capacity) and partly subjective (e.g. excavation difficulties) and assessment should be carried out for all the components of the subfactors.

(ii) Each subfactor was given a weighting factor on a scale from 0 to 1 and is considered a measure of increasing importance. The weighting factor is based on a number of assumptions the most important of which are: (a) the relative importance of the sub-factor among all other subfactors (this must necessarily be obtained from the analysis of nuclear power stations in operation); (b) the technological development and future constraints, such as environmental protection and quality assurance; (c) standards and regulations.

(iii) The grading number (G) was then multiplied by the weighting factor (W) and the products were added for each site representing a 'factor rating'.

(iv) The factors were graded under their 'factor rating' resulting from the foregoing stages, and then multiplied by a new weighting factor, on the scale 0 - 10 to show the relative importance of the factor among other factors. The resulting products were added for each site representing a 'complex rating'. An example is shown in Table I.

(v) The same process should be followed for the six complexes taking the 'complex rating' as the grading number. The final products are added for each site. The site with the highest rating may be considered the most promising.

TABLE I.  EXAMPLE OF COMPARATIVE EVALUATION FOR
THREE SITE LOCATIONS

| | Factor | Weighting factor W | Site I | | Site II | | Site III | |
|---|---|---|---|---|---|---|---|---|
| | | | $G_1$ | $G_1 \times W_1$ | $G_2$ | $G_2 \times W_2$ | $G_3$ | $G_3 \times W_3$ |
| 1. | COOLING WATER | | | | | | | |
| a | Sources | 7 | 5 | 35 | 5 | 35 | 5 | 35 |
| b | Availability | 9 | 4 | 36 | 5 | 45 | 4 | 36 |
| c | Depth of water | 4 | 4 | 16 | 5 | 20 | 4 | 16 |
| d | Distance | 5 | 4 | 20 | 5 | 25 | 3 | 15 |
| e | Characteristics | 6 | 4 | 24 | 5 | 30 | 2 | 12 |
| f | Temperature | 3 | 4 | 12 | 5 | 15 | 4 | 12 |
| | $\Sigma\, G \times W$ | | | 143 | | 170 | | 126 |
| 2. | FOUNDATION | | | | | | | |
| a | Load of reactor | 7 | 4 | 28 | 4 | 28 | 4 | 28 |
| b | Bearing capacity | 8 | 3 | 24 | 5 | 40 | 5 | 40 |
| c | Depth of foundation | 5 | 3 | 15 | 4 | 20 | 4 | 20 |
| d | Difficulties | 4 | 2 | 8 | 5 | 20 | 1 | 5 |
| e | Types | 6 | 4 | 24 | 2 | 12 | 5 | 30 |
| f | Design | 10 | 4 | 40 | 3 | 30 | 5 | 50 |
| | $\Sigma\, G \times W$ | | | 139 | | 150 | | 173 |
| 3. | SOIL | | | | | | | |
| a | Classification | 7 | 3 | 21 | 4 | 28 | 4 | 28 |
| b | Characteristics | 7 | 3 | 21 | 3 | 21 | 3 | 21 |
| c | Properties | 9 | 3 | 27 | 4 | 36 | 4 | 36 |
| d | Bearing capacities | 10 | 3 | 30 | 4 | 40 | 4 | 40 |
| e | Fixation of radioactive waste | 9 | 4 | 36 | 2 | 18 | 4 | 36 |
| f | Groundwater | 6 | 2 | 12 | 1 | 6 | 4 | 24 |
| | $\Sigma\, G \times W$ | | | 147 | | 149 | | 185 |

TABLE I (cont.)

| Factor | | Weighting factor W | Site I | | Site II | | Site III | |
|---|---|---|---|---|---|---|---|---|
| | | | $G_1$ | $G_1 \times W_1$ | $G_2$ | $G_2 \times W_2$ | $G_3$ | $G_3 \times W_3$ |
| 4. | CONSTRUCTION | | | | | | | |
| a | Land | 9 | 4 | 36 | 5 | 45 | 3 | 27 |
| b | Construction materials | 6 | 3 | 18 | 2 | 12 | 4 | 24 |
| c | Labour | 7 | 3 | 21 | 1 | 7 | 4 | 28 |
| d | Access | 6 | 3 | 18 | 2 | 12 | 4 | 24 |
| e | Types and methods | 5 | 3 | 15 | 2 | 10 | 4 | 20 |
| f | Difficulties | 4 | 2 | 8 | 1 | 4 | 3 | 12 |
| | $\Sigma G \times W$ | | | 116 | | 90 | | 135 |
| 5. | SERVICES | | | | | | | |
| a | Electricity | 7 | 2 | 14 | 1 | 7 | 3 | 21 |
| b | Gas | 5 | 1 | 5 | 1 | 5 | 3 | 15 |
| c | Drainage and sewerage | 5 | 3 | 15 | 4 | 20 | 3 | 15 |
| d | Location of services | 8 | 2 | 16 | 1 | 8 | 4 | 32 |
| e | Capacities | 7 | 4 | 28 | 4 | 28 | 4 | 28 |
| f | Transmission lines | 9 | 1 | 18 | 1 | 9 | 3 | 27 |
| | $\Sigma G \times W$ | | | 96 | | 77 | | 138 |
| 6. | DOMESTIC WATER | | | | | | | |
| a | Availability | 9 | 1 | 9 | 1 | 9 | 3 | 27 |
| b | Characteristics | 5 | 3 | 15 | 3 | 15 | 3 | 15 |
| c | Size, depth and pressure of mains | 8 | 2 | 16 | 1 | 8 | 4 | 32 |
| d | Consumption | 7 | 4 | 28 | 4 | 28 | 3 | 21 |
| e | Distribution systems | 6 | 2 | 16 | 1 | 6 | 3 | 18 |
| f | Sources and location | 7 | 2 | 14 | 1 | 7 | 4 | 28 |
| | $\Sigma G \times W$ | | | 94 | | 73 | | 141 |

TABLE I.  (cont.)

| COMPLEX | | Weighting factor | Site I | | Site II | | Site III | |
|---|---|---|---|---|---|---|---|---|
| Engineering | | W | G | G × W | G | G × W | G | G × W |
| 1 | Cooling water | 10 | 143 | 1430 | 170 | 1700 | 126 | 1260 |
| 2 | Foundation | 8 | 139 | 1112 | 150 | 1200 | 172 | 1373 |
| 3 | Soil | 6 | 147 | 882 | 149 | 894 | 185 | 1110 |
| 4 | Construction | 7 | 116 | 812 | 90 | 630 | 135 | 945 |
| 5 | Services | 5 | 96 | 480 | 77 | 385 | 138 | 690 |
| 6 | Domestic water | 3 | 94 | 282 | 73 | 219 | 141 | 423 |
| $\Sigma$ G × W | | | | 4998 | | 5028 | | 5801 |

G = grading number          W = weighting factor

It should be noted that the totals which emerge have no absolute
significance; what is aimed for is the relative order of site rating.


## 8.   CONCLUSIONS

In this paper an attempt has been made to show that there is a dynamic
relationship between the factors influencing site selection for nuclear power
plants, and that decision-making models depend not only on safety and
economic aspects but also on a proper understanding of the impact of nuclear
energy on the environment.

While the study of the early methods of site selection does not prove
that a balanced output of all the factors involved is necessarily guaranteed
(e.g. the economic factor of remote areas); yet study of the integrated
approach reveals that the possibility of an optimum output is always
achievable.  A further deficiency of earlier site selection studies is the
failure to assemble the wide number of the factors and their interrelationships in a model, which is necessary since a good site can only be chosen
by including all the subfactors into an integrated model.

The advantages of this integrated approach are numerous since the
model can be expressed in mathematical form, containing the subfactors,
the factors and the complexes, which can be handled by a computer.  This
is essential because of the vast quantities of data involved, their complex
relationships and the fact that most of them are dependent in their solutions
on other factors.

This integrated approach can only be carried out by an interface study
team.  In this way the effort of all concerned can be co-ordinated and
problems of information and optimization can be solved independently and

in an integrated way.  The advantages of computer techniques in planning
can then be well employed.

It must be re-emphasized that the integrated approach presented here
is only a step along the difficult path to the final development of a planning
programme.  It is not an ultimate solution but its further development will
certainly yield much insight into the solution of nuclear power plant siting
problems.

## ACKNOWLEDGEMENTS

I wish to thank Professor F.N. Morcos-Asaad, Professor of Architecture
University of Strathclyde, Glasgow, and Dr. K.E.A. Effat, Deputy Director,
Atomic Energy Establishment, Cairo, for their help and encouragement
in the preparation of this paper.

## REFERENCES

[1] FILLENBAUM, S., RAPOPORT, A., Structures in the Subjective Lexicon, Academic Press, New York
    (1971) 15.
[2] EDER, W.E., "Definitions and methodologies", The Design Method (GREGORY, S.A., Ed.), Butterworth,
London (1966).

## DISCUSSION

J.J. DiNUNNO:  The integrated approach you have described is very
similar to the method used in the two case studies which I discussed in
paper SM-188/42.  Have you applied your model to actual case studies
in your country?

E.M. HASSAN:  Yes, it has been applied to a case study in Egypt
involving comparative evaluation of three sites — but only for one complex
(engineering), as shown in Table I.

A.J. GAUVENET (Chairman):  Have you applied your method to
conventional power plants?

E.M.A. HASSAN:  No, the integrated approach described here was
developed specifically for the siting of nuclear power plants.  Nevertheless,
it could be applied to conventional plants if one replaced the nuclear hazards
complex by a new complex of possible hazards specific to conventional
power plants.

K.B. JANSSON:  It is not the responsibility of the power industry to
cover all the fields of knowledge which you have dealt with in your paper.
I agree with you that all these things must be considered at some stage
by someone; but this raises an important question.  Planning for nuclear
power stations requires effective collaboration between those responsible
for various planning activities in society.  How should the responsibility
for nuclear plant siting be divided between different sectors of society?
In other words, what is the role of the power industry, and what should be
left to other sectors?

E.M.A. HASSAN:  Obviously, such programmes must be carried out
by the nuclear energy authorities, the power industry and other planning
authorities which can co-ordinate all efforts towards the common goal of
supplying energy in a way which is consistent with maintaining the quality

of the environment.   This clearly means close co-operation between governmental agencies,  the power industry and the public bodies concerned with
the environment.   They must be encouraged to work together with a view
to applying this programme,  establishing new interagency agreements as
required.

H. SCHNURER:   In Table I you have indicated several site factors as
grade 1,  which corresponds to unsatisfactory site properties.   Does this
mean that you inevitably have to renounce such a site,  or would you allow
balancing of such unsatisfactory factors by others that are satisfactory?

E.M.A. HASSAN:   The objective of this approach is to obtain an optimum
balance between the numerous factors affecting site selection.   The site
would therefore be selected according to the overall evaluation;   the site
with the highest total rating would be chosen,  possibly in spite of having
certain less than wholly satisfactory features.

J.B. BURNHAM:   How did you assign weighting?

E.M.A. HASSAN:   The various factors were programmed for analysis
by computer.   Since a large number of standard programmes are available,
a computer programme was used to find the optimum relationships between
32 factors,  to assign distinct priorities among these factors and to determine their relative importance (on the scale 0 - 10).

J. XIMENEZ DE EMBUN RAMONELL:   Since the programme you have
outlined can be repeated or reviewed,  could you please indicate what the
territorial scope of the programme is,  and whether there are different
regions with different weighting factors or whether the same programme
holds good for the whole of a country?   Second,  is there provision for
review of these factors?   If so,  at what intervals?   Third,  have sensitivity
studies been performed to find out how the results would vary if the starting
hypotheses (factors or weights) were not fully accurate?

E.M.A. HASSAN:   To your question about territorial scope,  I can reply
that the approach has been applied to a case study on a regional basis.

As regards your second question,  I have mentioned in the paper (see
Section 4.3(c)) that the model should be regarded as a dynamic model; hence
it will be necessary to re-collect and update the files periodically.

Last,  the information flow diagram (Fig.6) shows the site selection
process and the steps involved in providing for a consistent set of decisions
to be made at the various stages,  whereas the feedback loops act to affect
the site selection process.

L. VENKATESH:   In the technique used for relative evaluation of sites,
by and large,  only the site factors seem to have been considered.   Have
you not considered the relative economics of power distribution from a
station at alternative sites to load centres which may develop by the time
the station could be expected to come on line?

E.M.A. HASSAN:   The economic complex in Fig.4 includes a 'power
factor',  which consists of several subfactors: energy demand and future
growth,  site proximity to load centre,  the energy supply sector,  electricity
prices and availability,  integration of nuclear power with existing power
systems, and the cost of transmission lines to the existing grid.   Consideration has to be given to all these subfactors if one is to have a thorough
comparative evaluation.

# RADIOACTIVE EFFLUENT RELEASES
# AND THE PUBLIC ACCEPTANCE
# OF NUCLEAR FACILITY SITES*

A.P. HULL
Health Physics and Safety Division,
Brookhaven National Laboratory,
Upton, N.Y.,
United States of America

**Abstract**

RADIOACTIVE EFFLUENT RELEASES AND THE PUBLIC ACCEPTANCE OF NUCLEAR FACILITY SITES.

A public controversy about the risks from radioactivity in effluents from nuclear power plants in the United States of America arose in the late 1960s, as their utilization was growing towards large-scale commercial basis. Several scientific critics alleged variously that the existing plants had occasioned excess infant mortality in their vicinities and that the growth of nuclear power would produce large increases in the cancer death rate in the general population. The controversy caused by these allegations led to the Biological Effects of Ionizing Radiation Committee's review of the effects of exposure to low levels of ionizing radiation. It also appeared to underlie the USAEC's 'Appendix I' proposals for numerical design limits for nuclear power plant effluents. These proposals are intended to limit the dose of any nearby individual to 5 mR/a. This critical review indicates that the critics' allegations were either without substance or irrelevant. Using effluent release data from recent years, the population health risk from power-reactor effluent radiation exposures appears to be far smaller than that from fossil-fuelled plant effluents, as well as within the range of those otherwise considered as negligible. Such comparisons are suggested as more hopeful towards achieving public acceptance than 'as low as practicable' measures, which serve to exaggerate the risks of radiation.

# INTRODUCTION

Until the late 1950's, only a few large nuclear facilities existed in the United States. Most were governmentally operated or funded, and many had intentionally been located at isolated sites, some under the initial cover of wartime secrecy. Thus, their establishment had occasioned little public reaction. Their routine radioactive effluents were generally accepted as relatively innocuous by the small nearby populations, and otherwise seemed of little public concern.

During the decade after the startup in 1957 of the first commercial U.S. nuclear power plant (the Shippingport Atomic Power Station), some eleven additional nuclear power reactors and one fuel reprocessing plant were built and put into operation. The prospective risk occasioned by their radioactive effluents was not an apparent public issue in their siting, even though many were within 50 km of large population centers, in contrast to their generally more remote predecessors.

Late in the 1960's, the era of general public acceptance of or indifference to the siting of nuclear facilities came to an end. Almost without exception, from then on, applications to the U.S. Atomic Energy Commission

---

* Research carried out at Brookhaven National Laboratory under contract with the USAEC.

for reactor construction permits or operating licenses were strongly con-
tested.  In the early phase of this opposition, their routine radioactive
effluents were frequently adduced as a grave and unacceptable risk to the
surrounding populations.

This opposition was seemingly generated by a combination of influences.
These included a growing public disenchantment with large impersonal governmental and private institutions, including those which seemed to be promoting nuclear power, at a time when the quality of the environment was becoming a matter of public concern, especially to some of the more idealistic
elements within the U.S.  This occurred concurrently with the first widely
disseminated warnings of the limits of growth, causing much of this environmental concern to focus on energy and its impact, and especially on nuclear
power reactors and related nuclear facilities.

That all exposure to radiation has some risk has been widely perceived
by the U.S. public as an established fact, rather than as a conservative
assumption for setting radiation protection standards.  This was particularly evident during the public controversy on the effects of fallout from
the atmospheric testing of nuclear weapons.  Armed with this unassailable
'fact', and filled with moral zeal, most of the environmental movement in
the U.S. seems to have uncritically assumed that being pro-environment
meant being anti-nuclear.

Thus, much of the opposition by U.S. environmentalist groups during
the late 1960's and early 1970's to the siting of nuclear power reactors
focused on their radioactive effluents.  It was alleged that these would
constitute a significant and unacceptable public health risk.  Seeming
evidence to buttress these contentions was forthcoming from several
scientist-critics, who appeared at public information meetings, at reactor
licensing proceedings, and even at congressional hearings, to testify in
opposition to citing the hazards of low-level radiation.

Principal among these scientist-critics were Drs. John W. Gofman and
Arthur R. Tamplin, who were at that time involved at the AEC's Lawrence
Livermore Laboratory in the assessment of the biological risks of the
Plowshare (peaceful uses of nuclear devices) program.  In their initial
paper[1] on this topic, within the context of the 'burgeoning program for
the use of nuclear power for electricity and other uses', they expressed
reservations about the adequacy of the Federal Radiation Council (FRC)
Guide of 170 mrem/yr for exposure of the general public.  Using the concept of 100 rads as a doubling-dose, they calculated that should the entire U.S. population be exposed at this rate from birth to age 30, some
16,000 'extra' cases of cancer per year would be induced.  During the next
two years, in a succession of papers[2] and a book[3] critical of nuclear
power, they extended their arguments and increased their estimates to as
many as 104,000 'extra' cancer deaths per year.

Another of the more prominent scientist-critics was Dr. Ernest
Sternglass, a radiation physicist at the University of Pittsburgh.  In
1970, he extended his previous much publicized allegations connecting fallout with 'excess' infant mortality, by claiming[4] that the radioactive
effluents from the Dresden I nuclear power reactor were related to excess
infant mortality in nearby counties in mid-Illinois.  In many subsequent
public appearances as a scientific expert on the effects of low-level radiation, he presented studies which made similar claims of 'excess' infant
mortality in the vicinity of most of the commercial nuclear facilities in
the U.S., as well as a number of small university research reactors[5].

Although their arguments were strongly disputed by the knowledgeable scientific community[6,7], these three scientist-critics had a large public impact. It seems more than coincidental that early in 1970, the FRC undertook a review of the scientific information on the risks associated with low levels of radiation, which led to the Biological Effects of Ionizing Radiation (BEIR) Committee report[8]. Concurrently, the United Nations Scientific Committee on the Effects of Atomic Radiation (UNSCEAR) reviewed much of the same information in its 1972 report[9] to the U.N.

Additionally, late in 1970, the USAEC formally adopted an 'as low as practicable' provision into its regulations governing reactor effluents (10 CFR 20 and 10 CFR 50). This was followed up in June 1971 with proposed numerical guides for design objectives and limiting conditions for operation to meet the criterion 'as low as practicable' for radioactive material in light-water-cooled nuclear power reactors, the 'Appendix I' amendment[10] to 10 CFR 50. This guidance was based on a nearby individual dose limit of 5 mrem/yr, or 1% of the previously applicable limit (10 CFR 20). A site boundary exposure limit of 10 mrem/yr from airborne radioactive effluents was adjudged consistent with this dose limit. The Commission noted that as a result, total population exposure would be less than about 400 person·rems/yr per 1,000 MW(e), installed capacity. Following a protracted period of consideration, the Regulatory Staff of the Commission has more recently proposed modifications[11], including an increase to 15 mrem/yr for the calculated dose to the thyroid and 30 mrem/yr for beta exposure of the skin. The essential details of the current proposal are contained in Table I.

These 'as low as practicable' proposals seem to have diminished the radiation controversy insofar as it was related to routine effluents. However, many of the critics have shifted the focus of their opposition to other issues, such as the adequacy of emergency core-cooling systems and of provisions for the storage of radioactive waste, the environmental risk of plutonium and the threat of nuclear sabotage.

Such acceptance for nuclear facility siting as may have been gained appears to be at a significant cost for additional effluent control features, as much as $1,000,000 per year per reactor and up to thousands of dollars per person·rem (or thyroid·rem) of calculated dose reduction[12,13]. Therefore, it seems appropriate to examine both the factual basis of the allegations of these scientist-critics, as well as the cost-effectiveness of the proposed Appendix I implementation of the long-standing 'as low as practicable' recommendations by standards-setting bodies[14].

S C I E N T I S T - C R I T I C   A L L E G A T I O N S

The Gofman-Tamplin inference that nuclear facility effluents would lead to a general U.S. population exposure of 170 mrem/yr is contradicted by the evidence to date. Knox[15] has shown if an exposure limit of 500 mrem/yr due to airborne gaseous effluents were reached at a reactor site boundary, then the average exposure to the population within 100 km would be about 1 mrem/yr. In fact, the average site boundary exposure at U.S. power reactors due to airborne effluents has been less than 1/100 of the above exposure limit[16]. No other routes of comparable population dose have been revealed in detailed surveillance studies at typical facilities[17-20].

Additionally, the BEIR Committee specifically examined the dose-effect estimates postulated by Gofman-Tamplin (Ref. 8, p. 183-188). It concluded that they had overestimated the relative risk of tumor induction following

TABLE I
Numerical Guides for Design Objectives and Limiting Conditions
for Operation to Meet the Criterion 'As Low As Practicable'
for Radioactive Material in Light-Water-Cooled Nuclear
Power Reactor Effluents†

| Liquid Effluents | Annual Quantity | Annual Dose Limit (mrems) | Remarks |
|---|---|---|---|
| (1) All | Calculated (per site) | 5 | |
| (2) Except HTO and Dissolved Gases | $\leq$ 5 Ci per reactor | | May be exceeded if 'base-line' control installed and provision (1) met. |

| Gaseous Effluents | | | |
|---|---|---|---|
| (1) All* | Actual | 10 | Calculated $\gamma$ air dose. |
| (2) All* | Actual | 20 | Calculated $\beta$ air dose. |

*May be reduced or exceeded so that total quantity will not
  result in annual doses $\leq$ 5 mrem/yr to body or 15 mrem/yr to skin.

| Radioactive Iodine & Particulate Radioactive Material | | | |
|---|---|---|---|
| (1) All** | Calculated (per site) | 15 (any organ) | Portion due to food may be evaluated on actual pathways. |
| (2) $^{131}$I** | $\leq$ 5 Ci per reactor | | |

**May be exceeded, provided 4x provision (1) not exceeded,
  if 'baseline' in plant control measures installed.

† As proposed in 'Concluding Position of the Regulatory Staff',
  Docket RM-50-2 (2/20/74).


irradiation by factors of 4-10 (depending on age) and had made the unreason-
able assumption of a life-long plateau (of additional cancer incidence)
following in-utero irradiation.  BEIR estimated that an additional exposure
of the U.S. population to 5 rem per 30 years (170 mrem/yr) would lead to
from 3,000 to 15,000 cancer deaths annually, and considered approximately
6,000 annually to be the most likely estimate.  For a calculated average
power reactor effluent related dose of about 5 x $10^{-2}$ mrem/person to a
nearby population of 2,000,000 persons[15] the BEIR dose-effect estimate
leads to the anticipation of 0.02 cancer deaths annually (which may be
viewed in the context of the current U.S. cancer mortality rate, about
325,000 per year).

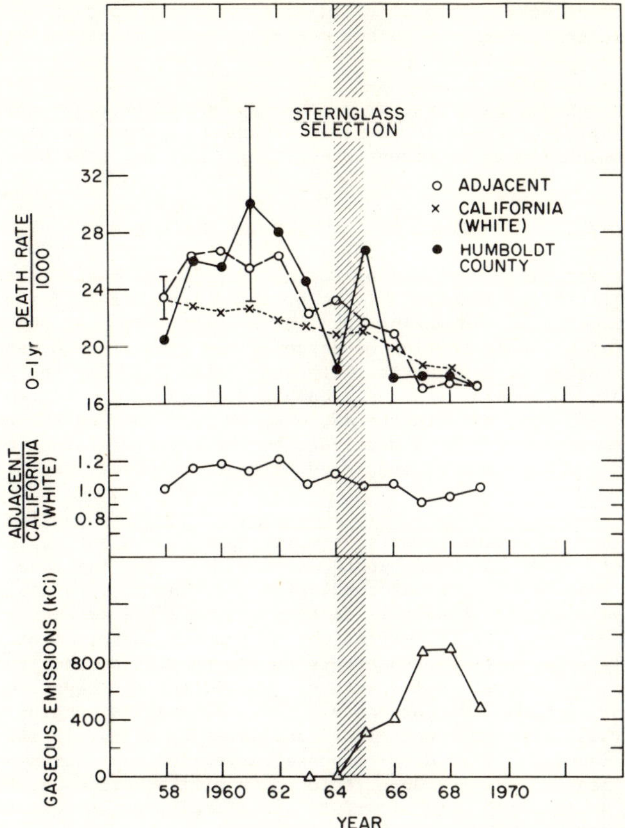


FIG. 1. Infant mortality trends and gaseous emissions in the vicinity of the Humbolt Bay Boiling-Water Reactor (Ref. [5]).

Sternglass has employed time-series correlations of increases of effluent releases with increases in infant mortality rates in surrounding regions, as the principal evidence for a relationship between them. Perhaps the most comprehensive examination to date of these allegations has been made by Hull and Shore[5] who find them unconvincing. They indicate that these correlations appear throughout to be selected instances of statistical fluctuations in infant mortality rates. When examined over a longer time span, the trends in these rates showed as many or more instances of negative correlations. A comparison of the Sternglass selection with the longer-term trends of effluents and infant mortality in the vicinity of the Humbolt Bay Boiling Water Reactor (which has released the largest amounts of airborne gaseous effluents of any in the U.S.) is shown in Figure I.

Additionally, the authors comment that when reported dose-effect relationships are considered, nuclear facility effluents cannot reasonably be held accountable for the indicated differences in infant mortality rates. They also note that Sternglass has not established a plausible biological

basis for such effects.  They also indicate other significant medical and
biological factors related to infant mortality rates which Sternglass has
not considered.

Thus, it appears that insofar as the Appendix I proposal was evoked
by these scientist-critics, it seems unwarranted by the factual strength
of their arguments--however attention-getting these may have been.

## C O S T - B E N E F I T   A S S E S S M E N T

The risk estimates utilized by the AEC Regulatory Staff in assessing
the cost-benefit effectiveness  of the proposed Appendix I limits are those
of the BEIR Committee.  In making them, BEIR admittedly followed the current
practice in radiation protection of assuming the linear no-threshold dose-
effect relationship in conjunction with the effect data for the exposures of
human beings to many rads of X- or A-bomb radiations.  The cost-effectiveness
of the Appendix I proposals is therefore quite dependent on the extent to
which the linear assumption is pertinent to the actual relationship at the
low doses and low dose-rates occasioned by routine nuclear facility effluents.

In a number of recent reports, a multi-factor or event process in the
development of cancer has been hypothesized[21-26].  The indicated models are
conflicting, some leading to the conclusion that the uniform application of
linear extrapolation for available data to estimate the effects of the low-
level, low-dose rate, low LET radiations of concern in this instance may be
insufficiently conservative, and others to the conclusion that such extrapolations may be excessively conservative.  The latter is suggested by some
recently reported human exposure data[27-29].  Basing its conclusions principally on available animal data (not considered by BEIR), in its 1972 report UNSCEAR concluded that dose-response curves for most human neoplasms
will be nonlinear in the low dose region.  On similar evidence, UNSCEAR
specifically anticipated with regard to genetic effects  a reduction of
one-third or more for chronic low-level radiations.

In their analysis of the Gofman-Tamplin estimates, Frigerio, et al.[27]
found an apparent inverse correlation between U.S. state-by-state background radiation levels (~100-200 mrem/yr) and their individual malignancy
rates between 1950 and 1967, as shown in Figure 2.  Additional data from
this analysis are contained in Table II.  They are grouped for comparison

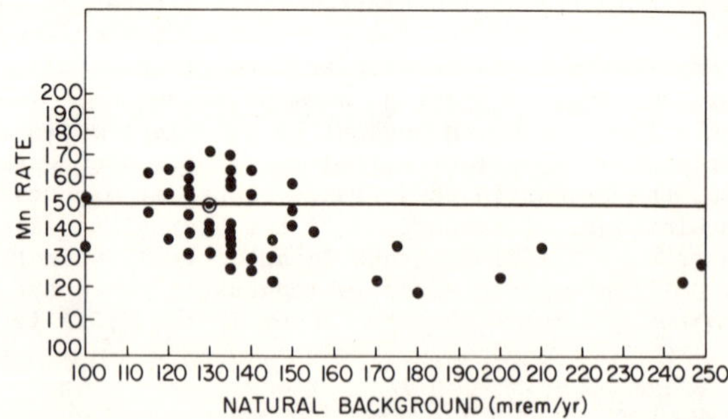


FIG. 2.  Malignant mortality rates (Mn) for the white population of the USA, 1950-1967, by state and natural
background.  The horizontal line and open circle indicate the rate and background for the USA as a whole.
(From Ref. [27].)

TABLE II

U. S. Low and High Background White Populations, 1950-1967

| No. | Characteristic | A | B | U.S. | C |
|---|---|---|---|---|---|
| 1 | Natural background, mrem/yr | 210 | 170 | 130 | 118 |
| 2 | White population, thousands | 5,735 | 16,897 | 158,051 | 59,683 |
| 3 | r, Mn[a] 140-159[b] (Alimentary system) | 42.9 | 45.6 | 52.4 | 50.3 |
| 4 | r, Mn 160-164 (Respiratory system) | 15.8 | 16.9 | 22.3 | 23.4 |
| 5 | r, Mn 170-181 (Reproductive system) | 36.8 | 38.2 | 41.5 | 40.1 |
| 6 | r, Mn 190-205 (Nervous, Lymphatic and Circulatory systems) | 30.8 | 31.5 | 33.3 | 33.0 |
| 7 | r, Stomach, 151 | 11.7 | 11.6 | 11.8 | 11.0 |
| 8 | r, All G.I., 150-159 | 40.7 | 43.0 | 49.0 | 46.7 |
| 9 | r, Lung, 163-164 | 14.5 | 15.5 | 20.4 | 21.5 |
| 10 | r, Breast, female, 170 | 21.5 | 22.6 | 25.3 | 24.4 |
| 11 | r, Thyroid, 194 | 0.055 | 0.054 | 0.057 | 0.054 |
| 12 | r, Bone, 196 | 0.92 | 1.03 | 1.12 | 1.07 |
| 13 | r, Leukemia, 204 | 7.03 | 7.23 | 7.13 | 6.91 |
| 14 | r, All malignancies | 126.3 | 132.2 | 149.5 | 146.8 |
| 15 | Residence altitude, ft. | 4510 | 2650 | 900 | 730 |
| 16 | Urbanization, % | 63 | 57 | 69 | 74 |
| 17 | Per capita personal income, $ | 2021 | 1922 | 2215 | 2255 |
| 18 | Physicians/1000 population | 1.27 | 1.25 | 1.49 | 1.49 |
| 19 | Hospital beds/1000 population | 8.24 | 8.82 | 9.49 | 8.76 |
| 20 | Poor diet households, % | 16.5 | 21.2 | 19.1 | 19.1 |
| 21 | Life expectancy, male | 67.7 | 67.7 | 67.6 | 67.5 |
| 22 | Life expectancy, female | 74.5 | 74.7 | 74.2 | 74.3 |
| 23 | Urban air, particulates, $\mu g/m^3$ | 129 | 119 | 115 | 116 |
| 24 | Urban air, benzene soluble, $\mu g/m^3$ | 9.9 | 8.8 | 9.5 | 9.6 |
| 25 | Urban air, beta, $pCi/m^3$ | 5.3 | 5.0 | 4.4 | 4.2 |
| 26 | Mortality rate, all causes | 892.0 | 893.2 | 928.5 | 903.9 |

[a] r, Mn = malignant mortality rate.
[b] International Classification of Disease Code Numbers.

as follows:  A - the seven U.S. states with natural backgrounds >165 mrem/yr,
B - the fourteen states with background above 140 mrem/yr, and C - the fourteen states with the lowest background; and the 50 U.S. states.  The indicated malignancy rates are per 100,000 persons.  It is evident that these
are uniformly lower in the high background states than in those with the
lowest background radiation levels.  While the data used for this analysis
do not make it possible to separate natural background radiation from other
potentially carcinogenic agents and thus to provide dose-effect estimates,
they document its secondary role in the current incidence of most forms of
cancer.  From this analysis, it appears that the BEIR estimate of the risk
for radiation-induced thyroid cancer is excessive or that there must be a
threshold for its induction, in that application of their estimate leads to
an estimated natural incidence greater than the observed rates shown in
Table II.  Mays[30] has recently suggested other evidence that the linear
hypothesis may not be applicable in the case of low-dose thyroid exposures
to radioiodine.

Much of the costs occasioned by Appendix I are for the reduction of
$^{131}$I in both airborne  and liquid effluents.  In this connection, it has
also been suggested elsewhere[31] that the BEIR estimates of the risk for
the production of thyroid cancer from $^{131}$I in nuclear facility effluents
also seem excessive, insofar as they are predicated on data from X-ray exposures which are demonstrably more efficient than those from $^{131}$I per
g · rad of dose[32].  Thus, the actual cost-benefit ratios related to the
control of $^{131}$I in effluents seem far less favorable than those estimated
by the AEC Regulatory Staff.

Even if the cost-benefit effectiveness of Appendix I is questionable
on technical grounds, it has been argued that they will occasion only a
small increment in the cost of nuclear power.  This is seen by Wilson[33]
as a reasonable price to pay if it will satisfy the critics, however
dubious their arguments, and thus secure public acceptance.  This rationale
runs up against the reality that however much the critics may have been
satisfied by the Appendix I proposals, they seem to have gone on to seek
out other seeming 'Achilles heel' issues over which to continue the attack.
Furthermore, although the costs may be small in the current instance, the
application of protective measures on the basis of popular clamor and
'doomsday' suppositions, rather than hard evidence, can lead to large and
even more questionable expenditures.

Those occasioned by the current thrust in the U.S. for the near-
universal application of cooling towers to power generating plant condenser
streams seem illustrative.  Parker[34] has recently observed that whereas
the U.S. industry will have to invest an additional $8 \times 10^9$ to meet the
requirements of 'best practicable' water pollution control, the cost of
thermal discharge limitation will be $9.5 \times 10^9$.  Considering the known
incidence of disease from the discharge of heavy metals, carcinogens,
mutagens, etc., with its known incidence from thermal discharge (zero),
he is appalled by the relative emphasis being given to thermal pollution
abatement.


R I S K   C O M P A R I S O N S

The risks associated with nuclear facility effluents, even at the conservative BEIR dose-effect estimates, seem small in comparison to those
from other sources of population radiation exposure, including medical exposures and natural background.  The component that they represent of overall population exposure, based on  projections[35] made by the U.S. Environ-

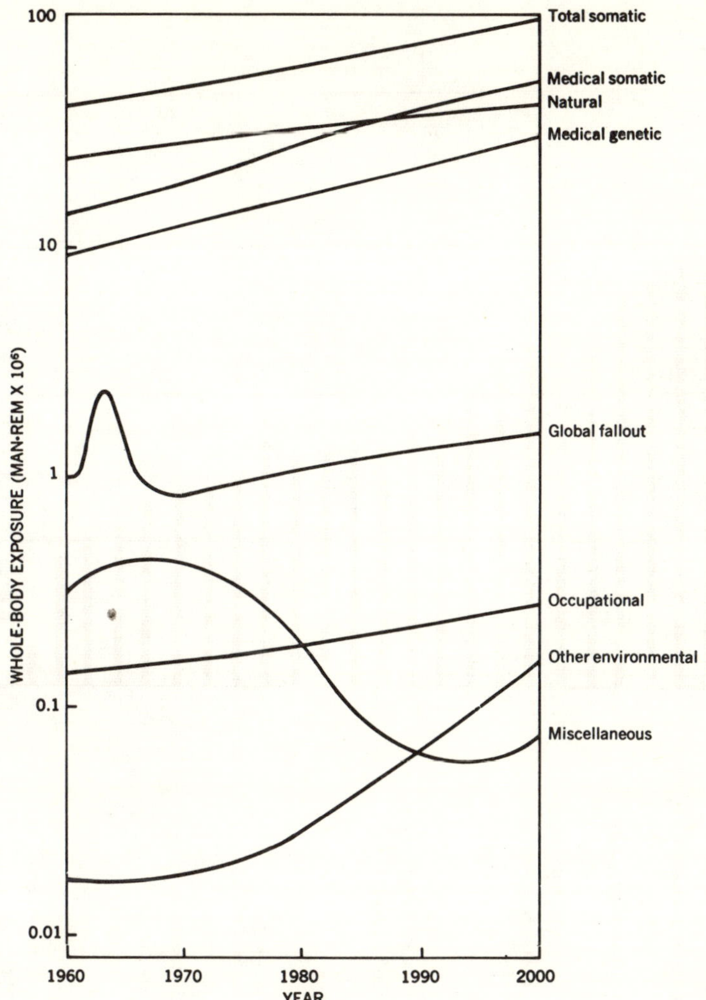



FIG. 3.   Estimated radiation doses in the USA, 1960–2000 (Ref. [35]).

mental Protection Agency (EPA), is shown as 'other environmental' in Figure 3. From this, the attention currently being devoted to reduction of these effluents seems all out of proportion to their relative contribution to population exposure.

The allocation of resources and manpower to the further reduction of nuclear effluents seems even more preposterous, when the associated theoretical risks are compared to a variety of demonstrable everyday risks to the abatement of which they might more beneficially be directed. Accepting the BEIR dose-effect estimates, the average risk to individuals within 80 km of a nuclear power plant, if exposed to an average whole-body dose of $5 \times 10^{-2}$ mrem/yr, may be calculated to be about $1 \times 10^{-8}$/yr.

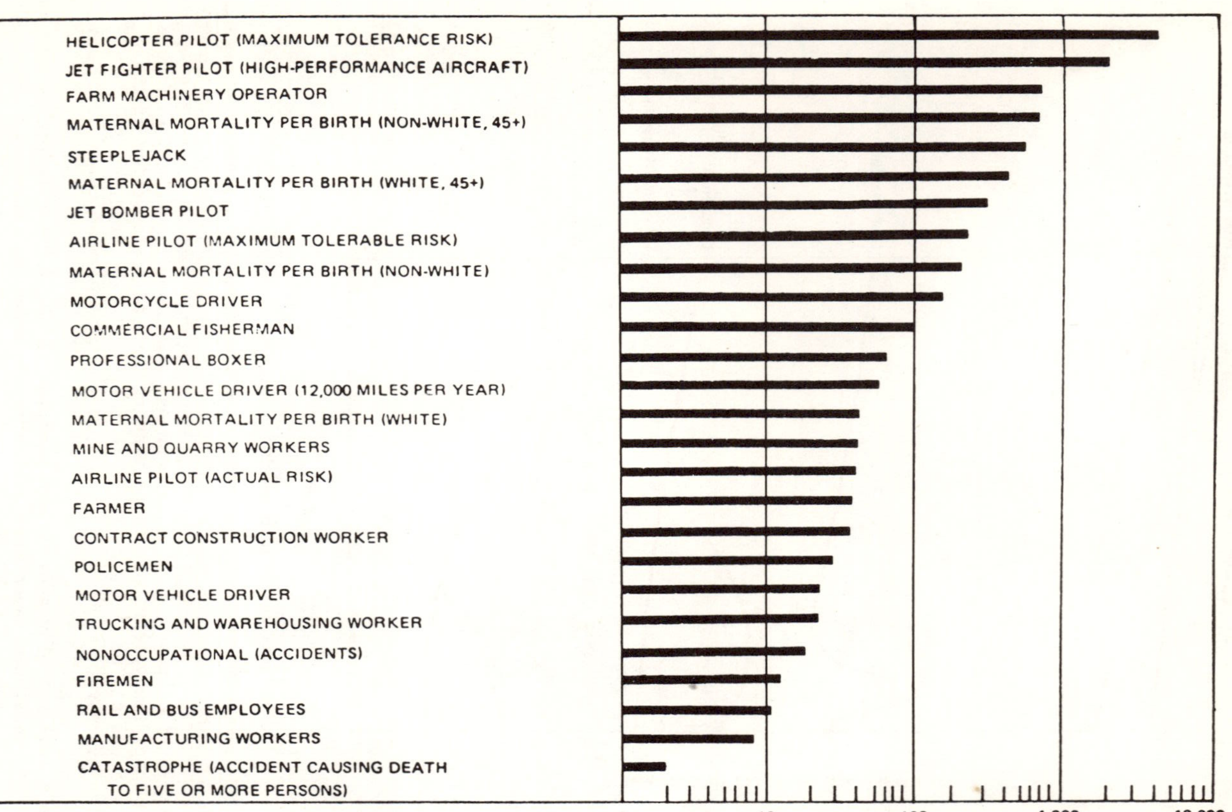


FIG. 4.  Risk to life per year as related to various socio-economic endeavors (Ref. [40]).

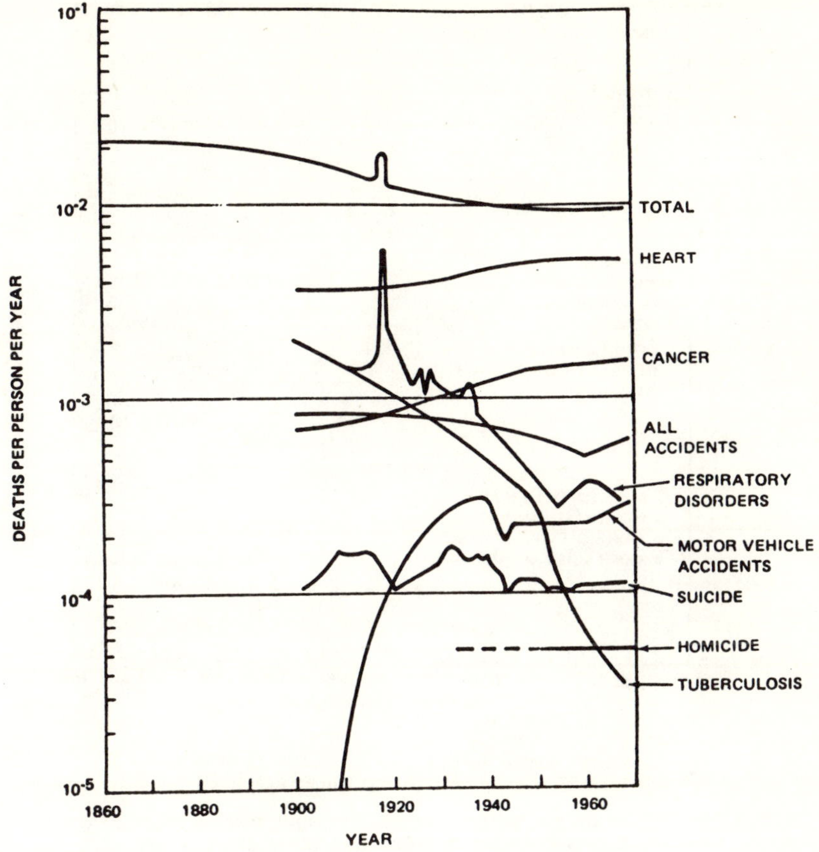



FIG. 5.  Death rates in the USA (Ref. [40]).

A selection of annual risks in the U.S. to life associated with various socio-economic endeavors is shown in Figure 4, and the risk of death from various diseases and accidents is shown in Figure 5. Additionally, from estimates provided by Wilson[33], the risk of death to individuals in the public attributable to fossil-fuel plant air pollution appears to be about $5 \times 10^{-5}$/yr. This the mid-range of the risk, $1 \times 10^{-5}$ - $1 \times 10^{-4}$, which may be derived from the number of deaths in the U.S. associated with the production of electric power estimated by Hamilton, et al.[36]. It can be shown[37] that with current airborne effluent technologies, the EPA's annual air pollution standards for general population exposures to $SO_2$ and $NO_2$ would be approached or exceeded at the site boundary of a 1,000 MW(e) coal- or oil-fired power plant. The health consequences of such exposures are suggested by the summary on data for effects of $SO_2$ pollution shown in Figure 6.

By comparison, it seems difficult to justify further expenditures in the name of 'as low as practicable' to reduce the hypothetical risk to an individual immediately adjacent to a comparable nuclear plant to less than $1 \times 10^{-6}$/yr, and that to the general nearby population to less than $1 \times 10^{-8}$/ year. To propose this is to suggest that there must be something extra-

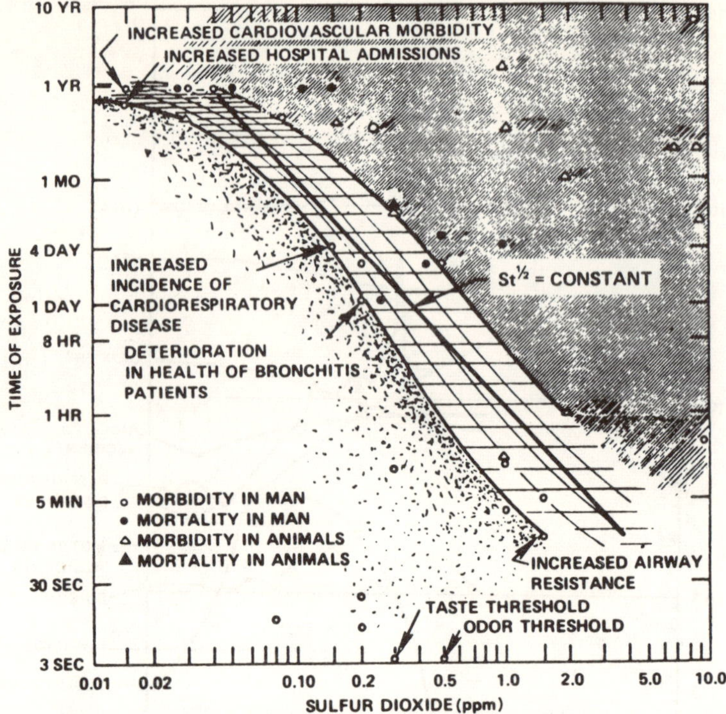



FIG. 6.  Effects of sulfur–dioxide pollution on health (Ref. [40]).

ordinarily obnoxious about radioactivity, so that to be exposed to even small concentrations of (or dose rates from) it is more hazardous than to be exposed to much larger concentrations (relative to air pollution standards) of fossil-fueled plant effluents.

In part, the excessive attention to the risks of exposure to radiation appear to stem from the emphasis on radiation protection standards on the hypothetical individual at the fence-line of a nuclear facility.  This seems inconsistent with other current provisions for the protection of the public from conventional technological activities.  Individuals living immediately adjacent to almost any facility in which toxic, explosive or flammable agents are either stored or utilized appear to be at greater risk than do persons at a greater distance, but still nearby.  Providing that the risk to these most potentially exposed persons is deemed small relative to their general life risks, it appears to be accepted whether or not they benefit directly

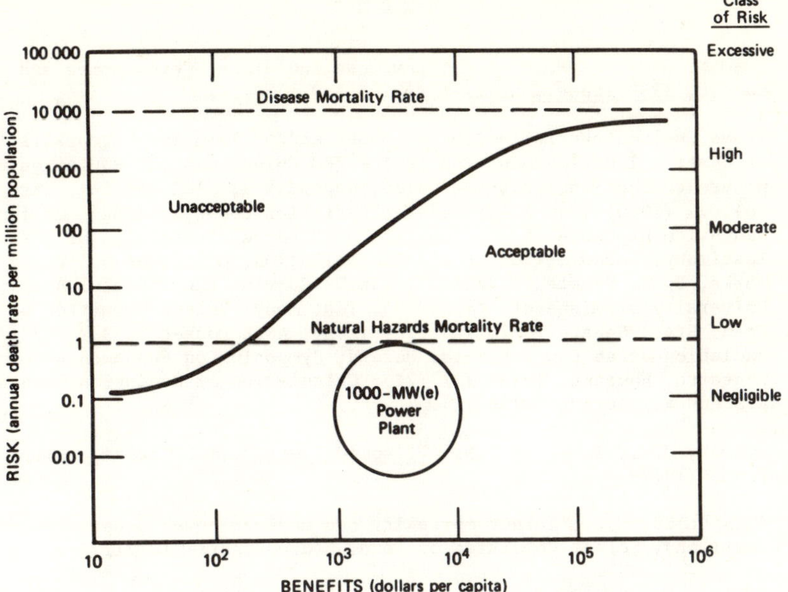



FIG. 7. Benefit-risk pattern for involuntary exposure (Ref. [40]).

from the activity. Radiation is made to appear especially hazardous by attempts to reduce the risk from it for any individual, however proximate to a nuclear facility, to one acceptable for the general population.


## CONCLUSIONS

Popular apprehension concerning specific nuclear facility sites, as well as of the overall nuclear enterprise, appears to contain a large psychological component of dread of the little understood effects of radiation. The nuclear critics appear to have exploited this dread quite skillfully in their efforts to forestall the implementation of nuclear power and other nuclear facilities in the U.S.

As long as it is considered sui generis, rather than within an overall context of potentially deleterious environmental agents, any related risk from radiation is likely to be considered detrimental and therefore unacceptable. However, it can be shown that the near-term practicable fossil-fueled alternatives for producing power contain a greater environmental insult[38], a larger occupational hazard[39], and a more significant public health risk[31,33,36-40].

Indeed, as indicated in Figure 7 from Starr[40], the risks associated with routine nuclear facility releases are within the range of those otherwise generally considered not only acceptable but negligible. Comparative presentations of this nature seem more hopeful toward putting risks in a reasonable perspective and thus securing public acceptance, than do measures such as 'Appendix I' to reduce radioactive effluents and environmental doses to such arbitrary 'as low as practicable' level, however small.

## R E F E R E N C E S

[1]  GOFMAN, J.W., TAMPLIN, A.R., Low Dose Radiation, Chromosomes and
     Cancer, *IEEE Trans.*, NS-17-1 (1970) 1-9.

[2]  These include the following by Gofman and/or Tamplin:  Proposals for
     at Least a Tenfold Reduction in the FRC Guidelines for Radiation Ex-
     posure to the Population at Large, Appendix 3, JACE Hearing, Part 2,
     Vol. II (1970) 1946-2144; Federal Radiation Council Guidelines for
     Radiation Exposure of the Population at Large: Protection or Disaster?,
     Testimony, Subcommittee on Air and Pollution, Committee on Public
     Works, U. S. Senate (11/18/69); Can We Survive the Peaceful Atom?,
     University of Minnesota (4/22/70); Testimony, Select Committee of
     Penn. State Senate (8/20/70); Testimony, ASLB Docket 50-222 (9/21/70);
     Radiation as an Environmental Hazard, Symposium on Fundamental Cancer
     Research, Houston, Texas (3/3/71); Calculation of Radiation Hazards,
     APS Annual Meeting (4/26/71).

[3]  GOFMAN, J.W., TAMPLIN, A.R., *Poisoned Power*, Rodale Press, Emmaus,
     Penn. (1971).

[4]  STERNGLASS, E.J., Infant mortality and nuclear power generation,
     Testimony, Select Committee of Penn. State Senate (10/21/70).

[5]  HULL, A.P., SHORE. F.J., Sternglass:  A Case History, BNL-16613 (1972).

[6]  HOLCOMB, R.W., Radiation Risk: A Scientific Problem?, *Science* 167 3919
     (1970) 853-855.

[7]  BOFFEY, P.M., Radiation Standards: Are the Right People Making
     Decisions?, *Science* 171 3973 (1971) 780-783.

[8]  ....., The Effects on Population to Low Levels of Ionizing Radiation,
     Report of the Advisory Committee on the Biological Effects of Ionizing
     Radiation, NAS-NRC (1972).

[9]  ....., Ionizing Radiation: Levels and Effects, Report of the U.N.
     Scientific Committee on the Effects of Atomic Radiation, Volume II,
     United Nations (1972).

[10] ....., Technical Specifications on Effluents from Nuclear Reactors,
     50.36a, Appendix I, 10 CFR 50, Licensing and Production Facilities,
     *Federal Register*, 36 111 (1971).

[11] ....., Concluding Statement of the Regulatory Staff, ASLB Docket
     RM 50-2 (2/20/74).

[12] HULL, A.P., Reactor Effluents: As Low as Practicable or As Low as
     Reasonable?, *Nuclear News* 15 11 (1972) 53-59.

[13] ....., Reply of Consolidated Utility Group to Concluding Statement of
     the Regulatory Staff, ASLB Docket RM 50-2 (3/7/74).

[14] ....., Implications of Commission Recommendations that Doses be Kept
     as Low as Readily Achievable, ICRP Pub. 22, Pergamon Press, Oxford (197.

[15] KNOX, J.B., Airborne Radiation from Nuclear Power Plants, *Nuclear News*
     14 2 (1971) 27-32.

[16]   ....., Calculations of Population Doses and Potential Health Effects
       Due to Atmospheric Releases of Radionuclides from U. S. Power
       Reactors in 1972, Radiation Data & Reports 15 8 (1974) 477-481.

[17]   KAHN, B.,et al., Radiological Surveillance Studies at a Boiling Water
       Nuclear Power Reactor, BRH/DER 70-1 (1970).

[18]   KAHN, B.,et al., Radiological Surveillance Studies at a Pressurized
       Nuclear Power Reactor, RD-71-1 (1971).

[19]   MC CURDY, D.R., RUSSO, J.J., Environmental Radiation Surveillance of
       the Oyster Creek Nuclear Generating Station, New Jersey Department
       of Environmental Conservation (1973).

[20]   SHLEIEN, B., An Estimate of Radiation Doses Received by Individuals
       Living in the Vicinity of a Nuclear Fuel Reprocessing Plant in 1968,
       BRH/NEHRL 70-1 (1970).

[21]   ALBERT, R.E., ALTSHULER, B., Considerations Relative to the Formulation
       of Limits for the Unavoidable Population Exposure to Environmental
       Carcinogens, Radiation Carcinogenesis, CONF-720505 (1973).

[22]   BAUM, J.W., Multiple Simultaneous Event Model for Radiation Carcinogen-
       esis, 5th International Congress Radiation Research, Seattle, Washing-
       ton (7/14-20/74). Also submitted to Science.

[23]   BUSTED, L.K., The Problem and Paradox that is Cancer, Radiation
       Carcinogenesis, CONF-720505 (1972).

[24]   CHADWICK, K.H., LENHOUTS, H.P., A Common Molecular Mechanism in
       Radiobiology? Its Implications in Radiological Protection, Assn.
       Euratom-IAL, Wagerinsen, Neth. (1973).

[25]   MAYNEARD, W.V., Radiation Carcinogenesis, Brit. Jnl. Radiol., 41 484
       (1968) 241-250.

[26]   ROSSI, H.H., KELLEHER, A.M., Radiation Carcinogenesis at Low Doses,
       Science 175 4018 (1972) 200-202.

[27]   FRIGERIO, N.A., ECKERMAN, K.J., STONE, R.S., Carcinogenic Hazard
       from Low-Level, Low-Rate Radiation, ANL/ES-26, Part I (1973).

[28]   MAYS, C.W., LLOYD, R.D., MARSHALL, J.H., Malignancy Risk to Humans
       from Total Body Radiation, Proceedings 3rd International Congress of
       IRPA (1973).

[29]   ROWLAND, R.E., The Risk of Malignancy from Internally Deposited Radio-
       isotopes, Center for Human Radiobiology, Argonne National Laboratory,
       Argonne, Illinois (1974).

[30]   MAYS, C.W., Cancer in Man from Internal Radioactivity, Health Physics
       25 6 (1973) 585-592.

[31]   HULL, A.P., As Low As Practicable: A Critical Reappraisal of Popula-
       tion Health Risks from Power Plant Effluents, Symposium on Population
       Exposures, CONF-741018 (1974).

[32]   COLE, R., Inhalation of Radioiodine from Fallout: Hazards and Counter-
       measures, ESA-TR-72-01 (1972).

[33]  WILSON, R., Examples in Risk-Benefit Assessment, Conference on Advanced
      Energy Systems, Denver, Colorado (6/21/73).  Available from author,
      Physics Department, Harvard University, Boston, Massachusetts.

[34]  PARKER, F.L., Water Pollution, Letter to Science 185 4151 (1974)
      568.

[35]  KLEMENT, A.W. et al, Estimates of Ionizing Radiation in the United
      States, 1969-2000, ORP/CSD 72-1 (1972).

[36]  HAMILTON, L.D., MORRIS, S.C., Health Effects of Fossil Fuel Power
      Plants, Symposium on Population Exposures, CONF. 741018 (1974).

[37]  HULL, A.P., Comparing Effluent Releases from Nuclear and Fossil-
      Fueled Plants, Nuclear News 16 4 (1974).

[38]  FINKEL, A.J.,(Ed.), Energy, the Environment, and Human Health,
      Publishing Sciences Group Inc., Acton, Mass. (1974).

[39]  LAVE, L.B., The Health Effects of Electricity Generation from Coal,
      Oil and Nuclear Fuels, Nuclear Safety 14 5 (1973) 409-428.

[40]  STARR, C., GREENFIELD, M.A., Public Health Risks of Thermal Power
      Plants, UCLA-ENG-7242 (1972).  Also summarized in Nuclear Safety
      14 4 (1973) 267-274.

# DISCUSSION

A.P.V. MODING:  How does public acceptance of reactor sites compare
with the attitude to reprocessing plant sites in the United States of America?

A.P. HULL:  Since there have been so few proposals for fuel reprocessi
facilities (and the sites that have been considered are remote), they have
not attracted much attention.  I think this is mostly a matter of numbers
rather than any perception of the relative hazards.

W. VINCK:  Have you analysed the problems of normal effluent releases
from large nuclear energy centres (parks)?  Would you not agree that, in
order to keep the consequences for the surrounding population within
acceptable limits, one would require supplementary retention equipment
(for noble gases and halides) at such installations?  It seems to me that the
conclusions you have reached in your paper concerning forecasts for the
future are too optimistic, having regard especially to the local effects of
normal effluents.

A.P. HULL:  Since the controversy in the United States of America
has centred largely on reactor siting, I have directed my attention to their
effluents.  The limited data available to us (from one commercial plant
in the USA) suggest that the inclusion of such a facility in a nuclear park
should not create an unmanageable problem.  An analysis of reported power
reactor effluents in the USA which I made last year suggests that the
amounts per MW·h of electricity generated were decreasing for pre-
Appendix I 'second-generation' plants.  Although it may be difficult to meet
Appendix I limits at multiple reactor facilities, their boundary exposures
due to routine releases should be small in relation to the standards
recommended by the ICRP (i.e. 500 mrem/a or 170 mrem/a).

E. TABET:  Do you not think that the small increase in nuclear energy cost required to reduce nuclear power plant releases is well worth accepting, especially in view of the large uncertainties (mentioned in your paper) in the evaluation of radiological risks?

A.P. HULL:  Although the cost increment in this instance is small, in principle any expenditure to reduce further an already insignificant risk (of the order of $10^{-7}$–$10^{-8}$/a) is difficult to justify.  I would call your attention to the example of the relative expenditures for control of water pollution and of thermal effluents indicated in my paper as an illustration of the excessive cost of obtaining public confidence in this fashion (in other words, large expenditure in relation to the problem).

A.J. GAUVENET (Chairman):  Could you please specify the statistical errors corresponding to each point in Fig.2, which gives malignant mortality rates as a function of natural background?

A.P. HULL:  Each point represents a large number of cases, of the order of hundreds of thousands.  The errors are therefore smaller than the size of the points themselves.

G. HAKE:  Would you hazard an opinion as to whether the arguments of Goffman and Tamplin would have had the impact they did if the USA regulations had from the start included an integrated population dose limit?

A.P. HULL:  Such a limit would have eliminated a proposition essential to their argument.  In my judgement, the radiobiological data and the risk estimates derived therefrom do not support anything beyond such a population dose limit when individual exposures are only a few mrem, i.e. the risk to the individual is trivial, but the collective risk warrants consideration.  The explanatory material issued with Appendix I indicated — although this is not mentioned explicitly in the Appendix itself — that its application could be expected to limit the population dose to 400 man·rem per 1000-MW(e) power reactor.

P. THOMAS:  In Fig.3 you show the whole-body exposure due to natural radiation increasing with time.  Would you mind explaining why this should be so?

A.P. HULL:  These are projections of the integrated population exposure in the United States of America, in $10^6$ man·rem, to the year 2000.  The increase simply reflects the anticipated growth in population during the next three decades.

S.O.W. BERGSTRÖM:  The collective dose of 400 man·rem mentioned in Appendix I is related, as far as I know, to an integration limit of 50 miles from the site.  We have calculated the collective dose to this limit, compared it with estimates of the aggregate collective dose for several actual sites, and found gross variations in the ratio.  What is the justification for the 50 mile limit?

A.P. HULL:  Since I had no part in the choice of the 50 mile integration, I cannot comment on its appropriateness.  However, if we consider a larger area, as we should do for tritium and noble gases, we must recognize that the significance of a given number of man·rem is somehow related to the unit size delivered to individuals.

D. BENINSON:  I should here add a comment in line with that of Dr. Bergström.  If we want to assess the impact of a source (and not the risk to individuals), we have to assess the collective dose resulting from the source, and there is no reason to stop the integration at 50 miles or any other arbitrary distance.

A.P. HULL:  Perfectly true.

# UN CRITERIO PARA LA EVALUACION
# DE LA SEGURIDAD NUCLEAR

A.J. GONZALEZ*
Comisión Nacional de Energía Atómica,
Buenos Aires,
Argentina

*Presentada por Ambretta Migliori de Beninson*

Abstract—Resumen

A CRITERION FOR NUCLEAR SAFETY ASSESSMENT.
The criterion presented has been developed extrapolating to the nuclear safety field the ICRP basic concepts and philosophies on radiological protection. The criterion postulates the use of a hyperbolic control curve basically similar to the one proposed by F.R. Farmer for probabilistic evaluations. The postulated control curve differs from Farmer's curve in both application range and characteristics of its mathematical function. A range of application is proposed from a minimum severity level (the $^{131}$I authorized discharge limit) to a severity level as large as the reactor's $^{131}$I inventory. The exponent value of the proposed hyperbolic function is variable with both the siting and $^{131}$I inventory of the reactor, and it also changes with the extrapolated radioprotection concept, as considered. The methodology to evaluate the control curve exponent is also presented. Three evaluation methods are described with the following objectives: (a) to limit expectation of individual dose in order to limit individual risk; (b) to limit expectation of collective dose in order to limit population detriment to justifiable levels; and (c) to optimize the installation in order to obtain a dose expectation as low as readily achievable. Three figures present the control curve exponent plotted versus dosimetric factor for $10^6$, $10^7$ and $10^8$ Ci of $^{131}$I inventories. The following conclusions can be deduced: (a) the expectation of individual dose changes very little with different inventories; (b) in sites where the collective dose per unit of activity released is high, only large reactors are justifiable; and (c) optimization analysis is generally less restrictive than the justification one. Finally, a criterion is suggested for limitation of collective dose commitment, to limit the future dose rate arising from accidents that could occur at present.


UN CRITERIO PARA LA EVALUACION DE LA SEGURIDAD NUCLEAR.
El criterio que se presenta ha sido desarrollado extrapolando al campo de la seguridad nuclear los conceptos básicos y las filosofías de protección radiológica desarrolladas por la CIPR. El criterio postula el uso de una curva de control hiperbólica básicamente similar a la propuesta por F.R. Farmer para las evaluaciones probabilísticas. La curva de control postulada difiere de la curva de Farmer en el campo de validez y en las características de la función matemática que la define. Se propone que el campo de validez se extienda desde un nivel mínimo de severidad (el límite autorizado de descarga para $^{131}$I) hasta un nivel de severidad tan grande como el inventario de $^{131}$I del reactor. El valor del exponente de la curva de control propuesta es variable con el emplazamiento y el inventario de $^{131}$I del reactor, y también se modifica con el concepto de radioprotección extrapolado. También se presenta la metodología para evaluar el valor del exponente de la curva de control. Se describen tres métodos de evaluación con los siguientes objetivos: a) limitar la expectancia de dosis individual para limitar los riesgos individuales; b) limitar la expectancia de dosis colectiva para limitar el detrimento de la población a niveles justificables; y c) optimizar la instalación para lograr que la expectancia de dosis sea tan baja como resulte razonablemente obtenible. En tres figuras se representa el valor del exponente de la curva de control en función del factor dosimétrico para los inventarios de $^{131}$I de $10^6$, $10^7$ y $10^8$ Ci. De las figuras pueden deducirse las siguientes conclusiones: a) la expectancia de dosis individual se modifica poco con el inventario del reactor; b) en los emplazamientos donde la dosis colectiva por unidad de actividad evacuada es alta, sólo se justifican los grandes reactores; y c) el análisis de optimización es generalmente menos restrictivo que el análisis de justificación. Finalmente se sugiere un criterio para limitar la dosis colectiva comprometida con el objeto de limitar la tasa de dosis futura debida a los accidentes que puedan ocurrir en el presente.

---

* Dirección actual: Atomic Energy Organization of Iran, P.O. Box 12-1198, Teheran, Irán.

## 1.  INTRODUCCION

El objetivo de este trabajo es proponer un criterio para la evaluación
de la seguridad de los reactores nucleares.  Este criterio está basado
principalmente en la extensión al campo de la seguridad nuclear de los principios generales de protección radiológica recomendados por la Comisión
Internacional de Protección Radiológica (CIPR), con la adecuación conse-
cuente a su campo de aplicación.

Hasta el presente no existen criterios internacionalmente aceptados
para evaluar la seguridad nuclear, y diversos utilizadores de la energía
nuclear, fabricantes de reactores y componentes nucleares y autoridades
licenciantes han empleado diversos criterios de seguridad para especificar,
construir o licenciar reactores nucleares.  Sin embargo, en los últimos
años se ha destacado el empleo de dos filosofías o métodos de evaluación
de la seguridad:  los denominados «determinístico» y «probabilístico».

Los criterios de evaluación basados en la filosofía determinística han
sido utilizados en muchos análisis de seguridad e incluso han sido regulados
por las autoridades licenciantes de diversos países.  El método de evaluación
se basa en suponer que las instalaciones nucleares seguras son aquellas
que cumplen ciertos criterios fijos predeterminados.  La facilidad de la
evaluación es posiblemente la principal causa de la popularidad de la filosofí
determinística.  No obstante, generalmente se efectúan diversas objeciones
a los criterios determinísticos;  las principales son las siguientes:  a) los
criterios determinísticos no están fundamentados científicamente sino que
se encuentran basados principalmente en la experiencia profesional de los
diseñadores de instalaciones y en los conceptos generales que la opinión
pública tiene respecto de los términos «seguridad» y «accidente»;  y b) en
ciertos casos, el uso de algunos criterios determinísticos puede ser causa
de altas inversiones sin un incremento significativo de los niveles de segurida

La filosofía probabilística se basa en el concepto de que ninguna instalación es absolutamente segura o insegura, dado que siempre existe un grado
de inseguridad asociado con la probabilidad de ocurrencia de accidentes y
acepta que la instalación resulta suficientemente segura cuando el riesgo
derivado de la ocurrencia de esos accidentes es razonable.  El origen de
este criterio se suele ubicar en varios trabajos presentados en el Simposio
sobre Confinamiento y Ubicación de Plantas Nucleares organizado por el
Organismo Internacional de Energía Atómica [1] y en especial en una memor
de F.R. Farmer [2].  En esta última fue presentada una curva de control que
relaciona la probabilidad y la severidad de los accidentes nucleares, determinando un límite superior de riesgo que sería aceptable para el público.
Este criterio asume que todos los reactores son conformados con un criteri
particular de excelencia en cuanto a las situaciones de fallas que involucren
liberación de productos de fisión, y que la probabilidad de ocurrencia de
accidentes nucleares está en relación inversa con la severidad de dichos
accidentes representada por el escape de actividad de $^{131}$I.  La curva de
control propuesta por Farmer establece un criterio para la ubicación de los
reactores nucleares determinando el riesgo de la población debido a situaciones accidentales en los mismos.  El criterio probabilístico basado en la
curva de control de Farmer ha servido para desarrollar métodos de evaluación de riesgos accidentales [3,4] y, recientemente, para proponer una
norma para riesgos debidos a escapes accidentales de gran magnitud [5];
también se ha extrapolado el criterio para evaluar la expectativa de escape

accidental promedio de determinados reactores nucleares [6]. Sin embargo,
no se ha encontrado en la bibliografía un método de evaluación de la seguridad
nuclear que, basándose en la filosofía probabilística, extrapole a las situaciones accidentales cada uno de los principios de protección radiológica
aceptados internacionalmente.

Tales principios, postulados por la CIPR, pueden resumirse en lo
siguiente:

a)  Las dosis recibidas por los individuos, debido a una determinada práctica
    u operación que involucre el uso de material radiactivo o radiaciones
    ionizantes, no deben exceder límites de dosis apropiados [7].

b)  El detrimento total producido por cualquier práctica u operación debe
    ser justificable en relación con el beneficio obtenido de dicha práctica
    u operación.  (La CIPR ha definido como «detrimento» la esperanza
    matemática del daño debido a las dosis de radiación teniendo en cuenta
    no sólo las probabilidades de cada tipo de efecto deletéreo sino también
    la severidad de cada uno de ellos.) [8]

c)  Las dosis debidas a una exposición justificable deben ser tan bajas como
    resulte razonablemente obtenible teniendo en cuenta consideraciones
    sociales y económicas [8].

Por otra parte, recientemente se ha sugerido fijar un límite de «dosis
colectiva comprometida por unidad de práctica», para controlar la dosis
promedio futura que recibirá la población debido a las aplicaciones nucleares
actuales [9], como un criterio adicional de protección radiológica.  El concepto
«dosis colectiva» ha sido definido por la CIPR [8] para ser empleado como
una medida de la exposición total de una población dada y, por extensión, se
ha utilizado el concepto «dosis colectiva comprometida» como una ampliación
del de «dosis comprometida» definido por el Comité Científico de las Na-
ciones Unidas para el Estudio de los Efectos de las Radiaciones Atómicas
(UNSCEAR) [10].

Los principios de protección radiológica expuestos en el párrafo anterior
han sido desarrollados conceptualmente para ser aplicados durante la operación normal de las instalaciones y, en consecuencia, no serían utilizables en
situaciones accidentales.  No obstante, el criterio para evaluar la seguridad
que se presenta en este trabajo postula la extensión de dichos principios al
campo de la seguridad nuclear relacionándolos con los criterios básicos
de la filosofía probabilística.  Para poder aplicarlos a las situaciones accidentales se debe reemplazar el concepto de «exposición debida a la operación
normal» por «expectancia de exposición debida a situaciones accidentales»
(en este caso se está usando el término «exposición» para significar «dosis
equivalente» recibida por los individuos o «dosis colectiva» o «dosis colectiva
comprometida» para toda la población).  Es decir que los principios extra-
polados podrían resumirse como sigue:

a)  La expectancia de dosis individual debida a los escapes accidentales de
    los reactores no debería exceder límites de dosis apropiados.

b)  El detrimento esperado para la población debido a los escapes acciden-
    tales de los reactores debería ser justificable, teniendo en cuenta el
    beneficio que se obtiene de la operación de las centrales nucleares.

c)  La expectancia de dosis debida a los escapes accidentales de una instala-
    ción justificable debería ser tan baja como resulte razonablemente
    obtenible teniendo en cuenta consideraciones sociales y económicas.

d)   La expectancia de dosis colectiva comprometida por unidad de práctica
     debida a las situaciones accidentales, debería estar limitada de manera
     de poder controlar la dosis individual promedio para la población del
     futuro.

El método propuesto tiene una triple utilización:  a) la evaluación del
nivel de seguridad de un emplazamiento donde será ubicado un determinado
reactor;  b) el análisis de la seguridad nuclear de un reactor ubicado en un
determinado sitio;  y  c) el diseño de los sistemas de seguridad de un reactor
teniendo en cuenta su posible emplazamiento.

La metodología sugerida consiste en la determinación de la pendiente de
tres curvas de control que tengan en cuenta un límite de expectancia de dosis
individual, un límite de expectancia de dosis colectiva que asegure que la
expectancia de detrimento será justificable y una expectancia de dosis optimi
zada.  Una vez determinados los valores de las tres pendientes mencionadas,
para la evaluación, el análisis o el diseño respectivo se utilizaría la curva de
control que tenga como pendiente el mayor valor encontrado; también se
presenta un criterio para limitar la expectativa de «dosis colectiva comprometida» debida a escapes accidentales producidos en centrales nucleares.


## 2.   LA CURVA DE CONTROL PROPUESTA PARA LA EVALUACION
## DE SEGURIDAD Y LA EXPECTANCIA DE ESCAPE RESULTANTE

La curva de control originalmente propuesta por Farmer (figura 1) y las
variantes de la misma, son básicamente líneas rectas representadas en un
plano de coordenadas logarítmicas probabilidad/severidad que pivotan en el
punto definido por las coordenadas $\{F(C) = 10^{-3}$ año$^{-1}$. reactor$^{-1}$ ;  C $=10^3$Ci $^{131}$I
Por razones de eficiencia operacional se ha propuesto que las líneas se haga
asintóticas a la ordenada $F(C) = 10^{-2}$ año$^{-1}$. reactor$^{-1}$.  La función matemática
que define la curva de control cuando las abcisas son mayores a C $=10^{-3}$ Ci $^{13}$
es una hipérbola del tipo

$$F(C) = A.C^{-a} \tag{1}$$

donde:

a es un exponente fijo determinado por el criterio de excelencia con el
que son diseñados los reactores nucleares, y
A es un factor que queda automáticamente determinado cuando se fija a.

Sin embargo, si lo que se pretende conseguir mediante el empleo de una
curva de control es limitar u optimizar la expectancia de exposición, el
exponente de la función hiperbólica (1) no debiera ser un número fijo.  La
expectancia de exposición depende, entre otras cosas, del escape potencial
máximo de radiactividad y de los parámetros meteorológicos y dosimétrico
que permiten evaluar la exposición por unidad de actividad evacuada.  Es
decir que, en principio, el exponente debería ser variable con el inventario
radiactivo del reactor y la ubicación elegida para el mismo.  Por otra parte
el exponente debe ser distinto si lo que se pretende es limitar la expectanci
de dosis individual (para asegurar que los riesgos individuales se mantendr
acotados), o limitar la expectancia de dosis colectiva (para asegurar que el

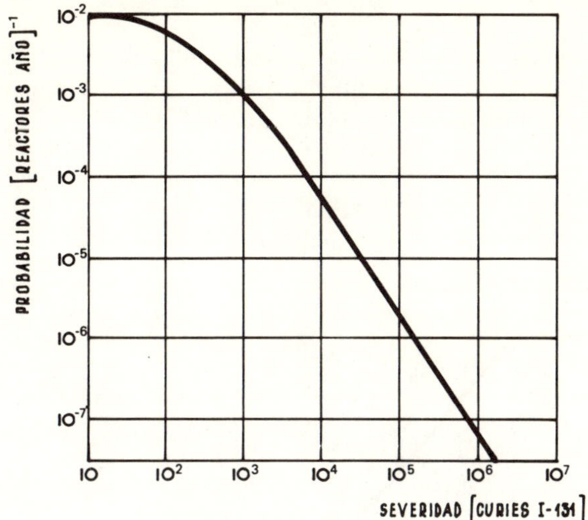


FIG.1.  Curva de control propuesta por Farmer.

detrimento a la población no será injustificable teniendo en cuenta los beneficios que dará el reactor), u obtener una expectancia de exposición «tan baja
como resulte razonablemente obtenible...» (de manera de optimizar los
sistemas de seguridad nuclear).

Por otra parte, el campo de aplicabilidad debería extenderse desde el
mínimo nivel posible de severidad hasta la mayor severidad potencial, es
decir aquélla de un nivel igual al inventario radiactivo de la instalación.
Si se considera que toda evacuación de radiactividad de magnitud mayor al
límite de descarga autorizado de material radiactivo es por definición una
«situación accidental», entonces el valor mínimo de severidad desde el cual
tiene validez la curva de control debiera ser igual al valor de ese límite de
descarga autorizado.   Debe destacarse que el límite de descarga autorizado
no es una fracción fija de la capacidad radiológica estipulada [11] para el
medio ambiente en el que ha sido ubicada la instalación, sino que es el valor
óptimo de descarga anual determinado en función del criterio que las dosis
debidas a la operación normal deben ser «tan bajas como resulte razonablemente obtenible...» [8].   Es decir que, dada la filosofía involucrada en la
determinación de dicho límite de descarga, existe la certeza que se produci-
rán evacuaciones de radiactividad iguales al mismo.

La curva de control propuesta por Farmer tiene un máximo
nivel de probabilidad de $10^{-2}$ año$^{-1}$. reactor$^{-1}$. Sin embargo, durante el año
fiscal 1973, en los Estados Unidos de América fueron señalados 680 incidentes
en los reactores nucleares de potencia [12]. Si bien muchos de esos incidentes
fueron menores y algunos no involucraron descarga de material radiactivo, el
elevado número de anormalidades aconseja adoptar una hipótesis cauta y
postular que la curva de control se debe extender hasta la certeza para una
severidad (expresada en Ci $^{131}$I) igual al límite de descarga anual de $^{131}$I(L).
Es decir, que un punto de la curva de control postulada que resulta fijo y
mandatorio es el definido por las coordenadas {F(C) = 1 año$^{-1}$.reactor$^{-1}$;

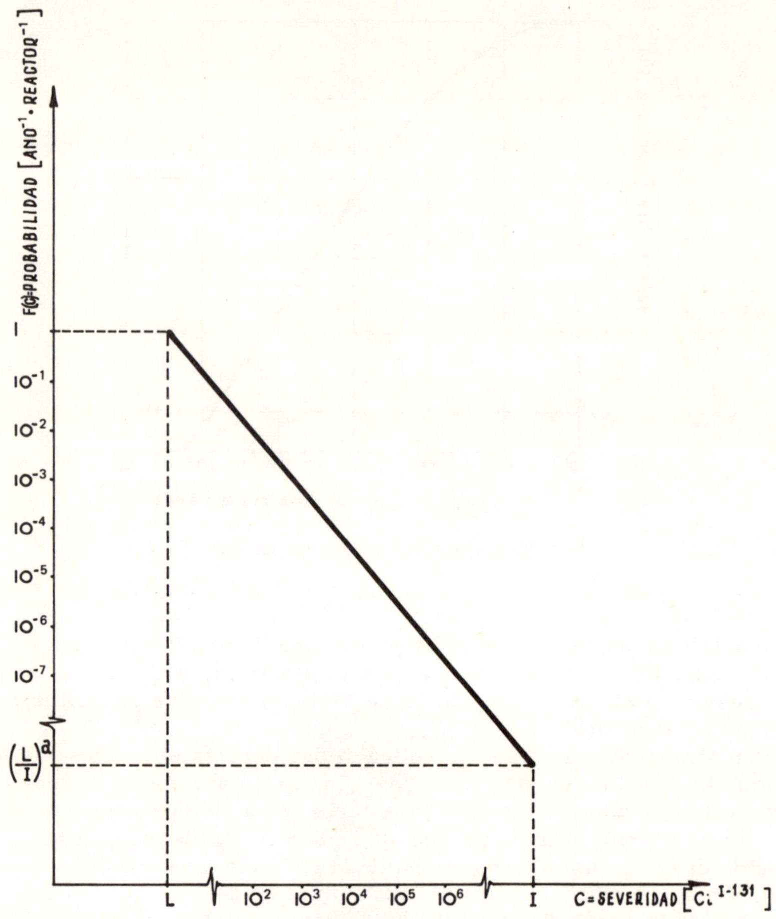



FIG.2.   Curva de control propuesta en este trabajo.

$C = L(Ci\ ^{131}I)\}$, de donde resulta que el factor A de la ecuación (1) debe ser $A = L^a$.

En consecuencia, la curva de control postulada en este trabajo, que se presenta en la figura 2, relaciona la probabilidad y la severidad de los accidentes nucleares mediante una función hiperbólica similar a la propuesta por Farmer, de la forma:

$$F(C) = (L)^a . C^a \qquad (2)$$

donde:

F(C) es la función de Farmer;

L    es el valor límite de descarga anual autorizada expresado en curies de $^{131}I$

C    es la variable de la curva de control expresada en curies de $^{131}I$

a       es el ponente de la curva de control que determina la confiabilidad que deben tener las distintas cadenas de un árbol accidental de modo que para una severidad dada exista una probabilidad determinada.

Resumiendo, la curva propuesta tiene las siguientes características: a) tiene un punto fijo, que corresponde a la unidad en el eje de probabilidades y a una severidad de L en el de abcisas, de manera que exista certeza que se producirá una evacuación de iodo $^{131}I$ igual al límite de descarga autorizado; b) su aplicabilidad se extiende hasta una severidad equivalente al inventario radiactivo del reactor (I), y c) el exponente de la función es variable con el inventario de la instalación y con las características radiológicas del sitio de emplazamiento, siendo ese el parámetro a determinar en cada caso para evaluar el nivel de seguridad nuclear de la instalación.

Por otra parte, la expresión general de la expectancia de escape de $^{131}I$ deducida de la curva de control es [3, 6]:

$$R = \int \frac{F(C)}{\ln 10} \cdot dC \tag{3}$$

Luego, integrando la curva de control propuesta en (2) entre L e I según la (3), la expresión de la expectancia de escape resulta ser:

$$R = \frac{L^a\left[I^{(1-a)} - L^{(1-a)}\right]}{(1-a)\,\ln 10} \tag{4}$$

En los capítulos siguientes se presentan los métodos propuestos para la determinación de los tres valores típicos del exponente. El primero define una función que determina una curva de control límite desde el punto de vista de la expectancia de dosis individual. El segundo determina una curva de control límite desde el punto de vista de la justificación socioeconómica de la instalación. Finalmente, el tercero determina una curva óptima para el diseño de los sistemas de seguridad de la instalación. Debido a que la función propuesta es similar a la función correspondiente a la curva de control presentada por F.R. Farmer, en ciertos casos particulares, el exponente resultante para alguna de las curvas de control puede ser equivalente al postulado por Farmer.


3.  METODO LIMITANTE DE LA EXPECTANCIA
    DE EXPOSICION INDIVIDUAL

En general, se acepta un riesgo máximo para las prácticas con radiaciones ionizantes siempre que esas prácticas sean justificables. Este riesgo máximo deriva de los «límites de dosis individuales» regulados por las autoridades nacionales para la operación normal. En consecuencia, parece razonable que el riesgo individual máximo debido a situaciones accidentales sea del mismo orden de magnitud que el aceptado para operaciones normales, dado que los efectos deletéreos en ambos casos son equivalentes siempre y cuando se tomen las previsiones necesarias para que no ocurran exposiciones agudas. Si la preevaluación de seguridad del sitio donde será ubicada la

instalación se efectúa siguiendo algún criterio determinístico apropiado [12,1?
es sumamente improbable que puedan ocurrir sobreexposiciones accidentales
que den origen a efectos agudos en el público. En ese caso parece razonable
fijar un límite para la expectancia de dosis individual debida a las situaciones
accidentales con el fin de acotar los riesgos emergentes de la misma forma
que los límites de dosis determinan un nivel superior de los riesgos derivado
de las operaciones normales.

La limitación de la expectancia de dosis individual debida a las situaciones accidentales puede lograrse limitando la expectancia de escape
accidental de material radiactivo. Si se utiliza el límite de dosis recomendado por la CIPR y teniendo en cuenta que para el $^{131}$I el órgano crítico es
la tiroides, puede establecerse la siguiente relación:

$$\frac{1,5 \ (\text{rem/año})}{\text{FD}(\text{rem/Ci})} \geqq R = \frac{L^{a}\left[I^{(1-a)} - L^{(1-a)}\right]}{(1-a)\ \ln 10} \tag{5}$$

donde FD es un factor dosimétrico expresado en rem/Ci ($^{131}$I) que tiene en
cuenta el parámetro meteorológico del sitio para el grupo crítico y el
parámetro dosimétrico de ese grupo crítico.

La ecuación (5) permite despejar el valor de «a» que define la curva de
control límite desde el punto de vista de los riesgos individuales máximos

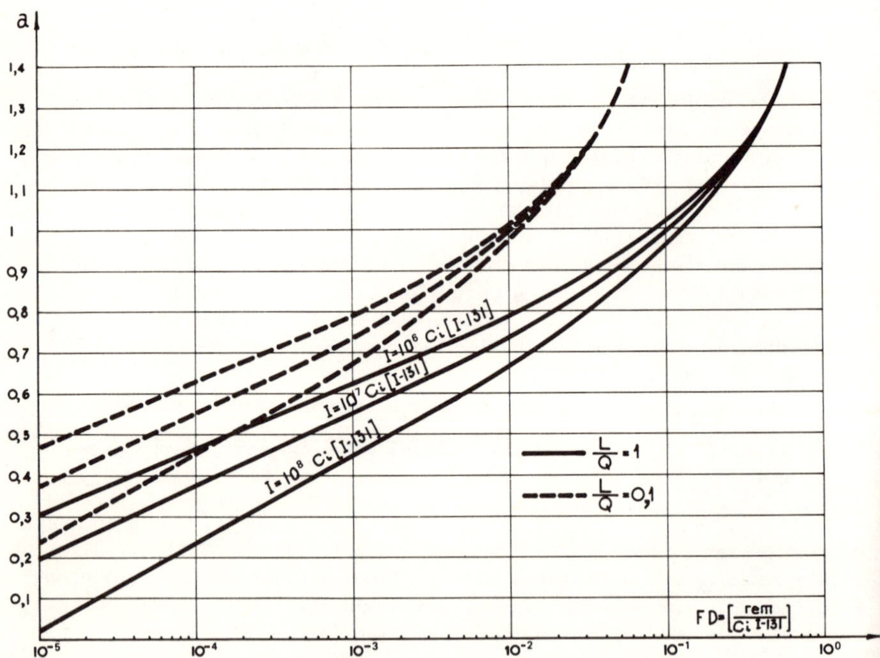


FIG.3.   El exponente «a» de la curva de control limitante desde el punto de vista de la expectancia de dos
individual en función del factor dosimétrico para el grupo crítico. (Se han diseñado juegos de curvas considerando L/Q = 1 y L/Q = 0,1, para distintos inventarios (I) de $^{131}$I.)

en función del inventario de $^{131}$I de la instalación y del factor dosimétrico del
sitio de ubicación. Sin embargo si se quiere ser coherente con la filosofía
empleada en la determinación de L, el valor de R no debiera ser de distinto
orden de magnitud que L. En consecuencia podría resultar razonable multi-
plicar el segundo miembro de la (5) por la relación L/Q donde Q es la capacidad radiológica limitante del medio ambiente [11], es decir, la cantidad de
radiactividad descargada por año de la cual resulta una dosis comprometida
igual a los límites de dosis recomendados por la CIPR.

La figura 3 permite encontrar el valor «a» de la pendiente de la curva de
control que limita los riesgos individuales máximos para diversas instalaciones (representadas por sus inventarios de $^{131}$I) y para varios sitios de
ubicación (representados por el factor dosimétrico); también se grafican
curvas para un valor de L/Q = 0,1. Las curvas han sido representadas para
un valor del límite autorizado de descarga igual a 2 Ci/año.


4.  METODO PARA EVALUAR LA JUSTIFICABILIDAD
    DE LA EXPECTANCIA DE EXPOSICION

Como se ha mencionado, la CIPR ha recomendado [8] que el detrimento
total debido a cualquier práctica u operación que involucre el empleo de
material radiactivo o radiaciones ionizantes no debe ser injustificable con
respecto al beneficio que la sociedad espera de dicha práctica u operación.
Si bien esta recomendación ha sido efectuada para las exposiciones debidas
al funcionamiento normal de las instalaciones, nada impide extender el
concepto a las situaciones accidentales de dichas instalaciones. En ese caso,
para determinar el beneficio esperado por la sociedad de la operación que se
realiza en la instalación, podría plantearse una ecuación similar a la empleada
para la evaluación de las exposiciones normales, es decir:

$$B = V - P - G \tag{6}$$

donde:

B   es el beneficio neto obtenido de la práctica u operación;
V   es el valor total de producto obtenido;
P   es el costo total de la práctica u operación, incluido el costo para
    obtener un nivel dado de seguridad, y
G   es la expectancia de detrimento total sobre la población debido a los
    eventuales accidentes de la instalación.

Para efectuar el análisis costo-beneficio planteado, la expectancia de
detrimento causado por las situaciones accidentales sobre la población puede
ser definida por analogía con la definición de la CIPR. En este caso se deno-
minará expectancia de detrimento a la esperanza matemática del daño debido
a una expectancia de dosis de radiación, teniendo en cuenta no sólo las probabilidades de cada tipo de efecto deletéreo sino también la severidad de cada
uno de ellos. Es decir que si $P_i$ es la probabilidad de que ocurra el efecto i,
cuya severidad es expresada por un factor de peso $g_i$, la expectancia de
detrimento G en un grupo compuesto por N personas es $G = N\Sigma_i P_i . g_i$.
Por otra parte, también por analogía con la definición de la CIPR, la
expectancia de dosis colectiva será una medida de la expectancia de exposi-

ción total del cuerpo entero o de un órgano específico de los miembros de una
población. Si el número de personas que tienen expectancias de dosis com-
prendidas entre H y H + dH es N(H).dH, la expectancia de dosis colectivas
está dada por S = ∫ H.N(H).dH donde la integración debe hacerse sobre la
distribución de la expectancia de dosis total en la población mundial. Si se
supone la linealidad de la relación dosis-riesgo, la expectancia de dosis
colectiva puede ser utilizada para evaluar la esperanza matemática del
número de efectos deletéreos en una población dada. En consecuencia, si se
supone que la gravedad de los efectos estocásticos es independiente de su
frecuencia, la expectancia de dosis colectiva puede considerarse proporcional
a la expectancia de detrimento sobre la población debido a las eventuales
situaciones accidentales.

Para poder efectuar el análisis costo-beneficio planteado en la ecuación
(6), la expectancia de detrimento debe ser expresada en las mismas unidades
(por ejemplo monetarias) en que han sido expresados los otros términos de
la ecuación. Se han publicado diversas estimaciones del costo de la unidad
de dosis colectiva (rem.hombre), que fluctúan entre las decenas y centenas de
dólares de los Estados Unidos de América [15-20].

La aplicación del criterio de que los beneficios netos deben ser positivos
implica que la ecuación (6) puede expresarse de la siguiente forma:

$$V - P - G \geq 0 \tag{7}$$

Para las centrales nucleoeléctricas V y P son proporcionales, con suficiente
aproximación, a la potencia del reactor y, en consecuencia, al inventario de
$^{131}$I en equilibrio, y G es directamente proporcional a la expectancia de
escape medio de $^{131}$I. En consecuencia, la ecuación (7) puede escribirse de
la siguiente forma:

$$k_1 I - k_2 I - FDC \cdot C_{r\text{-}h} \cdot T_v \left[ \frac{L^B \left[ I^{(1-a)} - L^{(1-a)} \right]}{(1-a) \ln 10} \right] \geq 0 \tag{8}$$

donde:

I       es el inventario de $^{131}$I de la instalación;

$k_1$     es igual al producto del valor de la unidad de energía nucleoeléctrica
        por la vida útil de la instalación, dividido por la actividad de $^{131}$I
        por unidad de potencia;

$k_2$     es igual al costo total (inversion, operación, mantenimiento, etc.)
        unidad de potencia, dividido por la actividad de $^{131}$I por unidad de
        potencia;

FDC es el factor dosimétrico para dosis colectiva correspondiente a la
        ubicación de la instalación;

$C_{r\text{-}h}$ es el costo estimado para la unidad de dosis colectiva.

$T_v$     es la vida útil de la instalación.

Teniendo en cuenta que $k_1$, $k_2$ y $C_{r\text{-}h}$ son valores económicos que resultan
constantes en un análisis dado, de la ecuación (8) se puede despejar el valor
del exponente «a», que será límite desde el punto de vista de la justificabilidad, para los distintos inventarios de $^{131}$I y para diversos sitios de
ubicación representados por el factor dosimétrico FDC.

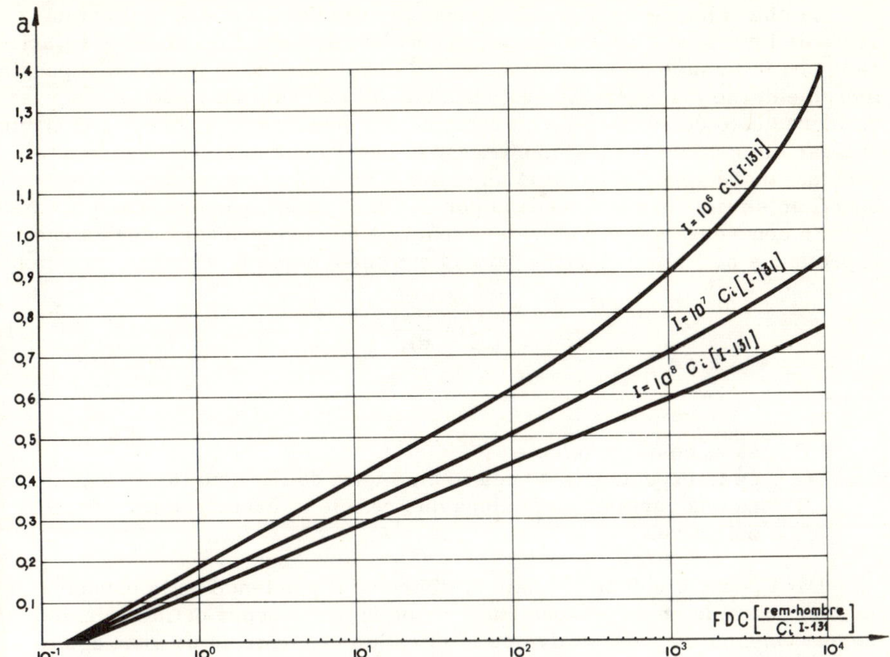



FIG.4.   El exponente de «a» de la curva de control limitante desde el punto de vista de la justificabilidad en función del factor dosimétrico de dosis colectiva.

La figura 4 presenta los valores de «a» para un valor fijo del límite autorizado de descarga.  Para hacer el gráfico se han supuesto valores de $L = 2$ Ci/año;  $C_{r-h}$ = US$\$$100/rem.hombre;  $k_1$ = US$\$$260/Ci ($^{131}$I);  y $k_2$ = US$\$$100/Ci ($^{131}$I).  Para evaluar las constantes $k_1$ y $k_2$ se ha supuesto un valor de la energía nucleoeléctrica de US$\$$ $10^{-2}$/kW(e)·h;  un costo de la potencia nucleoeléctrica instalada de US$\$$ $10^{6}$/MW(e);  un inventario de $^{131}$I en equilibrio de $10^4$ Ci ($^{131}$I)/MW(e);  y una vida útil de las centrales nucleares de 30 años.  Resulta evidente que cualquier otro análisis económico determinará valores diferentes de dichas constantes, sin invalidarse por ello el concepto de su aplicación.


5.   METODO PARA EVALUAR LA OPTIMIZACION
     DEL DISEÑO DE LA SEGURIDAD NUCLEAR

Los valores del exponente de la curva de control, resultantes de los criterios expuestos en los capítulos 3 y 4, son valores límites que tienen en cuenta los riesgos máximos aceptados por el público y la justificabilidad de las prácticas con radiaciones ionizantes.  Sin embargo, ninguno de los valores de «a» así obtenidos serán, en principio, los que determinen un diseño optimizado.  Si se emplea un criterio de optimización similar al recomendado para evaluaciones de radioprotección [8], la expectancia de exposición a la población debe ser tan baja como resulte razonablemente obtenible teniendo en cuenta consideraciones sociales y económicas.

La obtención de un nivel apropiado de seguridad nuclear implica un costo de inversión y operación que generalmente está directamente relacionado con el tamaño de la instalación. El costo total de la instalación podría incrementarse considerablemente sin que tal aumento de costo involucrase un incremento de seguridad equivalente. En consecuencia parece razonable evaluar el costo de las instalaciones y el beneficio que se obtiene de su eventual incremento, utilizando un criterio de análisis diferencial costo-beneficio similar al recomendado por la CIPR para las operaciones normales.

En consecuencia, el valor de diseño optimizado corresponderá a una expectancia de dosis colectiva para la cual se cumple la siguiente relación:

$$\frac{dP}{dE} = \frac{-dG}{dE} \tag{9}$$

donde:

P     es el costo definido en (6);

G     es la expectancia de detrimento total definida en (6);

E     es una variable cualquiera que refleje la expectancia de dosis colectiva.

Para poder efectuar el análisis diferencial planteado en la ecuación (9), las funciones de costo y expectancia de detrimento deben definirse en función de una variable representativa de E. Resulta conveniente utilizar como variable a la expectancia (R) de escape de $^{131}$I definida en (4). En esas condiciones, el análisis diferencial permitirá evaluar el valor «a» de la pendiente de la curva de control que asegure que el diseño ha sido optimizado. En este caso, de la ecuación (8) sólo interesan los valores de P y G:

$$P = k_2.I \tag{10}$$

$$G = FDC.C_{r\text{-}h}.T_v.R \tag{11}$$

Por otra parte, de la ecuación (4) se puede deducir la ecuación que relaciona al inventario de $^{131}$I con la expectancia de escape R.

$$I = \left[ \frac{\ln 10.(1-a).R}{L^a} + L^{(1-a)} \right]^{(1-a)^{-1}} \tag{12}$$

Luego la expresión del costo (10) puede escribirse:

$$P = k_2 \left[ \frac{\ln 10.(1-a).R}{L^a} + L^{(1-a)} \right]^{(1-a)^{-1}} \tag{13}$$

Derivando las ecuaciones (11) y (13) respecto de R, igualándolas y reemplazando el valor de R por el de la ecuación (4), se obtiene la relación siguiente:

$$\ln 10.k_2 \left(\frac{I}{L}\right)^a = -FDC.C_{r\text{-}h}.T_v \tag{14}$$

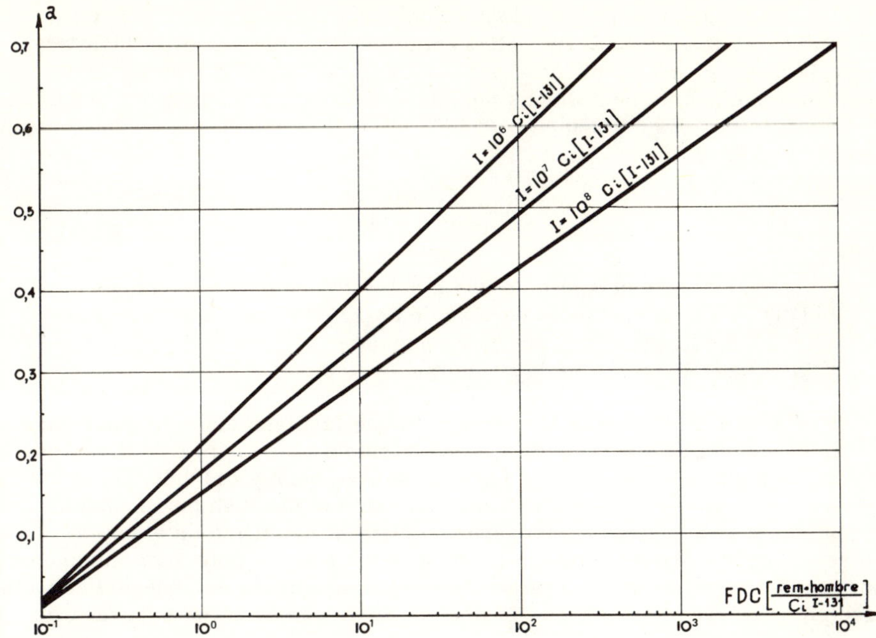


FIG.5.  El exponente «a» de la curva de control optimizada en función del factor dosimétrico de dosis colectiva.

Teniendo en cuenta que $k_2$, $C_{r-h}$ y $T_v$ son valores constantes para una instalación dada, de la ecuación (14) se puede despejar el valor de «a» óptimo para la instalación, para distintos inventarios de $^{131}$I y para diversos sitios de ubicación representados por el factor dosimétrico FDC.  La figura 5 presenta los valores de «a» suponiendo valores de L = 2 Ci/año; $C_{r-h}$ = US$100/rem.hombre y $k_2$ = US$100/Ci ($^{131}$I).

## 6.  PROPUESTA DE UN CRITERIO PARA LIMITAR LA EXPECTANCIA DE LA DOSIS COLECTIVA COMPROMETIDA POR UNIDAD DE PRACTICA

Recientemente se ha sugerido utilizar, como criterio suplementario de protección radiológica, la limitación de la dosis colectiva comprometida debida al funcionamiento normal de las instalaciones actuales, de manera tal de acotar la contribución de esas instalaciones a las dosis susceptibles de ser recibidas por el público en el futuro.  El objetivo de este criterio es evitar cargar a las generaciones venideras con costos de protección adicionales para mantener una cota de riesgos dada [9].  Dicho criterio puede ser extendido a la seguridad nuclear limitando en este caso la expectancia de dosis colectiva comprometida, de manera tal que la expectancia de dosis al público debida a los accidentes de las instalaciones actuales no sobrepase un determinado valor.

La expectancia de dosis colectiva comprometida puede definirse por
extensión del concepto «dosis comprometida» definido por el UNSCEAR,
como la integral temporal infinita de la expectancia de la tasa de dosis
promedio que podría recibir una población dada, multiplicada por el tamaño
de la población. Es decir que:

$$S_c = \int \overline{\dot{H}}(t).N(t).dt$$

donde:

$S_c$      es la expectancia de dosis colectiva comprometida;

$\overline{\dot{H}}(t)$      es la expectancia de tasa de dosis promedio a tiempo t de
comenzada la práctica u operación;

$N(t)$      es el tamaño de la población al tiempo t.


Por otra parte, se ha demostrado [9] que la dosis anual en una futura
situación de equilibrio es igual a la sumatoria de las dosis colectivas com-
prometidas por año de práctica (en condiciones normales) de todas las
instalaciones que utilizan radiaciones ionizantes, dividida por el tamaño de
la población. Se sugirió en consecuencia que si se utiliza un valor de
10 rem.hombre/MW(e).año, como límite para la dosis colectiva comprometi
por unidad de práctica, cuando se alcance un equilibrio de generación nucleo
eléctrica de 1 kW(e) per capita, la tasa de dosis promedio no superará el
valor de $10^{-2}$ rem/año.

Utilizando el mismo criterio este concepto podría ser extrapolado para
las situaciones accidentales, fijando un valor de 10 rem.hombre/MW(e).año,
como límite de la expectancia de dosis colectiva comprometida por unidad
de potencia eléctrica instalada, para asegurar que la expectancia de dosis
anual en el equilibrio no superará un valor de $10^{-2}$ rem/año. En consecuenc
podría escribirse la siguiente relación:

$$\frac{10 \text{ rem.hombre}}{MW(e).\text{año}} \geqq \frac{T_v \sum R_i.FDCC_i}{P(e).T_v} \tag{15}$$

donde:

$R_i$      es la expectancia de escape del radionúclido i debida a los
accidentes de la instalación nuclear;

$T_v$      es la vida útil de la instalación;

$FDCC_i$      es la tasa de dosis futura por unidad de dosis colectiva
comprometida del radionúclido i;

$P(e)$      es la potencia eléctrica instalada.


El valor de R puede calcularse a partir de una expresión similar a la (
utilizando como «límite de descarga» e «inventario» los valores correspondientes al radionúclido en consideración. De esta forma podría despeja
un juego de valores de la pendiente «a» de la curva de control para los
distintos radionúclidos involucrados en la evaluación. El mayor valor de
«a» sería límite desde el punto de vista de la expectancia de dosis colectiv
comprometida.

El valor de la sumatoria de la ecuación (15) debería computarse para
todos los radionúclidos que contribuyan substancialmente a la dosis colecti

comprometida.  Se ha estimado[9] que la mayor contribución derivada de la
operación normal se deberá al $^{85}$Kr, al tritio y al $^{129}$I.  La mayor parte de la
radiactividad de estos núclidos ingresará al medio ambiente aunque no ocurra
ninguna situación accidental debido a que aparecen como residuos de la
reelaboración de los combustibles nucleares;  en consecuencia, la contribución a la tasa de dosis futura debida al $^{85}$Kr, al tritio y al $^{129}$I debe ser
tenida en cuenta en la evaluación de la operación normal de las instalaciones,
por lo que resulta innecesario hacerla en los análisis accidentales.

Los radionúclidos que interesa evaluar en las situaciones accidentales
son aquellos que tienen un período de semidesintegración lo suficientemente
corto como para que su actividad sea insignificante en el momento de la
reelaboración, pero no tanto como para contribuir poco a la dosis colectiva
comprometida.  Se estima que algunos gases nobles, en especial el $^{133}$Xe,
deben ser, por sus características fisicoquímicas, los mayores contribuyentes
a la expectativa de dosis colectiva comprometida derivada de situaciones
accidentales en los reactores nucleares.


7.  CONCLUSIONES

Se ha propuesto un criterio para evaluar la seguridad nuclear de un
reactor de potencia basado en una metodología que se estima evita costos
innecesarios y garantiza un nivel adecuado de seguridad nuclear.  El criterio
se basa en la determinación del valor del exponente de una curva de control
similar a la empleada en las evaluaciones probabilísticas y determinada por
extrapolación de los principios básicos de la protección radiológica al análisis
de las situaciones accidentales.  En la determinación se supone que el reactor
ha sido diseñado de acuerdo a ciertos criterios determinísticos de manera de
asegurar que la instalación ha sido conformada con un criterio particular de
excelencia en cuanto a las situaciones de fallas que involucren liberación
de productos de fisión.  En consecuencia, puede evaluarse la seguridad del
reactor mediante una curva de control cuya relación probabilidad/severidad
pueda representarse con una función hiperbólica.

Se han presentado métodos de evaluación que determinan tres curvas
de control:  con el primero puede obtenerse el exponente limitante desde el
punto de vista de los riesgos máximos para los miembros individuales de la
población;  con el segundo, el exponente limitante desde el punto de vista del
beneficio que la población en su conjunto espera del reactor nuclear;  y
finalmente, con el tercero, el exponente que define la curva de control
óptima.

Se ha graficado el exponente de la función para diferentes factores
dosimétricos llevando como parámetro el inventario de $^{131}$I.  Del análisis de
esas figuras pueden sacarse las siguientes conclusiones:

1)  La expectancia de dosis individual varía muy poco con el inventario
    radiactivo del reactor en el orden de potencias habituales.
2)  Cuando la dosis colectiva por unidad de actividad de $^{131}$I es elevada
    para un emplazamiento dado, un reactor es tanto más justificable cuanto
    mayor es su potencia.
3)  En la mayoría de los casos el análisis de optimización es menos restric-
    tivo que el de justificación.

Se presentan también las bases de cálculo para determinar el exponente de una curva de control que limitaría la tasa de dosis futura debida a los accidentes en las instalaciones actuales. En un próximo trabajo se evaluarán los radionúclidos que contribuyen significativamente a la expectancia de dosis colectiva comprometida por situaciones accidentales, para hacer utilizable este criterio.


## AGRADECIMIENTOS

El autor agradece al Dr. D.J. Beninson la discusión de las ideas sobre las cuales se basa este trabajo, a la Dra. A. Migliori de Beninson y al Dr.A. Carrea las sugerencias que aportaron al mismo, y al Ing. A.E. Placer la discusión y revisión del texto final.


## REFERENCIAS


[1]   OIEA, Containment and Siting of Nuclear Power Plants (Actas Simp.Viena, 1967),OIEA, Viena (1967).

[2]   FARMER, F.R., «Siting criteria — a new approach» Containment and Siting of Nuclear Power Plants (Actas Simp.Viena, 1967),OIEA, Viena (1967) 303.

[3]   BEATTIE, J.R., BELL, G.D., EDWARDS, J.E., Methods for the evaluation of risk, UKAEA, AHSB (S) R-159 (1969).

[4]   BELL, G.D., Risk evaluation for any curie release spectrum and any dose-risk relationship, UKAEA, AHSE (S) R-192 (1970).

[5]   BEATTIE, J.R., BELL, G.D., «A possible standard of risk for large accidental releases», Principles and Standards of Reactor Safety (Actas Simp.Jülich.(1973),OIEA, Viena (1973) 11.

[6]   MELEIS, M., ERDMANN, R.C., The development of reactor siting criteria based upon risk probability, Nucl.Saf. 13 (1972), 22; FARMER, F.R., Letters to the editor, Nucl.Saf. 13 (1972).

[7]   CIPR, Recommendations of the International Commission on Radiological Protection, CIPR, Publication Pergamon Press, London (1965).

[8]   CIPR, Implications of Commission recommendations that doses be kept as low as readily achievable, CIPR, Publication 22, Pergamon Press, London (1973).

[9]   BENINSON, D.J., «Population doses resulting from radionuclides of worldwide distribution», Population Dose Evaluation and Standards for Man and his Environment (Actas Semin.Portoroz, 1974), OIEA, Vien (1974) 227.

[10]  UNITED NATIONS, A report of the United Nations Scientific Committee on the Effects of Atomic Radiation to the General Assembly, Vol.1 (Levels), Offic.Rec. General Assembly, 27th Sess.Suppl. N° 25 (A8725), Nueva York (1972).

[11]  SLANSKY, C.M., Principles for limiting the introduction of radiactive waste into the sea, At.Energy Rev. 9 4 (1971) 854.

[12]  COTTRELL, W.B., AEC testimony at the 1973 JCAE hearings on reactor safety, Nucl.Saf. 15 (1974) 132.

[13]  OIEA, Risk Evaluation for Protection of the Public in Radiation Accidents, Colección Seguridad N° 21, OIEA, Viena (1967).

[14]  USAEC, General Environmental Siting Criteria for Nuclear Power Plants, USAEC Rules and Regulations (Sept.1974).

[15]  DUNSTER, H.J., McLEAN, A.S., The use of risk estimates in setting and using basic radiation protecti standards, Health Phys. 19 (1970) 121-22.

[16]  HEDGRAN, A., LINDELL, B., POR, A special way of thinking, Health Phys. 19 (1970) 121.

[17]  OTWAY, H.J., BURNHAM, J.B., SOHRDING, R.K., Economic versus biological risk as reactor design criteria, Presented at the IEEE Nuclear Science Symposium (1970).

[18]  LEDERBERG, J., Squaring an infinite circle: radiobiology and the value of life, Bull.At.Sci. 27 (1971) 43-45.

[19]  COHEN, J.J., Plowshare: new challenge for the health physicist, Health Phys. 19 (1970) 633-39.

[20]  SAGAN, L.A., Human costs of nuclear power, Science 177 (1972) 487-493.

## DISCUSSION

Pamela M. BRYANT:  Would Mrs. Beninson please clarify whether the author included exposure to inert gases in his study?

Ambretta MIGLIORI DE BENINSON:  No, the author analysed only the exposures due to $^{131}$I, which was assumed to represent the highest risk.

Pamela M. BRYANT:  I understood you to say that the ICRP dose limit applicable to normal operation had been chosen as the limiting dose in the accident situation envisaged in the study.  This seems unduly restrictive. In a new study in the United Kingdom, we have used twice the annual occupational maximum permissible exposure, i.e. 10 rem to the whole body, as the emergency reference level of dose for guidance in instituting counter measures.

Ambretta MIGLIORI DE BENINSON:  The ICRP limits for normal operations have been chosen as a limit to dose expectation $(\sum \text{prob.} \times \text{dose})$ and not to actual doses.

A.P. HULL:  Calculations in the paper appear to be based on an estimated worth of US \$100 per man·rem.  Since this is a value derived on the basis of whole-body exposures, logically a somewhat smaller amount would be appropriate for thyroid·rem exposures due to $^{131}$I.

Ambretta MIGLIORI DE BENINSON:  I fully agree with you. As the risk of death from thyroid cancer is lower by an order of magnitude than that from any other radiation-induced cancer, the value is probably too conservative by that order of magnitude.

S.O.W. BERGSTRÖM:  Has the author considered only late effects on the thyroid (malignant tumours) or also acute effects (functional disturbance)?

Ambretta MIGLIORI DE BENINSON:  Only late effects were considered. But a very conservative figure is used for the cost of the man·rem to the thyroid, and this may to some extent cover the functional effects.

S.O.W. BERGSTRÖM:  The conclusion concerning the superiority of large reactor sizes, if based exclusively on late effects, may have to be changed when acute effects are also considered.  Some population distributions may be such that acute consequences will predominate over late consequences when more than a few per cent of the iodine inventory is released.

Ambretta MIGLIORI DE BENINSON:  I agree that the conclusion might have to be changed, but only for certain population distributions and for a very substantial release:  in our opinion it would have to be more than 10% of the inventory.  Plume rise will, in very large releases, tend to mitigate the effects to a substantial extent.

S.O.W.  BERGSTRÖM:  I doubt whether one can, without knowledge of population distribution, activity composition and release time, make general statements about the changes in the consequences.  A very large iodine inventory is sufficient to give high doses over very large distances.

E. IANSITI:  I think all these criteria which are based on releases should, for the moment at least, be used with caution;  in the first place constant diffusion characteristics independent of the size of the release cannot be accepted, because the heat transmitted to the cloud during a large release may have important effects, and, second, there are many kinds of release, releases from the core and from the containment, releases to the environment in gaseous form, and so on.

I have, of course, no comments on the general criteria but, according to the latest safety theories, it may be necessary to perform more elaborate calculations and formulate more detailed definitions.

# SITE SELECTION PROCEDURE FOR NUCLEAR POWER PLANTS IN DENMARK

L. HANNIBAL
National Health Service of Denmark,
State Institute of Radiation Hygiene, Copenhagen

J. SCHULTZ-LARSEN*
University Institute of Medical Genetics,
Copenhagen, Denmark

## Abstract

SITE SELECTION PROCEDURE FOR NUCLEAR POWER PLANTS IN DENMARK.
Although no political decision to introduce nuclear power in Denmark has yet been made, appropriate action has been taken to make land reservations that will be necessary for prospective nuclear power plants. As it is wished to treat the problem on the basis of long-range national planning, an interdepartment committee has been appointed to deal with the matter. The committee includes representatives from a number of ministries, the Danish Atomic Energy Commission, the National Health Service, various government advisers, and the Danish Electricity Council. This paper describes how the committee arrived at the selection of suitable sites in the western part of the country. For a radiological safety assessment of the sites, a number of dose-related parameters were calculated, based on a loss-of-coolant-accident $^{131}$I release and the pertinent population distribution. In accordance with the general recommendation by the International Commission on Radiological Protection that, economic and social considerations being taken into account, all doses should be kept as low as reasonably achievable, those sites were selected which had the lowest parameter values and which were acceptable from other points of view to the committee. Further, the paper mentions how a radius of 2-3 km was arrived at for an inner zone where very few new dwellings will be permitted, and reference is made to the choice of 10 km for the radius of an outer zone in which the population density will not be allowed to increase significantly.

## 1. INTRODUCTION

A political decision to erect nuclear power plants in Denmark has not yet been made, but Danish authorities have already started taking steps to reserve areas that will be necessary for the purpose. The problems are being treated by an interdepartmental committee headed by the Ministry of Environment. The committee is working under the following mandate:

"... to put forward one or several proposals as to the siting of necessary nuclear power plants, on the basis of present proposals and viewpoints and considering other existing interests in the areas, and to put forward proposals as to how the necessary areas be secured, including which instructions or directives this imposes upon local authorities, and which legislation may prove necessary."

The committee has included representatives from the following authorities: the Ministry of Environment, the Ministry of Public Works, the Danish Atomic Energy Commission, the National Health Service, the Secretariat for National Planning, the Government Adviser on Urban Planning, and the

---

* Consultant to the National Health Service of Denmark.

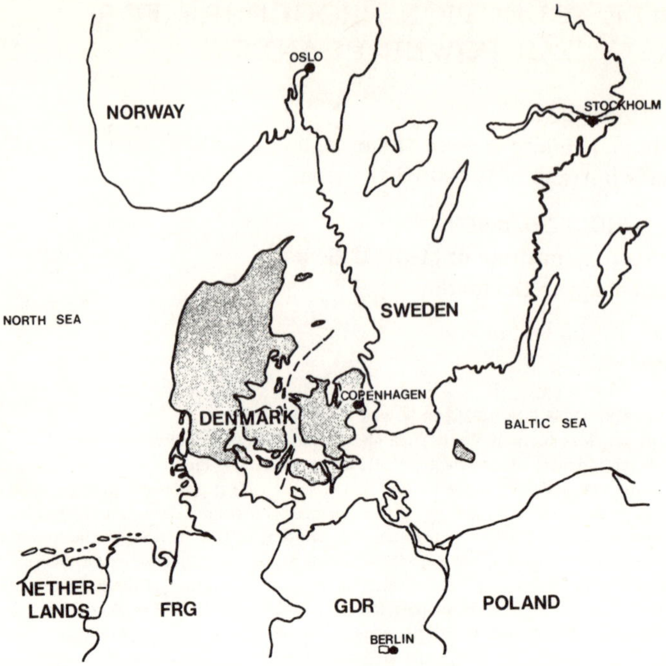


FIG. 1.   Map of Denmark and neighbouring countries.   Dashed line indicates the border between the two electricity supply regions.

Government Adviser on Conservation of Nature and Landscape Planning. Finally, the Electricity Council, an advisory body on power current under tl Ministry of Public Works, has also been represented on the committee.

The Danish electricity companies have united into two completely separ groups, one for the eastern islands (including the city of Copenhagen) and one for the western part of the country, i. e. the peninsula of Jutland and the island of Funen.   Figure 1 gives a rough map of Denmark and surrounding countries.   This paper describes how the committee arrived at the selection of suitable sites in the western region.

The committee is at present working on the selection of sites, according to similar principles, in the eastern part of the country.

The background for the task has thus been the intention of making land reservations now at adequate sites, independent of an eventual political decision to introduce nuclear power in Denmark.   In this way, when the tim comes to erect nuclear power plants, a situation may be prevented in which later arrangements for these areas would necessitate choice of inferior site The reason for this procedure is largely to be found in the fact that Denmar is quite a small country, homogeneously and densely populated, with an average distance of some 30 km between major towns and with a population density in rural areas of $\sim 60$ persons/km$^2$.   Many urban areas are develop rapidly and there is an ever increasing nation-wide demand for occupation land for various private and public purposes.   Within the procedure descril these trends may be incorporated into rational, long-term national plannin

At the beginning of the work, there were already legal provisions permitting land reservations.  According to the Danish Urban and Rural Zones Act, the Minister of Environment may issue a regulation to the effect that administration of matters of more far-reaching importance are to be transferred from local authorities to the Ministry.  Furthermore, the Minister may, under the terms of the Act on National and Regional Planning and upon negotiations with other ministers involved, decide that the basis of the regional planning is to be made according to the lines laid down in detail by the Ministry, and he may issue directives to the local authorities on how to administer the provisions.  Thus it is possible to ensure that the siting of prospective nuclear power plants in the areas suggested forms part of local as well as national planning.

## 2.    GENERAL CONSIDERATIONS

The basic principle in the safety assessment of possible sites has been the general recommendation expressed by the International Commission on Radiological Protection (ICRP):  "... that all doses from justifiable exposures be kept as low as is reasonably achievable, economic and social considerations being taken into account" [1].

Considering the situation in Denmark, it seems rational to employ this principle.  If it is found necessary on social grounds to introduce nuclear power in Denmark, and if in this connection the utilization of sites and security zones of a nature not compatible with the country's population structure are insisted on, then this would not be "reasonably achievable, economic and social considerations being taken into account".

In other words, the committee's task has been to perform a relative assessment of a number of possible sites originally proposed by the electricity companies in the western part of the country, and to lay down the provisional extent of security zones.  The committee was assigned the task to point out those sites at which accidental releases would result in minimum consequences to the neighbouring population.  In choosing the sites, due consideration had to be given to overall environmental consequences, urban planning trends, the existence of airfields and of military installations, seismic conditions, etc.  Also, major consideration was to be given to the preservation of natural amenities.

## 3.    RELATIVE ASSESSMENT OF SITES

The point of departure was a total of 49 sites, porposed by Elsam, the electricity company group in western Denmark, i. e. Jutland and Funen. The sites were all considered usable on operating and economic grounds, but the committee found it mandatory to add a number of sites in order to secure a representative choice.  For reasons of safety of supply and of service and economy, the committee decided that land reservations be undertaken within certain geographic regions: the west coast of Jutland, four parts of the east coast of Jutland, and the island of Funen.  The sites investigated are all situated on the coast or, in four cases, on a small off-shore island.  The reason for this is that the country has no large rivers and that the utilization of cooling towers is deemed undesirable.

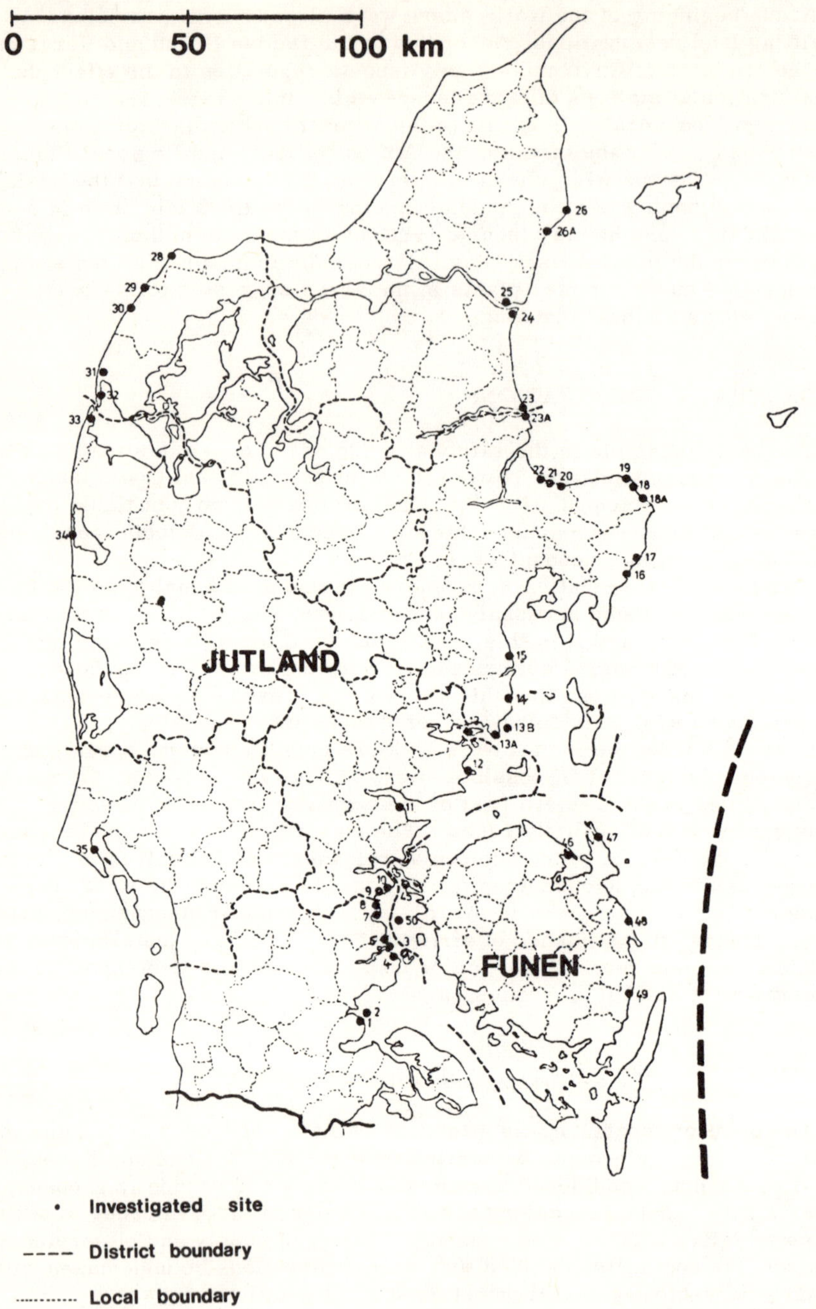


FIG. 2.   Key map of investigated nuclear power plant sites in the Elsam region, Denmark.   Heavy dashed lin
indicates the border between the two electricity supply regions.

For the radiological safety assessment of the sites, a number of dose-related parameters were calculated by the Health Physics Division of the Danish Atomic Energy Commission's Research Establishment Risö. Altogether 45 sites were considered. Some sites were soon dropped by the committee, being unacceptable for one or several reasons. Figure 2 is a map showing the sites investigated.

In the relative evaluation, the event considered is a release of $^{131}$I, and the consequences are taken to be represented by thyroid doses. Given the predominance of $^{131}$I in an actual accident release, this simplified procedure is considered to give a reasonable basis for a relative evaluation. In other words, one imagines a power plant on each site, and the same reference accident is applied. The accident regarded is a design-basis loss-of-coolant accident, as employed by the United States Atomic Energy Commission in their Safety Guides 3 and 4 [2]. Of the iodine component of the core inventory, 25% is considered to be immediately available for leakage from containment. Of this 25%, 5% is taken to be particulate iodine, 10% methyl iodide, and 85% elemental iodine. The leak rate from containment is assumed to be 0.1% per 24 hours, the release height is set at 0 metres, and no deposition is considered. The leak is assumed to last for two hours under the worst conceivable meteorological conditions (inversion, stability class F (Pasquill) [3]). The wind speed is assumed to be 2 m/s. For the dose calculation, the pessimistic assumption is made that elemental iodine is transformed into particulate, aerosol-borne iodine.

A polar mesh with twelve 30° sectors is centred in each site (Fig. 3). Somewhat arbitrarily, an inner zone is defined as the area within the radii of 0.5 km and 5 km; the area between radii of 5 km and 50 km is called the outer zone.

The release is imagined to proceed successively along any of the twelve sectors. For the centre of each sector segment, the adult thyroid dose caused by inhalation of the $^{131}$I during the two-hour passage of the cloud was calculated.

The parameters considered were the following:

(1)  The collective dose in the most densely populated 30° sector within the inner zone. The quantity is calculated as

$$\sum_i x_i$$

where $x_i$ is the collective adult thyroid dose in the $i^{th}$ sector segment of the sector in question, the summation being extended over all sector segments within the inner zones.

(2)  The collective dose in the most densely populated 30° sector in the area within 50 km (referred to as the total zone). Calculated in the same manner as (1).

(3)  The collective dose within 50 km, i.e. the sum of all sector contributions within the area 0.5–5 km.

Also, a couple of more sophisticated dose-based indices have been given consideration. These are:

(4)  An inner zone index, calculated as

$$\sqrt{\sum_j \left(\sum_i n_{ij}\, y_{ij}^2\right)^2}$$

where $n_{ij}$ is the number of persons and $y_{ij}$ is the individual adult thyroid dose in the $i^{th}$ segment of the $j^{th}$ sector, the summation being extended over all segments within the annulus 0.5-5 km.

(5) A total index, calculated in the same way as the inner zone index, the summation this time being extended to a distance of 50 km.

This procedure of squaring the doses in each partial area is cognate to the procedure proposed originally by Farmer [4]. By weighting the sector segment doses, according to the above root-square principle, around the compass card, greater importance is seen to be attached to persons living in the proximity of the site as well as to dense population clusters even at more remote distances.

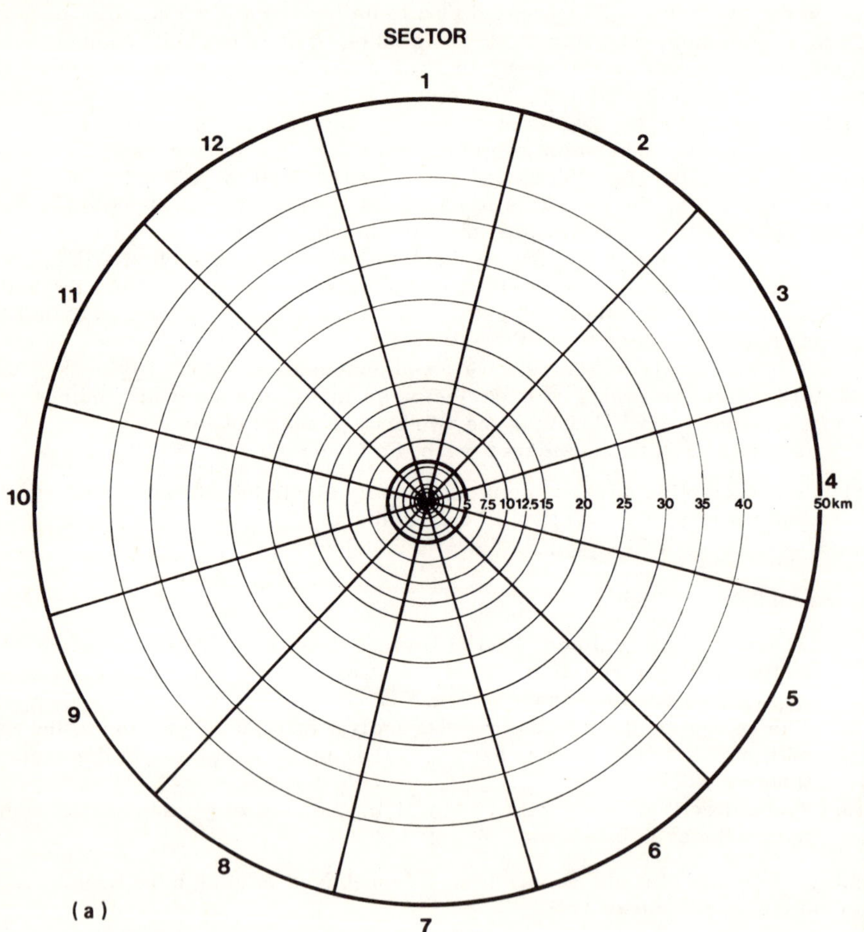


FIG.3. (a) Polar mesh for dose calculations. Both inner and outer zone shown. (b) Inner zone polar mesh for dose calculations.

In the dose calculations, seasonal and day-night fluctuations in the
population in summer houses, public institutions, etc., have been taken into
account and have been taken to represent the worst cases.  The size of the
population in the inner zone has been fixed upon the best available local
estimate of the present number of persons.  For the outer zones, allowance
was given to the anticipated population increase trends up to the time when
the first plants are believed to start operation, by making use of the population structure as predicted by existing forecasts for the year 1985.

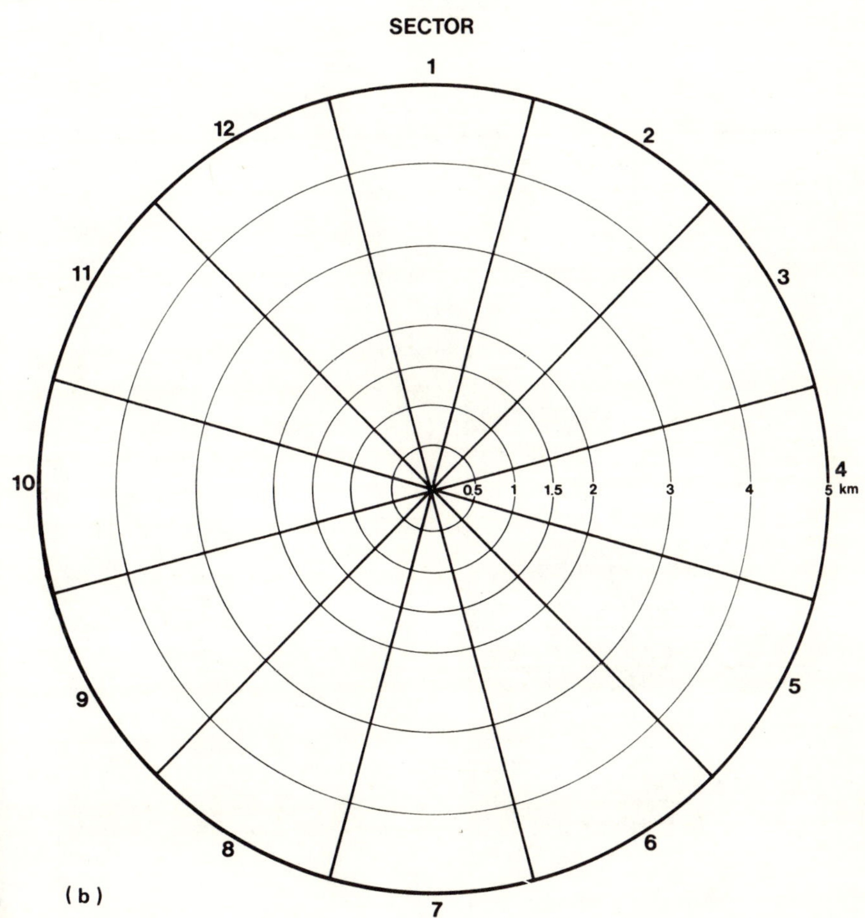

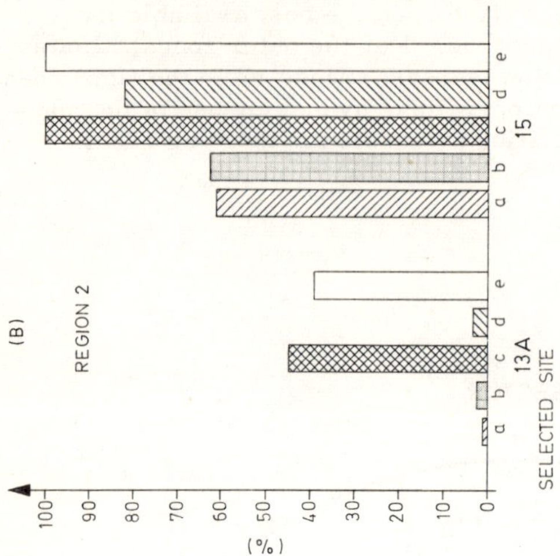
(A)
REGION 1
(%)
100 90 80 70 60 50 40 30 20 10 0
a b c d e
5
a b c d e
7
a b c d e
2
SELECTED SITE

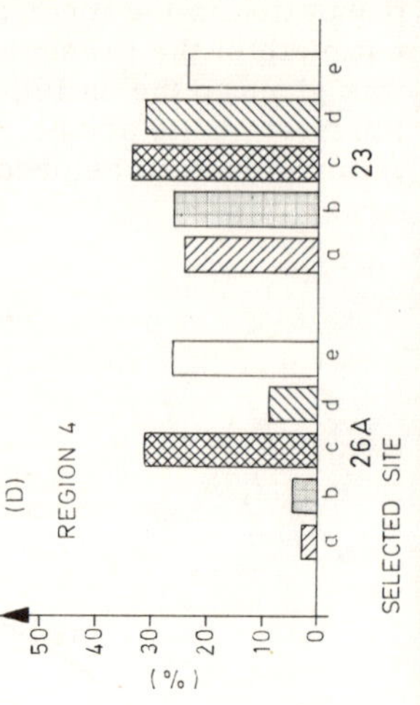
(B)
REGION 2
(%)
100 90 80 70 60 50 40 30 20 10 0
a b c d e
13A
a b c d e
15
SELECTED SITE

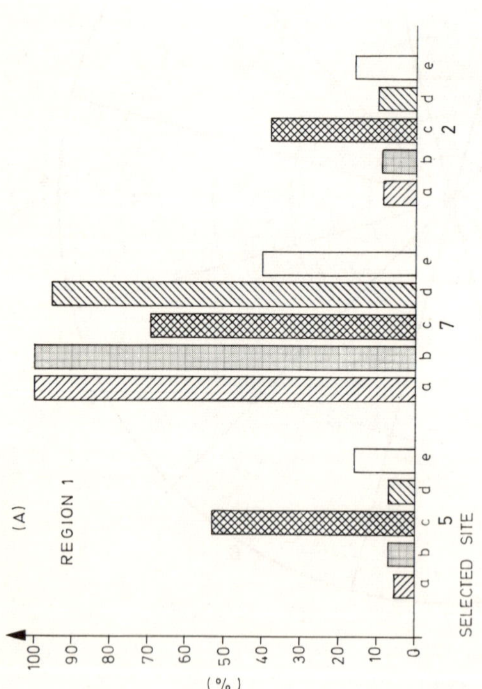
(C)
REGION 3
(%)
50 40 30 20 10 0
a b c d e
17
a b c d e
22
a b c d e
19
SELECTED SITE

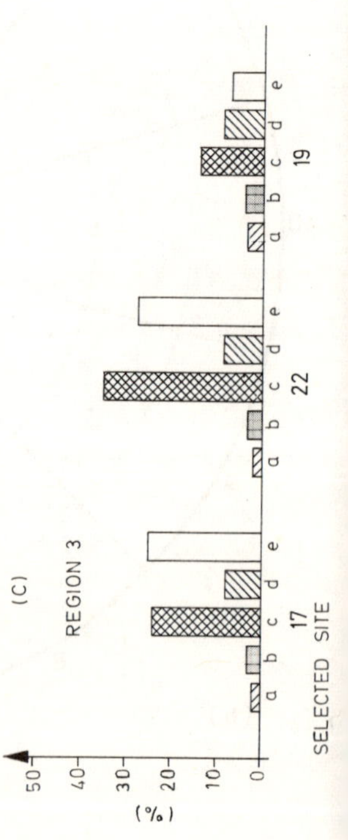
(D)
REGION 4
(%)
50 40 30 20 10 0
a b c d e
26A
a b c d e
23
SELECTED SITE

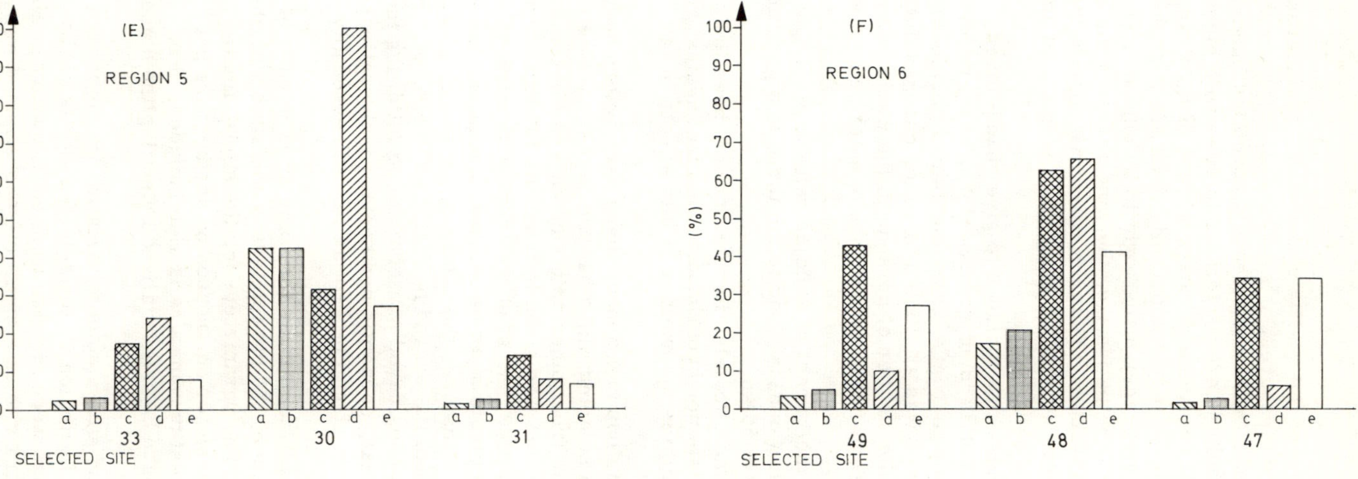



FIG.4.  (A)  Region 1.  5: Selected site.  7: Site with largest parameter values.  2: Site with lowest parameter values.

(B)  Region 2.  13 A: Selected site.  15: Site with largest parameter values.

(C)  Region 3.  17: Selected site.  22: Site with largest parameter values.  19: Site with lowest parameter values.

(D)  Region 4.  26 A: Selected site.  23: Site with largest parameter values.

(E)  Region 5.  33: Selected site.  30: Site with largest parameter values.  31: Site with lowest parameter values.

(F)  Region 6.  49: Selected site.  48: Site with largest parameter values.  47: Site with lowest parameter values.

Shaded blocks — a: inner zone index;  b: total zone index;  c: collective dose in total zone;  d: collective dose in most densely populated inner zone sector;  e: collected dose in most densely populated total zone sector.

## 4.   SELECTION OF SITES

In the committee's evaluation of the quality of the sites, considerable attention was given to emergency measures.

Sites with low collective doses in the most densely populated inner zone sector and with low inner zone indices are preferable to sites with higher values because fewer people will receive high doses in case of a release and because evacuation action will provide the possibility of an efficient dose reduction.  It was thought that emergency plans ought to be organized according to the worst conceivable situation, i. e. the wind carrying the release along the most densely populated sector.  Thus the collective dose in this sector gives a relative measure of demand for emergency plans, and major importance was attributed to this parameter.

The use of the inner zone index was included to give space for an evaluation of the impact on individuals in adjacent sectors, to allow for changes in the wind direction, and to treat a larger number of accidents statistically.  No great importance was attributed to minor changes in the index, as it was presumed that the number of persons in the whole inner zone (a few thousand people) is not greater than what adequate emergency measures are capable of coping with.

To limit as far as possible the impact of a major release on larger population groups, consideration was also given to the three last parameters including doses up to a distance of 50 km.

In other words, the aim of the site selection has not only been the protection of persons in the immediate neighbourhood of the plant against acute illness or fatality but also to minimize the number of late effects in the population at greater distances.

In each of the six coastal regions mentioned, these parameters have been compared, and those sites were selected which had the lowest parameter values and which were acceptable from other points of view to the committee. For instance, considerable attention was given to interests in regulations for preservation of natural amenities.  Most of the areas with highest amenities are exactly where the plants must be located, i. e. along the coast. If, in deference to conservation of nature, the choice of the radiologically preferable site was prohibited, then the second best site was selected.  By following this practice, we have considered ourselves to be following the general ICRP recommendation that, economic and social considerations being taken into account, all doses should be kept as low as reasonably achievable.

Figure 4 illustrates the parameter values for some sites in each of the six coastal regions.  For reference, the parameters of the sites with the highest parameters are given.  The parameters for the site(s) finally selected are shown, and when these for one reason or another are not the lowest, the site with the lowest parameters is also included.  For clarity the values for each parameter have been normalized by setting the highest parameter value equal to 100%.

Roughly speaking, there was quite a smooth variation in the parameters which facilitated the selection.  In Region 1, the selected site is really a two site, as the two adjacent locations were found to have equal quality in all respects.  In this region, an extra site on a small off-shore island was chosen at the request of the Ministry of Environment.  In Region 2, an alternative site was ultimately selected. Figure 5 shows the positions of the nine sites that were finally selected.

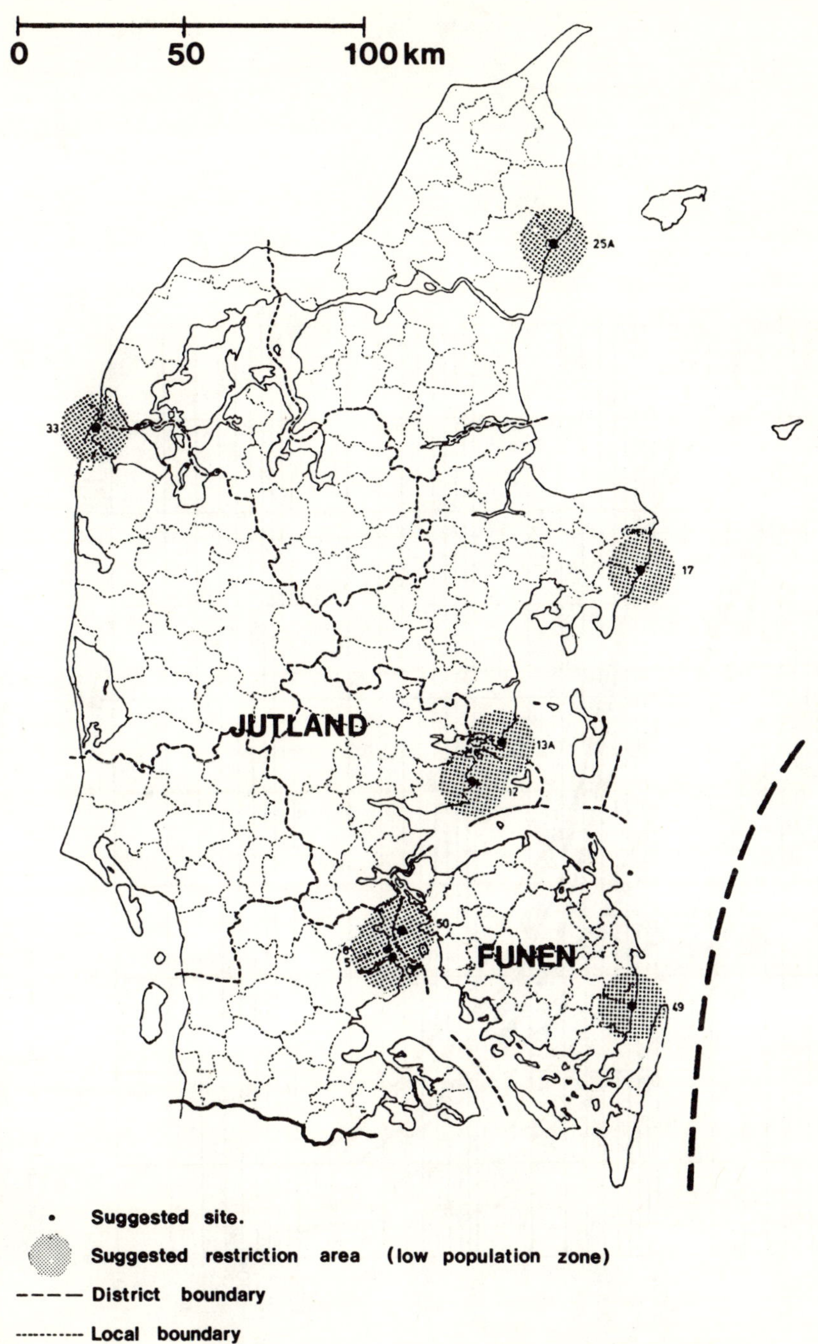


FIG.5.  Key map of the nine sites suggested for nuclear power plants in the Elsam region, Denmark.  Heavy dashed line indicates the border between the two electricity supply regions.

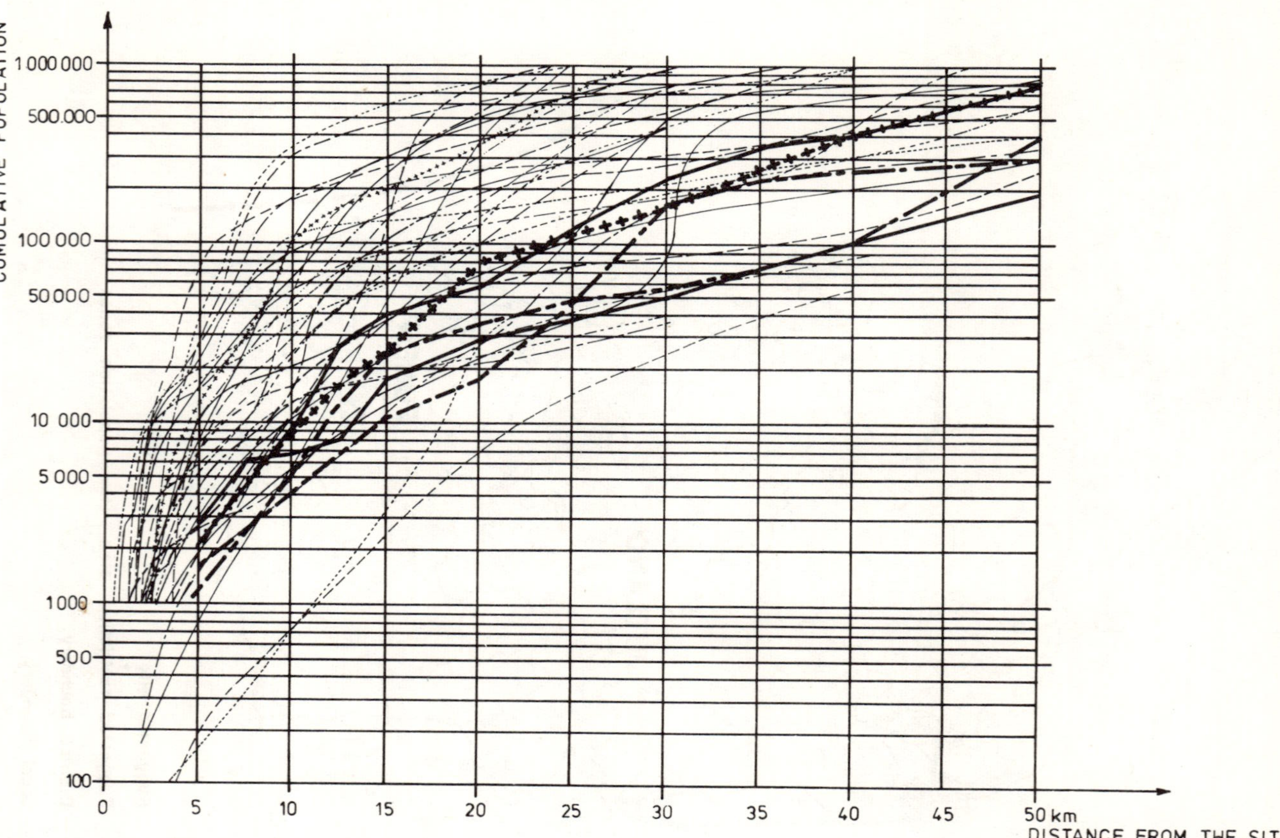


FIG. 6.    Cumulative population as a function of distance from site.   Curves with heavy lines represent typical Danish selected sites.   Thin lines represent existing sites in various European countries.

Figure 6 shows typical graphs of the cumulative population round some of the selected sites.  Comparison with the curves for some typical nuclear power plants in other European countries reveals that, despite Denmark's relatively high population density, sites have been successfully chosen that are not inferior to the average foreign site.


## 5.   EXTENT OF SECURITY ZONES

The evaluation of the necessary extent of security zones has been based on the following considerations.  Taking into account the character of the Danish landscape, with many medium-size and small towns and other population centres, the concept of a population centre distance has been avoided.  The choice was restricted to the adoption of an inner zone, in which emergency measures will provide the main protection to the public in case of an accident, and an outer zone,at whose outer border an individual is not likely to suffer an acute radiation syndrome even in the case of a massive accidental release.

The power of the reactor is taken to be 1000 MW(e), corresponding to 3000 MW(th).  Thus, the $^{131}$I inventory, after 1000 days irradiation time, will be $9.1 \times 10^7$ Ci.  Again, a design-basis loss-of-coolant accident is assumed to occur.  25% of the halogens inventory is assumed to be immediately available for leak from containment.  As previously, the leak rate from containment is stipulated to be 0.1% in 24 hours, and the weather conditions are taken to be given by stability class F.  Consistently, there is supposed to be no deposition.

To establish the extent of the inner security zone, the dose to a child's thyroid via inhalation of halogens as a function of the distance and of the duration of the release was considered.  The data were derived from Beattie and Bryant [5], who include iodine isotopes other than $^{131}$I, and tellurium isotopes.

Assuming an emergency reference level for evacuation of 300 rem to the thyroid, it is seen from Table I that at a distance of 2 km from the reactor, approximately 10 hours will be available for evacuation.  From the data, it can also be seen that a release ten times larger can still be managed by evacuation within the lapse of 2 hours at a distance of 3 km.  In view of these results and of the character of the country's rural areas, the limit was set at 2-3 km.  In these considerations it was borne in mind that design-basis accidents which may lead to thyroid doses of concern in the near neighbourhood are not likely to result in significant whole-body doses from noble gases.

To obtain a measure for the extent of the outer security zone, a severe accident is imagined to strike the previously mentioned 3000 MW(th) reactor. The accident is taken to be a core melt-out (or near melt-out), leading to the release of 100% noble gases and 25% volatiles.  Using the data in Ref.[5], the following values are obtained for the whole-body dose to an adult in class F weather, and assuming no plume rise:

| Distance (km) | Whole-body dose (rem) |
|:---:|:---:|
| 1 | 5000 |
| 2 | 1000 |
| 5 | 240 |
| 10 | 80 |

TABLE I.   DOSE TO CHILD'S THYROID FROM INHALATION OF I AND Te
ISOTOPES FOLLOWING DESIGN-BASIS ACCIDENT WITH 25% RELEASE
OF HALOGENS INTO CONTAINMENT AND CONTAINMENT LEAK RATE
OF 0.1% PER 24 HOURS.

| Distance from reactor (km) | Duration of release (h) | | | |
|---|---|---|---|---|
| | 2 | 4 | 8 | 16 |
| 1 | 180 | 360 | 720 | 1440 |
| 2 | 57 | 114 | 228 | 456 |
| 5 | 14 | 28 | 56 | 112 |
| 10 | 5 | 10 | 20 | 40 |

Reactor power 3000 MW(th).
Weather conditions:  stability class F [5].

These dose values include external radiation from the passage of the
cloud and internal doses from the inhalation of volatile $^{137}$Cs and iodine
isotopes.

In view of the pessimistic nature of the assumptions governing the accid
and with regard to the character of the Danish landscape, a radius of 10 km
was chosen as a realistic value for the outer zone border, making it extrem
improbable that any individual outside this area could ever suffer an acute
radiation syndrome as a result of the existence of the nuclear power plant.

## 6.   CONCLUSION

Out of a total of forty-five sites suggested for prospective nuclear power
plants in the western half of Denmark, the interdepartment committee called
attention to nine sites, recommending that land reservations be carried out
by administrative means, despite the fact that the political decision to
introduce nuclear power in Denmark has not yet been made.  The selection
the sites was made on the basis of a number of dose-related parameters and
with due regard to other points of view of the committee.

Within a distance of 2-3 km from the selected sites, the committee
recommended that permission for construction of further housing be given
solely in cases where it is commercially necessary for already existing far
and fisheries or for possibly inevitable dwellings for the staff of the plant,
assuming that an efficient emergency plan will be set up for the area.

In order to be able to cope as far as possible with accidental releases c
a drastic nature, which during rare meteorological conditions might have
serious radiological consequences, including acute radiation illness, up to a
distance of some 10 km, the committee by way of precaution advised that
urban districts within a radius of 10 km should not be allowed to develop int
densely built-up areas, and that the foundation of institutions not easily
evacuated should be avoided.

## REFERENCES

[1]   ICRP Publications 9 and 22, Pergamon Press, Oxford (1966 and 1973).
[2]   USAEC Safety Guide 3 (4):   Assumptions Used for Evaluating the Potential Radiological Consequences
      of a Loss-of- Coolant Accident for Boiling (Pressurized) Water Reactors    (1970).
[3]   PASQUILL, F. , The estimation of the dispersion of windborne material, Met. Mag. $\underline{90}$ (1961) 33.
[4]   FARMER, F. R. , The Evaluation of Power Reactor Sites, UKAEA Rep. DPR/INF/266 (1962).
[5]   BEATTIE, J. R. , BRYANT, Pamela M. , Assessment of Environmental Hazards from Reactor Fission
      Product Releases, UKAEA Rep.   AHSB(S)R 135 (1970).

## DISCUSSION

D. DAGAN:  Do you make allowance for an increase in population during the lifetime of the plant after 1985?

L. HANNIBAL:  We assumed the population predicted for 1985, because by that time the first power plants are expected to be in operation — provided the appropriate political decision is taken.  As I said in my conclusion, the construction of further housing would be severely restricted within a distance of 2-3 km from any site and no densely built-up areas would be permitted within a radius of up to 10 km.

D. DAGAN:  Could you please explain why the idea of wet cooling towers was given up?

L. HANNIBAL:  The utilities themselves did not want to use cooling towers, nor would the authorities have been very enthusiastic about such installations, for aesthetic and other environmental reasons.  Besides, excellent conditions for cooling water supply exist at the coastal locations in Denmark — far better than in most other countries.

Consuelo PEREZ DEL MORAL:  Have you assumed the application of any special safety devices associated with containment, such as sprayers, filters, penetration chambers or the like?  You mention in the paper that in defining the design-basis accident you used USAEC Safety Guides 3 and 4. These guides have been revised in the sense that the fraction of organic iodine assumed has been reduced; and this seems to be a more realistic hypothesis than the one adopted earlier.  I think this should be taken into account.

L. HANNIBAL:  Your point is probably an essential one in that the composition of the escaping iodine may have a significant influence on the doses obtained.  However, we used the original USAEC Safety Guides 3 and 4 because the revised versions were not available at the time.  Moreover, we made the pessimistic assumption that elemental iodine would be transformed into particulate, aerosol-base iodine, because we thought it wisest to take a conservative view.

Gloria CAMPOS VENUTI:  Could you please explain why the political decision to erect nuclear power plants in Denmark has not yet been taken? What is the public attitude to the use of nuclear energy in your country?

L. HANNIBAL:  In reply to the question why nuclear power has not already been introduced in Denmark, I can only say that until the advent of the energy crisis the country was well supplied with energy from fossil fuels.  The utilities, of course, are planning for nuclear power to meet the anticipated demand for energy.  Recently the political situation in Denmark has been somewhat chaotic, and the expected parliamentary debate on the subject will probably be postponed.

As regards the acceptability of nuclear power to the general public, I think the situation in Denmark resembles that in many other countries: there is some organized opposition, but the great majority of the population, I believe, take a positive attitude, especially in view of the doubtful prospects of obtaining cheap and uninterrupted supplies of fossil fuel. I am sure most people expect the Government to take the necessary steps to secure the energy supply.

Christina GYLLANDER: In Table I you have used the same diffusion factor for 2 h and 16 h. This implies a safety factor of about 3.

L. HANNIBAL: I quite agree with you. Again, we may have been somewhat too pessimistic.

E. IANSITI: Your hypothesis that 25% of the halogen would be free for dispersion to the gaseous environment is taken today as a basis for defining the safety features of a plant. But it cannot serve as a basis for evaluating the real consequences of an accident — at least not after the publication of the Rasmussen report — without a critical review of the Rasmussen calculation.

L. HANNIBAL: The committee worked out its evaluation about a year ago when the Rasmussen report had not yet come out. As stated in our paper it was decided to adopt a very conservative approach and assume that as much as 25% of the volatile inventory would be released to the containment, and even out of the containment if that was severely damaged.

# RISK AND POPULATION DOSE CALCULATIONS CARRIED OUT IN STUDYING NUCLEAR POWER PLANT SITES NEAR HELSINKI

O. J. A. TIAINEN, P. KORTELAINEN
Helsinki Electricity Works, Helsinki

I. SAVOLAINEN, R. TARJANNE
Technical Research Centre of Finland,
Otaniemi, Finland

Abstract

RISK AND POPULATION DOSE CALCULATIONS CARRIED OUT IN STUDYING NUCLEAR POWER PLANT SITES NEAR HELSINKI.

The combined production of electricity and district heat by large nuclear power plants was concluded to be economic in the Helsinki Metropolitan Area. It is also an advantageous alternative method of supplying energy from the point of view of air purity. The alternative sites for a nuclear power plant were compared, with the help of probable population doses caused by energy production and the consequences of radioactive releases due to reactor accidents. The latter comparison is more important in site selection because, in the normal operation of a nuclear power plant in the present stage of engineering, the individual and population doses round the nuclear power plant can be arranged to be well below the latest dose-limit guidelines also in densely populated areas. In analysing the accident consequences, both deterministic and probabilistic analyses were carried out. The probabilistic analysis performed showed that useful information for site selection was obtained even if the release spectrum was simplified. On the other hand, more data are needed for release spectrum before the probabilistic analysis can wholly displace the deterministic site evaluation.

## 1. INTRODUCTION

The Helsinki Metropolitan Area, dealt with in this paper, consists of the four cities, Helsinki, Espoo, Kauniainen and Vantaa, whose population is expected to be 850 000 in 1985. The energy service of the area is chiefly provided by municipal electricity works, which supply consumers with electricity and district heat [1].

On the basis of future energy demand estimates, the combined production of electricity and district heat by large nuclear power plants was concluded to be economic. The use of nuclear energy leads to about 180 million Fmk (1 Fmk = US $0.27) annual average saving compared with the most economic conventional energy production alternative. In the combined production of electricity and district heat, the nuclear power plant should be situated in the neighbourhood of a large population centre because the transmission of 2300 MW district heat (demand of the Helsinki Metropolitan Area in the first stage) leads to capitalized costs of about 15 million Fmk/km [1].

The environmental aspects of the nuclear power alternative for the Helsinki Metropolitian Area were studied by calculating risks and thermal pollution of various energy generation possibilities. It was concluded that somatic health risks caused by the nuclear power alternative were one to two decades less than in the coal- and oil-fired alternatives [1]. This was valid for all the alternative sites examined also when the effects of design-basis accidents were taken into account.  Only the use of natural gas led to a better result.  The effects of nitrogen oxides and trace-constituent emissions were lacking in the calculations of the risks caused by conventional energy production.  In the case of these releases it is difficult to carry out any risk analysis from present knowledge.

When power plants are used in combined electricity and district heat production, thermal losses are considerably smaller than in condensing electricity production.  On the basis of demand estimates for 1985 in the Helsinki Metropolitan Area, the average total efficiency of a nuclear power plant of $2 \times 1000$ MW equivalent condensing power would be $\sim 50\%$ instead of 32% in pure condensing electricity production.  Accordingly, the plant may also be allowed to cause higher population doses in combined production compared with pure condensing production [2].  This is due to the higher benefit of the plant.

Several alternative nuclear power plant sites, three of which are dealt with as examples in this paper, were compared, with the help of probable population doses caused by energy production.  The long-distance heat transmission lines are 19 km in site (a), 33 km in site (b) and 49 km in site (c).

## 2.    RADIOLOGICAL ASPECTS OF SITE SELECTION

The population dose caused by normal operation and probable disturbance releases from a $2 \times 1000$-MW(e) plant with light-water reactors during 40 years was estimated to be 3000, 2000 and 1500 man·rem for sites (a), (b) and (c), respectively [3, 4].  In this case a high stack (about 150 m) was assumed to be used.  The radioactive effluent release rates were chosen so that they can be considered typical for both pressurized-water and boiling water reactor units.  It can be concluded that population doses are well below the latest dose limit guidelines [2] in the case of all the site alternatives.  This was also valid for individual doses.  It can further be found that grounds for removing the plant from a nearer to a more distant site are economically not sufficient when it is assumed that society is ready to pay US $200-$250 for reducing collective doses by 1 man·rem [5].  Consideration of the effect of design-basis accidents led to the fact that all the site alternatives were acceptable.  Accordingly the effects of hypothetical accidents were studied and the sites were also compared using probabilistic methods.

## 3.    ACCIDENTS

The consequences of radioactive release due to reactor accidents were assessed by using both deterministic and probabilistic analyses.  In the deterministic analysis the values of accident parameters were chosen pessimistically, which led to conservatism.  In the probabilistic case the accident parameters were described by probability distributions.  Owing to inaccurate data and the roughness of the calculation model both methods included uncertainties.  Accordingly, results should be mainly considered as relative quality factors for comparing various nuclear power plant sites.  This work was restricted to direct radiological consequences, such as radiation effects of submersion and inhalation of radioactive material due to large accidents [6].

## 3.1.  Deterministic analysis

In the deterministic analysis, the consequences of a maximum hypothetical accident were evaluated.  It was assumed that because of the reactor
core melt-down together with leakage from the damaged containment, 100%
of the noble gas inventory, 25% of the iodine and some per cent of other
fission-product inventories of a 1000-MW(e) reactor core was released to
the environment.  The fraction of methyl iodine was taken to be 10%.

The effective height of the releases was taken to be the worst possible
including no plume rise.  These assumptions, together with the release
conditions used, are contradictorily pessimistic, because the rapid release
of 25% of iodine would be hot and thus cause plume rise.

Both acute and late health effects of thyroid, lung and the whole body
were studied.  The acute whole-body risk gave the most distinct differences
for comparison between the three alternative sites.  As large doses are fatal,
evaluation of the whole-body casualties gives important information for the
selection of the reactor site.

The number of acute health damage casualties was calculated by using
a 'detriment function' that describes the effect of radiation dose.  It has a
threshold $D_1$, under which the individual probability of undergoing the
considered detriment is zero, and another dose limit $D_2$, over which the
probability is one; between these limits the probability is assumed to be
linear.  The values of $D_1$ and $D_2$ were, for the whole body, 200 rem and
500 rem and for the thyroid 1000 rem and 7000 rem [7].

The weather conditions were assumed to be unchanged during the release
and dispersion, which leads to a conservative estimate.  The wind velocities
were assumed to be 1 m/s and 3 m/s in Pasquill F and D categories,
respectively.  The iodine deposition velocities were $1 \times 10^{-2}$ and $2 \times 10^{-2}$ m/s
in Pasquill F and D categories, respectively.  The doses were calculated by
using infinite cloud approximation.  Using the actual population distribution
round the reactor site, the number of casualties due to the release was
calculated in weather categories F and D, the wind direction being most
unfavourable.

The number of whole-body casualties were, in F category, 20 000,
500 and 600 for alternative sites (a), (b) and (c), respectively.  The
corresponding values for D category were 1400, 10 and 10.  The results
show that an acute whole-body risk has a clearly limited extent, and it
therefore causes great relative differences.

It should be emphasized that these numbers represent an extremely
pessimistic hypothetical case where no factors limiting the consequences of
the release were taken into account.  However, this analysis gives information on the comparison of the three site alternatives.


## 3.2.  Probabilistic analysis

In the probabilistic analysis the magnitude of release and the dispersion
conditions (weather category and wind direction) were described by probability
distributions.  Due to the lack of data, and to simplify the calculation model,
other parameters were taken into account with a single value as in the
deterministic analysis.  The most important parameters of this kind are the
fraction of methyl iodine and the effective height of the plume.

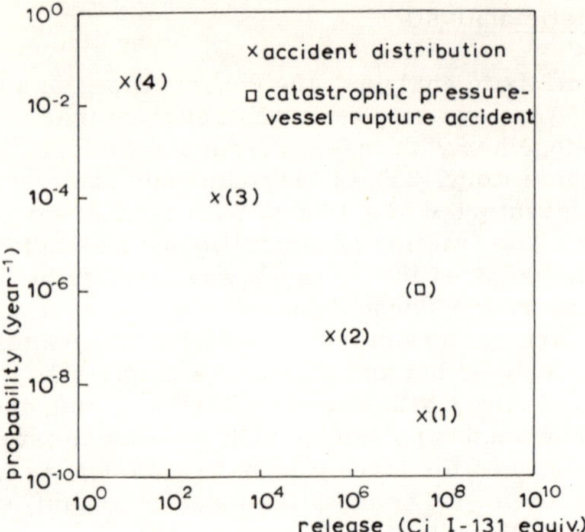


FIG.1.   Assumed discrete accident distribution.   The point describing the catastrophic pressure-vessel rupture accident is also shown.

In the probabilistic safety analysis the greatest difficulty is to generate a reasonable probability spectrum of radioactive release (called 'release spectrum' in the following).   Very little data for construction of a relevant release spectrum are available in the literature.   Thus the probabilistic analysis made was based only on a modified release spectrum of the loss-of-coolant accident caused by the primary pipe rupture of a 1000-MW(e) pressurized-water reactor.

The acute casualties and population thyroid dose caused by the inhaled iodine were considered.   The release spectrum used was discrete, consisti of the four points shown in Fig. 1.   Points 1, 2 and 3 are rough mean values based on various event tree analyses of primary pipe-rupture accidents of a PWR referred to in the literature (e.g. Ref.[8]).   The fraction of methyl iodine was assumed to be 10% in all releases.   The points are described as follows

Point 1: Most emergency systems fail, and the reactor core melts down. The containment is damaged and the equal quantity of radioactivity as in the maximum hypothetical accident is released rapidly to the environment.
Point 2: The core does not melt, but the containment isolation valves do no close.   The release is rapid.
Point 3: The emergency systems function correctly and the release takes place from the stack during a longer period.
Point 4: Not due to the primary pipe rupture, but takes into account smalle incidents during the reactor life.   The radioactive release is led through the stack.

The greatest incompleteness in the release spectrum is that the point corresponding to the catastrophic pressure-vessel rupture is not taken into consideration.   As can be seen from Fig. 1, this point would be dominant and make the results of the probabilistic analysis almost equal to those of t deterministic analysis.   The probability of this point could, however, be

decreased by those special safety arrangements presented in the connection
of the urban siting of power reactors (e.g. a double pressure vessel, a
prestressed concrete vessel or underground siting).  It should be noted that,
on the whole, the real release spectrum is individual for each reactor plant,
but using the model presented here the main aspects of the probabilistic
analysis can be obtained.

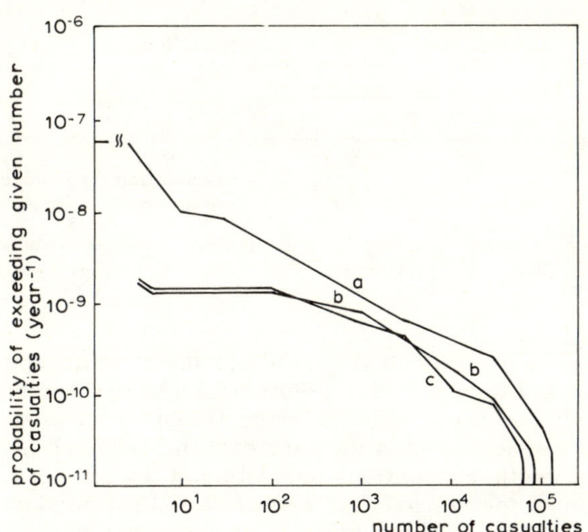


FIG. 2.   Annual probability of exceeding a given number of acute thyroid casualties due to the accident
distribution of Fig. 1 in three alternative plant sites.

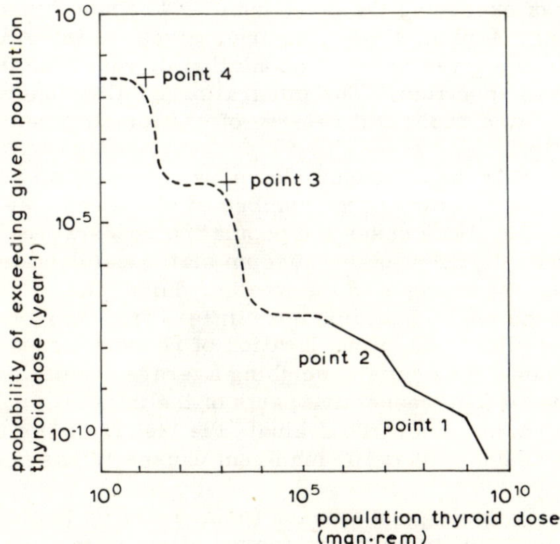


FIG.3.   Probability of exceeding a given population thyroid dose due to the iodine release spectrum of Fig.1,
in site alternative (a).

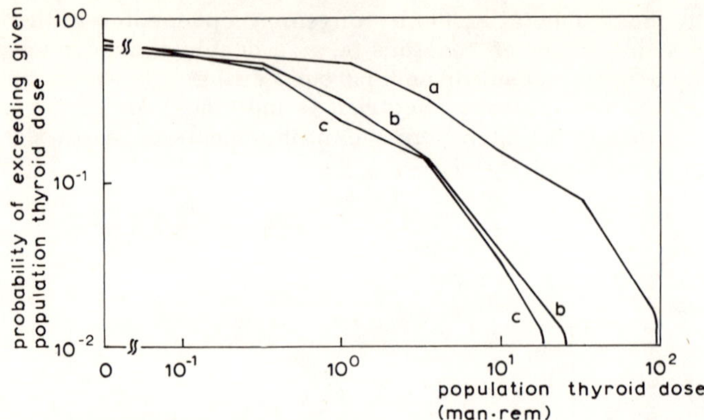

FIG. 4.    Probability of exceeding a given population thyroid dose due to a unit release of iodine (1 Ci [131]I equivalent) in the site alternatives (a), (b) and (c).


To calculate meteorological spreading, the Pasquill statistic dispersion model [9] was used together with a distribution of weather stability categorie and wind directions based on the Helsinki Airport weather of the years 1965-1969. It was assumed in the case of rapid releases (points 1 and 2 in Fig. 1) that the weather conditions prevailing at the time of the accident will persist throughout the release (the release should occur during a period of about one hour). For each combination of dispersion parameters, the numbe of acute thyroid effects and population thyroid doses was calculated in all three site alternatives. By accumulating the probabilities of combinations from the most serious to the least case, the curves presented in Fig. 2 can b calculated. The curves (called here 'integral probability functions') express the probability of exceeding the given number of acute thyroid casualties in the three alternative plant sites. Figure 3 gives the integral probability function to exceed a given value of population thyroid dose in site (a), due to the whole release spectrum. The integral probability functions of population thyroid dose due to a rapid unit release of $^{131}$I for all three site alternatives are shown in Fig. 4.

Because of their small magnitude, slow releases (points 3 and 4 in Fig. cause only late health effects, the number of which can be estimated from th population dose. In these cases the population doses were calculated by usin the whole weather distribution because no statistical data are available to take into account the changes in the weather during the release. The possibility that the real population dose differs from the average dose decreases quite quickly when the duration of release increases [10].

Figure 3 shows the points describing average population doses caused b the slow releases. The respective parts of the integral probability function are shown as a dashed line. To evaluate the risk from the population thyroi dose, it can be assumed that $10^6$ man·rem causes 40 cases of thyroid cancer [11].

In Fig. 2 only the biggest release (point 1 in Fig. 1) gives a contribution to the integral probability functions for the alternative sites (b) and (c). In the curve for site (a) the influence of release point 2 dominates under about $10^3$ casualties and the influence of point 1 dominates over about $10^3$ casualt:

In Fig. 3, the curves for sites (b) and (c) should be very close to and under the curve for site (a) because of the scale used. In Fig. 4 the differences of the same kind as in Fig. 2 are shown for the population thyroid dose; in both figures the alternative site (a) differs clearly from (b) and (c).

## 4. CONCLUSION

On the basis of both deterministic and probabilistic analyses, the alternative sites (b) and (c), which can be considered almost equivalent, are clearly preferable to site (a) as regards safety aspects.

The probabilistic analyses performed show that further information is obtained for site selection, even if the release spectrum is simplified. On the other hand, it can be concluded that a large amount of data for generation of real release spectrum should be collected before the probabilistic analysis is reliable enough to displace wholly the deterministic analysis. The extensive studies published recently in the United States of America [12] and Sweden [13] offer valuable new information for carrying out probabilistic analyses for the risk assessment of nuclear power plants.

## REFERENCES

[1]  MIKOLA, J., SEPPÄ, M., TIAINEN, O.J.A., TOIVIAINEN, E., HAAVISTO, H., NEVANLINNA, L.,
     "A study concerning different energy supply alternatives of the Helsinki Metropolitan Area in Finland",
     9th World Energy Conference, Detroit, 1974, paper No. 6.1-10.

[2]  RADIATION PROTECTION INSTITUTES, DENMARK, FINLAND, ICELAND and SWEDEN, Basic Principles
     for the Limitation of Releases of Radioactive Substances from Nuclear Power Stations, A Joint Statement,
     Copenhagen, Helsinki, Reykjavik, Oslo and Stockholm (1974).

[3]  HÄKKINEN, H.H., The Effect of Loviisa Nuclear Power Station on the Radioactivity of Air and Radiation
     Dose under Normal Conditions and Accidents, Finnish Meteorological Inst., Tech. Note 25 (1972).

[4]  TIAINEN, O.J.A., SEPPÄ, M., MIKOLA, J., KORTELAINEN, P., "Economic and safety aspects of a
     nuclear power plant used in combined electric and district heat supply", Czechoslovak Atomic Energy
     Committee, Conf. on Nuclear Heat Engineering and on Containments, Brno, 1973.

[5]  LINDELL, B., LÖFVEBERG, S., Kärnkraften, Människan och Säkerheten, Allmänna Förlaget, Stockholm
     (1972) (ISBN 91-38-01305-3).

[6]  SAVOLAINEN, I., Assessment of population risks due to accidents in a semi-urban-sited nuclear power
     plant by using probabilistic methods (in Finnish), Diploma thesis in Technical University of Helsinki,
     Dept of Technical Physics (1974).

[7]  FITZGERALD, J.J., Applied Radiation Protection and Control, Gordon and Breach, New York,
     Vol. I (1969), Vol. II (1970).

[8]  LINDACKERS, K.-H., STOEBEL, W., Probability analysis applied to light water reactors: loss-of-coolant
     accidents, Nucl. Saf. 14 (1973) 14.

[9]  VOGT, K.J., Umweltkontamination und Strahlenbelastung durch radioaktive Abluft aus Kerntechnischen
     Anlagen, Kernforschungsanlage Jülich GmbH Rep. Jül-637-ST (1970).

[10] HÜBSCHMANN, W., NESTER, K., Eine Neubewertung der atmosphärischen Diffusion bei Reaktorstörfällen
     in deterministischer und probabilistischer Sicht, Atomkernenergie 20 (1972) 315.

[11] UNSCEAR, Ionizing Radiation: Level and Effects, Vol. I, Levels; Vol. II, Effects, UN, New York (1972).

[12] USAEC, Reactor safety study, An assessment of accident risks in U.S. commercial nuclear power plants,
     Rep. WASH-1400 (1974).

[13] INDUSTRIDEPARTEMENTET, Närförläggning av Kärnkraftverk (The Urban Siting Study), SOU 1974:56,
     Stockholm (1974) (ISBN 91-38 01579-X).

# DISCUSSION

G. HAKE (Chairman): Could you elaborate on how you arrived at the figure of US $200-250 for the value of a man·rem?

R. TARJANNE: The value of a man·rem is based on the data of Lindell and Löfveberg [5]. However, it is not essential in our comparison because the extra cost of transmission lines is greater by two orders of magnitude than the cost of population dose in normal operation.

H. JAMMET: I must point out that Lindell's data are not isolated figures; they tally with the data published by several authors and are considered to be relatively acceptable, at the present stage of knowledge, by the ICRP Committee 4.

J.B. BURNHAM: The value of 1 man·rem of dose has been calculated by many investigators. This calculation includes a number of factors, one of which is the value of life, and this is probably one of the better known numbers in the health physics business: it ranges from US $100 to US $300 per man·rem, as calculated by a number of investigators (on the basis of United States of America data).

D. BENINSON: I should like to add that man·rem values have been derived by two procedures: (a) by multiplying the risk of death per rad by the 'cost' of life, as derived from insurance experience and court cases; and (b) by inquiring about the amounts which knowledgeable people have been willing to pay to reduce some exposures (see e. g. the Lindell-Hedgran study). Both types of assessment give values ranging from about US $20 to US $350 per man·rad (whole body).

A.P. HULL: I must emphasize that, whatever the worth of a man·rem where whole-body exposure is concerned, it seems reasonable that a somewhat smaller worth should be applied to a thyroid·rem, since thyroid cancer is seldom a fatal affliction.

R. TARJANNE: The values taken for man·rem are based exclusively on whole-body doses.

H. SCHNURER: You have compared three sites, and the closest one mentioned is 19 km. Is this distance measured form the centre of Helsinki?

R. TARJANNE: The distance of 19 km for site alternative (a) is a heat transmission distance. Its physical distance from the centre of Helsinki is about 17 km.

T.P. HAIRE: When you are considering possible nuclear power station sites, is any thought given to arrangements for evacuation of the local population near the station? What is the reaction of the public to nuclear power in Finland and what measures are taken to obtain public acceptance?

R. TARJANNE: No evacuation was taken into account in the calculation.

In Finland, too, there are opponents of nuclear power, and they wield some influence. The four towns included in the Helsinki Metropolitan Area are forming a company to build the nuclear power plants. Helsinki itself is the only one which has not yet decided to join the company. Its population (about 500 000) is the largest of the four towns. An intensive public debate going on, and representatives in the City Council are discussing the new company.

# QUANTIFIED SOCIAL AND AESTHETIC VALUES IN ENVIRONMENTAL DECISION MAKING

J.B. BURNHAM
Planning and Assessment,
Battelle Pacific Northwest Laboratories,
Richland, Wash.

W.S. MAYNARD
Battelle Human Affairs Research Center,
Seattle, Wash.

G.R. JONES
Jones and Jones, Landscape Architects,
Seattle, Wash.,
United States of America

Abstract

QUANTIFIED SOCIAL AND AESTHETIC VALUES IN ENVIRONMENTAL DECISION MAKING.
A method has been devised for quantifying the social criteria to be considered when selecting a nuclear design and/or site option. Community judgement of social values is measured directly and indirectly on eight siting factors. These same criteria are independently analysed by experts using techno-economic methods. The combination of societal and technical indices yields a weighted score for each alternative. The aesthetic impact was selected as the first to be quantified. A visual quality index was developed to measure the change in the visual quality of a viewscape caused by construction of a facility. Visual quality was measured by reducing it to its component parts — intactness, vividness and unity — and rating each part with and without the facility. Urban planners and landscape architects used the technique to analyse three viewscapes, testing three different methods on each viewscape. The three methods used the same aesthetic elements but varied in detail and depth. As expected, the technique with the greatest analytical detail (and least subjective judgement) was the most reliable method. Social value judgements were measured by social psychologists applying a questionnaire technique, using a number of design and site options to illustrate the range of criteria. Three groups of predictably different respondents — environmentalists, high-school students and businessmen — were selected. The three groups' response patterns were remarkably similar, though businessmen were consistently more biased towards nuclear power than were environmentalists. Correlational and multiple regression analyses provided indirect estimates of the relative importance of each impact category. Only the environmentalists showed a high correlation between the two methods. This is partially explained by their interest and knowledge. Also, the regression analysis encounters problems when small samples are used, and the environmental sample was considerably larger than the other two.

## ANALYTICAL METHOD

In the United States today there exists a major dilemma in environmental decision making. For example, diverse factors must be considered in the selection of nuclear facility sites, but many of these elements are abstract and unquantified. Social values should be factored into the decision-making process, but there is no consideration of community attitudes until an ad-

verse  situation arises in which inflexible positions and attitudes have been established.  In the face of these methodological
deficiencies, planners and regulators must make difficult choices
of major environmental import.

This familiar problem is most applicable to a nuclear plant,
but with the proper substitution of titles it could apply to any
major construction project.  In essence, there is a major time la
between the sponsor's and the public's awareness of the environmental aspects of the project, and Figure 1 illustrates the exter
of this mismatch in time and level.

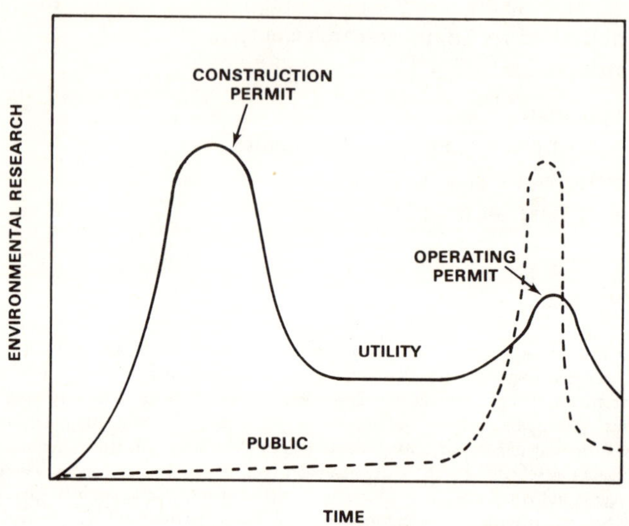


FIG. 1.   Typical time mismatch in environmental awareness of the public and the project sponsor.

Since the community's social values are seldom measured and
rarely factored into project design, an adverse  situation is
often generated.  This may result in a stalemate, a last-minute
compromise, or, worst of all, a court ruling in which one individual must make a judgment on a highly complex technical and
social matter.  Certainly, none of these are satisfactory solutions to the problem.

To resolve these undesirable alternatives, Battelle Pacific
Northwest Laboratories (BNW), sponsored by the U.S. Atomic Energ
Commission, has devised a method for environmental decision maki
that combines both societal and technical judgments.  In this
method, social values are measured in the affected community.
These relative values are used as weighting factors in a decisio
matrix.  Community judgment of alternatives is measured both
directly and indirectly on eight siting factors.  These same cr
teria are also analysed by experts using techno-economic method
The combination of these societal and technical indices yields
weighted score for each criterion.

## TABLE I.  COMPOSITE SOCIAL-TECHNICAL DESIGN ANALYSIS

| Criteria | Societal Weighting Factor | Techno-Economic Scores Design A | Design B | Composite Scores Design A | Design B |
|---|---|---|---|---|---|
| Aesthetic Impact | 0.3 | 20 | 90 | 6 | 27 |
| Land Use | 0.7 | 40 | 50 | 28 | 35 |
| Water Impact | 0.8 | 40 | 50 | 32 | 40 |
| Air Impact | 0.5 | 30 | 40 | 15 | 20 |
| Economic | 0.4 | 40 | 50 | 16 | 20 |
| Cultural/ Recreational | 0.3 | 40 | 80 | 12 | 24 |
| Health/Safety | 0.7 | 90 | 20 | 63 | 14 |
| Ecological Impact | 0.7 | 70 | 30 | 49 | 21 |
| Technical Total | | 370 | 410 | | |
| Composite Total | | | | 221 | 201 |

The methodology employed is illustrated in Table I.  When
the societal weighting factors are used to modify the independently derived technical evaluation, a weighted score results for
each criterion.  The sum of these weighted scores is a composite
score for each of the alternatives.  In the example shown, the
use of the composite score leads to a different conclusion than
afforded by the technical analysis alone.  Since a high score
indicates large impacts, hence unfavorable design, techno-economic
analysis would alone have indicated the selection of Design A.
The composite score, though, favors Design B.  The combination
of the weighted score can be pictured mathematically as a vector
in multidimension space.  In this case a root mean square sum
would be a preferable treatment.

The criteria used above were derived through a careful review
of environmental statements and intervenors' contentions.  The
goal of the evaluations was to quantify each of the criteria independently by a techno-economic method that can be as sophisticated
as necessary since it will be performed by technologists.  The
relative social values of the criteria will be measured independently in the affected community.

Pieces of this methodology have been developed by others.
Dee   et al.[1] at Battelle-Columbus have devised methods for
quantifying many individual environmental effects.  As can be
expected, the combination of a number of factors into a single
dimensionless index which represents the technical value for a
criterion is a challenging task.

This can be illustrated by the example of water quality.
The 'quality' of a given body of water is a complex combination
of many factors, including such diverse items as coliform count,
content of various elements and compounds, and dissolved oxygen.
Consider for now only the single measure of dissolved oxygen.
The saturation limit of dissolved oxygen is dependent on the temperature and pressure of the water.  For most common stream conditions, this theoretical limit is about 8.5 parts per million

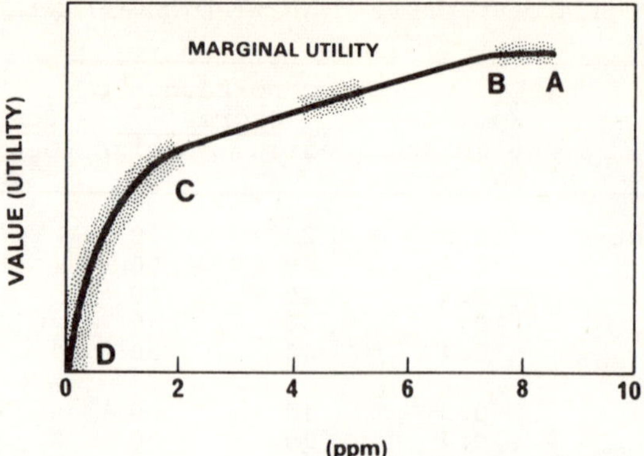



FIG.2.    Value of water quality as a function of dissolved oxygen content.

(ppm) of dissolved oxygen.  A very high quality natural water
body could be expected to have a dissolved oxygen content of
7.5 ppm or less.  To raise this body to the theoretical maximum
of, for example, 8.5 ppm would be a costly procedure involving
the control of temperature and pressure of the water body and
the injection of oxygen gas.  To maintain this quality would be
a most expensive venture; and since the water at a quality of
7.5 ppm dissolved oxygen (considering only this one measure) is
excellent for the support of aerobic life, the 'digestion' of
organic materials, etc., the utility of this additional expenditure presumably would be very low.  This region of the utility
curve is illustrated by the Section A-B of Figure 2.

Now consider the lower range of this curve.  Most water qua
ity standards require a dissolved oxygen content of at least 2 p
(C on the curve).  As the dissolved oxygen decreases below this
level, the water's ability to support aerobic life diminishes ra
idly and is, obviously, nil at a zero oxygen content.  The value
of this water body, by this measure, might be expected to drop
rapidly from C to D.  If it is assumed that the quality is linea
between B and C, the utility function is complete.

The marginal utility of one extra unit of water quality can
be measured by taking the slope of this curve.  If it is measure
in the A-B region, it shows a very low marginal utility, in the
C-D region a very high one, and an intermediate value in the lin
ear range.  One could argue that this strong dependence of value
on the overall level of water quality rules against the use of
any single value (utility) number.  In practicality this reason-
ing does not hold.  Economics eliminates from serious consideration any designs in the ultra-high-quality region, A-B.  The
natural levels of a high-quality water supply are such that ther
will never be proposed a design which is to operate in this regi
of 7.5 to 8.5 ppm dissolved oxygen for reasons previously discussed.  The other end (C-D) of the curve is ruled out in practi
by the adherence to water quality standards.  In theory, a plant
could be proposed using water already polluted below acceptable
levels, but this is highly unlikely.  If this rationale is correct, it should be possible to measure with reasonable accuracy

respondent's assessment of his marginal value of water quality.
While it is debatable that the B-C portion of the curve is truly
linear, it is demonstrably more likely to be linear (and, hence,
more tractable to measurement) than its extremities.

     The assumption that the values, even confined to region B-C,
will be linear is a gross one.  Einhorn[2] demonstrated that,
given the same multidimensional data, three decision makers each
used different cognitive processes to arrive at their overall
judgments.  Averaged over a number of judges, though, data are
best explained by a linear model in which the overall value of
a multidimensional alternative is the sum of the values of its
components.  Yntema and Torgerson[3] demonstrated that the linear model provides an excellent approximation of data even when
interactions or nonlinear relationships exist.  Thus, even though
individual respondents may not weight all criteria in a linear,
additive fashion, when their cumulative judgment is analysed, a
linear compensatory model will yield the best results in terms
of predicting their judgments.

     Although it is not possible in this phase of the project
to synthesize methods for quantifying each of the eight criteria,
one of the most highly subjective criteria--aesthetic impact--
was quantified.

QUANTIFICATION OF AESTHETICS

     The purpose of this phase of the study was to predictively
evaluate any changes in the visual quality of a landscape resulting from the physical introduction of a nuclear facility.  Pre-
ferably the methodology would:(1) allow the evaluation of a
range of potential facility types (varying principally by cooling option); (2) require relatively small numbers of evaluators
to be used effectively; and (3) be applicable to any landscape-
facility combination.

     Two major issues were faced from the outset.  First, a
workable definition of 'visual impact' had to be made.  It was
decided that the expression 'visual impact' could best be defined
as the change in visual quality over time, resulting from the
introduction of a facility into a landscape setting as viewed
from the surrounding area.  The second major problem was invent-
ing a reliable procedure to measure visual quality.  Through testing, research [4,5] and intuition and by synthesizing selected
techniques pioneered by other workers in the field,[6,7] it was
concluded that three components of visual quality are most important:  (1) the memorability of a scene; (2) its wholeness; and
(3) the harmony of its parts.  Hereinafter these components or
criterion values are referred to as <u>vividness</u>, <u>intactness</u> and
<u>unity</u>.  By carefully defining and scaling these values, it is
possible to objectively evaluate the visual quality of a given
scene.

     In order to select a workable number of viewpoints that
would represent the visual impact of a facility, the distance
between an observer and the facility is classified into three
zones:  foreground (0 to 1/2 mile); middleground (1/2 to 5 miles);
and background (beyond 5 miles).  Figure 3 illustrates this

FIG. 3.   Observer distance zones.

TABLE II.  IDEALIZED VIEWPOINT DISTRIBUTION:
NATURAL DRAFT COOLING-TOWER ALTERNATIVE
(12 final viewscapes required)

| Distance | Observer Inferior | Observer Normal | Observer Superior |
|---|---|---|---|
| Foreground (0-1/2 mile) | 1 | – | – |
| Middleground (1/2-5 mile) | 1 | 6 | 1 |
| Background (>5 mile) | – | 2 | 1 |

distance relationship.  In addition it is important to consider
the viewing height of the observer in relation to the facility,
since looking down upon an object in a scene is quite different
from seeing it on the horizon or looking up at it.  These positions are called:  observer inferior, observer normal and
observer superior [8].  In the background the facility appears
as a two-dimensional pattern or shape.  In the middleground the
facility becomes a strong visual element seen clearly in its
relationship to its setting; however, its true size and distance
may not be fully appreciated in this zone.  In the foreground,
texture, detail and size of the facility become readily apparent
and the facility clearly dominates the view.

A sight-line analysis procedure was used to delimit that
surrounding area of land or water from which an alternative
facility is visible, and hence which may be visually impacted
by the facility's introduction.  This surrounding area, termed
the facility's viewshed [5,7] is the key to the spatial location
of a potential viewing population and to viewpoint selection.
Visibility decreases with distance, and population usually
increases with distance, suggesting that the middleground zone
is the most significant for viewpoint selection.  Table II illustrates an idealized viewpoint distribution for facilities with
natural-draft cooling towers.

To evaluate visual quality, the components previously proposed of vividness, intactness and unity are most useful.  When
comparing a view of a pristine alpine lake to that of an
unsightly marsh landfill it may be noted that while both scenes
consist of land, vegetation, water and sky, one scene is strikingly vivid and the other mundane and nondescript; that while
one is intact and bears little or no trace of man's imprint, the
other is severely encroached upon; and that while one conveys
an overall visual harmony, balance and compositional integrity,
the other is merely chaotic, confused, and lacking strong
visual unity.

It was concluded that by breaking a viewscape into its
component parts and evaluating the vividness and intactness of
each part as well as considering the overall unity or compositional harmony of the parts, it was possible to write a simple
formula for visual quality as illustrated in the following:

$$VQ = 1/3 \ (I + V + U)$$

where VQ = Visual Quality
     I = Intactness
     V = Vividness
     U = Unity


Vividness, or memorability of the visual impression received from the viewscape, is dependent on contrast, mutual accentuation, distinctiveness and prominence of form, line, color and texture in a scene.  Eight considerations embrace this concept of vividness:

1.  Skyline boundary definition (Figure 4a).  For example, irregular and strongly defined definition between land and sky.

2.  And similarly, a strong sense of spatial enclosure (Figure 4b) or the overlapped weaving of sloping planes tends to hold an observer's interest longer than scenes of unenclosed spaces or those of uniform undissected surface.

3.  In many cases, scenes of extreme topographic relief (Figure 4c), such as mountains or deep canyons, are highly striking while flatter lands must depend upon other visual characteristics to maintain an observer's interest.

4.  The pattern of contrasting vegetation (Figure 4d) types and barren lands may be highly interesting, while seasonal changes in color and texture may also be very striking and memorable.

5.  Prominent natural landmarks (Figure 4e), such as peaks and strikingly colored rock outcrops, as well as less prominent features such as falls or solitary tree group ings, may be so memorable as to have been given names by earlier viewers.

6.  Water (Figure 4f) nearly always enhances the vividness and visual quality of a scene.  Variations in color, texture and motion of water through its various seasonal moods, the diversity of its shoreline edge, and its reflectivity or brilliant clarity in different settings may dramatically increase the memorability of a place.

7.  The sky (Figure 4g) is also an important visual element of a scene, casting its mood upon the observer while transmitting changes in light conditions and weather, sunrises and sunsets, cloud patterns and fogs.

8.  Man-made elements (Figure 4h) may also be highly vivid in a scene, depending upon their prominence, distinction, diversity and contrast with the other visual elements present.

Hence, the vividness of a scene or its elements may be evaluated as a summation of these eight considerations, and the vividness score inserted into the first term of the equation for visual quality.

The second indicator of visual quality, the intactness of a scene, is defined as the apparent degree of natural condition as judged by the level of development and the degree to which encroachment is present (Figure 5).  The level of urbanization may range from natural and undeveloped through rural and sub-urban development to urban areas which appear to have had their natural settings eradicated.  Similarly, a scene may be free of encroachment or nearly so, or may include major road cuts, trans-mission-line corridor scars, junk and a profusion of confusing signs and billboards.  These encroaching elements normally degrade the visual quality of a scene.

The third major visual criterion is unity (Figure 6), which is a measure of the combined intercompatability of a scene's components in forming a coherent harmonious visual unit.  Over-all unity may be achieved by similarity in form, line, color and texture, or by an organized balance between contrasting dominant and subordinate visual elements; compositional integrity is of prime importance in the achievement of unity.  The unity between man-made and natural elements is also measured in a scene, and this more specific rating is averaged together with the score for overall unity.

A photographic method was devised by which slides could be prepared (from actual site photographs) illustrating the viewscape with and without the facility.  Slides and illustra-tions were used to test the method with a special-competence group of landscape architects, environmental planners and urban designers using three views of the Trojan and Rancho Seco facilities.

A questionnaire was designed for this test.  Three ana-lytical methods were tried.  Each of these used the same aes-thetic quality principles, but they differed in the number of elements considered and in the subjectivity of the judgments required.  Table III shows the results of these three tests. In the first method respondents were asked only to grade the overall quality of a view with and without a plant.  In the second method they were asked to separately rate the intact-ness, vividness and unity of the same overall view before and after the plant appeared.  In the third method  the overall view or viewscape was broken into its major visual elements (landforms, waterforms, sky, etc.) and each of these elements was then rated for its intactness, vividness  and unity before and after the plant was added.

In analysing the three viewscapes, two at the Trojan nuclear plant and one at the Rancho Seco, the third method, Elemental Visual Attribute, was found to be the most reliable.  It consis-tently approached the mean of all measurements and exhibited the lowest standard deviation of values.  This result is certainly not unexpected since in this method the decision elements are more precisely defined and the subjectivity of the judgments

BURNHAM et al.

b

d

a

c

FIG.4.　Elements of vividness.

FIG. 5.  Intactness of a scene: increasing man-made encroachment.

minimized.  This objective evaluation approach is very desirable
since the research has shown that objective visual evaluations
show more normative results across different interest groups tha
do preferential ratings [7].

        Although environmental purists do not agree, our research
judged it correct to evaluate the visual impact as a function o

FIG.6. Unity.

TABLE III.  ANALYSES OF THREE VIEWSCAPES
BY SPECIAL-COMPETENCE GROUP

|  | Visual Quality Before | After | Ratio of Visual Change |
|---|---|---|---|
| **Viewscape A** | | | |
| 1.  Pre-Evaluation Measurement | 32 | 63 | 0.96 |
| 2.  Overall Visual Attribute | 38 | 62 | 0.63 |
| 3.  Elemental Visual Attribute | 34 | 64 | 0.91 |
| Average of Measurement Levels | 35 | 63 | 0.80 |
| **Viewscape B** | | | |
| 1.  Pre-Evaluation Measurement | 39 | 60 | 0.53 |
| 2.  Overall Visual Attribute | 46 | 55 | 0.20 |
| 3.  Elemental Visual Attribute | 35 | 51 | 0.45 |
| Average of Measurement Levels | 40 | 55 | 0.37 |
| **Viewscape C** | | | |
| 1.  Pre-Evaluation Measurement | 25 | 48 | 0.92 |
| 2.  Overall Visual Attribute | 35 | 47 | 0.34 |
| 3.  Elemental Visual Attribute | 29 | 47 | 0.62 |
| Average of Measurement Levels | 30 | 47 | 0.56 |

the number of visual contacts.  Functions were derived analyti-
cally for including the annual viewer contacts, both resident
and transient.  It was decided that because of the human accom-
modation factor, resident impacts (per contact) should not be
weighted as heavily as transient contacts.  These factors clearly
need field validation before they can be used with any degree of
confidence.

Thus, total aesthetic impact of a proposed facility is the
sum of the visual impacts measured at each of the selected viewpoints, with impacts before and after plant installation
described in terms of percentage change and modified by population contact.

QUANTIFICATION OF SOCIAL VALUES

The objective of this portion of the study was to develop
and test a method for assessing public values toward nuclear
generating stations.  Specifically, this work focused on the
measurement of the relative importance which the public attached
to the social and environmental impacts of nuclear power plant
construction and operation.

A set of social and environmental impact criteria were
identified which would serve as a basis for both (1) communicating necessary information to respondents, and (2) providing
a framework for obtaining questionnaire responses.  The goal
was to select a set of factors which were (1) comprehensive in
describing the effects of nuclear power plant siting, (2) meaningful both to the general public and to decision makers, and
(3) relatively independent of one another, so that there would
be little overlap among the factors.  A review of intervenor
actions and several 'workshop' discussions with researchers
engaged in studies of nuclear plants produced the following
criteria:  aesthetic, economic, water quality, air quality,
land use, animal/plant, cultural/recreational, and health/
safety.  These criteria were then used as a basis for describing the impacts created by various nuclear plant options.

Six examples of hypothetical nuclear plant sitings were
used to convey information about the eight criteria listed
above.  The six examples were constructed by presenting three
cooling options, each located on two different geographical
sites.  The rationale for presenting these six options was to
illustrate the differential effects on the eight criteria as
a function of nuclear plant design and location.  For example,
once-through, cooling pond and natural-draft cooling towers
illustrate wide variation on aesthetic, water quality, economic, and land-use criteria.

For each of the six hypothetical plant options, 'mini-
environmental impact statements' were prepared.  These statements consisted of eight paragraphs describing each plant
option with respect to the eight criteria.  Overhead slide
projections were also prepared to communicate the aesthetic
impact of each nuclear plant option.  In preparing these written and graphic descriptions, emphasis was placed on (1) rendering accurate descriptions of actual or likely environmental

impacts of the design/site plant options, and (2) minimizing
technical jargon so that the descriptions would be meaningful
to a broad spectrum of the public.

The eight environmental-social criteria and the six mini-
environmental impact statements were incorporated into a four-
part questionnaire.  Part I of the questionnaire consisted of
the impact statements for the six design/site alternatives.
Respondents were asked to judge each plant alternative by rating its acceptability on each of the eight environmental-social
criteria.  Acceptability was measured using 7-point rating
scales in which '1' represented 'very unacceptable' and '7'
'very acceptable'.  Finally, respondents were asked to judge the
general or overall acceptability of the option.  In analysing
the results of Part I, ratings of the acceptability of the
eight environmental impact categories were treated as predic-
tors of the overall rating.  Correlational and multiple regression techniques were used to obtain indirect estimates of the
relative importance of each impact category.  These analytic
techniques allow for estimates to be made of the relative importance of each environmental-social criterion indirectly, i.e.
the respondent is not required to assign importance estimates.
Rather, the pattern of responses to the eight criteria are
examined in relation to the overall acceptability rating for
each plant option.  The nature of this relation reveals the
importance of each criterion in affecting the overall rating,
if the eight criteria are mutually exclusive.

Part II of the questionnaire consisted of a ranking procedure in which respondents compared the eight environmental-
social criteria and rank-ordered them according to their
relative importance.  Compared to the technique used in Part I,
the ranking procedure could be described as a direct approach
to importance measurement, since the respondent was presented
the task of comparing and ranking the criteria.

Part III of the questionnaire consisted of 54 attitude
statements dealing with nuclear energy and related topics such
as the perceived urgency of the energy crisis and consumption
versus conservation orientation of respondents.  Respondents
were asked to rate the extent of their agreement or disagree-
ment with each statement.

Part IV contained a 7-item multiple choice 'test' of knowledge about nuclear power and a brief list of biographical questions to describe each respondent.  The results of Parts III and
IV were used to (1) characterize the background and orientation
of respondent groups, and (2) assist in interpreting the results
from Parts I and II.

The questionnaire was presented to three groups of respondents.  Participation in the study was by invitation and on a
voluntary basis.  The three groups and the number of analys-
able responses from each were:  environmentalists (91); high-
school students (57); and businessmen (29).  These three samples were drawn, not to reflect a representative sample of the
population, but rather to serve as validation groups.  These
three groups might be expected to differ from one another with
respect to interests, attitudes  and background, and conse-

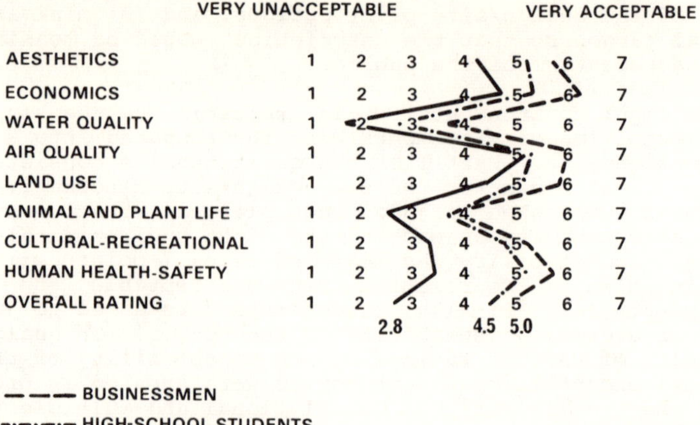


FIG. 7.   Mean responses, once-through cooling design, river location.

TABLE IV.  COMPARISON OF IMPORTANCE OF ENVIRONMENTAL ISSUES AS INFERRED FROM CORRELATION (INDIRECT METHOD) AND MULTIPLE RANKING (DIRECT METHOD)

| | High School Students | | Environ- mentalists | | Businessmen | | Combined Samples | |
|---|---|---|---|---|---|---|---|---|
| | Ind. | Dir. | Ind. | Dir. | Ind. | Dir. | Ind. | Dir. |
| Water | 4 | (8) | (7) | (8) | (7) | (7) | (6) | (8) |
| Animal/Plant | (7) | (7) | (8) | (6) | 4 | (6) | (8) | (7) |
| Air | 3 | (6) | 3 | 4 | 1 | 3 | 3 | 4 |
| Health/Safety | (6) | 5 | (6) | (7) | 5 | (8) | (7) | (6) |
| Economics | 1 | 3 | 1 | 2 | 3 | 4 | 1 | 3 |
| Land | 2 | 4 | 2 | 5 | 2 | 5 | 2 | 5 |
| Aesthetic | 5 | 2 | 5 | 3 | (8) | 1 | 4 | 2 |
| Cultural/ Recreational | (8) | 1 | 4 | 1 | (6) | 2 | 5 | 1 |
| Rank Order Correlation | -0.095 | | 0.643 | | -0.095 | | 0.560 | |

1 = Least Important
8 = Most Important

quently to differ somewhat in their responses to the survey instrument.

Summary statistics, calculated for the three groups across the six plant options, revealed significant differences among the three respondent groups in their 'acceptability' ratings for environmental-social criteria and for the overall ratings assigned to each plant option.  Figure 7 illustrates a typical response pattern.  The sample of businessmen tended to be more

accepting of nuclear plants than did environmentalists; support
for this finding was also found in the analyses of Part III,
the attitude survey.

The results of the indirect (Part I) and direct (Part II)
procedures for obtaining importance estimates on the eight criteria are summarized in Table IV.

Weights from the indirect procedure were averaged across
the six plant options to permit easy comparison with the results
of the direct-ranking procedure.  For the indirect procedure
(i.e.  ratings of mini-environmental impact statements were used
to predict overall acceptability), the criteria Water Quality,
Aesthetics, and Cultural-Recreational considerations received
the highest weights.  For the direct procedure (i.e.  direct
ranking by respondents), Water Quality, Animal-Plant Life and
Health-Safety considerations were judged highest in importance.
The numbers in parentheses show the three top-ranked criteria
for each group, for the two methods.  The rank-order correlations
show the extent of agreement between the two procedures; of the
three groups sampled, only the environmentalists showed high
agreement between the two methods.  From a methodological standpoint, this disparity of results between the two procedures may
have been caused by:  small sample sizes; need for more balanced
descriptions of the impacts of nuclear plants; and need for more
careful selection of criteria to ensure that each criterion is
mutually exclusive of the others.  On the other hand, the disparity of results may suggest that people do not always use or
weigh information in the same way that they state they would.
Field testing of the method with larger groups of respondents
will be required before more definitive explanations can be
made.

From a broader perspective, the approach presented here
illustrates an initial effort to develop a technique for obtaining community input to decision making on complex  technological
issues.  With refinements in the questionnaire and broader sampling of respondents, this approach may have considerable utility as an aid to environmental decision making.

ACKNOWLEDGEMENT

This work was done under the auspices of the U.S. Atomic
Energy Commission under Contract AT(45-1)-1830.


REFERENCES

[1]  DEE, N., et al., Environmental Evaluation System for Water Resource Planning, U.S. Bureau of Reclamation
     Research Paper, Jan. 1972.
[2]  EINHORN, H.J., The use of nonlinear, non-compensatory models in decision making, Psychol. Bull. 73
     (1970) 221.
[3]  YNTEMA, D.B., TORGERSON, W.S., Man-computer cooperation in decisions requiring common sense,
     IRE Trans. 2 1 (1961).
[4]  JONES, G.R., The Nooksack Plan: An Approach to the Investigation and Evaluation of a River System,
     Jones & Jones, Seattle, Wash. (1973).

[5]  JONES, G.R., "Techniques for assessing and quantifying environmental aesthetic and recreation values
     and impacts", Proc. of Conf. on Recreation Planning for Federally Licensed Hydroelectric Projects,
     Seattle 1974, Bureau of Outdoor Recreation, USDI, the Pacific Northwest Utilities Conference Committee,
     and the Federal Power Commission.
[6]  LITTON, R.B., Jr., et al., An Aesthetic Overview of the Role of Water in the Landscape, National
     Water Commission, Univ. of California, Berkeley (1971).
[7]  ZUBE, E.H., Scenery as a natural resource: implications of public policy and problems of definition,
     description, and evaluation, Landscape Architecture 63 2 (1973).
[8]  LITTON, R.B., Jr., Forest Landscape Description and Inventories — A Basis for Land Planning and
     Design, USDA, Forest Service, Pacific Southwest Forest and Range Experiment Station Research Paper
     PSW-49, Berkeley, Calif. (1968).

# DISCUSSION

H. JAMMET:  I must congratulate Mr. Burnham on his efforts at quantification where quantification is difficult.  However, what I find amazing is
that there seems to be greater concern for plants and animals than for huma
health.  Now, if we consider somatic effects in the context of the quantities
of radioactivity released, there cannot be any such effects on animals and
plants.  And, as to genetic effects, I often wonder whether public opinion is
aware of an important fact.  Here two distinct problems have to be considere
namely the induction of mutations and selection.  While the mechanisms
by which mutations occur are essentially the same among animals and huma
beings, selection is quite a different matter.  It follows a natural course
among animals, ruthlessly eliminating what is disadvantageous.  Among
human beings, on the other hand, disadvantageous characters are preserved
and reproduced.  So, among animals all the disadvantageous genetic effects
will be eliminated by nature and only the advantageous mutations will sur-
vive.  Biologically speaking, therefore, genetic mutation is desirable in
nature but would be disastrous among humans. I find it hard to understand
why the different consequences of radiation exposure are mixed up.  I think
a substantial part of Government efforts should be directed not to atomic
energy but to genetics.

J.B. BURNHAM:  Logically speaking I could not agree with you more,
especially as regards the need for a sense of proportion in the concern
expressed for animals and human beings and in public attitudes on the subje
However, the fact is that in the United States of America stiff opposition
is encountered from local groups if the habitat of any species is likely to
be disturbed; this was true, for example, of the debate about Steelheads in
the Columbia river.  The Portland General Electric Company was compelle
to build a cooling tower for its very first power station on that river, where
once-through cooling would have had no effect on the fish, considering the
extremely low temperature and the discharge of the river. So, what we
come up against is social values.  These values may be real or they may
represent the views of a small but vocal minority.  If we do not measure
this and make our findings known, the opinion of the minority will pass for
that of the whole people.  We are trying to find some logic to defend these
decisions.

Gloria CAMPOS VENUTI:  Could you please explain whether your test
to establish the societal weighting factor is based only on the three groups
consisting of 91 environmentalists, 57 high-school students and 29 businessmen, or on some other groups of different types which are more significant
from the statistical point of view?

J.B. BURNHAM: The tests cited in the paper were designed only to
gauge the sensitivity of the psychometric instrument. They were not intended
to measure in a meaningful way the values of the nation as a whole, or of a
single community. Distinct groups with predictable biases were tested.
If the instrument had not shown the sensitivity to reveal these biases statistically (as a matter of fact it did), it would clearly have needed to be redesigned.

G. HAKE (Chairman): I understand that in the United Kingdom problems
have been encountered with some species of birds. Would someone from
that country like to comment on the usefulness of the kind of work
Mr. Burnham has described?

E.F.F.W. USHER: As we reported in our paper (SM-188/39), in 1956
the Ministry of Power held a public inquiry because objectors to the Bradwell
power station feared that the oysters would be destroyed, the eel grass
would die and the Brent geese would disappear. When he gave his consent,
the Minister gave instructions that the marine ecology of the region should
be kept under review. Surveys carried out since 1959 have shown no change
due to the operation of the station; the oysters are still being fished and
the Brent geese have become so numerous that they are a nuisance to
farmers.

Objections to the Dungeness station were also received because of the
effect it might have on the Bird Reserve and Observatory. As it happens,
the station has prevented the public from wandering over the area, and the
Central Electricity Generating Board has helped to provide a warden.
Counts show that the birds are more numerous than ever before, so much
so that they are sometimes a nuisance at the station. Independent surveys
by wildfowl associations have shown very large increases in waders and
ducks along the foreshore at some estuarial power stations.

D.E. ANDERSON: Are your surveys of the public, along the lines
indicated in your paper, capable of determining the legitimacy of the environmental protest? It is not unknown for what are styled conservationist protests
to be in fact a convenient vehicle for achieving other ends, such as high
property values for the right-of-way of transmission lines.

J.B. BURNHAM: There is no doubt that some who intervene are
motivated by interests other than those they profess. At present, we do
not know how to differentiate between their professed opinions and their true
aims. The part of our analysis concerned with public attitudes is a first
attempt at such a differentiation.

A.P. HULL: For the benefit of participants from countries other than
the United States of America, I can vouch for the fact that on the Eastern
Seaboard the Striped Bass is just as sacred as the Steelhead is in the North-
West. The problems of siting on the Hudson River are illustrative, extending
even to a pumped storage project. We must recognize what people believe,
including the belief that radiation from nuclear effluents presents some
extraordinary hazards, because this affects the political process. In this
connection, does your technique take into account the relative influence
of the respondents?

J.B. BURNHAM: We are at present studying methods for sampling a
community in a meaningful manner. The optimum method would be a random
stratified sample of the community's power structure.

J.G. GROS: I should like to ask a technical question on Fig.2. Is this
a utility function, in the sense meant by von Neumann and Morgenstern, or
a value function? In common usage, a utility function can be used for uncer-

tain and certain situations, whereas a value function can be used only for
certain situations. How does one obtain such a function?

J.B. BURNHAM: This is a rough value function. Such functions will
have to be constructed and combined in some value to yield a dimensionless
index for each of the criteria.

W. KROEGER: In what way, in your opinion, would public response be
influenced by underground siting and/or by a change of reactor type (for
instance, a switch to the HTGR)?

J.B. BURNHAM: This is hard to predict. Judging by the public's
highly irrational response to the LMFBR, as compared with the LWR, it
is probably impossible to make any forecast at all.

B. OBERBACHER: It would be a mistake to overrate the aesthetic sense
of the public. In the 1950s a water power plant was planned at Schaffhausen,
at the Rhine Falls. All environmentalists were against it. Now, twenty
years later, 90% of these former opponents say that the utility did well to
build the plant!

In benefit analyses which compare technical-economic and social parameters, the most difficult thing is to fix the weighting factors. How did
you decide on the weighting factors in Table I?

J.B. BURNHAM: These weighting factors are only theoretical, and
were included solely to illustrate the principle of the method. We recognize
the change of human values with time and are working on techniques which
will enable us to obtain a quantitative assessment of them.

R.R. BOUSSARD: Do you not think that when your method is applied to
a community, the weighting of the different parameters will change as time
goes on and more information is made available to the public? Can you
estimate the sensitivity of the different parameters to public information?

J.B. BURNHAM: We do in fact anticipate that community values will
change with time. Hence we plan, in a later phase of this work, to assess
a community in which a reactor is under construction. We intend to return
after the reactor has operated for several years and repeat the measuremer

# ETUDES GENERALES DE L'ENVIRONNEMENT DES SITES NUCLEAIRES EN VUE DE L'EVALUATION DE LA PROTECTION DES POPULATIONS

G. BRESSON, R. COULON, D. MECHALI
Département de protection,
CEA, Centre d'études nucléaires
   de Fontenay-aux-Roses,
Fontenay-aux-Roses, France

**Abstract–Résumé**

GENERAL STUDIES OF THE ENVIRONMENT OF NUCLEAR SITES WITH A VIEW TO EVALUATING PROTECTION OF THE POPULATION.
   Of the tasks facing those responsible for a nuclear plant, that of selecting the site is necessarily of great importance, because it determines, apart from anything else, the impact on the population and the environment of the release of effluents both during normal operation and as the result of an accident. It is the task of radiological protection specialists, starting in the very first stages of the project and continuing until after commissioning, to make estimates of that impact and of its consequences, in order to supply the essential basic data for rational decision making. Such estimates can take various forms depending on the precision required. The principles developed by international organizations, which are sometimes adopted in national regulations, involve estimating the individual exposure of the critical group, the exposure of the population as a whole, and the radiological consequences of a possible accident. The studies necessary to achieve these objectives are aimed at establishing the characteristics of the environment and of transfers of radioactivity by the collection of existing data and by simulation, as well as the characteristics of the population, in particular its distribution, activities and habits. These data can be synthesized in a more or less elaborate fashion to determine the limiting radiological capacity on a regional level, to bring to light possible exclusion criteria or to make a comparison between short-listed sites.

ETUDES GENERALES DE L'ENVIRONNEMENT DES SITES NUCLEAIRES EN VUE DE L'EVALUATION DE LA PROTECTION DES POPULATIONS.
   Parmi les préoccupations des responsables d'une installation nucléaire, celle qui concerne le choix du lieu d'implantation doit revêtir une grande importance car il conditionne notamment l'impact sur la population et l'environnement d'une libération d'effluents tant en fonctionnement normal qu'à la suite d'un accident. Il appartient aux spécialistes de la protection radiologique de procéder, dès les premiers stades du projet et jusqu'après la mise en fonctionnement, à l'estimation de cet impact et des conséquences résultantes, afin de fournir les éléments indispensables à la prise de décisions rationnelles. Cette estimation peut revêtir différentes formes selon la précision souhaitable. A terme, les principes développés par les organisations internationales, et parfois repris dans certaines réglementations nationales, impliquent l'estimation de l'exposition individuelle au niveau du groupe critique, de l'exposition de l'ensemble de la population et enfin des conséquences radiologiques de l'accident éventuel. Les études nécessaires pour atteindre ces objectifs sont orientées vers la connaissance des caractéristiques de l'environnement et des transferts, par recueil des données existantes, ou inventaire, et par simulation, ainsi que des caractéristiques de la population, notamment sa répartition, ses activités et ses habitudes. La synthèse de ces informations peut s'effectuer d'une façon plus ou moins élaborée, soit pour définir la capacité radiologique limite au niveau régional, soit pour faire apparaître des critères d'exclusion éventuels, soit enfin pour établir une comparaison entre différents sites déjà retenus.

## 1.  INTRODUCTION

Pour assurer de façon efficace la protection de la population et de l'environnement contre les risques potentiels qui peuvent résulter du fonctionnement d'une installation nucléaire, il importe d'apporter lors de la préparation du projet la plus grande attention à deux aspects fondamentaux qui sont:
— le choix d'une implantation présentant les caractéristiques les plus favorables en ce qui concerne les facteurs susceptibles de limiter l'atteinte du milieu et de la population;
— l'utilisation de dispositifs visant à réduire au niveau de l'installation elle-même les conditions créant le risque.

Il est en fait pratiquement impossible, pour de multiples raisons, de parvenir à une situation idéale correspondant à un risque nul: c'est pourquoi les choix qui doivent être opérés se fondent sur le principe du risque «acceptable» compte tenu des avantages retirés par ailleurs de la pratique considérée.

Il est donc impératif d'être en mesure d'apprécier convenablement le risque éventuel et ses conséquences et c'est en cela que consiste tout particulièrement un des aspects du rôle de la protection radiologique.

L'estimation du risque repose notamment sur une étude de l'environnement de la future implantation. Il faut souligner que l'environnement, outre son rôle de récepteur et de vecteur des radionucléides rejetés, peut également exercer certaines contraintes sur l'installation, accroissant ainsi la probabilité que soient créées les conditions d'une situation accidentelle, par exemple du fait d'une sismicité élevée, d'un risque d'inondation ou de chute d'engin. Il s'agit là d'aspects relevant de la sûreté de l'installation dont il ne sera pas traité ici.

Dans la pratique, tous les éléments et considérations déjà évoqués sont interdépendants. Il est probable, par exemple, que le choix d'un site peu favorable entraînera la nécessité d'une meilleure épuration des effluents libérés dans le milieu, afin que soient respectées les limitations imposées. C'est donc une optimisation générale, basée sur l'ensemble des critères, qui doit être réalisée: les aspects relatifs à la sûreté et à la protection doivent bien entendu y trouver leur place.

La nécessité de la prise en compte des conséquences radiologiques que la présence d'une installation nucléaire est susceptible d'entraîner doit aussi apparaître très tôt dans les préoccupations des responsables. Bien entendu, la précision avec laquelle cet objectif doit être atteint dépend largement de la finalité qui lui est accordée: le développement d'études très élaborées au niveau du choix d'un site n'est certainement pas raisonnable alors que d'autres critères d'exclusion sont encore susceptibles d'intervenir. Il s'agit essentiellement d'accompagner le passage d'une simple phase de prospection à la phase finale de réalisation, par une intensification progressive des investigations. Afin de dégager les critères qui serviront de base aux différentes étapes, il est commode de considérer d'abord les objectifs finaux et la méthode permettant de les atteindre.

Ces objectifs peuvent être ramenés au nombre de trois:
— évaluer l'exposition individuelle subie par le groupe de population le plus exposé (groupe critique) et vérifier le respect des limites de dose réglementaires;

- évaluer l'exposition subie par l'ensemble de la population (dose collective);
- évaluer les conséquences de l'accident de référence.

Quant à la méthode généralement utilisée, elle repose sur trois types d'action:
- détermination des caractéristiques du site et de l'environnement ainsi que du comportement des radionucléides impliqués;
- détermination des caractéristiques de la population susceptible d'être affectée;
- synthèse de l'ensemble des informations et estimation des conséquences résultantes.


## 2.   ENVIRONNEMENT ET TRANSFERTS

L'impact susceptible d'être exercé sur l'environnement par les rejets d'effluents radioactifs ne constitue généralement pas un facteur limitatif; il en va toutefois différemment s'il s'agit d'effluents chimiques ou thermiques.

Par contre, intermédiaire entre la source de nuisances et la population, l'environnement, c'est-à-dire l'ensemble des éléments physiques (atmosphère, eaux, sol, etc.) et biologiques (végétaux, animaux, etc.) dans lesquels l'homme vit, peut intervenir comme un vecteur des éléments radioactifs qu'il reçoit. Il est donc nécessaire de bien connaître, d'une part les caractéristiques qu'il présente quant à son rôle de vecteur, d'autre part les relations que l'homme entretient avec lui, notamment par les diverses utilisations qu'il peut en faire.

Les actions permettant d'acquérir cette connaissance relèvent de deux domaines: l'inventaire et la simulation.


L'inventaire consiste à recueillir tous les renseignements existants utiles ou nécessaires pour estimer l'exposition de la population.

Les différents domaines où seront puisées ces informations varient selon les conditions particulières au problème traité, notamment la nature des rejets envisagés et les caractéristiques de l'implantation prévue. D'une façon générale, ils relèvent de:
- La météorologie et la climatologie. Les données obtenues sont destinées essentiellement à l'étude de la dispersion et du dépôt pour des rejets atmosphériques de gaz et d'aérosols.
- L'hydrographie, l'hydrologie, l'hydrogéologie. Dans le cas de rejets d'effluents liquides, ces informations permettent de connaître la dispersion, les échanges et le transfert des polluants dans le milieu aquatique ainsi que leur passage éventuel au niveau des nappes phréatiques.
- La géologie et la pédologie. Elles fournissent les éléments relatifs au comportement des radionucléides dans les sols.
- L'utilisation des eaux: alimentation des réseaux de distribution, irrigation, pêche (professionnelle ou récréative), loisirs, etc.
- L'utilisation agricole des terrains: agriculture et productions d'origine végétale, élevage et productions d'origine animale.
- La transformation et la distribution des productions agricoles.
- Les caractéristiques de la flore et de la faune naturelles.

L'ensemble des informations doit permettre une description de
l'état actuel du site et de son evironnement, mais aussi de l'évolution
prévisible durant les quelques décades qui suivront la mise en fonctionnement de l'installation. Les grands développements que l'on connaît
actuellement, aussi bien sur le plan de la démographie que de l'urbanisa-
tion, de l'industrialisation, des échanges, etc. sont en effet susceptibles
de modifier profondément la physionomie de certains bassins ou certaines
régions.

La phase d'inventaire comporte également la recherche de données
relatives au comportement dans l'environnement des éléments dont le
rejet est envisagé (dispersion, élimination, transfert). La littérature
scientifique offre un très grand nombre de modèles de transfert et de
valeurs de paramètres: le comportement des radionucléides peut en effet
varier considérablement selon les conditions particulières au milieu dans
lequel ils sont présents. C'est la raison pour laquelle il convient d'être
particulièrement vigilant dans le choix des valeurs utilisées.

Ainsi, lorsque l'on dispose des informations relatives à la nature,
l'importance et les conditions des rejets d'effluents, il est possible
d'évaluer la contamination des différents maillons des chaînes de transfert
et des vecteurs de la contamination.

Il ne s'agit bien entendu que d'une estimation préliminaire, de
caractère quelque peu théorique, dont le degré de précision dépend de la
qualité des informations disponibles. Elle devra être complétée par une
étude plus approfondie basée notamment sur la simulation.

Dans l'étude préliminaire, l'absence de certaines données est très
souvent compensée par des hypothèses, généralement conservatoires,
qui demandent à être vérifiées; les processus de dispersion y sont traités
par modélisation. Enfin certains points particuliers peuvent présenter
une importance justifiant un examen plus approfondi.

Le rôle de la simulation consistera donc à recréer, soit in situ,
soit en laboratoire, des situations identiques à celles qui se présenteront
à la suite des rejets d'effluents, et à caractériser l'action de tel ou tel
paramètre: étant donné la complexité des mécanismes qui interviennent
et le particularisme des situations écologiques, la simulation s'avère très
souvent indispensable pour garantir le bien-fondé des estimations.

Enfin, il ne faut aucunement négliger tout l'intérêt que peut présenter
une observation sur le terrain effectuée par une équipe de radioécologistes
expérimentés.


## 3.   LA POPULATION

L'objectif de la protection radiologique étant presque exclusivement
d'assurer le respect des impératifs sanitaires en ce qui concerne la
population, il est essentiel de procéder à un examen détaillé et précis des
caractéristiques qu'elle présente au regard d'une éventuelle atteinte.

En tout premier lieu, elle doit être située géographiquement par
rapport à l'installation concernée: densité par secteur, recensement des
centres de population, font partie des renseignements indispensables. En
général, cette répartition est fondée sur le lieu de résidence des familles,
alors que le séjour réel des individus peut être très différent. Aussi
convient-il de prêter attention aux activités, la plupart du temps profes-

sionnelles, qui tiennent écartée du périmètre concerné une partie de la population résidante ou qui, à l'inverse, drainent dans ce même périmètre des individus non résidants. Cet aspect peut, de même que la présence de groupements d'individus plus ou moins permanents (écoles, industries, organismes publics ou privés, hôpitaux, etc.), revêtir une importance particulière dans l'éventualité d'un accident.

Un autre élément essentiel est constitué par l'alimentation. Si la nature, l'importance et le devenir des productions agricoles sont des éléments qui figurent au niveau de l'inventaire, encore faut-il localiser les groupes de population qui en sont les consommateurs. La dilution par les circuits de distribution peut en effet être du même ordre de grandeur que celle qui se manifeste au niveau du milieu récepteur.

Tant qu'il s'agit de l'atteinte directe de la population par irradiation externe ou par inhalation, il est relativement aisé de cerner, en fonction du modèle de dispersion, les groupes d'individus qui seront exposés. Le problème est plus complexe lorsque l'atteinte est due à l'ingestion de produits alimentaires contaminés. L'attention se porte tout naturellement vers la population avoisinante, déjà concernée par les risques d'atteinte directe: l'hypothèse concervatoire de l'autoconsommation permet souvent de définir le groupe critique. Encore faut-il vérifier la réalité d'une hypothèse susceptible d'être infirmée par une dilution due à l'importation de produits alimentaires d'origine exogène. D'autres considérations peuvent parfois amener à définir un groupe critique géographiquement éloigné du lieu de rejet, principalement dans le cas d'effluents liquides libérés dans une rivière.

On doit également veiller, lors de la recherche du groupe critique, à respecter le principe d'homogénéité qui est à la base de sa définition: il peut parfois en résulter quelques difficultés dans la fixation d'une taille minimale ou le choix d'une taille optimale.

Lorsqu'il s'agit de l'expression de la dose collective, le problème de la recherche des individus exposés ne se pose en principe pas, puisqu'il suffit de connaître les niveaux de contamination des produits alimentaires et la quantité de ceux-ci.

Toutefois, la dose collective est souvent considérée à l'échelle régionale: la distribution de la production doit alors être prise en compte.


## 4.  SYNTHESE

L'exploitation des informations recueillies ou obtenues par simulation doit permettre d'effectuer une évaluation aussi précise que possible des conséquences entraînées par les rejets d'effluents en fonctionnement normal ou en cas de libération accidentelle: c'est ce qui est, à terme, généralement demandé aux responsables de l'installation.

Cependant dans les étapes préliminaires, et tout particulièrement lorsqu'il ne s'agit que de la sélection de lieux d'implantation, cette exploitation peut être beaucoup plus limitée, parfois même revêtir un aspect uniquement qualitatif.

Lorsque le problème posé initialement consiste à définir la capacité radiologique limite d'un bassin fluvial, d'un secteur côtier ou d'une région naturelle, c'est-à-dire estimer la quantité d'effluents pouvant être rejetée sans dépassement de la limite de dose réglementaire au niveau du groupe critique, il est nécessaire d'établir la relation qui existe entre l'activité rejetée et la dose d'irradiation résultante. Cette estimation peut cependant être effectuée d'une façon très globale qui ne requiert qu'une précision relative dans les données utilisées.

Lorsqu'il s'agit de sélectionner un certain nombre de sites présentant a priori une aptitude convenable à recevoir une installation nucléaire, il est possible de se contenter d'un jugement fondé sur des critères qualitatifs ou semi-quantitatifs. Une estimation précise de l'exposition résultant des rejets d'effluents normaux pourra paraître superflue, alors qu'il suffira de situer l'ordre de grandeur des conséquences dues à la présence proche d'une source importante de production d'une denrée alimentaire et de vérifier la comptabilité des contraintes qu'un tel avoisinage ferait peser sur l'installation.

La prise en compte des conséquences d'un éventuel accident peut s'avérer beaucoup plus déterminante. L'atteinte de la population s'effectuant alors essentiellement durant les premières heures par irradiation externe et par inhalation, la présence d'un important centre de population dans une zone où les niveaux d'activité entraîneraient des doses d'irradiation très élevées peut constituer un critère d'exclusion du site. Dans le cas de l'accident, il convient d'associer à la recherche d'un environnement favorable à une minimisation des conséquences, celle de la facilité de mise en œuvre des contre-mesures éventuelles.

Enfin, les sites acceptables, choisis généralement en nombre très supérieur aux réels besoins, sont classés par ordre préférentiel. Les conséquences des rejets d'effluents normaux prennent alors une plus grande importance puisque, toutes choses étant égales par ailleurs, le choix définitif portera sur les conditions correspondant à l'exposition la plus faible.

Il est d'ailleurs toujours possible de recueillir dans un premier temps les informations qui suffisent pour une estimation préliminaire, puis de compléter et améliorer par étapes cette première estimation.


## 5.  CONCLUSION

En plus des nombreux et importants critères économiques, sociaux, techniques qui conditionnent les actions liées aux étapes d'avant-projet, de projet et de réalisation d'une installation nucléaire, les responsables de ces actions doivent à tout moment demeurer vigilants quant aux conséquences pour la population du fonctionnement de celle-ci.

Il est du rôle de la protection radiologique d'estimer l'importance de ces conséquences et de fournir tant aux exploitants qu'aux autorités publiques responsables, les informations qui leur permettront d'orienter le choix des solutions à partir de principes rationnels de justification et d'optimisation.

C'est la raison pour laquelle la prise en compte des problèmes de protection doit s'effectuer au plus tôt en associant certains spécialistes aux équipes chargées de la conception et de la réalisation.

Ces éléments de conclusion peuvent servir d'introduction à l'étude d'un problème soulevé par de nombreux participants et autorités

techniques qui ont en charge le choix du site: celui de l'acceptation du
risque par l'homme.  Pour cela il serait souhaitable d'examiner plusieurs
questions:

- Quelle est l'idée que l'individu du public se fait de la notion de risque
  radioactif?
- Quelles sont les attitudes que l'individu du public adopte vis-à-vis de
  ce risque en fonction des informations dont il dispose?
- Comment caractériser les différentes attitudes observées et les
  classer d'une manière logique?
- Quelles sont les attitudes que la population ou les groupes de populations
  adoptent vis-à-vis de ce risque?
- Quels sont actuellement les moyens d'information utilisés et les
  méthodes mises en œuvre?
- Après avoir reçu ces informations, comment réagit l'individu, le
  groupe ou la population dans son ensemble?

# CHOIX DES SITES DE
# CENTRALES NUCLEAIRES ET
# PROTECTION DE L'ENVIRONNEMENT

H. JAMMET*, A. BELLIN**, G. LACOURLY*,
M. BEAU+, M. MORLAT++, *
Commissariat à l'énergie atomique
et
Electricité de France,
France

**Abstract—Résumé**

NUCLEAR POWER STATION SITE SELECTION AND PROTECTION OF THE ENVIRONMENT.
In France by the end of the century the installed nuclear capacity will be of the order of 200 GW,
which means that about forty sites will be needed. The selection of sites for nuclear power stations is
based on a number of diverse considerations. For convenience, the principal factors considered have been
classified here into three main groups: general criteria, safety criteria and radiological protection criteria.
The selection of sites also involves political, economic and social aspects of which account is taken in
the preliminary studies. Some of the selection criteria can be determined with good precision, some
remain subject to more or less subjective estimations and others are of a strictly political character. The
authors examine to what extent it is possible to reach as rational an analysis as possible with which to explain and
justify the selection of sites.

CHOIX DES SITES DE CENTRALES NUCLEAIRES ET PROTECTION DE L'ENVIRONNEMENT.
En France, à la fin du siècle, la puissance électro-nucléaire installée sera de l'ordre de 200 GW, ce
qui nécessitera la mise en œuvre d'une quarantaine de sites. Le choix des sites destinés à l'implantation
de centrales nucléaires prend en compte de nombreuses préoccupations dans des domaines très divers.
Pour la commodité de l'exposé, les principaux facteurs pris en considération ont été classés en trois grands
groupes: les critères généraux, les critères de sûreté et les critères de protection radiologique. Le choix
des sites comporte en outre des aspects politiques, économiques et sociaux dont il est tenu compte au stade
des études préliminaires. Parmi tous les critères de choix, certains sont mesurables avec une bonne précision, d'autres restent par contre soumis à des estimations plus ou moins subjectives, d'autres enfin ont un
caractère strictement politique. Les auteurs examinent dans quelle mesure il est possible de progresser
vers une analyse aussi rationnelle que possible permettant d'éclairer et de justifier le choix des sites.

## INTRODUCTION

En 1973, la consommation française d'électricité a été de 171 TW·h.
On prévoit 400 TW·h pour 1985 et 1000 TW·h pour la fin du siècle. Ces
chiffres correspondent à un rythme de croissance de l'ordre du double-
ment tous les 10 ans pour les prochaines années avec un ralentissement
à la fin du siècle.

---

* Département de protection, CEA, Centre d'études nucléaires de Fontenay-aux-Roses, Fontenay-
aux-Roses.
** Direction de l'équipement, Electricité de France, Paris.
+ Direction de la production et du transport, Electricité de France, St-Denis.
++ Direction des études et recherches, Electricité de France, Clamart.

Pour satisfaire de tels besoins, compte tenu de la crise pétrolière, la part de production d'origine nucléaire va croître très rapidement et passera de 8% en 1973 à 30% en 1980 et aux environs de 70% en 1985.

Le programme de construction des centrales nucléaires est donc un programme important, aussi bien par son volume (l'équivalent de 200 tranches nucléaires de 1000 MW d'ici à la fin du siècle) que par la traduction d'une volonté politique: la part grandissante donnée à l'énergie nucléaire dans les approvisionnements en énergie du pays.

La réalisation de ce programme implique au départ que soit résolu de façon aussi rationnelle que possible le problème posé par le choix des sites. On sait qu'il s'agit d'un problème très complexe, dont les données sont très nombreuses et dont la solution ne peut être qu'un compromis.

Parmi les critères à prendre en considération, certains sont plus importants que d'autres, et le problème qui se pose est de savoir comment parvenir à un schéma méthodologique, permettant d'éclairer et de justifier ces choix.

Quels sont ces critères?

## 1.   CRITERES GENERAUX: STRATEGIES D'IMPLANTATION

### 1.1.  La source froide

L'une des contraintes essentielles que rencontre le choix des sites est la source froide. Une centrale comportant quatre tranches de 1350 MW nécessite, si elle est refroidie en circuit ouvert, un débit d'eau compris entre 150 et 250 m$^3$/s échauffé de 10°C à 15°C. Cette contrainte conduit EDF à envisager deux techniques de source froide:
- le circuit ouvert sur rivière et surtout en bord de mer,
- le circuit fermé avec réfrigérants atmosphériques humides.

Pour la première technique, les calories non transformées en électricité et rejetées au condenseur sont évacuées à la rivière ou à la mer et passent progressivement à l'atmosphère par convection, par rayonnement et par évaporation. Pour la seconde technique, les calories du condenseur sont directement rejetées à l'atmosphère.

Pour les centrales refroidies en circuit ouvert et implantées sur rivière, c'est le débit d'étiage qui conditionnera la puissance équipable. Parmi les fleuves français, seul le Rhône a un débit d'étiage suffisamment important pour refroidir encore quelques unités en circuit ouvert.

Pour les centrales refroidies en circuit ouvert et implantées en bord de mer, les meilleurs sites seront ceux pour lesquels un bon mélange des rejets produira les échauffements les plus faibles dans toute la zone influencée par la centrale. Ce seront les sites où il y a des masses d'eau importantes, brassées et renouvelées par les courants de marées, les courants de dérive ou les courants de la circulation générale induite par les vents. Pour un site donné, la puissance équipable dépend de l'aptitude à assurer ce mélange et aussi de la configuration locale, en particulier des facilités ou difficultés qu'il y aura pour éviter que les eaux rejetées ne recirculent dans la centrale avant d'avoir été bien mélangées.

## 1.2.  Les critères liés au terrain

Une centrale est un ouvrage lourd nécessitant des fondations de
bonne qualité.

Les ouvrages d'eau en bord de mer qui transitent des débits importants
et qui doivent résister aux agressions nombreuses du milieu constituent
autant de cas particuliers qu'il y a de sites.

Les ouvrages d'évacuation de l'énergie constituent aussi un ensemble
ayant un impact important sur le paysage.  La proximité du réseau
d'interconnexion apportera une facilité de raccordement.

Les accès, notamment pour le transport des pièces lourdes, doivent
aussi pouvoir être exécutés simplement.

## 1.3.  Les critères liés à l'aménagement du territoire

La France ne figure pas encore sur la liste des pays pour lesquels
le territoire disponible est rare.  Cependant, certaines régions où la
densité d'occupation des sols est grande (la région parisienne, le Nord,
certaines vallées, certains estuaires) n'ont plus de terrains convenables
disposant de quantités d'eau suffisantes et de surfaces assez vastes
pour y aménager correctement une centrale.

Le littoral français, qui s'étend sur une longueur de plus de 5000 km,
est déjà très exploité: pêche, conchyliculture, tourisme, sont autant
d'activités peu favorables à l'accueil des centrales nucléaires.  De plus,
la volonté affirmée du Gouvernement français de préserver certains
espaces et de conserver leur cadre naturel ne facilite pas la réservation
des sites.

## 1.4.  Les critères liés à l'environnement

Les effets d'une centrale sur son environnement sont de nature très
variée.

Outre les effets radiologiques étudiés dans la section 3, on peut
schématiquement ranger les effets en trois catégories:
- les effets sur le milieu physique
- les effets sur le milieu vivant
- les effets sur le milieu socio-économique.

Sur le milieu physique, les effets les plus importants seront ceux
de la source froide: ce seront les échauffements pour le circuit ouvert,
les risques de panache pour le circuit fermé.  Ce seront aussi les
perturbations apportées à l'équilibre côtier par les courants ou par les
obstacles créés.

Sur le milieu physique, ce seront aussi les modifications de paysage
induites par la centrale qui interviendront.  Dans tous les cas, une
recherche architecturale sera nécessaire.  Cette recherche devra tendre
soit vers une insertion dans le paysage lorsque celui-ci a au préalable une
valeur, soit vers une création lorsque le paysage ne présente aucun
caractère.

Sur le milieu vivant, ce seront surtout les conséquences des modifications physiques qui vont intervenir.  Bien qu'il s'agisse d'un domaine
très vaste encore peu exploré, il est possible dès aujourd'hui d'affirmer

que les changements détectables n'interviendront que sur des zones très
localisées. Généralement les défenseurs de l'environnement considèrent
ces changements comme néfastes: en fait il y aura des changements
négatifs mais il y aura aussi des changements positifs et c'est en termes
de bilan qu'il faut raisonner. En attendant que ces bilans puissent être
dressés correctement, le responsable du choix des sites doit observer
vis-à-vis du milieu vivant une très grande prudence.

Sur le milieu socio-économique, la centrale apporte aussi des
transformations, notamment lorsqu'elle est implantée en zone rurale.
Sans être un facteur d'aménagement du territoire, car l'investissement
important qu'elle représente est surtout dépensé en dehors de la région
où elle est construite, la centrale constitue cependant un élément important. Elle apporte un véritable coup de fouet à l'économie régionale au
moment du chantier (en provoquant souvent une profonde mutation) et
ensuite elle apporte des ressources importantes pendant l'exploitation
(impôts et taxes). Sans être déterminants dans les choix, ces éléments
doivent être soigneusement examinés et les pouvoirs publics s'y attachent.

### 1.5. Les critères «liés au public et aux collectivités»

Une centrale fait l'objet d'un décret d'utilité publique pris par le
Ministre de l'Industrie et de la recherche au terme d'une enquête
administrative et d'une enquête publique.

En outre, le décret du 11 décembre 1963, modifié par celui du 27 mars
1973, a institué une procédure spéciale comportant un examen minutieux
des dispositions techniques prévues tant en matière de sûreté qu'en
matière de radioprotection. Cet examen constitue l'élément de base de
l'autorisation de création, puis de l'autorisation d'exploitation données
par les pouvoirs publics.

Dans la mesure où ces procédures sont longues (environ 2 ans) le
demandeur se doit de proposer des sites dont les chances d'aboutir sont
quasi certaines, c'est-à-dire des sites ayant au préalable fait l'objet
d'études d'avant-projet précisant notamment les effets de la centrale sur
son environnement et ayant fait l'objet de concertation avec les collectivités
locales et le public.

Cela conduit à engager des études d'avant-projet sur plusieurs sites
d'une même région: ces études, qui sont avant tout des études de factibilité
permettent de procéder sur le site à tous les relevés utiles (sondages,
courantologie, données météorologiques, hydrologie, etc.) tout en engagean
une discussion avec les collectivités locales. Le lancement d'une étude
d'avant-projet n'implique nullement que la décision de construire une
centrale sur le site sera prise. La concertation avec les collectivités
suppose que des contre-propositions puissent être faites et étudiées.


## 2.   CRITERES DE SURETE ET CRITERES DE PROTECTION

Tous les critères énumérés précédemment sont à quelques nuances
près ceux d'une installation importante. Ils s'appliquent aussi bien à une
centrale à fuel qu'à une centrale nucléaire, à une raffinerie qu'à une
usine de produits chimiques.

Si les critères qui suivent sont spécifiques des installations nucléaires, ils auraient leurs homologues dans beaucoup d'installations industrielles si l'on se donnait la peine de faire à l'égard de ces installations une analyse aussi fine que celle à laquelle les centrales nucléaires sont soumises.

## 2.1. Critères de sûreté

La sûreté recouvre l'ensemble des dispositions techniques imposées au stade de la conception de la construction, puis de l'exploitation et finalement du déclassement des installations pour assurer le fonctionnement normal, prévenir les accidents et en limiter les effets. Elle fait appel à la compétence technique des ingénieurs: ceux de l'exploitant et des constructeurs d'une part, ceux de l'Administration et des organismes techniques qui l'assistent d'autre part.

La sûreté implique deux types de liaisons entre la centrale et son environnement, conduisant à deux catégories de risques:

a) Les risques dus à des défaillances internes de l'installation. Ces risques ont une probabilité extrêmement faible, d'autant plus faible que les dommages qui en résulteraient pourraient être très graves.

b) Les risques liés à l'influence de l'environnement sur l'installation. Ces risques peuvent aussi bien être consécutifs à des cataclysmes naturels tels que séismes et phénomènes associés, tempêtes, tornades, inondations, etc. qu'à des accidents tels que ruptures de barrage, chutes d'avions et de missiles, agressions de l'environnement industriel, etc. Aucun de ces risques ne doit être à l'origine de situations accidentelles conduisant l'exploitant de la centrale à perdre la maîtrise des éléments radioactifs confinés dans l'installation.

## 2.2. Critères de protection

La radioprotection correspond aux dispositions prises pour protéger les travailleurs et le public contre les risques éventuels présentés par l'installation, tant pendant son fonctionnement qu'en cas d'incident. Elle concerne l'homme et son milieu.

Indépendamment des dispositions intrinsèques de sécurité de l'installation prises au titre de la sûreté nucléaire, la radioprotection se préoccupe de protéger la santé de l'homme. Elle effectue donc la surveillance de la contamination radioactive éventuelle de l'environnement, ainsi que de l'irradiation de l'homme qui pourrait en résulter, qu'il s'agisse des travailleurs ou de la population. La radioprotection passe donc par la protection et la surveillance de l'environnement. A cet égard, on peut distinguer:

a) Les critères de protection pour le fonctionnement normal de l'installation. Ils dépendent des paramètres qui influent sur l'atteinte des populations par les substances radioactives présentes dans les effluents normaux.

b) Les critères de protection en cas d'accident. Ces critères diffèrent peu de ceux qui sont pris en compte pour l'irradiation des populations à partir d'effluents normaux.

Pour mieux cerner ces critères, il est utile d'examiner les démarches qui permettent d'apprécier les conséquences des rayonnements ionisants émis par les effluents sur les populations.

## 3.  LA PROTECTION DES POPULATIONS ET DE L'ENVIRONNEMENT

Pour apprécier l'aptitude de l'environnement d'un site à recevoir les
substances radioactives présentes dans les effluents rejetés par une installation donnée, il est utile de disposer d'un critère global, intégrant les
caractéristiques de l'environnement et les populations concernées.

On sait que, suivant les recommandations de la Commission internationale de protection radiologique, l'activité rejetée dans le milieu par
les installations nucléaires doit être limitée de telle façon que:
- les limites de dose fixées pour les individus du public ne soient pas
  dépassées
- l'exposition des populations soit maintenue aussi basse que possible.

Tenant compte de ces deux impératifs, l'étude d'un site est faite selon
deux approches:
- estimation de la dose individuelle délivrée au groupe de population
  critique
- étude de l'exposition des populations concernées.

Ces deux approches permettent, la première, de vérifier l'application
du premier principe, la seconde, de définir le niveau d'exposition des
populations pour chacun des sites.


### 3.1.  Etude de la dose individuelle délivrée au groupe critique

Pour un même rejet unitaire, la dose individuelle varie suivant les
sites.  Elle est la résultante de la variabilité de différents facteurs
propres à chaque site et de la variabilité des caractéristiques et des
habitudes de vie (profession, alimentation, etc.) des populations concernées
Les rejets de la centrale sont habituellement divisés en deux catégories

a)  Les effluents atmosphériques

Les effluents atmosphériques peuvent atteindre les populations par
trois voies distinctes, d'importance variable, suivant les substances
radioactives rejetées et suivant les sites: l'irradiation externe, l'inhalation et l'ingestion de produits contaminés.

L'exposition des populations par irradiation externe et par inhalation
est directement liée à la concentration des substances radioactives dans
l'air au niveau du sol.  Pour l'évaluer, on a donc besoin de connaître
la délimitation géographique des zones d'isoconcentration autour du point
d'émission et la répartition des individus de la population dans chacune
de ces zones.

Outre les facteurs météorologiques, qui sont déjà nombreux et présentent des fluctuations importantes, certains facteurs géographiques tels
que la présence de reliefs montagneux, de vallées plus ou moins profondes,
d'étendues d'eau plus ou moins grandes, qui affectent la direction des
vents et les conditions de stabilité jouent un rôle considérable dans la
dispersion.

Les doses dues à l'ingestion de produits contaminés à partir du
dépôt sur le sol et sur la végétation des substances radioactives présentes
dans les effluents atmosphériques sont déterminées par l'activité ingérée
par la population.

La dose individuelle délivrée au groupe de population critique à la suite de l'ingestion de produits contaminés tient compte des radionucléides critiques et de la voie critique.

Les considérations développées ici pour les rejets d'effluents atmosphériques normaux sont susceptibles de la même application pour l'évaluation des conséquences des accidents, celles-ci étant évaluées sur la base de la composition isotopique des rejets accidentels présumés résultant de l'accident de référence.

b)    Les effluents liquides

D'une façon générale, les doses résultant des rejets d'effluents liquides sont essentiellement déterminées par la quantité d'activité ingérée, la contribution de l'irradiation externe étant beaucoup moins importante. Les facteurs à prendre en considération sont très nombreux, ils concernent notamment:
− les paramètres de dispersion: débits, courants, convection, etc.
− les paramètres de concentration soit dans les organismes aquatiques
   soit dans les produits irrigués.

On conçoit, dans ces conditions, que les conséquences des rejets peuvent varier d'un site à l'autre dans de très larges proportions suivant les caractéristiques locales qui déterminent la dispersion des substances radioactives suivant les habitudes de pêche locale et suivant les productions agricoles et piscicoles.

Finalement, qu'il s'agisse d'effluents atmosphériques ou d'effluents liquides, de rejets normaux ou de rejets accidentels, la dose individuelle délivrée au groupe critique peut varier beaucoup d'un site à l'autre: on conçoit aisément qu'un site mal ventilé en bordure d'une rivière à faible débit présente à cet égard des caractéristiques moins bonnes qu'un site en bord de mer où les masses d'eau sont perpétuellement brassées et mélangées et où les vents sont en général puissants.

Dans la pratique, on se borne jusqu'à présent à vérifier que la dose individuelle délivrée aux individus du groupe critique ne dépasse pas la limite de dose recommandée. Un concept différent consiste à définir, pour chacun des sites envisageables, l'activité des effluents qu'il serait possible de rejeter et qui entraînerait pour les individus du groupe critique une dose engagée égale à la limite de dose recommandée. Il s'agit de la capacité radiologique limite (CRL) du milieu récepteur. Pour un site donné la CRL, qui s'exprime en curies par an, représente donc une limite de rejet au-delà de laquelle la nuisance radiologique qui en résulte n'est plus considérée comme acceptable.

La valeur de la CRL est obtenue à partir de la relation qui existe à l'équilibre entre l'activité rejetée dans le milieu et la dose délivrée aux individus du groupe de population critique, en donnant à cette dose la valeur de la limite de dose recommandée. C'est donc une caractéristique de l'environnement et des populations. Elle intègre en particulier tous les facteurs qui jouent un rôle important dans la dilution et la reconcentra-tion des radionucléides dans le milieu récepteur et les chaînes alimentaires. Elle tient compte également d'un certain nombre de caractéristiques des populations locales.

Basée sur le critère de la limite de dose, cette notion permet de s'assurer que le milieu récepteur peut accepter les rejets projetés sans que la limite de dose fixée pour les individus du public soit atteinte. La comparaison avec la CRL du niveau de rejet projeté ou autorisé mesure automatiquement le degré de protection assuré au niveau du groupe critique. Utilisée à l'échelle locale ou régionale, la CRL permet en outre de comparer utilement entre eux plusieurs sites et éventuellement plusieurs points de rejet.

### 3.2. Etude de l'exposition de la population

La définition du groupe le plus critique tient compte de la dispersion des effluents de la centrale (irradiation directe) et de l'ingestion des produits contaminés (irradiation indirecte). Mais cette notion ne tient pas compte du nombre des individus concernés.

Pour la population qui vit autour du site, l'étude des doses relève de la même démarche que celle qui conduit à l'évaluation de la dose individuelle de groupe critique cité ci-dessus. Pour la population concernée par les productions alimentaires affectées par l'installation, une étude simplifiée peut être faite en tenant compte des activités rejetées et des voies de transfert. Pour un site donné, cette population est beaucoup plus difficile à identifier.

Des concepts tels que la «dose collective» ont été proposés par la CIPR dans le but de globaliser les effets des installations nucléaires sur la population. La définition de ce concept donne lieu à des études sur le plan national et international. Son exploitation nécessitera des études complémentaires appliquées à chaque cas particulier.

Si ce concept permet de comparer entre eux plusieurs sites, il apparaît aujourd'hui plus difficile d'établir une relation entre dose collective et conséquences sanitaires qui en résultent. En particulier les incertitudes actuelles sur les «relations dose-effet» font que cette évaluation ne pourrait conduire qu'à une valeur que l'on doit considérer comme limite supérieure, la valeur réelle étant comprise entre zéro et cette limite.

## 4.  LES CRITERES DE CHOIX:  ANALYSE ET PONDERATION

Est-il possible pour chaque critère de porter un jugement de valeur et ensuite, grâce à une pondération adéquate, d'établir une synthèse de tous les critères permettant de formuler un choix?

Les critères sont nombreux, mais aussi de nature très variée. Aux critères quantifiables (localisation des besoins d'électricité, source froide, fondations, accès, réseau, etc.) s'ajoutent des critères nouveaux plus difficiles à évaluer, tels que l'environnement, la sûreté, les conséquences radiologiques, l'opinion publique, qui font appel aux «sciences approchées».

Pour l'instant, aucun dénominateur commun n'existe entre ces critères nouveaux et les critères technologiques et économiques; la prise en compte de manière encore approchée de ces critères nouveaux suppose que les spécialistes se regroupent de plus en plus au sein d'équipes pluridisciplinaires ayant en charge l'étude d'un site sous ses multiples facettes.

S'il n'existe pas d'étalon de mesure commun à ces différents
critères il est encore bien plus difficile de donner à chacun de ces
critères nouveaux un poids.  Par exemple, est-ce que la prise en
considération d'un accident grave dont la probabilité d'occurrence est
infiniment petite a une importance inférieure, égale ou supérieure aux
risques d'atteinte à un paysage sauvage que l'on veut conserver comme
témoin des choses de la nature?

Comment évaluer en termes monétaires des données telles que la
dégradation de la nature, les atteintes à l'esthétique, les conséquences
sanitaires des rejets d'effluents radioactifs pour les populations, ce qui
implique une évaluation de la vie humaine.

L'analyse décisionnelle doit cependant rester l'objectif vers lequel
on doit tendre, étant entendu qu'il est possible dans certains cas de
tourner la difficulté en opérant un changement de variable.  C'est ainsi
par exemple que, dans le domaine de la protection, si deux sites sont
équivalents sur le plan des autres critères, la comparaison peut se faire
sur la base des critères de radioprotection explicités ci-dessus sans
qu'il soit nécessaire d'apprécier la valeur des conséquences sanitaires.

Il n'est pas sûr cependant que l'on parvienne à éliminer toutes les
difficultés et on risque d'être amené à comparer l'incomparable.  Faute
d'une unité de mesure commune, l'analyse rationnelle n'est plus possible
et le choix restera uniquement affaire de bon sens, à moins qu'il ne
résulte finalement d'une option politique fortement influencée par
l'opinion publique.

## BIBLIOGRAPHIE

LACOURLY, G., «La capacité radiologique de l'environnement, Base des études prospectives de
protection»,   Environmental Behaviour of Radionuclides Released in the Nuclear Industry (C.R. Coll.
Aix-en-Provence, 1973), AIEA, Vienne (1973) 603.

GARNIER, Arlette, LACOURLY, G., «Prévision des conséquences radiologiques des rejets normaux
d'installations nucléaires à l'échelle régionale», Population Dose Evaluation and Standards for Man and
his Environment (C.R. Journées d'études Portorož, 1974), AIEA, Vienne (1974) 183.

# DISCUSSION

(on the previous two papers)


C.J. van DAATSELAAR: May I ask Mr. Jammet whether, in the
accident analysis (fault tree analysis), there is a range above which one
can say that the probability is so small that such accidents need not be
taken into account any further?

In evaluating the radiological consequences of accidents, do you think
that for the calculation of the collective dose an 'insignificant dose' could
be applied to limit the integration?  And could a comparable 'insignificant
dose' be introduced in assessing effects on individuals?

H. JAMMET: The French bodies responsible for protection consider
only realistic accident hypotheses, i.e. those which will occur with a
probability greater than near-zero.  This follows from considerations of
reactor technology.  There are particular circumstances, such as an
aircraft crash on a reactor site and sabotage, which are considered in the
general context; we take account of all these factors.  But one must not
confuse actual deaths with hypothetical deaths from cancer at low doses.

A calculation aimed at determining whether there are limits on
significant dose is a difficult exercise.  As regards the collective dose,
I said earlier that it is sufficient to multiply the doses by the number of
individuals.  One must here know the pattern of dose distribution among
the population, because the usual simplistic hypotheses would give quite
a different result.  The hypotheses of linearity, i.e. of the absence of a
threshold, are valid only for very low exposures − of the order of rem.
If higher exposures are involved, one has to use man·rad.  The effects
will then no longer be stochastic phenomena and the calculations will be
wrong.  Then there are degrees of uncertainty.  The number of persons
is known very well, but the dose estimate is uncertain, so if we multiply
a small dose by a very large number of persons, the result will tend to
be uncertain, mathematically.  This means that there is a threshold below
which integrations cannot be performed over the full number of persons
because the uncertainty for this number is too high.  We take this point
into account so as not to upset the public.  In each case, it is sufficient
to take the released effluents, form a nice equation on world distribution
and multiply it by so many thousand million persons.  A computation of
this kind yields very high figures.  They may have some significance
where nuclear fall-out is concerned but are devoid of any meaning in
relation to nuclear reactors releasing iodine.

E.H. HUBERT: I should like to give some replies to the questions
raised by Mr. Bresson.  These replies are quite subjective and not based
on systematic inquiries.

Some four years ago the public in general were rather distrustful
owing to a lack of information.  The promoters of nuclear energy tried to
disseminate information by the usual means: the press, radio and television interviews, lectures, visits to facilities and distribution of very
simple brochures to a very large public.  What has been the result?  On
the whole, the public appear to be progressively more favourable (this is
also perhaps due partly to the oil crisis), whereas real opposition has
hardened and the opponents of nuclear energy have considerably improved

their actual knowledge of the phenomena involved, even though their numbers have clearly declined.

G. BRESSON: I am grateful to Mr. Hubert for these interesting observations, which support my idea of doing work in this direction. These activities must be carried out at the national level, but the studies obviously need to be compared internationally.

J.B. BURNHAM: Mr. Jammet deserves to be congratulated on his recommendation of the use of regal wisdom in the handling of nuclear matters. How would the King handle a situation like the one we have in California, where there is a real danger from combustion products, yet there are almost as many cars as people; where the risk from earthquakes is almost three orders of magnitude greater than that from reactors, yet the laws passed with a view to providing protection against earthquake hazards are universally ignored; while Proposition 9, seeking to ban not only nuclear reactors but even the manufacture of their components, was put to the vote in 1970? It failed in that election, but will come up again. How would the King handle this?

H. JAMMET: The wisdom of the Kings of France was brought to bear on many things, but their problems were not the same as ours. What is most striking today is that public concern with respect to nuclear power is based on false ideas. I have referred to genetic effects on plants and animals. People are totally ignorant of these. They believe that little is known about the effects of ionizing radiations, while the scientific truth is just the opposite. In cases where cancer is induced by ionizing radiations, the dose-effect relationship is known very well, much better than what is known of other carcinogenic agents. The dose-effect correlation for genetic mutations has also been determined most satisfactorily. At a recent meeting on virology, I was surprised to learn how few were the precautions taken in work with highly pathogenic viruses. But nobody is worried about them. The public often believe that radioactivity never disappears or disappears very slowly. But what we know for sure is that it does not increase exponentially like viruses or other pathogenic organisms. We must demonstrate to the public that the assertions of sensational journalism are very far from the scientific truth. What one learnt in school twenty years ago is simply not true today. People should worry more about earthquakes, chemical pollution and bacteriological hazards. The public is not concerned at hundreds of deaths every year from electrocution and gas explosions, but they are asked to worry about hypothetical cancer which may occur in ten or twenty years. Such attitudes do not stand up to scientific analysis. It is amazing that the public should be worried about plants and animals in the wild but not about their own health.

A. MERTON: I wonder if anybody can explain why public acceptance of nuclear power plants differs from one country to another.

H. JAMMET: I cannot explain that, but what I can say is that the public, ignorant of atomic problems and hence distrustful at first, changes its stand radically in a few years after coming into contact with the atomic realities.

Gloria CAMPOS VENUTI: I would be interested to know in what way the universities in France are involved in the problems of nuclear energy and nuclear siting, for a right choice is dependent on proper knowledge. Also, how are members of the public made aware of the health problems connected with fossil-fuelled plants?

H. JAMMET: In France, relationships with the universities are of two kinds. In most cases a constructive relationship is maintained with university men specializing in particular problems (e.g. ecology, cancerology and physical chemistry). In some cases, however, we have to deal with contentious men whose erroneous beliefs are impossible to change.

G. BRESSON: Mrs. Campos Venuti has struck an interesting topic, that of enlightening the public through improvements in general education. I am glad to say that the French authorities are giving attention to the matter, and studies bearing on questions of safety and risks, notably those relating to radiation, are included in the educational curricula, even at the higher levels.

# SITING AND REPROCESSING PLANTS

## (Session VI)

Chairman:

N.P. DERGACHEV (USSR)

# RADIOLOGICAL IMPACT ON THE ENVIRONMENT BY INTERMITTENT GASEOUS EFFLUENTS FROM FUEL-REPROCESSING PLANTS


W. HÜBSCHMANN, P. THOMAS
Abteilung Strahlenschutz und Sicherheit,
Kernforschungszentrum Karlsruhe,
Karlsruhe, Federal Republic of Germany



**Abstract**

RADIOLOGICAL IMPACT ON THE ENVIRONMENT BY INTERMITTENT GASEOUS EFFLUENTS FROM FUEL-REPROCESSING PLANTS.

Nuclear fuel-reprocessing plants are now, and will be in the future, main sources of radioactive emissions into the atmosphere. The radiological burden to the environment is calculated for different modes of release. This may help in site selection and licensing procedures. The annual local doses at several points in the vicinity of the plant are assessed and compared for continuous, periodic, and randomly changing activity release. Most unfavourable short-time emission is also considered. All doses refer to the same activity of $^{85}$Kr released annually. The emission rhythm and rate are extrapolated from the small Karlsruhe pilot plant. A 24-hour periodic or pseudoperiodic operation yields the lowest doses, if the time of the beginning of emission is suitably chosen.


## 1. INTRODUCTION

In the past, public opinion has been principally concerned with radioactive emissions to the environment from nuclear power plants. In the future, when additional facilities will be needed to reprocess the increasing amount of burnt fuel elements, emissions from fuel-reprocessing plants will also exercise public opinion. At the beginning of the 1980s a large reprocessing plant with an annual capacity of 1500 tonnes will start operation in densely populated western Europe [1]. This facility will serve forty nuclear power plants of the 1200 MW(e) power level. If filter and retention technology is not improved, the amount of radioactive effluents to the environment will be much larger than that from nuclear power plants.

At present $^{85}$Kr (100%), tritium (10%), radioiodine (1%), and aerosols ($10^{-6}$%) are released through the stack. The numbers in parentheses indicate the fractional releases [2]. In accordance with the process, the fission products are released in batches.

Because of public resistance to nuclear facilities, fuel-reprocessing plants will certainly not be allowed to charge the environment with higher radiation levels than modern nuclear power stations do. Site selection may therefore be much more difficult for reprocessing plants. To facilitate site selection and licensing procedures, the dose distributions in the vicinity of the plant have been assessed, taking into account the discontinuous process.

At present 100% of $^{85}$Kr is released to the atmosphere. For inhalation and submersion it is the critical isotope, even if only the skin dose produced by the soft $\beta$-radiation is considered. This fission product is therefore chosen for assessment of the radioactive burden to the environment. For the emission pattern, experience gained at the small Karlsruhe reprocessing plant (WAK) is applied in the calculations.

## 2. EMISSION PATTERNS

Different modes of release are considered, leaving unchanged the total annual activity emission.

### Case A: Continuous long-time emission

This case is calculated only for comparison. According to the dose distribution in the vicinity of the plant, ten characteristic locations were selected. These locations are situated in the two main wind directions, and one is exposed to the maximum dose.

### Case B: Periodic emission

In a 24-hour period the effluents are released over a constant length of time and at a constant emission rate. Twenty-four cases are calculated with releases at intervals of one hour (Fig.1).

### Case C: Randomly changing emission

The rate and interval of the emission vary at random between given limits (Fig.1). The cumulative frequencies of the doses are assessed. To demonstrate the differences between a pseudoperiodic release (case C1)

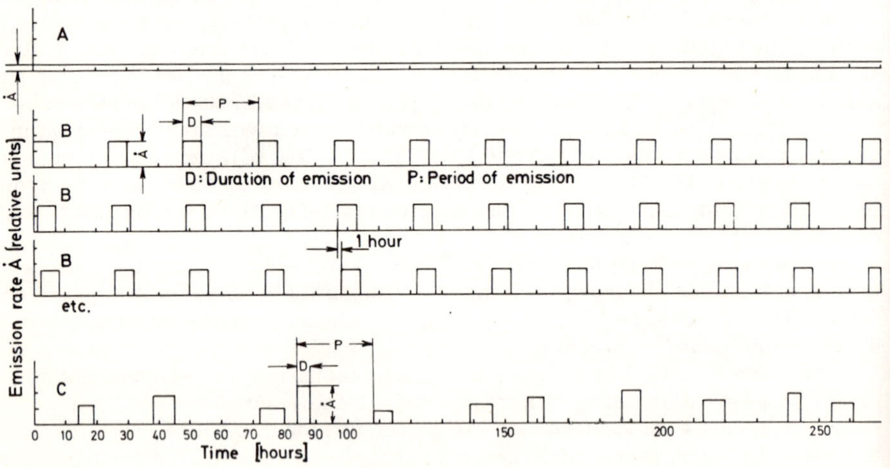


FIG.1.  Timing of emission.

and a rather random one (case C3), the limits of the emission rate and interval are changed.

## Case D: Short-time emission

For licensing procedures in the Federal Republic of Germany the most
unfavourable weather conditions are selected for a short-time emission
pattern [3]. But these weather conditions are rare. Even in case C3 the few
hours in which these weather conditions prevail are not sufficient to emit the
total krypton activity theoretically. The following model was therefore chosen
to calculate the upper limit of exposure:

The total annual duration of emission may be $x \cdot \Delta t$ hours. The total
annual activity is released by x equal shares at x time intervals of
duration $\Delta t$. These x intervals coincide with the annual hours yielding
the highest dose for a given location in the vicinity.

Of course this case is rather improbable, but it is theoretically possible. It
corresponds to the short-time emission model [3].


## 3.   METHOD OF EVALUATION

The meteorological input data are hourly mean values of the wind
direction measured at 60 m and of the wind velocity measured at ten different
heights of the 200-m tower erected on the site of the Karlsruhe Nuclear
Research Centre. The Pasquill diffusion categories are derived from the
vertical wind profile [4].

The hourly mean concentration $C_i$ at location i is calculated for each
hour n of two years, comprising the period from December 1967 to
November 1969 [5, 6]:

$$C_{i,n}(x_i, y_i) = \frac{\dot{A}}{\pi u \, \sigma_{y,n}(x_i) \, \sigma_{z,n}(x_i)}$$

$$\times \exp\left[ -\frac{H^2}{2\sigma_{z,n}^2(x_i)} - \frac{y_i^2}{2\sigma_{y,n}^2(x_i)} \right]$$

where

| | | |
|---|---|---|
| $\dot{A}$ | = | emission rate (Ci/s) |
| u | = | wind velocity at 60 m height at hour n (m/s) |
| H | = | stack height (m) |
| $\sigma_{y,n}(x_i)$, $\sigma_{z,n}(x_i)$ | = | diffusion parameters at point i and hour n (m) |
| $x_i, y_i$ | = | co-ordinates of location i |

The submersion dose:

$$D_{i,n}(x_i, y_i) = g C_{i,n}(x_i, y_i)$$

due to the concentration $C_{i,n}$, is proportional to the dose factor g for $^{85}$Kr [7]:

$$g = 0.073 \ \frac{rem \cdot m^3}{Ci \cdot s}$$

Case A:

In continuous emission all doses $D_{i,n}$ of location i are summed to yield the annual dose $D_i$:

$$D_i = \frac{1}{2} \sum_{n=1}^{17\,544} D_{i,n}$$

The time interval of two years equals 17 544 hours. The factor $\frac{1}{2}$ refers to the two years' integration period.

Case B:

For periodic emission the doses $D_{i,n}$ are multiplied by the ratio of period to emission time. Only doses corresponding to the periodic release intervals are summed.

Case C:

For randomly changing emission the summation is carried out analogousl The length of release intervals and the multiplication factor are integer random numbers homogeneously distributed between given limits.

The annual doses are calculated 100 times with continuously generated random numbers giving 100 sets of dose values $D_{i,j}$ (j = 1,2,...,100). The mean square error:

$$\sum_i{}^2 = \overline{D}_i^2 - \overline{D_i^2}$$

of the annual dose is assessed from the mean value $\overline{D}_i$ and the mean value of the square $\overline{D_i^2}$ obtained from the summation over j. Increasing the number to 1000 did not significantly change the results.

## 4. ACTUAL EMISSION PATTERN IN A REPROCESSING PLANT

The timing and amounts of activity release are chosen in conformity wi the pilot plant WAK at Karlsruhe. Its annual capacity is 40 tonnes uranium burnt up to about 20 000 MW·d/t. The authorized annual release of $^{85}$Kr amounts to 350 000 Ci.

According to the so-called chop-leach technique, the fuel elements are chopped by scissors and are subsequently dissolved. During the dissolution step lasting about six hours, the gaseous fission products are released. The timing of the $^{85}$Kr release during dissolution is shown in Fig.2 [8]. The

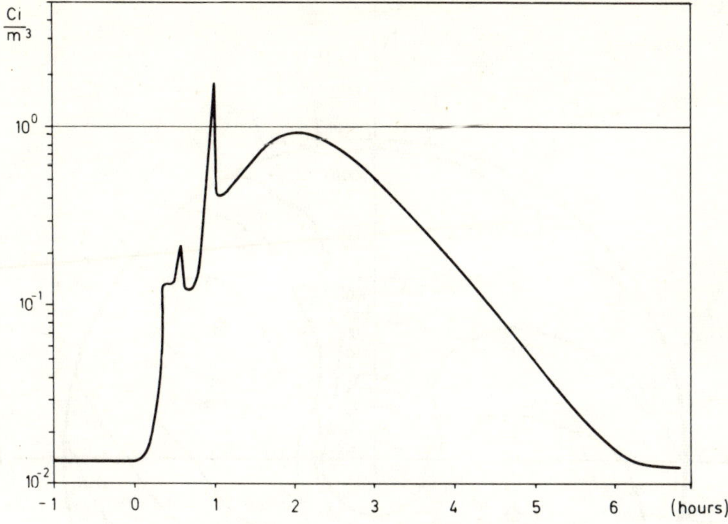

FIG.2.   Timing of $^{85}$Kr release, beginning of dissolution at t = 0.

TABLE I.   CHARACTERISTIC EMISSION FIGURES USED
IN THE DIFFERENT MODELS

| Case | Multiplication factor | | Period (h) | | Duration, D (h) | Mean released activity per batch, $\overline{A}$ (Ci) | Mean emission rate per batch, $\overline{\dot{A}}$ (Ci/s) |
|---|---|---|---|---|---|---|---|
| | $E_1$ | $E_2$ | $P_1$ | $P_2$ | | | |
| A.  Continuous | 1 | 1 | 1 | 1 | 1 | 39.95 | 0.0111 |
| B.  Periodic | 1 | 1 | 24 | 24 | 2 | 958.90 | 0.133 |
| C.  Random: | | | | | | | |
|     C1 | 1 | 3 | 13 | 17 | 2 | 599.32 | 0.0832 |
|     C2 | 1 | 8 | 13 | 45 | 2 | 1158.7 | 0.161 |
|     C3 | 1 | 8 | 13 | 240 | 2 | 5054.2 | 0.702 |

peaks on the left-hand side of the maximum indicate the addition of solvent.
The period of nearly constant release lasts for about two hours.

During the period from 27 Dec. 1971 to 25 March 1973 a total of
162 batches comprising three different types of fuel elements were reprocessed. The shortest interval between two batches was 13 hours. Experience
gathered during this period led to the selection of three different release
patterns, C1, C2, C3, whose characteristic data are compiled in Table I. The
annual activity release remains constant for each case.

Although the stack height of the pilot plant is 60 m, a 100-m stack was
chosen for a big plant and hence  also for the calculation.

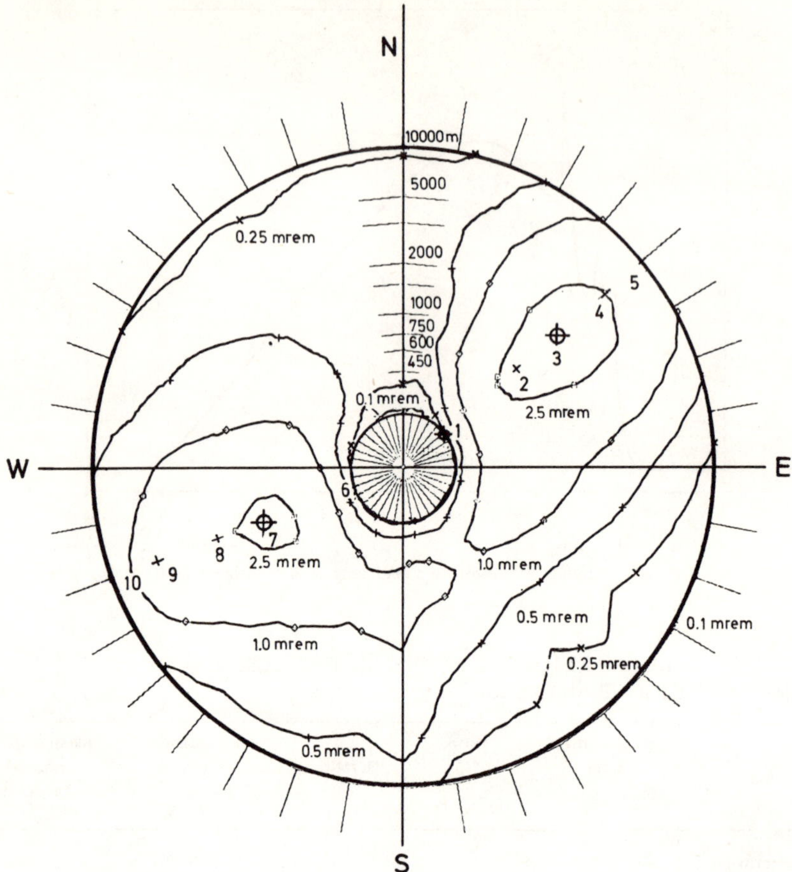



FIG.3.  Dose distribution in the vicinity of the plant.  At points 1 to 10 the doses are calculated for the different emission patterns.

## 5.  DISCUSSION OF RESULTS

Figure 3 shows the dose distribution in the vicinity of the plant for continuous emission.  At the ten locations indicated in the map, the doses are calculated for the different emission patterns discussed in Section 2.

Figure 4 shows the variation of the annual dose at locations 2, 3 and 7 i the release starts at the same hour every day.  The annual dose depends strongly on the starting time.  The horizontal lines indicate the continuous emission doses.

The differences between minimum and maximum doses vary greatly from location to location.  At some locations the doses hardly depend on the beginning of dissolution.  No significant difference for continuous and perio emissions is detectable.  At other locations the maximum doses for periodi emission may be larger by a factor up to 5 than the continuous emission do

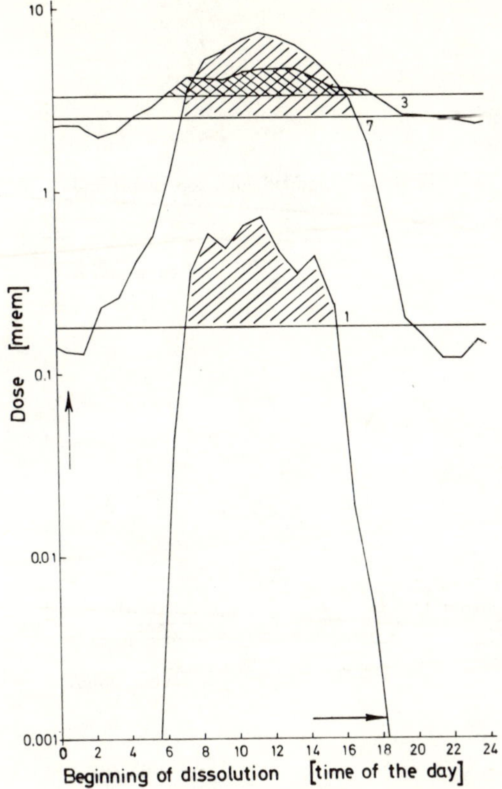



FIG.4.   Doses at points 1, 3 and 7 for periodic emission as a function of the beginning of dissolution.

The moment of the beginning of dissolution corresponding to a dose minimum
is different at each location.  If it is possible to warrant an almost periodic
operation and to choose the beginning of dissolution without restraint,  this
beginning should be chosen so that no maximum dose is produced in inhabited
areas.

Figure 5 shows the cumulative frequency for randomly intermittent
emission, again at the three locations 1, 3 and 7.  Case C1 demonstrates a
pseudoperiodic operation of the plant;  C3 an operation with an extremely
random character.  The random discontinuity of case C2 lies between the
two cases C1 and C3.

The mean values $\overline{D}_i$ of the three cases C1 to C3 agree well with the annual
doses for continuous emission.  This trivial result demonstrates that the
different models are comparable.  The mean values are smaller than the
maximum doses for periodic emission.  But for some locations there are
relatively high probabilities of significantly higher doses.  These high doses
are more probable when plant operation has a more random character.  The
square root of the mean square error is increased by a factor of 3 if the

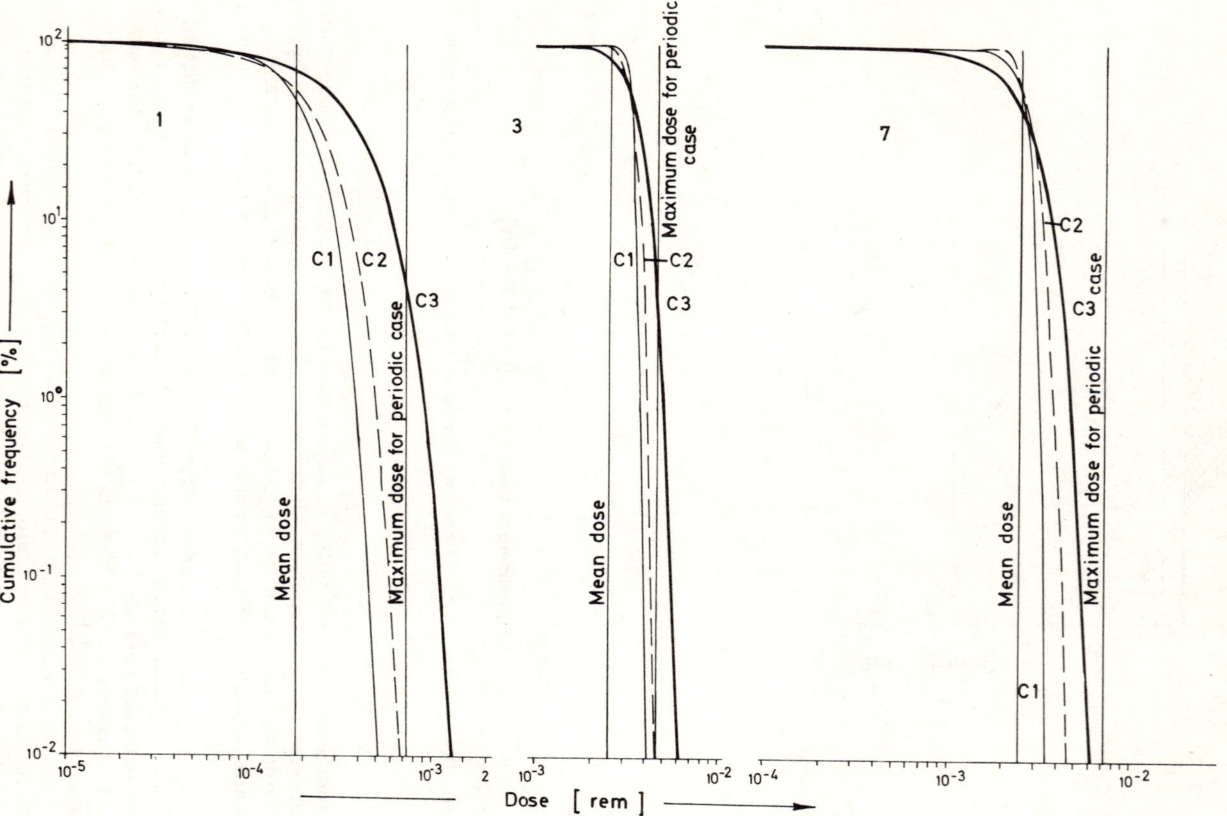



FIG.5.   Cumulative frequency for randomly intermittent emission at three points 1, 3 and 7.

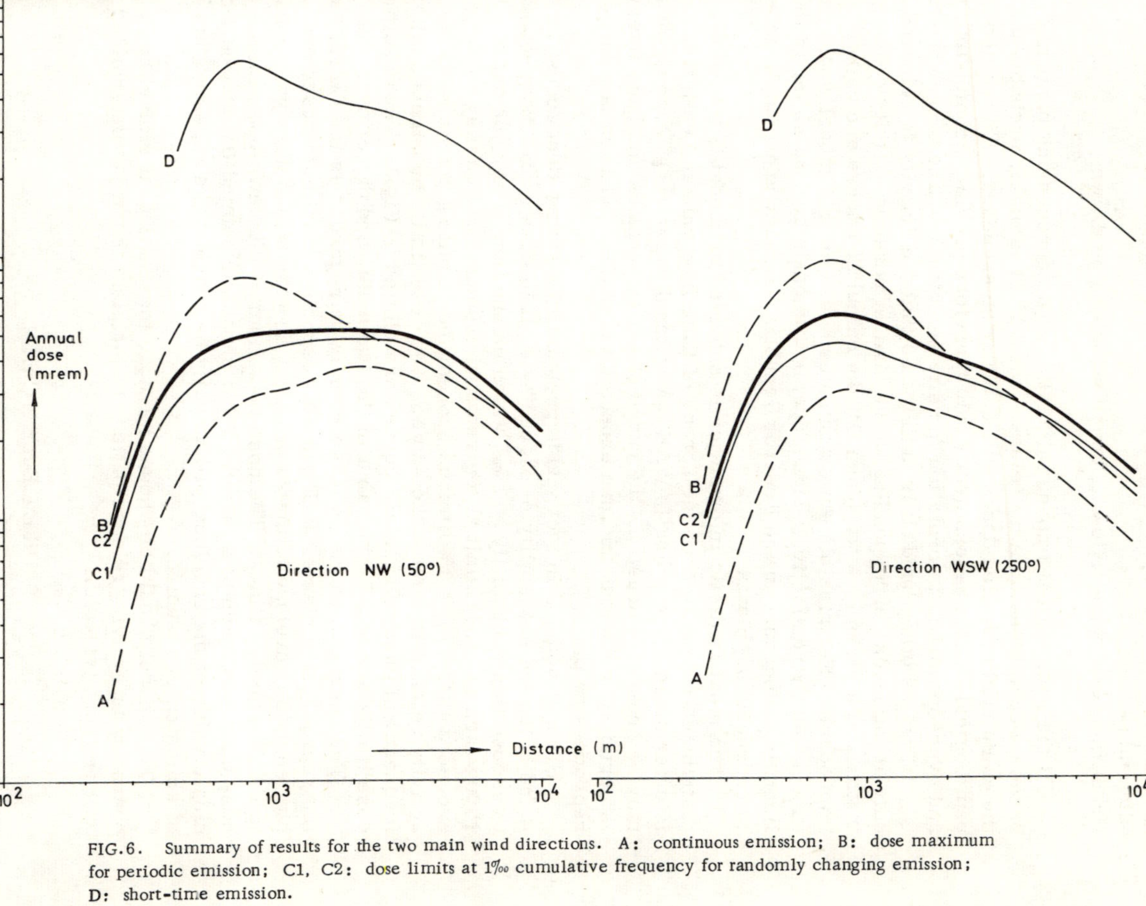


FIG.6. Summary of results for the two main wind directions. A: continuous emission; B: dose maximum for periodic emission; C1, C2: dose limits at 1‰ cumulative frequency for randomly changing emission; D: short-time emission.

operation of the plant is changed from pattern C1 to pattern C3.  So the
maximum dose of the periodic emission has a probability frequency of about
4% at location 1 or 3 for the extremely irregular operation of case C3.

If it is feasible, the reprocessing plant should be operated in a pseudoperiodic manner.

From the results the following conclusions can be drawn:

If the radiological impact calculation of a reprocessing plant is
based on continuous activity release (case A), it is highly probable that
the calculated radiation doses will actually be exceeded.  This method of
radiological impact assessment would therefore provide incomplete
and over-optimistic information for site selection and licensing procedure

If, on the other hand, it is anticipated that the activity is released
only at intervals corresponding to the highest possible activity concentrations at a given location (case D), the calculated doses are rather
improbable.  Permissible emission rates based on this model would be
unnecessarily restrictive.  Therefore, both models applied so far in the
licensing procedure in the Federal Republic Germany are inapplicable
to reprocessing plants.

The most realistic information is provided by model C (randomly
distributed release), once the spread of time intervals and emission rate
is determined.  According to WAK operational experience, case C1 or C
may be chosen.  In case C3 the range of emission rates and time interva
is certainly much greater than expected in routine operation of a large
reprocessing plant.

But the model C in fact yields probabilities rather than doses, i.e.
it indicates the probability of exceeding a given dose limit.  We propose
to choose this probability at 1‰ , which means that in any one out of
1000 years the dose limit at a given environment location may be
exceeded.  (Even if a 1% limit is chosen, the resulting permissible
activity release is almost the same.)  The dose limits at 1‰ cumulative
frequency of cases C1 and C2 curves are plotted in Fig.6.  The resulting
dose limits are nearly the same in the two main wind directions.

True periodic emission (case B) may be regarded as the 'worst cas
for reprocessing plants.  In Fig.6 the respective maximum doses are
plotted as an envelope of the 24 different dose distributions correspondi
to the 24 starting times.  The dose maximum of this envelope exceeds th
case C2 doses by up to a factor 1.7.  It should be pointed out that only o
single value of the envelope will actually be reached if a regular start
of emission is assumed.

This model should be chosen for dose forecasting  if either the
operating or activity release pattern of the plant is not yet determined o
will be truly periodical.


R E F E R E N C E S

[1]   ZÜHLKE, P., Atomwirtsch.-Atomtech. 7 (1974) 346.

[2]   LASER, M., BEAUJEAN, H., FILSS, P., MERZ, E., VYGEN, H., "Emission of radioactive aerosols from
      reprocessing plants", Physical Behaviour of Radioactive Contamination in the Atmosphere (Proc.Symp.
      Vienna, 1973), IAEA, Vienna (1974) 99.

[3]   BUNDESMINISTER FÜR BILDUNG UND WISSENSCHAFT, Schriftenreihe Kernforschung 6, "Emissionstärke
      von Kernkraftwerken" (1972).

[4]   NESTER, K., Kernforschungszentrum Karlsruhe Rep.KFK 1606 (1972).
[5]   SLADE, D.H., Meteorology and Atomic Energy 1968, USAEC Rep.TID-24190 (1968).
[6]   VOGT, K.J., Kernforschungsanlage Jülich GmbH.Rep., Jül-637-ST (1970).
[7]   COMPER, W., Kernforschungszentrum Karlsruhe Rep.KFK 1615 (1972).
[8]   FESSLER, H., KÖNIG, L.A., NESTER, K., WINTER, M., "Preliminary experience gained in monitoring
      85Kr received in the neighbourhood of the Karlsruhe reprocessing plant", Environmental Behaviour of
      Radionuclides Released in the Nuclear Industry (Proc.Symp.Aix-en-Provence, 1973), IAEA, Vienna
      (1973) 663.

# DISCUSSION

W. HEINZ:  The authors have collected some data on WAK (the Karlsruhe reprocessing pilot plant) but those data cannot simply be extrapolated to a big plant.  Owing to the vicinity of the Nuclear Research Centre (Gesellschaft für Kernforschung), there are some restrictions on the dissolving of fuels.  A big plant will have more fuel to dissolve and the operations will be overlapping, i.e. continuous.  Furthermore, WAK has reprocessed widely different fuels with varying burn-ups.

At the present stage, we are unable to agree that the gaseous effluents released from reprocessing plants will have to be reduced to the levels applicable to nuclear power plants.  The first generation of reprocessing plants treating LWR fuels will certainly release most of the gaseous effluents except iodine through the stack.  We fully concur in this view held by our British and French colleagues.

We are well aware of the problems, and research and development work has begun in many countries.  But to develop a retention technology which is as reliable as the PUREX process will take time.

However, as to $^{85}$Kr and $^{3}$H, we have a feeling that the solution of the retention problem will create a new problem, that of safe handling and storage of these concentrated gases.

In conclusion, I should like to inquire what medical and biological data are available which could clearly show the impact of the activity released by nuclear facilities on the population in the past.  It was very interesting to hear from Mr. Hull (SM-188/45) and Mr. Jammet (SM-188/10) that all such data are disputable and the assumption on which their interpretation is based not quite correct.

P. THOMAS:  A future big reprocessing plant will certainly operate in a less discontinuous manner than one might judge by extrapolation from the pilot facility WAK at Karlsruhe, but even with overlapping dissolutions the process nevertheless remains discontinuous, giving rise to periodically or randomly varying releases. Considering other input data, our method of assessing dose distributions in the vicinity of a reprocessing plant is acceptable;  a less discontinuous process would in fact reduce the maximum doses. However, the capacity of a big plant will be higher by a factor of 40, giving rise to higher activity releases and hence to higher dose values.

At present, scientists disagree as to the impact of activity releases on the population.  For the moment, therefore, radiation doses must be kept 'as low as possible'.  If there have been no perceptible consequences in the past, this is probably due among other things to the fact that only a few relatively small nuclear facilities have been distributed all over the world. In the future, the nuclear industry will grow exponentially, and the doses from individual facilities may even be reduced.

D. BENINSON:  It seems doubtful that emissions from reprocessing
plants should be limited to values similar to those for reactors.  One difficult
is that for individual doses (critical group) $^{85}$Kr would then be restricted by
skin limits (ICRP).  As to the application of the concept 'as low as reasonably
achievable', the differential cost-benefit analysis for a reprocessing facility
would be quite different from that of a reactor installation.

P. THOMAS:  In my opinion, emissions from reprocessing plants will
certainly be limited to values similar to those established for power reactors
At present, people will not accept higher dose limits for other nuclear
facilities than those already adopted for nuclear power stations.  With regard
to the 'as low as reasonably achievable' concept and the cost-benefit analysis
the higher charges of reprocessing plants consequent on low dose limits
will simply have to be shifted to the power stations and ultimately incorporat
in the price of electric power.

M. LASER:  I should like to make a short comment on the $^{85}$Kr problem.
The release of $^{85}$Kr from a big reprocessing plant is relatively high.  Nevertheless, it is possible to find a site with good meteorological conditions, such
that the whole-body dose will be lower than 30 mrem/a.  This means that it
is possible to run a reprocessing plant without krypton separation.  However
the Federal Republic of Germany Radiation Protection Ordinance requires
that the radiation dose be kept as low as possible;  economic considerations
are not taken into account.  Therefore we have to separate krypton, if we are
able to do so.  At present we have no proven technique but there are several
processes under development.  However, if we are speaking about the state
of the art, we must not forget the technology of storage and final disposal of
the separated krypton, nor the resulting risks.  In this respect I completely
agree with Dr. Heinz.

W. VINCK:  The releases from fuel-reprocessing plants should, in my
opinion, be considered on their own merits and not necessarily be compared
with releases from a nuclear power plant.  I could imagine an apportionment
of the individual dose to the critical groups of people which might be differer
from the 30 mrem whole-body dose for power plants (Federal Republic of
Germany) on the basis of a cost-benefit analysis.

P. THOMAS:  I do not believe that government agencies could raise the
30 mrem dose limit for nuclear facilities other than power plants on the
basis of a cost-benefit analysis.  Unless scientists can prove that there are
no genetic hazards from low doses, the 30 mrem value will stick as a limit
and the costs will have to be raised.

W. VINCK:  Is it possible that in future the xenon isotopes will become
critical, rather than $^{85}$Kr?  Indeed, it may be of economic interest to have
shorter cooling times, say of the order of 30 days, for future fuel cycles
(e.g. fast breeder fuels).  I am referring, of course, to future large-scale
reprocessing plants.

P. THOMAS:  These short cooling times will not be of interest for
reprocessing plants.  The activity released by xenon and other short-lived
fission products would be too high.  Moreover, radiolytic decomposition of t
solvent would become more severe because of stronger radiation from the
spent fuel elements.

# RADIOLOGICAL SAFETY CONSIDERATIONS IN THE SITING OF FUEL-REPROCESSING PLANTS

S. KRISHNAMONY, S.D. SOMAN
Health Physics Division,
Bhabha Atomic Research Centre,
Bombay, India

*Presented by L. Venkatesh*

## Abstract

RADIOLOGICAL SAFETY CONSIDERATIONS IN THE SITING OF FUEL-REPROCESSING PLANTS.
Fuel-reprocessing plants are characterized by a large inventory of fission products and plutonium present in the form of highly concentrated solutions in a mobile and often dispersible form. When selecting sites for plants of this type due consideration has therefore to be given to the potential exposure of the nearby population to radiation from accidentally released radioactive materials in addition to routinely released effluents. However, in contrast with reactor siting practices, where the concept of the maximum credible accident is widely used, a number of credible and relatively minor incidents will need careful consideration in evaluating the hazards of radiochemical plants. These factors are examined in this paper with the objective of formulating suitable guidelines for the siting of such a facility.

## 1. INTRODUCTION

Site selection criteria for nuclear power stations have been fairly well established in quantitative terms by many national authorities and have been the subject of several international symposia. Similar criteria do not appear to have been spelt out in detail in the case of fuel-reprocessing plants, possibly due to the very few such plants operating in the world today and the corresponding paucity of operating experience in this field. Fuel-reprocessing plants will increase in number and capacity in the coming years and the siting of such plants should take into consideration certain characteristics peculiar to plants of this type. Further, the recent requirement that the highly active liquid wastes arising from fuel reprocessing should be converted into solid form within five years of their generation makes it imperative to have the waste solidification plant as an adjunct to the reprocessing facility. For the purpose of site evaluation, therefore, both the plants must be considered as a single entity.

## 2. CHARACTERISTICS OF A TYPICAL REPROCESSING PLANT

For the purposes of this study, a reprocessing plant capable of processing 1000 t/a LWR fuel irradiated to 20 000 MW·d/t and cooled for 130 days is considered as a typical example. The annual throughput of fission products and plutonium in a plant of this size will amount to $3 \times 10^9$ Ci and 5000-6000 kg, respectively, present in the form of chemically reactive and corrosive solutions of high specific activity. In addition, the plant will generate in five years about one million gallons of highly active

liquid waste containing about $1.5 \times 10^{10}$ Ci of mixed fission products. The
inventory of long-lived fission products stored at the site is therefore
considerably larger than that present in a reactor core, with the difference,
however, that the potential energy available for initiating a dispersal of
radioactive material in the event of an accident seems to be considerably
less in the case of fuel reprocessing and waste calcination operations.


## 3.  RADIATION EXPOSURE UNDER NORMAL OPERATIONS

### 3.1.  Gaseous effluents

·  The gaseous effluents released through the stack during normal operation of the plant comprise:

(a)  The noble gases $^{85}$Kr and $^{133}$Xe;
(b)  Fission-product tritium;
(c)  Fission-product $^{131}$I and $^{129}$I;
(d)  Particulate activity composed of mixed fission products and
     plutonium;  and
(e)  $^{106}$Ru from the waste solidifaction process.

The quantity of gaseous effluents likely to be released from the typical
reprocessing plant under consideration and the resulting radiation exposures
to the public are shown in Table I [1, 6]. The release of $^{85}$Kr is not expected
to be a limiting factor in the sizing or siting of a fuel-reprocessing plant
so far as local exposures are concerned. $^{133}$Xe with a half-life of 5.27 days
will have to be considered while processing short-cooled fuel, about 1300 Ci
of the activity being present per MW·d of thermal power after 30 days of
cooling. This is about 4000 times as much as the $^{85}$Kr that would be present
and some interim hold-up will be required to reduce the $^{133}$Xe releases to
acceptable levels.

Radiation exposure from tritium released through the stack is also
not of concern, since the major portion of the tritium appears in the low-
level liquid effluent, as will be discussed later.

The real problem with $^{131}$I is encountered during the processing of
short-cooled fast-reactor fuel. Such short-cooled fuel with cooling time
of 30 days will contain about $1.5 \times 10^5$ Ci $^{131}$I per tonne of fuel processed
and will call for a plant retention factor in excess of $10^7$, which is four
to five orders of magnitude higher than that available in existing LWR
fuel-reprocessing plants [7]. The corresponding high inventory of $^{131}$I
in the liquid-waste handling system will also pose several problems. The
presence of large quantities of iodine in the off-gas will demand more
stringent containment and air-cleaning systems for the plant than are now
being provided [7]. Failure of iodine-trapping systems or accidents
involving such systems will need consideration in the siting of such a plant
since they provide a potential for releasing large quantities of iodine to the
environment.

$^{129}$I with a half-life of 16 million years is present to the extent of 20 Ci
per 1000 tonnes of fuel [8]. In view of its long half-life, the nuclide builds
up in the environment, a fact that was brought to light recently by measurements carried out round the Nuclear Fuel Services Plant, West Valley,

TABLE I.   GASEOUS EFFLUENTS FROM A TYPICAL REPROCESSING
PLANT AND THE RESULTING RADIATION EXPOSURES TO THE
SURROUNDING POPULATION
(1000 t/a of LWR fuel irradiated to 20 000 MW·d/t and cooled for
130 days)

| Radionuclide | Quantity released per year (Ci) | Exposure at point of maximum concentration downwind of the stack |
|---|---|---|
| $^{85}$Kr | $7 \times 10^6$ | 0.14 mrad (whole-body dose)[a]<br>20.4 mrad (skin dose) |
| $^3$H (HTO) | 14 000 [b] | 0.2 mrem whole-body dose to a child |
| $^{131}$I | 60 [c] | 3 mrem inhalation dose and<br>120 mrem ingestion dose to a child's thyroid [d] |
| $^{144}$Ce - $^{144}$Pr | 55 [e] | Insignificant |
| $^{106}$Ru - $^{106}$Rh | 25 [e] | Insignificant |
| $^{95}$Zr - $^{95}$Nb | 15 [e] | Insignificant |
| $^{90}$Sr - $^{90}$Y | 2 [e] | Insignificant |
| Total plutonium | (80 g) [e] | Insignificant |
| $^{106}$Ru - $^{106}$Rh from waste solidification plant | 700 | About 7 mrem/a to a child's lungs |

(Stack height 120 m;   annual weighted mean stack dilution factor $\chi/Q = 4.39 \times 10^{-8}$ s/m$^3$)[f]

[a] The dosimetry conversion factors used were: 1 pCi/cm$^3$ at 15°C and 760 mm Hg delivers $1.41 \times 10^{-2}$ rad/a whole-body dose and 2.08 rad/a to the skin [1].

[b] Experience at the Nuclear Fuel Services Plant at West Valley, New York, has shown that about 0.7 mCi of tritium is released through the stack per MW·d of fuel processed [2].

[c] Out of the 6000 Ci of $^{131}$I present in 1000 t of fuel after 130 days of cooling, 0.1% - 1% can be released through the off-gas system [3]. The conservative figure of 1% is assumed here. No credit is given for the amount of iodine that may be lost by plate-out in the exhaust ventilation ducts since it is likely that a significant portion of the airborne iodine may be in the form of unreactive methyl iodide [4].

[d] Dosimetry data used are those recommended by Bryant [5,6], modified to suit Indian conditions.

[e] Particulate activity release estimated for the plant under consideration was based on measurements carried out at the Trombay Fuel-Reprocessing Plant and the use of appropriate scale-up factors.

[f] The stack dilution factor used is the one calculated for the site of the second Indian fuel-reprocessing plant, located at Tarapur, on the west coast of India.

New York [9].   It appears prudent, therefore, to have an iodine clean-up
system in the off-gas system of a LWR fuel-reprocessing plant even though
the more abundant $^{131}$I may not be present in amounts that would result in
any significant exposure.

Radiation exposure from particulate activity released through the stack
is insignificant and not of concern.   Table II lists, for comparison, the

TABLE II.   DERIVED WORKING LIMIT VALUES FOR RELEASE OF
PARTICULATE ACTIVITY

| Radionuclide | DWL inhalation route (Ci/d) | DWL milk route (Ci/d) |
|---|---|---|
| $^{90}$Sr | 56 | 5 |
| $^{137}$Cs | 9000 | 300 |
| $^{106}$Ru, $^{144}$Ce | 425 | - |
| $^{239}$Pu (soluble) | 0.27 | - |
| $^{239}$Pu (insoluble) | 2.75 | - |

(Stack height 120 m;   annual weighted mean stack dilution factor $\chi/Q = 4.39 \times 10^{-8}$ s/m$^3$ )

derived working limits (DWL) for the release of particulate activity through
the 120-m stack of the 1000 t/a reprocessing plant considered here.

In view of the volatility of $RuO_4$, significant quantities of this nuclide
could be released in the waste solidification off-gas on account of the high
temperatures involved in the process.   The inventory of $^{106}$Ru in the high-
level waste from a 1000 t/a plant after five years of cooling will be about
$0.14 \times 10^4$ Ci.   Using a decontamination factor of 2000 [10], as reported
in the literature for the  waste calcination facility at the Idaho Chemical
Processing Plant, the annual release of $^{106}$Ru from the waste solidification
plant amounts to 700 Ci for the plant under consideration.   The annual
inhalation dose for an Indian child on account of this source works out to
7 mrem at the point of maximum concentration.

3.2.  Liquid effluents

The reprocessing of irradiated fuel gives rise to large volumes of low-
level liquid wastes ($\sim 10^{-4} \mu$Ci/mlitre) which have to be disposed of to a
nearby water body capable of providing adequate dilution.   The radionuclides present will be predominantly $^{144}$Ce, $^{137}$Cs, $^{90}$Sr, $^{106}$Ru and
isotopes of plutonium and transplutonics.   Detailed ecological investigations are required to assess the impact on the environment from the release
of these radionuclides prior to and after commissioning of the plant.   The
critical nuclide will depend on the characteristics of the local environment.
Ruthenium has been reported as the critical nuclide at Windscale [11] and
$^{137}$Cs at Trombay [12].

Most of the fission-product tritium (65%) [13] finds its way to the aquat
environment via the low-level liquid effluents.   The 1000 t/a plant will rele:
about 325 000 Ci of HTO via this route, a quantity that would require a river
or canal with a flow rate of $10^4$ litres/s to dilute the tritium activity to the
maximum permissible concentration of $10^{-3}$ $\mu$Ci/mlitre, applicable to

individual members of the public. Significant quantities of ruthenium and strontium will be present in the final condensate of the waste solidification plant also, and it may be necessary to recycle the effluents to the reprocessing plants rather than release them to the environment [14].

## 3.3. Solid wastes

A variety of solid wastes containing low to medium level fission-product radioactivity are now being disposed of by land burial in controlled sites. Reprocessing of high-burn-up fuels on a commercial scale will give rise to plutonium-bearing solid wastes that cannot be disposed of in this manner in view of its long half-life. The question of containing such wastes from man's environment for long periods needs careful consideration from the long-term point of view.

## 4. RADIATION EXPOSURE UNDER ACCIDENT CONDITIONS

Upper limit accidents that can be considered as credible in fuel-reprocessing operations are generally thought to be:

(a) A chemical explosion resulting from an uncontrolled or runaway exothermic reaction;
(b) A criticality incident;
(c) A fire involving solvent or ion-exchange resin loaded with plutonium; and
(d) A hydrogen-air explosion in a highly active waste-storage tank.

Of these, event (d) would be so violent that large quantities of vaporized radioactive material could be released at ground level. Such an eventuality must be precluded by suitable design and operation of the waste-storage tanks.

Table III lists the amount of radionuclides estimated to be released as a result of the upper limit accidents and the consequent radiation exposure to an individual located at the point of maximum cloud dosage downwind of the 120-m stack.

For the chemical explosion incident, the first-cycle raffinate evaporator is assumed to contain fission-product raffinate from 1 t of irradiated fuel having an activity of $2.5 \times 10^6$ Ci. It is assumed that $10^{-4}$ of this (250 Ci) becomes airborne and enters the cell ventilation air, 1% of which penetrates the HEPA filter bank which is credited with an efficiency of 99% (the efficiency will normally be maintained at not less than 99.95%). For the purpose of this discussion, the entire quantity is assumed to be $^{90}$Sr. Out of the total of 36 000 Ci of $^{106}$Ru present, it is assumed that $10^{-3}$ of the inventory escapes into the cell ventilation air as $RuO_4$ and is not removed by the particulate filters. Further, no credit is taken for ruthenium lost by deposition in the ventilation ducts.

For criticality incident, a total of $10^{18}$ fissions occurring over a short period is assumed. Nuclear excursions in reactor fuel reprocessing plants reported in the literature [15] have, with the exception of the 1959 incident at the Idaho Chemical Processing Plant, been about $10^{18}$ fissions or less.

TABLE III.   RADIONUCLIDES RELEASED IN UPPER LIMIT ACCIDENTS AND THE RESULTING RADIATION EXPOSURE DOWNWIND OF THE STACK

| Accident | Radionuclide released | Activity (Ci) | Radiation exposure |
|---|---|---|---|
| Chemical explosion in first cycle raffinate evaporator | Mixed fission products | 2.5 | Inhalation dose of 4 mrad/a and ingestion dose due to milk of 120 mrad/a and that from other foodstuffs of 5 mrad/a assuming the entire quantity to be $^{90}$Sr. |
| | $^{106}$Ru | 36.0 | Lung dose of 40 mrad and ingestion dose to GI tract of 0.3 mrad |
| | | | Inhalation dose to 6-months-old child's thyroid (mrad): |
| Criticality incident of $10^{18}$ fissions | $^{131}$I | 0.15 | 0.22 |
| | $^{132}$I | 18.90 | 0.16 |
| | $^{133}$I | 3.6 | 2.3 |
| | $^{134}$I | 90.0 | 3.6 |
| | $^{135}$I | 9.6 | 2.0 |
| | | | Ground deposition at point of maximum air concentration works out at 0.5% emergency reference level for ingestion route via milk |
| | $^{83m}$Kr | 13.5 | The external dose due to fission gases is less than 1 mrad |
| | $^{85m}$Kr | 18.5 | |
| | $^{87}$Kr | 112.0 | |
| | $^{88}$Kr | 69.5 | |
| | $^{133}$Xe | 2.7 | |
| | $^{135m}$Xe | 395.0 | |
| | $^{135}$Xe | 36.4 | |
| | $^{138}$Xe | 1050.0 | |
| Solvent fire | $^{239}$Pu | 0.62 | Insignificant |

(Stack height 120 m;  maximum cloud dosage $10^{-6}$ Ci·s/m$^3$ per Ci released)

This incident consisted of a series of excursions which, due to the particular design of the system, allowed a critical mass to re-assemble several times after the fissile material had been dispersed.  It is assumed that all the noble-gas fission products are released but the release of iodine isotopes is only 20% of that generated, 80% being deposited on the ventilation ducts.

For the solvent fire incident, it is assumed that 100 litres of solvent containing 1 kg plutonium at a concentration of 10 g/litre catches fire and

is converted into airborne particulate matter.  The quantity of plutonium
that would escape into the atmosphere through the stack works out at 0.62 Ci,
assuming a filter bank efficiency of 99%.  The exposure downwind of the
stack is insignificant.

## 5.  OFF-NORMAL CONDITIONS

Apart from the consequences of the upper limit accidents referred to
in Section 4, a number of off-normal situations in plant operations need to
be considered.

Although containment of the primary streams carrying the bulk of the
fission products and plutonium by 'successive envelopes' is a cardinal
principle in the design of a reprocessing plant, such containment in practice
may be more apparent than real unless close attention is paid to the details
of the design.  The possibility that process or waste streams containing
high or medium level activity may find their way into the low-level effluent
system by breaches in the primary or secondary containment has to be
analysed by a careful scrutiny of the plant design and operation.  Can such
effluents be recycled to the treatment system of the plant if such an event
occurs and is the design such that leakages can be contained and corrective
action taken sufficiently in advance for the surrounding community not to be
exposed to any undue risk?

A major assumption made in the analysis of the consequences of the
upper limit accidents is that the radioactive materials are invariably
released only through the 120-m-high stack.  It is necessary to make
certain that the design of the containment system ensures that such would
be the case.  The stringent concept of a containment building with an
acceptably low leak rate that is now being applied to reactor installations
has not so far been applied to any reprocessing plant.  The provision of a
suitable exclusion distance to a reprocessing plant would depend on its
containment capability.  In any case, a containment concept more similar
and closer to that of a nuclear power reactor will have to be evolved for
the short-cooled fast-reactor fuel-reprocessing plants of the future.

## 6.  TRANSPORT OF IRRADIATED FUEL

Apart from the safety-related questions already discussed, the siting
of reprocessing plants is closely tied to the technology of transporting
irradiated fuel long distances by rail or road or both, especially in a
developing country like India.  The answer to the question of whether a
large central reprocessing facility catering to a number of power stations
is preferable to small plants coupled to the reactor at the same location
will be determined largely by the economics and safety of irradiated fuel
transport vis-à-vis unit processing costs versus plant capacity.  A previous
study of this problem has shown that for conditions obtainable in India,
small plants (not more than 1 t/d) coupled to the nuclear power stations will
be cheaper and at the same time reduce the risk to the public from a transport accident [16].  The safety problems associated with the transport of
large quantities of plutonium-bearing fast-reactor fuel cooled with liquid
sodium are sufficiently serious to deserve careful analysis before siting
large centralized reprocessing facilities.

## 7.    CONCLUSIONS

No special problems which impose undue restrictions on site selection
are encountered in a reprocessing plant capable of processing 1000 t/a of
LWR fuel irradiated to 20 000 MW·d/t and cooled for 130 days insofar as
radiation exposure of the local population from gaseous effluents is concerned, both during normal operation of the plant and during credible
accidents, assuming that containment systems function as designed. Liquid
effluents, however, can cause problems if not properly controlled.  The
operation of the waste solidification plant which needs to be set up as an
adjunct to the reprocessing plant at the same site will contribute significant
amounts of radioactivity in the gaseous and liquid effluence unless properly
controlled.  The presence of large quantities of $^{131}$I in the short-cooled
fast-reactor fuel-reprocessing plants will call for more stringent containment of the radioactive materials processed than is currently being provided
in LWR fuel-reprocessing plants.  The siting of such a plant would therefore
depend on the degree of containment of the radio-iodine isotope in plants
of this type.  Considering the economic problems and risks involved in the
transport of irradiated fuel, especially in a country like India, it seems preferable to have small (not more than 500 t/a) reprocessing plants coupled
to the reactor at the same location rather than a large central reprocessing
facility catering to a number of power stations.  This is especially true
when one considers the magnitude of the safety problems involved in the
transport of large quantities of plutonium-bearing short-cooled fast-reactor
fuel.

REFERENCES


[1]   DUNSTER, H.J., et al.,  The disposal of noble gas fission products from the reprocessing of nuclear
      fuel, UKAEA Rep. AHSB(RP)R 101 (1970).
[2]   COCHRAN, J.A., et al.,  Characterisation of tritium stack effluent from a fuel reprocessing plant,
      Trans. Am. Nucl. Soc. 15 1 (1972).
[3]   BRYANT, P.M., WARNER, B.F.,  "Control of radio-iodine release from fuel reprocessing plants",
      Control of Iodine in the Nuclear Industry, IAEA  Technical Report 148, IAEA, Vienna (1973) 29.
[4]   HALLER, W.A., PERKINS, R.W., Organic $^{131}$I compounds released from a nuclear fuel reprocessing
      plant, Health Phys. 13 (1967) 733.
[5]   BRYANT, P.M., Data for assessment concerning controlled and accidental releases of $^{131}$I and $^{137}$Cs
      to atmosphere, Health Phys. 17 (1969) 51.
[6]   BRYANT, P.M., Derivation of working limits for continuous release rates of $^{131}$I to atmosphere in
      a milk-producing area, Health Phys. 10 (1964) 249.
[7]   YARBORO, O.O., et al.,  "Iodine behaviour and control in processing plants for fast reactor fuel", Proc.
      11th USAEC Air Cleaning Conference, Washington, 1970, USAEC Rep. CONF-700816.
[8]   COCHRAN, J.A., et al., An investigation of airborne radioactive effluents from an operating nuclear
      fuel reprocessing plant, Rep. BRH/NERHL-70-3, USPHS (1970).
[9]   RUSSELL, J.L.,  et al.,  Public health aspects of $^{129}$I from the nuclear power industry, Radiol. Health
      Data Reports 12 4 (1971).
[10]  COMMANDER, R.E., et al.,  Operation of the Waste Calcining Facility with Highly Radioactive Aqueous
      Wastes, Idaho Chemical Processing Plant Rep. 1DO-14662 (1966).
[11]  PRESTON, A., et al., "U.K. experience of radioactive waste release to the environment and expected
      waste management in fuel cycles in the 1980s", 4th Int. Conf. Peaceful Uses Atomic Energy (Proc.
      Conf. Geneva 1971), UN, New York, and IAEA, Vienna (1972) 415.
[12]  PATEL, B., et al., "Radio-ecology of Bombay Harbour", Proc. Bhabha Atomic Research Centre Seminar
      on Pollution and Human Environment, 1970, p.294.

[13] PETERSON, H.T., Jr., MARTIN, J.E., WEAVER, C.L., HARWARD, E.D., "Environmental tritium
     contamination from increasing utilization of nuclear energy resources", Environmental Contamination
     by Radioactive Materials (Proc. Seminar Vienna, 1969), IAEA, Vienna (1969) 35.
[14] McELROY, J.L., et al., Status of the waste solidification demonstration program, Nucl. Technol.
     12 (1971) 69.
[15] NICHOLS, J.P., Nucl. Saf. 3 (Sept.1961).
[16] SETHNA, H.N., et al., "Fuel Reprocessing in India", Technology and Economics of Nonaqueous
     Processing (1967).

# DISCUSSION

R.R. BOUSSARD: Where reprocessing plants are concerned, strict adherence to international standards on dose limits and to the general policy of keeping releases 'as low as practicable' ultimately means storage of radioactive solid waste for a long time (several centuries) under permanent surveillance. Gaseous effluents such as $^{129}$I and $^{131}$I as well as $^{85}$Kr are trapped and end up as solid waste. The fission products in liquid effluents have to be solidified and stored in pits, while other effluents are fixed on resins (thus producing more solid waste) for containment of $^{90}$Sr, $^{137}$Cs, $^{144}$Ce, and so on. There are yet other solid wastes (such as fuel cladding, filters, etc.) associated with the operation and maintenance of a plant.

Thus waste burial grounds are inevitably going to be established, and it is important that they should not be scattered all over. Europe, in particular, could not tolerate the use of much of its soil for such a purpose. The public everywhere, however, will be concerned at the prospect of receiving waste for burial. Careful public relations will accordingly be vital, and it is just possible that solidified waste may in some instances be returned to the sender.

Lastly, it will obviously be very important to prevent any diversion of plutonium when it is in a derivative form in the plant or during transport. For this reason it will be advisable to place side by side the facilities for reprocessing and fabrication of plutonium fuels used in the LWR and FBR fuel cycles.

L. VENKATESH: I appreciate your view that it may not be advisable to set up small reprocessing plants at a large number of locations in view of the close and constant surveillance which the storage of highly active wastes would call for. However, our attitude on the question of plant size in India is conditioned by the problems and risks involved in the transport of irradiated fuel over long distances. The existence of two different gauges of railway and the already heavy goods and passenger traffic on the existing lines impose a severe limitation in some sectors on the transport by rail of irradiated fuel. Movement by road will be even more difficult in view of the high frequency of such transport requirements.

W. HEINZ: The cooling time of 30 days, as mentioned in the paper, is already unrealistic today. The United Reprocessor Company is going to take 220 days and fast breeders will certainly not achieve less.

L. VENKATESH: The 30 day cooling time has been considered for fast breeder reactor fuel in order to reduce the inventory of fissile material; this is important for the economics of power generation by fast breeders.

J. PELSER: Can a 500 t/a reprocessing plant, as indicated in your paper, be considered a 'small plant', since it serves PWR-type reactors with a capacity of about 20 000 MW?

L. VENKATESH: It cannot be considered large, since it has to deal with fuel irradiated to about 8000 MW·d/t in the Candu-type stations.

# CRITERIA FOR HTR FUEL-REPROCESSING PLANT SITE EVALUATION

H.J. RIEDEL, M. LASER,
E. MERZ, H. SCHNEZ
Institute for Chemical Technology,
Kernforschungsanlage Jülich GmbH,
Jülich, Federal Republic of Germany

Abstract

CRITERIA FOR HTR FUEL-REPROCESSING PLANT SITE EVALUATION.
No site-independent criteria exist at present for reprocessing plants, but these plants must be designed
in such a manner that a suitable site can be found in nearly all important regions. Taking the criteria into
account, four typical areas in the Federal Republic of Germany are considered for siting a HTR reprocessing
plant.

## 1. INTRODUCTION

The siting of nuclear facilities in densely populated areas will become
increasingly difficult. Increasing restrictions on radioactive effluents as
well as public acceptance of the nuclear risk are the most important factors
influencing the licensing of nuclear facilities.

Several criteria have been established for evaluating nuclear facility
sites. The weight of the particular criteria, however, may change with the
safety philosophy, with technical progress and with the type of facility [1].
In the United States of America the advance in reactor technology has made
it possible to establish site-independent criteria for reactor design, characterized by the limitation of radioactive effluents independent of the real
impact to the population. Similarly, maximum exposure rates (e.g. a
hypothetical whole-body burden of 30 mrem/a at the most unfavourable
location offsite the plant) are recommended by the Deutsche Atomkommission.

No site-independent criteria for reprocessing plants exist at present,
but it should be taken into account that these plants must also be designed
so that in nearly all important regions a suitable site can be found, because
in the near future there will be a run on nuclear facility sites and site
selection will become increasingly difficult. However, this paper shows
that a completely site-independent design for reprocessing plants would be
impossible with present technology. An additional radiation burden will
result from the discharge of $^{14}$C from HTR reprocessing plants. Factors
which will influence site evaluation for such a  HTR reprocessing plant
are discussed in this paper.

## 2. CRITERIA FOR SITE EVALUATION

Economic considerations call for a big plant with an annual throughput
of about 365 t uranium and thorium corresponding to about 50 000 MW
installed electric power. The gaseous effluents will contain about 0.1 - 1 MCi

**TABLE I.  CRITERIA FOR SITE EVALUATION FOR HTR REPROCESSING PLANT**

---

A.    GENERAL CRITERIA

1. Geology:
   geological strata;  earthquake

2. Hydrology:
   groundwater table;  underground water source;  rivers;  inundation

3. Meteorology:
   wind direction and wind speed;  atmospheric precipitation;  humidity

4. Availability of ground:
   property;  use of ground;  farming;  agriculture;  forest;  topography

5. Traffic:
   railways;  highways;  waterways;  airways;  airports

6. Decommissioning of obsolete or worn-out plants


B.    RADIATION EXPOSURE DEPENDENT CRITERIA

7. State of the art of the technology:
   fractional release of tritium, $^{14}$C, $^{85}$Kr, $^{129}$I, $^{131}$I, aerosols;
   fractional release of fission products under accidental conditions;
   technical means to retain fission products under accidental conditions

8. Radiation exposure of the public by routine emission considering different exposure paths (inhalation, submersion, ingestion)

9. Radiation exposure of the public under accidental conditions

10. Population density


C.    ECONOMIC CRITERIA

11. Investment costs, shipping costs, available technology, operating costs

---

tritium, 10 MCi $^{85}$Kr, 1 - 10 kCi $^{14}$C per year, as well as $^{129}$I, $^{131}$I and aerosols.  The dose rate resulting from these emissions under the most pessimistic assumptions may be well above the recommended whole-body dose rate of about 30 mrem/a.  Therefore it must be shown that the real radiation exposure of the critical group of the population in a selected area will be appreciably below this limit.

The high fission-product and uranium inventory represents a relatively high potential danger.  Therefore in the Federal Republic of Germany, with an average population density of about 250 inhabitants/km$^2$, the inherent safety of such a plant must be very high because no site can be found which is safe by distance.  Safety against aircraft crash, earthquakes and explosi is essential anywhere and is therefore a general criterion for siting.  Precautionary measures must also be provided against sabotage.

These facts taken into account, the most important siting criteria are divided into three groups (Table I).  The first group (A) consists of general criteria which influence the design of the facility (e.g. earthquake stability, stack height corresponding to local meteorological conditions, safety against flooding, etc.) or others which can determine the final site in the selected region (e.g. water supply, railway and highway connection, ground stability).

The location of a plant should also take into account the need for its eventual decommissioning.  Lifetimes of reprocessing plants are estimated to be in the order of 30 - 40 years.  Present trends in decommissioning practices of complete removal of the installations and re-use of the plain land lead to estimated costs which will probably be prohibitive to the power economy of future generations.  A better approach to solving the problem would be either (a) to develop techniques of conditioning a radioactive contaminated facility 'in place' so that radionuclides would be fixed and kept excluded from the human biosphere long enough for all radionuclides to have decayed, or (b) to design the facility from the beginning in a way that permits essentially unlimited re-use of the facility and its site.

The decommissioning starts with partial decontamination of the plant. The obsolete equipment is dismantled and processed for final disposal. The shielded cells and fuel-storage basins may be remodelled and restored for re-use in the new reprocessing plant complex.  In case this is not practicable, the shielded cells, large waste-storage tanks as well as fuel-storage basins will be prepared for use as long-term storage vaults. Additional physical and chemical barriers against radioactive contamination of the environment may be provided by filling up the cell surrounding compartments and rooms with suitable materials such as highly adsorptive clays, gravel, etc.  It is also possible to use knocked-down building structure material as non-radioactive fill for the safety barrier.

Where possible, the superstructure of cells and buildings will either be recovered or, if permanently contaminated, placed in the storage vaults. The sealed, filled cells and storage vaults will be covered with 3 - 5 metres of water-impervious material and the grounds secured with a fence.

The second group (B in Table I) comprises the more important criteria which may be site-determining because they cannot be influenced by reasonable means.  The state-of-the-art technology, the radioactive discharges and the resulting radiation burden are criteria belonging to this group. Economic considerations dealt with in the third group (C in Table I) are also important for several items described in groups A and B.

With due regard to the different factors in evaluating the relevant criteria, we shall look at four typical extended areas in the northern part of the Federal Republic of Germany, where a location for a  HTR reprocessing plant would be imaginable (Fig. 1).

I.  An area near conglomeration centres of chemical and metallurgical industries

An accumulation of high-temperature reactors is expected to take place in this region because they can be employed for both power generation and production of process heat.

In this connection it will be possible in the most favourable case to select an area with 70 - 80 inhabitants/$km^2$ within a periphery of about 4 km. However, at a distance of about 50 km the population may already increase

FIG.1.   The Federal Republic of Germany with power reactor stations and the four areas discussed.

to 1000 and more inhabitants per km². This area in its extension represent
a variscitic zone of molasses; there are faults with markedly quarternary
movements and the seismographic values are limited to an intensity of
5 to 6. Gradations from loamy sand to sandy loam with heavy subsoil are
being encountered.

There are only a few minor river courses and the plain of the river
Rhine itself may present a risk of inundation. Large areas of this region
offer a good groundwater resource. The development of the traffic system
in this area is considered to be sufficient, whereas electricity supply is
inadequate in large parts of the area.

In this region as well as in the other three extended areas the air traffi
lines should be taken into account in site selection, and the execution of
construction work for the relevant plants should be based on those
requirements.

Only minor areas within the extended region are landscape-protection
areas or nature reserves. Agricultural utilization mainly covers root crop
cultivation, grain growing and cattle keeping. The forest plantation area
is widely divergent from region to region but, on the whole, of minor significance. As to the conveying distances, shorter distances for the spent

fuel elements would find themselves counterbalanced by longer distances
for the transport of the radioactive waste.

Meteorologically this area belongs to the North German Low Plain.
South-west winds prevail with a wind direction frequency of 35%. The
average wind speed is about 3 m/s. Strong inversion conditions are not to
be expected. The maximum long-time dispersion factor is in the order of
$1 \times 10^{-7}$ using a 200-m stack.

## II.  An area which would lie approximately half-way between the potential HTR fuel element supplier and the location of the ultimate waste disposal

The density of population within a periphery of up to 10 km amounts
to 70 - 80 inhabitants/km$^2$.

In this region, variscitic zones of molasses are encountered in the
north of a variscitic and prevariscitic formation. This area is nearly
aseismic, and harmful earthquakes are very rare. The nature of the soil
extends from stony loam over slightly loamy sand up to silty clay.

Minor river courses are being encountered and groundwater resources
are not estimated to be very high. Development of traffic and infrastructural
requirements is considered to be sufficient, all the more since motorways
cross or touch the area. There is also an adequate power supply. Mountains
and valley slopes in the southern part of this extended area contain large
tracts of woodland reserved as national parks. The agricultural utilization
of soil largely involves grain growing and root crop cultivation with prevalent
forest plantation areas. It should be noted that there is a military training
ground in the north-east of the area considered.

The wind direction frequencies here are determined by the nearby
highlands. The main wind directions are west, south-west, south and south-
east, with a frequency of about 70%. The average wind speed is about
3 - 4 m/s. Inversion conditions are rare. With a 200-m stack, the maximum
long-time dispersion factor is about $6 \times 10^{-8}$ in the north-east direction.

## III.  An area near the Baltic coast

The distance from both the fuel-element supplier and the ultimate waste
disposal would be great. Transport costs would accordingly be high unless
adequate salt stocks are found in this area for ultimate waste disposal.

In this region, there would be a population density of about 90 - 100
inhabitants/km$^2$ within a radius of 20 km. The eastern part is limited by
the open sea. In a zone of calcareous moraines there is a predominant
occurrence of loamy sand up to clay with heavy subsoil. Widely spread
permian salts are encountered. The area is aseismic. Away from the
coast, small lakes and rivers are encountered. In general there are good
prospects for groundwater resources. Possibilities of floodings cannot
be ruled out at unfavourable coastal locations.

The traffic system is sufficiently developed, unlike the electricity
supply which appears to be inadequate. Parts of the western region are
landscape-protection areas, and extensive coastal regions are typical
recreation areas. Agricultural utilization of land involves cultivation of
cereals and fodder whereas forestry plays a secondary part. Meteorologically this site is characterized by a high average wind speed of about 4 m/s.
The main wind directions are west and south-west. Inversion conditions

are rare.  A decisive dependence on the wind direction of the long-time
dispersion factor, however, does not exist.  With a 200-m stack its maximum
is about $6 \times 10^{-8}$ s/m$^3$.

### IV.  An area with subterranean salt formations that may be used for ultimate deposit of radioactive waste

Population density in this region amounts to 130 - 150 inhabitants/km
within a periphery up to 20 km.  Within an area of widely spread permian
salts, silty clay with heavy subsoil and slightly loamy sand predominate.
This zone is aseismic.

River courses of minor significance occur.  Groundwater resources
are expected to be different in the individual areas, but mostly of low pro-
ductivity.  The region is well developed in traffic and infrastructural aspects.
The same applies to electricity supply.  Smaller areas in this region are
landscape-protection areas.  In the northern part of the region sugar-beet
cultivation is the main component of agricultural land use, whereas large
areas in the south are wooded, and uninhabited districts are kept as national
parks and nature reserves.

The meteorological conditions are very similar to those of site B becaus
highlands are nearby.  The mean wind speed is about 3 m/s.  The prevailing
wind direction is south-west.  Inversion conditions are rare, if deep valleys
are excluded.  The maximum long-time dispersion factor is about $1 \times 10^{-7}$
north-east of the site.


## 3.  RADIOACTIVE EMISSIONS FROM HTR REPROCESSING PLANTS

At present there are no commercial HTR reprocessing plants in operation.  Experience can only be gathered from laboratory experiments and
pilot plant operations as well as from correlation with LWR reprocessing
plants.  Therefore the radioactive discharges through the stack of the HTR
plant must be estimated.  However, these estimates are good enough for
siting considerations.

According to the present state-of-the-art technology we have to assume
that the HTR fuel elements are stored about 180 days or longer before
reprocessing.  During reprocessing the total inventory of $^3$H, $^{14}$C and $^{85}$Kr,
1% of the iodine inventory and about $10^{-9}$ of the non-volatile isotope inventory may, if no additional measures are taken, be discharged to the atmosphere through the stack.

To keep the radiation exposure of the population as low as practicable,
processes for the separation of tritium, krypton, iodine and aerosols are
under development [1, 2].  Therefore, in the near future the $^3$H and $^{85}$Kr
discharges may be reduced to about 1%, and that of iodine to 0.1% of the
total inventory.

Accidental releases may result from severe malfunction of the plant.
According to the high inherent safety of the reprocessing plants, however,
these releases are small.  For a more detailed safety analysis [3, 4] a
critical excursion with about $10^{18}$ fissions and a fire of the uranium load
solvent are to be considered.  In both cases the hot cells and at least one
filter system remain intact, so that the radioactivity is released via the
stack.  These are maximum credible accidents.

TABLE II.  CALCULATED RELEASE FROM A 50 000-MW(e) HTR REPROCESSING PLANT

|  | Routine release (Ci/a) | Reduced routine release (Ci/a) | Criticality accident (Ci/accident) | Solvent fire (Ci/accident) |
|---|---|---|---|---|
| $^3$H | $1.6 \times 10^6$ | $1.6 \times 10^4$ | | |
| $^{14}$C | $5 \times 10^3$ | $5 \times 10^3$ | | |
| $^{85}$Kr | $2 \times 10^7$ | $2 \times 10^5$ | | |
| $^{129}$I | 0.5 | 0.05 | | |
| $^{131}$I | 1 | 0.1 | 0.75 | |
| $^{132}$I | | | 3.3 | |
| $^{133}$I | | | 18 | |
| $^{134}$I | | | 450 | |
| $^{135}$I | | | 48 | |
| $^{138}$Xe | | | 1050 | |
| $\alpha$-aerosols | 0.2 | 0.2 | | 0.1 |

TABLE III.  CALCULATED MAXIMUM DOSES RESULTING FROM RELEASES OF A 50 000-MW(e) HTR REPROCESSING PLANT

|  |  | Whole body | Skin | Thyroid gland | Lung |
|---|---|---|---|---|---|
| Routine emission | (mrem/a) | 26 | 140 | 27 | 19 |
| Reduced emission | (mrem/a) | 16 | 2 | 2.7 | 19 |
| Critical accident | (mrem) | | | 20 | |
| Solvent fire accident | (mrem) | | | | 40 |

Long-time dispersion factor $\quad \chi = 1 \times 10^{-7}$ s/m$^3$
Short-time dispersion factor $\quad \chi = 1 \times 10^{-6}$ s/m$^3$

Accidents with damage to the confinement are very improbable.  They should not be discussed here  because in such a case the exposure of the population in the immediate neighbourhood of the plant must be taken into account.  This, however, is a local rather than a regional problem.

Table II shows the calculated present state-of-the-art technology and future reduced routine emissions as well as accidental releases from a 50 000-MW(e) HTR reprocessing plant.

TABLE IV.  DOSE CONVERSION FACTORS (g) $[(\text{rem} \cdot \text{m}^3)/(\text{Ci} \cdot \text{s})]$

|  | Whole body | Skin | Thyroid gland | Lung |
|---|---|---|---|---|
| Routine release: | | | | |
| $^3$H | $1.5 \times 10^{-1}$ | $1.5 \times 10^{-3}$ | | |
| $^{14}$C | 31 | | | |
| $^{85}$Kr | | $5.7 \times 10^{-2}$ | | |
| $^{129}$I | | | $2.2 \times 10^5$ | |
| $^{131}$I | | | $1.6 \times 10^5$ | |
| $\alpha$-aerosols | | | | $3.8 \times 10^5$ |
| Accidental release: | | | | |
| $^{131}$I | | | $1.1 \times 10^3$ | |
| $^{132}$I | | | $3.7 \times 10^1$ | |
| $^{133}$I | | | $3.0 \times 10^2$ | |
| $^{134}$I | | | $1.9 \times 10^1$ | |
| $^{135}$I | | | $9.0 \times 10^1$ | |
| $^{138}$Xe | 0.83 | | | |

## 4.  EXPOSURE OF THE POPULATION TO RADIATION

The maximum radiation doses to critical organs outside the reprocessing plant with quantitative tritium and krypton release, assuming a dispersion factor ($\chi$) of $1 \times 10^{-7}$ s/m$^3$, are shown in Table III.  The concentration to dose conversion factors used in this calculation are given in Table IV [5]. Table III shows that the maximum whole-body dose reaches nearly the 30 mrem/a limit recommended by the Deutsche Atomkommission;  that of the skin dose, however, exceeds the recently recommended 100 mrem/a limit for organ doses.  If the calculated doses must be kept well below the said recommendation limits, the maximum dispersion factor in the neighbourhood of a reprocessing plant should not exceed about $5 \times 10^{-8}$ s/m$^3$. This can be achieved by avoiding sites with a low average wind speed or preferential wind direction (e.g. deep valleys).  If tritium and krypton separation processes are in operation, these siting restrictions can be omitted.

Most of the accidental releases do not contribute appreciably to the radiation exposure of the population.  They are less important than accidental releases from power reactors.  This will become clearer if the probability of such severe accidents are taken into account [3].  Therefore routine releases are decisive in siting considerations.

At all four sites considered in this paper for the erection of a  HTR reprocessing plant, the maximum dispersion factor is well below $1 \times 10^{-7}$ s/m$^3$, if a 200-m stack is used for the discharges.  So the recommended maximum dose limits can be achieved.

TABLE V.  COLLECTIVE DOSES TO THE POPULATION IN THE SITING AREAS

| Area | Exposure (man · rem/a) | | |
|---|---|---|---|
| | Whole body | Skin | Thyroid gland |
| I   (a) | 3 000 | 12 000 | 3 500 |
| (b) | 1 000 | 5 | 135 |
| II   (a) | 2 700 | 9 000 | 1 500 |
| (b) | 600 | 2 | 55 |
| III  (a) | 1 000 | 3 500 | 700 |
| (b) | 170 | 2 | 35 |
| IV  (a) | 1 800 | 9 000 | 1 700 |
| (b) | 460 | 3 | 100 |

(a): Routine emission (without $^3$H and $^{85}$Kr separation).
(b): Reduced emission ($^3$H and $^{85}$Kr separation, improved iodine retention).

TABLE VI.  WEIGHT · DISTANCE PRODUCT FOR TRANSPORT OF FUEL ELEMENTS AND FISSION PRODUCTS

| Area | Fuel-element transport | | Fission-product transport | | $\Sigma$ (t · km) |
|---|---|---|---|---|---|
| | Mean distance both directions (km) | Weight · distance (t · km) | Mean distance both directions (km) | Weight · distance (t · km) | |
| I | 350 | $3.0 \times 10^7$ | 700 | $2.5 \times 10^7$ | $5.5 \times 10^7$ |
| II | 380 | $8.3 \times 10^7$ | 600 | $2.2 \times 10^7$ | $5.5 \times 10^7$ |
| III | 1040 | $8.9 \times 10^7$ | 600 | $2.2 \times 10^7$ | $11.1 \times 10^7$ |
| IV | 880 | $7.5 \times 10^7$ | 0 | 0 | $7.5 \times 10^7$ |

For siting considerations, however, the collective dose to the population should also be taken into account, so far as biological effects of small doses cannot be completely excluded.  Table V shows these data for the whole-body, skin and thyroid exposure.  They differ by a factor of about 3.  The most favoured area from this point of view is the Baltic Sea site.  The highest collective doses are found near the industrial areas (areas I and II in Table V). The calculations cover all people who receive a maximum annual dose of 0.1 mrem, not exceeding, however, a distance of 150 km from the plant.

In addition, the transport distances between the power reactors and the reprocessing plant and between the reprocessing plant and the final disposal should be considered from the economic as well as from the safety point of view. Under normal conditions the shipping risk is relatively low, so that economic considerations prevail. However, transport of spent or refabricated fuel elements or of highly radioactive waste is extremely sensitive to sabotage, so that short or zero distances should be preferred if possible.

For comparison of the different sites, the product of shipping weight and shipping distances (t·km) may be used. The mean distances are estimated assuming that 70% of all HTR power plants are accumulated in the industrial area near the rivers Rhine and Ruhr. At a 50 000-MW(e) HTR reprocessing plant, $855 \times 100$-t shipping casks loaded with depleted fuel elements will arrive annually. Refabricated fuel elements can be shipped in the same casks. The weight of the shipping casks loaded with solidified fission products is estimated at about 36 500 t, corresponding to a shielding/heavy-metal ratio of 100. Table VI shows the products of distances and weights for fuel-element transport and for transport of fission products to the final disposal. The sum of these two products differs by a factor of about 2 with a maximum for the area at the Baltic Sea and a minimum for the industry near the areas discussed.

The relatively small differences of the areas in respect of the collective doses as well as of the t·km numbers neither favour nor exclude one of the four areas in a special way.

## ACKNOWLEDGEMENTS

D. Brenk and K.-J. Vogt (KFA-Jülich) made their extensive meteorological data available to us and computed the radiation exposure at the different sites. We gratefully acknowledge their important contribution.

## REFERENCES

[1] BEAUJEAN, H., BOHNENSTINGL, J., LASER, M., MERZ, E., SCHNEZ, H., "Gaseous radioactive emissions from reprocessing plants and their possible reduction", Environmental Behaviour of Radionuclide Released in the Nuclear Industry (Proc. Symp. Aix-en-Provence, 1973), IAEA, Vienna (1973) 63.

[2] LASER, M., BARNERT-WIEMER, H., BEAUJEAN, H., MERZ, E., VYGEN, H., in Proc. 13th USAEC Air Cleaning Conference, 1974.

[3] LASER, M., BRÜCHER, H., MERZ, E., WOLF, J., Jahrestagung des Fachverbandes für Strahlenschutz unter Beteiligung der Vereinigung Deutscher Strahlenschutzärzte (1974).

[4] USAEC Rep. Docket 50 268-26 (1972).

[5] BRENK, D., VOGT, K.-J., personal communication.

## DISCUSSION

A.P.V. MODING: Does Fig.1 mean that in the four areas indicated the siting of a LWR reprocessing plant would also be conceivable? The third area which you mention is near the Baltic Sea. May I ask why you have not chosen an area near the North Sea in addition to, or instead of, that near the Baltic?

H.J. RIEDEL: Our studies relate only to HTR reprocessing plants. They show that a site near the North Sea would not be suitable.

M. LASER: I should perhaps amplify Mr. Riedel's reply. The siting of a reprocessing plant near the Baltic Sea may give the impression that large, highly radioactive discharges were contemplated. But that was not what we had in mind. The radioactive emissions from the reprocessing plant, as indicated in the paper, would not affect the population to any significant extent. Better ground stability is one of the reasons why the site on the eastern shore has been considered.

R.R. BOUSSARD: What date do you think you can reasonably anticipate for the construction of a 365 t/a HTR reprocessing plant, corresponding to an installed capacity of 50 000 MW(e)? So far as I know, the reprocessing technology for this fuel has not yet been perfected on a full industrial scale, nor has the gaseous effluent retention equipment. Surely the methods ultimately adopted will have a bearing on your future studies of gaseous and liquid effluent releases to the environment and on the permissible dose limits?

H.J. RIEDEL: We reckon that a 50 000 MW(e) reprocessing plant will go into operation in 1995. The first stage, however, is to be a pilot plant corresponding to about 5000 MW(e).

The special head end needed for HTR elements is under development, and the THOREX solvent extraction process seems to present no insoluble problems.

S.O.W. BERGSTRÖM: Liquid effluent releases into the Baltic may be more critical than into the Atlantic. Do you not foresee any such releases into the sea in alternative III?

H.J. RIEDEL: No. We really do not foresee any liquid effluent releases.

W. HEINZ: I must point out that in the Federal Republic of Germany there are no plans to release active effluents into the sea either from LWR reprocessing plants or from thorium-treatment plants.

# NEW APPROACHES
# TO A SITING POLICY
# AND SPECIAL PROBLEMS

(Sessions VII, VIII, IX)

# A SYSTEMS ANALYSIS APPROACH
# TO NUCLEAR FACILITY SITING*

J. G. GROS, R. AVENHAUS
Energy Systems Project,
International Institute for Applied
   Systems Analysis, Laxenburg

Joanne LINNEROOTH, P. D. PAHNER**
Joint IAEA/IIASA Project,
International Institute for Applied
   Systems Analysis (at IAEA),
Vienna

H. J. OTWAY
Joint IAEA/IIASA Research Project,
Division of Nuclear Safety
   and Environmental Protection,
International Atomic Energy Agency,
Vienna, Austria

## Abstract

A SYSTEMS ANALYSIS APPROACH TO NUCLEAR FACILITY SITING.

An attempt is made to demonstrate an application of the techniques of systems analysis, which have been successful in solving a variety of problems, to nuclear facility siting. Within the framework of an overall regional land-use plan, a methodology for establishing the acceptability of a combination of site and facility is discussed. The consequences (e.g. the energy produced, thermal and chemical discharges, radioactive releases, aesthetic values, etc.) of the site-facility combination are identified and compared with formalized criteria in order to ensure 'legal acceptability'. Failure of any consequences to satisfy standard requirements results in a feedback channel which works to effect design changes in the facility. When 'legal acceptability' has been assured, the project enters the public sector for consideration. The responses of individuals and of various interested groups to the external attributes of the nuclear facility gradually emerge. The criteria by which interest groups judge technological advances reflect both their rational assessment and unconscious motivations. This process operates on individual, group, societal and international levels and may result in two basic feedback loops: one which might act to change regulatory criteria; the other which might influence facility design or site selection. Such reactions and responses on these levels result in a continuing process of confrontation, collaborative interchange and possible resolution in the direction of an acceptable solution. Finally, a Paretian approach to optimizing the site-facility combination is presented for the case where there are several possible combinations of site and facility. A hypothetical example of the latter is given, based upon typical preference functions determined for four interest groups. The research effort of the IIASA Energy Systems Project and the Joint IAEA/IIASA Research Project in the area of nuclear siting is summarized.

---

* The views expressed in this paper are those of the authors and do not necessarily reflect those of the organizations with which they are affiliated.
  ** Present address: Naval Regional Medical Center, Camp Pendleton, Calif., United States of America.

## INTRODUCTION

In recent years there has been a growing tendency in science to conduct
multidisciplinary studies of large-scale systems.  These studies include
the entire spectrum of economic, technological, environmental and societal
factors which characterize the complex problems of advanced industrialized
societies.  One of the more promising ways of addressing these problems
is the broad research strategy of applied systems analysis.  Basically this
is a rational approach to problem-solving which attempts to identify and
model interactions between the systems under study and all other systems.
This results in a thorough understanding of the system being studied which
may then serve as an aid in decision-making.

## 1.    PRESCREENING

The literature on prescreening of potential nuclear facility sites is
reviewed as a background for a proposed model for optimizing a combination
of site and facility.  The properties of the model are discussed in detail
and a hypothetical example is presented.

A rigorous systems analysis of a nuclear plant siting decision requires
that the number of options be finite.  Therefore, before analysis can start,
a finite number of possible sites must be selected in the region where a
plant is to be established.  As this prescreening is done well over ten
years before a nuclear plant comes on line, long-term aspects such as
regional development have to be taken into account.  Prescreening includes
the following steps:

(1)   Consideration of large regions as a whole;
(2)   Determination of acceptable zones in these regions;
(3)   Specific site studies.

Step 1 is done with the help of extended data bases including population
densities, meteorology, geology, seismology, water flows, etc., and may
even include aerial surveys (see e.g. Carlbom et al. [1] for a nation and
Hunt [2] for a region).  The end product of this kind of analysis is either
a number of acceptable regions or a list of specific sites.  Note that not
only the service area of the utility must be considered but also the total
region and even off-shore sites.

In step 2 one first identifies zones which are completely unacceptable
in one or more respects (e.g. the site being in an earthquake zone).  Thereafter one considers all the other consequences.  One common method of
analysis is the use of overlays coloured according to the value of specific
consequences:  where colours are too strong the area is eliminated from
consideration.  These non-quantitative techniques are usually supervized
by experienced decision-makers.

Step 3 is treated in a similar way to step 2 in the sense that unacceptable
sites are excluded.  One may then arrive at a finite number of acceptable
sites, which is the starting point for a quantitative analysis.

## 2.  SYSTEMS MODEL

### 2.1.  Literature

The importance of the problems of appropriate selection of a site for
a nuclear facility is best illustrated by the frequent international conferences
devoted to this subject (e.g. those of the International Atomic Energy Agency
and of the American Nuclear Society [3-6];  in fact this conference is just
another example.

Let us assume that a limited number of sites have been selected by
the prescreening procedure described in Section 1.  Then any systematic
study of a siting problem starts with a list of consequences important in
selecting a specific site.  This list has been established many times.  The
authors have found close agreement among these lists, irrespective of the
nationality of the authors (see e.g. Refs [7-10]).

A natural next step would be to devise an evaluation matrix where one
simply puts plus and minus signs against the desirability of a site with
respect to the criteria being considered (see Hill and Altermann [11]).
A more sophisticated way would be to put values on the attributes (e.g.
Beer [12]).  Sites unacceptable in any one respect could thus be eliminated.

After these preliminary steps, cost figures must be analysed, including
cost of land needed, equipment and maintenance for meeting environmental
standards (see e.g. the United States Environmental Protection Agency
work [13]), power transmission, etc.  Cost benefit may be considered without
explicit attention to other benefits.  In many cases the fact that cost figures
are not known accurately must be taken into account.  Initially these studies
were made only from the point of view of the utility companies;  examples
are given by Anderson [14].  However, with growing public interest in
nuclear plants, a siting decision becomes a matter in the public domain;
therefore it became necessary to model the impact of different interest
groups on these decisions.  An approach using decision analysis has been
outlined by Keeney and Nair [15];  previously, a similar approach and
extensions had been done by Gros [16].  In Section 4 an example is discussed
which is based on the work of Gros.  Work is continuing along these lines
at the International Institute for Applied Systems Analysis (IIASA), where
a critical review of nuclear facility siting techniques is near completion [17].

### 2.2.  The model

Figure 1 is a schematic representation of the flow of information
involved in judging the acceptability of a site-facility combination.  This
diagram intends to plot the real, practical flow of information rather than
the perceived flow, and is highly rationalized and simplified in order to
serve as a discussion aide.

Box 1, Fig.1, represents the possible combinations of site $S_k$ and
facility $F_j$ which might be proposed.  A decision to propose any combination
of site and facility carries with it a number of implied consequences which
include, in addition to the primary benefits intended by the sponsor (e.g.
electrical power), a number of side effects.  These side effects, which may
be adverse or beneficial, include:  radioactive, chemical and thermal discharges;  accident hazards;  aesthetic effects;  noise, etc.

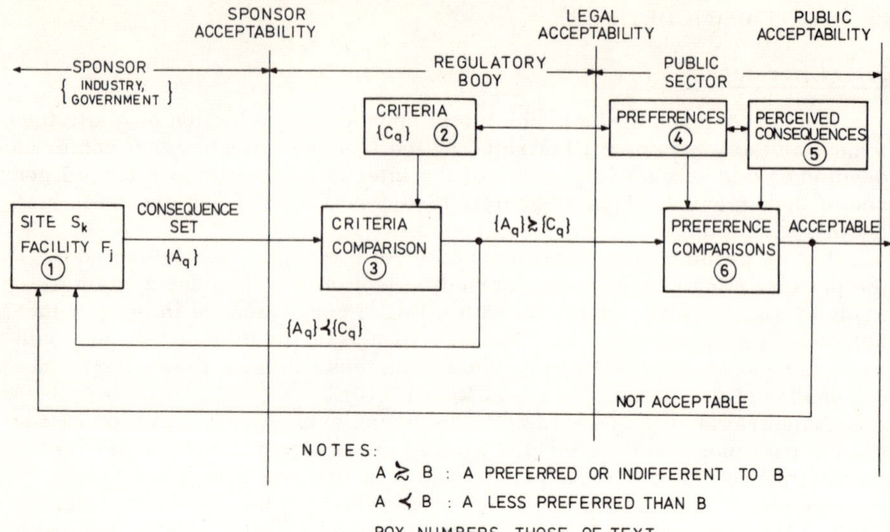



FIG. 1.  Siting process.

In addition to the set of primary and secondary consequences being satisfactory to the sponsor (industry or governmental agency, depending upon the country), they must also satisfy any formal regulations governing these effects.  This comparison (box 3) would be made by some regulatory body which uses the criteria shown in box 2.  If the consequences of the site-facility combination compare favourably with the respective criteria, then the proposal can be said to meet the requirements of 'legal acceptabilit' Failure to satisfy the criteria results in the postulated feedback loop shown in Fig.1, which acts to change the site-facility characteristics until the criteria can be met.  If the changes required make the proposal unacceptabl to the sponsor, it would be abandoned and a new proposal submitted to the regulatory body.

When the line of 'legal acceptability' has been crossed, we can say that the proposal, for practical purposes[1], first enters the public sector, where acceptability is judged on a quite different basis.

Here the responses of individuals and of various interested groups to the consequences implied by a site-facility combination gradually emerge. Interest groups use multiply determined criteria to judge, and perhaps challenge, technological advances;  that is, their preferences (box 4) reflec their rational assessment of the facts (consequences) as they know them, their perception of these consequences (box 5), and the effects of factors which may be buried deep in the nature of the groups themselves — i.e. unconscious fears and motivations, group dynamics.  These preferences are formed, in part, in the light of the past experiences of the individual and group.

[1] 'For practical purposes' because it is clear that the public can start forming their opinions before legal acceptability has been determined.  However, unless regulatory criteria are satisfied, any judgements made in the public sector may be considered premature.

It is important to note that this complex process operates on many organizational levels: the individual, the group, the societal or national, and perhaps even the international. In Fig.1 a number of interest groups are postulated which include all these levels of interest. Box 6 shows the process of comparison which has been described. Here, the acceptability of the site-facility combination to each group must be considered. In addition, by some unspecified process, the individual group's preference comparisons influence the final decision.

Here two feedback loops have been suggested: one which might act to change facility design or site selection if the comparison is unfavourable; another which might affect the regulatory criteria. These reactions and responses, on these many levels, result in a continuing process of confrontation, collaborative interchange and eventual resolution. In the following sections the consequences of site-facility combination and the individual boxes of Fig.1 will be discussed in more detail.

## 3. DISCUSSION OF FIG.1.

### 3.1. Selection of site and facility

We have said that the first box of Fig.1 represents possible site, $S_k$, and facility, $F_j$, combinations from which the sponsor may choose. These choices are to some degree interdependent and for this reason appear on the diagram together. It is assumed that the prescreening has previously narrowed the site selections to include only those candidates which have no 'unacceptable' characteristics; a further comparison of these candidates is now made to determine the 'best' site. In a similar fashion the facility designs are compared according to specific criteria set by the sponsor. To illustrate this process, a few of the possible considerations involved in the siting and design of a nuclear power plant will be discussed. The analysis is easily generalized to include fuel-fabrication plants, fuel-reprocessing plants and other nuclear facilities.

After prescreening, the sponsor must determine the 'most preferred' of the candidate sites. The usual basis of this decision is the sponsor's ranking of the site characteristics. Listings of these characteristics are well documented in the literature and generally include the locational requirements, site-related economic factors, and environmental considerations.

The characteristics of a potential site include the ease of land acquisiton, zoning, the ease of transporting major reactor components to the site, the proximity of transmission facilities, and distance to the load. The site geology and hydrology — including foundation conditions, location of faults, stability of subsurface materials, and velocity and availability of a cooling-water source — are important considerations. The possibility of flooding, landslides, and severe storms should also be considered.

Economic considerations include the costs of land acquisition, site preparation, material transport and energy transmission. The location of the site has a direct impact on specific costs of the facility, such as the extent of safeguards and the type of cooling system.

Included in the environmental impact considerations are the population density and the proximity of population centres. Meteorological factors, such as the influence of the topography and special atmospheric conditions

on the different paths of effluents, are also important. In addition, regional
land use must be considered. Some of the more sensitive areas include
park lands, wilderness and recreation areas, historical sites, wildlife habitat
and military installations. A few of the many other significant factors are
the aesthetics of the site, the use of natural resources, and the disruption
of local communities.

In choosing a facility design, power-plant sponsors will endeavour to
ensure reliable service at minimum per-unit power cost, given constraints
imposed by regulatory authorities and the site itself. More specifically
(avoiding any attempt at comprehensiveness), the considerations involve a
choice of reactor type and peripheral equipment: cooling system, intake
and discharge systems, chemical and sanitary waste handling systems,
biocide treatment system, and so forth. There are usually several available
technologies. For example, in the choice of cooling system the technological
alternatives are once-through cooling, mechanical and natural draught wet
and dry cooling towers, cooling ponds, spray ponds and spray canals.

## 3.2. Consequences

Various consequences, labelled $A_q$ (see Fig.1), result from the selection
of the site $S_k$ and the facility $F_j$. Since the nuclear power plant, at this
point in the analysis, has not yet been constructed, most of these consequence
are perceptions of some future occurrence. The accuracy of these perceptio
depends on the available information and will differ among the interested
individuals. But since continuation of the project will depend in part on
these perceptions, they should be fully anticipated. Only a brief discussion
of a few of the more important perceived consequences of the siting and
facility decisions can be included in this paper. The purpose is only to
introduce the reader to the sort of factors that should be included at this
stage of the analysis.

Consequences evolve from three stages in the development of the nuclea
power plant: site selection, plant construction, plant operation. They might
include, for example, land speculation after the siting decision, community
disruption during the construction period, and regional development during
plant operation. They impinge on local residents, users of the power and,
in a vague sense, regional and national interests.

Local communities are most directly affected by a siting decision,
and immediately so where there is dislocation of local residents. Then
there is disruption from plant construction. During operation of the plant
the primary impacts include radioactive discharges, thermal pollution and
aesthetic degradation of the landscape. In addition, local residents will be
anxious about the possibility of a nuclear accident, and they might perceive
a threat of further industrialization of the area as a result of the nuclear
power plant. The extent to which these impacts are perceived by the local
residents as real dangers, and thus lessen the desirability of locating in
an area, could be reflected in lowered property values. However, the
actual direction of the change in property values is unclear, since lowered
local property tax rates due to the increase in the tax base could increase
property values.

In some cases more specific impacts of a consequence, often referred
to as attributes, should be considered. (Finding the set of attributes is a
very important and difficult step in the analysis.) For example, the impact

of a radioactive release includes the possibility of gene mutations and an
increased cancer rate in human beings, and of danger to local agricultural
and fishing interests.  The extent of this potential damage is dependent
upon a complicated set of factors, e.g. type of isotope release, atmospheric
conditions for effluent transmission, pathways for radioactive effluent
absorption, and effects of this absorption on various forms of life.  Another
possibility, damage from thermal pollution, depends upon the type of abatement equipment used.  The most serious pollution results from the once-
through system, where the condenser cooling water is taken from a nearby
river, lake, estuary or ocean and then (usually) returned to the same source.
This procedure results in some modification of the aquatic environment;
an increase in water temperature changes the physical properties of water
such as density, viscosity and gas solubility, which can affect such phenomena
as the vertical migration of plankton and the mobility of higher organisms.
Also any change in the eco-system equilibrium can have other effects on
higher organisms.

## 3.3.  Criteria

As already mentioned, the levels of the consequences discussed must
satisfy certain criteria which may exist for regulating these consequences.
There are two different types of criteria:  those specifying the limits
within which certain consequences have to be kept, and those of a qualitative
nature which represent only general guidelines.  Examples of the first type
are regulations governing the normal operational releases of radioactive
isotopes into air or water, or the outlet temperature of cooling water into
a river in the case of once-through cooling of a power station.  Examples
of the second type are acceptable upper limits for the radioactive releases
in accidental situations, or population densities around nuclear plants;  in
these cases values have been traditional rather than stated in formal
standards.
Determining the values of the criteria requires a careful analysis of
the consequences:  for example, in the use of normal operational radioactive releases one must analyse what ambient dose rates result from what
emissions, taking into account biological pathways, etc.  Such studies, made
with great care (see e.g. Pochin [18] ), have resulted in standards for all
kinds of radioactive isotopes recommended by the International Commission
on Radiological Protection (ICRP).  With increasing knowledge of the
biological effects of radionuclides, these standards have changed:  whereas,
in 1930, 100 rem/a were thought permissible, in 1957 this value was reduced
to 5 rem/a.  The regulatory bodies inversely take these ICRP recommendations
as a basis for determining the emission standards for a specific plant, where
site characteristics such as geology  and main wind speed and direction now
have to be taken into account.
Regulations also exist for the outlet temperature of the cooling water,
but there are no international standards:  the local authorities limit the
maximum temperature increase as a function of the total amount of water
available, weather conditions, etc.
For accidental radioactive releases the situation is not so clear:
consequences are being considered even in monetary terms (see Beattie [19] ),
but no standards have yet resulted.  Unofficially, the value of 25 rem per

accident and human lifetime is sometimes stated.  Many of the issues are
covered in a paper by Majone [20].

Only general rules exist for permitted population densities around
nuclear facilities.  So far, in the Federal Republic of Germany, for example,
the following average values have been observed for nuclear power stations
(see the Institut für Reaktorsicherheit work [21]):

    Up to  5 km:   10 000 -  15 000 people
    Up to 10 km:   30 000 -  50 000 people
    Up to 20 km:  100 000 - 200 000 people

Already, however, in one case (Biblis [22]) these values have not been
observed;  in general, nuclear plants are tending to come closer to cities.

## 3.4.  Interest groups

Once the consequences of a siting decision meet the standards of the
regulatory agency, the siting process moves into the public sector.  Here
the perceived consequences are measured against the criteria of various
'interest groups'.  This is probably the least understood stage of the siting
decision, and it is here that behavioural scientists can contribute to the
analysis.

The psychologist views the interest group as a confluence of social
systems including individual responses, societal-cultural factors, political-
economic influences, and the input of the scientific community.  The interest
group per se represents the focal point of the interactions of these various
systems.

A unique feature of nuclear energy is its tremendous potential for both
constructive and destructive utilization.  It is perhaps this factor that in
part accounts for the difficulties sometimes encountered in the public
response to the siting of such facilities.  The population may respond on
an emotional-irrational level, with fear of nuclear holocaust and annihilation
worry about genetic effects and future generations, anxiety due to lack of
adequate knowledge and conceptualization of the power of the atom.

Assuming, then, that nuclear energy is a fear-provoking stimulus, we
can examine the response of the individual personality.  We owe much of
our understanding of the dynamics of personal responses to in-depth
psychology.  A helpful generalization is that external dangers lead to fear,
which in turn leads to a variety of healthy or unhealthy defenses against
this fear.  The two most primitive are flight and fight — apathetic withdrawal physically and emotionally or denial that any threat exists, or a
readiness to retaliate.  It is the latter response that accounts in part for
the opposition of the group to what is perceived as an external threat.

External dangers are not the only dangers in life for the human being;
there are internal dangers as well.  As the internal security and intrapsychi
balance of an individual is eroded, fear and anxiety mount.  Thus resistance
to the siting of nuclear power facilities may be due in part to projection of
our internal fears onto a symbolic external object, the facility.  As these
fears are expressed, the individual finds others who think and act similarly.

The social psychologist now provides further insights — based on
observations of group dynamics and laboratory studies — into the nature
of group responses.  An interest group reflects to varying degrees elements

of its members' individual responses, characteristics of the larger societal-
cultural group of which it is a part, and an indication of the information it
has obtained from the scientific-technological community;  and the political
milieu will influence the character of the resistance.  The group per se,
however, has its own unique characteristics.  A large body of literature
supports the following conclusions:

(a)  Interest groups tend to emerge and crystallize around affect-laden
     social-environmental concerns;
(b)  Groups tend to be solution-oriented rather than problem-oriented,
     i.e. they gravitate towards a dialectic-adversary position rather
     than engage in collaborative exchanges;
(c)  The constitution and cohesiveness of a group is likely to be directly
     related to the degree to which its members share similar values
     and attitudes;
(d)  Communication patterns are often distorted, especially in groups
     with a vertical hierarchy of status and power;
(e)  New information is accepted or rejected contingent on the support
     it provides for the beliefs and values of the group;
(f)  Behavioural responses of members are influenced by those of
     other members, and the strength and integrity of individual values
     are weakened.

Thus the interest group opposing the siting of a nuclear facility is
likely to be a well-organized, firmly entrenched, emotional body of persons
committed to their position and screening factual information according
to the utility it has for their position.  Of interest of course is that their
counterparts are frequently matched feature for feature.

On the final level, the interest group is viewed in its larger context,
the societal-cultural milieu from which it springs.  The anthropologist
and historian are now consulted for their insights.  Certainly the fervour
and diversity of movements in recent years have been remarkable.  Perhaps
this is evidence of a larger-scale, more generalized response to the ever
expanding technologies.  Analysis of previous technological revolutions
suggests that there gradually evolves a social structure which begins to
put the brakes on the process.  In that sense the interest group becomes a
culturally determined, expected response which emerges at the interface
of the individual faced with survival and the entire technological-societal-
political-cultural maze.

We have seen that the siting of a nuclear power facility is a complex
technological, economic, socio-political, environmental and psychological
issue.  Hopefully, collaboration and interchange among those involved in
the above disciplines will facilitate the process of making rational decisions
in the best interest of the community.


## 4.  THE PARETIAN APPROACH

### 4.1.  Introduction

In the last section we discusssed many of the impacts important in
the siting of nuclear power plants, and the interest groups that influence

the final decision.  Now we shall show how all these inputs can be integrated into one mathematical analysis.  The underlying assumption is the concept of Pareto optimality.  A Pareto-admissible decision is one from which any technologically feasible change would make at least one of the interest groups worse off;  in other words, if one makes a Pareto-admissible decision, it is impossible to make one interest group better off without making another worse off.  The mathematical analysis that finds the set of Pareto-admissible decisions is often called Paretian environmental analysis.  The basic concepts were pioneered by Dorfman and Jacoby [23].  Gros [24] extended the work to nuclear power plant siting decisions and showed how utility function analysis could be integrated into the framework of Paretian analysis.  In addition to describing the mathematical analysis, we shall discuss how these results can be used for descriptive, predictive or prescriptive purposes.

## 4.2.  Mathematical model

Let us assume that the required number of new generating units in a region $R_1$ is known for each year of a planning horizon, $\ell = 0, 1, ..., M$. Further, let there be K sites available, each of which can support a certain number of generating units.  The problem is then to find the set of Pareto-admissible unit designs and unit deployments.

The first step, after identifying the problem, is to determine which interest groups influence the decision.  Once the groups have been identified, a set of attributes for each group should be found which describes what is important to that group.  Let $X_{ni}$ be the $n^{th}$ attribute for the $i^{th}$ group. For instance, for those groups concerned about radiation effects, one or more of their attributes should cover this effect.  For each interest group, a utility function should be estimated which describes the relative preference for different unit designs and unit deployments.  In addition, a set of technological relations should be found which relate unit design parameters (such as size and type of cooling system, reactor type, etc.)  to the attributes.

If certain reasonable axioms are satisfied, it is possible to find a utility function which is a monotonic function of a group's preferences with the property that the expected value of the utility function is a guide for decision-making.  Let $U_{k\ell}^{imn}(X_{ni})$ be the utility function for the $n^{th}$ attribute for the $i^{th}$ interest group for site k and year m, given that a unit of some design was commissioned in year $\ell$.

Each single-attribute utility function is found from a set of indifference questions, and each function is generally scaled from 0, for the least-preferred value of the attribute, to 1 for the most-preferred value:

$$0 \leq U_{k\ell}^{imn}(X_{ni}) \leq 1$$

But each group's preferences depend on more than just one attribute.  Therefore, let $\underline{X}_i = (X_{1i}, X_{2i}, ..., N_{Ni})$ be the set of attributes of the $i^{th}$ interest group which describes what is important to this group in siting decisions. For this set of attributes, a multi-attribute utility function $U_{k\ell}^{im}(\underline{X}_i)$ should be estimated.  Based on the group's preference structure, this multi-attribute utility function could have one of several special forms.  For instance, if certain independence properties of preferences  hold, then the multi-attribute utility function can be written in the pure product form as follows [25]:

$$1 + dU_{k\ell}^{im}(\underline{X}_i) = \prod_{n=1}^{N} (1 + dd_n\, U_{k\ell}^{imn}(X_{ni}))$$

where d and the $d_n$'s are scaling constants which satisfy

$$0 \leq d_n \leq 1$$

for

$$1 \leq n \leq N, \quad -1 \leq d, \quad \sum_{n=1}^{N} d_n \neq 1$$

and

$$1 + d = \prod_{n=1}^{N} (1 + dd_n)$$

Naturally, if other preference structures are to be modelled, then the multi-attribute utility function will have a different, and generally more complicated, form.

In addition to the preference structure for these attributes, an interest group is concerned with when a unit is commissioned. Therefore, a multi-time-period utility function $U_{k\ell}^i$ should be obtained for each interest group. A common form for this function, given the multi-attribute utility function for each time period, is as follows:

$$1 + cU_{k\ell}^i = \prod_{m=0}^{M} (1 + cc_m\, U_{k\ell}^{im}(\underline{X}_i))$$

where c and the $c_m$'s are scaling constants with

$$0 \leq c_m \leq 1 \text{ for } 0 \leq m \leq M, \quad -1 \leq c, \quad \sum_{m=0}^{M} c_m \neq 1$$

and

$$1 + c = \prod_{m=0}^{M} (1 + cc_m)$$

A reasonable criterion for the power plant siting problem when more than one interest group is involved is to find the set of Pareto-admissible solutions. To find such solutions, objective function values of the following form can be used:

$$U_{k\ell} = \sum_{i=1}^{I} w_i U_{k\ell}^i$$

$$0 \leq w_i \leq 1 \text{ for } 1 \leq i \leq I, \quad \sum_{i=1}^{I} w_i = 1$$

The values $w_i$ are referred to as political weights since they give some indication of the distribution of benefits from a siting decision. (If more than one technology can be used at a site, then the technology that maximizes $U_{k\ell}$ is the Pareto-admissible unit design.) The objective function for the development decision can be expressed as follows:

$$\sum_{k=1}^{K} \sum_{\ell=0}^{M} U_{k\ell} Y_{k\ell}$$

where $Y_{k\ell} = 0,1$; 0 if the site is not committed in year $\ell$, and 1 if the site is committed in that year. This objective function should be maximized subject to the constraint that the number of sites committed in each year should equal the number needed, $R_\ell$, and that each site is committed only once. Mathematically, this can be written:

$$\sum_{k=1}^{K} Y_{k\ell} = R_\ell \quad \text{for } 0 \le \ell \le M$$

$$\sum_{\ell=0}^{M} Y_{k\ell} = 1 \quad \text{for } 1 \le k \le K$$

Everything was certain in the model just described; for instance, the number of sites that must be committed in each year was exactly known. But in real life, many quantities are not known exactly but can only be predicted in terms of a probability density. For instance, an electric utility company may need either three or four new nuclear units in 1985, with an equal probability of either value being correct; or regulatory standards cannot be predicted with certainty, but a distribution of possible standards may be predictable. One advantage of using utility functions is that these uncertain situations can be modelled easily, since the expected value of a utility function is a guide for decision-making. Details of modelling these uncertain situations, along with information on the utility function assessment process, can be found in the papers by Gros [16,24].

### 4.3. Hypothetical example

We shall illustrate the ideas outlined above with a hypothetical example based on a study of nuclear power plant siting in New England [16]. Let us assume that the technological alternatives are fixed (in the example, once-through cooling and spray canals for coastal sites). Based on a careful analysis of the political, societal and economic situation in the region under consideration, the following broad interest groups which have a major influence on siting decisions have been identified:

(1)  Electric utility companies;
(2)  Regulatory bodies;

(3)  Groups concerned with environmental, aesthetic and similar
     problems (not necessarily local groups); and
(4)  Local interest groups.

The last interest group, for instance, is concerned about the general welfare
of the region around each site.
     To obtain a multitime-period utility function for each of these groups,
a series of interviews is necessary. The first several would try to establish
qualitatively what is of concern to the groups, try to obtain the set of
attributes, and try to identify someone who can quantitatively describe the
preferences of the group when the utility function is assessed. For instance,
for the electric utility companies interest group, the following four attributes
cover many impacts of importance in the choice of sites:

(1)  Number of units at a site;
(2)  Cost;
(3)  Population within 10 km of a site;
(4)  Incremental water temperature at peak ambient water temperature
     period of year.

A set of typical single-attribute utility functions is shown in Fig.2. The
first attribute, number of units at a site, covers the perceptions of the
electric utility companies on the ease of obtaining regulatory approval
for a different number of units at a site; it also covers effects on system
reliability of having a different number of units at a site. To illustrate
the type of indifference question used in the assessment, let us consider
the second attribute, cost. As can be seen from Fig.2, the most preferred
cost, US $500 million, has a utility value of 1; the least preferred cost,
US $650 million, has a utility value of 0. Let us look at a lottery in which
there is a 50% chance of obtaining a US $500 million unit and a 50% chance
of obtaining a US $650 million unit. The expected utility value for that
lottery is 0.5 U (US $500 million) + 0.5 U (US $650 million = 0.5 × 1 + 0.5 × 0 = 0.5).
Now let us assume that the assessor was indifferent to the choice between
having a unit costing US $590 million for certain and having the lottery.
Then the utility value of US $590 million is 0.5, since this cost is indifferent
to a lottery with an expected utility value of 0.5. (Remember that the
expected value of the utility function is a guide for decision-making.) In
this situation, the value the assessor is willing to take for certain,
US $590 million, is less preferred than the expected value of the lottery
(US $575 million). This behaviour is common and the assessor is said
to be risk-averse over monetary costs. The analyst would ask similar
indifference questions, building up the utility function from its component
parts.
     The result of the analysis is a set of Pareto-admissible decisions.
The model does not identify one best decision, for there is no one best
decision from a political decision-making viewpoint. The results can be
used from a descriptive, predictive and prescriptive viewpoint. As a
descriptive tool, the Paretian model can be used to organize information
and to help identify what is happening during decision-making. The advantage
of this type of analysis is that it forces the analyst to consider explicitly
the important interest groups that influence the decision. To test the model
as a predictive tool, the analyst determines whether the actual decision is

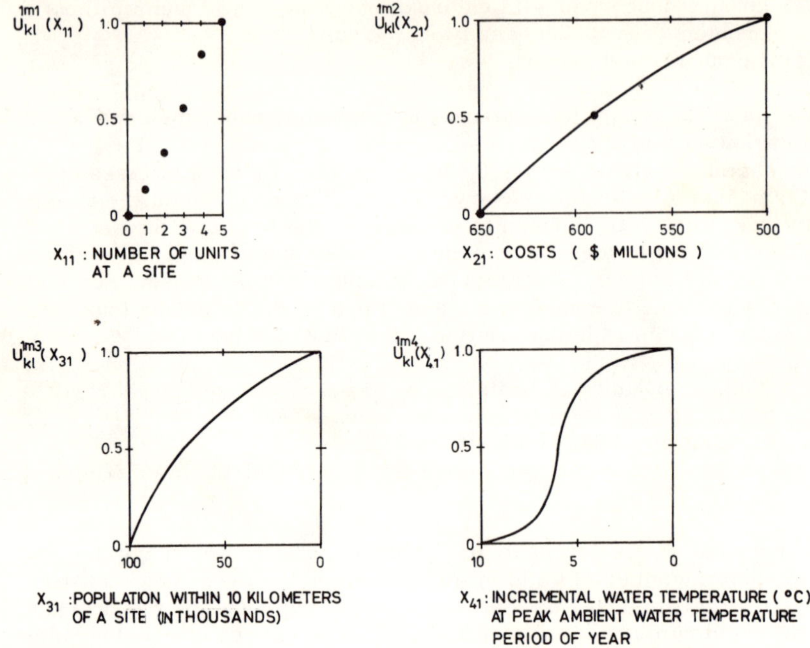

FIG. 2.  Single–attribute utility functions for electric utility companies interest group (i = 1).

Pareto-admissible.  The use of the model as a prescriptive tool is more
action-oriented than the other uses.  Here, the Pareto-admissible solutions
can be used in the actual decision-making process.  Alternatively, if a
Pareto-inadmissible decision is made, the analyst can offer for consideratic
an admissible decision which would make one group better off without
making any other group worse off.  We think that the Pareto-admissible
one would be chosen over the Pareto-inadmissible one.


## 5.   CONCLUDING REMARKS

Work is being done at the International Institute for Applied Systems
Analysis (IIASA) and by the Joint IAEA/IIASA Research Project on nuclear
facilities siting issues.  We have already mentioned that a thorough review
of contrasting siting methodologies with critical comparative comments is
being done at the Institute.  An aim of their larger project is to indicate
more clearly the interrelationships between decisions which must of
necessity be made at different levels and by so doing delineate clearly those
problems which are not part of the siting process per se.  The Institute's
work will extend Paretian analysis and will emphasize close co-ordination
between methodological studies and carefully planned and executed public
attitude assessment.  Their study of a specific siting problem will attempt
to incorporate rigorous risk perception work with decision methodologies
that include the broad spectrum of impacts.

The Joint IAEA/IIASA Research Project is an international, interdisciplinary group with the task of studying risk assessment principles and
their application in judging the acceptability of technological innovations.
The focus is on energy production systems, and more specifically on nuclear
energy as an interesting case study providing virtually all of the variables
which are of interest in risk assessment studies. The research interests
of the Joint Project  and its sponsors  lie primarily in the application of
risk assessment principles to standard setting, the study of the perception
of risks of a technological origin, methodologies for determining societal
preferences for risk acceptance and the group dynamics and information
transmission involved in making societal judgements on technological
applications. Within this framework the immediate research plans include:
(1) a study of the estimation procedures used in determining the probabilities
of certain outcomes or risks;  (2) the value society puts on these risks (in
the framework of modern utility theory);  (3) the relationship of these values
to standard setting procedures;  and, more generally, (4) the capacity of
societies, both traditional and modern, to cope with technological risks.
Longer term research plans include the development and application of
techniques to understand and estimate individual and group perception of
different types of risk, including those of a technological origin.

## REFERENCES


[1] CARLBOM, L., OBRADOVIĆ, J., MOJOVIĆ, L., "Analysis of criteria and their application in site
    selection for nuclear power plants in Yugoslavia", Containment and Siting of Nuclear Power Plants
    (Proc. Symp. Vienna, 1967) IAEA, Vienna (1967) 171.
[2] HUNT, F.R., "Power station site selection in England and Wales", Environmental Aspects of Nuclear
    Power Stations (Proc. Symp. New York, 1970), IAEA, Vienna (1971) 647.
[3] INTERNATIONAL ATOMIC ENERGY AGENCY, Siting of Reactors and Nuclear Research Centres
    (Proc. Symp. Bombay, 1963), IAEA, Vienna (1963).
[4] INTERNATIONAL ATOMIC ENERGY AGENCY, Containment and Siting of Nuclear Power Plants,
    (Proc. Symp. Vienna, 1967), IAEA, Vienna (1967).
[5] INTERNATIONAL ATOMIC ENERGY AGENCY, Environmental Aspects of Nuclear Power Stations
    (Proc. Symp. New York, 1970), IAEA, Vienna (1971).
[6] Conf. on Nuclear Power Plant Siting, Portland, 1974, Trans. Am. Nucl. Soc., Suppl. 1 (1974).
[7] MANDEL, H., Standortfragen bei Kernkraftwerken, Atomwirtsch. 16 (1971) 22.
[8] CANDES, P., Practice and experience of safety in nuclear plant siting in France, Conf. on Nuclear
    Power Plant Siting, Portland, 1974, Trans. Am. Nucl. Soc., Suppl. 1 (1974) 4.
[9] NORRIS, J.A.,"The guide to siting of nuclear power plants", ibid., 6.
[10] BOCK, H., Standortkriterien fur Kernkraftwerke, Elektrotech. Maschinenbau 91 (1974) 312.
[11] HILL, M., ALTERMANN, R., Power plant site evaluation: the case of the Sharon Plant in Israel,
    J. Environ. Management 2 (1974) 179.
[12] BEER, L.P., Quantification of siting factors for rational selection of alternative sites, Conf. Power
    Plant Siting, Portland, 1974, Trans. Am. Nucl. Soc., Suppl. 1 (1974) 12.
[13] UNITED STATES ENVIRONMENTAL PROTECTION AGENCY, OFFICE OF RADIATION PROGRAMS,
    Environmental Analysis of the Uranium Fuel Cycle; Part II — Nuclear Power Reactors  (1973).
[14] ANDERSON, D., Models for determining least cost investments in electricity supply, Bell J. Econ.
    Management Sci. 3 (1972) 267.
[15] KEENEY, R.L., NAIR, K., Decision Analysis for Siting of Nuclear Power Plants: The Relevance of
    Multiattribute Utility Theory, Tech. Rep. No. 96, Operations Research Center, MIT (1974).
[16] GROS, J., A Paretian Approach to Power Plant Siting in New England. Ph.D. dissertation, Harvard
    University (May 1974).
[17] McCUSKER, K., BAECHER, G., GROS, J., State of the Art of Multi-Objective Decision-Making
    Techniques for Nuclear Facilities Siting, IIASA Res. Rep., in preparation.

[18] POCHIN, E.E., "The development of quantitative bases for radiation protection", Environmental
     Aspects of Nuclear Power Stations (Proc. Symp. New York, 1970), IAEA, Vienna (1971) 119.
[19] BEATTIE, J., Rationale of reactor site selection for public safety, Conf. on Nuclear Power Plant
     Siting, Portland, 1974, Trans. Am. Nucl. Soc., Suppl. 1 (1974) 15.
[20] MAJONE, G., "On the logic of standard setting in health and related fields", Proc. IIASA Conf. on
     Systems Aspects of Health Planning, Mark Thompson, in press.
[21] INSTITUT FÜR REAKTORSICHERHEIT, Sicherheitskriterien für Kernkraftwerke, Rep. IRS-R-2 (1969).
[22] Series of articles on Kernkraftwerk Biblis, Atomwirtsch. 19 (1974) 405-438.
[23] DORFMAN, R., JACOBY, H.D., "An illustrative model of river basin pollution control", Models
     for Managing Regional Water Quality, (DORFMAN, R., JACOBY, H.D., THOMAS, H.A., Jr., Eds)
     Harvard Univ. Press, Cambridge, Mass. (1972).
[24] GROS, J., "Power plant siting: a Paretian environmental approach", Environmental Systems Program,
     Discussion Paper No. 74-4, Harvard Univ., August 1974.
[25] KEENEY, R.L., Multiplicative Utility Functions, Tech. Rep. No. 10, Operations Research Center,
     MIT (1972).

# DISCUSSION

**J.B. BURNHAM:** The difference between the shape of the water
temperature utility curve in Fig.2 and that of the water quality value function
which I showed in paper SM-188/43 is, I gather, due to the classical S-shape
of the Neumann-Morgenstern utilities. The major fish kill would then seem
to be in the 4-6°C range. The question I should like to ask, then, is how you
expect to make a physical assessment of the utility functions of a statistically
significant sample of the public.

**J.G. GROS:** You are right that the lethal temperature for the most
sensitive, economically important, fish species appears a little above 6°C
on the figure. In this connection, I would explain why this curve has what
you call the classical S-shape, while the other curves for continuous
attributes are concave in shape. The incremental temperature with a
single-attribute utility function value of 0.5 is 6°C, which, although still
safe for the fish, borders on the lethal range. The range between 6°C and
10°C is convex, which can be argued on the following grounds. Consider a
plant with the average temperature in this range, 8°C for certain. At this
temperature fish are killed. Then consider the lottery with a 50% chance
of having an incremental temperature of 6°C and a 50% chance of 10°C.
Now, most people would prefer the lottery with its 50% chance of no fish
kill (incremental temperature 6°C) and 50% chance of fish kill (incremental
temperature 10°C), instead of having a 100% chance of a fish kill, which
would happen if 8°C were chosen. Such behaviour would result in a convex
utility function for the range 6-10°C.

Now consider the incremental temperature range 0°-6°C. The average
here is 3°C, so let us consider having this value for certain as one alternative.
The second alternative is the lottery with a 50% chance of obtaining 0°C
and a 50% chance of obtaining 6°C. Many people, if faced with the choice
between these two alternatives, would choose the first alternative of 3°C,
since it is the average temperature of the lottery and avoids the uncertainty
of the lottery. This behaviour is consistent with the utility curve being
concave. It is also worth pointing out that utility functions can be used for
certain situations along with uncertain ones.

Interest groups can be defined in either of two ways.  All members of an interest group can share the same preferences (and thus have the same utility function), or the members can be grouped by the labels society gives to individuals, such as environmentalists.  Given the second situation, it may be that not all members of the group have the same preferences, so there is a need to test the range of opinion and get an average.  This is what I did to find these preferences.  I asked members of each interest group to identify one or several members of their group who had the mathematical ability to do the assessment, and who were knowledgeable about the preferences of the group.  These individuals were called 'knowledgeable observers'.  I asked these knowledgeable observers to assess the utility function for the group as a whole, and also their personal utility functions.  The functions of different knowledgeable observers for the group as a whole could be compared and presented to other group members for verification.  The personal utility functions of these and other members could be compared to find the variance of opinion within an interest group.

W. VINCK:  How do you incorporate or plan to incorporate accident considerations in your model, including the probabilities of equipment failure and its consequences (curves and doses)?

J.G. GROS:  Let me explain what was done in the study reflected in these figures, and also what we plan to do.  The study outlined here was for units of the same size:  1000 MW.  It was assumed that all were of the same type and had the same characteristics with respect to equipment failure probabilities.  Hence, the radiological consequences following upon a specific type of accident would depend only on the surrounding population.  Therefore, an assessment over population alone was needed.

In future work we expect to assess utility functions over all possible consequences, and then to assess a subjective utility function over the chance of having specific consequences.  These can be integrated to get the expected utility function value for consequences of accidents.

P.M. FISCHER:  Do the utility functions which you show in Fig.2 apply to only one unit at a site, and will they be different if there is more than one unit?  On what axioms is the utility theory you have used based?

J.G. GROS:  The utility function for a number of units at a site represents relative preferences for having different numbers of units at a site.  It can be used to find an optimum distribution of units if several units are to be distributed to more than one site.  If the relative preferences depend on site characteristics, then a utility function should be assessed for each site.

Before I state the axioms, it is worth mentioning that, regardless of what objective function form is used, we assume certain axioms to be satisfied.  Usually, there is no explicit consideration of them.  Whether the axioms are valid or not, for certain and uncertain situations, is a measure of the accuracy of the objective function.

Utility function theory and decision analysis depend on whether seven axioms are found reasonable.  Let us assume, for convenience, that we have one attribute x and use subscripts to designate specific values of the attribute.  In addition, let $x^*$ be the most preferred value of the attribute, and $x_*$ be the least preferred value.

Before formally stating the axioms, let us define some notations.  A simple lottery, to be written as $L(x_1,p,x_2)$, is the probabilistic event in which there is a chance p of obtaining $x_1$ and a chance $1-p$ of obtaining $x_2$.

This can be  pictured as

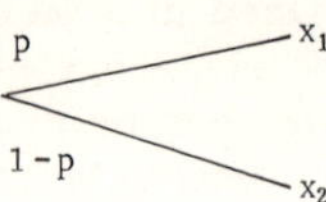


In monetary terms, a simple lottery is often called a wager.  The symbol $\succ$ means:  'is preferred to' ; $\sim$ means that the choice is a matter of indifference and $\succsim$ means:  'is not less preferred than'  (or, what is the same thing, that one alternative is either preferred to the other or that the choice is a matter of indifference).  Thus, we could write $x_1 \sim L(x_2,p,x_3)$ to mean that the decision-maker is indifferent to the choice between having $x_1$ for certain or having the lottery with a chance p of obtaining $x_2$ and a chance 1-p of obtaining $x_3$.

The first axiom is that of the existence of relative preferences.  It states that for every pair of values of the attribute, $x_1$ and $x_2$ for instance, the decision-maker will have preferences such that only one of the following will hold:  $x_1 \succ x_2$, $x_2 \succ x_1$ or $x_1 \sim x_2$.  What this means is that the decision-maker can say whether he likes one value or the other, or is indifferent to the choice between them.  Similarly preferences between simple lotteries can be stated.

The second axiom, transitively, states that if $x_1$ is preferred to $x_2$, and $x_2$ is preferred to $x_3$, then $x_1$ is preferred to $x_3$.  Similarly, for lotteries if $L_1 \succ L_2$ (lottery 1 preferred to lottery 2) and $L_2 \succ L_3$, then $L_1 \succ L_3$.  Also for indifference, if $L_1 \sim L_2$ and $L_2 \sim L_3$, then $L_1 \sim L_3$, and so on.

The third is called the axiom of the comparison of simple lotteries.  Let us assume that $x_1$ is preferred to $x_2$.  The axiom states than when $x_1$ and $x_2$ are placed in simple lotteries, the decision-maker prefers the lottery with the highest probability of getting the preferred value $x_1$.  This can be visualized as follows:  Lottery 1 is $L_1(x_1,p,x_2)$:

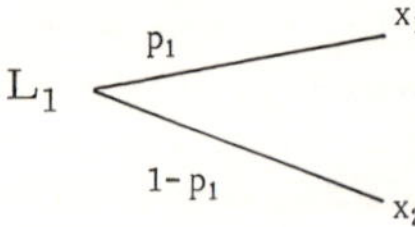


and lottery 2 is $L_2(x_1,p_2,x_2)$:

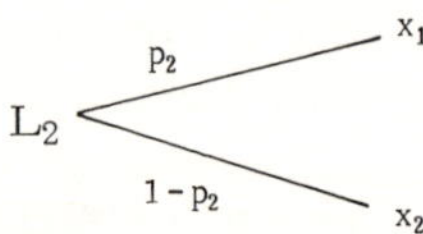


Since the decision-maker prefers $x_1$ to $x_2$, he would prefer the situation where he has the greatest chance of obtaining $x_1$.  Therefore, $L_1 \succ L_2$ if $p_1 > p_2$.  Formally, if for the decision-maker $x_1 \succ x_2$, then:

$$L_1(x_1,p_1,x_2) \succ L_2(x_1,p_2,x_2) \text{ if } p_1 > p_2$$

$$L_1(x_1,p_1,x_2) \sim L_2(x_1,p_2,x_2) \text{ if } p_1 = p_2$$

$$L_1(x_1,p_1,x_2) \prec L_2(x_1,p_2,x_2) \text{ if } p_1 < p_2$$

The fourth axiom is called the axiom of quantification of preferences.  It states that for any value of the attribute, say $x_1$, it is possible to find a probability for the simple lottery with outcomes $x^*$ and $x_*$ such that the decision-maker is indifferent to the choice between $x_1$ for certain and the lottery.  More formally, for attribute value $x_1$, the decision-maker can specify a number $\pi(x_1)$, where $0 \leq \pi(x_1) \leq 1$ ($\pi(x_1)$ is a probability), such that $x_1 \sim L(x^*, \pi(x_1), x_*)$.  This can be visualized as follows:

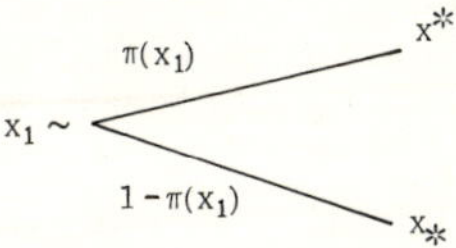

The fifth axiom states that for each possible event which may affect the decision it is possible to find a judgemental probability.  Moreover, the judgemental probabilities obey the axioms of probability theory.  This axiom is generally referred to as the axiom of quantification of judgemental uncertainties.  It is useful in those situations where a large sample survey on which to base the probabilities of various events is not available.  An example is the probability distribution for accidents at nuclear power plants. Since such accidents have rarely occurred, we do not have  enough data to specify a probability density, but we can specify the distribution using our judgement.  What the axiom states is that we can use the judgemental density in our analyses.

The sixth axiom, that of substitutability, states that if the decision problem is modified through replacement of one lottery or attribute value by another lottery or attribute value, and if the decision-maker finds the choice between the first lottery or attribute value and the replacement a matter of indifference, then he should be indifferent regarding the original and the modified decision problems.  Let us assume that a decision problem can be visualized as:

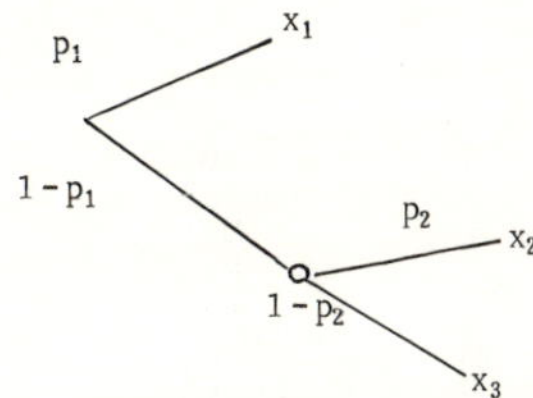

and let us suppose that $x_4 \sim L(x_2, p_2, x_3)$, i.e.

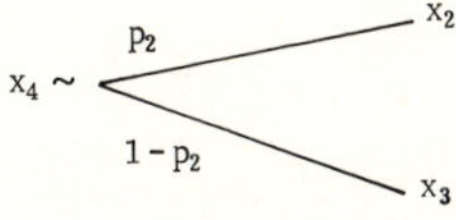

then the decision-maker is indifferent to the choice between the original problem  and $L(x_1, p_1, x_4)$, which can be illustrated as:

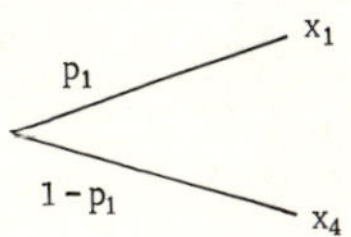

The last axiom is called the axiom of the equivalence of conditional and unconditional preferences. Let us assume we have two lotteries which are possible only if it rains tomorrow and let us rank the two lotteries today. Tomorrow we shall see if it rains, and if it does, we shall rank the lotteries again. For the axiom to be satisfied, today's ranking must be the same as tomorrow's. More formally, let $L_1$ and $L_2$ designate two lotteries which are possible only if some event E occurs. After it is known whether or not the even E occurred, the decision-maker must have the same preference between $L_1$ and $L_2$ as he had before it was known whether the event E had occurred.

Incidentally, the indifference probability $\pi(x_1)$ of axiom 4 is a measure of the relative preference for various values of the attribute x.

J.J. DiNUNNO: Your paper appears to contain a learned treatment of the subject, but the method is set out in terms so unfamiliar to many of us active in the field that we cannot at present understand it. Does the Institute plan to provide a primer or guide to translate the axioms and the approach into terms which the user community can better understand and apply?

J.G. GROS: As primers I would suggest Refs [16], [17] and [24] cited in the paper. These publications should be available from the Environmental Systems Program, Pierce Hall, Harvard University, Cambridge, Mass. 02138 USA. In addition, I would be most glad to provide any clarification.

# USE OF CRITICAL PATHWAY MODELS AND LOG-NORMAL FREQUENCY DISTRIBUTIONS FOR SITING NUCLEAR FACILITIES

D.A. WAITE, D.H. DENHAM*
Battelle Pacific Northwest Laboratories,
Richland, Wash., United States of America

*Presented by J.B. Burnham, Jr.*

**Abstract**

USE OF CRITICAL PATHWAY MODELS AND LOG-NORMAL FREQUENCY DISTRIBUTIONS FOR SITING NUCLEAR FACILITIES.

The advantages and disadvantages of potential sites for nuclear facilities are evaluated through the use of environmental pathway and log-normal distribution analysis. Environmental considerations of nuclear facility siting are necessarily geared to the identification of media believed to be significant in terms of dose to man or to be potential centres for long-term accumulation of contaminants. To aid in meeting the scope and purpose of this identification, an exposure pathway diagram must be developed. This type of diagram helps to locate pertinent environmental media, points of expected long-term contaminant accumulation, and points of population/contaminant interface for both radioactive and non-radioactive contaminants. Confirmation of facility siting conclusions drawn from pathway considerations must usually be derived from an investigatory environmental surveillance programme. Battelle's experience with environmental surveillance data interpretation using log-normal techniques indicates that this distribution has much to offer in the planning, execution and analysis phases of such a programme. How these basic principles apply to the actual siting of a nuclear facility is demonstrated for a centrifuge-type uranium enrichment facility as an example. A model facility is examined to the extent of available data in terms of potential contaminants and facility general environmental needs. A critical exposure pathway diagram is developed to the point of prescribing the characteristics of an optimum site for such a facility. Possible necessary deviations from climatic constraints are reviewed and reconciled with conclusions drawn from the exposure pathway analysis. Details of log-normal distribution analysis techniques are presented, with examples of environmental surveillance data to illustrate data manipulation techniques and interpretation procedures as they affect the investigatory environmental surveillance programme. Appropriate consideration is given these analytical methods to ensure that when contamination is released to the environment, due either to the planned disposal of wastes or to unplanned releases, siting selection will have been made to minimize the environmental consequences.

## 1. INTRODUCTION

Most installations which handle radioactive materials are designed and operated to contain any associated radiation and radioactive materials. Nevertheless, some release of radioactivity to the human environment may occur, due either to the disposal of low-activity wastes or to unplanned releases. Releases which do occur may end up in many different sectors of the environment, and the amount and composition of the releases will vary from installation to installation (even between installations of the same general type) and from time to time. Thus, in siting any facility, the potential environmental impact of plant emissions must be considered.

---

* Present address: Radiation Management Corporation, Philadelphia, Pa., United States of America.

In most situations where radioactive materials are introduced to the human environment, there are numerous and complex pathways by which each of the released nuclides may ultimately cause radiation exposure of man.  Experience has shown, however, that certain nuclides and exposure pathways are much more important than others, and hence these are designated as 'critical'. To evaluate critical exposure pathways, it is necessary to have:

1.   An estimate of expected releases from the plant.
2.   The location and age distribution of persons potentially exposed.
3.   Dietary habits, e.g. special foodstuffs and amounts consumed.
4.   Special occupational habits, e.g. the handling of fishing gear.
5.   Hobbies, e.g. sports, fishing  or sunbathing.

Key objectives in siting policies are:

1.   To maximize the economic advantage of the plant.
2.   To minimize the environmental consequences.

Secondary objectives of nuclear facility siting might be to:

1.   Reduce transport of released materials.
2.   Reduce the potentially exposed population.
3.   Eliminate or greatly reduce the plant effluents.

One can optimize these factors and determine their relative efficiencies to achieve the desired goal.  Balancing of environmental consequences and economics, which are often in direct conflict, should be addressed at the siting stage.  Our plan is to evaluate the advantages and disadvantages of potential sites for nuclear facilities through the use of environmental pathway and log-normal distribution analyses.

## 2.   APPLICATION OF PATHWAY ANALYSIS

Environmental considerations of nuclear facility siting are necessarily geared to the identification of media believed to be significant in terms of dose to man or to be potential centers for long-term accumulation of contaminants.  To aid in meeting the scope and the purpose of this identification, an exposure pathway diagram must be developed.  This type of diagram helps to locate pertinent environmental media, points of expected long-term contaminant accumulation and points of population/contaminant interface for both radioactive and non-radioactive contaminants.

The chart (Figure 1) was constructed using available data on the transport of plutonium through the environment.  The same pathways are involved where uranium is the contaminant; even though the values are not precisely applicable to uranium transport, they are sufficiently similar that the conclusions drawn are the same.  Pathways of important non-radioactive materials are included in the gas centrifuge plant-siting sections which follow.

Media appearing inside the diamond shapes on the figure are those commonly sampled in environmental surveillance programs.  Individual discrimination factors (IDF), defined as the recipient-to-donor concentration ratio or as the fraction of incident contaminant retained by recipients, are provided for each medium-to-medium transfer link.  Identification of the predominant exposure pathways was made by propagating the IDF's into combined discrimination factors (CDF) for each recognized potential transport route from effluent release to human exposure.  A CDF is defined as the product of all of the IDF's in an exposure pathway.

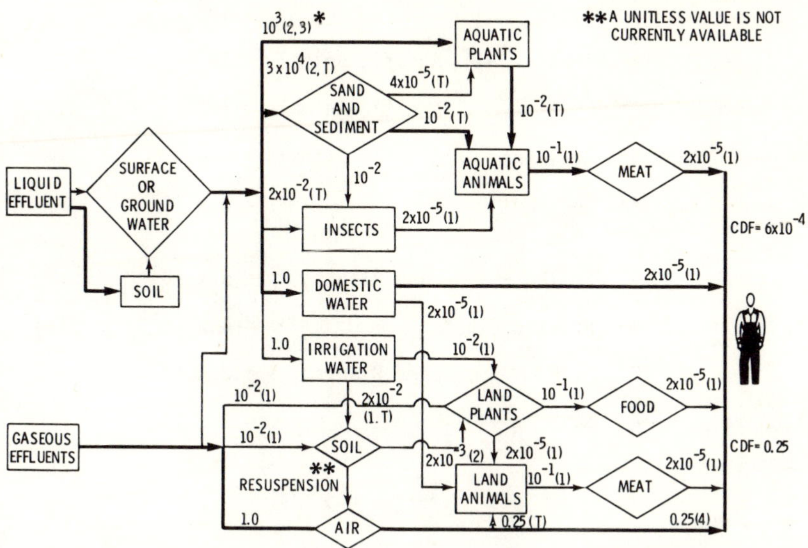



FIG.1. Exposure pathways and discrimination factors for plutonium in the biosphere.

Since identification of critical exposure pathways is the major objective of constructing pathway diagrams for radioactive or non-radioactive contaminants, only in special cases is it necessary to have all possible combined discrimination factors. Either of two accepted methods discussed here will generally be sufficient to generate IDF's necessary to eliminate less important pathways from the diagram. One method uses similarity assumptions; the other is based on log-normal distribution analysis.

Discrimination factors in Figure 1 which were not available from the literature were inferred by comparison as shown in Table I. Based on work by Langham [1], an IDF of $2 \times 10^{-5}$ was assigned to most animal ingestion processes, $10^{-2}$ for surface contamination mechanisms, and $10^{-1}$ for food processing. Other inferences indicated in Table I are based on work by Soldat [2] and Noshkin [3]. Exposure routes with CDF's less than $10^{-15}$ were considered insignificant and are not shown on the diagram. Four transport routes have CDF's greater than $10^{-5}$; these are indicated by heavy lines on the figure. The CDF's representing the most efficient transfer from source to man for both liquid and gaseous effluent exposure pathways are indicated at the right of the figure.

The relative importance of each potential critical pathway can be assessed by weighting the combined discrimination factor for each route with:

1.  The amounts of credible contaminants available to that pathway via expected environmental release of process materials.

2.  The extent to which human populations are exposed to each route.

TABLE I.   UNPUBLISHED DISCRIMINATION FACTORS DERIVATION

| | | | |
|---|---|---|---|
| Water → | Aquatic Plants → | Aquatic Animals = 10 | [2] |
| Water → | Aquatic Plants | = $10^3$ | [3] |
| | Aquatic Plants → | Aquatic Animals = $10^{-2}$ | |
| Sand and Sediment | → | Aquatic Animals = $10^{-2}$ | (a) |
| Water → | Sand and Sediment → | Aquatic Animals = 300 | [2] |
| Water → | Sand and Sediment | = $3 \times 10^4$ | |
| Irrigation Water → | Soil → | Land Plants = $4 \times 10^{-5}$ | [1] |
| Sand and Sediment → | Aquatic Plants | = $4 \times 10^{-5}$ | (a) |
| Soil → | | Land Plants = $2 \times 10^{-3}$ | [2] |
| Irrigation Water → | Soil | = $2 \times 10^{-2}$ | |
| Soil → Air | | = $10^{-2}$ | (b) |
| Air → Man | | = 0.25 | [4] |
| Soil → Air → Man | | = $0.25 \times 10^{-5}$ | |
| Soil → Air → Land Animals | | = $0.25 \times 10^{-5}$ | |
| Air → Land Animals | | = 0.25 | |

[a]Similarity of irrigation water → soil → plant and sand-and-sediment → plant
is assumed.
[b]Assuming uniform distribution 1 cm depth of soil.


By this process, the critical pathway of major importance for any released
contaminant can be clearly identified.

In simplest terms, selection of a nuclear facility site on a pathway evaluation basis  involves the comparison of hypothetical radiological or nonradiological doses calculated for the most critical pathways at each site under
consideration.  The site with the lowest calculated dose potential to individuals and/or to the population is therefore the optimum site from a pathway s
point.


3.  APPLICATION OF LOG-NORMAL ANALYSIS

Confirmation of facility siting conclusions drawn from pathway considerations must usually be derived from an investigatory environmental surveillanc
program.  Our experience with environmental surveillance data interpretation
using log-normal techniques indicates that this distribution has much to offe
in the planning, execution and analysis phases of such a program.

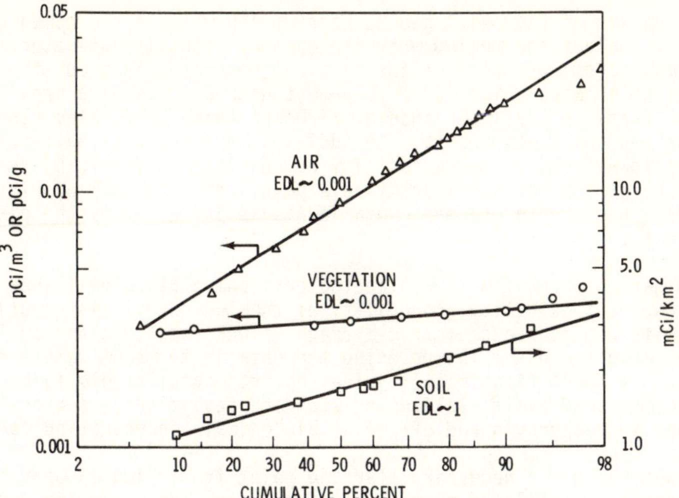


FIG.2.  Log-normal distributions of environmental concentrations of plutonium.  EDL = environmental detection
level.

To plan and conduct the preoperational environmental surveillance program
on a cost-effective basis, it is logical that media with the greatest contaminant-detection sensitivity and the most prominent role in the critical pathway
should be the media routinely analysed.  Evaluation of the pathways diagram
and the curves in Figure 2 or similar curves obtained at the site permit proper
selections to be intelligently made.

It has been demonstrated [5] that numbers of nuclide/medium combinations
exhibit log-normally distributed environmental concentrations when these are
plotted for a location over time or for one time over many locations.  In
fact, it has been said that when the applicable distribution is unknown, a
log-normal is the first alternative to try [6]. Figure 2 shows log-probability
plots of plutonium background concentrations in various environmental media.
Characteristics of special importance in the use of these curves are (1) their
linearity when plotted on log-probability paper, linearity denoting the presence
of a single distribution, and (2) their slope (or geometric standard deviation)
of between 1 and 3 which denotes a  weathered  distribution.   Comparison
of contaminant-detection sensitivities between media is a matter of identifying
the medium which exhibits the highest concentration distribution when exposed
to the background level of a contaminant in the biosphere.  Even though the
slope of the background concentration curves must be determined experimentally,
for many sets of media concentrations it has been found to have values near
2.  Greater slopes or nonlinearity usually denote the presence of a source-
related concentration distribution.

The log-normal method of determining needed IDF's, referred to earlier,
can be executed graphically from Figure 2.  It is necessary first to ensure
that distributions for all media are plotted on equivalent scales which generally
takes some conversion of units.  Once this comparability of data is achieved,
the intersection of each distribution with the 50 cumulative percent line yields
the relative IDF from source to that medium.

Examination of Figures 1 and 2, considering established human consumption
factors, to select the medium with the greatest contaminant-detection sensitivi
and the most prominent role in the critical pathway points out the importance
of air in this case, which is in agreement with results of a previous study [7]
In most situations the importance of airborne pathways is also consistently
shown for gaseous contaminants.  In addition, the media to be sampled should
include pathways media at the facility boundaries, at points of suspected
buildup and at the nearest identifiable points of potential population or
individual exposure to process contaminants as indicated by the diamond shapes
in Figure 1.

Examination of previously collected environmental surveillance data has
also shown the feasibility and utility of the log-normal distribution for
quoting with confidence geometric average values less than the analytical
detection levels and for interpreting the data in terms of sample representa-
tiveness, the identification of typical environmental levels (geometric means)
and expected upper limits (slopes or standard deviations), seasonal or yearly
variations in background and effective biological concentration factors.

Besides being the necessary starting point from which an operational
environmental surveillance program is formulated, the following details can
be studied during the preoperational phase:

1.   Background levels.
2.   Random statistical variabilities.
3.   Potentially exposed populations.
4.   Unusual exposure pathways.
5.   Pertinence of various environmental media.
6.   Operational surveillance program cost/benefit evaluation.

Once facility operation has commenced, obtaining some of these details with an
type of approach becomes impossible.


## 4.   SITING A CENTRIFUGE-ENRICHMENT PLANT

How these basic principles apply to the actual siting of a nuclear facilit
is  demonstrated for a centrifuge-type uranium enrichment facility.  Although
most detailed descriptions of current gas centrifuge technology are classifie
discussions of ultracentrifuge enrichment plant design are readily available [8, 12].

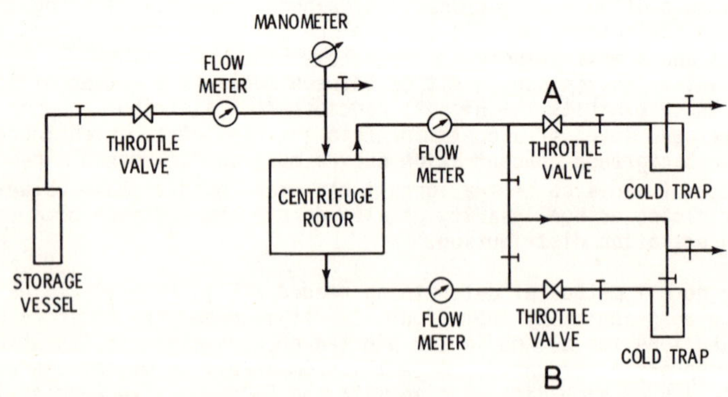


FIG.3.  Gas centrifuge plant flow diagram.

A simplified schematic of the centrifuge plant section which is unique to this type of enrichment facility is shown in Figure 3. Uranium hexafluoride is stored under high pressure in the storage vessel [10]. The enriched and stripped fractions, removed at A and B and in experimental apparatus, are condensed in traps [9]. In operating plants these product streams would be compressed into a liquid and the enriched fraction shipped to a fuel fabrication facility.

The signature of a tripartite agreement[11]in 1970 has significantly accelerated the schedule on which gas centrifuge plants are to be brought into the uranium fuel cycle. The plants now under construction are principally regarded as demonstration plants aimed at obtaining detailed information on centrifuge mass-production procedures and gaining plant operating experience on a plant scale. The product concentration will be variable between 2.5 and 4 percent $^{235}U$.

Since uranium hexafluoride is both the feedstock and the product of these enrichment plants, an examination of plant siting consideration warrants a discussion of $UF_6$'s chemical and physical properties [13].

Uranium hexafluoride is a highly reactive substance. It reacts readily with water, forming soluble reaction products $UO_2F_2$ and HF. It reacts with most organic compounds and with many metals. It does not react with oxygen, nitrogen or dry air.

Gaseous $UF_6$ when released to the atmosphere reacts with the atmospheric moisture to form HF gas and particulate $UO_2F_2$ which tends to settle on surfaces. The corrosive properties of $UF_6$ and HF are such that exposure can result in skin burns and temporary lung impairment. Water-soluble uranium compounds such as $UO_2F_2$, like most heavy-metal compounds, are toxic to the kidneys when inhaled or ingested in large quantities. For uranium, of less than 10 percent $^{235}U$ enrichment, the chemical toxicity is more important than the radio-toxicity.

Flora are frequently categorized in three classes as to sensitivity to atmospheric fluoride or tissue concentrations: susceptible, intermediate, and tolerant species. In a review [14] of current information on the concentrations of gaseous fluoride in air causing foliar markings, threshold levels were described as: (1) 3-4 $\mu g/m^3$ for the most susceptible species and 10 $\mu g/m^3$ or higher for species of intermediate susceptibility, for exposures of one day; and (2) for exposure periods longer than a month, about 0.5 $\mu g/m^3$ for susceptible and between 1-3 $\mu g/m^3$ for some intermediate species. Other considerations, especially that of effects of fluorides on animals via feed crops such as alfalfa, should be noted during the site evaluation process.

An examination of environmental pathway and concentration data in Figures 1 and 2 have revealed the general importance of the airborne pathway when uranium or plutonium are involved. Now a discussion of atmospheric properties of credible contaminants from a centrifuge enrichment plant re-emphasizes this point in this specific case. Since air is a medium common to any potential facility site, the determination of optimum site characteristics must resort to the effects of airborne contaminants.

As discussed earlier, offsite hazards of airborne contaminants are greater for HF than from uranium and greater to plants than to animals. Therefore, siting concerns should center on the presence or absence of environmental pathways which include plants in the susceptible category and should not overemphasize the accumulation of data on population fishing habits, milk consumption, protein sources or related matters.

Pollutant delivery to susceptible plants would probably be enhanced in climates with large annual rainfalls, because HF washout at source-deposition distances would be much smaller than under non-precipitation conditions. The very high reactivity of $UF_6$ with atmospheric moisture would indicate, however, that the quantity of HF available from the $UF_6$ to $HF + UO_2F_2$ reaction would be dependent upon the amount of $UF_6$ available for reaction and not upon relative humidity in the normally encountered relative humidity range. Therefore, rainfall data of prospective sites should not be of prime importance.

It can be seen in Figure 1 that instances might arise where liquid pathways are important if population habits involve the consumption of aquatic plants which concentrate uranium, again emphasizing the importance of pathway analysis in the recognition of potentially exposed populations and potential exposure pathways.

## 5.  SUMMARY AND CONCLUSIONS

The application of exposure pathway and log-normal distribution analysis techniques have been discussed and applied to siting criteria for a centrifuge enrichment facility. The same procedures are applicable for any contaminants, radioactive or non-radioactive, or facility type that one might encounter. Where many contaminants are present in gaseous and liquid effluent streams, the process of constructing exposure pathways and collecting environmental concentration data may be more laborious, but still informative and advisable. Use of existing regional environmental surveillance data and past knowledge of contaminant transport characteristics makes the process of eliminating non-critical contaminants and media relatively simple so that further investigations of critical pathways and contaminants can be emphasized. Thus when contamination is released to the environment, due either to the planned disposal of wastes or to unplanned releases, siting selection will have been made to minimize the environmental consequences.

R E F E R E N C E S


[1] LANGHAM, W.H., "Plutonium distribution as a problem in environmental science", Proc. Environmental Plutonium Symp., Los Alamos Sci. Lab. Rep. LA-4756 (1971).

[2] SOLDAT, J.K., ROBINSON, N.M., BAKER, D.A., Models and Computer Codes for Evaluating Environmental Radiation Doses, Battelle Pacific Northwest Labs Rep. BNWL-1754 (1974).

[3] NOSHKIN, V.E., Ecological aspects of plutonium dissemination in aquatic environments, Health Phys. 2 (1972) 537.

[4] INTERNATIONAL COMMISSION ON RADIOLOGICAL PROTECTION, The Metabolism of Compounds of Plutonium and Other Actinides, ICRP Publication 19, (1972).

[5] WAITE, D.A., DENHAM, D.H., Log-Normal Statistics Applied to the Planning, Collection and Analysis of Preoperational Environmental Surveillance Data, Battelle-Pacific Northwest Labs Rep. BNWL-SA-4840 (1974).

[6] MICHELS, D.E., "Log normal analysis for plutonium in the outdoors", Proc. Environmental Plutonium Symp., Los Alamos Sci. Lab. Rep. LA-4756 (1971).

[7] CORLEY, J.P., WAITE, D.A., Environmental Surveillance for Fuel Fabrication Plants, Battelle-Pacific Northwest Labs Rep. BNWL-1723 (1973).

[8] KISTEMAKER, J., LOS, J., VELDHYZEN, E.J.J., "The enrichment of uranium isotopes with ultra-centrifuges", 2nd Int. Conf. Peaceful Uses Atomic Energy (Proc. Conf. Geneva, 1958) 4, UN, Geneva (1958) 435.

[9] LONGON, L. (Ed.), Separation of Isotopes, Newnes, London (1961).

[10] GROTH, W.E., et al., "Enrichment of the uranium isotopes by the gascentrifuge method", 2nd Int. Conf. Peaceful Uses Atomic Energy (Proc. Conf. Geneva, 1958) 4, UN, Geneva (1958) 439.

[11]  AVERY, D.G., PARRY, J.V.L., BOGAARDT, M., JELINEK-FINK, P., "Centrifuge plants in Europe",
      4th Int. Conf. Peaceful Uses Atomic Energy (Proc. Conf. Geneva, 1971) 9, UN, New York,and IAEA,
      Vienna (1972) 53.
[12]  OLANDER, D.R., Technical Basis of the Gas Centrifuge, Lawrence Livermore Lab. Rep. UCRL-20506 (1971).
[13]  USAEC,OAK RIDGE OPERATIONS OFFICE, Uranium Hexafluoride: Handling Procedures and Container
      Criteria, Rep. ORO-651, Rev.3 (1972).
[14]  NATIONAL RESEARCH COUNCIL, Fluorides, a Report, Committee on Biological Effects of Atmospheric
      Pollutants of the Division of Medical Sciences, Washington, DC (1971).

# CHOIX DE LA METHODE D'EVALUATION DE LA CONTAMINATION RESULTANT DES REJETS ATMOSPHERIQUES D'UNE INSTALLATION EN FONCTIONNEMENT NORMAL, EN FONCTION DE L'ETUDE DES CARACTERISTIQUES DE L'ENVIRONNEMENT

Arlette GARNIER, A. BOUVILLE
Département de protection,
CEA, Centre d'études nucléaires
  de Fontenay-aux-Roses,
Fontenay-aux-Roses, France

Abstract–Résumé

CHOICE OF A METHOD OF EVALUATING CONTAMINATION RESULTING FROM ATMOSPHERIC DISCHARGES FROM A FACILITY DURING NORMAL OPERATION AS A FUNCTION OF ENVIRONMENTAL CHARACTERISTICS.
The contamination resulting from atmospheric discharges from a facility can affect the atmosphere, the soil and agricultural products, and the aquatic environment. Since deposition of atmospheric origin can be deduced directly from the concentration of the elements in the air, the choice of diffusion model is very important. Various theoretical or empirical methods are available for evaluating concentrations resulting from continuous discharges, each offering advantages and disadvantages according to the particular problem. It is desirable for the model to be simple enough to be general and easy to use. At the same time it must be able to allow for the environmental characteristics peculiar to a region or a site. This calls for a more or less detailed study depending on the ultimate aim of the evaluation and the degree of precision required. To determine maximum individual doses, it is sufficient to know the maximum concentrations and their location. On the other hand, the evaluation of collective doses calls for a much more elaborate study and, in the first place, the acquisition of data from which to calculate the mean concentrations in various zones situated in a large radius around the facilities; this radius will itself be determined by the limits of validity of the models.

CHOIX DE LA METHODE D'EVALUATION DE LA CONTAMINATION RESULTANT DES REJETS ATMOSPHE-RIQUES D'UNE INSTALLATION EN FONCTIONNEMENT NORMAL, EN FONCTION DE L'ETUDE DES CARACTERISTIQUES DE L'ENVIRONNEMENT.
La contamination résultant des rejets dans l'air émanant d'une installation peut concerner l'atmosphère, le sol et les produits agricoles, et les eaux. Les dépôts d'origine atmosphérique pouvant être déduits directe-ment de la concentration des éléments dans l'air, le choix du modèle de diffusion est essentiel. On dispose, pour l'évaluation des concentrations consécutives à des rejets continus, de différentes méthodes, théoriques ou empiriques, qui, selon le problème étudié, présentent chacune des avantages et des inconvénients. Il est souhaitable que le modèle soit suffisamment simple pour être général et d'un emploi commode. En même temps, il doit permettre de tenir compte des caractéristiques de l'environnement particulières à une région ou à un site. Celles-ci nécessitent une étude plus ou moins approfondie et détaillée selon le but final assigné à l'évaluation et selon le degré de précision souhaité. Pour la détermination de doses maximales individuelles, la connaissance des concentrations maximales et de leur localisation est suffisante. Par contre, l'évaluation des doses collectives nécessite une étude beaucoup plus raffinée, et, tout d'abord des informations permettant de calculer les concentrations moyennes en diverses zones situées dans un large rayon autour des installations; ce rayon sera lui-même déterminé par les limites de validité des modèles.

## INTRODUCTION

La contamination résultant des rejets atmosphériques d'une installation
nucléaire peut concerner l'atmosphère, le sol et les produits agricoles, et,
à un moindre degré, les eaux.  L'évaluation des quantités déposées et
transférées dans les divers maillons des chaînes biologiques reposant sur
celle des concentrations atmosphériques des éléments considérés, le choix
du modèle de diffusion est essentiel.  Or, il dépend non seulement des
caractéristiques du site influençant la diffusion, mais aussi du but que l'on
se propose, selon qu'il s'agit d'évaluer une limite supérieure de la dose
susceptible d'être délivrée à l'individu du groupe le plus exposé, ou bien de
la dose collective susceptible de concerner un groupe plus ou moins large
de population, ce qui nécessite des études plus raffinées,  et en premier
lieu, la détermination des concentrations moyennes en diverses zones plus
ou moins éloignées des points de rejets.  Le modèle utilisé doit donc
répondre à ces diverses exigences, ce qui conduit à rechercher parmi les
modèles existants celui qui est le mieux adapté aux conditions du cas à
traiter.

## MODES D'EXPRESSION DE LA DIFFUSION ATMOSPHERIQUE
## D'UN POLLUANT

De très nombreux travaux sur la diffusion dans un fluide ont permis
l'élaboration de modèles d'une très grande utilité pratique.

L'hypothèse d'une distribution gaussienne est à la base de différentes
expressions de la diffusion à partir d'une source ponctuelle, dans trois
dimensions [1].  Le calcul des dimensions du panache repose tantôt sur la
théorie de la diffusion turbulente, qui est à l'origine de différentes
expressions selon les auteurs, tantôt sur des résultats expérimentaux
mettant en évidence les valeurs numériques des paramètres.

Les valeurs numériques des paramètres de dispersion diffèrent assez
largement selon les conditions atmosphériques dont Pasquill a établi une
classification bien connue en fonction du degré de stabilité de l'atmosphère:
$\sigma_y$ augmente d'un facteur 5 entre les catégories F et A, à différentes
distances; $\sigma_z$ présente des variations plus importantes:

$$35 \text{ entre F et A à 1 km}$$
$$10 \text{ entre F et C à 10 km}$$
$$27 \text{ entre F et C à 50 km.}$$

Il en résulte, pour les concentrations correspondantes, une très
grande variabilité.  A des distances données de la source, les concentratio
maximales normalisées ($\chi \bar{u}/Q$) résultant de rejets brefs sont, en condition
plus élevées que celles en conditions C dans les rapports suivants:

à 10 km $\left\{ \begin{array}{l} \cong 30 \text{ dans l'axe du panache} \\ \cong 25 \text{ au sol, pour une hauteur effective de rejet de 30 m} \\ \cong \phantom{0}3 \text{ au sol, pour une hauteur effective de rejet de 100 m} \end{array} \right.$

à 50 km $\left\{ \begin{array}{l} \cong 100 \text{ dans l'axe du panache} \\ \cong 70 \text{ au sol } (h = 30 \text{ m}) \\ \cong 40 \text{ au sol } (h = 100 \text{ m}). \end{array} \right.$

Compte tenu des vitesses respectives du vent dans les conditions F et C, les rapports des concentrations ($\chi/Q$) pourraient être doubles des valeurs ci-dessus.

D'autres auteurs proposent des valeurs plus ou moins différentes des paramètres de dispersion, dont la détermination est influencée notamment par la durée d'échantillonnage, qui varie de dix minutes (Pasquill) à une heure (Le Quinio [2]), par le niveau auquel sont effectuées les mesures (dans l'axe du panache, ou au niveau du sol), ainsi que par les différences entre les conditions théoriques et les conditions réelles, telles que la rugosité des surfaces (sans parler du relief dont, par exemple, l'influence se traduit à Ispra par un facteur 10 sur les concentrations moyennes en certaines zones [3]). L'incertitude sur $\sigma_z$, qui est d'un facteur 2 pour les courtes distances, peut être de plusieurs unités pour des distances plus grandes.

C'est pourquoi Fletcher et al. [4] évaluent $\sigma_z$ par trois méthodes différentes: Pour des distances inférieures à 2 km, les valeurs de Pasquill sont retenues. Pour des distances variant de 2 km à environ 70 km, $\sigma_z$ est calculé selon la relation $\sigma_z = 1000\,a\,x^b$, dans laquelle x est exprimé en kilomètres, et où a et b varient respectivement de 0,45 à 0,015 et de 2,1 à 0,45 lorsque la stabilité va croissant. Enfin, pour des distances supérieures, $\sigma_z$ est pris égal à l'altitude de la couche limite.

Doury [5] a montré, dans une étude de synthèse, que l'on pouvait obtenir un bon ajustement de l'ensemble des résultats d'observations et d'expériences, en laboratoire ou in situ, en exprimant les paramètres en fonction du temps de transport, ce qui élimine les erreurs inhérentes à l'estimation des vitesses du vent: $\sigma_x$ et $\sigma_y$ vérifient la relation générale

$$\sigma_i = \left( A_i t \right)^{k_i}$$

où $A_i$ et $k_i$ prennent, pour une condition atmosphérique donnée, des valeurs numériques différentes par intervalles de temps successifs (tableau I).

Pour l'évaluation des concentrations, l'équation générale de diffusion basée sur l'hypothèse d'un modèle gaussien permet la résolution de nombreux cas particuliers:
— selon le niveau de la source par rapport au sol et les coordonnées du récepteur
— selon le caractère de l'émission (instantanée ou continue)
— selon le type d'évaluation recherchée: concentration maximale, concentration moyenne dans un espace donné ou un temps donné, etc.

La plupart des auteurs utilisent une série d'abaques donnant, pour chacune des six conditions de Pasquill, et en fonction de la distance à la source, la concentration atmosphérique rapportée à un rejet unitaire ($\chi/Q$) ou la même concentration normalisée par la vitesse du vent ($\chi\bar{u}/Q$), dans l'axe du panache, soit au niveau de la source, soit au niveau du sol pour différentes hauteurs effectives du point d'émission. De ces concentrations maximales on peut déduire les concentrations moyennes dans un secteur, consécutives à des rejets brefs ou continus, en utilisant les paramètres $\sigma_y$ et $\sigma_z$ appropriés à chacune des conditions.

L'application d'une méthode aussi complexe à un site particulier n'est évidemment possible que si l'on dispose d'informations suffisantes pour déterminer avec une bonne précision tous les paramètres numériques en

TABLEAU I.  ECARTS-TYPES DE LA DIFFUSION EN FONCTION DU TEMPS DE TRANSFERT, POUR UNE SOURCE PONCTUELLE
Relation pratique proposée par Doury [5]

$\sigma_i = (A_i t)^{k_i}$          $k_i$: sans dimension

$i = x, y, z$          $A_i$: pseudo-coefficient de diffusion ($L^{1/k_i} T^{-1}$)

Conditions «normales»

Diffusion horizontale (t en secondes)          $\sigma_x = \sigma_y = \sigma_h$

| | |
|---|---|
| 0 à 4 min | $\sigma_h = (0,405\ t)^{0,859}$ |
| 4 min à 1 j | $\sigma_h = (0,135\ t)^{1,130}$ |
| 1 j à 4 j | $\sigma_h = (0,463\ t)^{1,000}$ |
| 4 j à 15 j | $\sigma_h = (6,5\ t)^{0,824}$ |
| > 15 j | $\sigma_h = (2 \cdot 10^5\ t)^{0,5}$ |

Diffusion verticale

| | |
|---|---|
| 0 à 4 min | $\sigma_z = (0,42\ t)^{0,814}$ |
| 4 min à 40 min | $\sigma_z = (t)^{0,685}$ |
| > 40 min | $\sigma_z = (20\ t)^{0,5}$ |

Conditions «mauvaises»

Diffusion horizontale (t en secondes)

$\sigma_h$ (voir ci-dessus)

Diffusion verticale

$\sigma_z = (A_z t)^{k_z} = (0,2\ t)^{0,5}$

cause, ce qui ne peut se faire que dans un périmètre assez restreint. D'autre part, elle n'est justifiée que si les résultats obtenus varient d'un facteur appréciable d'un site à un autre.  A cet égard, les catégories de stabilité d'une part, et la distribution des vents d'autre part, n'influencent pas au même degré l'évaluation des concentrations résultant de rejets continus.

En effet, les coefficients de dilution pondérés diffèrent assez peu d'un site à un autre, pour une égale répartition des vents, comme le montrent quelques exemples.  En Grande-Bretagne [6], la concentration en un site particulier dans un rayon de 4 km du point d'émission varie de 0,76 à 1,38 fois la concentration moyenne calculée d'après les conditions pondérées sur huit sites d'Angleterre et du pays de Galles.  En France, les nombreuses mesures effectuées ont permis à Le Quinio [2] de proposer, pour des sites ne présentant pas de particularités orographiques, deux types de conditions de diffusion, «normales» et «mauvaises», ces dernières étant applicables notamment la nuit ou par temps de neige:  les résultats sont en effet moins

variables d'un site à l'autre qu'entre le jour et la nuit.  Comparons enfin
les coefficients de dilution moyens à des distances d'au moins 10 km pour
des régions très différentes, Grande-Bretagne et golfe de Tarante, où les
fréquences moyennes des catégories de Pasquill sont les suivantes:

|                        | A   | B    | C    | D    | E    | F    |
|------------------------|-----|------|------|------|------|------|
| Grande-Bretagne [6]    | 1,7 | 8,4  | 16,8 | 41,0 | 11,8 | 20,3 |
| Site de la Trisaia [7] | 5,1 | 19,4 | 20,2 | 37   | 14,8 | 3,5  |

Pour des cercles de 10 et 50 km de rayon, en supposant une égale
répartition du vent dans toutes les directions, les coefficients moyens de
dilution que l'on obtiendrait à partir d'une même série de coefficients
selon les catégories A à F et pour des hauteurs de rejet identiques
différeraient seulement d'un facteur 2.  Par contre, la fréquence des vents
dans un secteur déterminé variant de 0,4 à 28% sur le site de la Trisaia,
le coefficient de dilution pourrait varier d'un facteur 70 entre le minimum
et le maximum.  En fait, un calcul très détaillé, associant en chaque
secteur la fréquence et la vitesse des vents à chacune des catégories de
Pasquill, réduit ce facteur de variabilité à 50 à la distance de 10 km.  Entre
les concentrations obtenues par les deux méthodes (application des coefficients selon Bryant d'une part, ou calcul détaillé d'autre part), on trouve
un facteur au plus égal à 5 en un secteur donné, et voisin de 2 pour la
valeur moyenne.  Cette disparité, sans doute partiellement imputable aux
différences d'estimation de $\sigma_z$, est faible en regard de la variabilité des
données initiales.  Les valeurs moyennes et extrêmes apparaissent au
tableau II, où elles peuvent être comparées à des résultats calculés pour
les sites anglais et pour un site belge [8].
    Il résulte d'un tel examen que les données essentielles sont celles
concernant la fréquence, la vitesse et la persistance des vents, indispensables
à la fois au calcul des concentrations et à la prévision des trajectoires à
longue distance.  Par contre, la distinction de nombreuses catégories de
diffusion, qui peut être utile à la détermination des zones critiques dans
les environs immédiats d'une installation et à la protection de groupes
particuliers de population (bien qu'on puisse même dans ce cas lui préférer
une distinction en deux catégories, ou encore une présentation probabiliste
[9]) ne se justifie par au-delà d'une certaine distance: en cas de rejets
brefs, elle n'est plus valable en raison du manque de persistance d'une
condition atmosphérique donnée; en cas de rejets continus, elle apporte,
en outre, peu de variabilité sur le coefficient de dilution pondéré.  Pour
l'évaluation des concentrations à longue distance, il est donc préférable
d'appliquer un modèle général, d'une expression simple, mais qui intègre
la diversité des conditions possibles.  L'expression des paramètres de
dispersion en fonction du temps de transport, citée plus haut, apporte une
solution à ce problème, et a permis à Doury [5] de proposer des courbes
donnant, en fonction de la distance curviligne depuis la source, la concentration maximale dans un panache provenant d'une émission ponctuelle
continue, en cas de diffusion normale ou de diffusion faible.
    Il est intéressant de comparer ces estimations à celles de Cagnetti et
Pagliari ([10] fig. 6), qui proposent un modèle valable pour un intervalle
de quelques dizaines à quelques centaines d'heures, avec deux types de
conditions initiales, stable et neutre, et deux vitesses de vent (2 et 7 m/s).

TABLEAU II.   EXEMPLES DE VALEURS DU COEFFICIENT DE DILUTION $\chi/Q$

| Distance parcourue x (m) | Hauteur effective h (m) | Rejets brefs [6][a] | | Rejets continus | | | |
| --- | --- | --- | --- | --- | --- | --- | --- |
| | | $F$ $\bar{u}=2$ m/s | $C$ $\bar{u}=4$ m/s | Grande-Bretagne [6][a] | Mol [8][b] | Trisaia | |
| | | | | | | Calcul approché (voir texte) | Calcul détaillé [7][c] |
| $10^4$ | 0 | $1,2\cdot10^{-5}$ | $2\cdot10^{-7}$ | $4\cdot10^{-8}$ | - | - | - |
| | 30 | $10^{-5}$ | $2\cdot10^{-7}$ | $3,5\cdot10^{-8}$ | - | - | - |
| | $\simeq 60$ | $5,6\cdot10^{-6}$ | $2\cdot10^{-7}$ | $2,2\cdot10^{-8}$ | $\leq 2,2\cdot10^{-8}$ | $1,3\cdot10^{-8}$d  de $8,3\cdot10^{-10}$ à $5,6\cdot10^{-8}$ selon les secteurs* | $5\cdot10^{-9}$d  de $1,8\cdot10^{-10}$ à $1,1\cdot10^{-8}$ selon les secteurs* |
| | 100 | $1,3\cdot10^{-6}$ | $2\cdot10^{-7}$ | $1,2\cdot10^{-8}$ | - | - | - |
| $5\cdot10^4$ | 0 | $1,8\cdot10^{-6}$ | $10^{-8}$ | $4,5\cdot10^{-9}$ | - | - | - |
| | 30 | $1,5\cdot10^{-6}$ | $10^{-8}$ | $4\cdot10^{-9}$ | - | - | - |
| | $\simeq 60$ | $1,3\cdot10^{-6}$ | $10^{-8}$ | $3,4\cdot10^{-9}$ | $\leq 1,3\cdot10^{-9}$ | $1,5\cdot10^{-9}$  de $9,6\cdot10^{-11}$ à $6,7\cdot10^{-9}$ selon les secteurs* | $8,5\cdot10^{-11}$ et $2,2\cdot10^{-9}$ pour les secteurs extrêmes |
| | 100 | $8\cdot10^{-7}$ | $10^{-8}$ | $2,5\cdot10^{-9}$ | - | - | - |
| | | | | égale répartition de fréquence du vent | | * fréquences extrêmes du vent dans ces secteurs: 0,4% et 28% | |

[a] Réf.[6], fig.3 et 6 (rejets brefs) et fig.9 (rejets continus).
[b] Réf.[8], tableau 3,3; l'étude très détaillée du site de Mol a donné lieu à la définition de sept catégories de dispersion un peu différentes de celles de Pasquill.
[c] Réf.[7], fig.1 à 8 et tableau I.
[d] La différence entre les valeurs moyennes peut être imputable en partie à l'estimation de la dispersion verticale.

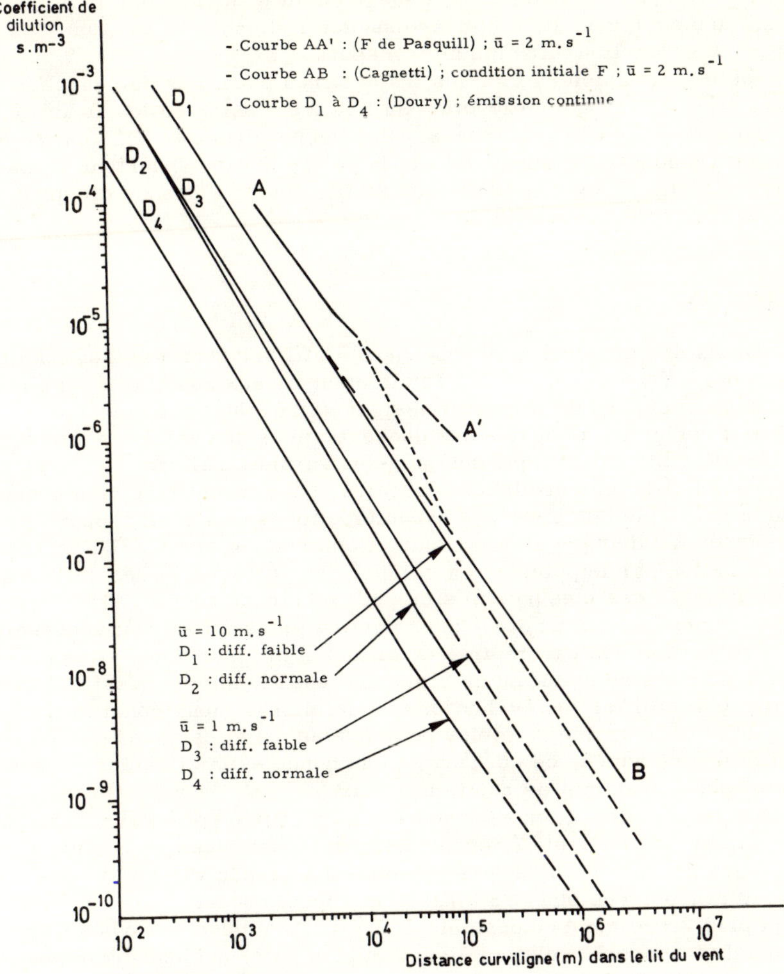


FIG.1. Comparaison de modèles de diffusion à longue distance.

Sur la figure 1 sont reproduites les courbes A et B dont l'ensemble est
proposé par Cagnetti et Pagliari pour fournir une limite supérieure du
coefficient de diffusion dans l'intervalle de 10 à 100 km, en vue de l'évaluation
de la dose consécutive à un rejet accidentel. Les valeurs ainsi obtenues
sont en effet légèrement supérieures à celles proposées par Doury dans
le cas le plus défavorable (diffusion faible, $\bar{u}$ = 10 m/s), et seraient trop
élevées pour l'évaluation des concentrations maximales résultant de rejets
continus pour laquelle il faut tenir compte de la répartition dans le temps
de conditions plus ou moins favorables.

Le modèle de Doury permet donc l'évaluation des concentrations
atmosphériques résultant de rejets continus, à longue distance, dans l'axe
du panache. Les valeurs présentées doivent évidemment être corrigées par

le coefficient d'appauvrissement du nuage dû au lessivage par la pluie, au
dépôt sur les surfaces, ou à la décroissance radioactive, qu'il est commode
d'exprimer en fonction du temps de transport.

Il reste enfin à déterminer les trajectoires possibles du panache, en
évaluant leur probabilité en fonction du lieu d'émission et des conditions
climatiques des régions traversées.  Une recherche de ce type est en cours
en Europe Occidentale pour l'étude de la pollution atmosphérique trans-
frontière (OCDE, Programme d'étude du transport, à grande distance, des
polluants atmosphériques).


CONCLUSION

Le choix des modèles mathématiques à utiliser pour l'évaluation de la
diffusion des effluents gazeux est étroitement lié aux possibilités d'obtention
des données et dépend de l'objectif assigné aux évaluations.

Pour l'évaluation détaillée des concentrations aux environs immédiats
d'une installation, on applique des modèles raffinés utilisant au mieux
toutes les données météorologiques locales, qui peuvent être bien connues;
une étude préalable complète des caractéristiques de l'environnement perme
de resserrer les marges d'incertitude et contribue à favoriser l'acceptation
de l'installation par le public.  La validité des résultats et des modèles doit
être vérifiée par des mesures in situ.  L'expérience acquise grâce à ces
mesures démontre que, si des études particulières demeurent nécessaires
à la détermination des paramètres de diffusion propres à des sites présenta
des particularités de relief ou de climat, il est néanmoins possible, bien
souvent, de simplifier les évaluations en définissant deux conditions moyenn
(favorable et défavorable) valables pour les autres sites, dans des régions
plus ou moins étendues, ce qui permet une bonne estimation des concentrations moyennes résultant de rejets continus.

Une telle simplification est encore plus justifiée pour l'évaluation des
concentrations de polluants à longue distance, pour laquelle l'extrapolation
des modèles précédents n'est pas valable.  La grande variabilité des
situations rencontrées oblige à utiliser un modèle général basé sur l'ajuste
ment du plus grand nombre possible de résultats d'observations en fonction
du temps de transport, celui-ci étant compris entre quelques dizaines et
quelques centaines d'heures.  Dans ce cas, le principal problème est celui
de la prévision des parcours possibles et de leur probabilité d'occurrence,
l'étude de l'environnement devrait avoir pour but de réunir en premier lieu
les données climatologiques en de vastes régions.


## REFERENCES


[1] SLADE, D.H. (Ed.), Meteorology and Atomic Energy, TID-24190 (mai 1968).
[2] LE QUINIO, R., Evaluation de la diffusion d'effluents gazeux en atmosphère libre à partir d'une source
    ponctuelle continue, Abaques et commentaires, Rapport CEA-R-3945 (février 1970).
[3] GAGLIONE, P., GANDINO, C., MARKOVINA, A., Valutazione dei rischi alla popolazione da
    scarichi radioattivi nell'atmosfera del CCR di Ispra, G. Fis. Sanit. Prot. Radiaz. 13 3 (1969) 208.
[4] FLETCHER, J.F., DOTSON, W.L., PETERSON, D.E., BETSON, R.P., «Modelling the regional transport
    of radionuclides in a major United States river basin», Environmental Behaviour of Radionuclides
    Released in the Nuclear Industry (C.R. Coll. Aix-en-Provence, 1973) AIEA, Vienne (1973) 449.

[5]  DOURY, A., Une méthode de calcul pratique et générale pour la prévision numérique des pollutions
     véhiculées par l'atmosphère, Rapport CEA-R-4280 (février 1972).
[6]  BRYANT, P.M., Method of Estimation of the Dispersion of Windborne Material and Data to Assist
     in their Application, UKAEA, AHSB (RP) R 42 (mai 1964).
[7]  CASSANO, G., FRASCHETTI, G., SICILIANO, F., Calcolo della diffusione atmosferica con il duplice
     peso delle frequenze della condizione di stabilità e degli elementi della matrice climatologica di
     effluenti aeroformi per il sito del CRN della Trisaia, G. Fis. Sanit. Prot. Radiaz. 17 3-4 (1973) 80-102.
[8]  BULTYNCK, R., MALET, L., SHARMA, L.N., VAN DER PARREN, J., Atmospheric Dilution Factors
     and Calculation of Doses in the Environment of SCK/CEN Mol for Short and Long Duration Stack
     Discharges, BLG-446 (1970).
[9]  LE QUINIO, R., Concentrations sur une heure de polluants dues à des émissions ponctuelles près du
     sol, Présentation probabiliste, Atmos. Environ. 7 (1973) 423-28.
[10] GAGNETTI, P., PAGLIARI, M., Il trasporto e la diffusione di effluenti gasossi su grandi distanze;
     considerazione e calcoli per un'analisi di protezione sanitaria, CNEN, RT/PROT (72)32 (1972).

## DISCUSSION

A. DOURY: I must first thank you for the interest you have shown in
my work.  You mentioned in particular the applications to continuous releases
at great distances.  However, the procedure referred to is derived from a
method of treatment of instantaneous emissions at short and long distances, any
other emissions being then treated as sequences of instantaneous emissions.
It has in addition the advantage of being highly flexible, so that it can be
adapted easily to all types of problems (stacks, explosions, etc.).  Finally,
this procedure, in which the dimensional parameters are a function of
transfer time and not of distance, combines a minimum of physical rigidity
with great ease of numerical treatment.

Arlette GARNIER: The general nature of the model naturally did not
escape me; that is one of its great merits.  I made particular mention of
the applications to continuous releases, because these were related to my
own problems.

E. IANSITI:  From the graph in Fig. 1 it appears that the diffusion
coefficient obtained by the Doury method would be very different from the
one given by the Pasquill method.  Could you please explain this difference?

Arlette GARNIER: The differences between the dilution coefficients
under Pasquill F conditions and under average diffusion conditions (qualified
as 'normal' and 'low') are due mainly to: (a) the roughness of actual
surfaces (since the Pasquill conditions postulate perfectly smooth surfaces);
and (b) the limited persistence of a particular meteorological condition —
the duration is not greater than a few hours under F conditions, and this
does now permit extrapolation of the curve.

P. CAGNETTI:  In connection with Mr. Iansiti's question I should like
to make a few comments.  In Fig. 1 of the paper, the curve AA' gives the
concentration values suggested by Pasquill for highly stable conditions.
It is well known that the estimates of dispersion coefficients used by
Pasquill rely on a set of data obtained from experiments on diffusion over
short distances.  Pasquill himself has never claimed otherwise.  In safety
analyses, however, the general tendency so far has been to extrapolate
these estimates to distances of the order of hundreds of km.

The concentration values suggested by Mr. Pagliari and me are based
on a simple model, described by Heffter, which uses all the horizontal
diffusion values obtained over long distances by different authors.  The

resulting curve (the second part of the curve AB in Fig.1) reflects an accidental release of short duration. I think the fact that these values for a distance of 100 km are lower by an order of magnitude than arbitrary extrapolations of the Pasquill curve does not justify any particular misgivings

A. DOURY: I should here point out that, considering our own experience and the probabilities we have found, our attitude is to regard Pasquill's category F as exceptionally severe and demanding.

Pamela M. BRYANT: I agree with Mme Garnier's conclusion. Although the calculated results vary according to the values selected for the parameters, this variation is small when compared with other variables in the calculation as a whole. As regards the estimation of dispersion at longer distances, it is worth mentioning that in the United Kingdom we have extended the Pasquill model, after discussions with him, to a distance of 1000 km. By this means and by means of a simple world-wide model, we have been able to calculate the effects of $^{85}$Kr discharges.

Christina GYLLANDER: A model for long-time releases based on trajectory statistics has been developed in Sweden. It is intended for computation of population doses during normal operation.

Arlette GARNIER: I am familiar with the important contribution Sweden has made to these studies, and I had this in mind when I referred to programmes for studying pollution across frontiers.

# IMPACT OF EXTERNAL EVENTS
# ON SITE EVALUATION
## A probabilistic approach

E. JACCARINO, P. GIULIANI, C. ZAFFIRO
Comitato Nazionale per l'Energia Nucleare,
Rome, Italy

### Abstract

IMPACT OF EXTERNAL EVENTS ON SITE EVALUATION: A PROBABILISTIC APPROACH.

A probabilistic method is proposed for definition of the reference external events of nuclear sites.
The external events taken into account are earthquakes, floods and tornadoes. On the basis of the
available historical data for each event it is possible to perform statistical analyses to determine the
probability of occurrence on site of events of given characteristics. For earthquakes, the method of
analysis takes into consideration both the annual frequency of seismic events in Italy and the probabilistic
distribution of areas stricken by each event. For floods, the methods of analysis of hydrological data and
the basic criteria for the determination of design events are discussed and the general lines of the hydraulic
analysis of a nuclear site are shown. For tornadoes, the statistical analysis has been performed for the
events which occurred in Italy during the last 40 years; these events have been classified according to an
empirical intensity scale. The probability of each reference event should be a function of the potential
radiological damage associated with the particular type of plant which must be installed on the site. Thus
the reference event could be chosen such that for the whole of the national territory the risk for safety and
environmental protection is the same.

## 1.   INTRODUCTION

In Italy there is not yet an official set of rules and criteria pertaining
to the protection of nuclear plants from the impact of natural external
events. The common practice so far has been to follow more or less
closely the USAEC regulations, since most important nuclear plants in
Italy were designed in the United States of America. Nevertheless, both
CNEN and other Italian organizations have been trying to define, from a
regulatory point of view, the natural events which must be taken into
account as a reference in the design of a nuclear plant. These events are:

(a)  Earthquakes
(b)  Floods
(c)  Tornadoes

The USAEC methodology is difficult to apply in Italy because, at least
in the case of earthquakes, it has been developed on the basis of observations and measurements made in California. As a matter of fact, it is
not always possible to detect faults, determine their length and activity
and then apply the empirical correlations between the length of faults
and the magnitude of earthquakes associated with them. In Italy tectonics
is not so evident as in California.

In the case of floods, the USAEC methodology is based upon probabilistic study of the precipitation data; these data are always consistent

enough for any statistical analysis.  So far as we know, there is not yet
an American methodology in the field of tornadoes.

In this paper a probabilistic method is proposed for defining the
characteristics of natural events which must be taken as a design basis
for nuclear plants.


## 2.   EARTHQUAKES

The design earthquake for a nuclear plant to be built on a proposed
site ought to be defined following an evaluation of the local and regional
geological and seismological characteristics of the tectonic province,
which includes the site, and of the adjacent provinces.  Furthermore, an
evaluation of the mechanical characteristics of the underlying materials
is indispensable.

It is known that for nuclear power plants this earthquake is the one
which causes the maximum potential vibratory motion on the site.  Should
this earthquake occur, a number of safety-important structures and
components of the plant must remain fully operational.

In Italy, seismic history goes back for two thousand years.  On the
basis of historical information a number of catalogues have been compiled  which are fairly complete, at least for events greater than the
VI[th] degree of the Mercalli scale.  Moreover, with the help of old
chronicles and records it has been possible to draw the isoseismal lines
for most of the known events.

From a statistical point of view, it has been possible to evaluate
for Italy:

(1)  The annual number of seismic events from the VI[th] to the XII[th]
     degree of the Mercalli scale which can occur in the part of the
     national territory deemed seismic (about one-third of the whole
     area);
(2)  The mean value of the epicentral area for each degree and the
     mean value of areas contained by isoseisms of a degree lower
     than the epicentral one;  and
(3)  The frequency of shocks of different intensities in each point of
     the national territory.

On the basis of these analyses it has been possible to establish the
validity of an identical statistical distribution law of epicentral areas of
any degree.  The radius of the average area is about 4 km.  Sufficient
events are available to enable statistical extrapolations to be performed
either by the Gumbel line method (extreme values) or by a method of
asymptotic interpolation of a variation law which is linear in the first
part.

On the basis of the available data and information and the results
of the statistical analyses, it is possible to evaluate the probability of
the reference earthquake.  This probability is of the order of $10^{-6}$ for
nuclear power plants; in fact, on the basis of historical data, the maximum event has a probability of $10^{-3}$ of occurring in any point of the
tectonic province, which includes the site.  Generally speaking, this
province has a surface one hundred times larger than the average area

of the epicentral isoseismal line. The extrapolation of at least a decade,
necessary to take into account that the reference earthquake is the
maximum possible on the site, allows us to reach $10^{-6}$.

For nuclear plants not of the power-generating type, whose associated
risk is lower, the probability associated with the reference event may be
higher than that used for nuclear power plants.

The same logic may be used if the seismic reference event associated
with a tectonic province is associated with a fault or with another tectonic
structure. In this case the causality of the event along the fault or the
tectonic structure may be defined by the ratio between the average length
of the diameter of the epicentral isoseismal line and the length of the
fault. In Italy this ratio is of the order of $10^{-1}$ to $10^{-2}$.


## 3.    FLOODS

The maximum possible flood in a river may be determined either by
a statistical analysis of precipitation data on the river basin or by a
statistical analysis of flow data if they are available for a section of
river close to the site and for a sufficient number of years. The analysis
of flow data may be employed when there are forty or fifty years of
reliable flow measurements for the river. A statistical analysis of these
data may allow the evaluation of a maximum flood flow to which it is
possible to associate a probability of occurrence. By an analysis of the
precipitation data, which are usually available for a substantial number
of years, it is possible to evaluate a maximum precipitation on a given
watershed and its probability of occurrence.

The rain or flood event can be chosen with a return time of one
hundred years, assuming a 2.5% risk that this event is exceeded in the
100-year period. This return period with the 2.5% risk corresponds to
a probability of the order of $10^{-4}$. Having determined a flood with a given
probability, one must then evaluate the maximum level that water can
reach at the site. Many different factors must be taken into account in
performing this evaluation: geological, geometric and hydraulic factors.

It is important to note that, in the vast majority of possible sites for
nuclear plants, the topographical and morphological situation is such
that a very small increase in water level corresponds to a large increase
in flow. It has been evaluated that, given a margin of 1 m over the maximum water level which has been determined for the site, the probability of
water reaching that level is decreased by about two decades. This may
be considered valid in the river valleys where large plants such as
nuclear power plants could be sited. Therefore, an overall flood probability of about $10^{-6}$ could be reached.


## 4.    TORNADOES

The design tornado for nuclear power plants is defined by the probability of occurrence on site and by the engineering parameters associated
with it. These parameters are translational and rotational speeds of the
wind, drop time of the pressure wave and velocity and type of missiles.

For statistical analysis, tornadoes have been classified into six classes (see Table I) on the basis of maximum rotational velocity, while the translational velocity has been assumed of the order of 30-60 km/h (in the United States of America 30-130 km/h).  This velocity corresponds to the average values of tornadoes which have occurred in Europe. Recordings and measurements of wind speed associated with historical tornadoes are not always available; on the other hand, information and data on their effects are plentiful.  Thus, as regards the subdivision of historical tornadoes according to Table I, the various classes have been correlated with the main effects they cause.  These effects go from the small damage caused by the smaller tornadoes — e.g. broken windows, partial destruction of roofs, etc. — to the great damage caused by large tornadoes — e.g. destruction of old buildings and damage to modern structures — and to the damage caused by disastrous tornadoes, e.g. destruction of modern buildings and of reinforced concrete structures; this last tornado is the disastrous American type.

TABLE I.  CLASSIFICATION OF TORNADOES ACCORDING TO MAXIMUM ROTATIONAL WIND SPEED

| Class | Maximum speed |
|-------|---------------|
| I     | $35 \leq V_r \leq 40$ m/s  |
| II    | $40 \leq V_r \leq 45$ m/s  |
| III   | $45 \leq V_r \leq 50$ m/s  |
| IV    | $50 \leq V_r \leq 60$ m/s  |
| V     | $60 \leq V_r \leq 80$ m/s  |
| VI    | $80 \leq V_r \leq 100$ m/s |

A collection of data on tornadoes which occurred during the last forty years has been made in Italy.  During this period the most ruinous event occurred in Venice in 1970; for this event good wind speed measurements are available.  This tornado was considered to be in class V.

Another event of comparable strength was the tornado which occurred in the Treviso-Udine region in 1930.  On the basis of the studies performed, this event is slightly lower than class V.

On the basis of the data so far available, the maximum event which can occur on Italian territory is of class V, which has occurred once in forty years.  Furthermore, it is possible to determine the probability of the event as a function of its class.

The probability of occurrence of a tornado of a given class on a given site is the product of the probability of the event and the probability that the event itself strikes that site.  This latter probability is given by the ratio of the surface swept by the tornado and the surface of the whole region in which the event may occur.

According to work done by the Italian Air Force Meteorological Service, the area swept by a tornado in Italy is of the order of 4 km$^2$. Thus, if for a nuclear power plant in the Po Valley (area ~ 25 000 km$^2$)

the design tornado has been chosen as class V (i.e. the maximum which
occurred in the region), the probability of such an event is of the order
of $10^{-6}$. For a plant located in another region the event with a probability
of the same order of magnitude will be of class V or less; in any case,
the applicability of the foregoing considerations will depend upon the
consistency of the available data. Thus, it may be necessary to extend
the analysis to a period longer than forty years.


5.   CONCLUSIONS

We have described a methodology for evaluating the probability of
occurrence of a reference event: an earthquake, a flood or a tornado.
This probability, for all natural events defined according to the practices
used so far, is of the order of $10^{-6}$. This value has no relation whatsoever
to the risk of nuclear accident which the event itself implies.
Studies on this important aspect of nuclear risk are still very far
from a final answer. Only for seismic events does it appear that a value
of $10^{-6}$ associated with the probability of the reference earthquake is
appropriate; this is based upon certain attempts to evaluate the probability
of the nuclear accident after a reference earthquake. On the other hand,
the financial cost associated with the seismic event is much greater than
that associated with other events; consequently, the definition of the
associated nuclear risk is more important for this event.
The methodology described suggests an important conclusion: when
the value of the probability associated with each natural event has been
decided upon, the reference event can be evaluated on a purely statistical
basis. The probability of the reference event will be the same for all
plants of the same type because it is a function of the nuclear risk
associated with the operation of the type of plant.


## BIBLIOGRAPHY

JACCARINO, E., Possibilità di rilevare un terremoto sul territorio italiano per mezzo di una rete di
accelerografi, CNEN RT/PROT(71)39 (1971).

CARROZZO, M.T., DE VISENTINI, G., GIORGETTI, F., JACCARINO, E., General catalogue of Italian
Earthquakes, CNEN RT/PROT(73)12 (1973).

JACCARINO, E., Probabilità della scossa di IX grado in Italia, CNEN RT/PROT(73)40 (1973).


## DISCUSSION

H.L. STRIEM: You mention that the radius of the average epicentral
area is about 4 km. Did you find a wide divergence from this mean value?
Is the radius a function of earthquake magnitude? At what distance does
the degree of epicentral intensity decrease by one and two degrees?
C. ZAFFIRO: There is a large divergence if we consider earthquakes over the whole territory of Italy. Within a given tectonic province
the divergence is of course not too large. In Italy few measurements of
magnitude are available, so the statistical analysis is based on intensity
data. We have found that the statistical law of distribution of epicentral

areas is independent of intensity (see the first reference cited in our paper).

So far as variation of intensity with distance is concerned, we have many attenuation curves. Some of these have been reported in the third reference cited.

H.L. STRIEM: Did you find quasicircular as well as elongated isoseismic areas? Was the form of the area a constant attribute of the location of the epicentre? If you also had elongated forms, were they associated with specific tectonic structures?

C. ZAFFIRO: Circular isoseismic lines may occur on nearly uniform soils, while elliptical forms are more likely to be found where the epicentres are located on faults. These forms can naturally be affected by particular local situations.

A. DE ACHA ARACAMA: Is there in Italy any official national seismotectonic map showing the seismotectonic provinces? Would you also be good enough to indicate what studies have been carried out on volcanism in relation to nuclear sites?

C. ZAFFIRO: The only official map at present is a geomorphological map prepared by the National Research Committee (CNR).

So far we have not performed any study of volcanism with reference to nuclear installations. We are inclined to exclude sites which might be affected by volcanic phenomena.

P.C. RIZZO: I have had the opportunity to conduct seismic risk analyses in various parts of the world, including Italy, and several other areas where the seismic risk is unquestionably lower than in Italy. Certainly, the results of probability analysis are highly dependent on the assumptions and techniques of analysis. But even with different methods and assumptions, we have never been able to justify values lower than $10^{-4}$, as opposed to your value of $10^{-6}$. Although I do not personally subscribe to the entire Rasmussen report, I note that they have calculated a value of the order of $10^{-4}$ for the probability of occurrence of extreme seismic events. This being so, your figure of $10^{-6}$ is most interesting; but I hope you do not plan to use it in a regulation. If you do, we would quickly find g values in the range of 0.6 to 1.0g, well above the current range for existing plants and those under construction.

C. ZAFFIRO: The differences do perhaps depend on the methods used. It would therefore be useful to compare methods and try to reveal the reasons for such differences. In any case, if the design-basis earthquake (DBE) were to be the maximum occurring in a period of 1000 years, it has a probability of $10^{-3}$ in a given tectonic province. This probability has to be reduced by at least two orders of magnitude if we assume that the event will occur at any given point in the tectonic province, since the ratio between the average epicentral area and the area of the tectonic province is $10^{-2}$. If the earthquake is associated with a fault whose length is about 10-100 times greater than the diameter of the epicentral isoseismic line, the yearly probability must be reduced at least to $10^{-4}$ or $10^{-5}$ in order to take account of the fact that the reference event may occur on any point on the fault.

The DBE as determined in current American practice is the maximum potential event which could occur at the site. It is greater than the maximum historical event, and therefore the probability of its occurrence in a tectonic province (or along a fault) is certainly less than $10^{-3}$. The

probability of occurrence of this event at the site (or on the point of the
fault closest to the site) must be at most $10^{-6}$ (or $10^{-5}$ if we consider
faults).

A. BARBREAU:  Did you derive the $10^{-6}$ probability for a seismic
event at a site on the basis of an earthquake of a given intensity (for
example, the strongest earthquake recorded historically in the tectonic
province in question during a period of 1000 years), or did you start with
the probability of $10^{-6}$ and then look for the corresponding seismic
phenomenon?  The $10^{-6}$ probability is often used in nuclear safety analyses.

C. ZAFFIRO:  We derived the probability of $10^{-6}$ on the basis of an
earthquake of a given intensity, but one determined by the seismotectonic
method used in the United States of America, not the strongest earthquake
encountered historically in the tectonic province.  The value of $10^{-6}$ could
be taken into account as the probability of reference earthquakes for
future nuclear buildings.

E. IANSITI:  I think some clarification would be in order here.  The
probability of the order of $10^{-6}$/a for a DBE in Italy was proposed by
experts using American criteria.  What Mr. Zaffiro is suggesting is simply
that if a DBE is evaluated by statistical methods, that probability must be
adopted for purposes of calculation.  If the probability of $10^{-6}$/a is too
small, we must bear in mind that it was the experts, not Mr. Zaffiro, who
calculated it.

D. STOIAN:  Were dam failures considered in your flood calculations?

C. ZAFFIRO:  Yes, of course.

W. VINCK:  Do the flood-effect criteria take into account a hypothetical
maximum snow melt?  I would imagine that the 2.5% risk of the statistically
assumed events being exceeded in a 100-year period includes such a
hypothetical event.

C. ZAFFIRO:  Yes, the flow-rate data used for the statistical analysis
take snow melt into account.

A.F. WYLDE:  Obviously the population living in the vicinity of a
nuclear plant would be at risk not just from the plant itself, if it were
damaged by an earthquake, but from the direct effects of such natural
phenomena.  Have you made any attempt to compare these hazards –
the hazard due to the plant itself (and associated with a reference earthquake having a probability of $10^{-6}$) and that attributable to the direct effects
of earthquakes and other natural phenomena?

C. ZAFFIRO:  No, we have not as yet done any evaluation of public
risk.

T.P. HAIRE:  I was extremely interested in the information provided
on certain external hazards.  However, there are other hazards such as
aircraft crashes, gas cloud explosions and sabotage.  I should be glad to
know if these subjects are under consideration in Italy and, if so, what
values are attached to the probability of such occurrences leading to
serious damage to the nuclear plant.

C. ZAFFIRO:  I agree that the other events you refer to are important
for safety analyses of nuclear plants.  In Italy, we are just beginning to
study events of this type.

# AN APPROACH TO SITING NUCLEAR POWER PLANTS: THE RELEVANCE OF EARTHQUAKES, FAULTS AND DECISION ANALYSIS

K. NAIR, G.E. BROGAN, L.S. CLUFF, I.M. IDRISS, K.T. MAO
Woodward-Clyde Consultants,
Oakland, Calif.,
United States of America

## Abstract

AN APPROACH TO SITING NUCLEAR POWER PLANTS: THE RELEVANCE OF EARTHQUAKES, FAULTS AND DECISION ANALYSIS.

The regional approach to nuclear power plant siting described in this paper identifies candidate sites within the region and ranks these sites by using decision-analysis concepts. The approach uses exclusionary criteria to eliminate areas from consideration and to identify those areas which are most likely to contain candidate sites. These areas are then examined in greater detail to identify candidate sites, and the number of sites under consideration is reduced to a reasonably manageable number, approximately 15. These sites are then ranked using concepts of decision analysis. The exclusionary criteria applied relate primarily to regulatory-agency safety requirements and essential functional requirements. Examples of such criteria include proximity to population centres, presence of active faults, and the availability of cooling water. In many areas of the world, the presence of active faults and potential negative effects of earthquakes are dominant exclusionary criteria. To apply the 'active fault' criterion the region must be studied to locate and assess the activity of all potentially active faults. This requires complementary geologic (including geomorphic), historical, seismological, geodetic and geophysical investigations of the entire region. Site response studies or empirical attenuation correlations can be used to determine the relevant parameters of anticipated shaking from postulated earthquakes, and analytical testing and evaluation can be used to assess the potential extent of ground failure during an earthquake. After candidate sites are identified, an approach based on decision analysis is used to rank them. This approach uses the preferences and judgements of consumers, utility companies, the government, and other groups concerned with siting and licensing issues in the ranking process. Both subjective and objective factors are processed in a logical manner, as are the monetary and non-monetary factors and achievement of competing environmental and design objectives.

## 1. INTRODUCTION

The cost and uncertain availability of fossil fuels have increased the urgent need to develop alternative energy sources. Most plans to meet energy requirements through the rest of this century project the increased use of nuclear energy. However, increased concern over safety, environmental, and social issues has often delayed the implementation of these plans.

Because of this concern, more organizations responsible for siting nuclear power plants are attempting to select and seek approval for several sites within a region. Such an approach is expected to improve siting of power plants in relation to safety, environmental and social issues.

This approach contrasts with the need to justify one preselected site and requires improved decision-making procedures. The procedures and methodology described by Keeney and Nair [1] for using decision analysis in site selection contribute significantly to the improvement of the decision-making process in site selection.

This paper discusses conceptual and concrete aspects of a regional approach to site selection and considers relevant decision analysis methodology.  The following paragraphs indicate the organization of subsequent sections.

Section 2 presents the concepts of the regional approach to siting and a brief description of a proven step-by-step procedure.  A critical aspect of this approach is the application of selected exclusionary criteria.  Among the most important of these are the presence of active faults and the effects earthquakes may have on the site.

Section 3 discusses a rational approach to the identification of active faults and the importance of conducting active fault studies early in the site selection process.

Section 4 outlines techniques for determining the parameters of anticipated shaking and the potential for ground failure which may be caused by earthquakes.

Section 5 outlines a procedure which uses decision analysis concepts to rank several candidate sites for nuclear power plants, while Section 6 presents conclusions.

## 2.  GENERAL CONCEPTS IN THE REGIONAL APPROACH

An important concept in selecting a nuclear power plant site in a large region, such as a large portion of a state or country, is that it is impractical to find the best site in an absolute sense.  Time and financial constraints make it necessary to concentrate on areas where the likelihood of finding candidate sites is greater than in excluded areas, even though the latter may also contain potential sites.  A number of candidate sites in the acceptable areas are identified and then ranked.

The approach presented here consists of a successive screening of maps of increasing scale.  Selected exclusionary criteria are used to identify acceptable areas within the region which have a high likelihood of containing candidate sites.  These areas are then examined in greater detail and candidate sites are identified.  The objective of this screening process is to reduce the number of candidate sites to a manageable number, approximately 1!  These sites are then ranked using concepts of decision analysis.

The exclusionary criteria are selected with certain concepts in mind.  All parties can agree on some general objectives in site selection:

Maximize human health and safety

Minimize environmental damage

Maximize desirable economic impact on owners

Provide quality service to consumers

Of course, different individuals and groups will attach varying degrees of importance to these objectives.

With these general objectives in view, regulatory agencies and decision makers establish specific criteria to ensure a minimum level of achievement on these objectives.  The nature of these criteria depends on the country, the region, and the current financial, social, political  and technological

conditions.  For example, the acceptable distance of a site from population
centers, wildlife sanctuaries, national parks, etc., will depend on existing
population densities and the cultural importance attached to preservation of
wildlife.  These criteria can change with time.  Furthermore, technological
breakthroughs, such as eliminating the need for a large supply of cooling
water, can change essential functional requirements.

A rational step-by-step procedure for regional site selection is discussed below.  Although the procedure is best described in a series of steps,
in practice there is considerable interaction among the steps.  Steps 2 and
5 are described in greater detail in Sections 3 and 5, respectively.

Step 1.  <u>Establish Regional Exclusionary Criteria</u>

Exclusionary criteria are generally based on regulatory agency requirements which are primarily related to safety, essential functional requirements, and unique environmental considerations.  Other exclusionary criteria
may be specified by those responsible for the siting decision.

Regulatory guides from appropriate government agencies and discussions
with agency staff to evaluate future changes in regulation form the basis of
regulatory criteria.  Essential functional requirements are determined in discussions with the designers and operators of the facility, while discussions
with appropriate government agencies determine environmental and land-use
criteria.  The owners and operators of the facility may specify certain
exclusionary criteria for economic reasons.  Finally, the criteria, once
established, must be 'ratified' by the decision makers, who may be representatives of a national government, the utility company  or a regulatory agency.

Examples of regulatory criteria which may be used in this step are the
presence of active faults, population centers  and national parks.  Essential
functional requirements include the availability of cooling water and topographic constraints.  Environmental considerations include the presence of
rare and endangered species and land-use considerations, such as the presence
of valuable mineral resources.  Exclusionary criteria specified by the decision makers may include distance from load centers and transportation facilities and postulated earthquake-induced ground acceleration above a certain
level.  The latter affects both safety and cost of the plant.

Step 2.  <u>Apply Exclusionary Criteria - Identify Candidate Areas</u>

These criteria are applied by a system of overlays prepared manually or
with computer graphic techniques.  The base map scales typically used in this
step are 1:500,000 and 1:250,000.  Computer graphic techniques have the
advantage of allowing easy examination of the effect of variations in criteria
and rapid retrieval of data for future use.

The application of these criteria excludes significant parts of the
region under study and identifies candidate areas which have a higher likelihood of containing candidate sites than the excluded areas.  Section 3
presents an example of the operation of this step.

Step 3.  <u>Establish Criteria to Identify Candidate Sites Within Candidate
         Areas</u>

Plant sites are usually identified in a two-stage process.  First,
approximately 50 candidate sites are identified.  In this stage, the cri-

teria used are primarily objective; for example, acceptable topographic
relief, accessibility to existing transportation, and appropriate local
land-use patterns.  Conceptually, there is no difficulty in working with 50
sites and ranking them in accordance with the procedure outlined in Step 5.
However, practical limitations in time and money usually require dealing
with fewer sites.

In the second stage, subjective judgments of experienced personnel on
environmental factors and design and construction conditions are used to
reduce the number of sites from 50 to a more manageable number, approximately
15.  The importance of using the judgments of qualified, experienced personnel
cannot be overemphasized.  Typical environmental factors include aquatic and
terrestrial ecology, air and water quality, and socioeconomics; design considerations include foundation conditions and design accelerations; construction conditions include proximity to building materials and construction
access.  The decision maker ratifies the criteria to be used.

## Step 4.  Identify Candidate Sites

At this stage, larger scale maps varying from 1:62,500  to 1:24,000 are
used.  Typically, overlays are used to identify approximately 50 sites.
These 50 sites are reduced to 15 by listing the criteria, bringing together
the experienced people, and obtaining a consensus on the best 15.

## Step 5.  Ranking the Candidate Sites

The candidate sites are ranked by a decision analysis approach as described by Keeney and Nair [1].  This approach considers the preferences and
judgments of the decision makers.  Section 5 describes this approach in
greater detail.

During the first four steps, except for possible required studies for
exclusionary criteria, no detailed field investigations are conducted.
Available information and reconnaissance site visits must be relied upon in
ranking the 15 sites.  A great deal of subjective judgment is involved.
Careful documentation and the use of experienced, knowledgeable personnel is
essential in reaching valid decisions and obtaining public acceptance.
Again, this is a practical matter of cost and time.  After the sites are
ranked, detailed site-specific information is collected on the most desirable
two or three sites, and a prime site is selected.

## 3.  AN EXAMPLE OF APPLYING EXCLUSIONARY CRITERIA:  THE IDENTIFICATION AND DELINEATION OF ACTIVE SURFACE FAULTS

Exclusionary criteria can be divided into two categories--those which
do not require a significant amount of investigation and interpretation in
application and those which do.  Typical among the former are the presence
of population centers, designated national parks, and government reserves.
Typical among the latter are the presence of active faults and ecologically
unique areas, and the availability of cooling water.  Of those criteria
which require investigation and interpretation, the criterion of dominant
interest will depend on the region.  For example, in the seismically active
areas of the world, earthquake-related exclusionary criteria tend to dominate the selection of candidate areas.  In some other regions, the availability of water for cooling purposes is the dominant criterion.

The investigative studies necessary to apply these criteria successfully can often delay the total siting study.  However, the consequences
of rejecting a site at a much later stage of the design and construction
process, which may occur if comprehensive initial studies are not conducted,
are a greater problem in terms of monetary loss than the delay the investigative studies may cause.

The presence of active surface faults is a particularly important
exclusionary criterion because they can be the source of earthquakes and
surface fault displacement.  Earthquakes can also be generated from faults
which have no surface manifestation.  The effects of an earthquake can be
divided into four categories:

      Fault displacement

      Strong ground motion (shaking)

      Ground failure

      Tsunami (seismic sea waves) and other water waves [2]

<u>Fault displacement</u>, the process by which the two walls of a fracture slip
relative to one another, is generally accepted as the mechanism that generates
most of the damaging earthquakes in the world [3].

When fault displacement occurs suddenly in the upper portion of the crust
of the earth (generally shallower than 30 km), the fault rupture may intersect
the surface of the ground.  When this occurs, the ground surface will be rup-
tured and structures located astride the fault will be displaced along with
the ground.

FIG. 1.   Oblique aerial photograph of the San Andreas Fault, looking north towards San Francisco.  The
fault extends vertically through the photograph and is marked generally by the contrast between the forest
to the west (left) and open fields to the east (right).

FIG. 2.  Aerial view of the Wasatch Fault scarp south of Provo, Utah.  The fault is near the base of the mountains.

FIG. 3.  View of the Fairweather Fault at Lituya Bay, Alaska.  The Fairweather Fault occupies the trench through the center of the photograph.  The fault is covered by glacial ice over approximately 80% of its length.  An earthquake in 1958 generated a landslide from the steep wall on the right (A).  The landslide, in turn, generated a sea wave that stripped vegetation as high as 500 m from the ridge on the left (B).

Fault displacement can have horizontal, vertical, or oblique components
that range from a few millimeters to several meters. The largest of these
are most likely to occur on those long faults that have a geologically young
history of displacements. Examples of such faults are the San Andreas Fault
of California (Figure 1), the Wasatch Fault of Utah (Figure 2), the Fairweather and Denali Faults of Alaska (Figures 3 and 4) [4], and the Boconó
Fault of Venezuela (Figure 5). These are a few representative examples of the
many active faults that exist throughout the world.

Small displacement along faults can be damaging to structures located
astride them. A displacement of only a few centimeters on a fault in
Managua destroyed many structures, including the Customs Building (Figure 6).
Even large fault displacements can occur across faults of narrow width.
The 1906 rupture along the San Andreas Fault is often extremely narrow
(Figure 7).

Studies of historic surface faulting show that ground surface ruptures
occur along pre-existing faults, although the prior existence of identifiable
evidence has often not been recognized until after the latest fault displacement. Evidence of repeated fault displacement along the most recently active
trace of a fault is apparent along many faults, including the San Andreas
(Figure 8).

Strong ground motion caused by earthquakes, unlike surface fault displacements resulting from active faulting, are not confined to well-defined
zones but affect large areas.

The entire downtown area of Managua, Nicaragua, was destroyed mainly
by shaking in the 1972 earthquake (Figure 9). The Caracas earthquake of
1967 showed that shaking damage can be selective (Figures 10 and 11).

Ground failure may cause damage in several forms, including landsliding
(Figures 12 through 14) [5], liquefaction (Figure 15), and settlement.

FIG. 4. Oblique aerial view of the Denali Fault, Alaska, showing a series of 'en echelon' surface ruptures
in alluvium that mark the most recently active trace of the fault. The ruptures are indicated by arrows
at either end of the series.

FIG. 5.   View showing the Boconò Fault, Venezuela.  The major valley marks the fault zone, which is longer than 500 km.  The view here is northeast from Lagrita towards Merida in the far distance.


Saturated sand layers located at shallow depth beneath the surface may be susceptible to liquefaction during an earthquake.  This phenomenon has been observed frequently in past earthquakes; notably in Niigata, Japan, in 1964, Chile in 1960, and San Fernando in 1971.  Shallow, loose, saturated sands appear to be the most susceptible to liquefaction; dense sands are less susceptible.

Tsunami are sea waves generated by large, rapid displacements of the sea floor resulting from earthquakes or other phenomena such as submarine land sliding.

In the open ocean, tsunami are difficult to recognize, because they have long wave lengths (usually hundreds of kilometers), long periods of oscilla- tion (about an hour), high velocities (more than 800 km/hr), and low wave heights (no greater than a few meters).  As they approach shallow waters, however, the velocity decreases and wave height increases; wave heights as great as 20 m have been recorded.  Their approach is generally heralded by a number of surges separated by low water.

FIG. 6.   Customs Building in Managua, Nicaragua, that collapsed during fault displacement associated with the 1972 earthquake.  This building was located astride the Chico Pelon Fault, which slipped only a few cm during the earthquake.

FIG. 7.   Fault displacement of a fence by the San Andreas Fault in 1906.  The width of active fault displacement is restricted to a few cm between formerly adjacent posts of the fence (indicated by A and B). The fault is indicated by a broken line.  The curvature of the fence on the far side of the fault shows warping adjacent to the fault displacement.

FIG. 8.   View across the San Andreas Fault in the Carrizo Plain, California.  The fault trace indicated
by arrows slipped approximately 10 m during a great earthquake in 1857.  Note that several streams that
cross the fault are displaced differentially, indicating that repeated fault displacements have occurred
along the same fault trace, although not along the other fault traces in the field of view.

Seiches are long-period oscillations of bodies of water generated by
earthquake vibrations or tilting of the basin in which the body of water is
located.  Typically, seiches are not as destructive as tsunami, and their
immediate effects are limited to the shoreline areas of bodies of water.
Overtopping of reservoirs and flooding of near-shore facilities may occur
during a period of seiching.

Locally destructive water waves generated by nearby submarine and surface
landslides in the sea, lakes, and reservoirs may reach tremendous heights; a
rockfall at Lituya Bay, Alaska, in 1958 generated a wave that stripped vegeta
tion to a height greater than 500 m (Figure 3).  Landslides in Port Valdez,
Alaska, appear to have generated destructive waves with run-ups in excess of
60 m.  The effects of earthquake-generated waves are significant for coastal
sites and are not discussed further in this paper.

The level of earthquake effects in terms of ground shaking and potential
ground failure have in some cases been specified as exclusionary criteria by
the decision makers.  These effects are discussed in Section 4.  The remainde
of this section deals with the location and delineation of active faults.

## Active Fault Definition

For the purposes of this paper, an active or potentially active fault is
defined as a fault on which there is sufficient evidence of displacement with
in the recent geological past to make it reasonable to expect that future dis
placements of engineering significance could occur along the fault.

FIG. 9. Damage from strong ground motion (shaking). View shows Managua, Nicaragua, shortly after the earthquake of 25 Dec. 1972. Faults crossing the city were marked by linear belts of damage due to surface faulting.

FIG. 10.   Damage from strong ground motion (shaking), San Jose Apartment Building in Caracas, Venezuela, showing total pancake collapse after the 1967 earthquake.  Note that the degree of damage appears to be selective, since surrounding structures are damaged but did not collapse.

FIG. 11.   Damage from strong ground motion (shaking), Mijagual Apartment Building in Caracas, Venezuela, showing total pancake collapse.  Note that individual floors can be counted.

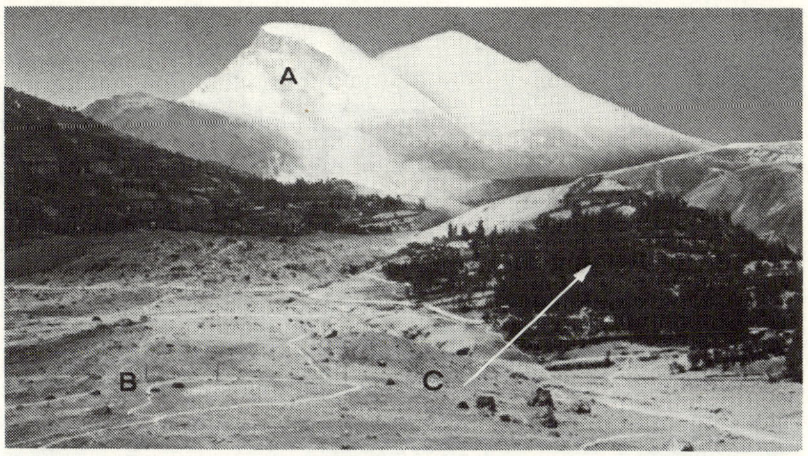

FIG. 12.   Damage from ground failure (landsliding), Yungay, Peru.   The source of the 31 May 1970 earthquake and debris avalanche was the mountainside marked by the dark scar through the snow (A). The landslide then flowed down the valley and into the townsite of Yungay (B) which was completely destroyed.   The large boulders in the right foreground are similar to boulders among the trees at C, signifying that similar landslides occurred in the recent geologic past.

FIG. 13.   Damage from ground failure (landsliding).   This view shows the Turnagain Heights landslide in Anchorage, Alaska, which resulted from the earthquake of 27 March 1964.   The landslide is believed to have been generated when near-surface sand lenses within a clay liquefied during the earthquake.

FIG. 14.   Damage from ground failure.   View shows the Government Hill landslide in Anchorage, Alaska, that resulted from the 1964 earthquake.   The Government Hill Elementary School, shown in the photograph, was destroyed.

FIG. 15.   Damage resulting from ground failure (liquefaction).   Buildings in Niigata, Japan, tilted during a 1964 earthquake there when a shallow sand layer liquefied beneath the tilted buildings.

According to the definition adopted by the USAEC [6] for nuclear reactor
siting, an active fault (now termed by the USAEC a 'capable' fault) has one
or more of the following characteristics:

> Displacement at or near the ground surface at least once in
> 35,000 years or movement of a recurring nature within the past
> 500,000 years;
>
> Instrumentally measured macroseismicity determined with
> sufficient precision to demonstrate a direct relationship
> with the fault;
>
> A structural relationship with an active fault having the two
> characteristics mentioned above, so that displacement on the
> one could be reasonably expected to be accompanied by displace-
> ment on the other.

The USAEC definition is considered to be a reasonably conservative one.

The chronology of recent displacements on most faults cannot as yet be
established precisely from objective data.  Therefore, the classification of
a fault as active depends on geologists' subjective interpretation of the
evidence as to how recently and how often the fault has slipped in the past
and may slip in the future.  For this reason, there will be differences of
opinion among geologists as to the activity of a fault.


Evidence for Recognizing Active Faults

Five general types of evidence are used to evaluate whether or not a
fault is active [7]:  _geological_, including geomorphic and stratigraphic;
_historical_,  including accounts of past historic earthquakes and surface
faulting; _seismological_, considering the preinstrumental historic record and
the instrumental record; _geophysical_, considering various remote techniques
used to obtain data that are difficult or impossible to obtain otherwise;
and _geodetic_, using data from two or more precise surveys.

Classification of a fault in terms of its activity is based on the
assumptions that a fault is likely to slip if it has slipped in the recent
geologic past, and is not likely to slip if it has not slipped for a
sufficient length of time.  For faults with no record of historic displacement, the time of the most recent surface displacement usually is evaluated
best from geomorphic and stratigraphic evidence.  The ages of unfaulted
deposits or surfaces that lie astride the fault provide a basis for estimating the time since the most recent fault displacement [8].

_Geological Evidence._  Geological evidence provides the most reliable long-term
record of the recent activity of a fault.  The geological record of surface
faulting is often tens of thousands to millions of years long.  The Quaternary
period, especially the late Quaternary (approximately a few thousand to a few
hundred thousand years), is generally considered sufficient for the evaluation
of recent fault activity.

Geological studies can be divided into two main types:  geomorphic and
stratigraphic.  If a fault has been recently active at the surface, geomorphic
features indicating active faulting are likely to occur at various places
along the fault.

Stratigraphic evidence of active faulting consists mainly of faulted
sediments or rocks that are geologically young, generally of late Quaternary

age.  This evidence includes natural or artificial exposures of the fault
within the young deposits or the displacement of them, or evidence from
subsurface investigations, such as drilling, trenching, or geophysical survey-
ing, that reveal faulted units.

It is usually easier and quicker to identify diagnostic geomorphic
evidence of active faulting than to search for stratigraphic evidence, though
both are often needed for conclusive evaluations, especially in showing that
no geologically young faulting has occurred [7].

<u>Historical Evidence</u>.  Historical evidence can be divided into two main cate-
gories:  accounts of the historical record of earthquakes and accounts that
describe historical evidence of surface faulting.  Evidence of past earthquake activity usually becomes less reliable with older accounts, so that
much of the early historic record is often unreliable in terms of identifying surface faulting.  In addition, evidence of past earthquakes may or may
not be indicative of past surface faulting.

Evidence of surface faulting may be described in some early accounts,
but these are generally rare.  Such accounts may describe differences in
elevation, open fissures, changes in spring activity, or other phenomena.
These must be separated by careful scrutiny from descriptions of landslides,
settlement of weak soils, effects of seiches or tsunami, and other phenomena
that may not actually represent surface faulting.

The value of the historical record is limited by its short duration in
relation to the geological record.

<u>Seismological Evidence</u>.  Seismological evidence may be considered in two ways
when evaluating active faulting.  First, evidence of macroearthquakes[1] generally denotes a seismically active fault that has slipped and may slip again
but the fault rupture may or may not have intersected the surface.  Second,
evidence of microearthquakes[2] may or may not indicate a fault that could
generate a large earthquake or surface faulting.  It is generally felt that
seismological evidence should be considered as a guide for studies of active
surface faulting, particularly in areas where a concentration of shallow-
focus earthquakes has been detected [9].

A major difficulty in using the seismological record as a guide to interpreting fault activity is that many faults classified as active on the basis
of geological criteria have no record of earthquake activity.  This apparent
inactivity may result from absence of earthquake activity, failure to record
such activity, or failure to locate the epicenters accurately.  Therefore, the
apparent lack of earthquakes in an area may be misleading, both in terms of
potential future earthquakes and in terms of future surface faulting.

Sherard et al. [8] discuss circumstances in which seismological data
might be most useful.  These include the following:

---

[1] Macroearthquake as used in this paper refers to earthquakes equal to or larger than a Richter
magnitude of approximately 4.5.

[2] Microearthquake as used in this paper refers to earthquakes smaller than a Richter magnitude
of 4.5.  Generally, it is possible to detect most earthquakes of magnitude 4.5 or greater; these are
generally assigned an epicentral location.  Other use of the term microearthquake may refer to small
earthquakes that are not generally felt; these are usually smaller than a Richter magnitude of
approximately 3.0.

A location at which it is very difficult to interpret the known
geology in terms of fault activity because of exceedingly heavy
vegetation, water cover, glacial ice cover, or river valleys
covered with extremely young flood plain deposits;

A project for which two to four years of seismic observation
can be allowed;

A project for which the installation of seismographic stations
is desired for other reasons, such as studying the possible
increase in seismic activity related to filling reservoirs.

In regional studies for siting, one method commonly used to display
seismicity is to plot epicenters of instrumentally located earthquakes on a
map.  The epicentral locations are often obtained from the records of a
seismological laboratory or government agencies.  These locations may be
only approximate, especially in regions where the recording stations are
widely spaced.  For this reason, caution must be exercised in interpreting
recent fault activity from plots of epicenters.

Another limitation in using seismicity data is that the instrumental
seismological record is of shorter duration than the geological and historical
records.

Geophysical Evidence.  Geophysical evidence for determining the recency of
faulting depends upon identifying anomalies within geologically young materials.  As such, it is an extension of geological criteria for which the
essential data can be obtained mainly through one or more of the following
geophysical surveys:  seismic refraction, seismic reflection, magnetics,
gravity, or resistivity.  Generally, the geophysical data must be interpreted
very carefully, and should be corroborated with other definitive evidence if
an active fault is inferred.

Geophysical surveys are often useful because they allow a large area to
be studied in a relatively short period of time, and they allow gathering of
data that may be difficult or impossible to obtain otherwise.  Many areas
have been surveyed for mineral and/or petroleum exploration by one or more
of the techniques listed above.  These existing surveys can be of great
benefit in guiding studies of fault activity.

Geodetic Evidence.  Strain data from repeated geodetic surveys may indicate
current fault activity.  This evidence generally shows the presence of either
aseismic slippage along a fault (fault creep), or slippage that is rapid and
accompanied by an earthquake.  Geodetic data may also indicate an accumulation
of strain.  Because geodetic data require thorough interpretation and precise
surveys repeated over decades, they are generally not as usable as other types
of criteria.  Also, they are not always available in areas of interest.

Conclusion.  Most of the reliable data for evaluating surface faulting are
derived from geological evidence.  Important evidence may also be derived
from studies of historic earthquakes, instrumental measurements of micro-
earthquakes, geophysics, and geodetic measurements of creep and crustal
strain.

Largely because of the recent increasing emphasis on the safety of
nuclear power plants, the number of fault studies conducted has greatly
increased.  Although no significant recent advances have been made in deter-

mining the exact mechanisms of surface faulting, accumulated knowledge of
techniques for evaluating faults and of the usefulness of extensive and
detailed geologic studies is increasing rapidly.  In earthquake-prone regions,
more geologists are specializing in evaluating active faults and are devoting
much of their practice to the field of earthquake activity and seismic geology, an area of study previously receiving the attention of only a few
investigators.

For these reasons, one can now have greater confidence in studies of
fault activity than was previously possible.  The approach discussed below
provides a high degree of reliability in identifying areas for nuclear reactor
siting that may be free from active faults.

## An Approach to Regional Studies of Fault Activity

In developing an approach to locating, identifying  and delineating
active faults, one must recognize that evidence of the most recent displacement along an active fault can often be identified at only a few locations,
even though the entire fault may be considered active.  Hence, conditions
existing along any limited length of a fault, such as in the immediate vicinity of a proposed site, will not always provide conclusive evidence of recent
fault activity.  Studies should be conducted along the entire length of known
faults.  The degree of confidence with which conclusions can be drawn for an
area depends greatly on the amount of effort expended during the regional
fault studies, and on the size of the area in which the investigations are
carried out.

The approach discussed in this paper is divided into two phases.  In
Phase 1, areas are classified in broad categories as acceptable, doubtful,
or excluded.  In Phase 2, the acceptable areas are studied in greater detail
to identify potential candidate areas.  The second phase requires a higher
level of investigation and, usually, a significant field effort.

The level of investigation undertaken depends on the degree of confidence
one can have that the studies can show the acceptable areas to be free of
active faults, on the level of such confidence acceptable to the decision
maker, and on the willingness of the decision maker to exclude areas from
further consideration.  In some cases the Phase 2 effort may not be necessary.

**Phase 1.  Classification Into Acceptable, Doubtful, and Excluded Areas.**  The
process of classifying involves the tasks described below.

Task A.  **Review of existing data.**  This task includes acquiring and
reviewing pertinent publications and unpublished data on the geology, seismic
history and seismicity, geodesy  and geophysics of the region; consulting
with experts in these fields of study who know the region; and locating,
acquiring, and interpreting aerial photographs and other remote-sensing
material, such as satellite (ERTS, Sky Lab) and radar (SLAR) images.  This
task provides a preliminary basis for assessing the regional geologic and
tectonic conditions related to active faulting.  These data will also be
useful at a later time when design earthquakes are being assigned to sites.

Task B.  **Photo interpretation.**  The interpretation of stereoscopic aerial
photographs by geologists specializing in active fault studies is essential
identifying active faults.  Geomorphic features characteristic of active

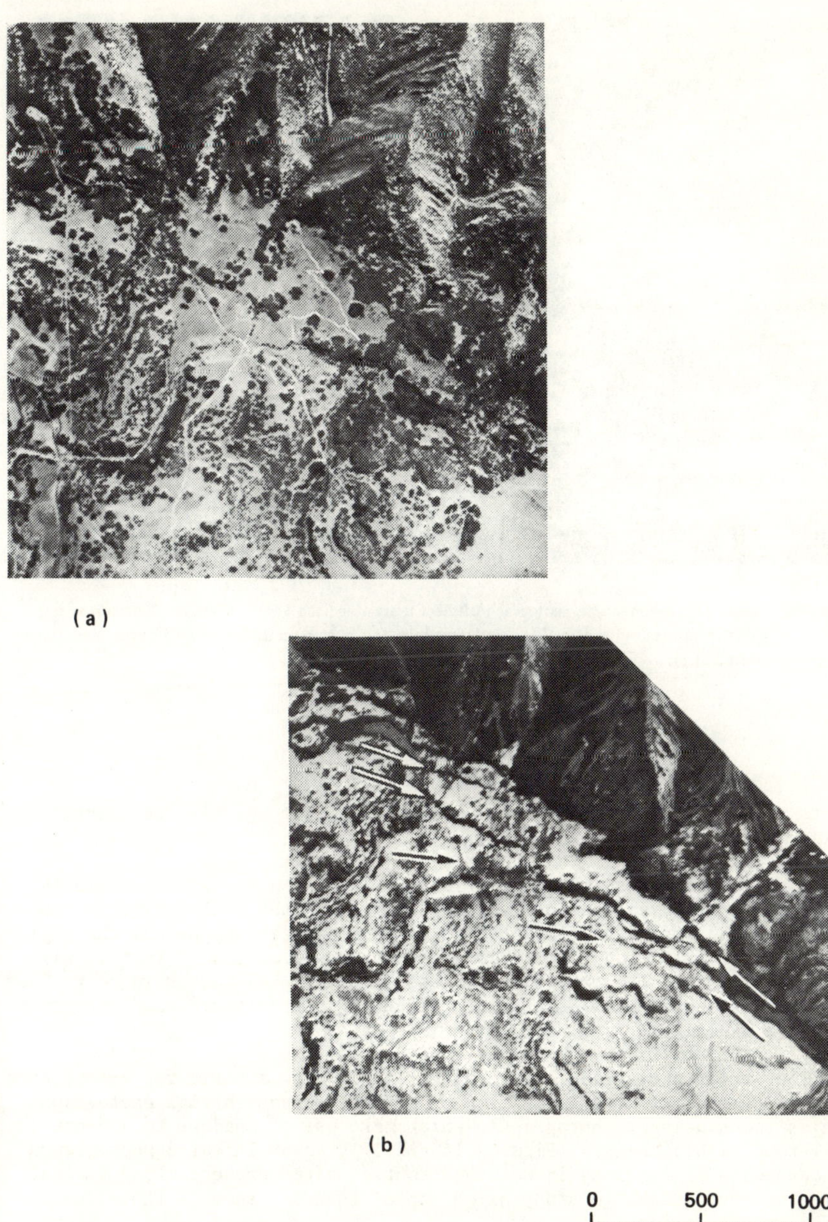



FIG. 16. Conventional aerial photograph (a) and low-sun-angle aerial photograph (b) of the same area
of the Wasatch Fault, Utah. The low-sun-angle photographic technique can delineate subtle topographic
features that may pass unrecognized on conventional aerial photographs. Note that individual fault traces
are difficult to recognize in (a) but are shown clearly as linear shadows (indicated by arrows) in (b).

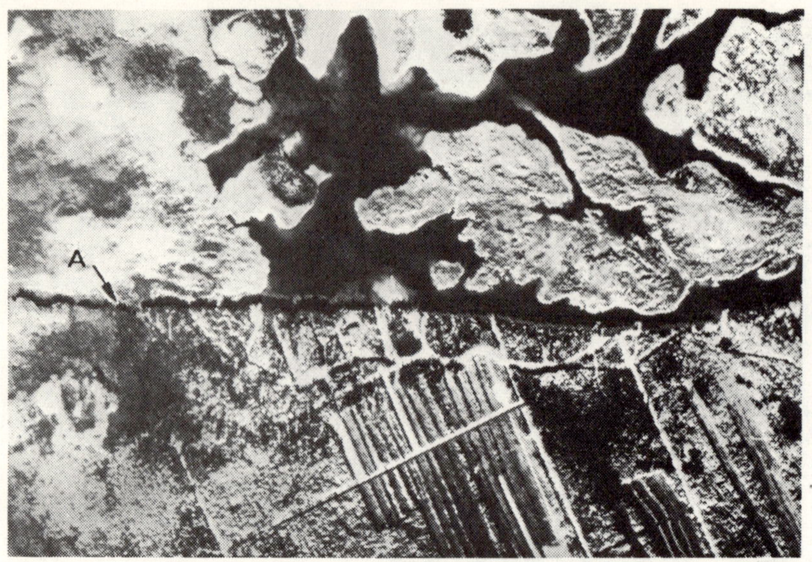

FIG. 17.    Vertical low-sun-angle photograph of the Clearwater Lake Fault, Alaska.  The fault scarp,
shown as a continuous line of shadow (A),  separates lake and swamp on the upper half of the photograph
from the forest and plowed fields of the lower half.

faults are often visible on aerial photographs, but may also be identified
during aerial reconnaissance [10].

     Natural processes, the activities of man, and urban growth commonly
modify surface indications of faulting along even the most active faults.
Therefore, the aerial photographic coverage should extend beyond the region
under consideration  and should cover a broad span of time.  In this way,
subtle features no longer visible or easily overlooked may be found to line
up with pronounced but widely scattered geological features characteristic
of active faulting.

     The aerial photographic interpretation is also a basis for determining
areas to be covered at a later time by special-purpose aerial photographs
such as low-sun-angle photographs, which make use of shadows to accentuate
small topographic changes (Figures 16 and 17), color infrared photographs
to accentuate differences in moisture content often present along faults,
and color photographs to study slight color changes, such as those due to
different rock types [10].

     Task C.  Regional aerial reconnaissance.  Aerial reconnaissance in a
high-wing aircraft allows the geologist to view a large region in a short
time and to gain an understanding of the geomorphology and topography.  Specific geologic and geomorphic features that may be significant to classifying an area can often be identified only from the air.  The locations of
such features can be plotted during the flight.

Task D.  <u>Classification of areas</u>.  This task involves compiling and synthesizing all of the data from the first three tasks.  The pertinent data are usually plotted on overlays to a base map of the region, with each item on a separate overlay.  These items may include the following:

> Faults and lineaments
>
> Geologic rock units
>
> Groundwater information
>
> Geomorphic features
>
> Aerial photographic interpretations
>
> Remote sensing imagery interpretations
>
> Earthquake epicenters
>
> Tectonic features
>
> Information from aerial reconnaissance
>
> Credibility of existing information for the various parts of the region

On the basis of this information and the subjective judgment of knowledgeable and experienced personnel, the region is divided into acceptable, doubtful  and excluded areas.


Phase 2.  <u>Selection of Candidate Areas From Acceptable Areas</u>.  The objective of Phase 2 is to select candidate areas for which the level of confidence in meeting regulatory agency criteria relating to active faults is considerably higher than for the acceptable area as a whole.  If, however, the decision maker is willing to exclude certain areas within the acceptable area from consideration, this phase may not be necessary.

The tasks listed below cover a general investigation.  The level of effort in each of these tasks can vary greatly depending on the region and the investigations conducted in Phase 1.


Task A.  <u>Data review</u>.  A review of all literature, maps  and other data collected in Phase 1 should be made and all material pertinent to the desirable area under study should be organized, catalogued  and used by the geologists. At this point, the scale of the study must be changed.  A scale allowing features as small as 200 m long to be accurately mapped should be selected.  All regional features located in the acceptable area should be plotted at suggested scales of 1:50,000 to 1:20,000.


Task B.  <u>Aerial and ground reconnaissance</u>.  The aerial reconnaissance undertaken at this time provides the geologist with an overall view of the area in which he is working.  Locating outcrops for later mapping, viewing prominent lineaments observed during photo interpretation in the regional study, and identifying prominent features mapped previous to or discovered during the regional study would be included.  The ground reconnaissance should be directed toward gaining a preliminary familiarity with the rock units to be mapped and studied later.

Task C.  <u>Detailed photo interpretation</u>.  Photo interpretation has the specific goal of scrutinizing the area under consideration within the region.

Also, special-purpose photography or imagery should be selected, obtained
and interpreted at this time.  The types of imagery generally used in evaluating fault activity are listed under Task A of Phase 1.

Because there is a large element of experience and judgment in the
interpretation of aerial photographs, and particularly in interpretation of
recent fault activity, images of any kind should be studied independently
by a team of at least two geologists experienced in active fault studies [7].
The many types of photographs and images available for such work have been
used and evaluated by Sherard et al. [8].

Task D.  <u>Geologic mapping</u>.  An adequate geologic map is of primary
importance in this step as the map depicts the distribution and ages of the
various geologic deposits and geomorphic surfaces exposed and provides a
means of estimating subsurface relationships.  Geologic mapping is normally
required for nuclear reactor sites, and mapping acceptable areas during this
stage of the site selection process would be of great importance in selecting
candidate areas.  The geologic mapping should use any past mapping of the
area, but the geologist should not accept past mapping without checking the
quality of that mapping in the field.  In selecting candidate areas, the
following three points are emphasized.  First, faults that are not covered by
a datable horizon within the acceptable area should be traced into an area
where the faulted rock is overlain by younger datable deposits.  If these
younger deposits show no evidence of faulting, it may be presumed that they
have been deposited after the time of faulting and that their depositional
age is a minimum date of the most recent fault displacement.  The tracing
procedures may require that the fault be followed outside the acceptable area

Second, field geologic mapping techniques should make use of stratigraphic studies, marker beds, pediments, or any other recognizable horizons,
zones, or geologic units which relate to the time of most recent faulting.

Third, time markers should be checked for the reliability of the date
assigned to them.  If the ages of markers are in question, samples should be
taken for radioactive dating or for other dating, where possible.

Task E.  <u>Subsurface exploration</u>.  Subsurface exploration is normally
used at a candidate site, but should also be considered during the site
selection process.  Three subsurface exploration methods commonly used to
evaluate the potential for surface faulting through a site are geophysics,
boring  and trenching [10].

Geophysical methods and exploratory borings are useful where pronounced
geologic contrasts are present on either side of a fault.  These contrasts
are typically due to differences in water table elevation, distinct changes
in material, or variation in depths to bedrock.  Such contrasts may be
detected by seismic refraction, electrical resistivity, or magnetic methods.
Geophysical methods and borings do not provide diagnostic proof that a fault
does or does not exist on a given site, but they provide targets for further
exploration [11].

The change in material or stratification across a fault in recent
alluvium is likely to be so minute that it can be detected only by the
closest possible scrutiny.  It is therefore necessary in most cases to
directly observe the subsurface conditions to verify the location and nature
of faulting.  Direct observation is possible by trenching.

Exploratory trenches are normally oriented at right angles to the suspected fault and excavated by a backhoe to a depth of about 4 to 6 m. For safety, trenches in alluvium must be supported by shoring. The length of a trench will depend upon the precision used to locate the fault initially. Therefore, length may vary within broad limits from 15 to more than 600 m [11].

Detailed examination of the trench walls is required to reveal most fault-related features. The material disturbed by the backhoe must be carefully removed by hand tools from one or both walls of the trench where the features revealed are examined in minute detail for evidence of faulting.

Trenching in alluvium to 4 to 6 m in depth is usually sufficient to expose a geologic section representing several thousand years of deposition. Age dating of material encountered in the trench walls can establish the total age of the deposits penetrated. The absence of fault features in the trench signifies that a fault is either absent or its recurrence interval is greater than the age of the deposits exposed.

Task F. <u>Classification of areas</u>. Once the data from Tasks A through E are completed and synthesized, potential candidate areas which satisfy this category of exclusionary criteria can be determined. These areas are large enough for one or more nuclear reactor sites, and they should meet all of the exclusionary criteria related to active faulting. Specifically, the site should be apparently free from active faults and should have continuous stratigraphic units that are apparently unfaulted and are of ages that demonstrate the lack of recent fault activity across the site.

Other exclusionary criteria must be applied in a similarly careful manner to determine a candidate site that meets all criteria.


## 4. OTHER EARTHQUAKE-RELATED EXCLUSIONARY CRITERIA


Among the major factors to be considered in site selection for a nuclear power plant are: level and characteristics of anticipated shaking associated with the controlling earthquake, and potential for ground failure during the assigned controlling earthquake. In some cases these factors may constitute safety-related exclusionary criteria for the siting of nuclear power plants. The decision maker may specify limits based on economic considerations which then also become exclusionary criteria.

The approach used to evaluate fault activity discussed in the previous section can be used in a regional study to identify potential earthquake sources that will determine the controlling earthquake. As a general rule the size of the region that should be studied to determine potential earthquake sources should extend 200 km beyond the region being considered for potential sites.

The major elements of the regional study to determine future seismic activity are:

> Evaluation of past regional seismic activity including all historic data, felt reports, and instrumentally determined epicenters;

> Evaluation of the seismic geology of the region to identify earthquake sources (active faults) based on geologic evidence.

The information from these evaluations should be combined to estimate past seismicity.  From this estimate, the potential future seismic activity can be estimated and will include the identification of the various earthquake sources.  These earthquake sources will aid in establishing the controlling earthquake.

The level and characteristics of shaking at a site are influenced by the magnitude and distance to the site of the controlling earthquake, and by the local site conditions.  Once the source and the magnitude of the controlling earthquake have been established, the level and characteristics of shaking can be evaluated by a variety of available techniques.  These techniques may include site response studies and empirical attenuation correlations.

Site response studies require detailed evaluations of several engineering properties of the local soils.  Such detailed evaluations, however, may not be possible or practical during the site selection process.  Therefore, it may be necessary to rely on empirical correlations to arrive at the level and characteristics of shaking.  These empirical correlations provide values of peak ground acceleration as a function of magnitude and distance.  Other characteristics, such as frequency content (spectra) and duration, can also be obtained using empirical correlations.  With the aid of these correlations, the design parameters (peak ground acceleration, spectra  and duration) associated with the postulated controlling earthquake can be selected.  Correlations such as those proposed by Housner [12], Schnabel and Seed [13], Donovan [14], and Tocher et al. [15] can be used to evaluate the peak ground surface acceleration.  Newmark et al. [16] presented generalized smooth response spectra that were based on the average, plus one standard deviation, of the spectral ordinates of 33 recorded motions.  These generalized spectra were proposed by Newmark et al. [16] for use in the analysis and design of nuclear power plants; they were adopted by the USAEC and issued in Regulatory Guide No. 1.60.  More recently, Seed et al. [17] analyzed the spectra for 104 recorded motions and also presented generalized smooth response spectra.  Because they had more data to use, Seed et al. [17] were able to construct smooth spectra for four different local site conditions:  rock, shallow stiff site conditions, deep cohesionless soils, and soft to medium clay and sand.  The spectra for these generalized local soil conditions differ from those of the regulatory guide (which is independent of local site conditions) over a wide range of frequencies.  During a site selection phase, only a general knowledge of the local conditions is available.  This general knowledge of local site conditions could be sufficient to help in selecting the most applicable - at least for a preliminary assessment - design spectrum.

In some parts of the world it may not be possible to assign the controlling earthquake to a specific fault.  Therefore, it may not be possible to determine the distance from the epicenter to the site.  In many of these cases, selection of the controlling earthquake is then based upon historic data and the concept of a seismotectonic province.  Usually this leads to assigning intensity data that would be associated with the controlling earthquake.  Empirical correlations, such as those proposed by Gutenberg and Richter [18], Hershberger [19]  and Neumann [20], relating peak ground acceleration and intensity have been widely used  but must be used with caution.

Ground failures such as excessive settlement, liquefaction  and slope instability have been observed during many earthquakes throughout the world.  Recent advances in analytical, testing, and evaluation procedures have improved the capability to assess the potential extent of ground failure during a postulated earthquake [21].  A complete assessment would require detailed site explorations, laboratory testing  analytical studies, and evaluations.

For site selection purposes, however, it may be sufficient to conduct a
limited assessment based on the results of a brief field exploration program.
Such an assessment should be made following the selection of the controlling
earthquake.

Extensive ground failures that may be caused as a result of the application of the controlling earthquake at a candidate site may constitute an
exclusionary criterion for reasons of safety.  Limited ground failures could
be alleviated by corrective measures in the site preparation and final construction of the plant, but may be specified as exclusionary criteria for
economic reasons.


## 5.  RANKING THE CANDIDATE SITES

At the conclusion of Step 4, as discussed in Section 2, the decision
maker is faced with the choice of selecting among several candidate sites.
Decision analysis provides a logical and systematic approach to ranking
these sites in accordance with the preferences and judgments of the decision
maker.  A detailed discussion of the decision analysis approach is presented
by Keeney and Nair [1].  The major problems that must be addressed in this
ranking process are discussed below.

Multiple objectives.  A number of conflicting objectives which include
safety, environmental, social  and economic considerations must be evaluated
in making the decision.  Some of these have direct measures of effectiveness
(e.g.  cost), and others are intangibles (e.g.  aesthetics).  For the intangible objectives, it will be necessary to develop subjective indices of
measurement.  The decision maker will be faced with the inevitable conflicts
in achieving competing objectives.

Uncertainty.  There is always a degree of uncertainty associated with the
possible impacts of an action such as the siting of a nuclear power plant.
Future effects on safety, environmental  and social conditions cannot be
predicted with certainty at the time the siting decision has to be made.
Public resistance to nuclear power plants is largely due to the uncertainty
associated with the possible consequences.

Several impacted and decision groups.  There are several identifiable
groups who are impacted by a siting decision.  The preferences of these groups
are different.  Not only can there be several groups involved in making the
decision, but there may also be some ambiguity and overlap concerning their
responsibilities.

Impact over time.  The decisions made today affect the future.  In
economic studies, this factor is considered by discounting to net present
worth.  However, when considering radioactive waste, for example, it does not
appear that discounting would be appropriate because at most moderate discount
rates (three to seven percent), society should effectively neglect all waste
generated after 50 years.  Such an implication in discounting does not appear
to be intuitively justifiable.  More appropriate procedures for addressing
impact over time must be developed and used.

The decision analysis approach advocated in this paper can effectively
address these problems.

<u>General Concepts of the Decision Analysis Approach</u>

The underlying idea in the decision analysis approach is that decisions
are made on the basis of the decision maker's judgments about the likelihood
of the occurrence of various consequences and his preferences for these consequences.  This approach breaks the problem into parts which are easier to
analyze than the whole and then puts the parts back together, using a logical
and systematic procedure.  The procedure for putting the parts together
depends on the individual decision maker.  The experience, professional
judgment, and knowledge of the individual responsible for making the siting
decision should be used.  Decision analysis provides procedures for formal-
izing the judgments and preferences of the decision makers and integrating
these elements, using <u>their</u> logic to evaluate alternative courses of action.


<u>The Ranking Problem in a Decision Analysis Framework</u>

To briefly illustrate some of the essential concepts in the decision
analysis approach, the problem of ranking three hypothetical candidate sites
is considered.  The 'do-not-build' alternative must also be considered and
the consequences of such a decision examined.  A step-by-step procedure for
analyzing this example is presented below in a series of sequential tasks.


Task A.  <u>Identification of the decision makers, their objectives  and the</u>
<u>measures of effectiveness (or attributes)</u>.  The decision makers must be iden-
tified.  In many cases, it may appear that a group is responsible for a
decision, but in actuality a single individual is responsible, even though
he may have delegated various aspects of the total decision to different
individuals.  True group decisions are more difficult to handle.  However,
an individual analysis for each member of the group can provide the basis
for resolution of conflicts and achievement of consensus.  For the purposes
of discussion, it is assumed that the decision maker is an individual.

The objectives and the measures of effectiveness should be chosen by
the decision maker with the assistance of experts in the various disciplines.
It is obvious that the general objective of the decision maker is to select
the best site.  This objective is too general to be of any use in analyzing
the problem.  A level below this general objective may include the following
more specific objectives:

> Maximize human health and safety
>
> Minimize environmental damage
>
> Maximize desirable economic impact on owners
>
> Provide quality service to the consumers

Even these objectives are too broad.  It is necessary to break these
down into subobjectives so that a measure can be assigned to them.  From a
practical point of view, unless achievement on an objective can be measured,
it is of little use in a formal analysis.  This does not imply that an
objective numerical measurement is required.  In many cases, it will be
necessary to develop subjective indices for measuring achievement on an
objective.

The concept of obtaining measures (attributes) and subobjectives from
general objectives is illustrated in Table I.  By formally analyzing a
problem with alternative measures of effectiveness for a specific objective,

TABLE I. CONCEPTS ON DEVELOPING OBJECTIVES, SUBOBJECTIVES, AND MEASURES OF EFFECTIVENESS (after Keeney and Nair [1])

| OBJECTIVE | SUBOBJECTIVE | MEASURE OF EFFECTIVENESS*<br>(Attributes) |
|---|---|---|
| Maximize Human Health & Safety | Radiation Exposure | Level of radiation per person |
| Minimize Environmental Damage | Aquatic Ecology | Fish population<br>Change in ambient temperature of water<br>Subjective indices in terms of other sites |
| | Terrestrial Ecology | Animal population<br>Change in ambient temperature of water<br>Subjective indices in terms of other sites |
| | Water Quality | Objective measures available** |
| | Air Quality | Objective measures available** |
| | Aesthetic and Scenic Values | Subjective indices |
| | Impact of Transmission Lines | Length of line |
| | Historic & Cultural Features | Subjective indices |
| Maximize Desirable Economic Impact on Owners | Profit to Owners:<br>Cost (Initial)<br>Operating revenues<br>Operating expenses<br>Cost of Delays in Licensing | Monetary measure |
| Provide Quality Service to Consumers | Service Interruptions<br>Cost to Consumer<br>Delays in Initial Operation | Days<br>Annual cost<br><br>Days |

*For each subobjective, there can be more than one measure of effectiveness. When no objective measure is apparent, a subjective index can be developed from the experience of the decision maker.

**These are usually based on government agency specifications.

to conduct a sensitivity analysis to determine if the choice
influences the decision.

discussion, let us assume that there are n measures of effective-
referred to as attributes, labeled $X_1$, $X_2$,..., $X_n$. The small
, will represent a specific amount of attribute $X_i$.

For each attribute, a range has to be established. This range represents the practical limits within which achievement on the objective will
be measured.

Task B. <u>Specification of consequences and associated probabilities</u>. At
the time of decision, the future consequences of an action cannot be predicted
with certainty. The explicit treatment of uncertainty is a feature of
decision analysis. To describe the uncertainty associated with the various
consequences of an action, a probability distribution function $p_j$ $(\underline{x})$
associated with action j and the resulting consequences $(\underline{x})$ = $(x_1,...,x_n)$ is
required. The subjective concept of probability is used in decision analysis.
It represents the decision maker's degree of belief in the likelihood that
various consequences will occur and his willingness to act on this belief.
It is necessary to assess the assumption of probabilistic independence
between attributes. If this assumption is valid, then the probability
distribution over the n attributes is equivalent to the product of the
probability distribution for each attribute. For each attribute over the
specified range, the probability density function must be assumed for each
action $p_j(x_i)$. That is, locating the plant at any site results in a probability being associated with every level of an attribute. For some attributes, data and prior experience may provide the basis for estimating
probabilities. However, such information is often not available and probabilities based on the judgment of the decision maker and his designated
experts would have to be assessed.

If probabilistic independence is valid, then the joint probability
distribution function over n attributes is given by:

$$p_j(\underline{x}) = p_j(x_1,...,x_n) = \left[p_j(x_1)\right]\left[p_j(x_2)\right]\cdots\left[p_j(x_n)\right]$$

If probabilistic independence is not valid, then the problem is more
difficult and analytical and simulation models have to be used to generate
probability distributions.

Task C. <u>Describing the preferences of the decision maker: development
of a utility function</u>. The issue of achievement on competing objectives is
addressed by assessing the decision maker's relative preferences for the
consequences. It is necessary to quantify the decision maker's subjective
preferences so that it can be clearly seen, for example, if

$\underline{x}' = (x_1',...,x_n')$ is preferred to $\underline{x}'' = (x_1'',...,x_n'')$, if he is indifferent,

or if the latter is preferred. Then, if $(x_1',x_2',x_3^o,...,x_n^o)$ is indifferent

to $(x_1'',x_2'',x_3^o,...,x_n^o)$, one can say that given $x_3 = x_3^o,...,x_n = x_n^o$, the

decision maker is willing to change $x_1'$ to $x_1''$ if $x_2'$ is changed to $x_2''$.

To quantify preferences, it is necessary to assess the decision maker's utility function which assigns a number u to each of the possible consequences $(x_1, x_2, \ldots, x_n)$. This function has two desirable properties:

(1) $u(x_1', \ldots, x_n') \geqslant u(x_1'', \ldots, x_n'')$ if, and only if, $(x_1', \ldots, x_n')$ is

preferred to $(x_1'', \ldots, x_n'')$, and

(2) in situations with uncertainty, the expected value of u is the appropriate guide to make decisions. For each alternative j, the probability distribution $p_j$ and the preference function u are necessary and sufficient to calculate the expected value of u. Then the alternative with the highest expected value is chosen. This last result follows from some fundamental behavioral assumptions specified in von Neumann and Morgenstern [22]. Raiffa [23] has a discussion of their appropriateness in decision making.

For a problem with multiple attributes, it would be very convenient if a simple functional form for u could be developed so that

$$u(\underline{x}) = u(x_1, x_2, \ldots, x_n) = f\left[u_1(x_1), u_2(x_2), \ldots, u_n(x_n)\right] \quad \text{where}$$

$u_i(x_i)$ denotes utility of the amount $x_i$ of the attribute $X_i$

The assumptions of preferential independence and utility independence should be validated with the decision maker. If they are valid, then the multiattribute utility function can be expressed either in additive or in multiplicative form.

Additive form: $\quad u(x_1, x_2, \ldots, x_n) = \sum\limits_{i=1}^{n} K_i u_i(x_i)$

Multiplicative form: $\quad 1 + K u(x_1, x_2, \ldots, x_n) = \prod\limits_{i=1}^{n} \left[1 + KK_i u_i(x_i)\right]$

In both the cases $K_i$'s are positive scaling constants. Thus, the problem of finding the n-attribute utility function reduces to the simpler problem of finding n single-attribute utility functions, $u_i$, and the trade-offs on achievement of various objectives as measured by K's. A detailed discussion of the assessment of the utility function is provided in Keeney and Nair [1].


Task D. <u>Synthesizing the information and sensitivity analysis.</u> All the information obtained in the previous steps is placed in a logical framework through a decision tree. The expected utility at each node of the tree can be calculated and the alternate actions (i.e. choice of sites) can be ranked on the basis of expected utilities. Typically, a computer program is written to perform these calculations. This facilitates the conduct of sensitivity analyses to study variations in preferences and the estimates of uncertainties associated with various outcomes, and including alternate measures of effectiveness or additional objectives.

To illustrate how the expected utility is calculated, it should be noted that a probability density function associated with each attribute for each action has been determined.

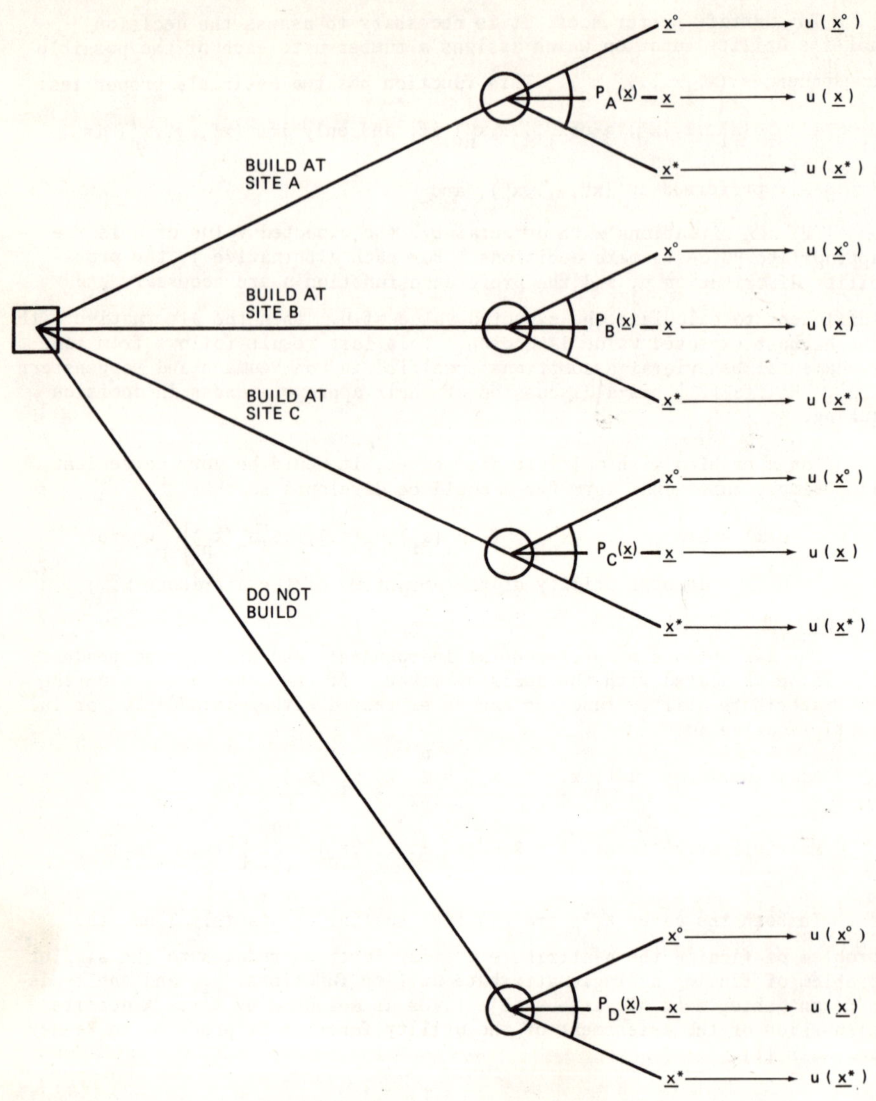


FIG. 18.   Simple decision analysis model of a nuclear siting decision (after Keeney and Nair[1]).

If probabilistic independence is assumed, then

$$p_j(\underline{x}) = p_j(x_1, x_2, \ldots, x_n) = \left[p_j(x_1)\right], \ldots, \left[p_j(x_n)\right]$$

Also, we have determined a multiattribute utility function for all the attributes over the appropriate range--

$$u(\underline{x}) = u(x_1, x_2, \ldots, x_n)$$

Therefore, the full range of consequences can be expressed in terms of the multiattribute utility function.

The expected utility of each action is the product of the probability density function associated with each action and the multiattribute utility function as follows:

If alternative j is taken, the expected utility =

$$\int_{x_1},\int_{x_2},\ldots,\int_{x_n} \left[ p_j(\underline{x}) \right] \left[ u(\underline{x}) \right] \, dx_1,\ldots,dx_n$$

The alternative actions can then be ranked according to the expected utility.

A picture of the decision analysis model in simple form is illustrated in Figure 18. The alternatives are evident. The notation $p_j(\underline{x})$ is the probability distribution if alternative j is taken, where $j = A,B,C$ and D, and D indicates 'do not build'. The possible consequences range from a worst, labeled $\underline{x}^0$, and a best, labeled $\underline{x}^*$; and the decision maker's utility function is u.

In some cases, the regulatory agencies and other concerned individuals have specified an upper limit in terms of probability of occurrence for certain attributes. In a decision analysis, this information can be readily made available by explicitly displaying the probabilities associated with each attribute.

## 6.  CONCLUSIONS

A systematic and logical procedure to identify candidate nuclear power plant sites on a regional basis has been developed. Further, decision analysis has been shown to be a sound approach for ranking the identified candidate sites. A formal analysis provides the basis for the public to evaluate such decisions, focuses discussion on substantive issues, and provides a framework for the resolution of conflicts.

The identification and application of exclusionary criteria is of special significance in the selection of candidate sites. As an example, an approach to the location of active faults is presented. Different levels of effort are required depending on the seismicity of the region under consideration and the level of confidence in the results that is acceptable to the decision maker. Adequate investigative techniques are available for identifying and locating active faults.

The effect of an earthquake at a site can be of sufficient magnitude to make construction of a plant uneconomical. The parameters that have to be considered have been identified, and suitable analytical techniques for determining the level of these parameters at a site are available.

An essential step in selecting sites is the ranking of several candidate sites. All the critical issues in the ranking problem can be addressed within the decision analysis framework. Of special significance is the explicit inclusion of the preferences of the decision maker and the treatment of the uncertainty associated with the consequences of a decision.

Techniques are available to structure and assess the utility function
for the decision maker over multiple attributes which include both tangible
and intangible factors.  Conflicts in achievement on competing objectives
are accounted for in developing a utility function.  Uncertainties in the
impact of an action measured in terms of the attribute can be expressed in
terms of a joint probability distribution.

Since this approach uses the preferences and judgments of the decision
maker and knowledgeable professionals selected by him, it is necessary that
there be close coordination between those implementing the approach and those
responsible for the decision.

## ACKNOWLEDGMENTS

The authors wish to acknowledge the contributions of Jefferson R.
McCleary, Leslie E. Oborn  and David J. Gross in preparing this paper and
the assistance of Carolyn Geiger and Patty Reitman in editing the manuscript.

## REFERENCES

[1]   KEENEY, R.L., NAIR, K.,"Decision analysis for the siting of nuclear
      power plants--the relevance of multi-attribute theory", Proc. IEEE
      Special Issue on Social Systems Engineering (in press, 1975).

[2]   CLUFF, L.S., SLEMMONS, D.B., WAGGONER, E.B.,"Active fault zone hazards
      and related problems of siting works of man", 4th Symposium on Earth-
      quake Eng. (Proc. Symposium, Univ. of Roorhee, India, 1970).

[3]   CLUFF, L.S., BOLT, B.A., Risk for earthquakes in the modern urban
      environment, with special emphasis on the San Francisco Bay Area,
      Urban Environmental Geology in the San Francisco Bay Region, special
      publ. of Assoc. Engineering Geologists (1969).

[4]   BROGAN, G.E., CLUFF, L.S., KORRINGA, M.N., SLEMMONS, D.B.,"Active
      faults in Alaska", Int. Symposium on Recent Crustal Movements (Proc.
      Int. Symposium, Zurich, Switzerland, 1974, in press).

[5]   CLUFF, L.S., Peru earthquake of May 31, 1970, engineering geology
      observations, Bull. Seismol. Soc. America 61 3 (1971) 511.

[6]   USAEC, Code of Federal Regulations, Appendix A, Pt. 100, Federal
      Register 36 228 (1971).

[7]   CLUFF, L.S., BROGAN, G.E.,"Investigation and evaluation of fault activi
      in the USA", 2nd Int. Congress of Int. Assoc. of Engineering Geology
      (Proc. Conf. São Paulo, Brazil, 1974, in press).

[8]   SHERARD, J.L., CLUFF, L.S., ALLEN, C.R., Potentially active faults in
      dam foundations, Geotechnique 24 3 (1974) 367.

[9]   OBORN, L.E.,"Seismic phenomena and engineering geology", 2nd Int.
      Congress on Int. Assoc. of Engineering Geology (Proc. Conf. São Paulo,
      Brazil, 1974, in press).

[10]  CLUFF, L.S., HANSEN, W.R., TAYLOR, C.L., WEAVER, K.D., BROGAN, G.E.,
      McCLURE, F.E., IDRISS, I.M., BLAYNEY, J.A.,"Site evaluation in seis-
      mically active regions, an interdisciplinary team approach"(Proc. Int.
      Conf. Microzonation for Safer Construction; Research and Application,
      Seattle, Washington, 1972).

[11]  TAYLOR, C.L., CLUFF, L.S., Fault activity and its significance assessed
      by exploratory excavation, Stanford Univ. Publication, Geol. Sci. 13
      (1973) 239.

[12]  HOUSNER, G.W., "Intensity of earthquake ground shaking near the causative
      fault", 3rd World Conf. on Earthquake Eng. (Proc. Conf. Auckland,
      New Zealand, 1965) 1.

[13]  SCHNABEL, P.B., SEED, H.B., Acceleration in Rock for Earthquakes in the
      Western United States, EERC 72-2, Earthquake Engineering Research
      Center, Univ. of Calif., Berkeley (1972).

[14]  DONOVAN, N.C., "A statistical evaluation of strong motion data including
      the February 9, 1971, San Fernando earthquake", 5th World Conf. on
      Earthquake Engineering, Rome, Italy (1973).

[15]  TOCHER, D., PATWARDHAN, A., "Attenuations of peak acceleration in earth-
      quakes", Seismological Society of America, Annual Meeting (Proc. in
      press, 1975).

[16]  NEWMARK, N.M., BLUME, J.A., KAPUR, K.K., Design response spectra for
      nuclear power plants, American Society Civil Eng.  (San Francisco, 1973).

[17]  SEED, H.B., UGAS, C., LYSMER, J., Site-Dependent Spectra for Earthquake-
      Resistant Design, 74-12, Earthquake Engineering Research Center, Univ.
      of Calif., Berkeley (1974).

[18]  GUTENBERG, B., RICHTER, C.F., Earthquake magnitude, intensity, energy
      and acceleration, Bull. Seismol. Soc. America 46 2 (1956) 105.

[19]  HERSHBERGER, J., A comparison of earthquake accelerations with intensity
      ratings, Bull. Seismol. Soc. America 46 4 (1956) 317.

[20]  NEUMANN, F., Earthquake Intensity and Related Ground Motions, Univ.
      of Washington Press, Seattle, Wash. (1954).

[21]  IDRISS, I.M., SEED, H.B., "Effects of local geologic and soil conditions
      on damage potential during earthquakes", 2nd Int. Congress of the Int.
      Assoc. of Engineering Geology (Proc.  São Paulo, Brazil, Conf., 1973, in
      press).

[22]  VON NEUMANN, J., MORGENSTERN, O., Theory of Games and Economic Behavior,
      2nd ed., Princeton Univ. Press, Princeton, New Jersey  (1947).

[23]  RAIFFA, H., Preferences for Multi-Attributed Alternatives, RM-5868-
      DOT/RC, The Rand Corporation (1969).


## DISCUSSION


L. VENKATESH (Chairman): I must congratulate Mr. Cluff on the
impressive catalogue of faults and their devastating effects, covering most
regions of the world.  In every case, I noted that there were surface indications of the existence of the fault.  But how do we detect deep-seated faults
which may not show up on the surface owing to heavy recent overlying deposits
and the long return time of earthquakes on such faults?  This points to the
need for intensive and prolonged microseismic studies covering a reasonably
representative area around a potential site as a matter of routine, before
any siting decision is taken.  I should be most interested to hear Mr. Cluff's
comments on the subject.

L.S. CLUFF: It must be clearly understood that there are two aspects of active fault behaviour which are quite distinct, although related: the fault may represent a hazard through surface fault rupture, or a hazard only from shaking. If it can be convincingly shown that a fault has not displaced geologically young deposits (half a million years by USAEC standards), then it need not be considered capable of surface faulting. However, some faults may be capable of generating small-to-moderate earthquakes at depth withou a causative fault rupturing at the surface. A fault which does this is accord-ingly a problem only as an earthquake source. It may be desirable to instru-ment and monitor nuclear reactor sites for microearthquakes, but this woulc depend on the region, local conditions and other factors. Each case would have to be considered on its own merits. As a general rule I see definite advantages in installing a good microearthquake monitoring network. It could be used, among other things, to determine the distribution of micro-earthquakes as a guide to the probable distribution of larger, less frequent earthquakes.

J.J. DiNUNNO: What are your views on the usefulness of earth satellite data for identifying geological faults and evidence of historical seismic activity?

L.S. CLUFF: We have utilized imagery from satellites in a large number of regional studies, and have found it useful in identifying some major faults and long lineaments that may be faults. However, this is only one of several types of information that should be utilized in regional studies. It is quite impossible to identify a fault as active by means of satellite imagery alone. Generally speaking, the most effective procedure in nuclear reactor siting studies is to use the interpretations of imagery as a guide for further studies aimed at defining fault activity. In this way, the examination of individual sites can benefit from the interpretations of satellite imagery completed during a regional study.

P.C. RIZZO: You cannot dismiss intensity data, as one might infer from your presentation. Fortunately, or unfortunately, depending on one's point of view, we are not all working in southern California, where faults are reasonably well defined and much strong motion data is available. Most of us are working in parts of the world where no strong motion data exist, only intensity data. Certainly one will always find anomalous data forming exceptions to the rule, for both magnitude and intensity, especially in the nea field where your photographs were taken. Therefore, it is obvious that one must use a conservative, logical approach in dealing with either type of data

L.S. CLUFF: The limited time allowed for the presentation of our pape did not allow a full discussion of the usefulness of earthquake intensity data; my comment about intensity data was meant as a caution regarding its use. Certainly one should always utilize all the data available, including intensity data from field reports, instrumental epicentre data and geological data. One of the main objectives of this presentation is to stress the value of data other than historic seismicity records. In most seismically active regions of the world, including California, we have found that a more realistic evaluation of the potential for seismic activity can be obtained by evaluating geologic information rather than relying strictly on the historic record of seismicity, which is likely to be incomplete. One must always take intensit and magnitude data into consideration, but they should be used with caution and shrewd judgement — otherwise misleading conclusions may be drawn. As to your comment on strong ground motion, copious data are in fact available in California but the record there, too, is far from complete.

A. DE ACHA ARACAMA:  In establishing seismotectonic zones, do you take account of deep tectonics in any way?  What is your opinion about the curves for seismic intensity versus horizontal acceleration of the ground, especially the curve of Coulter and co-workers (Coulter, H.W. et al., "Seismic and geological siting considerations for nuclear facilities", 5th World Conf. on Earthquake Engineering, Rome, 1973)?

L.S. CLUFF:  We are not worried about deep earthquakes (those deeper than 300 km) causing serious damage, as they are already 300 km or more from the site, even though they may be directly beneath it.  Even earthquakes of intermediate depth (between 71 and 300 km) are a long way from the site;  but of course the shallower the event, the greater is the likelihood of its causing damage.  By and large, it is the shallow earthquakes (0-70 km depth) which cause most damage, the very shallow events being the most likely to generate surface faulting.  Thus, the greatest emphasis in nuclear reactor siting should be placed on understanding shallow earthquakes and less on deep earthquakes.  Sometimes deep seismicity may serve as a guide to understanding shallow seismicity, especially along the boundaries of some plates.

The available intensity versus acceleration data indicate a significant scatter, and selection of a design acceleration based on intensity is at best a crude procedure.  We use intensity data mainly because they provide an extension of the historic record and increase our knowledge of the seismic activity in an area, thus giving us a better basis for assessing past seismicity and for estimating future activity.  In many parts of the world (e. g. the eastern region of the United States of America) most of the available seismic information is in the form of intensity data.  This makes it necessary to select design accelerations based on intensity.  The most widely used correlations for the purpose are those proposed by Hershberger [19], Gutenberg and Richter [18] and Neumann [20].  When used with judgement, exercised in the light of experience, these correlations can provide reasonable estimates of conservative design accelerations.

Coulter, Waldron and Devine attempted to incorporate local site conditions into acceleration versus intensity correlations.  A lack of sufficient data (especially in the high-intensity ranges) made these correlations inadequate.  I can say this  because two of the authors have expressed concern regarding the misunderstanding and misuse of these curves.

Our approach is to consider magnitude, distance and local site conditions as fully as possible.  When only intensity data are available, we use them with caution, relying on experience and good judgement to derive design accelerations from them.

R.B. ELLWOOD:  The attitude of the USAEC regulatory authorities is that sites which contain active faults are not suitable.  What is your interpretation of the term 'site'?  Does the site consist of the ground beneath Class I structures, or does it include the general construction area, the exclusion area or the area within, say, two miles of the reactor? Secondly, how close can a reactor be built to a known active fault?  One or more of your figures would suggest that as little as ten feet might not be so disastrous.

L.S. CLUFF:  I interpret the term 'site' as a more or less confined area where Class I structures are to be located.

To your second question there may be several answers.  The answer that counts has to do with being able to license the site through a regulatory agency, such as the USAEC.  The USAEC guidelines indicate that it is

desirable to locate nuclear reactors at least five miles from an active fault.
In my discussions with USAEC representatives, I find most of them would
not consider a site with an active fault closer than five miles.  To pursue
a site close to an active fault may take more time and cause expensive
delay, so it is generally avoided in the United States of America.  Other
regulatory agencies in other countries may decide that five miles is not
the correct distance.  My own opinion is that each active fault, and even
different segments along a single active fault, should be considered on its
own merits.  I believe that the risks from surface faulting depend on many
factors which vary from one active fault to another and even along the strike
of a single  long active fault.  Some of the most important factors are the
type of fault (strike slip, dip slip or oblique slip), the direction of displacement, the amount of displacement, the attitude of the fault plane and the
near-surface geologic conditions.  It is my belief that if studies were carrie
out which permitted a reliable determination of all these factors, it would
be possible to locate a nuclear reactor within a mile of some active faults.
On the other hand, for some other faults five miles might not be a safe
distance.  In short, we must consider each active fault and each site
individually.

G. B. BAECHER:  During the past five years quantitative approaches
to geological exploration strategies have been introduced.  These are based
on random process models, inductive philosophy and Bayesian decision
theory.  The models make it possible to optimize the allocation of exploratic
effort over time and space and to draw quantitative inferences from data
with associated levels of uncertainty.  This process yields information whic
could be directly incorporated in the quantitative decision methodologies
suggested in your paper.  However, in your paper you stressed the qualitative basis of geological exploration.  Have you used quantitative strategies
of exploration, and if so, with what success?

K. NAIR:  We are aware of the quantitative approaches you mentioned,
but so far they have not been accepted by regulatory agencies in the United
States of America.  For this reason we have not applied them to the problen
of siting nuclear power plants, but we have used similar concepts in certain
other branches of our work.  We believe, in fact, that these approaches
should be applied to the siting of nuclear power plants.

G. HAKE:  You assess the level of protection to be given to a nuclear
facility against earthquakes or other external hazards, but is this level
related in your methodology to the hazard to people from the event itself?
There must be a point of no return benefit.

K. NAIR:  This can readily be accounted for, since in the 'no build'
alternative we would consider the case of the occurrence of the earthquake
without a nuclear facility being present.  A comparison of this case with
that of the facility being present would enable us to assess the consequence,
depending on the degree of safety of the nuclear facility.

H. A. G. KOHLER:  Before we leave the subject of reactor siting in the
vicinity of active faults, I should like to ask whether you know the distance
between the main fault of the San Andreas system and the nuclear power
stations under construction in this region, for instance, those at San Onofre
or Diablo Canyon.

L. S. CLUFF:  The distance from the main trace of the San Andreas
fault to San Onofre is approximately 70 miles (110 km), but the most
historically active fault of the San Andreas system in the Los Angeles

area is the San Jacinto fault, located approximately 45 miles (70 km) east
of San Onofre.  Another major active fault, which is broadly considered
to be part of the San Andreas fault system, is the Newport-Inglewood fault,
which appears to pass approximately 5 miles (8 km) offshore from San
Onofre.  The Diablo Canyon nuclear station is located approximately
50 miles (80 km) west of the San Andreas fault and lies to the north of the
Newport-Inglewood and San Jacinto faults.

H. JAMMET:  I would like to comment on the fact that we are asked
to consider possible genetic effects on many generations as a risk connected
with nuclear power plants.  The seismic events described to us have claimed
tens of thousands of victims.  Has anybody considered how much genetic
loss these slaughters represented?  We are here dealing with real deaths,
while in the case of reactor accidents the evaluation of genetic risk is based
on hypotheses which, though cautious, are by no means certain.

L.S. CLUFF:  Professor Jammet makes a valid point, in that the
possibility of death from an earthquake without any nuclear accident must
be considered in conjunction with the consequences of a nuclear accident
resulting from an earthquake.  However, it is also important, in attempting
to obtain public acceptance, to consider the public's perception of these
consequences.  This perception is not always logical, but it must be
understood.

J.B. BURNHAM:  The risk the public is willing to accept varies a
great deal depending on its nature.  There is ample proof (this has been
discussed by Starr, Otway and others) that people will accept 'voluntary'
risks at least two orders of magnitude greater than 'involuntary' ones.
Nuclear hazards are in the latter category, while the natural hazards to
which Professor Jammet has referred are in the former.

K. NAIR:  Yes, but the analysis of acceptable risk in terms of
'voluntary' or 'involuntary' risk, presented in the oft-quoted work of
Starr, does not satisfactorily explain the concept.  Acceptable risk is
related to the ability to pay for reducing risk.  In cases where the cost
of reducing risk can be spread over a large part of the population, the
level of acceptable risk is much lower than when the cost is spread over
a few.  This point has been explained by Lave (Lave, L.B., Risk, Safety
and the Role of Government Perspectives on Benefit Risk Decision Making,
National Academy of Engineering, Washington, D.C. (1972)).

Acceptable risk should be examined in the context of utility theory,
where the cost of changing the expected utility of an undesirable consequence has to compare with the expected utility of other uses for the same
money.  This acceptable risk is dependent on the values of society and
will change with time.

E. IANSITI:  So far no assessment has been made of the consequences
of the collapse of a nuclear power plant as the result of an earthquake.
No evaluation of this kind seems to be included in the Rasmussen study,
and for this very reason one has to be conservative in selecting the design-
basis earthquake.

In many countries I have visited areas damaged by severe earthquakes
only a few hours after the event.  I have often wondered what would have
happened if there had been nuclear power plants in those areas.  I was
always convinced that a reactor designed with the safety margin used today
would have withstood the quake without suffering damage.  But those who
feel that these margins ought to be reduced must consider what would then

happen, bearing in mind, at least, the public apprehension about possible damage to the reactor.

L.S. CLUFF: I agree with your comments, having regard to the many devastated earthquake sites which I have visited. One must keep in mind, though, that it is not only the reactor itself which is vulnerable, and that facilities associated with the reactor must also be properly designed. The damage to a nuclear reactor facility designed in accordance with today's standards would probably not be noticeable, provided the reactor was not actually astride an active fault. This, of course, is the reason why so much attention is being focussed on active faults. They constitute one of the most critical factors in siting.

# RECHERCHE D'UN CRITERE D'EVALUATION DE SITE PAR LA PONDERATION DES DONNEES RADIOLOGIQUES, METEOROLOGIQUES ET DEMOGRAPHIQUES

A. DOURY, R. GERARD,
Département de sûreté nucléaire,
CEA, Centre d'études nucléaires de Saclay,
Gif-sur-Yvette, France

## Abstract–Résumé

FINDING A CRITERION FOR SITE EVALUATION BY WEIGHTING RADIOLOGICAL, METEOROLOGICAL AND DEMOGRAPHIC DATA.
The general problem of compatibility between given installations and their site, as well as more specific problems such as those associated with decisions to be taken in regard to demographic equilibrium in the environment, logically involve finding quantitative or at any rate semiquantitative criteria that will enable selection to be made under the least subjective conditions possible. Since the air is frequently the principal primary vector of effluents, especially in the case of an accident, atmospheric discharges are considered first separately. As the task is to find a "site security index" integrating the meteorological and demographic data with radiological data relating to the potential sources and the consequences of contamination, the authors first consider the possibility of using the man‧rem concept, which has the advantage of possessing the dimension of a collective dose. In view of the objections that have been raised, particularly in connection with the controversy over the existence of a threshold for the dose-equivalent effect relation, a number of other practices are described and discussed. The conclusions are complex and subject to revision. Several definitions of site security indices can be identified and, pending the final decision stage, actually put into use, in conjunction with specific maps or graphs serving to reveal potentially important features which might be obscured by excessive integrations or summations.

RECHERCHE D'UN CRITERE D'EVALUATION DE SITE PAR LA PONDERATION DES DONNEES RADIOLOGIQUES, METEOROLOGIQUES ET DEMOGRAPHIQUES.
Le problème général de la compatibilité entre des installations données et leur site, ainsi que des problèmes plus particuliers comme ceux qui peuvent être liés aux décisions à prendre en matière d'équilibre démographique dans l'environnement, impliquent logiquement la recherche de critères quantitatifs ou au moins semi-quantitatifs susceptibles d'orienter les choix dans les conditions les moins subjectives possibles. L'air étant souvent le principal vecteur primaire d'effluents, surtout en cas d'accident, les rejets atmosphériques seront seuls considérés dans un premier temps. Comme il s'agit de trouver un «indice de niveau de sûreté de site», intégrant les données météorologiques et démographiques avec des données radiologiques relatives aux sources potentielles et aux conséquences de la contamination, on a d'abord pensé à utiliser la notion d'homme‧rem, qui a l'intérêt d'avoir la dimension d'une dose collective. Compte tenu des objections liées notamment aux controverses sur l'existence de seuils d'équivalents de dose entraînant l'apparition de dommages, un certain nombre d'autres pratiques sont exposées et discutées. Les conclusions sont complexes et sujettes à révision. Plusieurs définitions d'indices de niveau de sûreté de site peuvent être retenues, et même conservées comme éléments de jugement jusqu'au stade des décisions finales, en liaison avec des cartes ou des graphiques spécifiques faisant apparaître des particularités éventuellement importantes qui risqueraient d'être masquées par des intégrations ou totalisations abusives.

## 1. ORIGINE ET POSITION DU PROBLEME

Malgré des difficultés de tous genres, l'augmentation et l'amplification des programmes de constructions de centrales nucléaires et d'usines associées, obligent les spécialistes de la sûreté à fournir aux autorités

responsables de l'orientation des choix en matière de classement de site
et d'équilibre démographique, des critères quantitatifs, ou au moins semi
quantitatifs, susceptibles de leur permettre de prendre leurs décisions
dans des conditions les moins subjectives possibles.

Comme il faut bien sérier les problèmes en tenant compte des urgence
et comme en cas d'accident l'air est généralement reconnu comme le vecteur le plus préoccupant (rapidité, problèmes de confinement et de contr
mesures, consommation permanente), les transferts par voie atmosphérique
seront seuls considérés dans un premier temps.

Parmi les données de base indispensables, devront naturellement figu
rer les données radiologiques, c'est-à-dire tout ce qui concerne les
rejets radioactifs, leurs probabilités et les seuils d'équivalents de
dose entraînant l'apparition de dommages. Les données démographiques,
considérées dans leur configuration tant actuelle que future, seront
présentées selon une distribution adéquate, compatible avec celle des
vents, en secteurs de 20 degrés dont l'un centré sur le Nord géographiqu
et en éléments de secteurs de 5 ou 10 kilomètres, sans limitation de dis
tance a priori jusqu'à une centaine de kilomètres.

## 2. CONDITIONS DE TRANSFERT ATMOSPHERIQUE

On utilisera la notion très commode de coefficient de transfert atmosphérique (CTA), considérée comme une fonction de transfert particulière réduite, au moins en l'absence de phénomènes secondaires comme les
dépôts et la décroissance, à un simple facteur multiplicatif entre le
terme source ou rejet et la réponse cherchée en chaque point du site
après la traversée du "filtre" atmosphérique. Comme on ne s'intéresse
finalement qu'aux concentrations intégrées ou aux équivalents de dose
correspondants, la seule connaissance du CTA (équivalent de la concentra
tion intégrée normalisée fréquemment utilisée), permettra toujours d'ac-
céder au résultat recherché, quelles que soient la nature et la forme du
rejet, instantané ou prolongé, permanent ou variable, pourvu toutefois
que la quantité totale sur l'intervalle de calcul (et de validité du
CTA) soit connue.

Le traitement du problème en secteurs centrés sur la source, entraî
ne bien entendu, avec les approximations que cela implique, une assimi-
lation plus ou moins valable des trajectoires d'effluents avec leur tan
gente à l'origine. Cette circonstance fait apparaître l'une des difficu
tés de la prise en considération de distances égales ou supérieures à u
centaine de kilomètres, ainsi que de sites au relief compliqué. Moyenna
ces réserves les directions de transfert seront assimilées aux directio
du vent convenablement mesurées au niveau de la source sur une durée su
fisante (plusieurs années). Ces directions seront distribuées en dix-
huit secteurs de 20 degrés (dont l'un centré sur le Nord géographique)
affectés chacun d'une probabilité de présence du vent.

Pour une direction donnée les conditions de transfert (ou CTA) dépe
dent de la vitesse du vent et de la diffusion turbulente. Leur réparti
tion est effectuée d'une manière optimale (nécessité de disposer de plu
sieurs points pour construire certaines courbes) en cinq lois (fig. 1)
de coefficients de transfert en fonction de la distance (LCTA). Pour
chaque distance la valeur de CTA retenue est éventuellement celle qui
correspond à la moyenne sur la largeur du secteur, celui-ci étant de
toute manière considéré comme centré sur la valeur maximale, et les valeurs extérieures étant négligées. Les valeurs de CTA sont déduites
du Rapport CEA-R-4280 /¯1_7 et déterminées en considérant deux classes

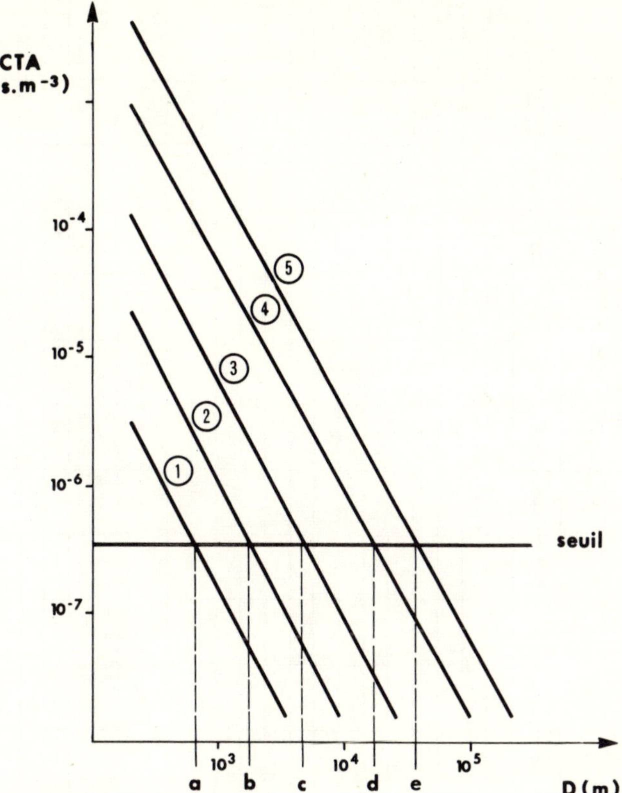



FIG. 1.   Lois de coefficients de transfert atmosphérique CTA en fonction de la distance D (représentation schématique).

de stabilité comprenant respectivement trois et deux classes de vitesses. Chaque LCTA est affectée d'une probabilité d'apparition dans chaque secteur géographique.

On a ainsi quatre-vingt-dix probabilités météorologiques, désignées dans la suite par PM directement déduites de tableaux de fréquences d'observations de vent par plages de direction et de vitesse associées pour différentes conditions de stabilité, tels qu'ils doivent figurer dans les rapports de sûreté (fig. 2).

3.   REGIME NORMAL DE ROUTINE - CALCULS POUR UN AN DE FONCTIONNEMENT

Les rejets de routine étant prévus pour n'avoir aucune incidence sur les populations, ils ne sont généralement pas retenus pour la recherche de critères de sélection. Pour la clarté de l'exposé et pour des raisons qui seront indiquées en conclusion, nous avons cependant choisi de les prendre en considération, au moins à titre d'entrée en matière.

Les différentes opérations proposées sont présentées dans leur ordre chronologique naturel.

**STATION :**
latitude :
longitude :
altitude :

**MESURES**
hauteur :
durée d'échantillonnage :
périodicité :

**FREQUENCES POUR 1000 DES VITESSES**

**DE VENT EN FONCTION DE LA DIRECTION**

**PERIODE**

**CRITERES DE SELECTION POUR LE TABLEAU**
sous-période (saison, mois...) :
classe de stabilité :
classe de précipitations :

**NOMBRE D'OBSERVATIONS**
total (sur la période) :
partiel (pour le tableau) :

| DEGRES m.s⁻¹ | 360 | 020 | 040 | 060 | 080 | 100 | 120 | 140 | 160 | 180 | 200 | 220 | 240 | 260 | 280 | 300 | 320 | 340 | direction inconnue | toutes directions |
|---|---|---|---|---|---|---|---|---|---|---|---|---|---|---|---|---|---|---|---|---|
| 30 | | | | | | | | | | | | | | | | | | | | |
| 25 | | | | | | | | | | | | | | | | | | | | |
| 20 | | | | | | | | | | | | | | | | | | | | |
| 15 | | | | | | | | | | | | | | | | | | | | |
| 14 | | | | | | | | | | | | | | | | | | | | |
| 13 | | | | | | | | | | | | | | | | | | | | |
| 12 | | | | | | | | | | | | | | | | | | | | |
| 11 | | | | | | | | | | | | | | | | | | | | |
| 10 | | | | | | | | | | | | | | | | | | | | |
| 9 | | | | | | | | | | | | | | | | | | | | |
| 8 | | | | | | | | | | | | | | | | | | | | |
| 7 | | | | | | | | | | | | | | | | | | | | |
| 6 | | | | | | | | | | | | | | | | | | | | |
| 5 | | | | | | | | | | | | | | | | | | | | |
| 4 | | | | | | | | | | | | | | | | | | | | |
| 3 | | | | | | | | | | | | | | | | | | | | |
| 2 | | | | | | | | | | | | | | | | | | | | |
| 1 | | | | | | | | | | | | | | | | | | | | |
| | | | | | | | | | | | | | | | | | | | | |
| calmes | | | | | | | | | | | | | | | | | | | | |
| toutes vitesses | | | | | | | | | | | | | | | | | | | | |

nb. le signe* signifie fréquence non nulle inférieure à 0,5 pour 1000
chaque fréquence correspond à la classe de vitesse de la ligne dans la classe de direction de la colonne
chaque fréquence est calculée par rapport au nombre total d'observations de la période d'acquisition, tous types de temps confondus

FIG.2.    Tableau de fréquences d'observations météorologiques.

3.1.   La carte de l'espérance mathématique des CTA est établie par éléments de secteur (somme des produits de chaque CTA par sa probabilité dans le secteur).

3.2.   La carte des produits de l'espérance mathématique des CTA par la population concernée est établie par élément de secteur.

3.3.   La transformation de la carte précédente en carte de concentration intégrée collective ou d'équivalent de dose collectif annuel (ou engagé) est éventuellement effectuée par multiplication par un terme "rejet", réel ou arbitraire (curie.homme.seconde.$m^{-3}$ ou homme.rem).

3.4.   La carte précédente est éventuellement étendue à d'autres rejets ou à tous les rejets envisageables.

3.5.   Les concentrations intégrées collectives ou les équivalents de dose collectifs sont éventuellement sommés, par secteur, par zone ou par région. La distance de sommation est soit concertée en fonction des particularités de chaque site, soit limitée automatiquement par le jeu de l'importance des différents rejets et de valeurs limites d'équivalent de dose individuel. C'est le résultat de cette somme, exprimé en curie.homme.seconde.$m^{-3}$ ou en hommes.rem qui pourrait constituer un premier "indice de niveau de sûreté de site".

## 4.   SITUATIONS EXCEPTIONNELLES ACCIDENTELLES

Par l'importance des rejets mis en cause, les situations accidentelles sont probablement les seules qui soient susceptibles de fournir des éléments de détermination de critères quantitatifs indiscutés pour le classement ou la sélection des sites. Il resterait cependant à démontrer que les classements par situations exceptionnelles sont toujours et systématiquement différents des classements par régime de routine.

Les situations exceptionnelles sont généralement traitées de deux manières: l'une est à rejet unique ou de référence, l'autre considère le spectre de tous les rejets en fonction de leur probabilité. Les différentes opérations sont toujours présentées dans leur ordre chronologique naturel.

### 4.1.   Prise en considération d'un seul rejet maximal, ou de référence, secteur par secteur, avec la LCTA la plus sévère et sa probabilité

4.1.1.   La carte des produits (CTA).(Population), est effectuée par élément de secteur.

4.1.2.   La carte précédente est transformée en carte de concentrations intégrées collectives ou d'équivalents de dose collectifs, par simple multiplication par le terme "rejet".

4.1.3.   Les concentrations intégrées collectives ou équivalents de dose collectifs sont éventuellement sommés dans chaque secteur sur des distances concertées en fonction des particularités de chaque site, ou limitées automatiquement par le jeu de l'importance du rejet et de la valeur de seuil d'apparition des dommages pour la dose correspondante. Une telle somme peut constituer un "indice de niveau de sûreté de site".

4.1.4.   Une carte du nombre de personnes susceptibles de recevoir, dans chaque élément de secteur, un équivalent de dose égal ou supérieur à celui du seuil d'apparition des dommages, peut aussi être établie. Les valeurs peuvent en être sommées par secteur dans les mêmes conditions que précédemment pour constituer un autre "indice de niveau de sûreté de site".

4.1.5.    Plusieurs cartes analogues peuvent être établies pour plusieurs accidents de référence traités séparément.


4.2.    Prise en considération de tous les rejets possibles en fonction de leurs probabilités respectives

4.2.1.    Pour chaque groupe "rejet-seuil" et pour chaque secteur, on établit, à partir de cinq LCTA de la figure 1, une courbe (fig. 3) de la probabilité météo PM d'atteindre ou de dépasser le seuil considéré, en fonction de la distance ou du nombre NC de personnes concernées (correspondance directe déduite de la carte de population).

4.2.2.    Pour chaque groupe "rejet-seuil", une carte de courbes "isoprobabilité météo" (iso PM) d'atteindre ou de dépasser le seuil considéré est établie à partir des courbes précédentes (fig. 4).

4.2.3.    Pour chaque groupe "rejet-seuil", une carte des produits (PM). (Population) est établie par élément de secteur.

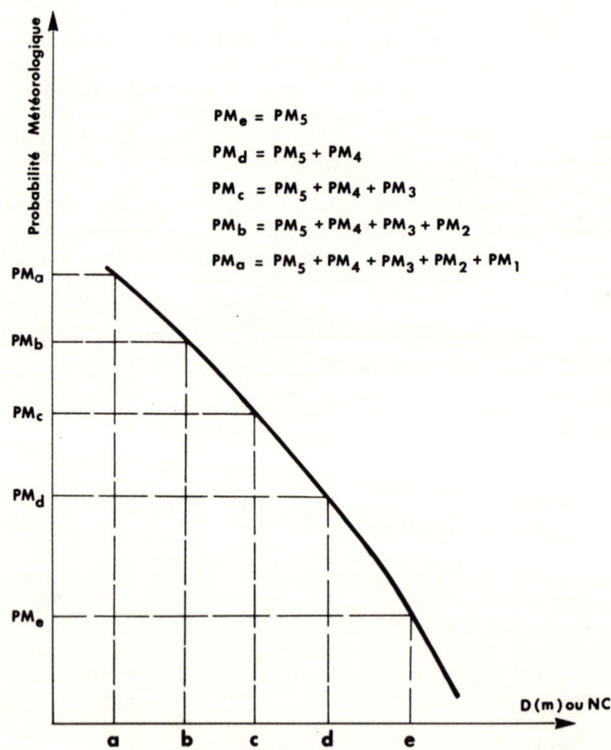


FIG. 3.    Probabilité météorologique PM d'atteindre ou de dépasser dans un secteur donné un seuil donné pour un rejet donné en fonction de la distance D ou du nombre d'individus concernés NC.

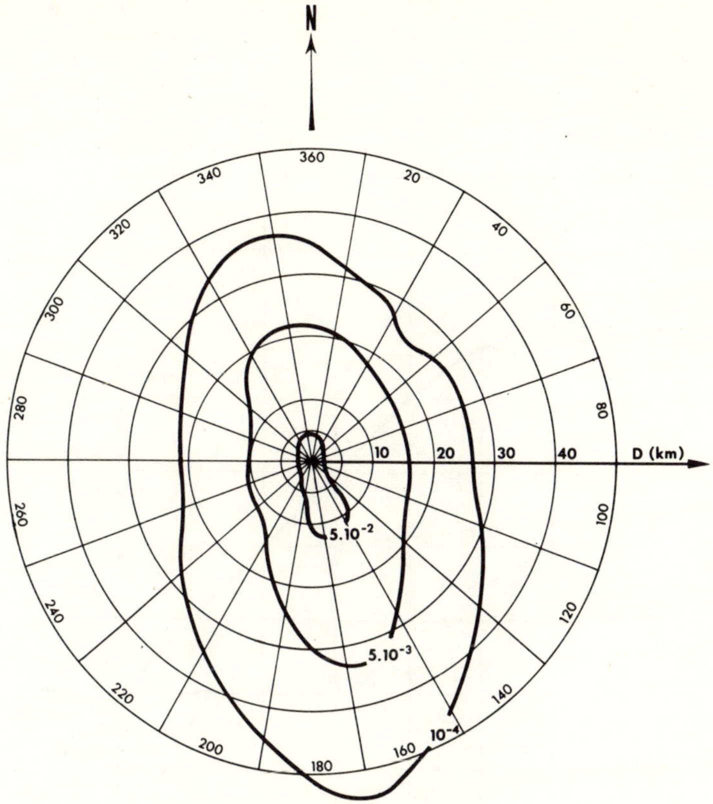



FIG.4.   Courbes isoprobabilité météo d'atteindre ou de dépasser un seuil donné pour un rejet donné.

4.2.4.   Pour chaque groupe "rejet-seuil", les produits (PM).(Population)
sont sommés par secteur ou pour tous les secteurs. Le résultat, somme
pondérée ou moyenne stochastique des individus susceptibles de recevoir
un équivalent de dose supérieur ou égal à la valeur de seuil d'apparition
des dommages, peut constituer un <u>"indice de niveau de sûreté de site"</u>
/⁻2_7 .

4.2.5.   A partir d'une courbe de correspondance Probabilité-Rejet (fig.5)/⁻3_7,
et de la relation entre le nombre de personnes concernées et l'importance
du rejet, un réseau de courbes de la probabilité de rejet, en fonction du
nombre de personnes concernées par la dose seuil (PR-NC), est construit
dans chaque secteur, à raison d'une courbe  par point considéré sur la
courbe PM-NC de la figure 3. Chacune des courbes de ce réseau (fig. 6)
est paramétrée par la probabilité météo PM correspondante, ce qui permet
le calcul (par le produit des probabilités) de la probabilité absolue en
tous points, et la construction de courbes de probabilité absolue en fonction du nombre de personnes concernées (fig. 7).

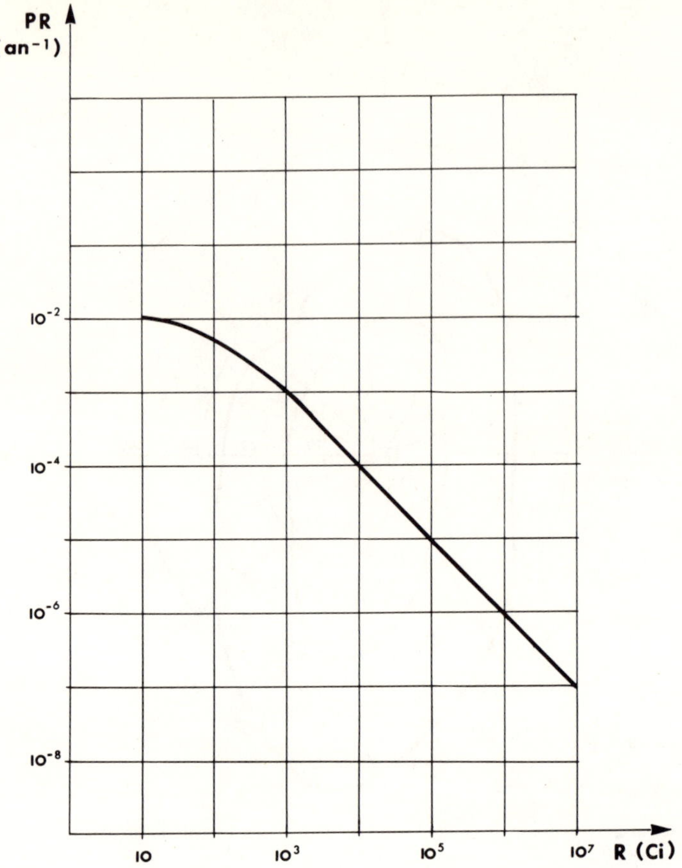

FIG.5.   Probabilité PR d'apparition des rejets en fonction de leur importance R [3].

4.2.6.   En passant par la considération de la moyenne des nombres de pe
sonnes concernées, pondérée par les probabilités météo, les courbes iso
probabilité météo du réseau précédent peuvent être transformées en une
seule courbe de correspondance PR-NC par secteur, entre le nombre de pe
sonnes concernées par la dose seuil, et une probabilité de rejet donné /

5. CONCLUSIONS

   Par la quantité même des différentes possibilités évoquées, on voit
qu'il est difficile de conclure sur un seul "nombre-magique" prétendant
représenter, dans des conditions susceptibles d'entraîner un certain
consensus, un facteur de qualité de site, ou "indice de niveau de sûret
de site", permettant de procéder à des classements ou à des choix sans
interventions de nature empirique (comparaisons de cartes de situation
ou de courbes de probabilités).

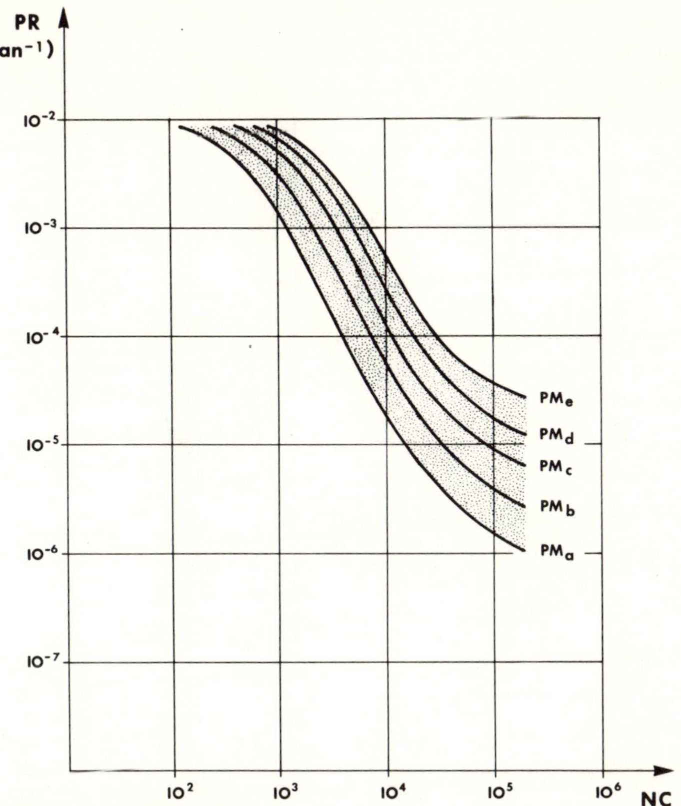



FIG.6.   Correspondance entre la probabilité d'un rejet PR et le nombre de personnes concernées NC par un seuil donné dans un secteur donné pour cinq valeurs de probabilité météorologique PM.

Le régime de routine, peu intéressant en soi, sauf peut-être s'il aboutit sensiblement aux mêmes classements que les régimes accidentels, reste en tout cas le plus facile à traiter, et le seul capable, sur des bases relativement simples, de fournir un indice de niveau de sûreté de site, et de permettre des comparaisons avec les doses dues aux sources naturelles.

Le cas du rejet unique maximal, ou de référence, se traduit par des conditions de traitement assez commodes, mais comporte de grandes parts d'arbitraire. Il est de plus incapable de fournir un indice unique pour un site donné par suite de l'impossibilité de principe d'effectuer la sommation finale des différents secteurs.

La solution la plus satisfaisante est aussi la plus exigeante en matière de données. Elle comporte essentiellement la prise en considération du spectre complet des accidents possibles en fonction de leur probabilité, ainsi que de valeurs de seuils d'apparition de dommages, pour aboutir à des calculs de détriments. Même en simplifiant à l'extrême l'introduction des paramètres météorologiques, cette solution restera encore, pendant quelque temps, assez utopique par sa complexité /¯5_7.

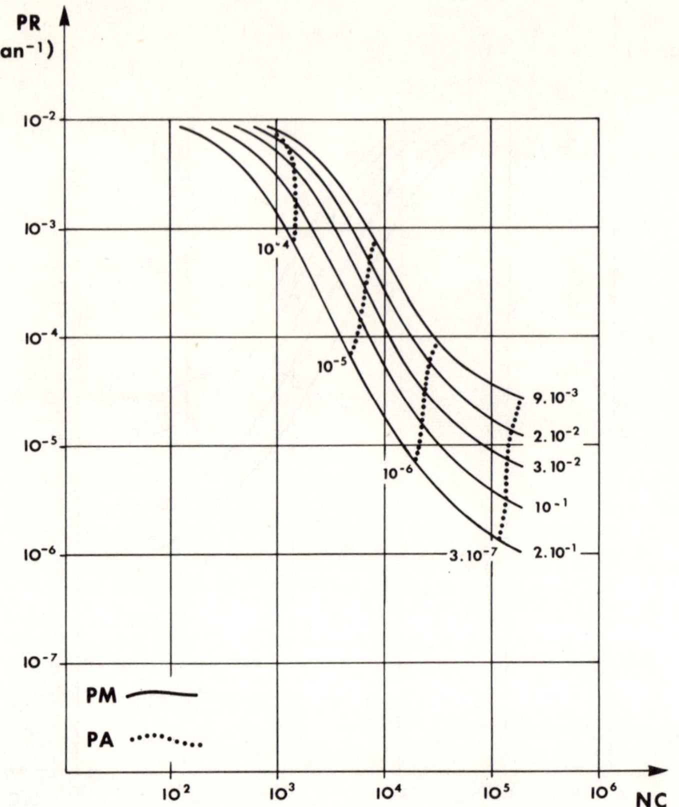



FIG.7.  Réseaux de courbes d'isoprobabilité (météo PM et absolue PA) d'atteindre ou de dépasser dans un secteur donné un seuil donné pour un nombre de personnes concernées NC et une probabilité de rejet PR.


        Parmi les pratiques les plus satisfaisantes et immédiatement appli-
cables, il semble que l'on puisse recommander, comme critères semi-quan
titatifs :

- des cartes de nombres de personnes susceptibles de recevoir un équiva
  lent de dose égal ou supérieur à celui du seuil d'apparition d'un dom
  mage donné pour un rejet donné, éventuellement arbitraire;

- des courbes de correspondance, pour chaque secteur géographique, entr
  des probabilités partielles ou absolues, et le nombre de personnes su
  ceptibles de recevoir un équivalent de dose égal ou supérieur au seu
  d'apparition d'un dommage donné, pour une série de rejets de probabil
  té donnée.


## REFERENCES


[1]  DOURY, A., Une méthode de calcul pratique et générale pour la prévision numérique des pollutions
     véhiculées par l'atmosphère, Rapport CEA-R-4280 (1972).
[2]  TVETEN, U., «A quantitative approach to site evaluation», Principles and Standards of Reactor Safety
     (C.R. Coll. Juliers, 1973), AIEA, Vienne (1973) 255.

[3]  BEATTIE, J.R., BELL, G.D., ≪A possible standard of risk for large accidental releases≫, Ibid. p.11.
[4]  CANDES, P., AUSSOURD, Ph., ≪Critères de sûreté nucléaire orientant le choix des sites — Pratiques
     françaises≫, ces comptes rendus, IAEA-SM-188/12.
[5]  LE QUINIO, R., ≪Influence de la puissance et de la distance sur les risques présentés par un réacteur
     nucléaire — Facteur atmosphérique de site≫, Population Dose Evaluation and Standards for Man and
     his Environment (C.R. Journées d'études Portorož, 1974), AIEA, Vienne (1974) 83.


## DISCUSSION

A. ALUJEVIČ:  In Fig.4 you use a sector of 20° instead of the more
commonly used 22.5° (which is, for example, the United States of America's
Standard Format), the latter being based on the 'compass rose' directions.
By departing from the 22.5° sector, you introduce an octadecimal sector
system in place of the more usual hexadecimal system, even though in your
results (Fig.4) it can be seen that the wind at that particular location blows
generally in the NNW-SSE direction.

A. DOURY:  We have adopted 20° sectors for two reasons. The first is
practical: the Meteorological Services with which we are co-operating use
weather vanes with 18 directions. The other is theoretical, and is based on a
study of the widths of plumes or clouds as a function of distance for different
meteorological conditions of wind and stability. We find, in particular, a
near-coincidence between an angular opening of 20° and the width of a cloud
for a wind velocity of 5 m·s$^{-1}$ under 'normal' diffusion conditions.

P. THOMAS:  My question is about the meteorological probability of
reaching or overstepping a given dose threshold. Have you studied, or
rather considered, the probability of the wind remaining in a particular
sector and not changing direction?

A. DOURY:  For the problem considered here, the persistence of the
wind in a given direction is automatically taken into account because the
statistics used are based on long series of close observations (made every
three hours over five to ten years).

# EXPERIENCES DE DIFFUSION
# DANS L'ATMOSPHERE ET
# DANS L'EAU DES REJETS
# DES INSTALLATIONS NUCLEAIRES

P. CAGNETTI
Laboratorio Radioattività Ambientale,
CNEN-CSN, Casaccia, Rome, Italie

**Abstract–Résumé**

EXPERIMENTS ON DIFFUSION INTO THE ATMOSPHERE AND INTO THE SEA OF DISCHARGES FROM
NUCLEAR INSTALLATIONS.
     The Environmental Radioactivity Laboratory (Casaccia Nuclear Research Centre, Rome) and the Marine
Radioactive Contamination Laboratory (Fiascherino, La Spezia) of the Italian National Nuclear Energy
Commission have developed various experimental methods for studying the diffusion of pollutants in the
atmosphere and in sea-water.  Micrometeorological techniques have been developed to study the thermodynamic behaviour of the lower layer of the atmosphere (0-500 m).  Various methods have also been developed
for determining the diffusion coefficients in the vertical and horizontal directions.  In addition, theoretical
and experimental studies have been performed with the aim of evaluating the mean concentrations in the
atmosphere for continuous discharges.  Numerous examples of the application of these techniques are cited in
the text.  The Fiascherino Laboratory has performed numerous experiments on the diffusion of liquid effluents
in sea-water in the Gulf of Taranto (Southern Italy), off the Trisaia Nuclear Research Centre (National Nuclear
Energy Commission).  The outlay necessary to collect all the various data is justified by the intention to study
the capacity of the Centre to accept large nuclear installations in the future.  These data have been used to
calculate the limiting radiological capacity for different lengths of discharge pipe and they will also provide
a means of evaluating the impact on the environment of releases from other facilities which might be
constructed on the same site in the future.

EXPERIENCES DE DIFFUSION DANS L'ATMOSPHERE ET DANS L'EAU DES REJETS DES INSTALLATIONS
NUCLEAIRES.
     Le Laboratoire de radioactivité ambiante (CEN de la Casaccia, Rome) et le Laboratoire de la
contamination radioactive de la mer (Fiascherino, La Spezia) du CNEN italien ont mis au point différentes
méthodes pour l'étude expérimentale de la diffusion des polluants dans l'atmosphère et dans l'eau de mer.
En ce qui concerne la diffusion atmosphérique, on a utilisé des techniques relevant de la micrométéorologie
pour étudier le comportement thermodynamique de la couche limite de l'atmosphère (0-500 m).  On a ensuite
mis au point des méthodes de détermination des coefficients de diffusion dans le sens vertical et horizontal.
Enfin des études théoriques et expérimentales ont été effectuées pour évaluer les niveaux moyens de concentration dans l'atmosphère, dans le cas des rejets continus.  De nombreux exemples d'application sont cités dans
le texte.  Dans le domaine de la diffusion des effluents liquides dans l'eau de mer, le laboratoire de Fiascherino
a conduit de nombreuses expériences dont l'objectif était l'étude de la diffusion dans le  golfe de Tarente
(Italie du Sud), en face du CEN de la Trisaia (CNEN).  L'effort financier nécessaire pour réunir les nombreuses
données a été justifié par l'intention d'étudier la capacité du Centre à recevoir de futures installations nucléaires
importantes.  Ces données ont non seulement permis de calculer la capacité radiologique limite pour
différentes longueurs de la conduite de décharge, mais elles serviront aussi, à l'avenir, à évaluer l'impact sur
le milieu des déchets provenant d'autres installations qui pourraient éventuellement être construites sur ce
même site.

## INTRODUCTION

     Lors du choix d'un site, les considérations socio-économiques ou
politiques sont en général primordiales.  Bien que l'importance de l'étude
du milieu soit souvent sous-estimée à ce stade, nous pensons que des

recherches écologiques préliminaires sont très utiles. Ces recherches
écologiques deviennent indispensables pendant et après la construction de
l'installation, soit pour permettre d'organiser rationnellement le réseau
des points de prélèvement pour le contrôle des niveaux de contamination
du milieu ambiant, soit pour faciliter le repérage de la voie critique et
du groupe critique de population.

Dans le passé, en Italie aussi les études préliminaires n'ont eu que peu
de poids dans le choix des sites; mais aujourd'hui, à la veille d'un développement intense dans le domaine des centrales nucléaires, une programmatic
territoriale ne peut être raisonnablement conçue que dans la mesure où une
analyse générale du milieu a été effectuée.

Sans que ce stade soit déjà atteint, à l'heure actuelle des analyses
préliminaires du milieu sont effectuées dans le but de réunir des informatio
qui seront utiles dans la phase de construction des installations pour le
traitement avant rejet des effluents radioactifs et leur rejet même, de
manière à réduire aux valeurs les plus faibles auxquelles l'on peut parvenir
sans difficulté («as low as reasonably achievable») les doses à la population
et en général la contamination du milieu naturel.

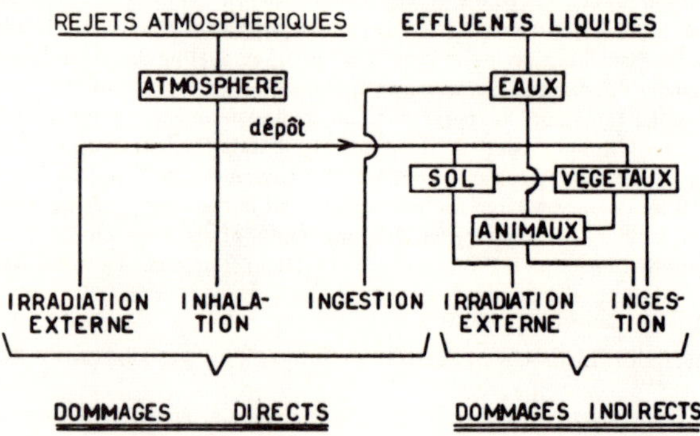


FIG. 1. Schéma des dommages directs et indirects produits par les rejets radioactifs.

On sait que l'atmosphère et l'eau sont les secteurs primaires à prendr
en considération dans l'évaluation des limites imposées par l'homme et le
milieu aux rejets de polluants radioactifs. En effet si l'on exclut les
effluents solides, qui sont généralement stockés et qui posent un problème
à part, les rejets d'une installation nucléaire se présentent sous forme
liquide ou gazeuse. Il est de ce fait indispensable de connaître le mieux
possible l'évolution des effluents dans ce premier stade, c'est-à-dire leur
diffusion dans l'eau ou l'atmosphère: les évaluations qui conduisent à
l'identification de la voie critique dépendent de ces premières informations
Enfin, comme il ressort du schéma très simplifié de la figure 1, la
connaissance de l'évolution de la contamination dans l'eau ou dans l'air
permet immédiatement d'établir la portée des dommages directs.

L'objectif de ce mémoire est de résumer ce qui a été fait dans ce
domaine par le CNEN au cours de ces dernières années.

## 1.  METHODES ADOPTEES POUR L'ETUDE DE LA DIFFUSION DANS L'AIR DES POLLUANTS RADIOACTIFS

En ce qui concerne l'étude de la diffusion atmosphérique, un des critères qui, d'une manière générale, est appliqué automatiquement consiste à proportionner les recherches sur le milieu soit à l'importance des rejets et aux résultats d'une enquête démographique préliminaire, soit aux caractéristiques locales de diffusion.  Par conséquent, si l'importance des rejets dans l'atmosphère est limitée et si le site est plat et éloigné de configurations qui favorisent les effets locaux, l'analyse préliminaire de l'atmosphère peut être limitée à la recherche de données suffisantes pour déterminer la distribution des vents, éventuellement en fonction de la stabilité de l'atmosphère.  Ceci a permis dans quelques cas d'effectuer une bonne étude de la diffusion simplement en utilisant les valeurs de la dilution proposées dans la littérature.  A cette fin, une étude générale sur tout le territoire national de la fréquence des catégories de stabilité [1] s'est révélée très utile, soit pour établir un terme de comparaison avec les données relevées directement sur le site, soit pour faciliter le choix de zones favorables à l'implantation d'installations nucléaires.  Les figures 2 et 3 montrent la fréquence des situations de neutralité et de stabilité pour les stations météorologiques au niveau de la mer.  Dans le tableau I on peut comparer les fréquences des différentes catégories de stabilité pour des sites en haute montagne, sur une côte, une île, ou dans la plaine du Pô [1].

Il faut dire qu'en général en Italie, en raison de la configuration particulière du relief et des côtes, il est très probable que le site soit choisi dans une zone qui ne peut pas être comparée à la situation idéale supposée dans les modèles de diffusion usuels.  Dans ces cas il est raisonnable de prévoir un programme de recherche plus important, qui peut aller de l'analyse expérimentale d'un simple cas de diffusion jugé des plus intéressants du point de vue de la protection, jusqu'à une série complète de campagnes d'études micrométéorologiques visant à déterminer les effets anémologiques locaux et leurs interactions avec les systèmes régionaux, ainsi que la structure thermique de l'atmosphère dans la couche limite (0-500 m).

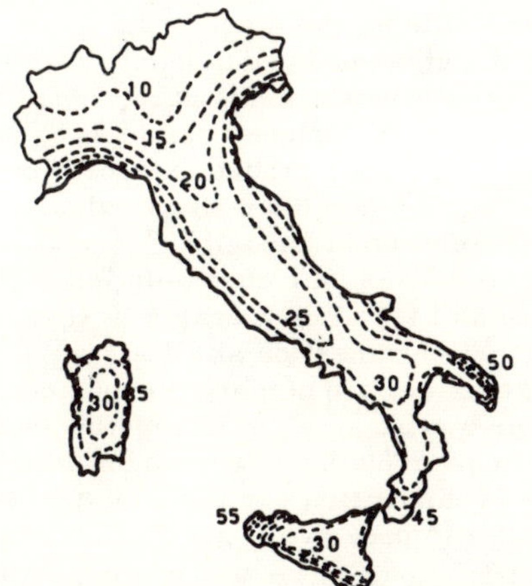

FIG.2.   Stations météorologiques au niveau de la mer: fréquences (%) des situations de neutralité.

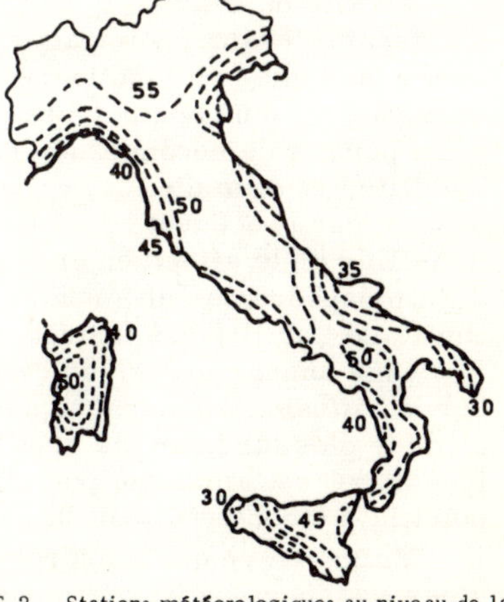

FIG.3.   Stations météorologiques au niveau de la mer: fréquences (%) des situations de stabilité.

TABLEAU I.  FREQUENCES DES CATEGORIES DE STABILITE POUR DIFFERENTS TYPES DE SITES

| Type de site (nombre de stations) | Moyenne des fréquences des situations | | |
|---|---|---|---|
| | instables (écart-type) | neutres (écart-type) | stables (écart-type) |
| Montagne (10) alt. > 800 m | 22,0  (3,0) | 50,0  (13) | 28,0  (10) |
| Ile (3) | 24,8  (0,2) | 41,6  (1,5) | 33,6  (1,7) |
| Côte (18) | 24,7  (2,4) | 36,2  (5,9) | 39,1  (4,5) |
| Plaine du Pô (7) | 32,6  (1,4) | 11,1  (2,8) | 56,3  (1,7) |

2.  QUELQUES EXEMPLES DE RECHERCHES DE DIFFERENTS DEGRES D'IMPORTANCE SUR LA DIFFUSION ATMOSPHERIQUE AUX ALENTOURS DU SITE

Le premier exemple est le cas simple d'un site dans la plaine du Pô, où se trouve une petite installation de retraitement du combustible (EUREX) [2].  Puisque l'installation était déjà en fonctionnement, il s'agissait dans ce cas d'estimer les niveaux de concentration dans l'air, pour rédiger le rapport de sûreté.  Disposant des informations météorologiques obtenues sur le site, complétées par des données sur la climatologie de la plaine du Pô publiées par l'Istituto Centrale di Statistica, il a été possible de limiter l'analyse à des calculs directs, en utilisant les évaluations de Pasquill. La figure 4 donne les courbes d'isoconcentration dans l'air au niveau du sol des distributions analogues ont été trouvées pour les doses d'irradiation externe dues aux rayons gamma du krypton-85, d'après la méthode de Gamertsfelder.

Le site du CEN de la Casaccia représente un cas à part.  Là aussi il s'agissait d'évaluer l'impact des rejets à l'atmosphère des laboratoires, après leur mise en fonctionnement.  Contrairement au cas précédent, on disposait ici d'un bon nombre de données météorologiques sur le site, ce qui a permis de reconstituer la rose des vents pour chaque catégorie de stabilité, et donc d'effectuer avec plus de précision les mêmes calculs que dans le cas précédent [3].  Un calcul complet de la capacité d'acceptation du milieu a été effectué; on a aussi examiné dans leur ensemble les rejets à l'atmosphère des différentes installations [4], et on a évalué la courbe limite de fiabilité des installations en fonctionnement du site [5].

On a aussi conduit sur ce même site un certain nombre d'expériences sur la diffusion atmosphérique, soit pour mettre au point des méthodes à utiliser plus tard sur d'autres sites, soit pour étudier directement les effet locaux.  C'est ainsi que pour l'étude de la dispersion verticale on a mis au point la méthode utilisant la photographie par pose prolongée des fumées [6

Enfin, en ce qui concerne la dispersion horizontale au niveau du sol et à des distances petites et moyennes, l'utilisation de la Fluorescéine comm traceur s'est montrée très efficace [7].  A ce propos il faut relever qu'une

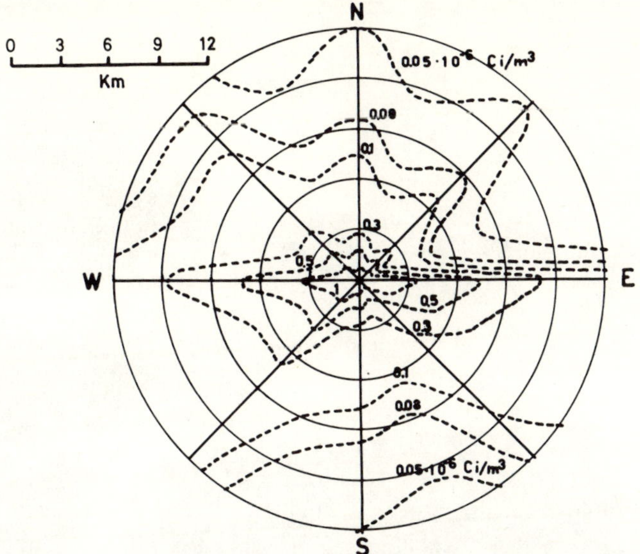



FIG. 4. Courbes d'isoconcentration dans l'air au niveau du sol normalisées pour un rejet de 1 Ci/s sur le site EUREX.

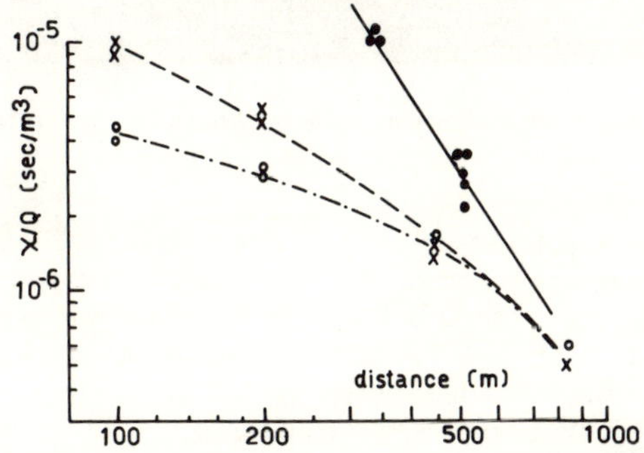



FIG. 5. Valeurs des concentrations dans l'air en fonction de la distance:

● durée de prélèvement 1 h, pas de bâtiments sous le vent
○ durée de prélèvement 1 h, bâtiment sous le vent
x durée de prélèvement ½ h, bâtiment sous le vent.

vingtaine d'expériences ont été effectuées sur terrain plat. Ces données ont été récemment comparées avec celles obtenues à proximité de constructions, pour évaluer la dilution additionnelle introduite par la turbulence induite par le bâtiment [8]. Cette étude s'est révélée utile puisqu'en cas de situation accidentelle les rejets à l'atmosphère d'un des laboratoires les plus importants auraient lieu en présence d'une brise de mer du moment que ce laboratoire opère seulement de jour. Dans ce cas, un bâtiment de 1700 m² de section transversale, situé à environ une centaine de mètres de distance, se trouverait dans l'axe du vent. Les résultats sont donnés dans la référence [8], dont a été tiré le graphique de la figure 5 qui résume les données principales.

FIG. 6.    Photographie de fumée avec pose de 5 minutes dans un cas de situation de neutralité.

Le dernier exemple est celui d'un site sur lequel les installations n'avaient pas encore été construites.  Il s'agit d'une zone dans les Apennins (Brasimone) où sera construit à l'avenir un réacteur expérimental.  En raison de la situation orographique particulière de la région et de l'emplace ment du site à l'intérieur d'une petite vallée qui débouche sur un bassin artificiel au bord duquel se trouve un pâté de maisons, il a été estimé utile de mettre à exécution un programme plus important de recherche sur la diffusion atmosphérique.  Ces recherches ont été orientées dans trois directions principales.  En premier lieu on a évalué la dispersion verticale sur de petites distances (1 km) à l'aide de la photographie de fumées [9]. La technique, légèrement modifiée par rapport à celle utilisée à la Casacci consiste à exposer la même plaque sensible à un photogramme toutes les 5 secondes;  on obtient ainsi des valeurs moyennes sur des temps beaucoup plus longs que dans le cas de la méthode précédente.  La figure 6 est un exemple de photographie exécutée avec un temps de pose de 5 minutes.

Une deuxième ligne de recherche a été conçue pour étudier la structure thermodynamique de l'atmosphère dans la couche limite.  De nombreux sondages thermiques avec des ballons captifs et, simultanément, des lancers de ballons pilotes ont permis de relever la température, l'humidité relative ainsi que la direction et l'intensité du vent, jusqu'à 500 m de hauteur [10, 11].  La figure 7 montre les valeurs des températures relevée au cours d'une série de 24 de ces sondages effectués pendant l'hiver, à des intervalles de 6 heures.  L'analyse des nombreuses données ainsi obtenues

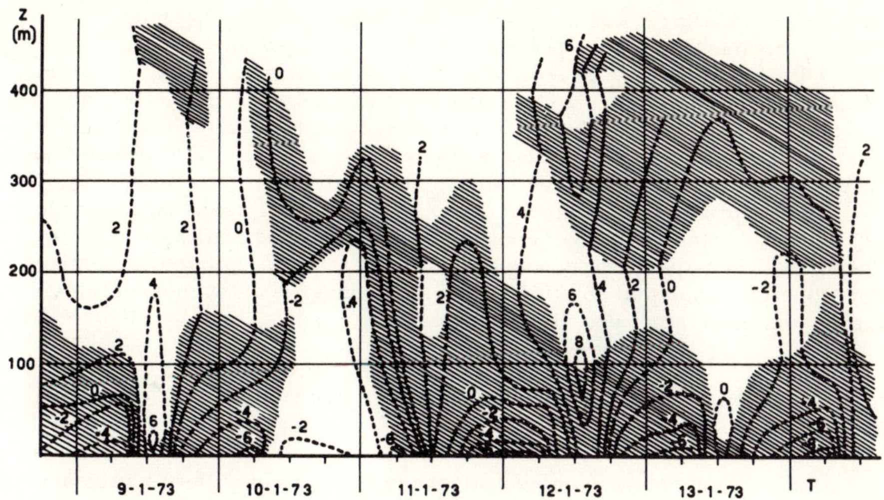

FIG. 7.   Courbes isothermes (°C) pour la période du 7 au 13 janvier 1973, selon la hauteur dans l'atmosphère
(site du Brasimone).   Les hachures représentent les couches stables (inversion).

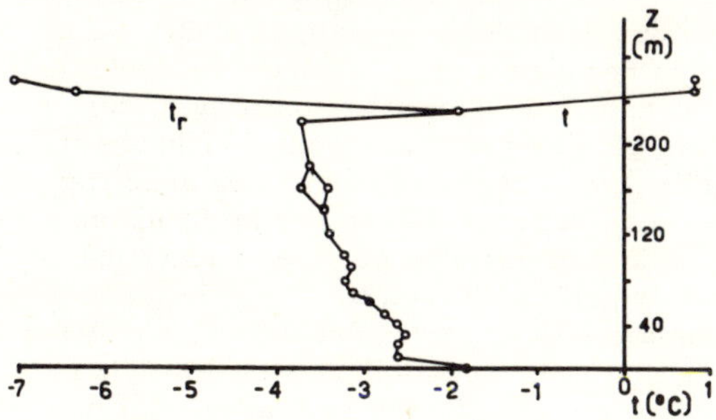

FIG. 8.   Sondage vertical de température (t) et température du point de rosée ($t_r$) en présence de brouillard.

a permis de proposer une hauteur optimale pour la cheminée qui représente le
meilleur compromis entre le coût d'une augmentation de la hauteur et l'avantage
d'une diminution des doses moyennes aux groupes de population avoisinants.
Pour établir ce bilan il est indispensable de posséder des informations sur
les hauteurs les plus fréquentes des couches d'inversion au sol, sur l'altitude
de la base des inversions en hauteur, etc. Enfin d'autres observations utiles
ont été faites sur le rôle de la vallée en ce qui concerne soit la canalisation
des vents dans les basses couches, soit la formation des inversions. On a
par exemple observé que, même si le vent est assez fort mais perpendiculaire
à l'axe de la vallée, pendant la nuit en l'absence de nuages il y a également
formation d'une couche stable jusqu'à une hauteur de 80 à 120 mètres, qui
est précisément la hauteur des collines qui bordent la vallée. Si le sol est
enneigé et le ciel serein, la couche de l'inversion nocturne subsiste après
le lever du soleil, bien qu'au cours de la journée son épaisseur se réduise
à quelques mètres au-dessus du sol. Au contraire, par brouillard, comme
d'ailleurs on pouvait s'y attendre, l'inversion se déplace au sommet de la

couche de brouillard, tandis qu'à l'intérieur le gradient thermique reste
neutre, favorisant ainsi le phénomène de la « fumigation » (figure 8).

A l'heure actuelle, un troisième type de recherches est en cours de
réalisation. Il s'agit d'une étude sur les trajectoires à l'échelle régionale
(10 à 100 km), à l'aide soit de ballons tétraédriques (« tetroons ») suivis
par radar, soit de photographies aériennes de fumée en situation de forte
inversion sur des distances de l'ordre de 5 à 10 km.

Les résultats de ces expériences, comme nous l'avons déjà souligné,
ont été très utiles, soit pour donner des informations réalistes sur les
doses absorbées par les personnes qui habitent aux alentours du centre,
soit pour proposer des solutions pratiques concernant les installations de
traitement et de rejet des effluents gazeux.


## 3. EXPERIENCE RELATIVE A LA DIFFUSION DES EFFLUENTS LIQUIDES DANS L'EAU DE MER

D'une façon générale on peut affirmer que les quantités de radioactivité
rejetées sous forme d'effluents liquides ont été jusqu'à aujourd'hui assez
faibles pour permettre une analyse approximative de la diffusion dans l'eau,
en faisant l'hypothèse de conditions sûres de dilution. Néanmoins, dans
le but d'acquérir une certaine expérience dans le domaine de la diffusion
des effluents liquides dans l'eau de mer, le CNEN a financé un programme
assez important de recherche dans la partie de mer située en face du CEN
de la Trisaia (Italie du Sud). En effet, on prévoit sur ce site un certain
développement des activités de traitement du combustible irradié, et de
conditionnement et de stockage des déchets radioactifs; on a donc estimé
utile de faire des études particulières sur la diffusion des effluents liquides
dans cette zone de la mer, et d'étendre les recherches sur le transport et
la diffusion à tout le golfe de Tarente.

A cette fin on a procédé: premièrement à des recherches sur la
diffusion à courte distance, aux alentours du point de rejet, dans un rayon
de l'ordre du kilomètre, en utilisant un traceur fluorescent (Rhodamine B)
[12, 13]; deuxièmement à des recherches sur le transport et la diffusion
à plus grande échelle (10 à 100 km), en étudiant les courants soit par la
méthode directe (mesure avec des courantomètres), soit par la méthode
indirecte, qui se fonde sur l'hypothèse géostrophique en océanographie [13]
Cette dernière méthode permet de décrire les courants de surface, à
partir de mesures de salinité et de température effectuées sur tout le volum
d'eau examiné. Les résultats des expériences avec traceurs ont permis
de choisir le modèle de diffusion qui se rapprochait le plus des données
expérimentales, en ce qui concerne la diffusion à de faibles distances du
point de rejet. L'étude des courants a révélé la présence d'un tourbillon
d'environ une dizaine de kilomètres de diamètre, juste en face du Centre
(fig. 9). Ce phénomène a été confirmé à plus petite échelle par les
expériences avec traceurs (fig. 10). Ses effets sur les concentrations
moyennes atteintes dans le cas d'un rejet constant et continu ont été évalués
à l'aide d'un modèle mathématique à compartiments.

A l'aide des modèles ci-dessus on a trouvé que lorsqu'on augmente la
distance entre le point de rejet en mer et le rivage, la quantité maximale
qu'on peut rejeter augmente à son tour, mais que pour une certaine valeur
de la distance la capacité de réception se stabilise, ou augmente beaucoup

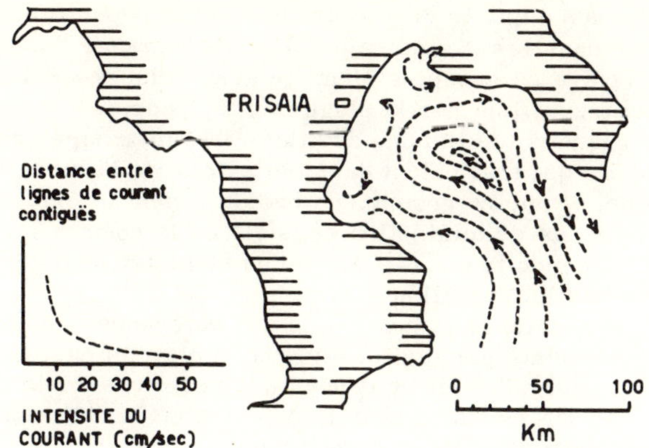



FIG. 9.    Courants dans le golfe de Tarente pendant le mois d'avril 1970.

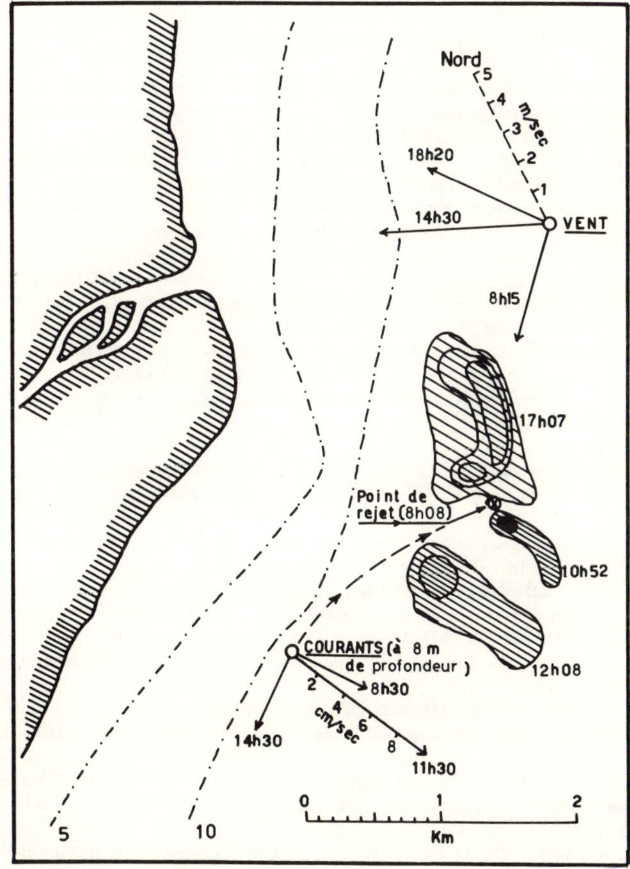



FIG. 10.    Evolution dans le temps d'un rejet en mer de 2 kg de Rhodamine B, juste en face du CEN de la
Trisaia (golfe de Tarente).

plus lentement, parce que le groupe critique de la population change. En effet, tandis que dans les cas de rejets à faible distance (quelques centaines de mètres) du rivage ce groupe s'identifie avec celui de quelques familles de pêcheurs qui travaillent sur la plage même, pour des rejets à des distances plus grandes (de l'ordre du kilomètre) le groupe critique devient celui des pêcheurs qui travaillent dans cette zone de la mer, à bord de bateaux de pêche. Pour ce groupe, un prolongement ultérieur de la conduite de rejet n'apporte pas de diminution sensible de la dose reçue, tout au moins tant que la conduite n'a pas des dimensions telles qu'elle sorte de la zone intéressée par le tourbillon.

Pour plus de détails nous renvoyons aux références, où on trouvera tous les renseignements particuliers que nous n'avons pas pu inclure dans ce texte, dont le seul but était de donner une vue d'ensemble de l'activité déployée par le CNEN en Italie dans le domaine de la diffusion des effluents radioactifs dans l'atmosphère et dans l'eau de mer.

# REFERENCES


[1] CAGNETTI, P., GIUDICI, G., ROSINI, E., « Condizioni medie di stabilità nei bassi strati dell'atmosfera per le stazioni meteorologiche italiane », présenté au: XVII Congresso dell'Associazione Geofisica Italiana, 1968; CNEN RT/PROT (70) 30 (1970) 68.

[2] CAGNETTI, P., FRITTELLI, L., Ricettività del sito circostante l'impianto di ritrattamento EUREX per rilasci continui nell'atmosfera di Kr 85, I 131, Sr 90, Pu 239 e Trizio, CNEN RT/PROT (70) 38 (1970).

[3] CAGNETTI, P., Determination of the equal-concentration curves in the air around a continuous release of a radioactive effluent, on the basis of the meteorological information collected on site, G. Fis. Sanit. Prot. Radiaz. 12 4 (1968) 239-51; CNEN RT/PROT (69) 37 (1969).

[4] CAGNETTI, P., FRITTELLI, L., NARDI, A., « La coordination des rejets aériformes radioactifs des différentes installations d'un centre nucléaire », présenté au 2$^e$ congrès Int. Radiat. Prot. Ass., Brighton, 3-8 mai 1970; CNEN RT/PROT (70) 41 (1970).

[5] CAGNETTI, P., FRITTELLI, L., NARDI, A., « Some considerations on the reliability function of a nuclear plant with reference to its site », Ibid; CNEN RT/PROT (72) 19 (1972).

[6] CAGNETTI, P., Valutazione dei parametri di diffusione atmosferica sul sito della Casaccia mediante l'uso della fotografia di fumate, G. Fis. Sanit. Prot. Radiaz. 5 3 (1969) 168-76; CNEN RT/PROT (70) 11 (1970).

[7] CAGNETTI, P., « Application of methods and first results of atmospheric diffusion at small and moderate distances around CSN Casaccia », présenté au 2$^e$ congrès Int. Radiat. Prot. Ass., Brighton, 3-8 mai 1970.

[8] CAGNETTI, P., Downwind concentrations of an airborne tracer released in the neighbourhood of a building, Atmos. Environ., sous presse.

[9] CAGNETTI, P., Valutazione della dispersione verticale di rilasci gassosi di breve durate sul sito del Brasimone, Rapporto di avanzamento n. 1, CNEN RT/PROT (73) 19 (1973).

[10] CAGNETTI, P., PALMIERI, S., SIMONINI, G., Misure meteorologiche nei bassi strati e loro interpretazione nello studio della diffusività dell'atmosfera sul sito del Brasimone, Periodo estivo, CNEN RT/PROT (73) 10 (1973).

[11] CAGNETTI, P., PALMIERI, S., SIMONINI, G., Misure meteorologiche nei bassi strati e loro interpretazione nello studio della diffusività dell'atmosfera sul sito del Brasimone, Periodo invernale (sous presse).

[12] BERNHARD, M., CAGNETTI, P., ZATTERA, A., La diffusione in acque basse di mare: prime esperienze con Rodamina B, G. Fis. Sanit. Prot. Radiaz. 16 2 (1972) 71-80; CNEN RT/PROT (72) 42 (1972).

[13] CAGNETTI, P., BERNHARD, M., ZATTERA, A., « Radioecological investigations in the Gulf of Taranto, 1. General circulation in the Gulf of Taranto and diffusion processes offcoast the Trisaia Center », Procs Int. Symp. Radioecology Applied to the Protection of Man and his Environment, Rome, 7-10 sept. 1971, EUR 4800 dfie (1972).

## DISCUSSION

D. BENINSON: In the last part of the paper, you describe your studies
on dispersion in the sea at short distances.  Would you be good enough to say
what relation you find between the amounts released and the time integral of
concentration (c.g. at about 1 km)?

P. CAGNETTI: After a series of experiments using Rhodamine B as
tracer, we have found that the theoretical dispersion model which gives
results closest to the experimental values obtained is that suggested by Joseph
and Sendner.

# AREAS AFFECTED BY GROUND DEPOSITION OF $^{137}$Cs AND DESCRIPTION OF THE COMPUTER CODE ARCON

U. TVETEN
Institutt for Atomenergi,
Kjeller, Norway

**Abstract**

AREAS AFFECTED BY GROUND DEPOSITION OF $^{137}$Cs AND DESCRIPTION OF THE COMPUTER CODE ARCON.
The paper deals with the combined probability analysis and parametric study of $^{137}$Cs deposition after a postulated nuclear reactor accident. The study is performed for a 3000 MW(th) reactor at a site typical of the Oslo Fjord area. No efforts have been made to determine definite accident sequences or amounts of release. It is simply postulated that an accident has taken place that will lead to release of a large amount of radio-activity to the atmosphere during a relatively short time period: on the order of one hour. The size of the area affected by a certain release of radioactive caesium depends upon the meteorological conditions during the release time period. Local weather data have been used to determine the probability of a certain magnitude of the area. The results are given in the form of conditional, cumulative probabilities. Conditional because they give the probability of a certain magnitude of the consequence if the release has taken place and the wind is blowing in the direction for which the weather data are valid. Cumulative because, e.g., the 5% probability means that the weather conditions may give these or worse consequences with a 5% probability or, expressed differently, they will give less severe consequences with a 95% probability. Since the calculations do not apply to a specific accident sequence, several pertinent release characteristics cannot be assigned a definite value. The most important are: amount of $^{137}$Cs released; initial heat at release point; heat generated during plume transport due to decay of radioactive materials; and deposition velocity. The last item is only partially a release characteristic. Besides depending upon particle size and chemical conditions, the deposition velocity depends upon characteristics of the land area over which deposition takes place. A para-metric study is performed by varying each of these characteristics over a certain range while keeping the others fixed at what it is reasonable to assume are typical values under extreme accident conditions. A computer code ARCON has been developed for this analysis, adaptable to a large range of calculations connected to atmospheric releases during accident conditions as well as normal operation, e.g. calculation of reference levels for milk consumption, health effects on specified populations, and the area within any kind of concentration limit or dose limit (except external $\gamma$ dose).

## 1.  INTRODUCTION

A large number of aspects go into the decision making related to site evaluation, and each one of these aspects is very complex and must be treated in a proper way for the evaluation to be of any value. The present paper deals with one such aspect: deposition of radioactive caesium on the ground.

A study has been performed recently in Norway in order to give the best possible estimate of the consequences of a postulated large accident at a nuclear power plant, and a computer program called ARCON has been developed in connection with this study. The study has been performed for a site typical of the Oslo Fjord area, using local weather and population characteristics. This paper deals with calculation of deposition in the environment of $^{137}$Cs released after a postulated accident.

An accident leading to core melt-down may under certain conditions lead to release to the atmosphere of a sizable fraction of the $^{137}$Cs inventory in the reactor core. The caesium will deposit on the ground, and the resulting radiation dose may be so high that it will be decided to evacuate the contaminated area. The effective halflife of the deposited caesium is of the order of years, and the size of the affected area is estimated at several time periods after deposition. The calculations do not assume any clean-up actions.

The calculations have been performed as a combined probability analysis and parametric study. The parameters chosen for parametric study are:

Amount of $^{137}$Cs released;
Initial heat at release point;
Heat generated during plume transport due to decay of radioactive materials; and
Deposition velocity.

The various parameter values are not assigned any specific probability of occurrence. It would, for instance, have been desirable to base the calculations on a probability distribution of the amount of caesium released. This has, however, not been possible. On the other hand, a probability analysis based upon weather data is performed for each parameter set. The weather data are assumed to be available in the six Pasquill classes and four wind speed groups.

The size of affected areas has been calculated. In addition, the size of affected population groups might have been determined, but this has not yet been done.

There are many alternative uses for the ARCON program. A reactor accident that leads to release of radioactive iodine might make it necessary to restrict consumption of milk produced in the neighbourhood. Calculations of such affected areas and other related problems may be performed with ARCON. The program may also be used for calculating the size of affected population groups as well as population doses of various types (excepting external radiation from airborne materials).

## 2.  ATMOSPHERIC DIFFUSION

The calculations are based upon local measured weather data, given as percentages of stability in each of the six Pasquill classes and of wind speed in each of four wind speed groups. Separate diffusion calculations are performed for each stability group and wind speed, using standard methods for short time release. The relative concentration is then given as:

$$\chi/Q = \frac{1}{2\pi\sigma_y\sigma_z u} \exp\left[-\frac{y^2}{2\sigma_y^2}\right]\left(\exp\left[-\frac{(z-h)^2}{2\sigma_z^2}\right] + \exp\left[-\frac{(z+h)^2}{2\sigma_z^2}\right]\right) \tag{2.1}$$

where

$\chi$        is concentration in the point $(x,y,z)$
$Q$        is release rate from the release point, which is at $(0,0,h)$

u          is average wind speed

h          is effective release height. (Origin of the co-ordinate system is at ground level directly underneath the release point.)

$(x,y,z)$   describe the point in the direction downwind, crosswind horizontal and vertical, respectively

$\sigma_y$ and $\sigma_z$ are the Gaussian diffusion parameters horizontally and vertically, at right angles to the plume axis.

The Gaussian diffusion parameters as functions of distance along the plume axis are given in various forms in source literature. The present program uses the graphic form given e.g. in Ref.[1], pp 102-3 with appropriate extrapolation.

In the type of problems solved with ARCON only the ground-level concentration is of interest, and Eq.(2.1) is reduced to

$$\chi/Q = \frac{1}{\pi \sigma_y \sigma_z u} \, \exp\left[ -\left( \frac{y^2}{2\sigma_y^2} + \frac{h^2}{2\sigma_z^2} \right) \right] \qquad (2.2)$$

This equation is, as mentioned before, valid for short-time release. A modification to the program to treat extended releases is possible, but cannot be done in the usual straightforward manner where the plume is assumed to be uniformly distributed over a certain angle, the size of the angle depending upon release time and stability. When the size of the affected area is to be calculated, knowledge of the ground concentration variation at right angles to the wind direction is imperative.

In connection with ground deposition of caesium and similar problems, however, the most severe consequences are expected from short-time releases, and the present calculations are limited to these.

## 3. PLUME RISE

An accident at a nuclear power plant that will lead to a large release of radioactivity within a short time of the accident-initiating event will be accompanied by the release of large amounts of hot gas and steam. The release will be hotter than the ambient air and will have a vertical motion. Furthermore, a release of $^{137}$Cs is accompanied by a much larger release of other radioactive materials. Radioactive decay gives rise to radiation, which is absorbed, mostly in the plume itself, and transferred to heat. Both of these effects give plume rise.

The temperature of the gases at the time of release depends upon the accident sequence, while the heat generation caused by radioactive decay depends upon the total amount and characteristics of the released radioactivity, as well as release rate. The amount of $^{137}$Cs released is not a determining factor for either of these two effects causing plume rise. For this reason release of a certain amount of $^{137}$Cs may take place under very varying conditions concerning plume rise, and initial heat at release point and heat generated during plume transport due to decay of radioactive materials are consequently among the parameters chosen for a parameter study.

Extensive experimental and theoretical work has been done on plume rise. Much of this work is summarized in Ref.[2], which considers the rise due to initial heat content only. In more recent work, reported in Ref.[3], the plume rise of a radioactive plume due to radioactive decay was treated theoretically. An experimental verification of the results is out of the question since only release of a considerable fraction of the gaseous and volatile fission products in a reactor core during a short time interval will give radioactive plume rise of consequence. The effect will not be noticeable for releases related to, e.g., the design-basis accident.

A plume that is released from a stack may have an initial plume rise due to 'jet effect' if the exit velocity is sufficiently high. The conceivable reactor accident sequences that might lead to large releases of radioactivity to the atmosphere are such that the release will not necessarily pass through the stack. For this reason the jet effect is disregarded in the following.

According to Ref.[2], the rise of most non-radioactive plumes is dominated by what is referred to as transitional rise. Transitional rise is described by the following equations, according to Ref.[3]:

Momentum-dominated plume:
$$z = c_1 \, Fm^{1/3} \, u^{-2/3} \, x^{1/3} \qquad (3.1)$$

Buoyancy-dominated plume:
$$z = c_2 \, F^{1/3} \, u^{-1} \, x^{2/3} \qquad (3.2)$$

Radioactivity-dominated plume:
$$z = c_3 \, F^{*1/3} \, u^{-4/3} \, x \qquad (3.3)$$

Here

$z$      is plume rise
$c_1, c_2, c_3$    are constants of proportionality
$u$      is average wind speed
$x$      is horizontal distance travelled by the plume

and we have

$$F_m = (T/T_0) \, w_0 \, r_0^2 \qquad (3.4)$$

$$F = \frac{gQ}{\pi \rho \, c_p \, T} \qquad (3.5)$$

$$F^* = \frac{gQ^*}{\pi \rho \, c_p \, T} \qquad (3.6)$$

where

$T$      is absolute air temperature
$T_0$     is absolute temperature of the stack gas
$w_0$    is plume exit velocity at stack
$r_0$     is stack radius
$g$      is gravitational acceleration
$Q$      is rate of heat emission from stack (cal/s)

$\rho$     is air density
$c_p$   is air specific heat at constant pressure
$Q*$  is rate of heat emission from fission-product decay $(cal/s^2)$

The constants $c_1$ and $c_2$ are known from theory and observation:

$$c_1 \approx 2.5 \tag{3.7}$$

$$c_2 \approx 2.0 \tag{3.8}$$

The constant $c_3$ cannot be determined from observation but may be determined theoretically as follows. Ref.[2] gives the following equations for transitional rise of a buoyancy-dominated plume:

$$\frac{d}{dt}(wur^2) = F \tag{3.9}$$

$$\frac{dr}{dt} = \epsilon w \tag{3.10}$$

where

    $w$   is vertical velocity of the plume
    $r$   is radius of the plume
    $\epsilon$   is entrainment constant

In Ref.[3] the radioactivity-dominated plume is treated in direct analogy to the above by letting $F^*t$ replace $F$, where $t$ is the time after emission from stack. This leads eventually to the equation:

$$z = \left(\frac{F^*}{2\epsilon^2 u}\right)^{1/3} t \tag{3.11}$$

Remembering that $x = ut$, and using $\epsilon = 0.445$ [4], it is found that

$$c_3 = 1.36 \tag{3.12}$$

In ARCON we wanted to include buoyancy as well as radioactivity in the transitional rise calculation. This was achieved by replacing $F$ in Eq.(3.9) by $F + F^*t$, leading to

$$z = \left(\frac{F + F^*t}{2\epsilon^2 u}\right)^{1/3} t^{2/3} \tag{3.13}$$

or

$$z = 1.36 \frac{\left(F + F^* \dfrac{x}{u}\right)^{1/3}}{u} x^{2/3} \tag{3.14}$$

and introducing Eqs (3.5) and (3.6) along with the parameter values [3]:

$$g = 980 \text{ cm/s}^2$$
$$c_p = 0.24 \text{ cal/g °C}$$
$$\rho = 1.2 \times 10^{-3} \text{ g/cm}^3$$
$$T = 300 \text{ °K}$$

we get

$$z = 2.78 \frac{\left(Q + Q^* \frac{x}{u}\right)^{1/3}}{u} x^{2/3} \tag{3.15}$$

In Eq.(3.15) the units have been converted, so that we have m and m/s for length and velocity and MW and MW/s for the heat terms.

Equation (3.15) for transitional rise is used out to a distance of

$$x_{tr} = 2.44 \, u \, \beta^{-1/2} \tag{3.16}$$

Here

$$\beta = \frac{g}{T} \frac{d\theta}{dz} \tag{3.17}$$

where $d\theta/dz$ is the vertical potential temperature gradient in the atmosphere.

For longer distances an equation [3] is used valid for plume rise due to radioactivity only:

$$z = \left(\frac{3F^* x}{\epsilon^2 \beta u^2}\right)^{1/3} \tag{3.18}$$

Transformed to units of m, m/s and MW, this gives

$$z = 5.07 \left(\frac{Q^* x}{\beta u^2}\right)^{1/3} \tag{3.19}$$

Plume rise for distances longer than $x_{tr}$ in Eq.(3.16) is calculated using a combination of Eqs (3.15) and (3.19).

The present calculations are representative of extreme accident conditions. The release is then expected to start one to two hours after the initiating event. Even though the fission process will terminate almost immediately, there will be strong heat production in the reactor core during the release. This heating is caused by decay of fission products in the fuel. Much of this heat will be carried away by gases and steam released together with the radioactive materials, and this effect is represented by Q in Eq.(3.15), which in the calculations is varied around 10 MW.

The radioactive material contained in the plume will heat the plume while undergoing decay. For release of a large amount of the fission products in the reactor core this heat rate may be several MW for a 3000-MW(th) reactor, distributed over the time period during which the release is taking

place.  In the present calculations 7 MW is taken as an upper limit, and it is
assumed that the duration of the release is one hour.  This gives 0.002 MW/s
as an upper limit for $Q^*$ in Eq.(3.15).

## 4.  DRY DEPOSITION

Dry deposition is calculated using the following equations [5]:

$$\frac{q(x)}{Q_0} = \exp\left[-\sqrt{\frac{2}{\pi}}\,\frac{v_d}{u}\,I(x)\right] \tag{4.1}$$

$$I(x) = \int_0^x \frac{dx}{\sigma_z(x)}\,\exp\left[-\frac{h^2}{2\sigma_z(x)^2}\right] \tag{4.2}$$

when

    $v_d$    is deposition velocity
    $u$    is wind speed
    $\sigma_z(x)$ is the Gaussian diffusion parameter in the vertical direction at
        distance x along plume axis
    $h$    is the height of the centre of the plume

    $\sigma_z(x)$ is of course a function of x, the distance from release point.  But
it should be remembered that if plume rise is important, the plume height h
will also be a function of x.

The deposition velocity depends on the surface upon which deposition is
taking place and on particle size and chemical nature.  Deposition velocities
of $10^{-3}$ to $2 \times 10^{-3}$ m/s have been given as reasonable values for caesium over
grassland [6].  Deposition on snow-covered agricultural land is a case of
special interest in Norway.  The deposition velocity would then be expected to
be lower, but no information on this has been found.

## 5.  EMERGENCY REFERENCE LEVEL AND DOSE FROM GROUND CONTAMINATION

If an accident leads to ground deposition of $^{137}$Cs, the persistent dose
rate from the deposited activity may be so high that it is decided to move the
inhabitants out of the area for a shorter or longer period of time.  In reality
such a decision will be taken on a case basis.  But for computational purposes
one must refer to a certain emergency reference dose limit to be able to
quantify the consequences.

The persistent dose rate of 0.5 rad/a has been referred to [7] as a
reasonable value, and will be used in this paper.

The radiation dose above ground after deposition will decrease with
time due to radioactive decay (halflife ~ 30 years), and because of weathering.
The caesium is gradually brought deeper into the soil mainly through the
action of rainwater.  These combined effects have been measured in the
United Kingdom [8].  Chosen plots of ground were contaminated with$^{137}$Cs and

the dose rate one metre above ground was measured at intervals over seven
years.  It was found that the dose reduction during the first four years was
fairly rapid and then grew much slower, approaching the reduction attributabl
to radioactive decay only.  These results indicate that the caesium becomes
permanently fixed to the soil after some years.  It does not ever penetrate
very deep into the ground — at the most some  5 - 10 cm.  The variation with
time of the dose rate above an infinite contaminated plain was found to
vary as

$$D(t) = 0.63\ e^{-1.155\,t} + 0.37\ e^{-0.0305\,t} \qquad (5.1)$$

This is a decay scheme constructed of two overlapping exponentials of
halflives 0.6 years and 22.7 years respectively.

It was also found that $^{137}$Cs contamination of 10 mCi/km$^2$ gave an initial
dose rate of 1.2 mrad/a.  This, combined with Eq.(5.1), has been used here t
calculate the integrated radiation doses for several individual years, and the
results are given in Table I.

From Table I it is of course possible to calculate back to the initial
deposit in Ci/km$^2$ that will give the integrated dose corresponding to the
emergency reference level (0.5  rad/a).  This has been done, and the results
are given in Table II.

It is assumed in this section and in the calculations referred to in the
following sections that there are no clean-up efforts.  Relatively simple
measures may reduce the doses considerably.  Furthermore no credit is
given for the fact that people, even in rural areas, are indoors most of the
time, and the radiation doses will at least be halved if this is taken into
account.

TABLE I.  INTEGRATED DOSES FOR INDIVIDUAL YEARS FROM A GROUND DEPOSITION OF 10 mCi/km$^2$ OF $^{137}$Cs

| Year | mrad |
|------|------|
| 1st  | 0.89 |
| 2nd  | 0.56 |
| 3rd  | 0.46 |
| 4th  | 0.41 |
| 5th  | 0.33 |
| 6th  | 0.31 |
| 11th | 0.27 |
| 16th | 0.23 |
| 21st | 0.20 |
| 31st | 0.15 |
| 51st | 0.08 |

TABLE II.  INITIAL GROUND DEPOSITION THAT WILL GIVE THE INTEGRATED DOSE 0.5 RAD OVER ONE INDIVIDUAL YEAR

| Year | Ci/km$^2$ |
|------|-----------|
| 1st  | 5.6  |
| 2nd  | 8.9  |
| 3rd  | 11.0 |
| 4th  | 12.0 |
| 5th  | 15.0 |
| 6th  | 16.0 |
| 11th | 19.0 |
| 16th | 22.0 |
| 21st | 25.0 |
| 31st | 33.0 |
| 51st | 63.0 |

TABLE III.   WIND SPEED AND STABILITY DISTRIBUTION

| Wind speed: | 2 m/s | 2-4 m/s | 4-6 m/s | 6 m/s |
|---|---|---|---|---|
| Stability | | Wind from north | | |
| A | 1 | 3 | 0 | 0 |
| B | 1 | 4 | 1 | 0 |
| C | 14 | 10 | 1 | 0 |
| D | 14 | 10 | 2 | 0 |
| E | 14 | 3 | 0 | 0 |
| F | 20 | 1 | 0 | 0 |
| | | Wind from south | | |
| A | 1 | 3 | 2 | 0 |
| B | 1 | 3 | 2 | 0 |
| C | 2 | 14 | 9 | 1 |
| D | 2 | 14 | 10 | 1 |
| E | 3 | 10 | 10 | 3 |
| F | 1 | 5 | 3 | 0 |

## 6.   WEATHER DISTRIBUTION

The probability analysis procedure described in Section 8 makes use of
weather distributions.  The weather distributions used in the present analysis
have been measured near a proposed site in the Oslo Fjord area.  Two
opposite directions were chosen for analysis:  north and south.  The difference
in distribution between the two directions is explained in part by the fact that
winds from the north are prevalent during the winter season, while winds
from the south dominate during the summer.  The distributions are given in
Table III.

Most of the present calculations were performed with 'wind from the
south' weather.  Only one case of 'wind from the north' has been run for
comparison.

## 7.   RELEASE OF CAESIUM

Deposition of caesium on the ground may represent a problem only in
the unlikely event of a very severe reactor accident.  Caesium is an alkali
metal and, as such, chemically very active.  It will react very strongly with
water and also with oxygen.  The melting point of the metal is 28.5°C and
the boiling point 670°C.  A certain amount of $^{137}$Cs will be present in the fuel
plenum, and can be released if the canning is damaged.  In a core melt-down
accident a considerably larger release of caesium from the fuel would be

expected. The amount of release to the environment, however, is dependent
upon a number of other factors, like absorption in cooling water, deposition
on surfaces, and wash-out by spray.

For a Mark III reactor of 2000 MW(th) the $^{137}$Cs plenum activity has been
given as 200 Ci [7] for the whole core. In the event of a core melt-down
accident, much larger releases might be possible. In a recent accident study [9]
50% of the core inventory was arrived at as an upper limit for the caesium
release. For a 3000-MW(th) reactor this is $2 - 3 \times 10^6$ Ci. In a United Kingdom
study of the consequences of a large release of $^{137}$Cs, the release was assumed
to be $10^5$ Ci [10]. No experimental data are available on caesium release
during actual accident conditions, and both of the above values are probably
conservative. In the present calculations the $^{137}$Cs release is chosen as one
of the parameters for parameter study, and it is varied around $10^5$ Ci.


8.   PROBABILITY AND CONSEQUENCE

The size of the affected area after a postulated $^{137}$Cs release will depend
upon the weather conditions. Unstable weather and high wind speed would
generally be expected to give a smaller affected area than stable weather and
low wind speed, though the dependence upon stability and wind speed is more
complicated when plume rise is taken into account.

The program ARCON calculates the size of the affected area for each
combination of stability and wind speed, and then the areas are sorted by
magnitude. The measured weather data supply the probability for each
combination of stability and wind speed, and consequently also for the corresponding magnitude of the affected area. The probabilities are most con-
veniently expressed as cumulative, e.g. the 5% probability is the probability
of the affected area being of a certain magnitude or larger.

It must be stressed at this point that the probabilities referred to here
and in the results are conditional probabilities. They are probabilities on
condition that a release of the specified magnitude has taken place, with
the properties otherwise assigned to it, and on condition that the wind blows
in the direction for which the weather data are valid. The probabilities refer
only to variations in atmospheric stability and wind speed. To estimate the
actual probability of a consequence of a certain magnitude, one would also
have to take into account the probability distribution of wind direction, of
magnitude of release and of other release characteristics.

It will be seen from Table III that the distribution pattern is rather
coarse. One single combination of wind speed and stability may have a
probability of occurrence as high as 20%. It is, however, reasonable to
assume that the probability distribution is flat over each wind speed group,
i.e. every wind speed inside one group has equal probability of occurrence.
This assumption has been used in the calculations. It has then also been
necessary to assign a lower and upper limit to the lower and upper wind-
speed group. They have been chosen at 0.5 m/s and 8 m/s respectively.


9.   RESULTS AND DISCUSSION

The results of the calculation are given in Figs 1-8. When not otherwise
indicated, the parameters have the following values:

| | |
|---|---|
| Amount of $^{137}$Cs released | $10^5$ Ci |
| Initial heat at release point | Q = 10 MW |
| Heat generated during plume transport | |
| due to radioactive decay | $Q^*$ = 0.001 MW/s |
| Deposition velocity | $v_g$ = 0.001 m/s |
| Conditional probability | 5% |

Most of the calculations (where not otherwise indicated) are for a conditional probability of 5%, that is, if the release has taken place and the wind is blowing in the direction for which the weather distribution is valid, there is a 5% probability that the affected area will have the magnitude indicated or larger.

It must be explained that the figures do not give the right impression of the shape of the areas, as the dimensions along the wind direction and at right angles to it are not the same. The actual areas are much narrower than appears from the figures. One figure with the same dimensions in both directions has been included for illustration (Fig.8).

Figure 1 illustrates the probability analysis scheme described in Section 8. The difference between the two is only the initial heat, where we have Q = 10 MW in Fig.1(a) and Q = 5 MW in Fig.1(b). It is apparent that the discontinuous stability classes of Pasquill cause abrupt changes in the shape of the area. A continuous stability scheme like the one proposed by Högström [11] is to be preferred, but weather data are usually available only in the Pasquill classes. The magnitude of the areas will be quite sensitive to the parameter values and to the accuracy of the calculation of areas; and the accuracy of the calculation of areas depends upon the number of points calculated on the isoconcentration curve. For these reasons the results shown in Fig. 1(b) must not be regarded as final answers. In cases like the one shown in Fig.1(b) where the 1% and 10% probability occurs under Pasquill E weather, while the 5% probability is Pasquill F weather, one must look upon the results as a whole rather than assign too much importance to each single result. Furthermore, these results show, as already mentioned, a weakness in the Pasquill type of weather classification.

It is evident from the results that the size of the affected area is very sensitive to the amount of caesium released and to the deposition velocity. The size is less sensitive to the heat conditions. They are also quite insensitive to the weather distribution characteristics of two opposite directions, although there is some difference. The difference due to weather distribution will, of course, be larger from one site to another.

Figure 7 illustrates how the area inside the emergency reference level decreases as a function of time, on condition that no action is taken to reduce the dose level.

The calculations are based upon the plume-rise model of Gifford, as described in Section 3. It is, however, evident that this model is not good enough for very large releases. It has in the present calculations frequently given rises of several hundred metres. For rises of this magnitude it is wrong to assume that the wind speed does not change with altitude. A better model is desired and will be introduced into ARCON as soon as it is available.

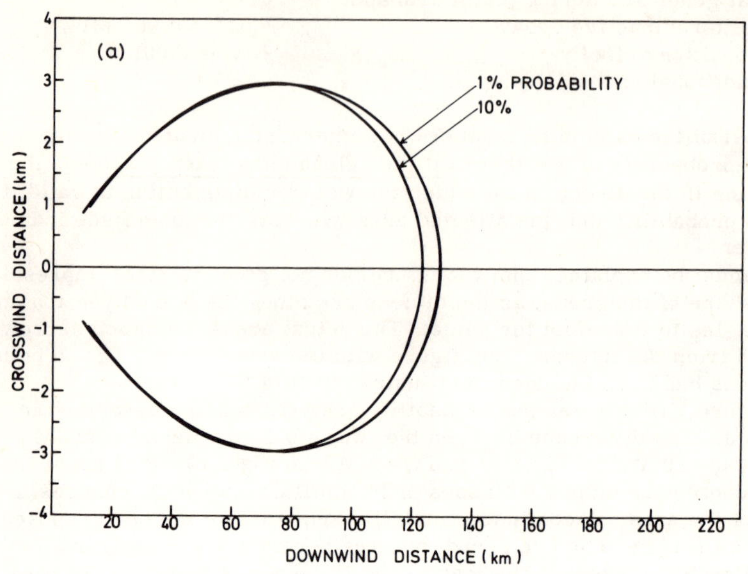


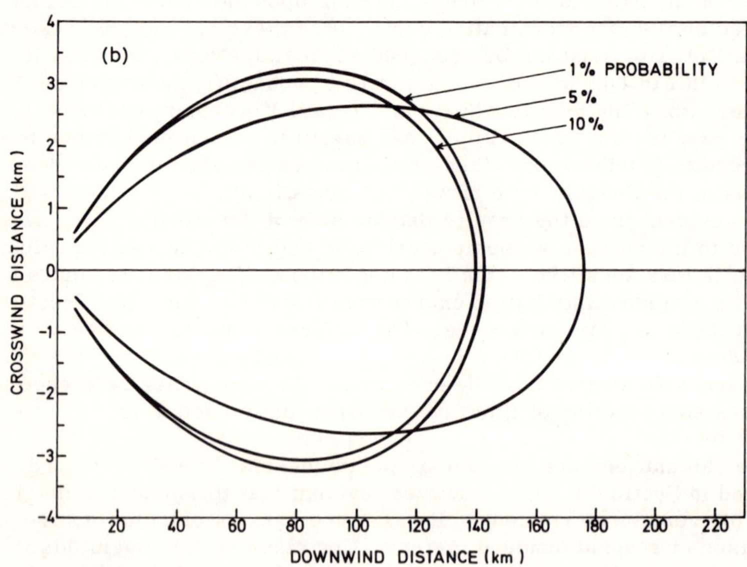


FIG.1.  Dependence on stability and wind speed.

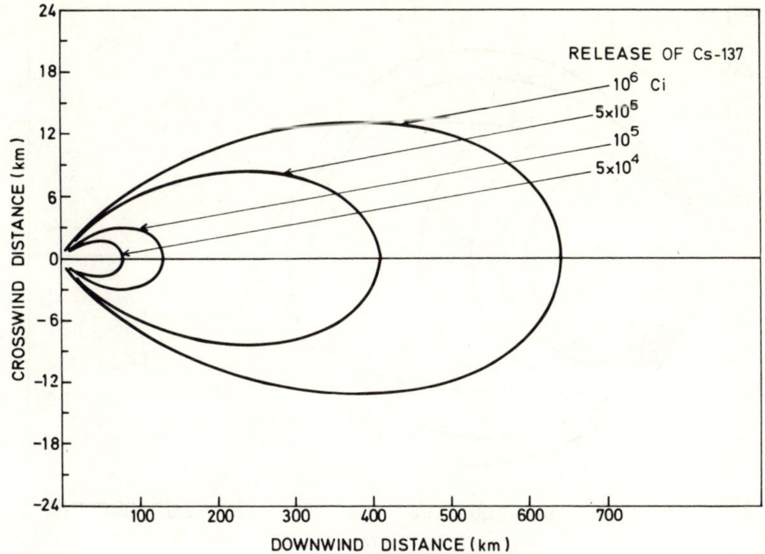


FIG.2.   Variation of amount of $^{137}$Cs released.

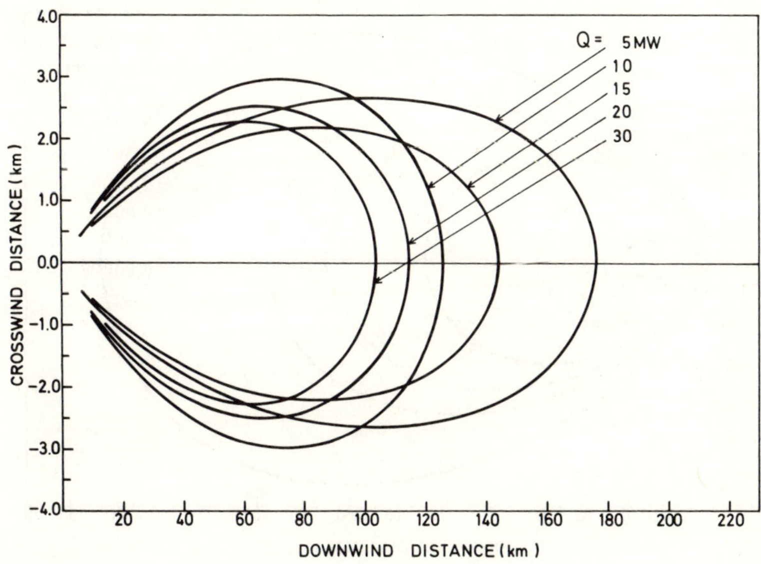


FIG.3.   Variation of initial heat.

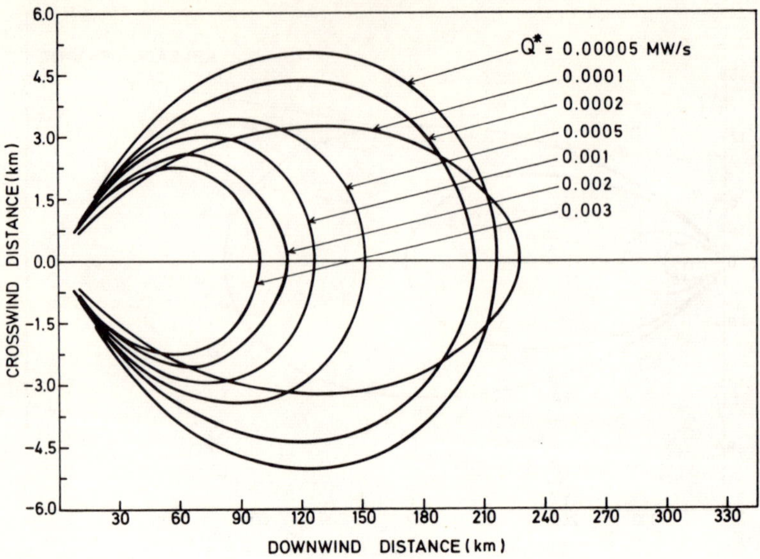



FIG.4. Variation of decay heat in plume.

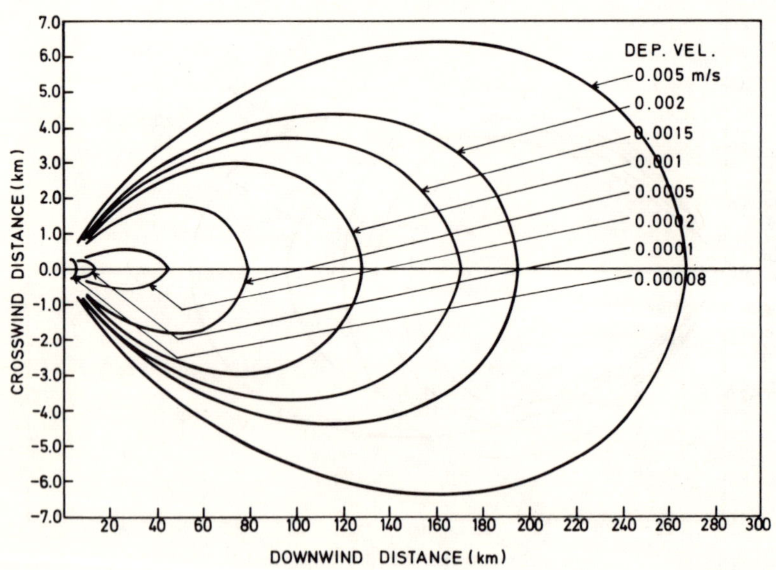



FIG.5. Variation of deposition velocity.

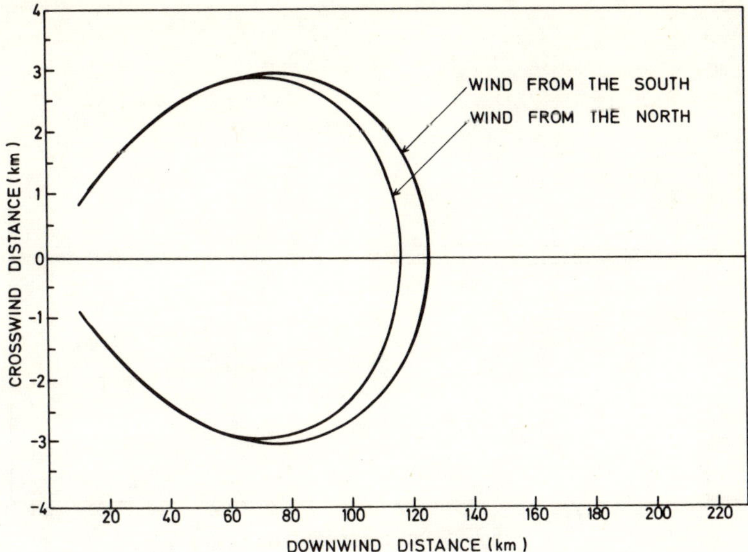



FIG.6.  Variation of weather.

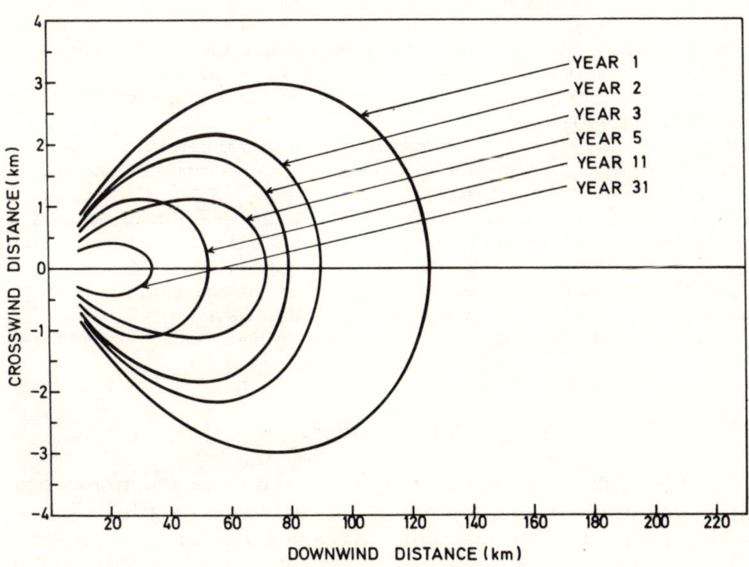



FIG.7.  Depletion with time after initial deposit.

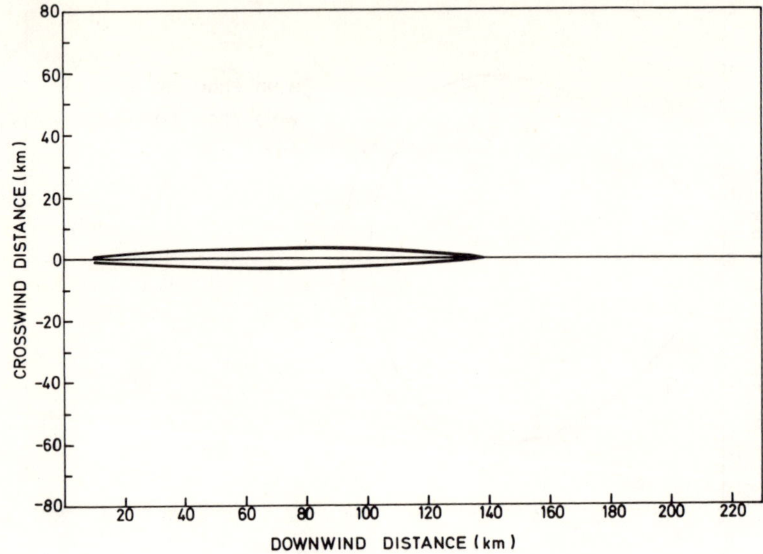



FIG.8.    Actual shape of a typical deposit area.


# REFERENCES


[1]   USAEC, Meteorology and Atomic Energy, Rep.TID-24190 (1968).

[2]   BRIGGS, G.A., Dimensional Analysis of Smoke Plume Rise, Thesis, Dept of Meteorology, Pennsylvania State University (1965).

[3]   GIFFORD, F.A., Jr., The rise of strongly radioactive plumes, J.Appl.Meteorol. 6 (1967) 644.

[4]   SCORER, R.S., The behaviour of chimney plumes, Int.J.Air Pollution 1 (1959) 198.

[5]   Van der HOVEN, I., "Deposition of particles and gases", Meteorology and Atomic Energy, Chapters 5-, USAEC Rep.TID-24190 (1968).

[6]   GIFFORD, F.A., Jr., PACK, D.H., Surface deposition of airborne material, Nucl.Saf. 3 (1962) 76.

[7]   BEATTIE, J.R., Radiological significance of caesium-137 releases from Mark II and Mark III gas-coole reactors, UKAEA Rep.SRD R 11 (1972).

[8]   GALE, H.J., et al., The weathering of caesium-137 in soil, UKAEA, Harwell, Rep.AERE-R 4241 (196

[9]   USAEC, Reactor Safety Study, An Assessment of Accident Risks in U.S. Commercial Nuclear Power Pla Rep.WASH-1400 (1974).

[10]  BEATTIE, J.R., BELL, G.D., A possible standard of risk for large accidental releases, Principles and Standards of Reactor Safety (Proc.Symp.Jülich, 1973), IAEA, Vienna (1973) 11.

[11]  HÖGSTRÖM, U., An experimental study on atmospheric diffusion, Tellus 16 (1964) 205.



# DISCUSSION

D.C. COOPER:  ARCON is highly sensitive to assumptions regarding deposition velocity and is therefore of limited use on a site-specific basis unless empirical data are available.  Have you attempted to use empirically derived data on a case-by-case basis under different conditions of meteorol topography and ground cover?  If not, this would seem to be the next step i evaluating the code's applicability.

U. TVETEN:  You are right in remarking that ARCON is very sensitive to deposition velocity, although the sensitivity actually lies in the area size. The values used in the paper are not experimental, but the basic value of 0.001 m/s is generally considered reasonable for grassland. When calculations for a specific site are performed, deposition velocity values have to be determined, and varying values may be assigned to the area around the site within a polar sector system.

J.L. WEEKS:  Do I understand correctly that ARCON is programmed to give the dose rate quite quickly after a release?  If so, how long a time elapses between the release and the receipt of information?

U. TVETEN:  I think you will be interested in a variation of the program which I have also developed:  it calculates area size for a specific combination of stability and wind velocity.  In this version you can also obtain a series of curves as output, but these will be isodose or isoconcentration curves requested by the user as input data for the program.  In the unlikely event of an accident taking place, one could then run the program with the actual weather conditions, and the area size will come out on a plotter, if one is available.  How quickly you can obtain the result depends more on the equipment you have than on the program itself.  Data preparation is fairly simple, the program is of moderate size and the running times are just a matter of seconds on a modern computer.

A. DOURY:  Your formulae for the calculation of deposition are rather complicated.  Is this perhaps because you have considered depletion of the cloud by deposition itself during transport?  If so, is the accuracy obtained by such a procedure not a little illusory in view of the very small fraction of the cloud removed by deposition?

U. TVETEN:  You are right;  depletion of the cloud due to deposition is included.  For small deposition velocities, this will probably be unimportant, but for average values the result will be affected.  I suppose that, in your opinion, the accuracy with which the deposition velocities can be determined does not justify a complex computational model.  There are at least two different approaches to questions of this type.  Personally, I am of the opinion that if a more exact calculational model does not unduly complicate or delay computations, it should be used, even if present data do not seem to justify it.

Christina GYLLANDER:  Is it ground wind velocity which you introduce into the plume rise formula?  If so, you may obtain unduly high plume levels. I should think it would be possible to introduce the actual wind velocity at plume level by iteration between the wind velocity determined from wind profile data and the plume level obtained by the plume rise formula.

Did I understand correctly that your statistics are based on the frequency of weather and not on the size of the contaminated area?

U. TVETEN:  The wind velocity used in the present calculations is ground-level velocity.  I am aware that Gifford's model is not good for very large releases, giving plume rises of the order of several hundred metres. The program was intentionally designed to allow substitutions to be made as better models become available.

As to your second question, the area calculation is performed for each combination of wind velocity and stability, and then adjusted according to area magnitude, so that in fact the 1% case represents a greater area than the 5% case, and so on.

W.N. LABLANS:  I should like to know the duration of the release to which Fig.8 refers.  It seems to me that for a release lasting an hour (or

longer), if one considers the time required to travel the distance shown here
the wind direction would seldom be so constant as to restrict contamination
to the narrow sector indicated on your figure;  in general, the contamination
resulting from a one-hour release would  spread  in an irregular pattern ove
a wider sector.

U. TVETEN:  The calculations are not considered valid for releases
lasting longer than one hour, and the wind direction is taken to be constant.
This may be somewhat conservative, but under stable conditions the fluctuations in the course of an hour are not so great.

Christina GYLLANDER: In an accident, much more severe than the
design-basis accident, the plume-rise effect will be quite important in the
consequence analysis.  Buoyancy forces will be created by sensible heat fro
the melting core, latent heat, and self-heating of the plume due to activity
decay during the transport.

We have  computed the plume-rise effect based on Gifford's guidelines,
supplemented by expressions for neutral and unstable conditions.  The basic
data for this computation are:

(a)  Core melt-down accident in a reactor of 1700 MW(th) effect;
(b)  Release to the atmosphere is assumed to be:  100% noble gases;
     10% I, Te, Cs;  and 1% of all the other less volatile fission products
     in total 40 nuclides.

Before making a choice of weather type, however, we made a paramete
study for three different stability categories with variation of the wind velocity in the interval 1-10 m/s.  The general diffusion formula implies that
the higher the wind velocity, the lower the concentration, but if the plume-ri
effect is included, a low wind velocity will allow the plume to rise to rather
high levels and the concentration at the ground will be low.

A very high wind velocity will 'blow off' the plume at the release point
but on the other hand promote good dilution conditions.  Thus, these effects
counteract and there is an intermediate range with a weak maximum in the
concentration.  In fact, we found that the most common weather in Sweden
(near neutral or unstable conditions with moderate winds) will cause the
maximum acute consequence to the environment  if plume-rise effects are
included.

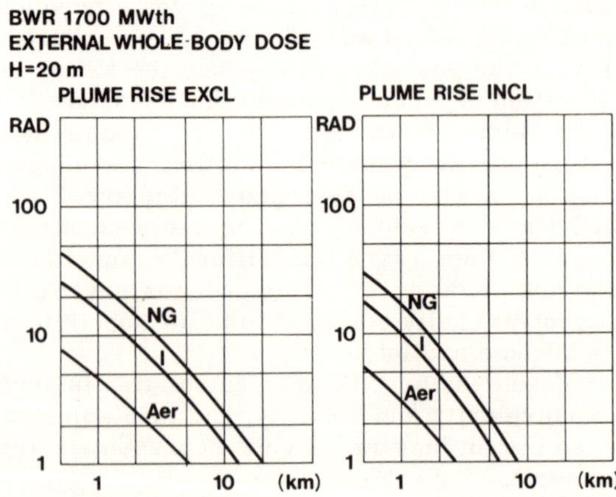


FIG.A.   External gamma dose from the plume.  NG: noble gases;  I: iodine;  Aer: aerosols.

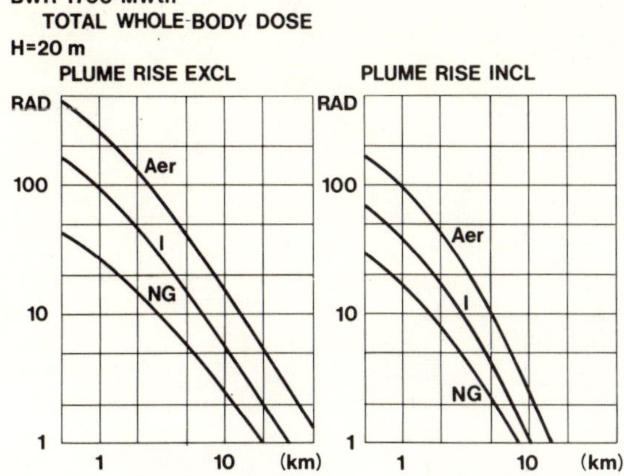


FIG.B.   Total whole-body dose from the plume, external and internal.   Aer:   aerosols;   I:   iodine;
NG: noble gases.

Based on these results we computed doses with and without plume rise
for the weather conditions:   neutral stability,   and wind velocity 5 m/s at the
starting level.   It was assumed to be a low-level release, and the starting
level was 20 m.   The dose was calculated separately for noble gases, iodine,
and other fission products.   Figure A shows the external $\gamma$ dose from the
plume.   Following the Urban Siting Study the shielding factor used is   0.3.
At 1 km, where the plume height is 35 m, the total $\gamma$ dose is reduced from
$\sim$ 50 rad to 30 rad, i.e. to about 60%.   Figure B gives the total whole-body
dose, external and internal, obtained by inhalation of iodine and other fission
products.

I will make a reservation about the conversion factors for these nuclides.
They are the values used by Rasmussen in Appendix 6 to his report of 1974
(WASH-1400),   An Assessment of Accident Risks in US Commercial Nuclear
Power Plants, and I have understood that these factors will be the object
of a special study.   However, they do not affect this comparison as the curves
will be equally displaced.

If we add the three contributions, the total whole-body doses will be the
following:

At a distance of 1 km:
     $\sim$ 400 rad will be reduced to $\sim$150 rad, i.e. a reduction to $\sim$ 40%.
At a distance of 10 km:
     $\sim$ 20 rad will be reduced to $\sim$3 rad, i.e. a reduction to $\sim$15%.

As to the acute consequences, this reduction will represent a significant
effect, as 400 rad would cause a large number of fatalities while 150 rad
would mainly give cases of radiation syndrome.   This is the effect of self-
heating only.   Even the effect of sensible heat from the melting core can be
included in the program, which will result in even greater dose reductions.
As a special study is in progress on the amount of heat released under
defined accident conditions, I have not included this effect.

# LES ETUDES SISMOLOGIQUES EFFECTUEES AU CEA DANS LE DOMAINE DE LA SURETE DES SITES NUCLEAIRES

A. BARBREAU, H. FERRIEUX, B. MOHAMMADIOUN
Département de sûreté nucléaire,
CEA, Centre d'études nucléaires de Saclay,
Gif-sur-Yvette, France

**Abstract—Résumé**

SEISMOLOGICAL STUDIES CARRIED OUT BY THE CEA IN CONNECTION WITH THE SAFETY OF NUCLEAR SITES.

In order to evaluate the seismic risk at nuclear sites, the Department of Nuclear Safety of the French Atomic Energy Commission (CEA) has been conducting a programme of seismological studies for several years past. This programme is aimed at acquiring a better knowledge of seismic phenomena, in particular the spectral distribution of the energy of earthquakes, considered to be the only correct approach to the problem of earthquake protection, as well as a better knowledge of the seismic activity of the areas surrounding nuclear sites. The authors propose defining the design spectrum of the site on the basis of the probable energy at the source, the distance from the epicentre and the transfer function of the geological formations. The need — for the purpose of defining this spectrum — to acquire data on the characteristics of French earthquakes and on regional seismicity led the Department of Nuclear Safety to set up a network of seismic stations. It now has an observatory at the Cadarache Nuclear Research Centre and mobile stations with automatic magnetic recording for studying aftershock sequences and the activity of faults in the vicinity of nuclear sites, and for making the measurements necessary to calculate the transfer functions. With this equipment it was possible to record six aftershocks of the Oleron earthquake on 7 September 1972 close to the epicentre, and to calculate the spectra therefrom. The latter contained a lot of high frequencies, which is in agreement with the data obtained from other sources for earthquakes of low energy. The synthetic spectra calculated on the basis of one magnitude and one distance are in good agreement with the spectra obtained experimentally.

LES ETUDES SISMOLOGIQUES EFFECTUEES AU CEA DANS LE DOMAINE DE LA SURETE DES SITES NUCLEAIRES.

Afin d'évaluer le risque sismique sur les sites nucléaires, le Département de sûreté nucléaire du Commissariat à l'énergie atomique développe depuis plusieurs années un programme d'études sismologiques. Ce programme est orienté vers la recherche d'une meilleure connaissance des phénomènes sismiques, notamment de la répartition spectrale de l'énergie des séismes, considérée comme permettant seule une approche correcte du problème de la protection parasismique, ainsi que vers une meilleure connaissance de l'activité sismique des régions entourant les sites nucléaires. On se propose de définir le spectre du séisme de référence du site en prenant en compte l'énergie probable à la source, la distance et la fonction de transfert des formations géologiques. La nécessité, pour définir ce spectre, d'acquérir des données sur les caractéristiques des séismes français et sur la sismicité régionale a conduit le Département de sûreté nucléaire à constituer un réseau de stations sismiques. Actuellement on dispose d'un observatoire installé au Centre d'études nucléaires de Cadarache et de stations mobiles automatiques à enregistrement magnétique destinées à étudier des séquences de répliques et l'activité des failles au voisinage des sites nucléaires, et à faire les mesures nécessaires pour calculer les fonctions de transfert. On a ainsi pu enregistrer, à proximité de l'épicentre, six répliques du séisme d'Oléron du 7 septembre 1972 et en calculer les spectres. Ceux-ci se sont révélés riches en hautes fréquences, ce qui est en accord avec les données obtenues par ailleurs pour les séismes de faible énergie. Les spectres synthétiques calculés à partir d'une magnitude et d'une distance sont en bon accord avec les spectres obtenus expérimentalement.

## 1. INTRODUCTION

Une pratique courante en matière de génie parasismique consiste à se référer à une échelle d'intensité macrosismique évaluée en degrés auxquels

on essaie d'associer des valeurs d'accélération. Il s'agit alors de connaître le degré d'intensité caractérisant le site concerné, degré déterminé à partir de données historiques (carte des intensités macrosismiques
observées) pour que soit défini le niveau d'accélération correspondant.
Malheureusement les valeurs d'accélération associées aux différents degrés d'intensité sont déterminées à partir de formules empiriques ou de
données fragmentaires et disparates. L'accélération maximale n'est pas
seule représentative des effets destructeurs d'un séisme et bien d'autres
facteurs entrent en ligne de compte. Il n'y a pas de valeurs d'accélération  que l'on puisse associer valablement aux différents degrés d'une
échelle d'intensité macrosismique $/\bar{\ }1\_/$. Seule la connaissance de la répartition spectrale de l'énergie permet une approche scientifique du problème.

Il est souhaitable de ne plus parler en termes d'accélération maximale
mais de se reporter plutôt au spectre du séisme caractéristique du site.

Le but des études entreprises au sein du Département de Sûreté Nucléaire
par la Section d'Etudes de Sûreté des Sites Nucléaires, est la recherche
des paramètres numériques destinés à définir ce spectre. Celui-ci, exprimé
en spectre de réponse, est déterminé à partir de la connaissance de la
magnitude maximale et de la distance du séisme compatibles avec les données
dont on dispose sur la sismicité de la région $/\bar{\ }2\_/$ $/\bar{\ }3\_/$ $/\bar{\ }4\_/$ $/\bar{\ }5\_/$
et les caractéristiques sismotectoniques régionales.

Les séismes qui se produisent en France sont habituellement de moyenne
ou de faible magnitude $< 6,5$, mais peuvent provoquer dans leur zone épicentrale des intensités parfois assez élevées, IX à X (MSK), par suite
de la faible profondeur de leur foyer. Ils constituent donc dans certaines régions un risque non négligeable pour la sûreté des installations
nucléaires.

Depuis 1960 on a développé un programme d'études sismologiques orienté vers la recherche d'informations utiles à la sûreté des installations
nucléaires. En 1961 fut créé, afin d'étudier l'activité sismique de la
Provence, l'observatoire sismologique de CADARACHE, Centre d'Etudes Nucléaires où se trouvent implantés plusieurs réacteurs expérimentaux. Par
ailleurs, la nécessité d'obtenir des données précises sur les séismes et
sur la sismicité des sites nucléaires a conduit à développer un matériel
constituant des stations sismologiques mobiles destinées à la surveillance de l'activité sismique autour des sites et à l'étude des répliques de
séismes importants.

## 2.   ETUDES POURSUIVIES

### 2.1. Etude de la sismicité régionale

Ces études font appel à un réseau de stations sismologiques mobiles à
enregistrement magnétique, placées de telle sorte qu'elles assurent la
surveillance de la sismicité de la région entourant le site. On peut ainsi surveiller l'activité des failles, le réseau de stations mobiles permettant une localisation précise des épicentres. L'emplacement des stations peut être modifié en fonction des premiers résultats obtenus.

### 2.2. Recherche directe d'enregistrements de séismes au voisinage de
#### l'épicentre

En matière de génie parasismique, il est essentiel de disposer d'enregistrements sismiques obtenus au voisinage de la zone épicentrale, là

où le séisme a des effets destructeurs. C'est en effet le seul moyen de
connaître directement les phénomènes physiques qui ont été à l'origine des
effets observés.

La principale difficulté réside dans la nécessité de disposer d'un
réseau extrêmement dense de stations couvrant les régions sismiques. On
peut, à défaut d'un tel réseau, chercher à enregistrer les répliques
qui suivent souvent les séismes importants. Certaines de ces répliques
peuvent avoir un niveau d'énergie assez élevé  pour être utiles au génie parasismique.

A cette fin, on peut installer immédiatement après le séisme un réseau
de stations mobiles permettant une étude précise et détaillée des répliques susceptibles de se produire par la suite.

Les enregistrements ainsi obtenus fournissent un certain nombre d'informations sur le  mouvement du sol, soit sous forme de fonctions temporel-
les en accélération, vitesse ou déplacement, soit sous forme fréquentiel-
le en spectre de FOURIER ou en spectre de réponse de résonateurs.

On peut aussi déterminer :

- la variation du spectre en fonction de la magnitude,
- la relation entre la vitesse maximale du sol et l'intensité spectrale,
- le rapport existant entre ces résultats et les données macrosismiques.

## 2.3. Etude de la loi de décroissance de l'énergie avec la distance

Dans une région considérée, la connaissance de cette loi est importante, car elle permet d'évaluer les effets du séisme de référence régional au droit du site considéré. Elle est déterminée soit à partir de
l'enregistrement des séismes locaux en des stations situées à
des distances différentes soit au moyen de tirs. L'atténuation en fonction de la fréquence, de l'énergie des ondes sismiques au cours de la
propagation est obtenue à partir du rapport des amplitudes spectrales
correspondant aux enregistrements effectués dans des stations alignées
avec les points de tir et sur une même formation géologique.

## 2.4. Etude de la fonction de transfert locale

Les conditions géologiques locales modifient le spectre des ondes
sismiques. La connaissance de la fonction de transfert (voir 4.3.)
permet ainsi de comparer des sites entre eux ou bien, sur un site donné,
de comparer des emplacements afin de déterminer ceux qui sont les plus
favorables. La fonction de transfert est obtenue à partir de l'analyse
comparative d'enregistrements, sur le substratum rocheux et sur le site, de séismes locaux ou de tirs.

## 3. DISPOSITIFS EXPERIMENTAUX

Pour appliquer le programme que l'on vient d'indiquer, on dispose
des moyens expérimentaux suivants :

- des stations mobiles,
- un observatoire sismologique installé en zone sismique (Cadarache).

## 3.1. Stations mobiles

### 3.1.1. Stations mobiles automatiques

Elles sont destinées plus particulièrement à l'étude de la sismicité régionale et des répliques de séismes.

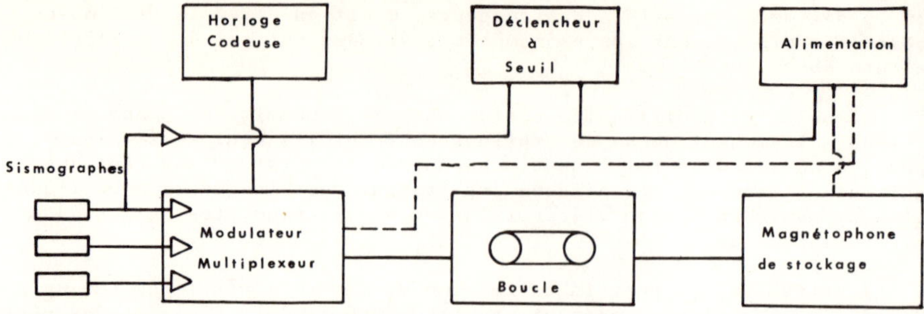

FIG. 1.    Schéma d'une station mobile.

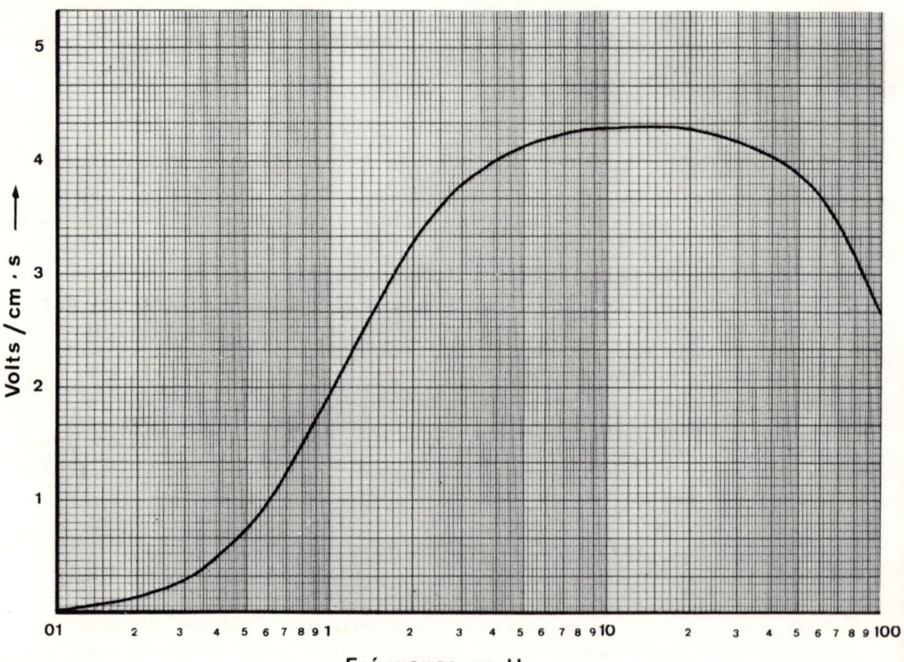


FIG. 2.    Courbe de réponse de l'ensemble de la chaîne d'enregistrement.

Le dispositif est représenté par la figure n° 1. Les sismographes 1 Hz sont orientés habituellement suivant la verticale et les deux com santes NS et EO. Le modulateur multiplexeur permet l'utilisation d'enr gistreurs magnétiques monopistes.

La figure n° 2 montre la courbe de réponse de la chaîne de mesure.

Rôle du déclencheur à seuil et des deux magnétophones

Seuls les signaux sismiques qui ont dépassé un certain niveau sont enregistrés sur le magnétophone de stockage, dont le démarrage est com mandé par le déclencheur à seuil.

Le seuil n'étant généralement pas atteint dès le début d'un phénomène, on utilise, pour ne pas perdre d'information, un enregistreur magnétique à boucle sans fin mis au point au CEA et tournant en permanence. La
durée de passage de toute la boucle est d'environ deux minutes mais peut
être modifiée selon les besoins. La bande rencontre dans l'ordre les tê-
tes de lecture, d'effacement et d'enregistrement. Un phénomène enregistré
sur la boucle est donc lu après le temps de passage complet de celle-ci.
La mise en marche de l'enregistreur de stockage s'effectue avant que le
début du phénomène ne passe devant la tête de lecture.

Horloge codeuse

C'est une horloge à quartz fabriquée par le CEA, pilotée par un récepteur horaire. Elle délivre chaque minute un signal codé représentant
le temps écoulé depuis un repère de temps connu. Ce signal est modulé
en fréquence et multiplexé avec les signaux sismiques modulés.

3.1.2.  Stations mobiles à démarrage manuel

Elles sont utilisées pour enregistrer des phénomènes attendus (ex :
explosions).

Le système est allégé par la disparition du magnétophone à boucle et
du déclencheur à seuil.

3.2.  Observatoire sismologique

Il comporte :

- un système classique à enregistrement photographique,
- un système à enregistrement magnétique (fig. n° 3), fonctionnant suivant le même principe que les stations automatiques mobiles. Trois
  étages d'amplification permettent de couvrir une dynamique d'amplitude
  de 120 dB. On utilise deux magnétophones 7 pistes.

3.3.  Disposition des stations mobiles sur le terrain

Différents types de dispositifs, correspondant à des études de natu-
res diverses, peuvent être envisagés. Citons-en quelques exemples.

3.3.1. Etude de la sismicité d'une région ou des répliques d'un séisme

La méthode de détermination des épicentres est basée sur l'étude des
temps d'arrivée des ondes aux différentes stations . Celles-ci doivent
être disposées de manière à ce que l'on puisse déterminer par triangulation l'emplacement des épicentres avec une bonne précision (fig. n°4).

3.3.2. Détermination de la profondeur de foyer d'un séisme

La méthode, là encore, utilise les temps d'arrivée des ondes. La
meilleure précision est obtenue si les stations sont disposées en pro-
fils à l'aplomb de la zone sismique.

3.3.3.  Etude de l'atténuation des ondes sismiques

Les stations sont alignées avec le point de tir (fig. n° 5), sur une
même formation géologique.

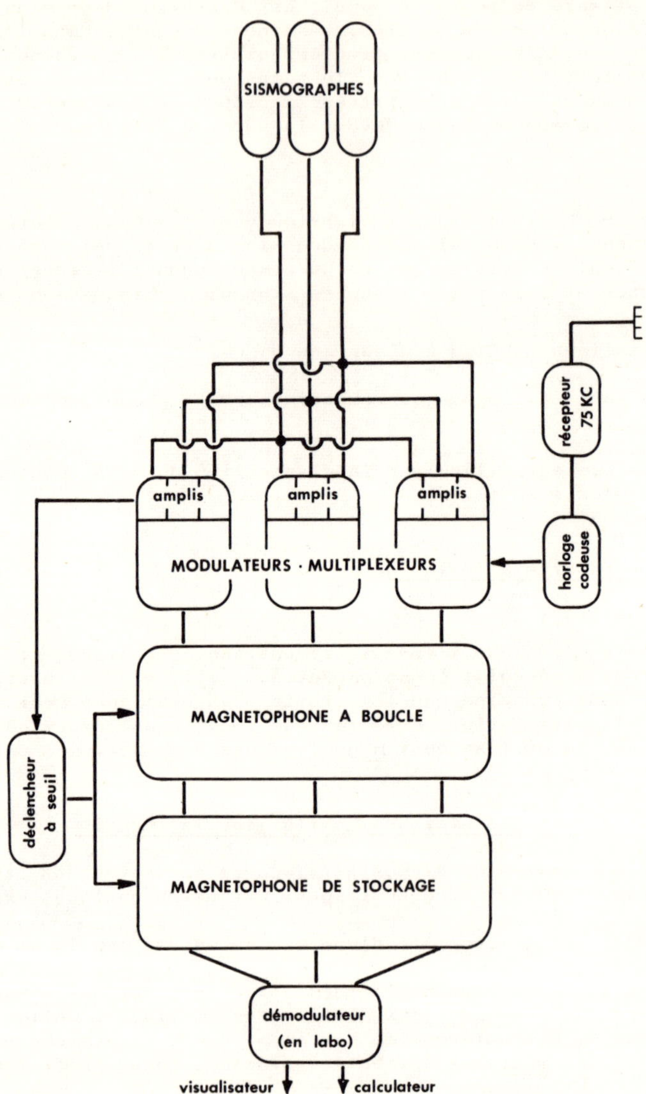


FIG. 3.   Station automatique à enregistreur magnétique de Cadarache.

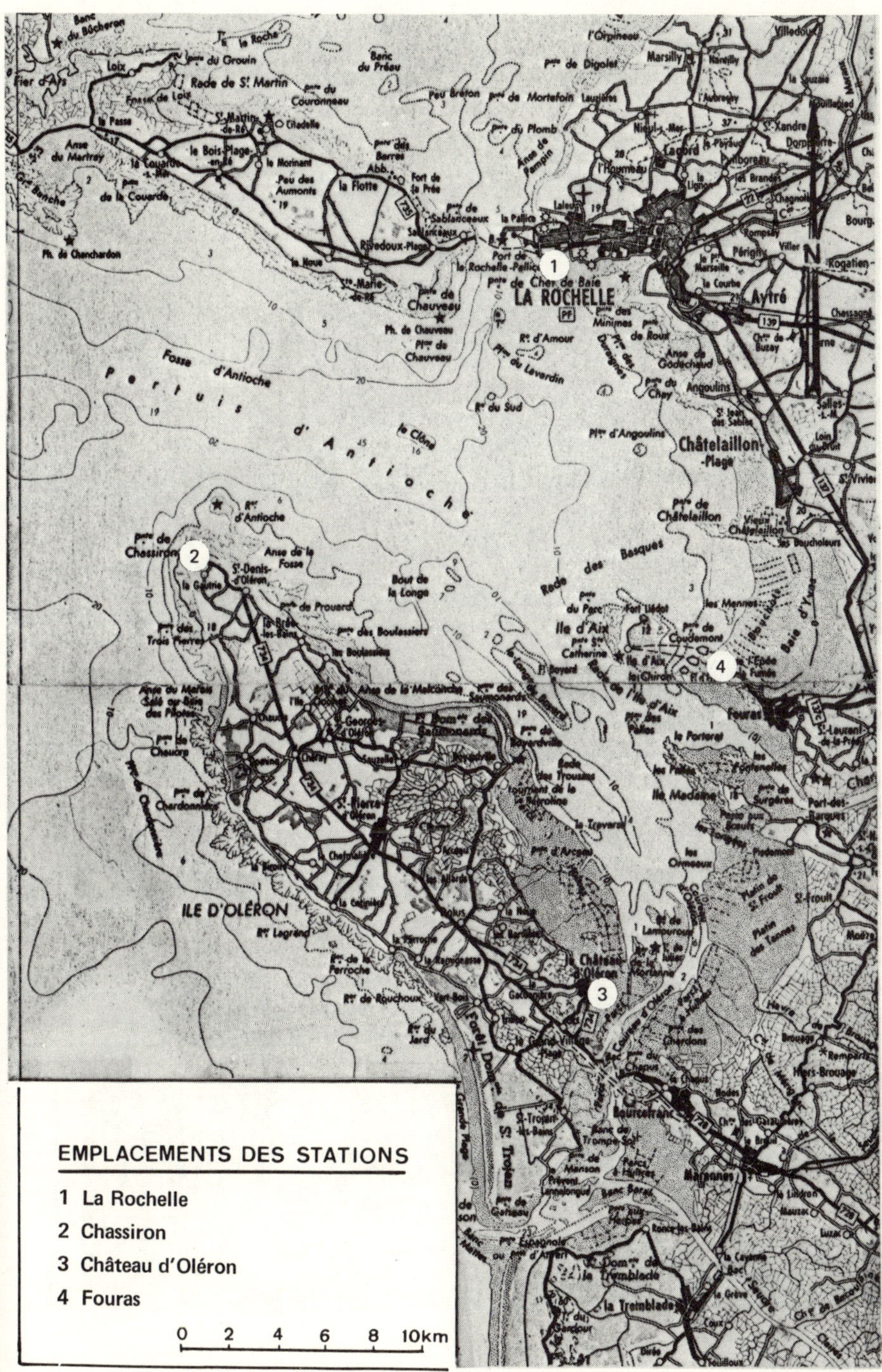


FIG.4.  Carte d'Oléron.  Emplacement des stations.

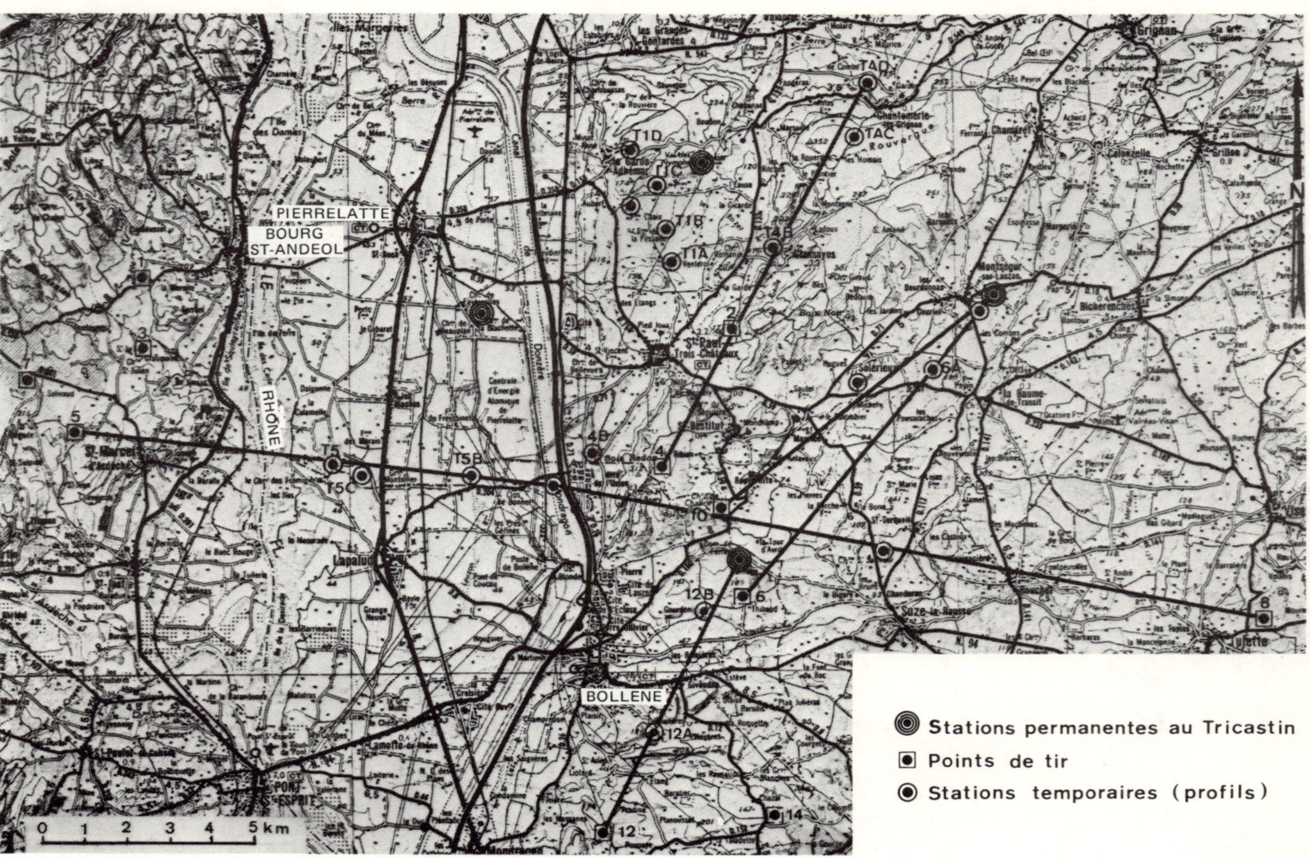


FIG.5. Campagne sismique au Tricastin. Etude des atténuations.

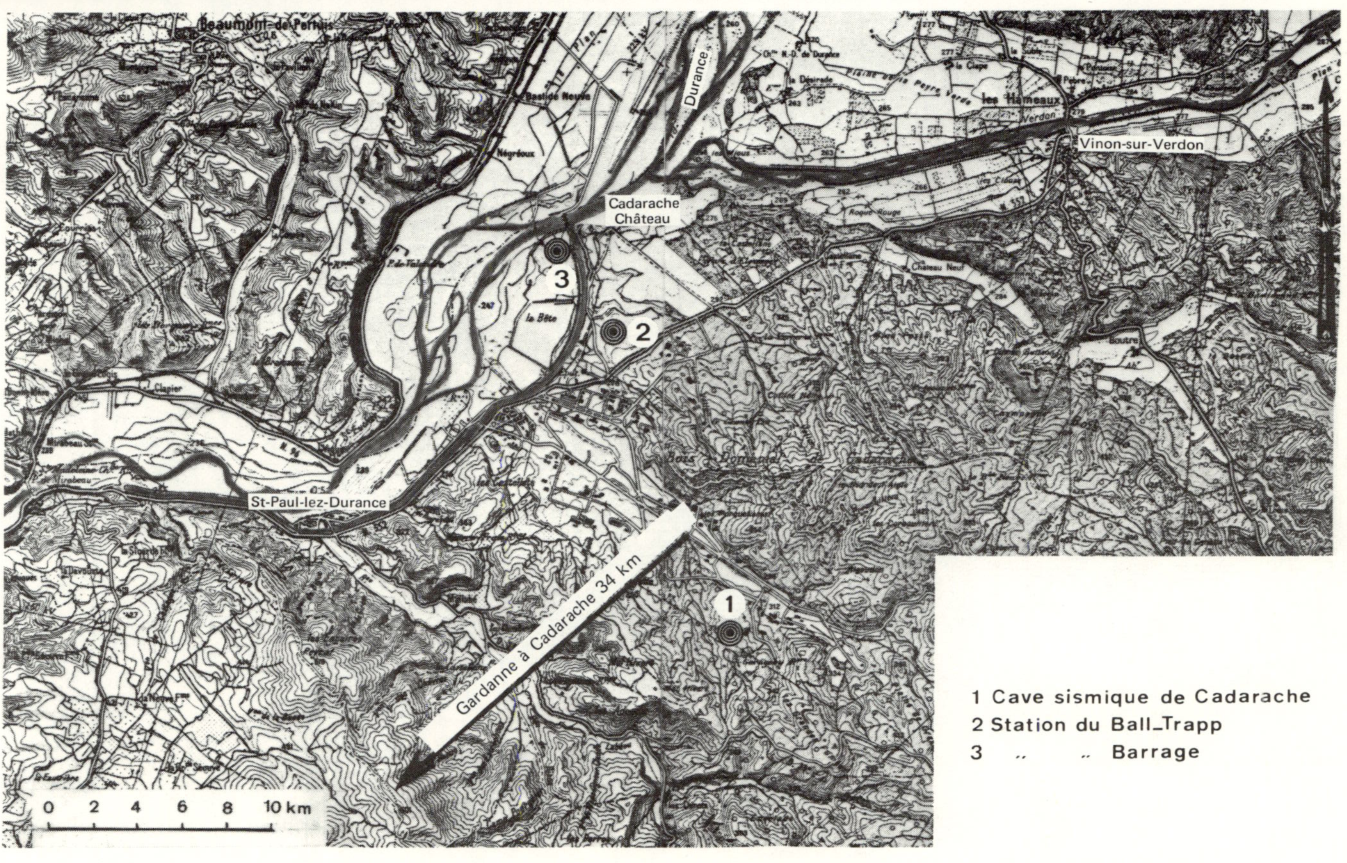


FIG. 6.  Mesures sismiques dans la région de Cadarache.  Etude des fonctions de transfert.

      BARBREAU et al.

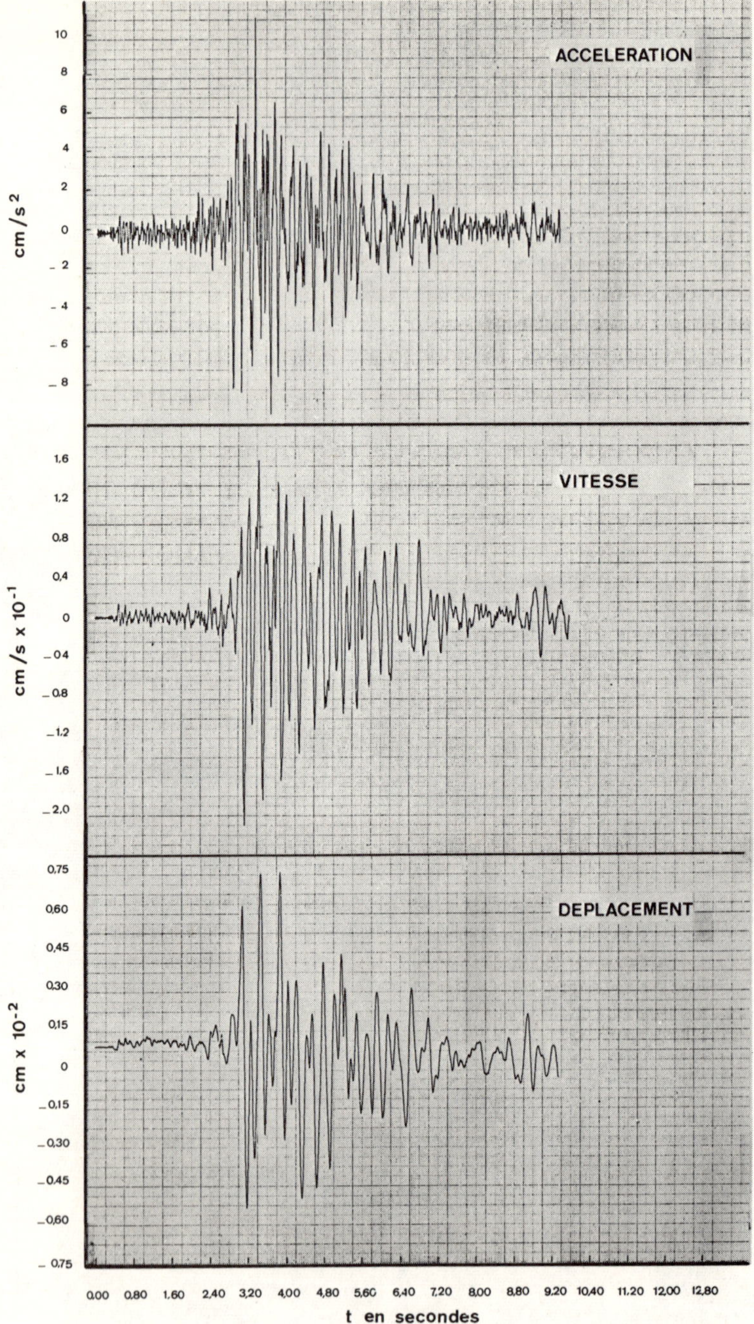


FIG. 7.  Spectres d'Oléron (6 janvier 1973) en accélération, vitesse, déplacement (trace est-ouest déconvoluée).

### 3.3.4. Etude des fonctions de transfert locales

Les stations sont situées sur des terrains de nature différente
(fig. n°6) à la même distance du point de tir, ou en forage au niveau de
la limite supérieure de différentes couches géologiques.


## 4. RESULTATS

### 4.1. Enregistrement de séismes naturels à proximité de l'épicentre

1) Spectres de réponse

L'accélération du sol est obtenue sous forme de fonction temporelle
à partir de l'enregistrement en éliminant l'effet de la chaîne de mesure
(déconvolution) $/^-6_7$ . Les figures 7 et 8 montrent deux exemples de traces déconvoluées.

On en déduit le déplacement relatif, en fonction du temps, d'un oscillateur de fréquence propre f et d'amortissement donné. La valeur maximale
du déplacement multipliée par la pulsation $\omega = 2\pi f$ de l'oscillateur défi-
nit la pseudo-vitesse relative . En faisant varier f, on détermine le
spectre de pseudo-vitesse relative (SPVR) correspondant à un amortissement
donné a , dont l'intérêt est de fournir une information pratique pour les
calculs de génie parasismique $/^-7_7$ .

Les spectres représentés dans la figure n° 9 se rapportent respectivement

1 - aux enregistrements (composante EO) de quatre répliques du séisme du
    7/9/1972 dans la région de l'Ile d'Oléron (Tableau 1A) obtenus à
    Château d'Oléron, sur une formation calcaire (courbes 1 à 4),

2 - à l'enregistrement (composante NS) d'un séisme du Tricastin (tableau
    1B) réalisé près du Centre de Pierrelatte, sur des alluvions (courbe
    5).

On voit que ces spectres de séismes ayant des magnitudes différentes
mais relativement faibles sont riches en hautes fréquences. Les différences entre les spectres des répliques du séisme d'Oléron et celui du séisme du Tricastin enregistré à une distance épicentrale comparable, peuvent
s'expliquer par la présence des alluvions au Tricastin et par la différence des magnitudes de ces phénomènes : le spectre devient plus riche
en basses fréquences quand la magnitude augmente.

2)  Intensité macrosismique et intensité spectrale

Comme il n'existe pas de relation simple entre l'intensité macrosismique et les paramètres du mouvement du sol, HOUSNER a proposé de considérer sous le nom d'"intensité spectrale" l'aire comprise sous le spectre de réponse dans une bande de période entre 0,1 et 2,5 secondes.

$$I_a = \int_{0,1}^{2,5} SPVR\,(a,T)\; dT$$

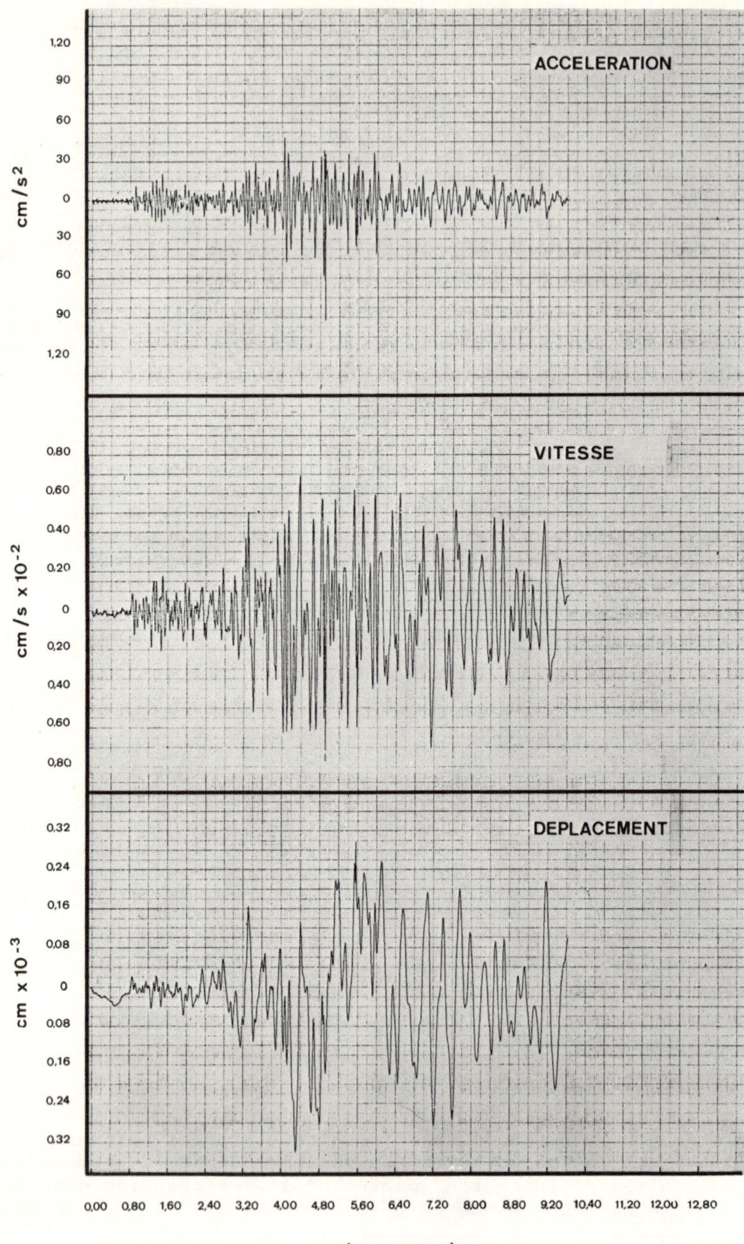


FIG. 8.   Spectres du Tricastin (10 mai 1974) en accélération, vitesse, déplacement (trace est-ouest) déconvoluée).

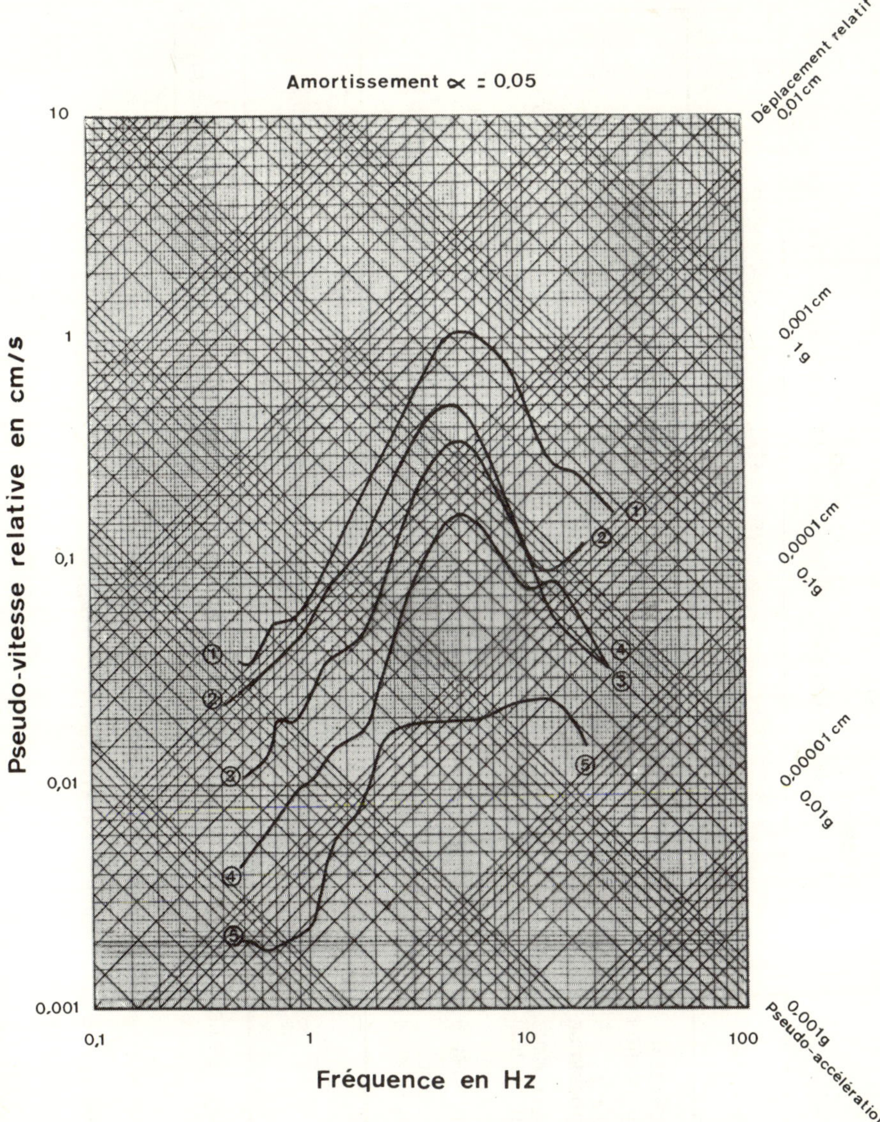



FIG. 9. Spectres de séismes enregistrés à proximité des épicentres. Courbes 1 à 4, répliques du séisme du 7 septembre 1972 enregistré à Château d'Oléron; courbe 5, spectre du séisme du 10 mai 1974 de Grignan enregistré au Château de Faveyrolle (Pierrelatte).

TABLEAU 1A : Caractéristiques des répliques du séisme du 7 septembre 1972 (OLERON)

| Dates | Station | Composante | Fréquence de la vitesse maximale (Hz) | Magnitude | Intensité spectrale ($I_s$) | $I_M$ | Distance épicentrale (km) | Profondeur h (km) |
|---|---|---|---|---|---|---|---|---|
| 6/01/1973 | CHATEAU D'OLERON | E | 5 | 4 | 0,39 | V | 20 | $\simeq$ 15 |
| 6/01/1973 | LA ROCHELLE | E | 5 | 4 | 0,17 | V | 34 | $\simeq$ 15 |
| 20/01/1973 | CHATEAU D'OLERON | E | 5 | 3,5 | 0,22 | IV | 20 | $\simeq$ 15 |
| 19/02/1973 (I) | CHATEAU D'OLERON | E | 5 | 3,0 | 0,12 | - | 21 | $\simeq$ 15 |
| 19/02/1973 (II) | CHATEAU D'OLERON | E | 5 | 2,6 | 0,06 | - | 21 | $\simeq$ 15 |

TABLEAU 1B : caractéristiques du séisme du 10 mai 1974 (GRIGNAN)

| | | | | | | | | |
|---|---|---|---|---|---|---|---|---|
| 10/5/1974 | FAVEYROLLES | N | 10 | 1,5 | 0,02 | - | 20 | Très superficiel |

$I_M$ = Intensité macrosismique à la station d'enregistrement

$I_S$ = Intensité spectrale pour un amortissement de 0,05

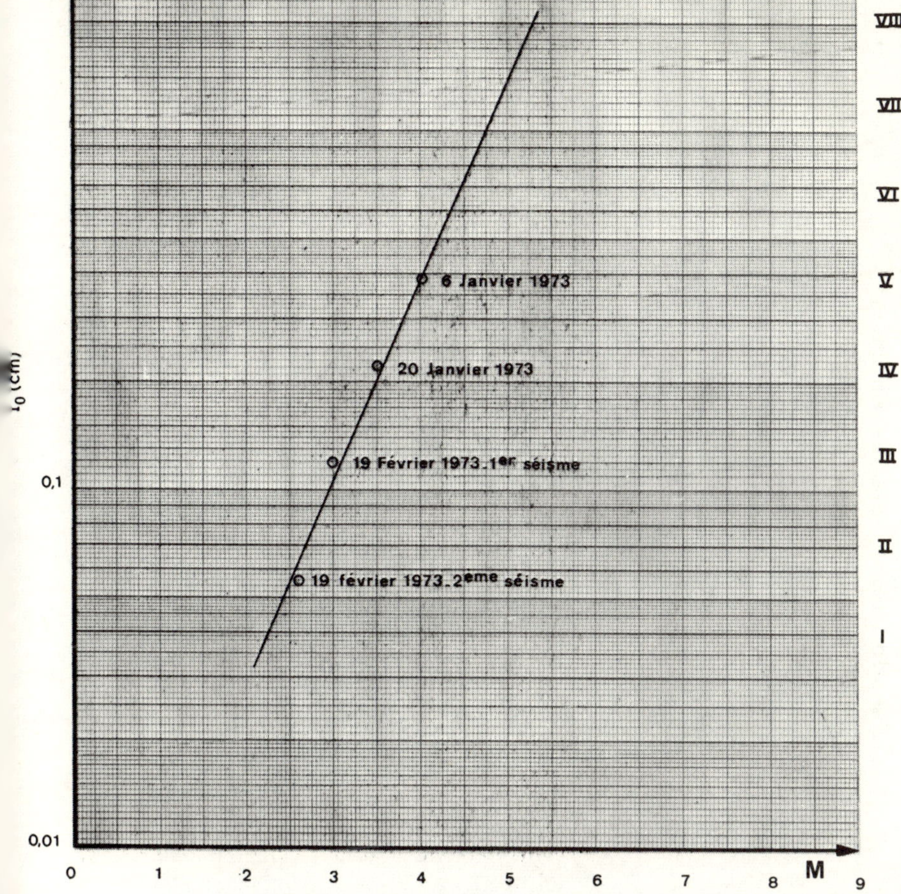



FIG. 10.   Variation de l'intensité spectrale en fonction de la magnitude.

La figure n°10 montre la variation de cette intensité spectrale calculée pour les répliques du séisme d'Oléron (tableau 1A) en fonction de la magnitude de ces répliques. A titre de comparaison on a porté à droite de la figure n° 10 les degrés d'intensité macrosismique d'après les observations faites sur les deux séismes du 6 et du 20 janvier 1973, qui sont les répliques les plus fortes du séisme principal. L'intensité macrosismique du séisme principal était évaluée à VII (MSK) sur l'Ile d'Oléron ; pour cette valeur l'extrapolation de la droite (fig. 10) correspond à la magnitude 5 et à une valeur d'intensité spectrale de 1,2.

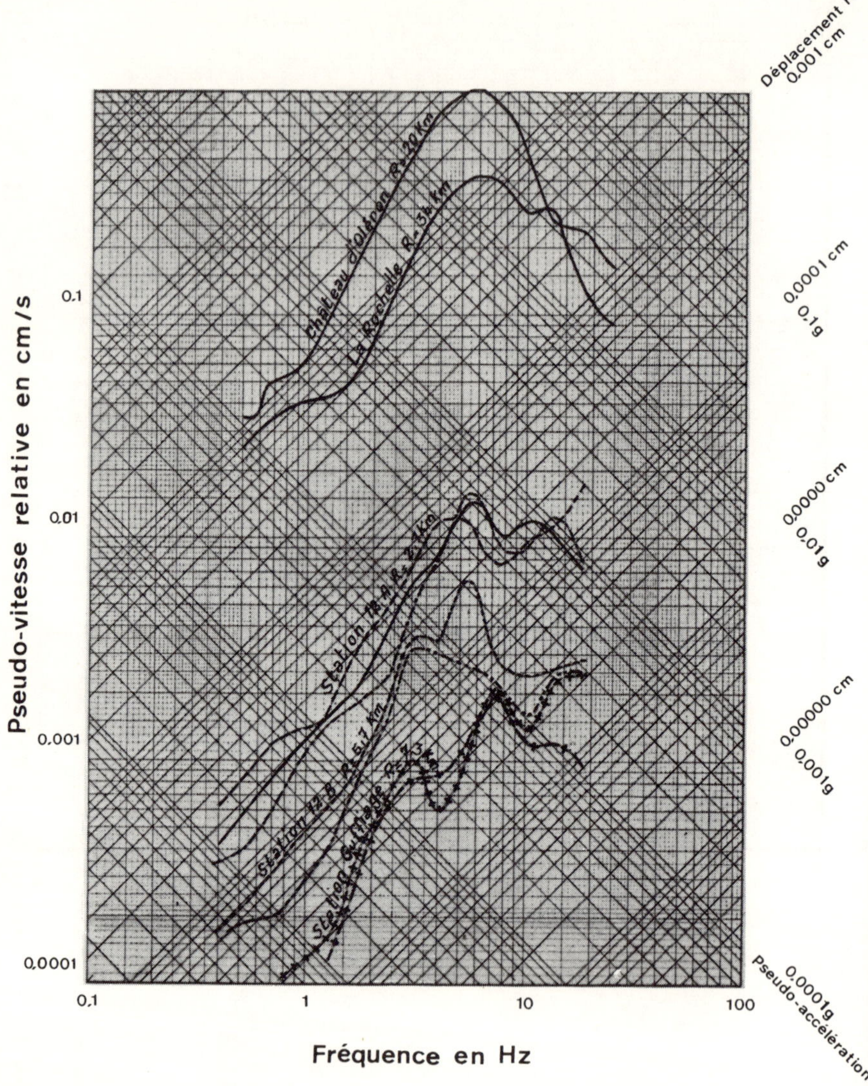


FIG. 11.   Variation du SPVR en fonction de la distance.

## 4.2.   Atténuation de l'énergie en fonction de la distance

La figure 11 représente les spectres des phénomènes suivants :

1 - Le séisme du 6 janvier 1973, précédemment cité, enregistré à la
Rochelle et à  Château d'Oléron (voir figure n° 4) (les deux
points de mesure étant sur calcaire, mais à des distance focales
différentes) (tableau 1A).

2 - Un tir dans la région du Tricastin enregistré sur le profil 12 (voir fig. 5) en trois points (12A,
    12B et Guffiage, à 3, 6 et 7 km de distance respectivement), tous sur une formation calcaire.

    La figure n° 12 montre l'atténuation de la valeur maximale du SPVR
en fonction de la distance :

    1) dans le cas du séisme (composante E)
    2) pour le tir (composantes verticale  et longitudinale). On constate
que pour ces deux sites les lois de décroissance du SPVR en fonction de
la distance sont très voisines (les deux droites sont sensiblement paral-
lèles).

## 4.3.  Fonctions de transfert

    Les couches superficielles ont une influence considérable sur le si-
gnal sismique. Il a été souvent constaté que les terrains mous amplifient
le mouvement du sol.

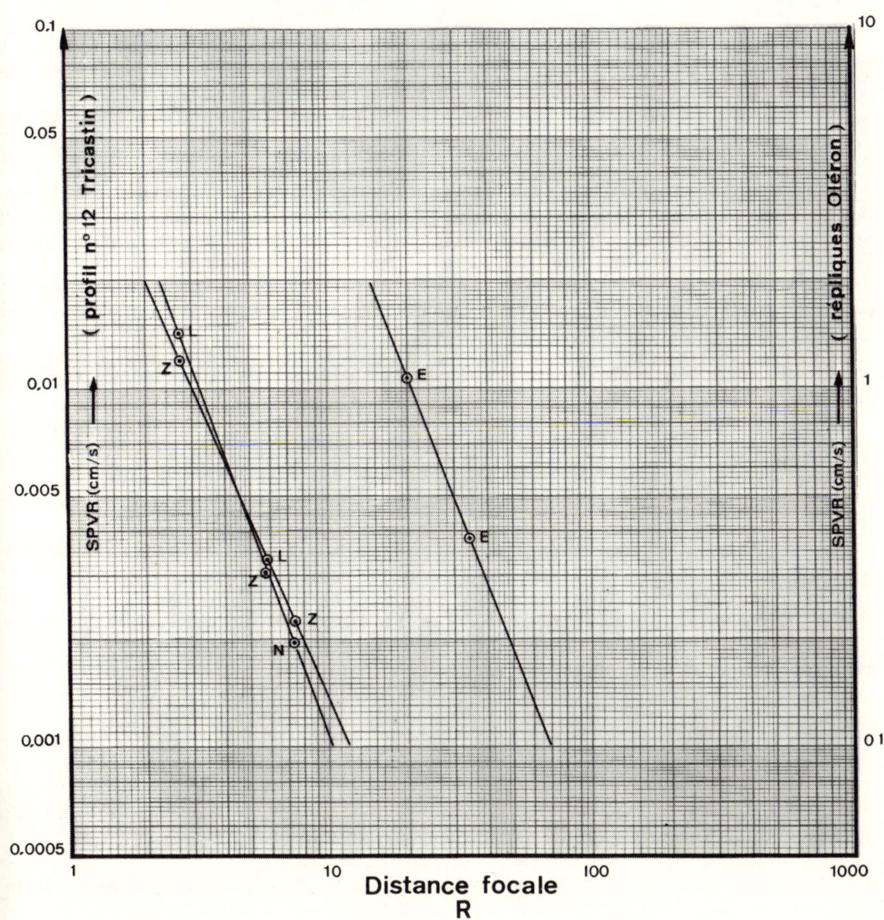


FIG.12.   Atténuation du SPVR en fonction de la distance.

Si B(t) est le mouvement incident à la base des couches stratifiée
et S(t) le mouvement à la surface du sol, on peut écrire :

$$S(t) = \int_{-\infty}^{+\infty} B(t)\, T\,(t-\tau)\, d\tau = B(t) * T(t)$$

$\tau$ est une variable d'intégration.

La fonction T(t) est la réponse temporelle caractéristique de ce
couches. La fonction de transfert est définie par :

$$T(f) = \frac{S(f)}{B(f)}$$

où S(f) et B(f) sont respectivement les spectres des fonctions tempore
les S(t) et B(t).

On peut calculer la fonction de transfert des couches superficiel
en partant de modèles théoriques. Il faut connaître les épaisseurs des
différentes couches (supposées homogènes et horizontales dans le cas
modèle simple), leurs densités et les vitesses de propagation des ond
longitudinales et des ondes transversales.

On peut également obtenir la fonction de transfert expérimenteleme
Dans le cas idéal, les géophones seront placés en forage, au sommet de
différentes couches. Une approximation du cas idéal précédent consiste
à placer les géophones sur le site et sur les affleurements les plus
proches du substratum rocheux. La validité des fonctions de transfert
repose, dans ce cas, sur la supposition que le substratum rocheux s'é
le de façon régulière sous les terrains du site.

Les figures 13 et 14 présentent :

- les spectres obtenus à partir des enregistrements à Cadarache d'un
  séisme (19 juillet 1973), composante E et Z, sur les alluvions anci
  nes d'une épaisseur de 15 à 20 m et sur une formation calcaire (voi
  la carte figure n° 6),

- les fonctions de transfert alluvions/calcaire qui en ont été déduit

La figure n° 15, analogue aux précédentes, montre les spectres ob
nus à partir des enregistrements sur le calcaire (Turonien) et sur le
alluvions (voir figure n° 5, carte Tricastin) d'un tir, ainsi que la
tion de transfert alluvions/calcaire.

## 4.4. Prévision d'un spectre de référence

On constate d'après ce qui précède que le SPVR varie en fonction
plusieurs paramètres, dont les principaux sont :

- l'énergie émise à la source sous forme d'ondes sismiques (magnitude
- la distance focale,
- la loi d'absorption de l'énergie en fonction de la distance,
- la fonction de transfert.

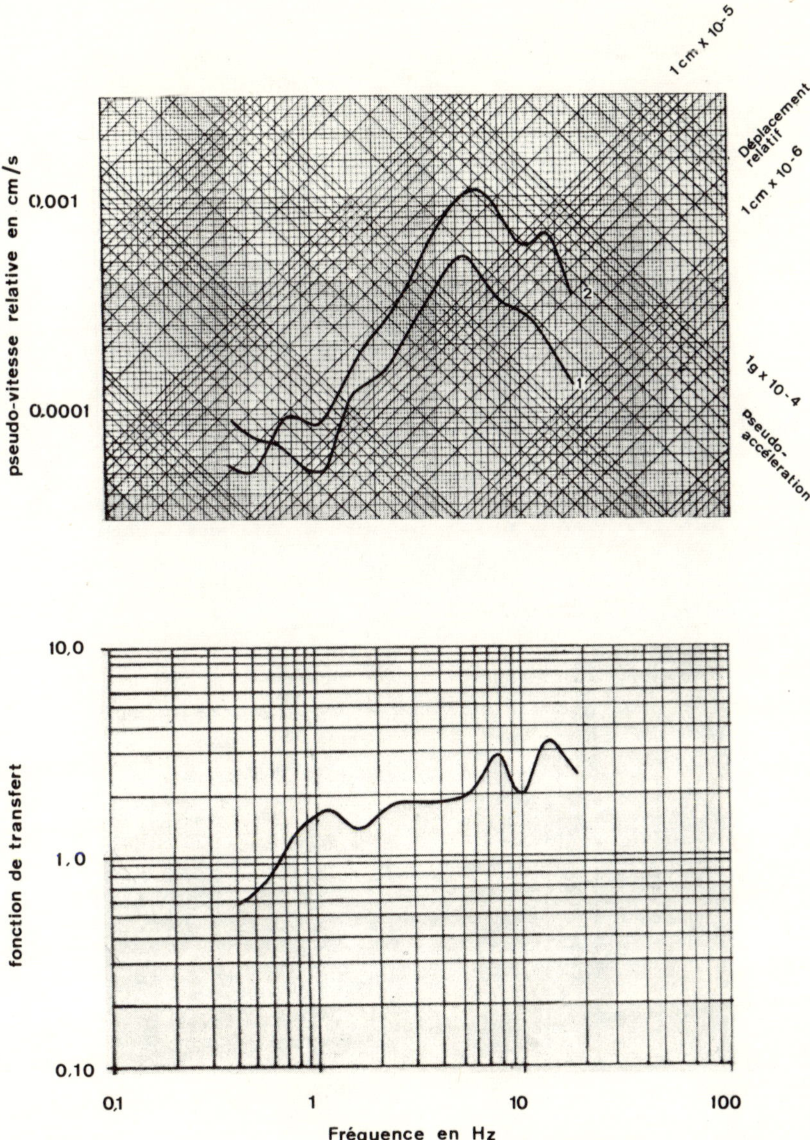


FIG. 13. Cadarache: Comparaison des SPVR (amortissement 0,05) du séisme du 19 juillet 1973, composante est-ouest, obtenus à partir des enregistrements sur des alluvions et sur du calcaire, et fonction de transfert alluvions/calcaire (1 - calcaire; 2 - alluvions).

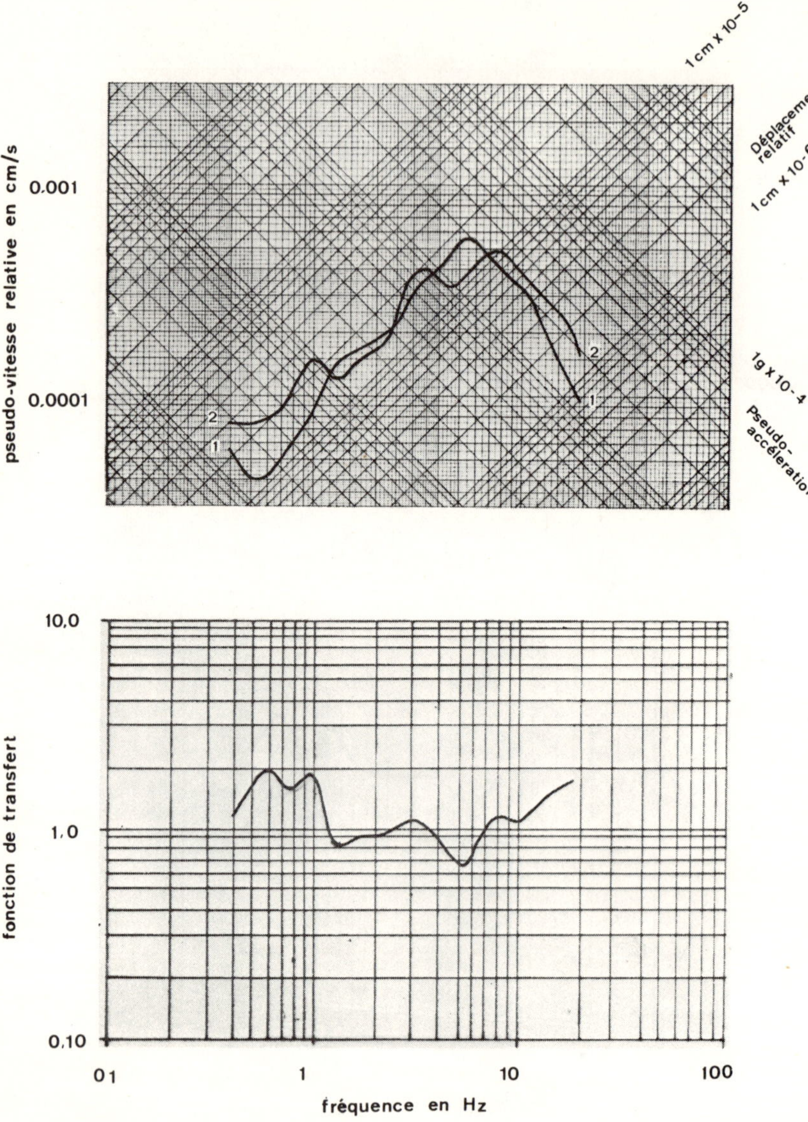



FIG. 14.　Cadarache: Comparaison des SPVR (amortissement 0,05) du séisme du 19 juillet 1973, composante verticale, obtenus à partir des enregistrements sur des alluvions et sur du calcaire, et fonction de transfert alluvions/calcaire (1 - calcaire; 2 - alluvions).

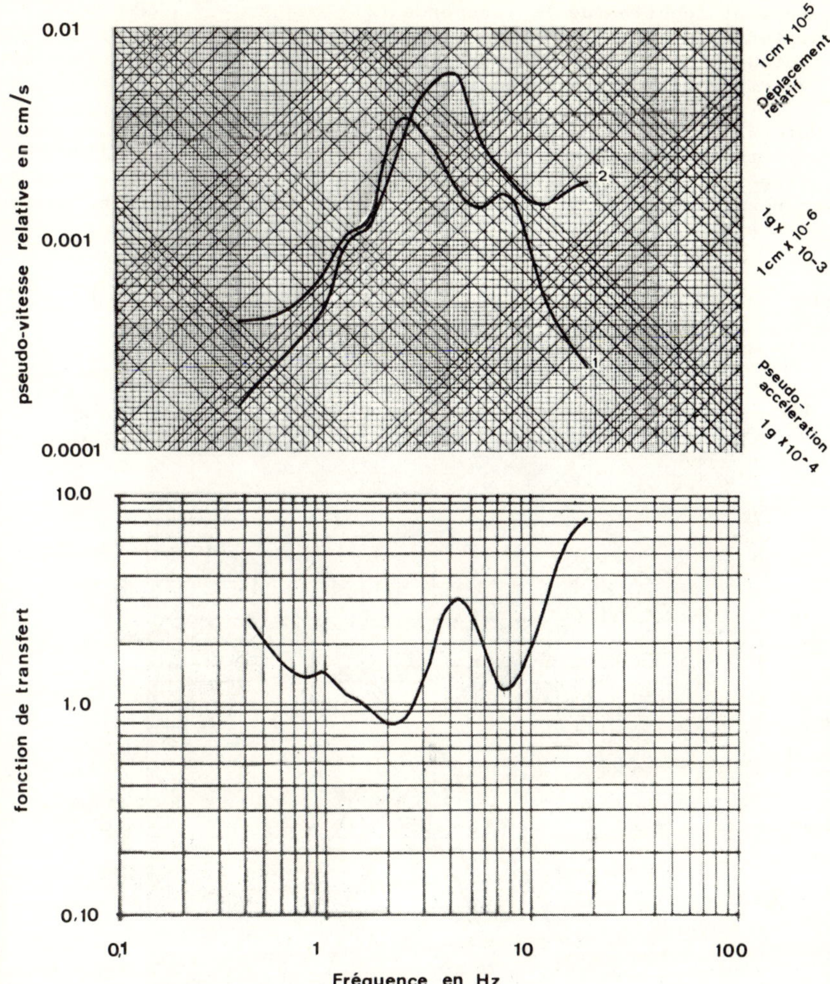



FIG. 15.  Tricastin:  Comparaison des SPVR d'un tir, composante est-ouest  (amortissement 0,05) obtenus
à partir des enregistrements sur des alluvions et sur du calcaire, et fonction de transfert alluvions/calcaire
(1 - calcaire;  2 - alluvions).

Des tentatives ont été faites par différents auteurs pour établir
des relations empiriques entre le SPVR et ces paramètres. La plus récen-
te est une étude statistique prenant en compte des tirs nucléaires /⁻8 7
/⁻9 7 /⁻10 7 du Nevada et des tremblements de terre de Californie qui
conduit à représenter le SPVR pour un amortissement de 5 % par la formule empirique suivante :

$$SPVR = C \; 10^{\alpha M} \; R^{n}$$

où :

C : constante pour une fréquence et un amortissement donnés
M : magnitude

α : exposant fonction de la fréquence
R : distance focale
n : facteur d'atténuation en fonction de la distance ; ce facteur est
    également variable avec la fréquence.

   Notons que les données statistiques américaines sont issues de mesu
res effectuées indifféremment sur la roche dure ou sur les alluvions.
Pour faire intervenir la fonction de transfert locale, il convient don
de séparer les données obtenues sur ces deux types de terrain. La form
précédente s'écrit alors :

$$\mathrm{SPVR} = \mathrm{C'}\ 10^{\alpha M}\ \mathrm{R}^{n'}\ \mathrm{A}$$

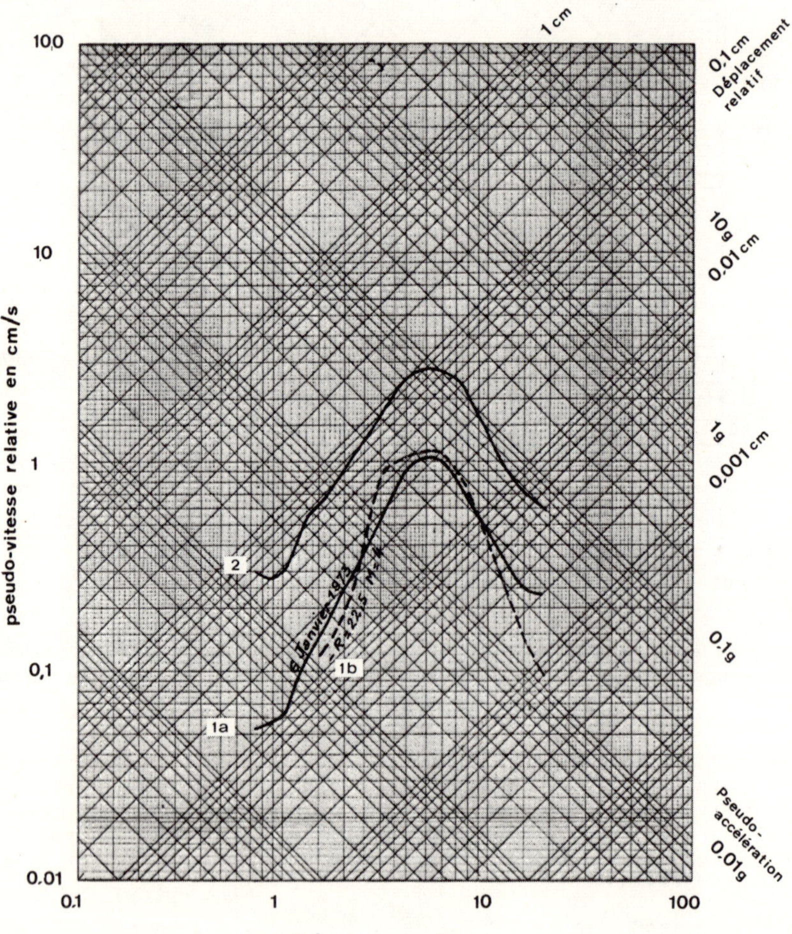


FIG. 16.   Région de l'île d'Oléron: Spectres de réponse (x = 0,05).  1a – Spectre de réponse du séisme du
6 janvier 1973.  1b – Spectre théorique avec M = 4 et R = 22,5.  2 – Spectre synthétique du séisme principal
du 7 septembre 1972 avec M = 5,4 et R = 22,5.

Il est donc intéressant de comparer les spectres synthétiques que l'on peut déduire de cette formule avec des spectres réels obtenus à partir d'enregistrements.

A titre d'exemple, on a présenté dans la figure n° 16 un groupe de trois spectres qui comprend :

1 - le spectre réel du séisme du 6 janvier 1973 (voir tableau 1A),
2 - le spectre du 6 janvier 1973 tel qu'on peut le calculer avec les coefficients C et α déduits des données statistiques américaines, n étant calculé pour la région considérée en utilisant le rapport des spectres obtenus à partir des enregistrements à Château d'Oléron et à la Rochelle (voir figure n° 11). On voit que les deux spectres sont très comparables et que le spectre synthétique est dans ce cas bien en accord avec les données expérimentales,
3 - un essai de reconstitution (en partant des hypothèses M = 5,4 et R = 22,5) du spectre du séisme principal du 7 septembre 1972 ressenti dans l'Ile d'Oléron avec l'intensité VII et dont on ne possède pas d'enregistrement.

## 5 - CONCLUSION

Les séismes de faible magnitude dont les foyers sont situés à faible profondeur (et dont la distance focale est, par conséquent, faible) sont riches en hautes fréquences. Il est donc nécessaire si on veut assurer la sûreté des installations nucléaires, de prendre en compte les spectres de ce type de séisme. Le programme d'étude actuellement en cours de développement a pour but d'améliorer nos connaissances en la matière par la récolte d'un maximum d'informations sur les séismes caractéristiques des styles tectoniques que l'on rencontre en France et de fournir les bases pour la définition d'un spectre de référence réaliste correspondant à ce genre de séisme. Les résultats obtenus jusqu'à maintenant sont en bon accord avec les spectres synthétiques calculés à partir des données statistiques déduites de tirs nucléaires et de séismes de Californie. Des compléments d'étude faisant intervenir d'autres sources d'informations sont toutefois nécessaires pour confirmer ces premiers résultats.

# REFERENCES


[1] AMBRASEYS, N.N., «Dynamics and response of foundation in epicentral regions of strong earthquakes», Proc. 5th World Conf. Earthquake Engineering, Rome, 1973, Edigraf, Rome (1974).

[2] ROTHE, J.P., DECHEVOY, N., Sismicité de la France de 1940 à 1950, Ann. Inst. Phys. Globe, Nouvelle série VII, Troisième partie, Géophysique (1954).

[3] ROTHE, J.P., DECHEVOY, N., La sismicité de la France de 1951 à 1960, Ann. Inst. Phys. Globe VIII (1967).

[4] ROTHE, J.P., La sismicité de la France de 1961 à 1970, Ann. Inst. Phys. Globe IX (1972).

[5] KARNIK Vit, Seismicity of the European Area, Parts 1 et 2, Dordrecht, D. Reidel Publ. Co., Dordrecht (1969, 1971).

[6] MAX, J., Les principales méthodes de traitement du signal (corrélation, analyse spectrale ...) et leurs applications aux mesures physiques, Rapport CEA-R-4018 (1970).

[7] HARRIS, C.M., CREDE, C.E., Shock and Vibration Handbook, McGraw-Hill, New York (1961).

[8] JOHNSON, R.A., An earthquake spectrum prediction technique, Bull. Seismol. Soc. Am. 63 4 (1973) 1255-74.

[9] LYNCH, R.D., Response spectra for Pahuta Mesa Nuclear Event, Bull. Seismol. Soc. Am. 59 6 (1969) 2295-2309.

[10] KEMMISH, W.B., Communication privée.

# DISCUSSION

E. IANSITI: Is it possible to extrapolate the response spectrum associated with a small earthquake and with a low maximum acceleration to more severe earthquakes and greater accelerations?

A. BARBREAU: The spectra corresponding to low energies cannot be extrapolated directly to high energies simply by changing their levels; they are different in shape. In general, a low-frequency enrichment is observed as the magnitude increases. It is therefore necessary to determine separately the spectra corresponding to different magnitudes. One can do so, because these magnitudes are directly involved in the calculation of spectra.

E. IANSITI: What do you think of the new theories which correlate the source mechanism with near-field response and far-field response spectra (the works of Trifunac, for example)?

A. BARBREAU: A spectrum can indeed be calculated from the source mechanism. The difficulty lies in knowing this source mechanism in advance. It depends, above all, on the characteristics of the fault concerned; but these are often little known, especially in regions where it is difficult to correlate earthquakes with particular faults.

# CONNECTIONS BETWEEN SITING, NORMAL OPERATING PRACTICES, STRUCTURAL LAY-OUT AND QUALITY ASSURANCE
## Present situation and future outlook for nuclear power plants

W. VINCK, F. LUYKX*, H. MAURER,
J. VAN CAENEGHEM
Commission of the European Communities,
Brussels, Belgium

**Abstract**

CONNECTIONS BETWEEN SITING, NORMAL OPERATING PRACTICES, STRUCTURAL LAY-OUT AND
QUALITY ASSURANCE: PRESENT SITUATION AND FUTURE OUTLOOK FOR NUCLEAR POWER PLANTS.
   The choice of reactor site must take into consideration, amongst other things, release of radioactive
effluents during normal operation and the potential of accident situations. The paper shows how, in the
European Community, the actual practice of effluent control has succeeded in limiting exposure around
nuclear power plant sites to extremely low values. Procedures to limit releases followed in the different
countries are described and ideas are given on the approach which would be the most appropriate in the
future, taking into account the increase in unit and the tendency to create multi-unit sites. In the context
of prevention of accident initiation, the paper discusses some of the efforts employed in the analysis of
design and construction problems, quality control, in-service inspection and testing of the thick-walled
primary components (pressure vessel). Emergency core cooling and containment systems are briefly
discussed for their role in reducing the consequences of accident situations.

## 1. INTRODUCTION

It should be understood that the scope of this paper is deliberately limited in several ways:

(a) It deals only with nuclear power plants as related to siting problems from the standpoint of nuclear health and safety and the environmental implications (excluding, for example, reprocessing plants).
(b) For this type of installation it deals only with practices and criteria applied in the general areas of release and control of normal effluents and in prevention and mitigation of accidents.
(c) In the latter area it only summarizes a few technological considerations for some primary structures and systems, and secondary containment designs for LWR-type reactors, from the point of view of assessing their integrity and/or performance characteristics as required to ensure safe operation or sufficiently limited accident consequences. It does not deal with numerous other design and accident analysis aspects which are also related to the nuclear plant and site; the considerations presented in this limited context summarize practices encountered in national safety reviews and those sponsored by the Commission of the European Communities (CEC).

---

* Present address: CEC, Luxembourg.

## TABLE I.  GENERAL CHARACTERISTICS OF NUCLEAR POWER STATIONS IN THE EUROPEAN COMMUNITY[a]

| Facility/Location | Type of reactor[b] | Maximum output capacity (MW(e)) | First link-up with grid | Water body receiving liquid effluents |
|---|---|---|---|---|
| **FED. REP. GERMANY** | | | | |
| VAK, Kahl (Bavaria) | BWR | 15 | 17.06.1961 | Main |
| AVR, Jülich (North-Rhine Westphalia) | GCR | 13 | 17.12.1967 | Rur |
| KRB, Gundremmingen (Bavaria) | BWR | 237 | 12.11.1966 | Danube |
| MZFR, Karlsruhe (Baden-Württemberg) | HWR | 57 | 09.03.1966 | Rhine |
| KWL, Lingen (Lower Saxony) | BWR | 174[c] | 20.05.1968 | Ems |
| KWO, Obrigheim (Baden-Württemberg) | PWR | 328 | 29.10.1968 | Neckar |
| KWW, Würgassen (North-Rhine Westphalia) | BWR | 640 | 18.12.1971 | Weser |
| KKS, Stade (Lower Saxony) | PWR | 630 | 29.01.1972 | Elbe |
| BIBLIS A, Biblis (Hessen) | PWR | 1146 | 10.1974 | Rhine |
| **BELGIUM** | | | | |
| BR-3, Mol (Antwerp) | PWR | 10 | 28.10.1962 | Molse Nete |
| DOEL (Antwerp) | PWR | 2 × 392 | 1) 09.1974<br>2) 1975 | Scheldt |
| TIHANGE (Liège) | PWR | 870 | 1975 | Meuse |
| **FRANCE** | | | | |
| EDF-1, Chinon (Indre-et-Loire) | GCR | 70 | 14.06.1963 | Loire |
| EDF-2, Chinon (Indre-et-Loire) | GCR | 210 | 24.02.1965 | Loire |
| EDF-3, Chinon (Indre-et-Loire) | GCR | 480 | 04.08.1966 | Loire |
| SENA, Chooz (Ardennes) | PWR | 280 | 03.04.1967 | Meuse |

TABLE I (cont.)

| Facility/Location | Type of reactor[b] | Maximum output capacity (MW(e)) | First link-up with grid | Water body receiving liquid effluents |
|---|---|---|---|---|
| FRANCE (cont.) | | | | |
| EL-4, Monts d'Arrée (Finistère) | HWR | 70 | 09.07.1967 | Ellez |
| SL-1, St-Laurent-des-Eaux (Loir-et-Cher) | GCR | 480 | 14.03.1969 | Loire |
| SL-2, St-Laurent-des-Eaux (Loir-et-Cher) | GCR | 515 | 09.08.1971 | Loire |
| BUGEY (Ain) | GCR | 540 | 15.04.1972 | Rhône |
| PHENIX, Marcoule (Gard) | FBR | 233 | 13.12.1973 | Rhône |
| ITALY | | | | |
| LATINA, Latina (Latina) | GCR | 153 | 12.05.1963 | Thyrrhenian Sea |
| GARIGLIANO, Sessa (Caserta) | BWR | 152 | 23.01.1964 | Garigliano |
| TRINO VERCELLESE, Trino Vercellese (Vercelli) | PWR | 247 | 22.10.1964 | Po |
| CAORSO (Piacenza) | BWR | 840 | 1975 | Po |
| NETHERLANDS | | | | |
| DODEWAARD, Dodewaard (Gelderland) | BWR | 53 | 25.10.1968 | Waal |
| BORSSELE (Zeeland) | PWR | 450 | 04.07.1973 | W. Scheldt |
| UNITED KINGDOM CEGB[d] | | | | |
| BERKELEY, Gloucestershire (England) | GCR | 276 | 06.1962 | Severn Estuary |
| BRADWELL, Essex (England) | GCR | 250 | 06.1962 | Blackwater Estuary |
| HINKLEY POINT A Somerset (England) | GCR | 460 | 02.1965 | Severn Estuary |
| TRAWSFYNYDD, Merioneth (Wales) | GCR | 390 | 01.1965 | Lake Trawsfynydd |
| DUNGENESS A, Kent (England) | GCR | 410 | 09.1965 | English Channel |

TABLE I (cont.)

| Facility/Location | Type of reactor[b] | Maximum output capacity (MW(e)) | First link-up with grid | Water body receiving liquid effluents |
|---|---|---|---|---|
| UNITED KINGDOM (cont.) | | | | |
| SIZEWELL, Suffolk (England) | GCR | 420 | 01.1966 | North Sea |
| OLDBURY, Gloucestershire (England) | GCR | 400 | 11.1967 | Severn Estuary |
| WYLFA, Anglesey (Wales) | GCR | 840 | 11.1971 | Irish Sea |

[a] From "Operation of Nuclear Power Stations during 1973" Statistical Office of the European Communities - April 1974.

[b] BWR:  Boiling-water reactor
PWR:  Pressurized-water reactor
GCR:  Gas-cooled reactor
HWR:  Heavy-water reactor
FBR:  Fast-breeder reactor

[c] Plus 87.5 MW(e) by conventional superheating.

[d] Central Electricity Generating Board.

This paper therefore covers some special (non-exhaustive) aspects which demonstrate:

(a)  For normal operations the release and control of radioactive effluents the present safety record and the outlook for future choice of sites and multi-unit development;  and

(b)  That there is a relationship between design of plant and of vital systems and long-term quality assurance on one hand, the initial acceptance of the site and subsequent protection of its environment on the other hand.

## 2.  CONTROL OF RADIOACTIVE EFFLUENTS IN NORMAL OPERATION

The release of radioactive effluents during normal operation of a nuclear power plant is the factor of primary importance in environmental siting.  The level of radioactive releases, the resulting radiation dose and significance in relation to the natural radiation exposure and to international recommendations are amongst the determining factors for the acceptability of nuclear plants at a given site.

A.  For many years, the International Commission on Radiological Protection (ICRP) has included in its recommendations the requirement that all radiation exposure of man be kept as far below the dose limits as

## TABLE II.  ANNUAL DISCHARGE OF GASEOUS RADIOACTIVE WASTE (NOBLE GASES)

| Country | Facility | Discharge limit (Ci/a) | Activity released (Ci/a) | | | |
|---|---|---|---|---|---|---|
| | | | 1969 | 1970 | 1971 | 1972 |
| FED. REP. GERMANY | VAK | $8.8 \times 10^4$ | 1 750 | 3 340 | 2 455 | - |
| | AVR | 51.6 | 18 | 32 | 27 | 30 |
| | KRB | $1.9 \times 10^6$ | 11 400 | 7 350 | 6 780 | 11 105 |
| | MZFR[a] | $3 \times 10^3$ | - | - | 526 | 955 |
| | KWL | $3.1 \times 10^6$ | 166 000 | 114 000 | 9 000 | 5 300[b] |
| | KWO | $8 \times 10^4$ | 5 560 | 7 700 | 1 456 | 3 202 |
| | KWW | $3.2 \times 10^4$ | - | - | - | 594 |
| | KKS | $6.1 \times 10^4$ | - | - | 3 | 2 445 |
| BELGIUM | BR-3 | | - | 26 680[c] | - | 252 |
| FRANCE | Chinon | $4 \times 10^{5d}$ | 12 300 | 8 085 | 4 225 | 11 515 |
| | SENA | $2.5 \times 10^{6d}$ | - | 3 | 4 500 | 31 342 |
| | EL-4 | $4 \times 10^{5d}$ | 46 | 72 | 53 810 | 144 450[e] |
| | St-Laurent-des-Eaux | $4 \times 10^{5d}$ | 1 900 | 305 | 3 425 | 3 863 |
| | Bugey | $4 \times 10^{5d}$ | - | - | - | 841 |
| ITALY | Latina | $5 \times 10^3$ | 1 500 | 2 500 | 2 470 | 3 660 |
| | Garigliano | $6.3 \times 10^5$ | 140 000 | 275 000 | 640 000 | 290 000 |
| | Trino | $5 \times 10^5$ | - | 19 | 585 | 1 031 |
| NETHERLANDS | Dodewaard | $3 \times 10^5$ | - | $\sim 3\ 000$ | $\sim 3\ 000$ | 8 400 |
| UNITED KINGDOM | CEGB Power Stations:  gaseous activity not measured routinely[f] | | | | | |

[a] In addition MZFR discharged the following amounts of tritium into the atmosphere:  in 1969 1300 Ci; in 1970 1190 Ci;  in 1971 1130 Ci;  in 1972 542 Ci.  The discharge limit is 4000 Ci/a.

[b] Shut-down of power station for 8 months.

[c] Exceptional discharge due to reactor operating experimentally with faulty fuel elements.

[d] At this discharge level, assuming an atmospheric dilution factor of $1.5 \times 10^{-5}$ s/m$^3$ and a probability of 20% that the wind is blowing in one direction, the maximum concentration in air at ground level is equal to the MPC to population in air.

[e] In addition, EL-4 discharged 83 Ci of tritium into the atmosphere.

[f] In the CEGB power stations, gas activity (Ar-41) is not systematically measured.  Occasional measurements have shown that at the Bradwell, Hinkley Point and Trawsfynydd power stations the annual discharges are approximately 40 000 Ci, 200 000 Ci and 130 000 Ci, respectively.

can reasonably be achieved and that any exposure not strictly necessary
be avoided.  This recommendation was taken over in the Euratom Basic
Standards on Radiological Protection and has consequently been included
in the legislation of the Member States.

The authorities responsible for radiation protection and nuclear power-
plant operators have applied this principle to discharges of radioactive
effluents from nuclear power stations by limiting discharges to levels as
low as readily achievable.

Table I shows the general characteristics of nuclear power stations
in the countries of the European Community from 1969 to 1972.  Tables II-V
give releases of gaseous and liquid effluents from these plants as well as
the discharge limits for the various effluents.

An evaluation [1], based on conservative assumptions, of the maximum
exposure to individuals living in the vicinity of nuclear power stations as
a result of the activities released (Tables II-VI) shows that:

(a)  Whole-body doses due to gaseous effluents are generally less than
     1% of the dose limit;  in the case of the newer power stations they are
     lower than 0.1%.
(b)  Doses to the thyroid of an infant consuming milk produced near the site
     are less than 0.3% of the dose limit;  and
(c)  Doses to critical groups of the population exposed to discharges of
     liquid radioactive waste are generally lower than 1% of the dose limits.

Table VII gives the annual discharge limits for gaseous and liquid
radioactive waste per 100 MW(e) output for the most recent Community
power stations with water-cooled reactors.  From this table it can be seen
how, without prior consultation between the different countries, the current
state of technical knowledge in waste treatment has brought about an
increasing convergence of the limits.  For $^{131}$I, in particular, all limits lie
within a factor of about 2.  However, practice regarding the control of
radioactive waste discharges differs considerably from country to country.

B.   In this connection a distinction must be made between approaches
followed for:  (a) the design of a new installation and (b)  the setting of
discharge limits for a nuclear plant.

(a)  Design of a new installation

Three different approaches are followed in the Member States:

(i)   The authorities require that doses due to releases from the plant do
      not exceed a stipulated fraction of the ICRP dose limits.  Calculating
      backwards, using conservative assumptions, the plant designer can
      work out the corresponding maximum rate of discharge of radioactive
      effluent.
(ii)  The authorities have drawn up general guidelines and numerical
      reference values for the release of radioactive effluents from nuclear
      facilities into the environment (e.g. the LAWA guidelines in the
      Federal Republic of Germany [2]).
(iii) The designer of a new installation is asked to plan discharge equipment
      as efficient as reasonably achievable and in line with the current state
      of technology.

## TABLE III. ANNUAL DISCHARGE OF RADIOACTIVE AEROSOLS

| Country | Facility | Discharge limit (Ci/a) | Activity released (Ci/a) | | | |
|---|---|---|---|---|---|---|
| | | | 1969 | 1970 | 1971 | 1972 |
| FED. REP. GERMANY | VAK | 88 | $8.8 \times 10^{-2}$ | 0.13 | $6.8 \times 10^{-2}$ | - |
| | AVR | | | | 0.152 | 0.21 |
| | KRB | 2 850 | $7.6 \times 10^{-3}$ | $7.5 \times 10^{-2}$ | $5.0 \times 10^{-2}$ | $1.4 \times 10^{-2}$ |
| | MZFR | | - | - | $<1.2 \times 10^{-3}$ | $1.2 \times 10^{-3}$ |
| | KWL | 15 800 | 0.25 | 0.67 | - | $< 1.4 \times 10^{-2}$ |
| | KWO | (a) | $< 3 \times 10^{-2}$ | $< 5 \times 10^{-2}$ | $5.8 \times 10^{-2}$ | $1.7 \times 10^{-2}$ |
| | KWW | 10.5 | - | - | - | $6.5 \times 10^{-3}$ |
| | KKS | 17.5 | - | - | - | $1.2 \times 10^{-2}$ |
| BELGIUM | BR-3 | | - | - | - | - |
| FRANCE | Chinon | $1 \times 10^{3b}$ | $< 1 \times 10^{-2}$ | $< 1 \times 10^{-2}$ | $18 \times 10^{-3}$ | $7.5 \times 10^{-2}$ |
| | SENA | $1 \times 10^{3b}$ | - | - | - | $5 \times 10^{-4}$ |
| | EL-4 | $1 \times 10^{3b}$ | $< 2 \times 10^{-3}$ | | $7.3 \times 10^{-2}$ | $6.2 \times 10^{-3}$ |
| | St-Laurent-des-Eaux | $1 \times 10^{3b}$ | $< 1$ | $< 1 \times 10^{-2}$ | $4.7 \times 10^{-2}$ | $7 \times 10^{-3}$ |
| | Bugey | $1 \times 10^{3b}$ | - | - | - | $4 \times 10^{-4}$ |
| ITALY | Latina | 0.1 | - | - | - | - |
| | Garigliano | 1 | $6.3 \times 10^{-2c}$ | $6.3 \times 10^{-2c}$ | $6.3 \times 10^{-2c}$ | $6 \times 10^{-2}$ |
| | Trino | 2 | - | $<1.2 \times 10^{-4}$ | $<1.4 \times 10^{-4}$ | $1 \times 10^{-5}$ |
| NETHERLANDS | Dodewaard | (a) | - | $2 \times 10^{-2}$ | $4 \times 10^{-2}$ | $2 \times 10^{-2}$ |
| UNITED KINGDOM | Berkeley | (d) | $6.6 \times 10^{-3}$ | $5.7 \times 10^{-3}$ | $5.2 \times 10^{-3}$ | $5.7 \times 10^{-3}$ |
| | Bradwell | (d) | $3.7 \times 10^{-3}$ | $5.6 \times 10^{-3}$ | $3.2 \times 10^{-3}$ | $3.0 \times 10^{-3}$ |
| | Hinkley Point A | (d) | $2.8 \times 10^{-2}$ | $1.7 \times 10^{-2}$ | $1.3 \times 10^{-2}$ | $4.3 \times 10^{-2}$ |
| | Trawsfynydd | (d) | $2.2 \times 10^{-2}$ | $2.0 \times 10^{-2}$ | $2.6 \times 10^{-2}$ | $3.4 \times 10^{-2}$ |
| | Dungeness A | (d) | $1.55 \times 10^{-1}$ | $1.25 \times 10^{-1}$ | $1.12 \times 10^{-1}$ | $9.4 \times 10^{-2}$ |
| | Sizewell | (d) | $1.6 \times 10^{-2}$ | $1.3 \times 10^{-2}$ | $1.1 \times 10^{-2}$ | $8.4 \times 10^{-3}$ |
| | Oldbury | (d) | $9.1 \times 10^{-3}$ | $2.0 \times 10^{-2}$ | $5.6 \times 10^{-3}$ | $3.2 \times 10^{-2e}$ |
| | Wylfa | (d) | - | - | - | $3.2 \times 10^{-3f}$ |

[a] No limit laid down in the operating licence.

[b] At this level, assuming an atmospheric dilution factor of $1.5 \times 10^{-5}$ s/m$^3$ and a probability of 20% that the wind is blowing in one direction, the concentration at ground level is equal to the MPC to population in air ($10^{-9}$ Ci/m$^3$).

[c] Average value for 1969, 1970 and 1971.

[d] Authorizations for discharge of radioactive gases and aerosols from CEGB power stations place no limit on the quantities but require that the best practicable means be used to minimize the amount of radioactivity to be discharged.

[e] In addition, Oldbury discharged 0.3 Ci of S-35 in 1972.

[f] In addition, Wylfa discharged $5.6 \times 10^{-2}$ Ci of S-35 and 194 Ci of H-3 in 1972.

## (b)  Setting of discharge limits

In all Member States discharges of radioactive effluents from a nuclear installation are subject to a system of prior authorization.  However, the approaches to limit discharges differ considerably.

## (i)  Limits for general application

Some countries apply or are envisaging the application of general limits for nuclear power plants or even for all types of nuclear installation. Two kinds of limitations are applied:

### Allocation of a stipulated fraction of the existing dose limits to exposure resulting from radioactive effluents

For instance, in the Federal Republic of Germany [3] the licensing authorities set discharge limits such that the whole-body dose due to internal and external irradiation, taking into account all activity releases relevant for the environment at the most unfavourable point of each exposure pathway, does not exceed 30 mrem/a;  the thyroid dose to small children due to radioiodine taken up through the food chain may not exceed 90 mrem/ calculated for a hypothetical person drinking milk produced at the most unfavourable point in the environment.

### Limitation of the population dose due to activity releases from a given plant

The nordic countries have published a joint statement on "Basic principles for the limitation of releases of radioactive substances from nuclear power stations" [4].  The Danish authorities intend to apply these principles when authorizing discharges for their future power plants.  One of the recommendations in this statement is that the collective whole-body dose from the full fuel cycle operations (exclusive of occupational exposure) be limited to 1 man · rem/a per MW(e) installed, of which no more than 50% should be due to releases from nuclear power stations.

## (ii)  Limits set on a case-by-case basis

In several countries discharge limits are set individually for each plant or for each site.  Usually the applicant has to propose a maximum discharge in his application to the authorizing body and to justify his proposal.  On the basis of the real discharge needs of the plant, allowing some reasonable margin for operational flexibility, and after considering the radiological consequences of the release, the authorities set the discharge limits.  In some Member States, the technical services of the authorizing bodies proceed at the same time to an assessment of the radiological capacity of the environment[1], and check that the proposed release does not exceed a stipulated fraction of this capacity.  This is, for instance, the case in Italy and the United Kingdom (for liquid effluents).

---

[1]  The radiological capacity of the environment can be defined as the maximum amount of activity which can be discharged yearly into a given environment without exceeding, through the critical pathway, the ICRP dose limits to the critical group of the population.

## TABLE IV.  ANNUAL DISCHARGE OF $^{131}$I INTO THE ATMOSPHERE

| Country | Facility | Discharge limit (Ci/a) | Activity released (Ci/a) | | | |
|---|---|---|---|---|---|---|
| | | | 1969 | 1970 | 1971 | 1972 |
| FED. REP. GERMANY | VAK | 0.61 | $7 \times 10^{-3}$ | 0.6 | $2.9 \times 10^{-3}$ | - |
| | AVR | | - | - | - | - |
| | KRB | 22 | 0.36 | 0.2 | 0.35 | 0.19 |
| | MZFR | | - | - | - | - |
| | KWL | 16 | - | 0.26 | 0.38 | 0.15 |
| | KWO | $15^a$ | $6.3 \times 10^{-2}$ | $4.4 \times 10^{-2}$ | $1.5 \times 10^{-2}$ | $6.2 \times 10^{-3}$ |
| | KWW | | - | - | - | - |
| | KKS | 0.21 | - | - | - | $4.0 \times 10^{-2}$ |
| BELGIUM | BR-3 | | $1 \times 10^{-4}$ | $6.3 \times 10^{-2}$ | $< 2 \times 10^{-5}$ | $< 1 \times 10^{-3}$ |
| FRANCE | Chinon | $1.5^b$ | - | - | - | $2.7 \times 10^{-2}$ |
| | SENA | $1.5^b$ | - | - | - | $2.3 \times 10^{-2}$ |
| | EL-4 | $1.5^b$ | - | - | - | - |
| | St-Laurent-des-Eaux | $1.5^b$ | - | - | - | $6.5 \times 10^{-2}$ |
| | Bugey | $1.5^b$ | - | - | - | $1 \times 10^{-4}$ |
| ITALY | Latina | $1 \times 10^{-3c}$ | - | - | - | - |
| | Garigliano | $1^c$ | - | $6 \times 10^{-2}$ | 0.13 | $6 \times 10^{-2}$ |
| | Trino | $0.5^c$ | - | $< 5.9 \times 10^{-4}$ | $1 \times 10^{-3}$ | $1 \times 10^{-6}$ |
| NETHERLANDS | Dodewaard | | - | $6.3 \times 10^{-3}$ | $6.3 \times 10^{-3}$ | $6 \times 10^{-3}$ |
| UNITED KINGDOM | CEGB power stations$^d$ | (e) | | | | |

[a] During the grazing period, the discharge of I-131 is limited to 2.5 mCi per week and 1 mCi per day.
[b] Annual permissible discharge based on the milk consumption exposure path.
[c] Limit for all halogens.
[d] The discharge of I-131 by British power stations is stated to be negligible.
[e] Authorizations for the discharge of iodine from the CEGB power stations place no limit on the quantities, but require that the best practicable means be used to minimize the amount of iodine to be discharged.

## (iii)  Discharge restriction without specification of discharge limits

In one country, the United Kingdom, authorizations for gaseous emissions from nuclear power stations do not include specific limits for the radionuclides to be discharged. Instead, the operator is required to use the best practicable means to minimize the amount of radioactivity discharged. The reason for this procedure is that the authorities fear discharge limits will be regarded as working limits by the plant operators.

### C.  Future practice

It is worth considering which of the above approaches to the limitation
of radioactive waste discharge would be the most appropriate in the future,
taking into account the predicted growth in nuclear power production, the
increase in the unit size of power plants and the tendency to create
multi-unit sites and to release liquid effluents from many plants into the
same water-body.  Indeed, fears exist that, without a good system of
discharge control, the cumulative effects of discharges from several units
would cause unacceptable exposure of the population living in the
neighbourhood.

Whatever the approach adopted, the two basic ICRP principles:

(a)  Respect for the dose limits;
(b)  Optimization of radioprotection, i.e. reduction of effluent discharges
     to levels as low as readily achievable.

will certainly have to be taken into account.

### (a)  Respect for the ICRP dose limits

A sound approach to site evaluation would be to determine beforehand
what fraction of these dose limits should be allocated to radioactive effluent
discharges from nuclear power plants.  For gaseous effluents a quite
straightforward calculation then allows conversion of the fraction derived
into maximum discharge rates from a given site.  For liquid effluents the
radiological capacity of the environment into which liquids will be released
should be determined and a fraction of that capacity allocated to each site
releasing into it.

### (b)  Optimization of radioprotection

To minimize releases of radioactive effluents, nuclear plants should
always be built according to the latest state of technical knowledge in the
field of emission control.  This means that a case-by-case analysis of
discharge control should be carried out during the design and construction
of a new unit and before setting discharge limits.  During this optimization
stage, use could be made of the concept of collective dose to compare the
merits of various possible solutions to effluent control.

### 3.  EXAMPLES OF CONSIDERATIONS IN PREVENTION AND MITIGATION OF POTENTIALLY SEVERE ACCIDENT CONDITIONS IN THE COUNTRIES OF THE CEC

Discussion of siting problems also takes into consideration possible
radioactive release in accident situations.  Analysis, control and tests of
the structural features of the plant are essential in preventing accident
initiation.  The considerations on prevention and mitigation of potentially
severe accident situations treated hereafter are limited to the thick-walled
components (pressure vessel) of the primary circuit of LWRs, and

particular emphasis is placed on the problems related to the initiation of accidents which could be caused by deficiencies in design, fabrication, inspection and material-failure modes.

Engineered safety features are the means of reducing the consequences of accidents. The problems related to the study of the evolution of accidents such as loss-of-coolant accidents and containment response are discussed briefly.

## 3.1. Reactor pressure vessel and primary pressure boundary

### 3.1.1. Design problems

Problems relating to accident initiation start with the design and stress analysis of the pressure components. The stress analysis is to be performed for the various operating conditions of the components during the lifetime of about 40 years.

For the reactor pressure vessel, about 400 occurrences of heat-up and cool-down are to be considered and about 18 000 occurrences each of unit loading and unloading at 5% of the full power per minute, about 2000 occurrences each of step load increase and decrease at 10% of the full power, and about 400 occurrences of reactor trip at full power. These are the most important reactor-coolant operating transients. Other operating transients to be considered are large step load decrease, loss of load (without immediate turbine or reactor trip), loss of power, loss of flow (including partial loss of flow), hydrostatic test conditions performed before initial start-up and post-operation tests [5].

Under all these transient conditions the core support structure must hold the core in position; there should be no over-pressurization of the pressure vessel and no undue crack development. The design and the stress analysis of the cylindrical vessel parts pose no real problems for the designer, but the connections to the nuclear pressure vessel (e.g. the pipework) necessitate structure discontinuities which may significantly alter the normal stress patterns in the vessel walls and introduce undesirable stress concentrations.

A vast amount of experimental and analytical information on stress distribution near nozzle penetrations in cylindrical and hemispherical vessel walls have already been made available. Most information has been concerned with the establishment of elastic stress concentration factors, while some has been directed towards defining the elasto-plastic behaviour in the intersection area. It is current practice in European countries to apply, for the design and the stress analysis of nozzle penetrations, the information on the elastic stress concentration factors or the stress indices stated in the ASME Boiler and Pressure Vessel Code, Section III. These indices are defined as the ratio between the highest stress peak in the nozzle-to-shell intersection area and the corresponding membrane stress in the undisturbed shell. For geometries of the nozzles which are exactly the same as described in the ASME code, the stress indices stated in the code were confirmed by model tests. It seems, however, unacceptable to apply such stress indices also in cases where the geometries of installed nozzles are slightly different in shape from the models given in the ASME code. So far as only primary membrane stresses are to be considered in the analysis, the maximum stress index of 3.1 given in the code will

## TABLE V.  ANNUAL DISCHARGE OF LIQUID RADIOACTIVE EFFLUENTS (EXCLUSIVE OF TRITIUM)

| Country | Facility | Discharge limit (Ci/a) | Activity released (Ci/a) | | | |
|---|---|---|---|---|---|---|
| | | | 1969 | 1970 | 1971 | 1972 |
| FED. REP. GERMANY | VAK[a] | 0.6 | $6 \times 10^{-3}$ | $6.4 \times 10^{-2}$ | $6.0 \times 10^{-2}$ | $2.3 \times 10^{-2}$ |
| | AVR[b] | | - | - | - | $9.7 \times 10^{-2}$ |
| | KRB | 14.4 | 1.65 | 1.52 | 1.89 | 1.54 |
| | MZFR[c] | | - | - | - | - |
| | KWL | 5.4 | 0.65 | 0.60 | 0.38 | 0.11 |
| | KWO | 18 | 10.5 | 3 | 4.4 | 3.33 |
| | KWW | 17 | - | - | - | 1.8 |
| | KKS | 5 | - | - | - | 0.63 |
| BELGIUM | BR-3[d] | | | | | |
| FRANCE | Chinon | 900[e] | 7.44 | 2.25 | 2.0 | 3.0 |
| | SENA | 100[e] | 3.8 | 6.4 | 34.4 | 12.4 |
| | EL-4 | 4[e] | $2.7 \times 10^{-2}$ | $6 \times 10^{-3}$ | 0.1 | 0.22 |
| | St-Laurent-des-Eaux | 800[e] | 2.71 | 0.77 | 2.25 | 9.4 |
| | Bugey | 680[e] | - | - | - | $3.9 \times 10^{-2}$ |
| ITALY | Latina | 20[f] | 29.6 | 10.2 | 1.5 | 16.5 |
| | Garigliano | 25[f] | 9 | 11.9 | 19.1 | 14.4 |
| | Trino | 15[f] | 3.09 | 2.96 | 19.07 | 6.0 |
| NETHERLANDS | Dodewaard | 2.6 | 0.5 | 2.33 | 1.6 | 2.03 |
| UNITED KINGDOM | Berkeley | 200[g] | 55.1 | 23.2 | 15.0 | 23.3 |
| | Bradwell | 200[g] | 113 | 129 | 82.9 | 118.9 |
| | Hinkley Point A | 200[g] | 194 | 128 | 161 | 147.4 |
| | Trawsfynydd | 40[g] | 10.8 | 13.5 | 22.8 | 31.4 |
| | Dungeness A | 200[g] | 181 | 83.7 | 29.4 | 29.0 |
| | Sizewell | 200[g] | 9.91 | 23.4 | 12.6 | 14.6 |
| | Oldbury | 100[g] | 2.32 | 7.74 | 2.54 | 5.2 |
| | Wylfa | 65[g] | - | 6.32[h] | 0.31 | 0.30 |

[a]  Including effluents from the experimental power station HDR-Grosswelzheim.

[b]  The liquid waste from AVR is transferred to the decontamination plant at Jülich Nuclear Centre (KFA), which discharged 0.41 Ci to the Rur in 1972, of which $9.7 \times 10^{-2}$ Ci originated from AVR.

TABLE V (cont.)

c  Liquid waste from MZFR is transferred to the decontamination plant at Karlsruhe Nuclear Centre (GfK), which discharges into the Rhine.

d  Liquid waste from BR-3 is transferred to the decontamination plant at Mol Nuclear Centre (CEN), which discharges into the Molse Nete.

e  Values inferred from a MPC to population in drinking water of $10^{-7}$ Ci/m$^3$ (any mixture of $\alpha$, $\beta$, $\gamma$ emitters from which Ra-226 and Ra-228 can be excluded) and the annual flow of water body; at SENA a discharge formula is applied.

f  Discharge formulae are applied:

at Latina: $\dfrac{\text{H-3}}{10^4} + \dfrac{\text{P-32}}{0.5} + \dfrac{\text{Sr-90}}{10} + \dfrac{\text{Cs-134} + \text{Cs-137}}{20} + \dfrac{\beta-\gamma}{3} + \dfrac{\beta}{100} + \dfrac{\alpha}{0.1} \leq 1$ Ci/a

at Garigliano: $\dfrac{\text{H-3}}{5 \times 10^3} + \dfrac{\beta}{1} + \dfrac{\text{Cs-137} + \text{Co-60}}{25} + \dfrac{2 \times \text{I-131}}{1} + \dfrac{\beta-\gamma}{2} + \dfrac{\alpha}{1} \leq 1$ Ci/a

at Trino: $\dfrac{\text{H-3}}{10^4} + \dfrac{\text{I-131} + \text{Cs-137}}{15} + \dfrac{\text{Sr-90}}{0.1} + \dfrac{\beta-\gamma}{10} \leq 1$ Ci/a

g  The discharge limits are laid down according to the waste discharge needs of each power station, but within the maximum permissible discharge limit as estimated by the critical path approach.

h  Mainly Br-82 used in radioactive tracer tests.

normally not be exceeded. Difficulties arise, however, if reaction forces of the pipework are to be taken into consideration. In the latter case a detailed stress analysis, including confirmation by a model test, should be the best way to handle this problem.

## 3.1.2. Fabrication and quality control problems

The most important contribution towards ensuring the material quality with reference to the design specifications is the establishment and continued enforcement of control of both the quality of the material and the fabrication procedures. To evaluate any unexpected structural deterioration or damage sustained in service, each fabrication step must be monitored so as to obtain a fabrication history of the vessel. Quality control is considered the best way to obtain an extra margin in the assurance of overall nuclear plant safety.

Guidance on quality assurance requirements is given, for instance, by the "Quality assurance criteria for nuclear power plants and fuel reprocessing plants", 10 CFR 50, Appendix B (and in supplementary detailed implementing documents). The general criteria are applied largely also in Europe. However, difficulties arise from time to time in translating 'general' requirements into an implemented programme. Because such a programme depends on the co-operation in the design (between designer, constructor, owner and engineering consultants), on the personnel, which is not always well trained before the start of the activities which are to be controlled, and on the organization responsible for quality assurance, which is not always independent of those departments whose work they have to review.

## TABLE VI. ANNUAL TRITIUM DISCHARGE IN LIQUID EFFLUENTS

| Country | Facility | Discharge limit (Ci/a) | Activity released (Ci/a) | | | |
|---|---|---|---|---|---|---|
| | | | 1969 | 1970 | 1971 | 1972 |
| FED. REP. GERMANY | VAK[a] | 480[b] | - | - | 1.43 | 3.45 |
| | AVR[c] | | - | - | - | 42.46 |
| | KRB | 432[b] | 17.8 | | 45.60 | 90.23 |
| | MZFR[d] | | - | - | - | - |
| | KWL | | 26 | 31.7 | - | 23.7 |
| | KWO | | 328 | - | 311.1 | 319.77 |
| | KWW | 300 | - | - | - | 4.59 |
| | KKS | 1600 | - | - | - | 101.4 |
| BELGIUM | BR-3[e] | | - | - | - | - |
| FRANCE | Chinon | | - | - | - | - |
| | SENA | | - | 340 | 706 | 1762 |
| | EL-4 | $1.5 \times 10^{5f}$ | - | - | - | 5 |
| | St-Laurent-des-Eaux | | - | - | - | - |
| | Bugey | | - | - | - | - |
| ITALY | Latina | $1 \times 10^{4g}$ | 25.2 | 16.7 | 13.0 | 16.9 |
| | Garigliano | $5 \times 10^{3g}$ | 7.0 | 5.0 | 5.0 | 3.0 |
| | Trino | $1 \times 10^{4g}$ | - | 135 | 1117 | 1078 |
| NETHERLANDS | Dodewaard | | - | 2.37 | - | 2.5 |
| UNITED KINGDOM | Berkeley | 1500 | 60.5 | 60.1 | 43.1 | 44.2 |
| | Bradwell | 1500 | 183 | 95.3 | 102 | 251 |
| | Hinkley Point A | 2000 | 35.3 | 18.6 | 24.9 | 38.6 |
| | Trawsfynydd | 2000 | 233 | 67.7 | 41.9 | 46.0 |
| | Dungeness A | 2000 | 72.2 | 18.6 | 35.5 | 28.9 |
| | Sizewell | 3000 | 9.56 | 20.9 | 77.0 | 53.2 |
| | Oldbury | 2000 | 16.4 | 17.3 | 64.4 | 15.0 |
| | Wylfa | 4000 | - | 0.551 | 30.2 | 82.7 |

[a] Including effluent from the experimental power station HDR-Grosswelzheim.

[b] Figure derived from the monthly limit.

[c] Liquid waste from AVR is transferred to the decontamination plant at Jülich Nuclear Centre (KFA), which discharged 50.23 Ci of H-3 in 1972, of which 42.46 originated from AVR.

[d] Liquid waste from MZFR is transferred to the decontamination plant at Karlsruhe Nuclear Centre (GfK).

[e] Liquid waste from BR-3 is transferred to the decontamination plant at Mol Nuclear Centre (CEN).

[f] Value inferred from the MCP in drinking water of $3 \times 10^{-3}$ Ci/m$^3$ and from the annual flow of the watercourse.

[g] See Table V(f).

TABLE VII.  ANNUAL DISCHARGE LIMITS FOR GASEOUS AND LIQUID
RADIOACTIVE EFFLUENTS PER 100 MW(e) OUTPUT
(for the most recent power stations with water-cooled reactors)

| | Noble gases (Ci/100 MW(e)·a) | I-131 (Ci/100 MW(e)·a) | Liquid effluents (Ci/100 MW(e)·a) | |
|---|---|---|---|---|
| | | | Exclusive of tritium | Tritium |
| **PWR** | | | | |
| KKS STADE | 9700 | $3.3 \times 10^{-2}$ | 0.79 | $2.5 \times 10^2$ |
| BIBLIS A | 7900 | $2.6 \times 10^{-2}$ | 0.87 | $1.4 \times 10^2$ |
| DOEL | 5100 | $2.6 \times 10^{-2}$ | 3.1 | $4.6 \times 10^2$ |
| TIHANGE | 4600 | $2.3 \times 10^{-2}$ | 0.92 | $4.6 \times 10^2$ |
| BORSSELE | 2700 | $5.3 \times 10^{-2}$ | 3.3 | |
| **BWR** | | | | |
| KWW WÜRGASSEN | 5000 | | 2.7 | $0.47 \times 10^2$ |

### 3.1.3.  In-service inspection

To demonstrate that continued operation of the nuclear power plant,
especially of the pressure components in the primary circuit, does not entail
the risk of rupture, a programme of periodic examination and inspection
is always foreseen.  The test methods almost exclusively applied are radiography and ultrasonics.

Radiography is considered to be a common technique for crack detection
but difficulties arise in large wall thicknesses.  The application is therefore
limited (at least in the Federal Republic of Germany) to wall thicknesses up
to 40 mm.  For larger wall thicknesses, ultrasonic tests are, in addition,
required for tests in the shop as well as the in-service inspection tests.
For in-service inspection of components the radiographic technique is
nearly always impaired by the radiation caused by contamination so that
the ultrasonic test method is the only one applicable.  However, in order to
apply ultrasonic test methods for in-service inspection, the designer must
take care that all parts to be examined are accessible for ultrasonic
scanning.

The following difficulties which were experienced recently may
demonstrate the problems of applying ultrasonic methods in the field:

(a)  There is a substantial variation in the operator's capacity to locate
     defects;
(b)  A threshold of sensitivity makes it possible that, under laboratory
     conditions, very small flaws would be detected whereas, under field
     conditions, flaws representing 1% or even 2% of the total thickness
     of the vessel may be missed.  The most detectable crack size was
     found to be a significant function of the crack orientation.

(c)  If enclosures or cracks are detected in a vessel and the decision is
     made to repair the flaw, one of the most difficult problems of the entire
     operation is the requirement for stress relief, because it is difficult
     to stress-relieve a local vessel region.

For permissible flaw dimensions, the Federal Republic of Germany's
acceptance standards for cracks up to 10 mm, for example, define both the
allowable defect length and the number of defects per metre of weld as a
function of the wall thickness.  If the wall thickness is $60 < S < 120$, ten
defects, each 10 mm long, or three defects, each 40 mm long per metre
of weld,are acceptable [6].  The ASME III code specifies the allowable
crack length and depth as a function of the wall thickness but independent
of the number of cracks.


3.1.4.  Periodic proof tests

In European countries, the periodic proof test is recognized as a useful
leak and strength verification test and it may often serve to detect improper
design or faulty materials.
The following conclusions were drawn in at least one country (France)
from the discussions on this test method:  the danger involved in carrying
out overpressure tests only becomes evident if a fault is present which
with a high degree of probability would otherwise have led to a catastrophic
fracture during operation.  Because of the variety of factors which may
cause weakening in a pressure vessel, it is difficult to express with absolute
certainty the cycling potential established by an overpressure test.
However, an estimate can be made.  It was found that a test at 1.3 times
the operating pressure gave rise to a propagation of the fault corresponding
to approximately three normal cycles.  Since the potential of the vessel is
at least 300 normal cycles, carrying out a dozen tests during the vessel's
lifetime would not represent more than about 12% of the total vessel potential
European countries are seriously examining whether more frequent
periodic proof tests, probably associated with non-destructive test methods
such as acoustic emission, would significantly improve safety without
excessive costs.   As an indication in this direction it should be mentioned
that a French regulation from February 1974 asks for a test pressure of
at least 1.20 times the design pressure, thirty months after first fuel
loading and every ten years.  In the Federal Republic of Germany the first
periodic overpressure test after eight years of operation was performed
recently at the boiling-water reactor KRB Gundremmingen at 115 bars,
corresponding to 1.3 times the design pressure.
The in-service system hydrostatic test, as required by the ASME
Section XI code, reflects the acceptance of the pressure test primarily
as a means to enhance leakage detection during the examination of components under pressure, rather than solely as a measure to determine the
structural integrity of the components.  The test pressure is about 1.1 times
the operating pressure and depends on the relationship of the yield strength
corresponding to the operating pressure.  The test time required by the
ASME code, Section XI, is, with about four hours, substantially longer than
necessary for pressure testing.  These pressure and time data characterize
the hydrostatic test, according to ASME, as a leak test.

## 3.2.  Engineered safety features

Some very general remarks now follow on the engineered safety
features which are provided to mitigate the consequences of major accidents,
e.g. the emergency core-cooling system and the containment system.

### 3.2.1.  Emergency core-cooling system

It is possible that in case of a loss-of-coolant accident (depending on
the break type and size) the whole core or at least part of it would be no
longer covered by the coolant.  The emergency core-cooling system has to
remove the stored energy in the fuel rods and the energy input by decay
heat and metal-water reactions from the core to prevent core damage
and/or partial melting.  A protection against core damage for the entire
spectrum of conceivable break types and sizes requires engineered safeguards consisting of several independent and overlapping core-cooling
systems.  Two main subsystems can be distinguished which show quite
different characteristics with respect to efficiency and mode of operation:
the high-pressure injection systems installed to cope with small ruptures
and the low-pressure injection system to cope with large ruptures including
the double-ended pipe break.  The effectiveness of emergency core-cooling
systems must be such that all the pressure-vessel internals are maintained
in position and stay essentially intact.  It is essential to perform quantitative
prediction of the blow-down effects on the pressure-vessel internals (and
especially of the pressure time history within the primary coolant system)
because these effects influence the starting point of the different subsystems.
Numerous tests on the effectiveness of the different spray and flooding
systems have already been performed and showed in general that the vessel
and core geometries play an important role.  Also, the heat transfer from
the fuel rods to the coolant depends on whether water is injected at the top
or at the bottom of the core.

If flooding from the bottom and spray from the top of the fuel elements
are applied simultaneously, it is difficult to assess the cooling efficiency
precisely by analysis or experiments [7].  The emergency core-cooling
problem is still the subject of important research work at present under
way in the European  as well as other countries.

The question of whether the systems installed in the newer plants are
fully capable of avoiding core melting or whether they have to be modified
has still to be solved case by case, taking into account existing criteria
and detailed independent reviews supported if necessary by analytical check
calculations.

One should in the future, however, also weigh the possible increased
(disadvantageous) complexity of such mitigating systems against reasonable
probabilities of the various initiating primary system-failure types which
should define the desired engineered safeguard requirements for reliability
and efficient thermohydraulic performance.

### 3.2.2.  Containment systems

The containment and its associated engineered safeguards are designed
to limit the consequences of nuclear accidents.  The containment is generally
considered to be the third barrier against uncontrolled fission-product

release from the core.  Actual containment systems may be divided into two different types:

(a)  Full pressure containments designed to withstand a maximum pressure and temperature calculated by steady-state calculations from an energy balance, taking into account the initial energy of the primary system and of the fuel, and possible chemical reactions;

(b)  Pressure suppression containments which are designed to withstand a pressure calculated on the assumption of thermodynamic equilibrium between air, released high-energy steam and water.

The containment is subjected to the short-term pressurization transient corresponding to the sudden release of the high-energy primary coolant and the long-term pressure and temperature effects.  These effects are influenced by the containment-cooling system and core-cooling systems as well as by their availability.

We should like to discuss more specifically the long-term behaviour which especially influences the containment leakage.  The relationship between leak-rate and fluid conditions is emphasized, as well as the containment-testing procedures.  After a loss-of-coolant accident, a steam-air mixture is present in the reactor containment.  The composition of this mixture strongly depends on the containment structural characteristics.  The amount of fluid released to the external atmosphere is related to the accident and the post-accident fluid conditions.  Normally a decrease in leak-rate due to natural heat-sink and cooling by engineered safeguards can only be assumed for the calculation of the radiological consequences if the relationship between leak-rate and internal fluid conditions is known.  This is essential, since the periodic leak-tests are performed at other than accident conditions and the extrapolation of the test results to the accident conditions underlies the effects of fluid-nature (air, steam, steam-air mixture) pressure and temperature.

The most correct way to obtain the leakage characteristics are leak-tests at different pressures up to the accident pressure.  If the leakage characteristics of a containment system are not determined in this way, the leak-rate at low pressure has to be extrapolated on a conservative basis, which means on the basis of laminar flow conditions.  For determination of the radiological consequences in these circumstances, a conservative basis means that the calculation is performed under the condition of maximum accident pressure.  However, if the leak-rate characteristics of the containment are well known (e.g. as outlined above), the decrease of the leak-rate in post-accident conditions can be taken into account.

Besides the periodical overall leak-rate tests, which should be performed at the highest possible pressure compatible with the internal equipment, local tests are to be performed on the various types of components which are part of the containment boundary.  Since the means for performing these tests depend strongly on the component type, it is recommended that in designing a containment system each of the components will be carefully examined and properly designed to allow the test performance.  The penetration test pressure should be the maximum containment accidental pressure because the leak geometry may be influenced by the stress conditions of the test area.

The major concern, however, is an important leak-rate increase arising from a valve failing to close or a door failing to shut properly.

Although containments for LWRs have been built and tested for two decades, difficulties still exist in the application of the actual leakage test results as basis for a conservative calculation of the release consequences after a loss-of-coolant accident.


# REFERENCES


[1] COMMISSION OF THE EUROPEAN COMMUNITIES, Radioactive Effluents from Nuclear Power Stations in the Community; Discharge Data, Radiological Aspects, Doc. V/1973/74 e (1974).

[2] LANDESARBEITSGEMEINSCHAFT WASSER (LAWA), Richtlinien für das Einleiten radioaktiver Stoffe aus kerntechnischen Anlagen in die Gewässer (1973).

[3] BUNDESMINISTERIUM DES INNERN, Sicherheitskriterien für Kernkraftwerke (1974).

[4] RADIOACTIVE PROTECTION INSTITUTES IN DENMARK, FINLAND, ICELAND, NORWAY AND SWEDEN, Basic Principles for the Limitation of Releases of Radioactive Substances from Nuclear Power Stations, a joint statement, Copenhagen, Helsinki, Reykjavik, Oslo and Stockholm (1974).

[5] MAGER, T.R., RICARDELLA, P.C., "Use of linear elastic fracture mechanics in safety analysis of heavy-section nuclear reactor pressure vessels", Proc. Conf. Practical Application of Fracture Mechanics to Pressure Vessel Technology, London, 1971.

[6] HAAS, "Definition of the maximum flaw size in base and weld metal that may remain undetected by non-destructive testing methods", discussion at 5th Mtg. of Working Group: Mechanical and Material Problems Relating to the Safety Aspects of Steel Components in Nuclear Plants, Brussels, 1974 (Summary Report).

[7] BROSCHE, D., "Emergency core cooling", Water-cooled Reactor Safety, an assessment prepared for the Committee on Reactor Safety Technology, OECD, Paris (1970).



# DISCUSSION

P. GRANDE: Are the Member States of the European Communities expected to use dose commitment calculations in the near future?

W. VINCK: I think the concept is already being used in cases of internal contamination by long-lived radionuclides.

P. GRANDE: What is Euratom's view on summation of very low doses to large populations, a procedure which yields a fairly high man·rem dose?

W. VINCK: Here I can only give my personal view. As we have mentioned in the paper, the collective dose concept can be a useful tool in comparing the relative merits of different solutions to waste-discharge control, with a view to optimizing the release conditions. I do not believe in its usefulness as an absolute value so long as integration limits are not well established.

J. DiNUNNO: I am sure it will not be out of place to make a few observations on dose commitment calculations, particularly on the accuracy with which population dose commitments can be calculated by current prediction techniques. To stimulate discussion, let me expound the position that, using the meteorological dispersion models and critical-pathway techniques available to us now, we cannot calculate population dose commitments to great distances with a degree of certainty that would justify taking the calculations very seriously. Such calculations are frequently done, and are even required by some regulatory authorities. However, they are fraught with uncertainties which should be recognized.

Earlier in this Symposium we heard mention of dispersion calculations made out to a great many km. A USAEC guide gives 50 miles as a frame of reference. But I doubt whether enough meteorological data will ever be gathered to allow dispersion calculations to be made with confidence to anything like these distances. Surely such data will not come from the on-site meteorological tower alone. There are real time techniques which have been applied and which involve trajectory predictions, but these require regional data inputs from multiple meteorological stations. Even then, the dispersion predictions out to distances of many km, which would frequently be required if one wished to encompass large accumulations of people, at best remain rough estimates. This view is supported by experience with measurements for radionuclides in the environment, which provide little evidence that one can use off-site environmental measurements to confirm model predictions of the dose commitment resulting from normal operation in a plant.

W.N. LABLANS: As regards Dr. DiNunno's doubts about the value of meteorological dispersion models, I might just point out that the largest concentrations of released substances are usually found within two or three km of the facility. For calculations pertaining to this relatively important range of distances, the meteorological models can be applied with reasonable accuracy.

D. BENINSON: Collective dose commitments can give a fairly good assessment in the case of some nuclides, e.g. $^{85}$Kr, because the contribution from the global distribution is much higher than the local or regional collective dose contribution. For other nuclides which deliver the dose through a given food, the collective dose commitment can be assessed from predictions of food contamination and of the amount of food produced. In the case of some nuclides of medium half-life ($^{133}$Xe), the assessment is more difficult. The collective dose commitment must be calculated, however, because by limiting it per unit of power production, one can limit the average dose rate to the population which will be reached when reactors are spread all over the world.

S.O.W. BERGSTRÖM: Following up Dr. Beninson's comment, I would add that so long as we profess to apply a linear relationship between dose and health effects we have to estimate collective doses. For mass conservation reasons, it does not seem possible that large errors would be made in the calculation of atmospheric exposures at large distances if trajectory methods are used.

E. TABET: As can be seen from Table II, the activity released by different plants differs widely. So we are faced with the problem of backfitting, i.e. we have to take appropriate action to reduce the discharges from old plants to the levels attainable with new reactors. What is your view of this problem, and of the related question of possible backfitting applied to certain specific safety features?

W. VINCK: In our paper we refer mostly to new plants. Backfitting has to be considered by the licensing authorities on an ad hoc basis, in relation to the merits of each plant; the benefit must be weighed in terms of safety, and whether this balances with the cost will be deduced from discussions with the utility. Depending on the type of backfitting, it is either the utility or society in general that should bear the cost.

J. EDWARDS: Referring to the integrity of thick-walled pressure vessels and the need for rigorous and long-term quality control and assurance, you posed the question what one should do if a defect were to be found. May I ask you to answer this question in relation to determining whether it should be repaired, and if so, how a successful repair would be guaranteed?

My other question is a corollary to this and relates to the small undetected crack which may grow during the lifetime of the plant.  Do you advocate frequent in-service monitoring inspections to guard against this possibility?

W. VINCK:  To the first question, there is no universally applicable answer.  The matter would have to be evaluated in each separate case — e.g. in the light of ASME code requirements.  Essentially, heat treatment would be applicable.

I do advocate more emphasis on non-destructive testing methods such as ultrasonic techniques  to improve the sensitivity of fault detection and obtain access where it would otherwise be difficult.

# NUCLEAR ENERGY IN A DENSELY POPULATED AND BUILT-UP COUNTRY SUCH AS THE FEDERAL REPUBLIC OF GERMANY

H. BONNENBERG, M. ESCHHAUS, W. HENSEL, J. KAYSER
Bonnenberg & Drescher Ingenieurgesellschaft mbH,
Aldenhoven, Federal Republic of Germany

## Abstract

NUCLEAR ENERGY IN A DENSELY POPULATED AND BUILT-UP COUNTRY SUCH AS THE FEDERAL REPUBLIC OF GERMANY.

Nuclear energy is absolutely necessary for the Federal Republic of Germany. By 1995, 75% of the energy used in that country for electricity production will be nuclear, which means an installed nuclear power of ~100 000 MW with all its nuclear infrastructure such as fuel fabrication, reprocessing, waste disposal and transport of a great deal of radioactive material between these services. Many sites are needed for these activities, including transport, which is, in fact, comprised of moving sites on special lines. Risk analysis is also needed for all this, the basis of which is the use of land in the western part of the country by population and industry. The Federal Republic of Germany is one of the most densely populated and built-up areas in the world. A computer program with a voluminous information system which stores all the relevant land-use data was written. A simulation program was also written to analyse the required regional capacity for transport of nuclear material in respect of different sites for power plants, reprocessing plants and waste disposal. This simulation program makes optimization with variable criteria. All the railway lines which can be used for heavy transport are stored in the computer as well as the traffic routes and rivers. The computer calculations show that, especially for selection of sites for reprocessing plants, the minimization of transport should be an important criterion in an intensely used country like the Federal Republic of Germany. The country will be forced to undertake detailed regional planning for the introduction of nuclear energy to achieve minimum risk to the population and minimum risk to the introduction itself.

## 1. INTRODUCTION

The growing energy demand and the necessity for diversification of energy supply require rapid installation of nuclear energy in the Federal Republic of Germany.

There is no doubt, and should be no discussion, that nuclear energy is absolutely necessary. Therefore the revised energy programme of the Federal Republic of Germany (FRG) of 30 Oct. 1974 postulates that in 1985 15% of the entire energy used and 45% of the energy used for electricity production will be produced by nuclear energy. For 1995 the energy percentage used for electricity production will be 75%; i.e. an installed nuclear power of 100 000 MW. Because of this necessity, the FRG is forced to discuss and analyse all the problems which could be connected with nuclear installation in the country. Problems of risk occur whenever radioactive material is handled: in nuclear power plants, in fuel fabrication and refabrication plants, in the reprocessing plant, in waste disposal, and in transport of nuclear material. This problem arises in all countries engaged in nuclear energy, but is it completely understood, especially in countries very densely built-up and populated? In such countries, the problems of siting nuclear plants and nuclear transport lines are very ticklish.

TABLE I.   POPULATION DENSITY AND ANNUAL INDUSTRIAL PRODUCTION

| Population (man/km$^2$) | | Production of: | | | | | | | |
|---|---|---|---|---|---|---|---|---|---|
| | | Iron (t/km$^2$) | | Synthetics (t/km$^2$) | | Cars/km$^2$ | | Electricity (MW·h/km$^2$) | |
| Bangladesh | 525 | Belgium | 476 | Netherlands | 49 | FRG | 16 | Netherlands | 1224 |
| Taiwan | 399 | Japan | 262 | FRG | 24 | Japan | 16 | Belgium | 1180 |
| South Korea | 338 | FRG | 176 | Japan | 19 | UK | 9 | FRG/Japan | 1106 |
| Netherlands | 326 | Netherlands | 137 | Belgium | 15 | Italy | 6 | UK | 1000 |
| Belgium | 318 | UK | 103 | UK | 9 | France | 5 | Italy | 441 |
| Puerto Rico | 316 | Italy | 66 | Italy | 8 | USA | 1 | France | 292 |
| Japan | 289 | France | 44 | France | 4 | Sweden | 0.7 | USA | 197 |
| Lebanon | 285 | USA | 13 | USA | 1.4 | USSR | 0.05 | Sweden | 157 |
| FRG | 248 | USSR | 6 | USSR | 0.1 | | | USSR | 38 |

TABLE II.   ANNUAL AGRICULTURAL PRODUCTION, USE OF
ENERGY AND GNP (industrialized countries only)

| Production of cereals (t/km²) | | Cattle (animals/km²) | | Use of energy (t coal/km²) | | GNP (10⁶ DM/km²) | |
|---|---|---|---|---|---|---|---|
| FRG | 81 | Netherlands | 334 | Belgium | 1770 | Netherlands | 3.8 |
| France | 66.1 | Belgium | 232 | Netherlands | 1446 | Belgium | 3.4 |
| UK | 61 | UK | 198 | FRG | 1190 | FRG | 3.1 |
| Belgium | 59 | FRG | 145 | UK | 1170 | Japan | 2.2 |
| Italy | 56 | Italy | 92 | Japan | 783 | UK | 2.0 |
| Netherlands | 44 | France | 82 | Italy | 428 | Italy | 1.2 |
| Japan | 41 | Japan | 29 | France | 323 | France | 1.1 |
| USA | 23 | USA | 23 | USA | 234 | USA | 0.4 |
| USSR | 8 | USSR | 14 | USSR | 45 | | |

## 2.   DENSITY SITUATION

The Federal Republic of Germany is one of the very densely populated
and built-up countries of the world.  In an area of 250 000 km², 60 million
people live and produce a gross national product (GNP) of $9 \times 10^{11}$ DM.  To
give an idea of that density, various typical values of population and economics
of the FRG are divided by the area (Tables I and II).  The average population
density of the FRG is nearly 250 persons/km².  At this figure, the country
ranks ninth in the world (excluding some very small states).  Only the highly
industrialized countries, the Netherlands, Belgium and Japan, have a higher
average population density.  Other highly industrialized countries, such as
the UK, Italy, France, the USA, Sweden and the USSR, are less densely populated than the FRG.  In spite of the higher GNP/person of the USA, that
country has only one-twelfth the population density of the FRG.  For industrial
use of the area available, the FRG is always in the world's foremost group.
Also, industrial use of land is a measure of density, so the FRG has the
third place in iron production average, second place in synthetics production,
first place in motor car production, and, together with Japan, third place in
electricity production.  All such averages are more than in the USA by a
factor of 6 to 17.  Besides this very dense industrialization, the FRG is also
intensively used for agriculture.  So the FRG has the highest cereal production average, being greater than that of the USA by a factor of 4.  In cattle
raising the FRG is behind the Netherlands, Belgium and the UK, while higher
than in the USA by a factor of 7.

Such high averages in all areas of economics and life give rise to all the
problems of environmental burden.  So the use of land, water and air postulates
more and more planning and organization, especially with improving standards
of life and production.  Misdevelopment of the environmental situation must
be prevented very early in order to obtain the consent of the population for
future growth, i.e. for building factories, refineries, power plants, etc.

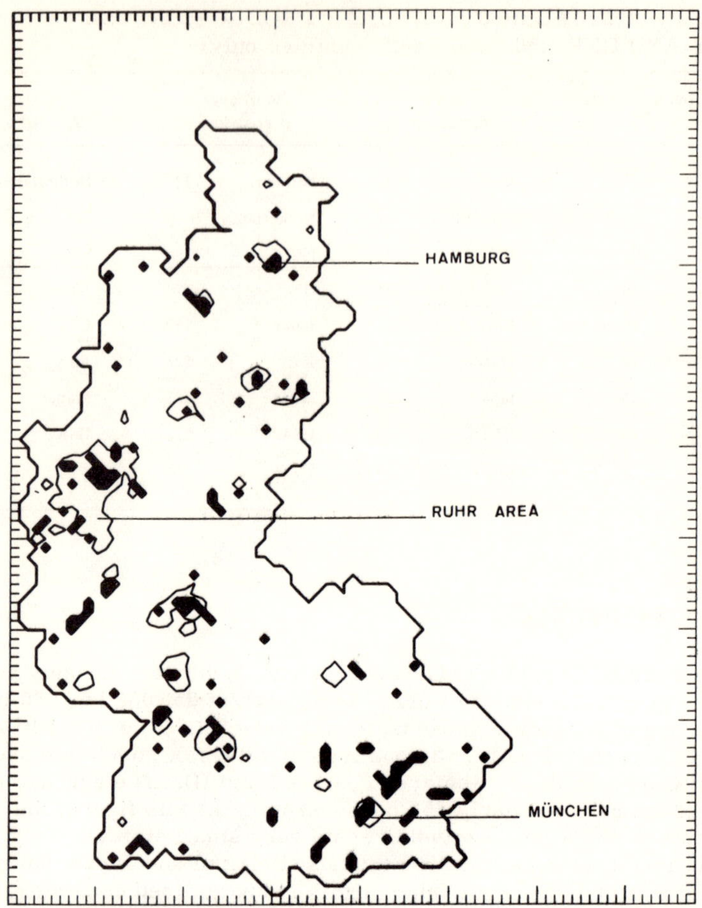



FIG.1. Regions of dense population (white enclosures) and power plant sites (black) in the Federal Republic of Germany; plotted by computer.

The above averages give only a first, rough idea of the situation in our densely populated and built-up country. In addition, there are certain areas which are much more intensively used than the others, the so-called Verdichtungsräume (areas of very high density). Among other things, a Verdichtungsraum is defined by a population density higher than $1250/km^2$. The FRG has 24 Verdichtungsräume so defined. In these very dense areas, 42% of the population live on 6.3% of the area; 60% of the total energy is used there and 80% of the total steel is produced there. The average density in these areas is higher by a factor of about 10 than in the other parts of the FRG, where the average population density is about $160/km^2$. By the way, this last 'low' population density is higher by a factor of 7.5 than the average population density of the entire USA. The average electricity production

figure in the FRG in 1990 is expected to be 0.8 MW/km$^2$; in the Verdichtungsräume this value will be 10-20 MW/km$^2$. In the cities, an electricity demand
of more than 100 MW/km$^2$ is expected. Figure 1 is a computer plot showing
these Verdichtungsräume and is the result of a very extensive computer
program written by us to study siting and transport problems in the FRG.

This computer program has a voluminous computer information system
which stores all the relevant data of energy such as the demand of the
different users (industrial traffic, residential, commercial, etc.), the kind
of energy supplied (coal, oil, gas, electricity, etc.), the environmental burden,
etc., all on the basis of well defined governmental districts (Kreise). The
FRG contains nearly 500 such districts, for which we make energy balances.
This information system also stores the population and the area of each
district. Moreover, all sites of power plants (thermal and hydro-plants) of
more than 10 MW and all sites of refineries are stored in this system. Power
plants sites can be seen in Fig.1. One black 'diamond' is one plant, irrespective of its power, and many 'diamonds' overlap. There are also areas with
high densities of power plant sites. Figure 1 shows that power in the FRG
is produced within or in the direct neighbourhood of the areas of demand,

TABLE III.   PROBABLE REGIONAL DISTRIBUTION OF
NUCLEAR POWER PLANTS IN 1995

| State in FRG | Electricity demand (MW) | Installed nuclear power (MW) |
|---|---|---|
| Schleswig-Holstein | 4 800 | 3 400 |
| Hamburg | 5 500 | 3 800 |
| Niedersachsen and Bremen | 16 200 | 17 500 |
| Nordrhein-Westfalen | 57 700 | 18 000 |
| Hessen | 12 500 | 14 000 |
| Rheinland-Pfalz | 8 100 | 5 000 |
| Saarland | 2 800 | 1 300 |
| Baden-Württemberg | 21 500 | 20 000 |
| Bayern | 20 900 | 19 000 |
| Total | 150 000 | 102 000 |

TABLE IV.   OUTPUT OF A 1000-MW NUCLEAR POWER
PLANT 6900 h/a

| | PWR (t) | BWR (t) | HTR (t) |
|---|---|---|---|
| Irradiated fuel | 33 | 42 | 8 |
| Cladding material | 9 | 11 | 94 |

TABLE V.   TRANSPORT CAPACITY IN 1995 (Case:  Emden-Asse)

| | | | BWR | PWR | BWR + PWR | HTR | BWR+PWR+HTR |
|---|---|---|---|---|---|---|---|
| **FROM NUCLEAR POWER PLANTS TO:** | | | | | | | |
| Reprocessing plant: | Irradiated fuel | (t) | 2 026.35 | 1 741.96 | 3 768.31 | 2 009.22 | 5 777.53 |
| | (with cladding) | (m$^3$) | 254.72 | 217.74 | 472.46 | 1 113.83 | 1 586.29 |
| | | (Ci/1000) | 4 865 531 | 6 241 873 | 11 107 405 | 1 161 068 | 12 268 474 |
| Waste disposal: | Operational waste | (t) | 2 547.20 | 2 752.62 | 5 299.82 | 1 397.21 | 6 697.02 |
| | | (m$^3$) | 2 927.38 | 3 163.46 | 6 090.84 | 1 574.32 | 7 665.15 |
| | | (Ci/1000) | 15 | 16 | 31 | 10 | 42 |
| **FROM REPROCESSING PLANT TO:** | | | | | | | |
| Nuclear power plants: | Refabricated fuel | (t) | 1 596.75 | 1 372.20 | 2 968.95 | 159.40 | 3 128.35 |
| | (without cladding) | (m$^3$) | 167.28 | 143.79 | 311.07 | 25.58 | 336.65 |
| | | (Ci/1000) | 3 | 4 | 7 | 1 | 9 |
| Waste disposal: | Fission products | (t) | 121.66 | 106.82 | 228.48 | 96.43 | 324.90 |
| | | (m$^3$) | 237.61 | 205.42 | 443.03 | 186.95 | 629.98 |
| | | (Ci/1000) | 337 598 | 435 489 | 773 088 | 80 683 | 853 771 |
| Waste disposal: | Cladding | (t) | 418.20 | 369.75 | 787.95 | 0.00 | 787.95 |
| | | (m$^3$) | 87.44 | 73.95 | 161.39 | 0.00 | 161.39 |
| | | (Ci/1000) | 7 603 | 7 970 | 15 573 | 0 | 15 573 |
| Waste disposal: | Operational waste | (t) | 4 991.75 | 4 383.65 | 9 375.40 | 944.59 | 10 319.99 |
| | | (m$^3$) | 8 287.90 | 7 189.68 | 15 477.58 | 1 534.96 | 17 012.54 |
| | | (Ci/1000) | 460 | 455 | 916 | 296 | 1 213 |

particularly the areas of high demand. Because of the historical development
of the supply system and of the organization of the energy suppliers, it can be
expected that nuclear power plants will also be in the neighbourhood of the
densely populated and built-up electricity demand areas.

We have now extended this computer program to the ~ 12 000 municipal
districts (Gemeinden) of the FRG and it is completed on this basis with all
data of recreation areas, the air-traffic situation, the seismic situation, the
natural radiological situation, transport systems (railway lines, roads,
rivers, electricity transmission lines, pipelines) and district-heating systems.
A selection program is being written with defined criteria to look for sites.


## 3.  CONSEQUENCES OF INSTALLING NUCLEAR ENERGY

In 1995, 100 000 MW of nuclear power are expected in the Federal
Republic of Germany. This means that about 40 sites for nuclear power plants
and nuclear power plant concentrations are necessary, i.e. a nuclear
system will operate at an average of every 79 km. This average gives a hint
of the high nuclear power density of the future. Nevertheless there will be a
regional distribution of nuclear power plants because of the regional distribution of the demand. Table III indicates the probable regional distribution of
nuclear power. There are ten states in the FRG, and the figure gives the
expected electricity demand which is in total 150 000 MW.

Because of the hitherto existing decisions in the field of nuclear plant
siting, there will probably be a development of future siting. This leads to
the column in Table III for 'installed nuclear power'. It is of special interest
that the state Nordrhein-Westfalen will have a comparatively smaller
proportion of nuclear power plants than the others; this is owing to the
coal and lignite resources in that state. Besides, the areas of the energy
suppliers are not the same as those of the state. The big energy suppliers of
Nordrhein-Westfalen run power plants in other neighbouring states. The
provision for siting nuclear plants is different in our different states; it is
very advanced in Baden-Württemberg and Bayern (Bavaria).

One of the future problems of nuclear energy is management of radioactivity outside the nuclear power plants. This problem must be solved with
a minimum of risk to the population and with maximum economy. The
annual output of a 1000-MW power plant with 6900 h/a is given in Table IV.
It can be seen that the HTR has very high fuel efficiency, but has a great deal
of cladding material (graphite) to transport. Besides the fuel output, a large
operational waste output from the nuclear power plants must be managed.
The reprocessing plant also has an output of much waste of different degrees,
e.g. fission products, used cladding material and operational waste. For our
calculations it was taken that in 1995 40% BWR, 40% PWR and 20% HTR will
be installed in the FRG. For transport from the nuclear power plants to the
reprocessing plant, this spectrum of nuclear energy leads, e.g., to a necessary annual transport capacity of $12 \times 10^9$ Ci. Table V gives the different
annual transport capacities required in 1995 in more detail. Transport
between power plants and reprocessing plant and vice versa is shown once
without cladding to obtain information about the proportion of transport needed
for cladding. The regional management of this waste is dependent on the
siting of the different nuclear power plants, the reprocessing plant, and waste
disposal. We have therefore written a computer program which analyses the

TABLE VI.   SHORTEST TRANSPORT ROUTE FROM NUCLEAR POWER
PLANT GUNDREMMINGEN TO REPROCESSING PLANT EMDEN.
Case (a) Emden-Asse.   Distance transported:  814.50 km.

| Starting point | Arrival point | Distance (km) |
|---|---|---|
| KRB Gundremmingen | Neuoffingen | 10.38 |
| Neuoffingen | Ulm | 28.87 |
| Ulm | Plochingen | 69.47 |
| Plochingen | Stuttgart | 21.76 |
| Stuttgart | Ludwigsburg | 9.96 |
| Ludwigsburg | Bietigheim | 8.87 |
| Bietigheim | Kirchheim | 11.05 |
| Kirchheim | Heilbronn | 15.21 |
| Heilbronn | Bad Friedrichshall | 12.15 |
| Bad Friedrichshall | Eberbach | 36.90 |
| Eberbach | Babenhausen | 58.45 |
| Babenhausen | Hanau | 19.83 |
| Hanau | Beienheim | 24.60 |
| Beienheim | Giessen | 35.37 |
| Giessen | Wetzlar | 11.05 |
| Wetzlar | Siegen | 58.36 |
| Siegen | Plettenberg | 54.63 |
| Plettenberg | Hagen | 39.71 |
| Hagen | Schwerte | 9.40 |
| Schwerte | Froendenberg | 15.63 |
| Froendenberg | Hamm | 18.70 |
| Hamm | Muenster | 38.10 |
| Muenster | Ibbenbueren | 37.22 |
| Ibbenbueren | Rheine | 3.97 |
| Rheine | Salzbergen | 7.93 |
| Salzbergen | Lingen | 24.22 |
| Lingen | Leer | 79.58 |
| Leer | Emden-West | 37.83 |
| Emden-West | Reprocessing Plant | 15.28 |

required regional transport capacity for the different sites mentioned and
for siting the possible transport.  This is a simulation and optimization
program.  All plant sitings and all plant outputs can be varied.  Also, the
optimization criteria can be changed.  In this computer program all the
railway lines which can be taken for heavy transport (120-180 t) are stored
in the computer;  the same is done with traffic routes and rivers.  For
example, 600 sections of railway lines are taken.  In this network of transport possibilities, the sitings of the plants are simulated.  It was worked
with the optimization criterion of the shortest transport distance.  In Table VI
an example is given of the shortest transport distance from the nuclear
power plant Gundremmingen to the simulated site of a reprocessing plant in
the neighbourhood of Emden.  Of course it is possible in the program to
prohibit transport on specific railway lines or to impose limitations on
their transport capacity.  The sites for nuclear power plants in the FRG are
fairly well known until 1985.  For our analysis of the situation of 1995, we
expanded some sites already used and simulated new ones.  This was done
in correlation with the numbers given in Table III.  In the state of Nordrhein-
Westfalen, in particular, we were forced to design new sitings not previously
discussed.  The different types of nuclear power plants are spread all over
the FRG, following some criteria, e.g. the cooling problem.

In addition to many other cases, we discussed in particular the following two:

(a)  Reprocessing plant and waste disposal situated at different places
     (Emden-Asse);
(b)  Reprocessing plant and waste disposal on the same site (Asse-Asse).

For both cases it was postulated that the waste disposal is in Asse II in the
north-eastern part of the country; for case (a) it was taken that the reprocessing
plant is at Emden in the north-west.  In Table V the required annual transport capacity of 1995 is given for case (a) (Emden-Asse);  for case (b) (Asse-
Asse) there is no transport from the reprocessing plant to the waste disposal.
Figure 2 shows the regional distribution of the required annual transport
capacity for 1995.  The broken lines are those railway lines which are not
taken for transport of radioactive material.  The total annual transport capacity
is given in the upper half of the figure.  The regional transport capacity
depends on the site of the reprocessing plant as well as the waste disposal
plant.  Four main areas of waste management can be considered:

<u>The south-west, between Lörrach and Frankfurt</u>
16% of the necessary overall capacity for transport of irradiated fuel.
<u>The south, below the line Stuttgart-Würzburg</u>
28% of the necessary overall capacity for transport of irradiated fuel.
<u>The west, from Saarland to the Ruhr area</u>
12% of the necessary overall capacity for transport of irradiated fuel.
<u>The north, between Göttingen and the northern border</u>
20% of the necessary overall capacity for transport of irradiated fuel.

In case (a) (Emden-Asse), fuel transport is concentrated in the western
part of the country, and the power plant waste transport is concentrated in the
eastern part.  Transport from the reprocessing plant to the waste disposal
plant only touches the area between both these plants and has, of course,

a) <u>IRRADIATED FUEL</u> (WITH CLADDING)
FROM THE REACTORPLANTS TO THE
REPROCESSING PLANT <u>E M D E N</u>
TOTAL : 5800 t/a

$1600\ m^3/a \qquad 12\times10^9$ Curie/a

b) <u>FISSION PRODUCTS</u>
FROM THE REPROCESSING PLANT <u>E M D E N</u> TO
FINAL WASTE DISPOSAL <u>A S S E II</u>
TOTAL : 324 t/a

$630\ m^3/a \qquad 8\times10^8$ Curie/a

<u>OPERATION-WASTE AND CLADDING WASTE</u>
FROM THE REPROCESSING PLANT <u>E M D E N</u> TO
THE FINAL WASTE DISPOSAL <u>A S S E II</u>
TOTAL : $11\times10^5$ t/a

$17\times10^5 m^3/a \qquad 10^6$ Curie/a

c) <u>OPERATION WASTE</u>
FROM THE REACTORPLANTS TO THE
FINAL WASTE DISPOSAL <u>A S S E II</u>
TOTAL : 6700 t/a

$7700\ m^3/a \qquad 4\times10^4$ Curie/a

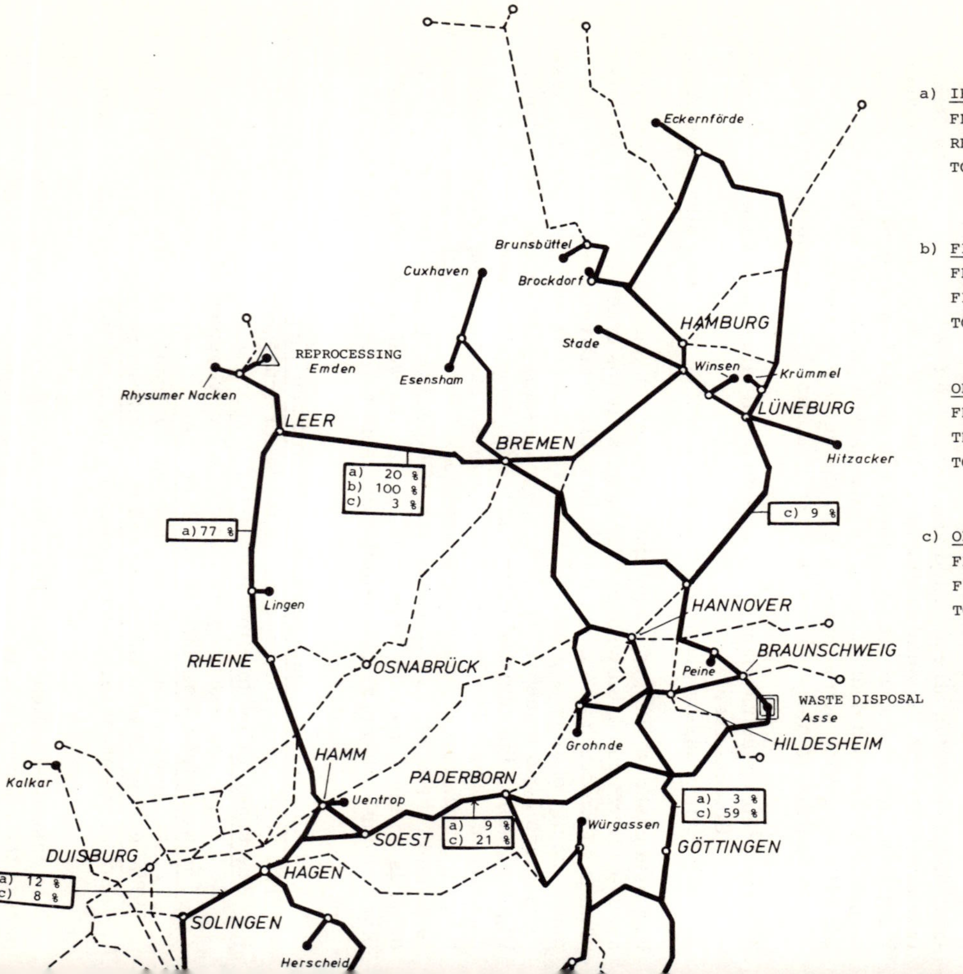

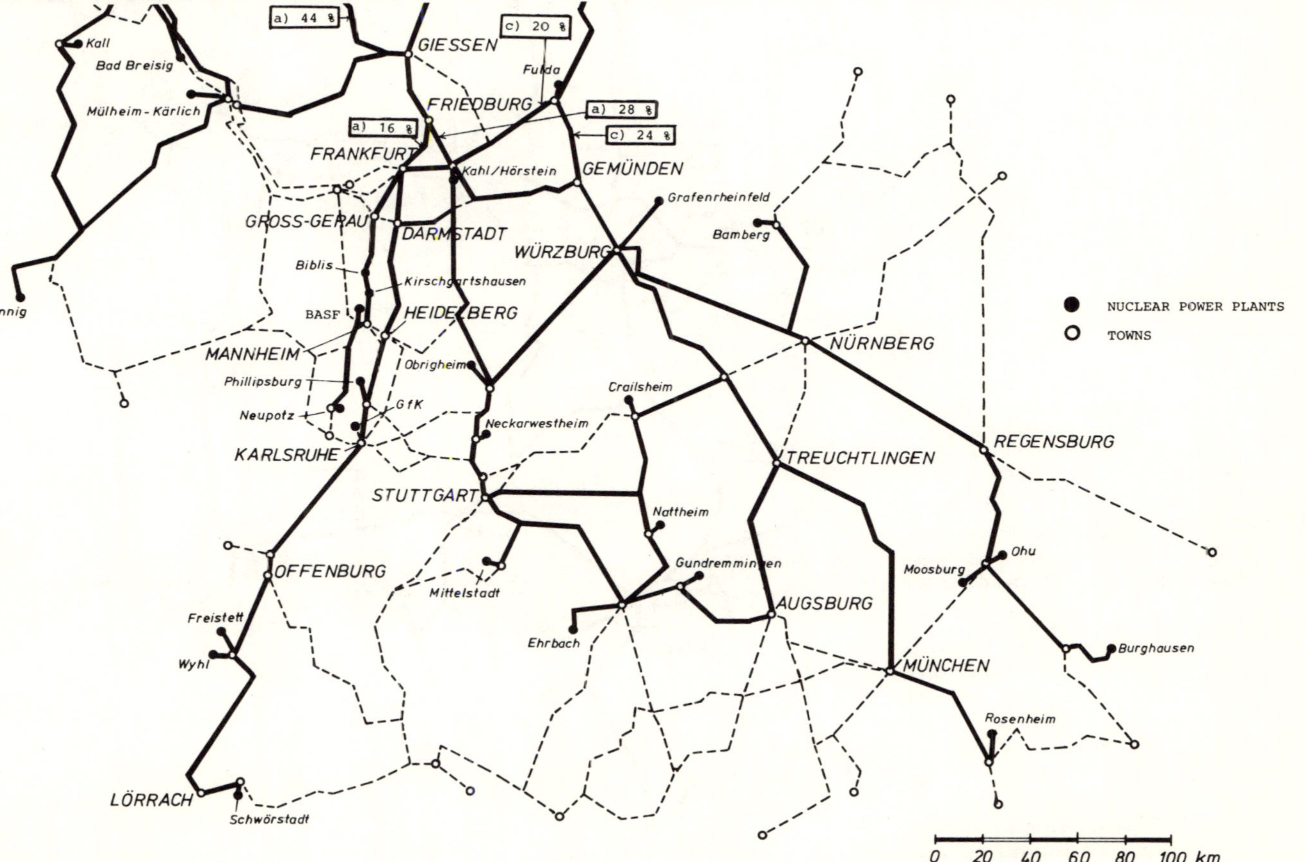


FIG.2.  Transport capacity in 1995;  case (a):  Emden–Asse.  Broken lines:  railway lines not taken for transport of radioactive material.

a) <u>IRRADIATED FUEL</u> (WITH CLADDING)
FROM THE REACTORPLANTS TO THE
REPROCESSING PLANT E M D E N
TOTAL : 5800 t/a

$$1600 \ m^3/a \qquad 12\times10^9 \ \text{Curie/a}$$

b) <u>FISSION PRODUCTS</u>
FROM THE REPROCESSING PLANT E M D E N TO
FINAL WASTE DISPOSAL <u>A S S E II</u>
TOTAL : 324 t/a

$$630 \ m^3/a \qquad 8\times10^8 \ \text{Curie/a}$$

<u>OPERATION-WASTE AND CLADDING WASTE</u>
FROM THE REPROCESSING PLANT E M D E N TO
THE FINAL WASTE DISPOSAL <u>A S S E II</u>
TOTAL : $11\times10^5$ t/a

$$17\times10^5 m^3/a \qquad 10^6 \ \text{Curie/a}$$

c) <u>OPERATION WASTE</u>
FROM THE REACTORPLANTS TO THE
FINAL WASTE DISPOSAL <u>A S S E II</u>
TOTAL : 6700 t/a

$$7700 \ m^3/a \qquad 4\times10^4 \ \text{Curie/a}$$

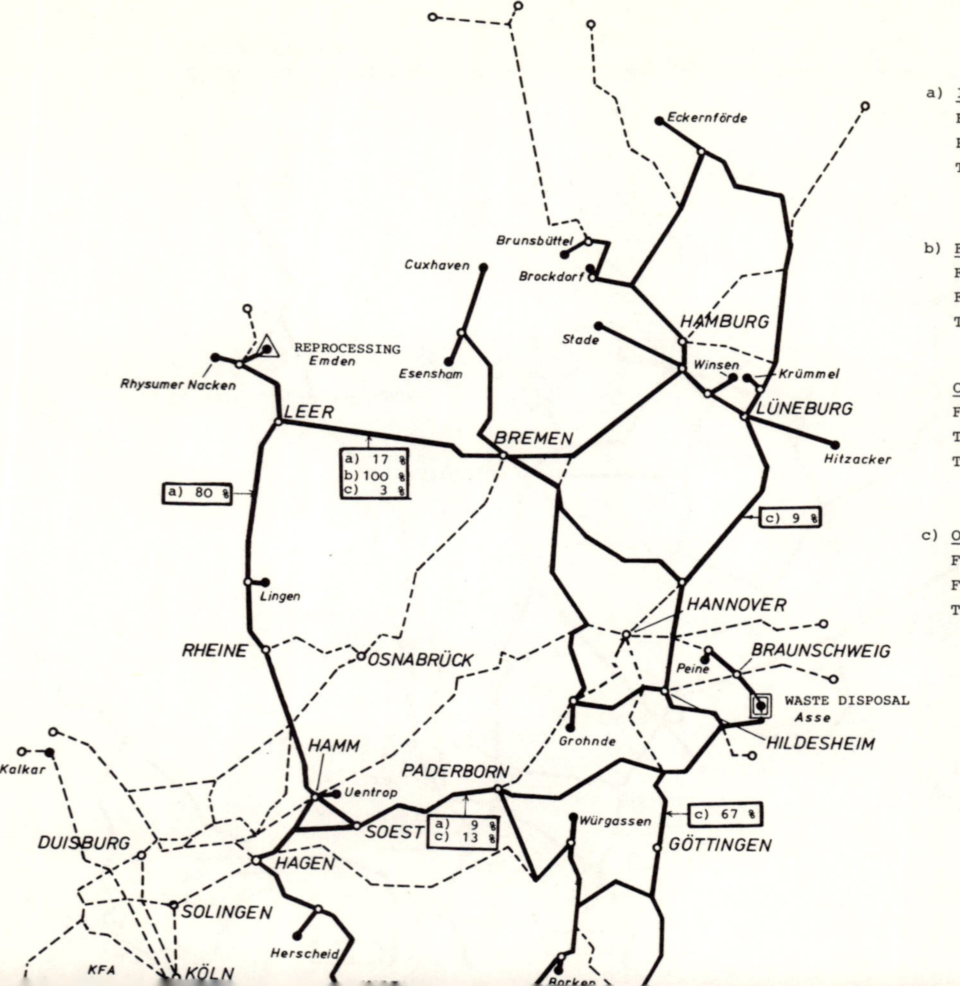

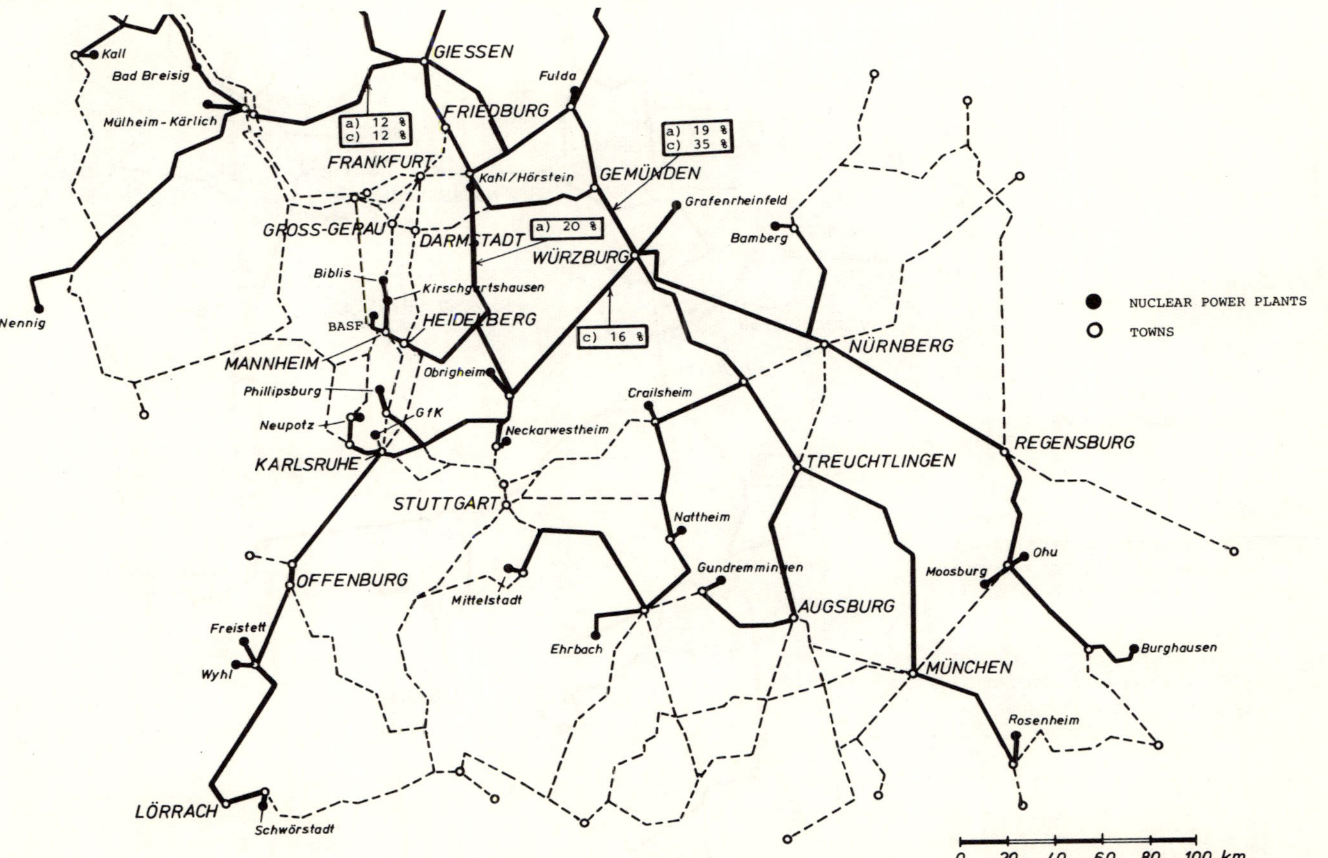


FIG.3. Transport capacity in 1995; prohibited lines.

a) <u>IRRADIATED FUEL</u> (WITH CLADDING)
FROM THE REACTORPLANTS TO THE
REPROCESSING PLANT <u>E M D E N</u>
TOTAL : 5800 t/a
$1600 \ m^3/a$ $\qquad 12 \times 10^9$ Curie/a

b) <u>FISSION PRODUCTS</u>
FROM THE REPROCESSING PLANT TO
FINAL WASTE DISPOSAL <u>A S S E II</u>
TOTAL : O

<u>OPERATION-WASTE AND CLADDING WASTE</u>
FROM THE REPROCESSING PLANT TO
THE FINAL WASTE DISPOSAL <u>A S S E II</u>
TOTAL : O

c) <u>OPERATION WASTE</u>
FROM THE REACTORPLANTS TO THE
FINAL WASTE DISPOSAL <u>A S S E II</u>
TOTAL : 6700 t/a
$7700 \ m^3/a$ $\qquad 4 \times 10^4$ Curie/a

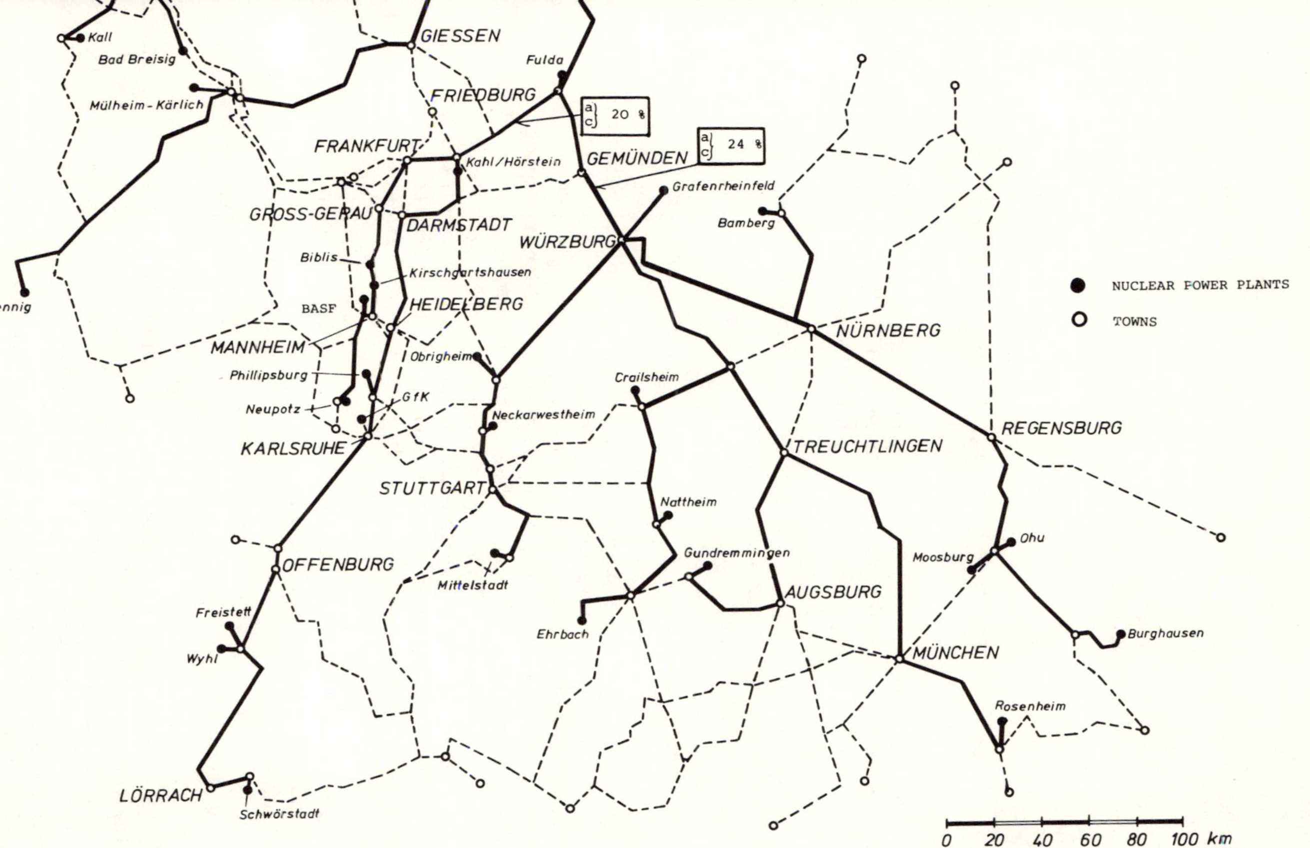


FIG.4.   Transport capacity in 1995; case (b): Asse–Asse.

no influence on all the other parts of the country.  The following results are
of great interest:

(a)  The heavily used railway lines in the very narrow Rhine valley
     between Frankfurt and Köln (Cologne) are used almost only by the
     nuclear power plants operating in that area.

(b)  The Ruhr area between Duisburg and Hamm is barely touched by
     transport of radioactive material.  Only the southern line between
     Köln and Hagen bears nuclear transport.

(c)  High transport capacity exists in the very dense area between
     Karlsruhe and Frankfurt.

(d)  Contact with nuclear transport is less in all the other intensively
     used areas such as Hamburg, Hannover, Nürnberg, Stuttgart and
     Munich.

The result of case (a) (Emden-Asse) is that the average distance by
railway from the reactor plant to the reprocessing plant is 545 km.  Multiplication of the transported tonnes by the transporting distances is:

| | |
|---|---|
| Irradiated fuel (with cladding) | $3.5 \times 10^6$ t·km |
| Transport from reprocessing plant to waste disposal | $3.7 \times 10^6$ t·km |
| Transport from reactor plants to waste disposal | $2.7 \times 10^6$ t·km |

Figure 3 shows the same case (a) (Emden-Asse) but with prohibited
lines.  Some railway lines in the regions of dense population are closed to
nuclear transport.  It has been found that the densely populated area between
Karlsruhe and Frankfurt can be made free of nuclear transport.  The same
applies to Hannover, Köln and Stuttgart.  Other densely populated areas such
as Hamburg, Bremen, Nürnberg and Munich cannot be taken out of the
nuclear transport zone  but they are not touched by much transport.  In this
case the average distance between reactor plant and reprocessing plant
reaches only 585 km.  The t·km multiplications are also somewhat higher:

| | |
|---|---|
| Irradiated fuel (with cladding) | $3.8 \times 10^6$ t·km |
| Transport from reprocessing plant to waste disposal | $4 \ \times 10^6$ t·km |
| Transport from reactor plants to waste disposal | $2.8 \times 10^6$ t·km |

The second case, (b) (Asse-Asse), shown in Fig.4, gives useful information
as the railway lines are changed with the change of the reprocessing plant
site.  Thus the whole transport capacity is needed in the eastern part of the
country.  The average distance between reactor plant and reprocessing and
waste disposal plant is 406 km.
The t·km multiplications for the different transports are now:

| | |
|---|---|
| Irradiated fuel (with cladding) | $2.7 \times 10^6$ t·km |
| Power plant waste | $2.7 \times 10^6$ t·km |

(The two numbers coincide by chance.)  If in this case the prohibited transport lines (Fig.3) are also considered, there is only transport distance of
5% more.

Other cases have been computed, e.g. if two reprocessing plants operate
in the FRG, one in Emden and one in the south.  All such examples show that
the transport problem would become more complicated.


## 4.  CONCLUSION

All the computer calculations have shown that the siting very strongly
influences the utilization of the transport lines.  Thus one criterion for siting
the reprocessing plant (or some reprocessing plants) should be minimization
of transport.  This seems very important in such a densely populated and
built-up country as the Federal Republic of Germany.  Of course, additional
computer programs should be written about the problems of risks of transport
and siting.  The two computer programs already mentioned (the information
system on population distribution, energy demand, regional use of the country,
and the simulation program on the problems of siting and transport) are the
basis for such a study about risk in our densely populated country.  At present
we are expanding our work and programs to that topic.  Furthermore it seems
to be necessary to integrate the whole problem into regional planning.  We
are poor in land, poor in water, poor in air.  We must also plan for the above
situation.


## ACKNOWLEDGEMENTS

This work was done for the Federal Ministry of Research and Technology
of the Federal Republic of Germany and for the Kernforschungsanlage Jülich.


## DISCUSSION

R.R. BOUSSARD:  A foreseeable programme of 100 000 MW(e) installed
power based on LWR plants in 1995 would involve the transport of 3000 tonnes
of oxides a year to reprocessing plants.  Do you think that the Asse site
alone would be sufficient for two plants with a capacity of 1500 t/a each, or
are you thinking of looking for sites other than Emden and Asse?  I suppose
that the capacity of 1500 tonnes of oxides per year represents the current
optimum size for a reprocessing plant.  And am I right in thinking that the
Asse site is your final choice for storage of high-level solid wastes, in view
of its geological and hydrogeological qualities (salt mines)?

H. BONNENBERG:  Ours is a simulation program, and we are not taking
decisions about the siting of reprocessing or waste disposal facilities.
However, I think that a single site for reprocessing and waste disposal would
provide the best solution for all environmental and organizational problems
encountered in a country where every inch of space and all resources are
used intensively.

I do not think that Asse has been finally chosen.

W. HEINZ:  In reply to Mr. Boussard's last question, I should like to
point out that the facilities provided by the Asse salt mine are envisaged for
use as a pilot plant;  at present if is not planned to use them for final storage.

I have a comment on Mr. Bonnenberg's paper. In his study, he has taken
into consideration only one or two reprocessing plants for 1995. However,
for approximately 6000 tonnes of irradiated fuel we should need four plants
because, judging by current experience, the maximum capacity of a plant is
1500 t/a. Besides, with four plants in the country the transport problem
which he has mentioned could be handled more easily.

H. BONNENBERG: The figure of 6000 t/a includes cladding and graphite
without which the amount would be about 3100 tonnes of heavy metal per year.
That volume can be handled in two plants. Contrary to your belief, I have the
feeling that transport to one big plant can be managed more easily and safely.
The problems of organization become more complex with a greater number
of plants.

R. MÜLLER: In Fig.1 you show the population density of the Federal
Republic of Germany. In one case at least, in the south-west of the country,
the population density on the other side of the border is greater than on the
German side. Did you consider population density across the border?

H. BONNENBERG: No.

A.P.V. MODING: I am a member of a Swedish Government committee
that considers problems of radioactive waste, including reprocessing and,
of course, the transport of nuclear material. We are studying essentially the
same type of transport problem as you have examined here. I should like to
ask you why you have chosen movement by rail as the only means of transport
in your analysis.

H. BONNENBERG: Road transport is not safe enough in our country.
Besides, the transport of nuclear material by road would hinder all other
traffic, the density of which is fearfully high. The railways are extremely
safe for moving this kind of material, and no other traffic will be impeded if
the trains move at a speed of 100-130 km/h.

# UNDERGROUND SITING OF NUCLEAR POWER REACTORS*

V.N. KARPENKO, C.E. WALTER
Lawrence Livermore Laboratory,
University of California,
Livermore, Calif.,
United States of America

## Abstract

UNDERGROUND SITING OF NUCLEAR POWER REACTORS.

A preliminary study of reactor siting has evolved a new concept of undergrounding that calls for excavation of an open pit. The reactor containment structure can be built using conventional construction techniques. After construction, the reactor vessel is covered with backfill material of defined permeability and porosity. Calculations show that because of this controlled permeability, radioactive seepage is confined in the backfill should an accident rupture the containment structure. The required depth of burial for complete confinement of the radioactive release is on the order of 100 m. Even a few metres of selected backfill, however, provide a dramatic reduction in the consequences of a transient release of radioactivity due to filtration and adsorption. In addition, backfill makes the reactor invulnerable to tornadoes and falling objects. Solving the containment issue in this economic manner makes close-in, metropolitan reactor siting a distinct possibility.

## INTRODUCTION

Numerous factors influence the world's need for fission as an energy source and the ultimate potential of fission technology. These factors include future growth rates of energy consumption, the forms in which energy will be needed, desires for national self-sufficiency in energy, and technological alternatives to fission. In the last case, the potential magnitude, economic and technical feasibility, environmental consequences, and availability time scales of fission vs. its alternatives strongly dictate a future need for fission power.

In the United States, we are currently making a major commitment to nuclear energy as our predominant source of electrical power. Authoritative sources predict that in the next twenty-five years, 900 1000-MW plants may have to be built to meet our projected energy consumption. However, greater public acceptance of nuclear power plants is necessary for more rapid use of fission. Fears that present designs cannot cope with the worst possible accident must be allayed. Actually, a detailed study just completed by the U. S. Atomic Energy Commission [1] indicates that the probability of a catastrophic reactor accident is small enough to fall into the category of a negligible risk. It remains to be seen, though, if the study will assuage the public's 'risk aversion' attitude.

To enhance public acceptance, underground nuclear power plants should be studied for possible commercial use. Preliminary results of a study made at LLL [2] show that underground nuclear reactors for large central plants are practical, economical, and offer several advantages in confining radioactivity from a major accident. In this paper we will explore

---

* Work done under the auspices of the USAEC.

the protection offered to the public and plant alike by the underground siting plan developed in our study.  We feel that it not only reduces the consequences of a major nuclear accident  but also better protects the plant
against tornadoes, falling objects  and sabotage.  While we do not mean
to imply that nuclear power plants are unsafe as presently constructed,
we do suggest that our underground plan offers more options in siting these
plants.

## LLL UNDERGROUND SITING PLAN

Previous proposals for underground siting had the reactor, and
sometimes the power conversion plant as well, buried deep in a competent
rock formation [3-5].  The LLL plan locates the reactor, built using conventional construction techniques, in an open pit excavated in almost any
geological strata.  The reactor containment structure is then covered
with backfill of defined permeability and porosity.  The conversion equipment and auxiliary structure can be located near ground level, as shown
in Fig. 1.

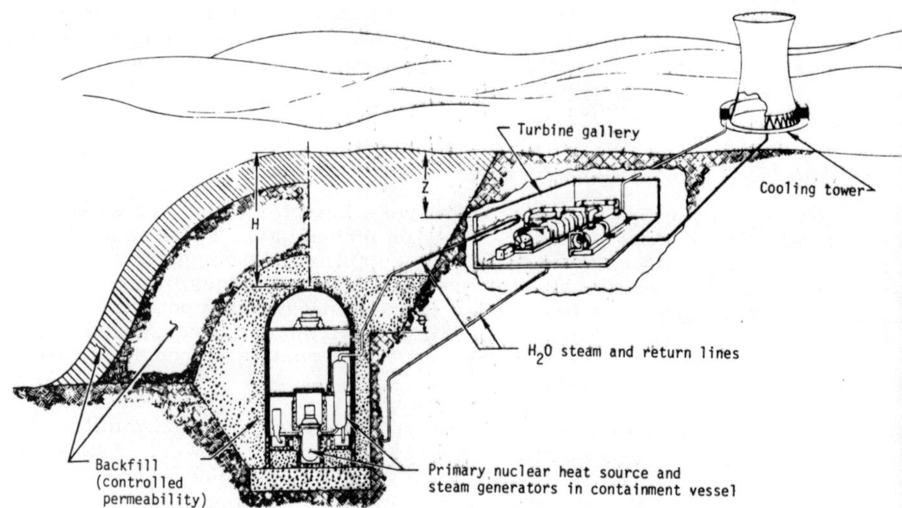


FIG. 1.    Underground siting of nuclear power plant.

We estimate the additional cost for such underground siting to be
less than 5%, and the plan may be equally applicable to LWR and GCR
systems.  Reactor burial depth is governed by (1) the degree of protection
required against tornadoes, hurricanes  and falling objects, (2) the
desired reduction in consequences of a major reactor accident, especially
one involving core meltdown, and (3) site characteristics such as population density.  A typical PWR reinforced-concrete containment structure
designed for an internal pressure of 0.31 MPa can withstand 40 m of dry
overburden with only minor design modifications.

## TORNADO PROTECTION

Recent observations indicate that the contiguous United States is subjected to about 600 tornadoes a year [6]. Most of these tornadoes are less energetic than the Design-Basis Tornado (DBT), defined as having an internal wind velocity of 483 km/h, a translational velocity of 97 km/h, and a pressure differential of 0.02 MPa in 3 s. The probability of a DBT striking a specific location is on the order of once in ten million years [6]. However, some areas in the United States might experience tornadoes with destructive force of 600 km/h combined with wind velocities.

While it is possible to design surface structures to withstand tornadoes to the extent of preventing catastrophic failure, an underground installation would have a greater chance of escaping unscathed. Flying missiles like telephone poles, automobiles, livestock, or parts of houses could easily incapacitate a reactor situated above ground for some time, but would have little effect on a suitably buried reactor. Our calculations in the next section indicate that missiles, such as flying telephone poles, generated by 644-km/h winds would probably penetrate the soil less than 6 m. Thus with 6 m of soil overburden, an underground structure can easily accommodate maximum wind velocities without our resorting to probabilistic considerations. Presumably, power lines and substations, though vulnerable, can be repaired in a short time.

## FALLING OBJECTS

There has been some concern over the consequences of objects falling on a reactor, as in an airplane crash. While the probability of an accidental crash is acceptably low, the probability of an intentional one is not. It has been postulated that in crashing, an airplane would penetrate the containment structure and its engine could severely damage the reactor's safety system.

The backfill over a buried reactor is excellent protection against these eventualities. The depth of cover necessary to prevent penetration can be found using the empirical correlation developed by Young [7], who assumed a functional relationship to describe the event:

$$X = f_1(n)\, f_2(A)\, f_3(W)\, f_4(V)\, f_5(S) \tag{1}$$

where
    n = nose-performance coefficient of a falling object
    A = cross-sectional area of the object in square metres
    W = object weight in kilograms
    V = terminal velocity in metres per second
    S = constants that lump together for soil properties
    X = penetration distance of object in metres

The values of constants n and S are listed in Tables I and II.

Evaluating numerous field tests, Young determined the functions in Eq. (1) and arrived at a prediction equation for terminal velocities above 61 m/s, which is of particular interest for our case:

$$X = 117 \cdot S \cdot n \sqrt{\frac{W}{A}} \times 10^{-6}\, (V - 30.48)$$

TABLE I.  Nose performance coefficient, n

| Nose shape | n |
| --- | --- |
| flat nose | 0.56 |
| 2.2 tangent ogive | 0.82 |
| 6.0 tangent ogive | 1.00 |
| 9.25 tangent ogive | 1.11 |
| 12.5 tangent ogive | 1.22 |
| cone, $\ell/d = 2$ | 1.08 |
| cone, $\ell/d = 3$ | 1.32 |

TABLE II.  Soil constant, S

| Description | S |
| --- | --- |
| Rock, soft to hard | 1.1 |
| Sandstone | 1.3 |
| Permanently frozen layer of earth | 3.8 |
| Desert alluvium | 4.4 |
| Clayey, silty sand, dry to moist | 5.0 |
| Nonuniform deposit of loose to medium-dense moist sand | 6.8 |
| Mud, soft estuarine saturated | 22.2 |

Thus, the vertical penetration of an object, such as an aircraft engine,
impacting alluviumlike soil at a terminal velocity of 335.28 m/s would be

$$X = 117 \ (4.46)(0.56)\sqrt{0.28 \times 10^4} \times 10^{-6} \ (335.28 - 30.48) = 4.7 \ m$$

Other calculations for a variety of postulated cases indicate that 6 m of
backfill would provide adequate cover.  Hydrodynamic Lagrangian calculations [8] show that the shock wave generated in the soil should dissipate
quickly and have little adverse effect on the integrity of an underground
structure.

The main disadvantage to Young's approach is that soil constants
are difficult to obtain for soil types not used in previous penetration experiments.

RADIOACTIVE RELEASE PROTECTION

The major issue of reactor safety is the consequence of an accident
involving core meltdown and a breach in the containment structure, thus
releasing radioactive material.  Whether such an event is a real possibility is moot; at this time, we cannot absolutely rule out its occurrence.
Again, we are not disputing the findings of the AEC study [1], but merely
want to show the value of underground siting for protection against accidents, however remote.

There are three principal advantages to underground siting from the standpoint of protection against radioactive release.

1. Because of its porosity, backfill provides an additional reservoir for escaping steam and uncondensable gases, thereby delaying any radioactive release into the atmosphere.

2. Backfill provides extra weight and therefore extra internal pressure and shock capability for the containment vessel.

3. Properly selected, backfill is an effective filter for removing aerosol particles and acts as an absorber.

The first and third points merit closer scrutiny.

<u>Backfill as a Reservoir for Radioactive Release</u>

To evaluate the effectiveness of underground siting in delaying or confining radioactive releases, we must examine a multitude of accident scenarios. Obviously a thorough analysis of even a few here is beyond our scope, but we can illustrate the principles and techniques of evaluation by applying them to specific cases.

The effectiveness of backfill as a confining reservoir will depend on the volume of radioactive release, pressure characteristics of the source, and composition of the backfill itself. Analytical techniques developed at LLL [9-11] can predict the penetration of radioactive fluid through porous media. They can solve two-dimensional flow for noncondensable gases and one-dimensional flow for condensable fluids.

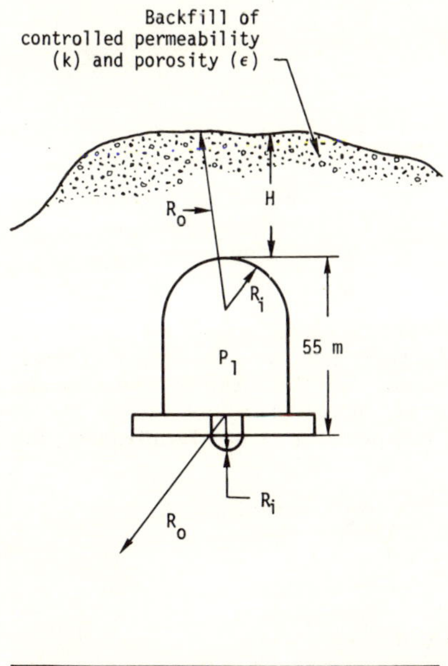


FIG.2. Configuration model for gas flow in porous media.

In this paper, most of our bounding calculations use normalized spherical geometries for noncondensable gases with limited comparison of the effect of condensable fluid on the final results. The geometries are shown in Fig. 2. A gas-release surface with a 3.2-m radius was used for the containment vessel foundation. In formulating a model for gas flow in porous media, we wanted one detailed enough to predict the salient features of the flow, yet simple enough for its results to have a broad range of application.

Assuming (1) isothermal flow of an ideal gas through soil with constant porosity, (2) the continuity equation for compressible flow, and (3) Darcy's law at low Reynolds number, the following basic relations apply [12]:

$$\frac{\partial^2}{\partial X^2}\left(P^2 + \frac{2P}{N-1}\right) + \frac{n-1}{X}\frac{\partial}{\partial X}\left(P^2 + \frac{2P}{N-1}\right) = 2\frac{\partial P}{\partial \tau} \; . \tag{2}$$

$$\frac{dM}{d\tau} = -\frac{n}{(1-X_o)^n}\frac{N-1}{2}\times\frac{d}{dX}\left(P^2 + \frac{2P}{N-1}\right)\Bigg|_{X=1} \tag{3}$$

$$M = -\frac{n}{(1-X_o)^n}\frac{N-1}{2}\int_0^\tau \frac{d}{dX}\left(P^2 + \frac{2P}{N-1}\right)\Bigg|_{X=1} d\tau \tag{4}$$

where

$$X = \frac{r}{R_o}$$

$$P = \frac{p - p_o}{p_1 - p_o}$$

$$\tau = \frac{k(p_1 - p_o)\,t}{\epsilon\,\mu\,R_o^2}$$

$$N = p_1/p_o$$

$$M = m/m_o$$

Using finite-difference techniques, we solve Eq. (2) with the aid of Eqs (3) and (4) to find the rate and amount of mass leaving the solid at a given outer radius. Also, for the case of one-dimensional multiphase flow involving air, water and water vapor, we can determine water saturation pressure and temperature as a function of one parameter, $\theta$ [10,12]:

$$\theta = \frac{r}{2}\left(\frac{\epsilon\mu_a}{k\,p_1\,t}\right)^{1/2} \tag{5}$$

For Eqs. (2) through (5):

$p_o$ = initial fluid pressure
$p_1$ = initial driving pressure
$\mu$ = fluid viscosity, $\mu_a$ = air viscosity
$k$ = permeability

   $r$ = spatial variable
$R_O$ = outer radius of medium
  $m$ = total mass that has left outer radius of the solid
$m_O$ = mass of gas originally in the pores of medium
  $\epsilon$ = porosity
  $t$ = time

The first set of calculations applied to the geometry of Fig. 2 and were made using spherical coordinates and holding the source pressure constant at 0.34 MPa. Surface arrival times of the gas released from the top of the containment structure are given in Table III. For noncondensable gases at constant pressure, the time delay for release of radioactivity is small. Appropriate changes in porosity and permeability, however, would increase the delay. If we consider the condensability of the radioactive mixture in the containment vessel, the arrival times of Table III will increase by two orders of magnitude, depending on the mixture ratio of steam and noncondensable gas [12].

A second set of calculations were made to show the effectiveness of backfill as a reservoir when source pressure is decaying. Under these circumstances and with a backfill 24 m deep, radioactivity never reached the surface, as shown in Table IV.

TABLE III.  Flow from reactor dome at constant
pressure of 0.34 MPa

| $R_i$ (m) | $R_O$ (m) | H (m) | k (Darcy) | $\epsilon$ | t (h) |
|---|---|---|---|---|---|
| 15.2 | 21.3 | 6.1 | 1 | 0.3 | 0.216 |
| 15.2 | 27.4 | 12.2 | 1 | 0.3 | 0.866 |
| 15.2 | 33.5 | 18.3 | 1 | 0.3 | 1.94 |
| 15.2 | 61.0 | 45.8 | 1 | 0.3 | 18.6 |

TABLE IV.  Flow from reactor dome with decaying pressure of 0.17 MPa

| $R_i$ (m) | $R_O$ (m) | H (m) | k (Darcy) | $\epsilon$ | t (h) |
|---|---|---|---|---|---|
| 15.2 | 21.3 | 6.1 | 1 | 0.3 | 0.64 |
| 15.2 | 33.5 | 18.3 | 1 | 0.3 | 5.89 |
| 15.2 | 39.3 | 24.1 | 1 | 0.3 | $\infty$ |

To model the geometry of a large power reactor in a two-dimensional ideal gas calculation (see Fig. 2), we started with our one-dimensional model and extended it to two dimensions. Thus, the transient flow is governed by the same equations except that the permeability takes on the form of a tensor. The equations are then solved for pressure in cylindrical coordinates using an explicit, finite-difference approximation with a constant temperature.

Variations of soil permeability between layers of sand and clay are permitted in this second model. The permeability is assumed to be isotropic within each soil layer and does not vary with time. The porosity represents the fraction of interconnected void volume available for gas flow in the porous media. It is assumed that the only effect of liquid in the pores is to reduce the porosity.

Pressure within the containment shell is assumed to decay with time as a result of mass flow from the shell into the surrounding soil. The containment shell is assumed to be impermeable except for a 3.13-m-radius cylinder representing the volume occupied by the molten core.

A set of calculations were made on this second model representing an accident sequence in PWR involving core meltdown, but with containment sprays and heat exchanger working. Leakage pressure for the source comes from hydrogen from the zirconium-water reaction, carbon dioxide generated from the decomposition of concrete during the core melt-through of the foundation, and some steam. The pressure at melt-through was specified as 0.34 MPa; pressure decay is governed solely by the amount of gas lost to the formation. With a backfill greater than 3 m, radioactivity never reached the surface, as shown in Table V.

TABLE V.   Flow from reactor foundation with
decaying pressure of 0.34 MPa

| $R_i$ (m) | $R_o$ (m) | H (m) | k (Darcy) | $\epsilon$ | t (h) |
|---|---|---|---|---|---|
| 3.1 | 9.1 | 0 | 1 | 0.3 | 0.2 |
| 3.1 | 30.4 | 0 | 1 | 0.3 | 8.6 |
| 3.1 | 58.0 | 3.0 | 1 | 0.3 | $\infty$ |

The conditions for a driving-pressure source will change as soon as the molten core penetrates the foundation. Decay heat from the core will then vaporize the water in the pores of the adjacent soil [12]. Estimates of the quantity of steam for fully saturated soil are given in Fig. 3. Obviously, dry soil conditions under the foundation will be more conducive to lowering the driving pressure.

The sequence of events in a nuclear reactor accident that leads to a driving source of constant pressure results in a reduction of radioactive release due to decay time and filtration of particles. It is estimated that melt-through might take from a half to almost a full day. Calculations of integrated radioactivity decay (Fig. 4) indicate that approximately 10 d will bring an order of magnitude reduction in the radioactivity released.

It is important to reevaluate safety systems that might contribute to a driving source of constant pressure when containment is breached. It is also important that any breach of a partially or fully buried containment structure occur at the bottom of the vessel. The effectiveness of backfill as a filter, discussed later, can then be fully exploited.

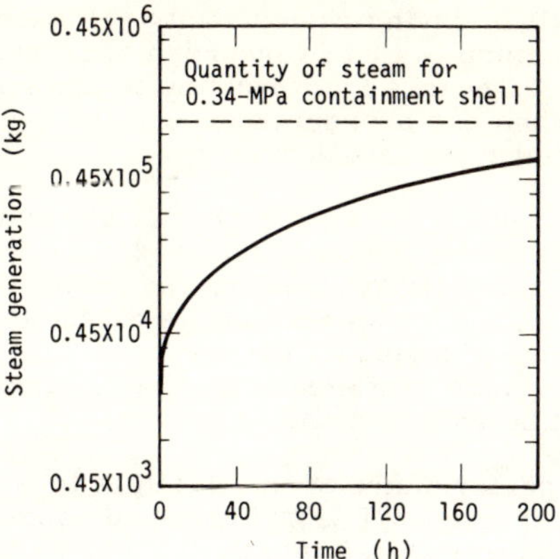



FIG.3.    Steam generation for core melt in contact with wet soil for a core after-heat of $10.3 \times 10^6$ W and a containment shell volume of $5.6 \times 10^4$ m$^3$.

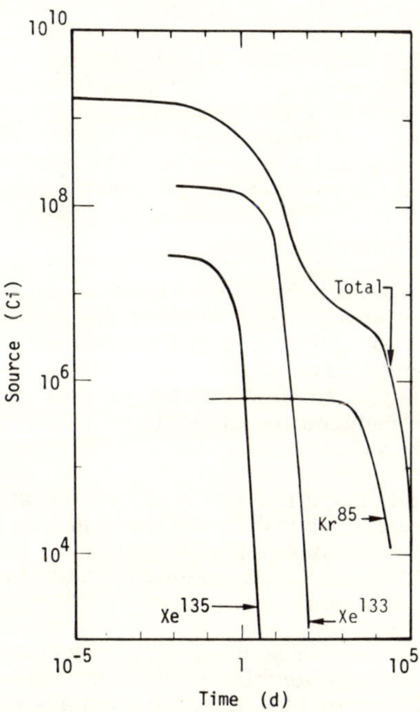



FIG.4.    Fission-product activity.

A second benefit in having radioactivity released from the bottom of a containment structure is that overburden pressure is often sufficient to prevent gas-initiated cracks from running to the surface [13]. In shallow depths and when gas pressures exceed the overburden pressure, layered media might prevent this occurrence.

The recent AEC study [1] should greatly alleviate public concern over nuclear reactor safety. Indeed, the study is a two-fold accomplishment. First, it quantifies the probability and consequences of a major accident. The risk from such an accident is calculated to be very low when compared with other naturally occurring or manmade catastrophes. Second, from it, safeguard systems whose failure increases the risk from an accident can be analysed and improved.

Post-accident performance of the HTGR is less troublesome than that of the LWR, because the HTGR failure mode does not require large amounts of water to be introduced into the reactor. Also, if the reactor is underground, pressure might be controlled by venting gases harmlessly through the overburden, which would then act as a filter.

## Backfill as a Filter and Adsorber

The initial composition of the radioactive source can be narrowed to 45 isotopes [1]. Because of short half-lives or limited availability, all other isotopes are eliminated from consideration. Integrated fission-product activity as a function of time is shown in Fig. 4.

Except for noble gases and halogens, most fission products released from a breached containment vessel will be solid particles. Since solid particles behave like aerosols, they would be effectively removed in flowing through the backfill. McFee and Sedlet [14] have shown that sand is a good filter for removing P-U-Mo alloy fumes from air. Their particles ranged from less than 0.02 to 4 $\mu$m, which is probably smaller than those encountered in our case. Filtering efficiency varies with particle size and velocity and with sand characteristics. For a 0.76-m bed of sand with 35% void space, they measured the penetration fraction as $4 \times 10^{-5}$ at 0.15 m/s, with a slight dependence on velocity above 0.1 m/s (see Fig. 5).

The velocities encountered in the release of radioactivity through backfill could vary from almost zero to about 0.5 m/s. It is reasonable to assume, therefore, that a penetration of less than $10^{-5}$ is possible for even a metre or so of properly engineered backfill. If we subtract the noble gases shown in Fig. 4 and account for particulate filtration, fission-product activity will be reduced to below 1000 Ci with minimal backfill and release in less than 24 h.

Effective filtration of particulates reduces the intensity of fission activity to that associated simply with the noble gases, $Xe^{135}$, $Xe^{133}$ and $Kr^{85}$. Since these noble gases are nonreactive, they will not be retained in the backfill by any known chemical reaction; however, the adsorption mechanism is of some significance.

Based on the difference in surface areas between ground charcoal and average soil and on the adsorption coefficient of $6.10 \times 10^{-5}$ g Kr/g charcoal, Tadmor and Cowser [15] postulated an adsorption coefficient of $10^{-7}$ g Kr/g soil. This important parameter must be experimentally verified for specific backfill materials before any firm conclusions can be drawn. However, we estimate that it would take about $4 \times 10^3$ m$^3$ of

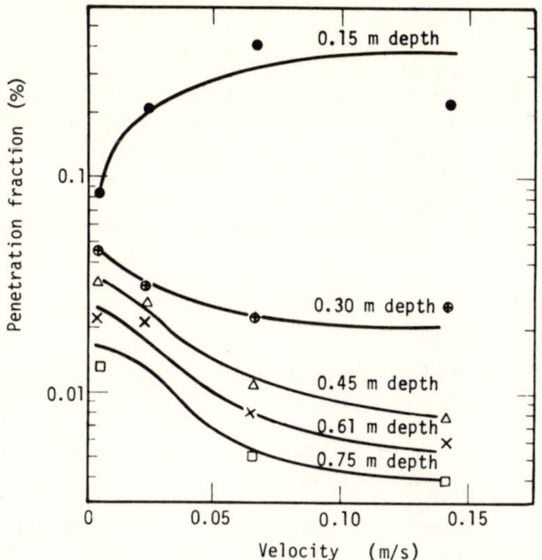


FIG. 5.   Penetration versus velocity through different depths of sand (from McFee and Sedlet, Ref.[14]).

backfill to adsorb $Kr^{85}$ with an activity of $6 \times 10^5$ Ci, and even the shallowest cover over a reactor considered here (6 m) has a factor of 3 more volume above the dome.

Xenon, especially $Xe^{133}$, would require $3 \times 10^3$ m$^3$ adsorbing backfill, assuming the same adsorption coefficient ($10^{-7}$ g Kr/g soil). The radioactive decay characteristics of $Xe^{133}$ make confinement of close to 100 d desirable with 10 000 d needed for $Kr^{85}$. To achieve adequate confinement time, we must also allow for molecular diffusion. The average travel time due to diffusion for $Kr^{85}$ can be approximated by the following equation [16]:

$$t = \frac{L^2}{2D}$$

where L is path length and D is the average diffusion coefficient of 0.015 cm$^2$/s for ground material. A path length of 31 m will result in a travel time of greater than 10 yr after the adsorption process has been completed.

## POPULATION DENSITY

Although siting reactors in areas with low population densities is commendable, it may not always be possible in the future. Also, since population density will likely change outside the site fences, it is a criterion of only temporary value. Fully exploiting underground siting techniques, however, will reduce or eliminate health hazards and property damage, and should give additional flexibility to site selection.

CONCLUSIONS

We have avoided probabilistic arguments over reactor safety in this paper not because we quarrel with the approach, but because we wanted to explore the merits of underground siting on a deterministic basis where possible.  Hopefully, our deterministic solutions are both verifiable by full-scale tests and economically feasible.

We see the following advantages to underground siting of nuclear reactors:

1.  It offers extreme protection against tornadoes and falling objects, accidental or deliberate.

2.  Backfill can delay or fully contain radioactive releases, allowing more time for evacuation or remedial action.

3.  As a filter for particulates, backfill greatly reduces the magnitude of radioactive releases.

4.  Backfill may also adsorb noble gases effectively, although adsorption coefficients need to be experimentally verified.

5.  Underground siting offers an economic alternative when placing nuclear power plants near populated areas.

6.  Many types of nuclear power plants, including the HTGR, can be adapted to underground siting.


REFERENCES

[1]   Reactor Safety Study:  An Assessment of Accident Risks in U. S. Commercial Nuclear Power Plants, USAEC Rep. WASH-1400 (1974).

[2]   BLAKE, A., et al, A Concept for Underground Siting of Nuclear Power Reactors, Lawrence Livermore Laboratory Rep. UCRL-51408 (1973).

[3]   OLDS, F. C., Power Eng. 75 10 (1971) 34.

[4]   WATSON, M. D., et al, Underground Nuclear Power Plant Siting, Environmental Quality Laboratory, California Institute of Technology, EQL Rep. 6 (1972).

[5]   Underground Nuclear Power Plants, A Preliminary Evaluation, United Engineers and Constructors Rep. UEC-UNP-710701 (1971).

[6]   BECKERLEY, J. G., Technical Basis for Interim Regional Tornado Criteria, USAEC Rep. WASH-1300 (1974).

[7]   YOUNG, C. W., J. Soil Mech. Foundation Div. 95 (1969) SM3.

[8]   McCAULEY, E., Study of Earth Penetration by Flying Objects, Lawrence Livermore Laboratory Rep. ENN 74-111 (1974).

[9]   MORRISON, F. A., Jr., Transient Gas Flow in Porous Column, Lawrence Livermore Laboratory Rep. UCRL-52929 (1971).

[10]  MORRISON, F. A., Jr., Transient Multiphase Multicomponent Flow
      in Porous Media, Lawrence Livermore Laboratory Rep. UCRL-74327,
      Rev. 1 (1972).

[11]  BOWMAN, B. R., TWIG, A Computer Code for the Calculation of
      Two-Dimensional Ideal Gas Flow in Porous Media, Lawrence
      Livermore Laboratory Rep. UCID-16320 (1972).

[12]  PITTS, J. H., et al, Diffusion of Radioactive Fluid through Soil
      Surrounding a Large Power-Reactor Station after a Core Meltdown
      Accident, Lawrence Livermore Laboratory Rep. UCRL-51494
      (1973).

[13]  PITTS, J. H., Gas Initiated Crack Propagation in a Porous Solid,
      A Progress Report, Lawrence Livermore Laboratory Rep.
      UCID-16236 (1973).

[14]  McFEE, D. R., SEDLET, J., J. Nucl. Energy 22 (1968) 641.

[15]  TADMOR, J., COWSER, R. E., Nucl. Eng. Des. 6 (1967) 243.

[16]  REIST, P. C., Krypton Disposal Studies, Internal Rep., Department
      of Industrial Hygiene, Harvard School of Public Health, Boston,
      Mass. (1964).

## DISCUSSION

M. NELKEN: Reactor manufacturers are reluctant to modify their
'standard' design. So, are you assuming a conventional Nuclear Steam
Supply System lay-out, space and volume? Secondly, would you consider
'standard' emergency core cooling and other protection systems? Lastly,
what is the additional capital cost for a 'standard' nuclear site?

V.N. KARPENKO: We have taken an average volume and a lay-out
close to the standard one with a few minor modifications from available
records. As I recall, our volume was about $2 \times 10^6$ ft$^3$.

The use of some standard protection systems would be in order, but let
me suggest that the balance between systems designed for plant protection
and protection of the population should be examined, as should the whole
approach to protection. If we take the published probabilities of major reactor
accidents at their face value, then from the economic standpoint the loss of
a plant would be so infrequent that it hardly needs to be considered. Therefore, the emphasis should be on systems which limit the consequences of
accidents for the public. The approach might, however, be somewhat different
in the case of underground plants.

B.K. GRIMES: I have a question about accidental releases. Every
reactor must have access penetrations — for equipment, for personnel,
for ventilation of equipment, for changing the air in the containment before
personnel entry and, of course, for transfer of steam from the reactor to
the turbines. Did you in your study consider the question of a release of
radioactivity resulting from failures in these paths which might bypass
the sand or other backfill?

V.N. KARPENKO: This is a very good question, and one which is asked
frequently. It has two aspects, one of which is technical, viz. whether closure

systems are available or not; our experience indicates that high-reliability closure can in fact be designed. We have done some work in this area. Closures can be designed to act in milliseconds.

The other aspect of the problem relates to surface facilities. If the closures on a surface plant fail, the radioactivity is released directly into the air with obvious consequences. In the case of underground conduits, if a break occurs, the radioactivity in many instances will be released to the backfill. We should be better off that way.

E.F.F.W. USHER: In the United Kingdom preliminary investigation of stations in underground caverns has shown the additional cost to be considerably higher than 5%. Have you assumed a site already excavated in a suitabl manner? Where appropriate ground features already exist, we have considered placing the turbine house on the surface where the steam circuit involves heat exchange. Have you considered this arrangement?

V.N. KARPENKO: We envisaged a site not already excavated. Our estimate for excavation is based on published current industrial costs in the United States of America. Cut and backfill costs range from US $1 to US $3 per cubic yard.

The location of the turbine house is not critical. The heat and pumping losses are calculated to be very small for the length in question. The turbine building was placed underground with only a shallow cover, for purposes of plant protection.

W. KROEGER: In the Federal Republic of Germany we are also studyin, underground siting of an HTGR, using an open pit (soil) with backfilling and covering. Our purpose is simply to analyse the safety potential. The buildir industry in the Federal Republic of Germany has estimated the cost of a pit 50 m in diameter and 50 m in depth at about 20 million DM, given average underground conditions.

V.N. KARPENKO: We have estimated that a pit 50 m deep with cover, prepared by cut and backfill techniques, would cost US $4 million in fairly dry locations. It will probably be necessary to increase the figure slightly to allow for accessibility during the construction of the containment vessel and so on. This figure is based on $1-3 per cubic yard. The problem of hydrology was not studied.

P. GRANDE: What is the average probability of a reactor site in the United States of America being hit by a tornado?

V.N. KARPENKO: The probability of a DBT striking any location is of the order of one in ten million years. Mr. Grimes of the USAEC should be able to give you more details.

R. MÜLLER: I think that when you go underground once, you will always have to go underground, as you will realize that other types of construction are less safe!

M. LASER: What do you think about the decommissioning of an underground power plant? Should the soil be removed to bring out the reactor vessel and other machinery, or would you fill up the whole site with concret

V.N. KARPENKO: Underground sites should have access openings to allow removal of the major components. This is justified by maintenance needs and economic considerations. The equipment could be left in place, however, because underground storage on the original site is more advantageous than building another storage place.

A participant from the United Kingdom mentioned earlier that they were considering piling a mound of earth round one of their old plants as part

of the procedure of decommissioning. This precaution is of course already
taken care of when one resorts to underground siting.

B. OBERBACHER: In the underground concept all activity is released
in the soil. Do you have any idea about dissipation from the soil into the
environment?

V.N. KARPENKO: I have discussed the question of diffusivity and
adsorption, and it should be possible to calculate the rate of release from
these considerations. We have not studied other mechanisms for all the
constituents in question. I should think that a perusal of the literature would
supply the necessary data.

J.B. BURNHAM: Some questions have been raised concerning the cost
of underground siting. But even if we assume a 10% additional capital cost —
or US $50 million for a 1000-MW(e) reactor — if this would allow siting
close to conglomerates of population and hence the use of district heating,
it would not take very long to recover the extra cost.

R.A. TARJANNE: Could you give us an approximate quantitative estimate
of the advantage of underground siting for a reactor located in or near an
urban area, assuming the most serious type of reactor accident — say,
core melt-down?

V.N. KARPENKO: In core melt-down accidents we would have all the
advantages described in the paper. The path length to the surface would
be considerable. One would have to be careful, however, about steam generation from soil moisture produced by heat from the melted core. The condensable flow of gases through porous media can be calculated. The moisture
content of the soil under the foundation might have to be reduced, but calculations should show whether that was necessary.

R.A. TARJANNE: When do you think reactor manufacturers are likely
to offer underground plants? In the Helsinki area in Finland, underground
siting has been considered as an alternative, but of course not for the first
power plant.

V.N. KARPENKO: More and more contractors are beginning to realize
that underground construction is not all that different. If I may venture a
guess, they may well have enough information to proceed with this type of
work in a couple of years.

# EVALUATION OF BLAST WAVE DAMAGE
# FROM VERY LARGE UNCONFINED
# VAPOUR CLOUD EXPLOSIONS

G. MUNDAY
Department of Chemical Engineering
  and Chemical Technology,
Imperial College, London

L. CAVE
Pollution Prevention (Consultants) Ltd,
Crawley, Sussex, United Kingdom

**Abstract**

EVALUATION OF BLAST WAVE DAMAGE FROM VERY LARGE UNCONFINED VAPOUR CLOUD EXPLOSIONS.
A mathematical model is described for estimating the damage potential from unconfined vapour
cloud explosions. An attempt has been made to cover the salient details of the explosive phenomenon
including finite flame accelerations and finite vapour cloud sizes. The model has been evaluated against
two industrial incidents and the results extrapolated to large-volume vapour clouds. The authors conclude,
on the evidence of this model, that great care must be taken in the evaluation of the explosion hazard
from the probable occurrence of very large unconfined explosions even at distances in excess of 1 km from
the centre of initiation.

## 1. INTRODUCTION

In assessing the possible effects of hydrocarbon vapour cloud explosions
on the safety of nuclear power stations the problems must be regarded as
twofold, with two fundamental questions to be considered. These are:

(a) How do the peak pressure and duration of the pressure pulse at the
stations vary with the size of the cloud and the distance of the point of
initiation from the stations?
(b) How would the vulnerable features of the stations, including personnel,
respond to the blast loadings?

Although the answers to these two questions are equally important, the
present paper is limited to consideration of the explosion phenomenon.
The potential hazards from the explosion of large unconfined clouds of
hydrocarbon vapours have been studied in the United States of America [1-4]
and in the Federal Republic of Germany [5,6] for some years, and more
recently the independent work described in this paper has been undertaken
in the United Kingdom.
The major problem in prediction of the effects of unconfined vapour
cloud explosions is the lack of fundamental understanding of the transient
behaviour of flames in this situation. All writers on this subject agree that,

if the flame-gas interactions were even partly understood, our present
knowledge of aerodynamics and our ability to handle many forms of
mathematical model would allow us to estimate the damage potential of
this form of explosion.  The accuracy of such predictions would be similar
to those from the standard analysis of damage from conventional high
explosives.

In the absence of this fundamental understanding of flame behaviour,
three techniques for damage estimation have been applied in practice:
estimates of blast wave strength, based on point source solutions for the
equivalent TNT explosions (see e.g. Ref.[7]); estimates from similarity
solutions integrated from the boundary conditions [8]; and the use of
analytical methods applied to a simplified constant velocity model [9].
All these techniques suffer from inherent disadvantages either within the
model or in the difficulties in forming an estimate of the scaling parameter
or the boundary conditions.


2.   MATHEMATICAL MODEL

In this paper, a mathematical model which is not subject to these
limitations is described.  It is assumed that the expanding gases in the
explosion act as a 'spherical piston', compressing the surrounding unburned
vapour and air with sufficient rapidity to create a blast wave.  Allowance
is made for the difference between the velocity of the flame which defines
the propagation of the explosion process and that of the solid piston used
in the model;  the piston is assumed to accelerate at a uniform rate from
rest, starting from a finite distance from the source of initiation.  The
piston is allowed to accelerate up to a constant maximum velocity but its
movement at this velocity is limited to the volume enclosed by the vapour
cloud;  the expansion of the cloud as a result of the explosion is taken into
account.  Thereafter, movements of the air outside the cloud are governed
by conditions of equality of pressure and velocity across the boundary
defining the moving edge of the burnt vapour cloud.

The position-time variation of the gas state and its velocity are obtained
numerically at points of intersection of a grid of lines in this plane by
solving the appropriate hydrodynamic equations for spherically symmetric
compressible flow using the method of characteristics.  This mathematical
method is described in all modern text books on compressible flow and
the basis of the present model was originally developed by Gibbs [10] for
estimating the blast wave hazard associated with the partial failure of high-
pressure vessels.

The boundary conditions are fixed by the second-order expression for
the piston trajectory, the maximum velocity of the piston and the limitation
imposed by the size of the vapour cloud.  The results can be mapped on a
radial distance-time diagram and depict quite clearly where the blast wave
occurs.  Appropriate relations are used to transfer the calculation procedure
across the shock associated with this blast wave.

With this knowledge, pressure loads and wind forces can be computed
for any point within the domain of the calculations and in this way the damage
to buildings and equipment can be established.

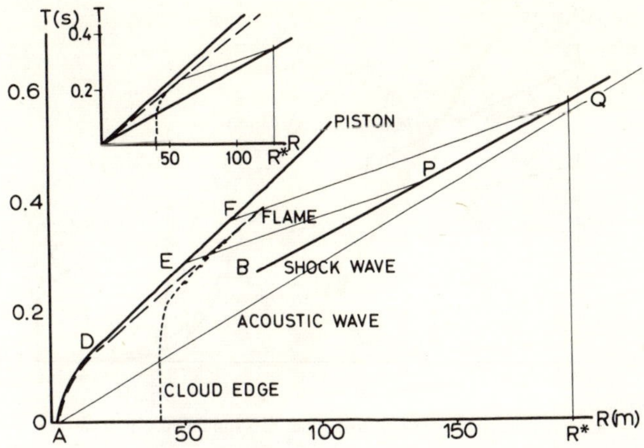


FIG. 1.  Radial distance–time diagram for a simulated vapour cloud explosion.

## 3.  SAMPLE RESULTS AND DETAILED PHENOMENOLOGICAL DESCRIPTION

The results from such a calculation are depicted in Fig.1.  In this example, it is assumed that the piston eventually moves relative to the observer at a constant velocity of about two-thirds of the speed of sound in air, accelerating from an initial sphere of 4 m diameter.  It should be noted that the flame velocity relative to the unburnt gas is about one-tenth that of the piston.

Referring to Fig.1, as the piston moves through the surrounding gas, accelerating from rest at the point A, compression waves are set up which coalesce to form a shock wave.  In this illustration  the shock is formed at the point B within the fan of compression waves which are bounded by the piston path ADEF on one side and the acoustic wave AQ on the other. In certain situations the shock may form at a point on the acoustic wave. It is important to realize that the piston motion at the point E in the diagram influences the shock propagation at the point P and beyond.

The piston reaches a constant velocity at the point D and continues to expand until the explosion exhausts its energy supply and the flame reaches the boundaries of the vapour cloud.  This occurs at the intersection of the flame trajectory and the particle path associated with the cloud edge (found at an initial radius of 40 m in this case).  The occasion corresponds to a piston position indicated by F and will affect the propagation of the blast wave at a radius R*.

The subsequent movement of the shock is then governed by the momentum of the expanding gas and the manner in which the burnt gases which originally formed the vapour cloud retard this motion.  This complex interaction may be handled by the same mathematical techniques but would involve lengthy numerical analysis to cover the distances of interest.  In the present instance

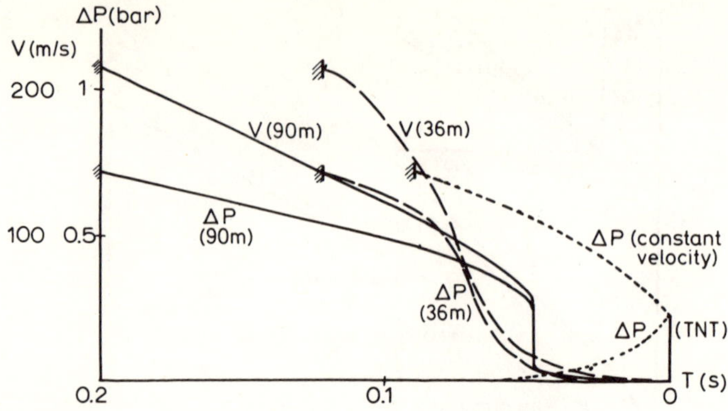


FIG. 2.   Pressure and wind velocity distributions in a simulated vapour cloud explosion.

a semi-empirical analytical approach, suggested by Brinkley [11] and based
on energy losses from a blast wave, has been used to describe in a quantitati
manner the characteristics of a decaying blast wave.

In Fig.1, the trajectories for the constant velocity piston model [9] are
included to illustrate the main differences between the two models.  The
influence of the size of the exploding vapour cloud on the blast wave in this
latter model is greatly reduced, as can be seen by a comparison of the two
values of R*.

Figure 2 depicts the pressure and wind velocity variations experienced
by observers at two radial distances from the explosion centre.  At a
distance of 36 m, the observer would experience a shockless compression
from the incidence of the leading disturbance (t = 0) to the arrival of the pisto
Quite large overpressures would occur and high wind speeds would be
encountered.  At 90 m from the centre, an observer would be subjected to
a discontinuity in pressure and wind speed a short time (t = 48 ms) after
the initial disturbance has passed him as the blast wave has been formed.
Consequently, the overpressures and wind speeds are now considerable.

Overpressure results for other models are also shown in Fig.2 for the
radial distance of 90 m.  The two models considered are the constant
velocity piston and the equivalent TNT high explosive.  In the former model,
equal maximum velocities enable a direct comparison to be made but in
the latter case the comparison is based on the same shock overpressure
at this radial distance.  Two main differences are immediately apparent:

(a)  Magnitudes of pressure are throughout greater for the present
model;  and
(b)  The pulse length in the present model is considerably greater.

Both these differences are the result of the finite time required for the
development of the shock and the resultant increased distances between
the shock and the compression source.  Both factors have an important
bearing on the long-distance effects of blast waves since shock decay rates
are most rapid for sharp pressure pulses.

## 4.   MODEL FITTING

Several accidental explosions involving vapour clouds have been
sufficiently well documented to evaluate the coefficients for the expression
governing the trajectory of the piston in the model.  However, in any attempt
to determine the best fit of damage distribution to the pressure and wind
velocity results for a given model, two levels of attack are available.  The
first involves a detailed analysis of individual items of damage in terms
of transient pressure and wind loads and their interaction with the dynamic
response of the mechanical structure involved;  the best example of this
approach has been given by Penney et al. [12].  Alternatively, the accumulated experience of experts in the assessment of damage due to high explosives
may be used to express the effects in the form of peak overpressures for
specific categories of damage.  This latter method is not very satisfactory
when dealing with vapour cloud explosions which are far from 'ideal' in
the sense of a high-explosive blast source.  However, insufficient detail in
the descriptions of damage to individual structures in past vapour cloud
incidents is available and the second approach must perforce be used.
The one exception involves the most recent major incident in the United
Kingdom but the lengthy analysis required for the former approach has not
yet been completed.

Two incidents have been chosen to provide data for model fitting.
The first occurred on 20 January 1968 at the Shell Pernis petrochemical
complex in the Netherlands [13].  At 0423 hours an explosion occurred,

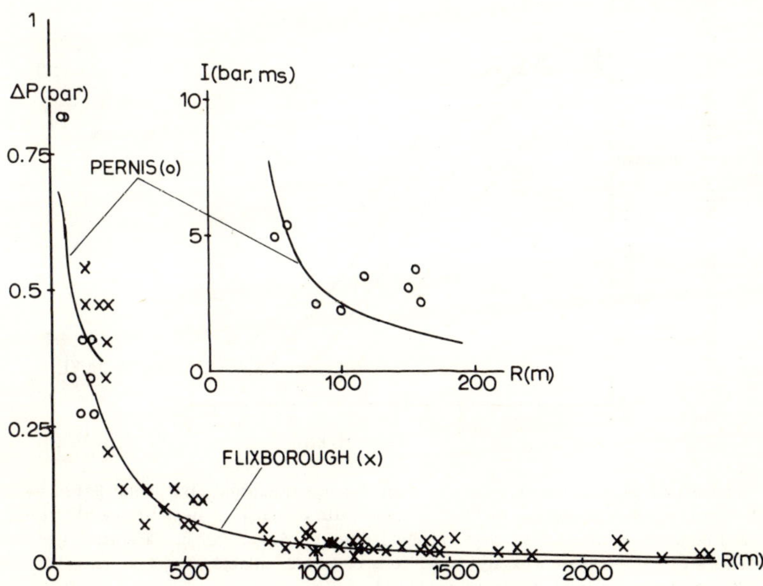


FIG. 3.   Peak overpressures and impulse as a function of radial distance for two industrial explosions.

killing two people and injuring seriously a further nine. The enquiry showed
that the cause was to be found in a storage tank for oil slops from which was
expelled a large quantity of hydrocarbons to form a cloud of an explosive
mixture of air and mist. The report of the enquiry provides sufficient
evidence for a distribution of peak shock overpressures to be deduced as
shown in Fig.3. Since these figures are for distances close to the centre
of the explosion, the more useful impulse-distance distribution is also shown
in Fig.3. These results are obtained on the basis of experience with conventional high explosives and with nuclear explosions.

The second incident considered is that at Flixborough in the United
Kingdom on 1 June 1974. The accident has received considerable publicity
and is at present the subject of a court of enquiry. Although the court has
not yet completed its investigations, a very extensive survey of the damage
has been undertaken by the Safety in Mines Research Establishment and
has been made available for a comparison with the model presented here.
However, as indicated above, the effects have, as yet, been interpreted
only in terms of a conventional high-explosive charge. The disaster occurre
between 1550 and 1600 GMT at the Flixborough works of Nypro Ltd. A
quantity of cyclohexane was expelled from the process equipment and
ignited some sixty seconds later. The subsequent explosion and fire caused
extensive damage. The data described here relates to damage caused by
this explosion but in areas outside the factory and not affected by the fire.
The results are also shown in Fig.3.

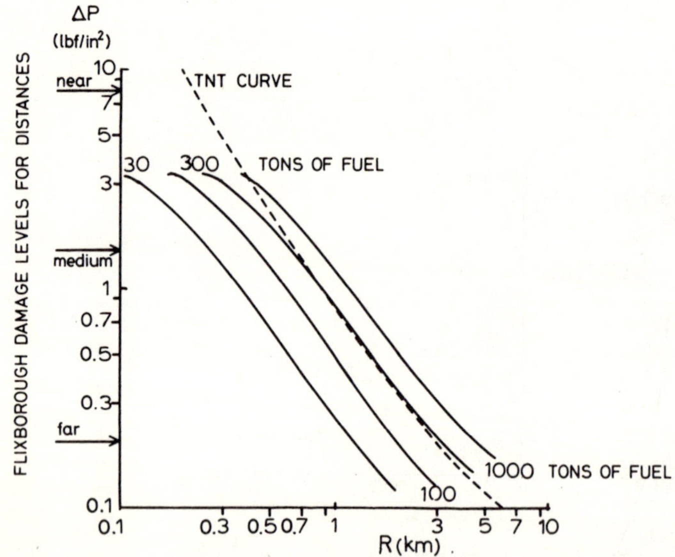


FIG. 4.   Extrapolated peak overpressure curves.   Near distance (Flixborough): Normal British house
completely destroyed; severe damage to 11-inch brick walls; collapse of steel framework; severe
distortion and collapse of oil storage tanks.   Medium distance: Slight roof displacement; roof tiles
removed; extensive window-glass breakage; steel posts deflected; asbestos cladding destroyed;
extensive damage to trees with appreciable number blown down.   Far distance: A few roof tiles displaced; 25% window glass broken; extensive damage to plastic windows; broken asbestos sheets;
distorted corrugated steel sheets.

Curves derived from the model using the appropriate factors relevant
to each incident and employing coefficients for the piston trajectory to give
the best fit are included in Fig.3.  The trajectory coefficients are the same
for both incidents although there are some minor differences in the scaling
parameters involving the cloud size.  The cloud radii are 50 m and 40 m
respectively and the piston trajectory represents a uniform acceleration
to a maximum velocity of two-thirds the speed of sound in air.

## 5.   EXTRAPOLATION OF THE MODEL TO VERY LARGE VAPOUR CLOUDS

Using these coefficients to model large-scale explosions with particular
reference to the blast wave strength at large distances of the order of
1 km, extrapolations have been performed.  From the results of these extrapolations it will be possible to evaluate the risks associated with the siting
of nuclear facilities in the vicinity of chemical and refining plant or large-
tonnage storage depots.

Peak overpressure curves for explosions involving 30, 100, 300 and
1000 tons of hydrocarbon fuel are shown in Fig.4 using logarithmic axes.
The figure includes the TNT point source curve for the largest cloud
assuming that the 'percentage equivalent yield' of the vapour cloud is 5%;
this is the value normally assigned to such explosions.  Further, hazard
levels are indicated in terms of peak overpressure values for various
categories of damage experienced in the Pernis and Flixborough explosions.

## 6.   CONCLUSIONS

The authors believe that Fig.4 clearly indicates the formidable nature
of the hazard presented by large vapour cloud explosions.  They would
like to emphasize that the damage from incidents involving fairly small
amounts of fuel is found to extend over a large area surrounding the
explosion and that simple extrapolation based on dimensional scaling
techniques is not adequate for determining the risks involved in siting
nuclear facilities near chemical or refining plants or large fuel terminals.

As a final warning we should like to quote from an article describing
the use of vapour cloud explosions for military purposes [14].  "Fuel air
explosives open a potential for increasing blast effects while reducing
weight!" ; and earlier:  "The fuel air explosive bomb was designed ... to
generate a more effective blast than standard munitions ... with blast
pressures of 300 $lbf/in^2$ within a 50-ft circle ...".  Perhaps it is ironic that
the test area where the United States Air Force perfected these devices
is to be used for a scientific investigation into the civil problems of
unconfined explosions involving large-scale experiments.

REFERENCES


[1]  STREHLOW, R.A., in Proc. 14th Int. Symp. on Combustion, Salt Lake City, 1972, Combustion Inst.
     (1973) 1186.
[2]  BRASIE, W.C., SIMPSON, D.W., Loss Prev. Bull. 3 (1966) 91.
[3]  WOOLFOLK, R.W., Stanford Res. Inst. Final Rep. (1971).

[4]   IOTTI, R.C., et al., in Proc. Topical Mtg. on Water Reactor Safety, Salt Lake City, 1973, USAEC Rep. CONF. 73-0304.

[5]   WAGNER, H.G., two of a set of papers from Bochum describing German studies on gas cloud explosions, issued as NSAC (73), 9 (1971).

[6]   GEIGER, W., 2nd Int. Conf. on Structural Mechanics in Reactor Technology, 1973, Battelle Inst., Frankfurt.

[7]   BURGESS, D.S., ZABETAKIS, M.G., US Department of the Interior, Bureau of Mines, Report of Investigation 7752 (1972).

[8]   FRIEDMAN, M.P., J. Fluid Mech. $\underline{11}$ (1961) 1.

[9]   KUHL, A.L., et al., in Proc. 14th Int. Symp. on Combustion, Salt Lake City, 1972, Combustion Inst. (1973) 1201.

[10]  GIBBS, J.C., Ph.D. Thesis, Imperial Coll., London (1974).

[11]  BRINKLEY, S.R., Loss Prev. Bull. $\underline{3}$ (1966) 79.

[12]  Lord PENNEY, SAMMUELS, D.E.J., SCORGIE, G.C., Phil. Trans. R. Soc. (London) $\underline{A266}$ (1970) 357.

[13]  STATE PUBLISHING HOUSE, HAGUE, Report on the Cause of the Explosion at Shell Pernis (1968), (in Dutch; Engl transl. available from publisher).

[14]  ROBINSON, C.A., Aviat. Week Space Technol. (Feb. 1973) 42.

# CHAIRMEN OF SESSIONS

| | | |
|---|---|---|
| Session I | R. GAUSDEN | United Kingdom |
| Session II | M. NASIM | Pakistan |
| Session III | H. SCHNURER | Federal Republic of Germany |
| Session IV | A.J. GAUVENET | France |
| Session V | G. HAKE | Canada |
| Session VI | N.P. DERGACHEV | Union of Soviet Socialist Republics |
| Session VII | L. VENKATESH | India |
| Session VIII | L.B. SZTANYIK | Hungary |
| Session IX | B.K. GRIMES | United States of America |

# SECRETARIAT OF THE SYMPOSIUM

| | | |
|---|---|---|
| Scientific Secretaries | J.D. McCULLEN | Division of Nuclear Safety and Environmental Protection, IAEA, Vienna |
| | K.B. STADIE | OECD/NEA, Paris |
| Administrative Secretary | Caroline DE MOL VAN OTTERLOO | Division of External Relations, IAEA, Vienna |
| Editor | Miriam LEWIS | Division of Publications, IAEA, Vienna |
| Records Officer | S.K. DATTA | Division of Languages and Policy-making Organs, IAEA, Vienna |

# LIST OF PARTICIPANTS

## ARGENTINA

Migliori de Beninson, Ambretta          Comisión Nacional de Energía Atómica,
Av. Libertador 8250, Buenos Aires

## AUSTRALIA

Hanna, G. L.          Permanent Mission of Australia to the International Atomic
Energy Agency,
Mattiellistrasse 2-4/III, A-1040 Vienna

## AUSTRIA

Binner, W.          Österreichische Studiengesellschaft für Atomenergie GmbH,
Lenaugasse 10, A-1082 Vienna

Cehak, K.          Zentralanstalt für Meteorologie und Geodynamik,
Hohe Warte 38, A-1190 Vienna

Frantz, Anny M. T.          Bundesanstalt für Wasserbiologie in Abwasserforschung,
Kaisermühlen, A-1223 Vienna

Hefner, A.          Österreichische Studiengesellschaft für Atomenergie GmbH,
Lenaugasse 10, A-1082 Vienna

Oszuszky, F.          Österreichische Elektrizitätswirtschafts AG,
Am Hof 6a, A-1010 Vienna

Powondra, F.          Österreichische Elektrizitätswirtschafts AG,
Am Hof 6a, A-1010 Vienna

Pusch, W.          Bundesministerium für Gesundheit und Umweltschutz,
Stubenring 1, A-1010 Vienna

Putz, F.          Österreichische Studiengesellschaft für Atomenergie GmbH,
Lenaugasse 10, A-1082 Vienna

Randa, K.          Brown-Boveri Vienna,
Pernersdorferstr. 95, A-1100 Vienna

## BELGIUM

Boulenger, R.          SCK/CEN,
200 Boeretang, B-2400 Mol

Collée, R.F.                           Université de Liège,
                                       19 rue Ponson, B-4500 Jupille-sur-Meuse

Dresse, H.M.                           Electrobel,
                                       1 place du Trône, B-1000 Brussels

Fieuw, G.H.M.                          SCK/CEN,
                                       200 Boeretang, B-2400 Mol

Gillon, L.                             Université catholique de Louvain,
                                       2 chemin du Cyclotron, B-1348 Louvain-la-Neuve

Hubert, E.H.                           Union des exploitations électriques en Belgique,
                                       4 Galerie Ravenstein, B-1150 Brussels

Planquart, J.                          SCK/CEN,
                                       200 Boeretang, B-2400 Mol

Schelfaut, G.F.                        Office de promotion industrielle,
                                       26 square de Meeus, B-1040 Brussels

Versele, Huguette.M.H.                 Institut d'hygiène et d'épidémiologie,
                                       14 rue J. Wytsman, B-1050 Brussels


## CANADA

Anderson, D.E.                         Ontario Hydro, c/o Atomic Energy of Canada Ltd.,
                                       Power Projects Division,
                                       Sheridan Park, Mississauga, Ontario

Duncan, R.M.                           Atomic Energy Control Board,
                                       107 Sparks Street, P.O. Box 1046, Ottawa, Ontario

Hake, G.                               Atomic Energy of Canada Ltd.,
                                       Chalk River Nuclear Laboratories,
                                       Chalk River, Ontario

Shah, J.                               Environmental Protection Service, Environment Canada,
                                       Ottawa, Ontario

Weeks, J.L.                            Whiteshell Nuclear Research Establishment,
                                       Atomic Energy of Canada Ltd.,
                                       Pinawa, Roe Ilo, Manitoba


## CUBA

Alfonso Bilbao, A.V.                   Centro de Investigaciones Energéticas,
                                       Havana


## CZECHOSLOVAK SOCIALIST REPUBLIC

Chochlovský, I.                        Chemoprojekt,
                                       Štěpánská 15, Prague 2

Dlouhy, Z.                             Nuclear Research Institute,
                                       Řež near Prague

Hejlek, R.                          Chemoprojekt,
                                    Štěpánská 15, Prague 2

Kříž, Z.                            Czechoslovak Atomic Energy Commission,
                                    Slezská 9, Prague 2

Namestek, L.                        Institute of Hygiene and Epidemiology,
                                    Srobárova 48, 10042 Prague 10

Sulkiewicz, M.                      Terplan, Czechoslovak Institute for Regional Planning,
                                    Platnéřská 19, 11000 Prague 1

Vesely, V.                          Institute of Nuclear Fuels,
                                    Prague


# DENMARK

Gjörup, H.L.                        Research Establishment Risö,
                                    D-4000 Roskilde

Grande, P.                          State Institute of Radiation Hygiene,
                                    378 Frederikssundsvej, DK-2700 Copenhagen Brh.

Hannibal, L.                        State Institute of Radiation Hygiene,
                                    378 Frederikssundsvej, DK-2700 Copenhagen Brh.

Henningsen, E.J.                    National Health Service of Denmark,
                                    St. Kongensgade 1, DK-1264 Copenhagen K

Jensen, A.                          Danish Atomic Energy Commission,
                                    Inspectorate of Nuclear Installations,
                                    Risö, DK-4000 Roskilde

Jensen, H.                          Danish Atomic Energy Commission,
                                    Inspectorate of Nuclear Installations,
                                    Risö, DK-4000 Roskilde

Øhlenschläger, N.                   National Health Service of Denmark,
                                    St. Kongensgade 1, DK-1264 Copenhagen K

Schultz-Larsen, J.                  University Institute of Medical Genetics,
                                    71 Raadmandsgade, DK-2200 Copenhagen N

Sörensen, T.F.                      Research Establishment Risö,
                                    D-4000 Roskilde


# EGYPT

Hassan, E.M.A.                      Nuclear Power Division, Atomic Energy Establishment,
                                    Cairo


# FINLAND

Åkesson, T.E.                       Institute of Radiation Physics,
                                    P.O. Box 268, SF-00101 Helsinki 10

Blomquist, L.G.                     Ministry of Commerce and Industry,
                                    Aleksanterink 10, SF-00170 Helsinki 17

Tarjanne, R.                        Nuclear Engineering Laboratory,
                                    Technical Research Centre of Finland,
                                    SF-02150 Otaniemi


# FRANCE

Barbreau, A.                        Département de sûreté nucléaire, CEN de Saclay,
                                    B.P. n° 2, F-91190 Gif-sur-Yvette

Bellin, A.                          Electricité de France, Direction de l'équipement,
                                    3 rue de Messine, F-75 Paris 8e

Boussard, R.R.                      Commissariat à l'énergie atomique, Centre de La Hague,
                                    B.P. n° 209, F-50107 Cherbourg

Bouville, A.C.                      CEN de Fontenay-aux-Roses,
                                    B.P. n° 6, F-92260 Fontenay-aux-Roses

Bresson, G.                         Département de protection, CEN de Fontenay-aux-Roses,
                                    B.P. n° 6, F-92260 Fontenay-aux-Roses

Candès, P.                          Service central de sûreté des installations nucléaires,
                                    Ministère de l'industrie et de la recherche,
                                    13 rue de Bourgogne, F-75 Paris 8e

Chapelle, A.                        Ministère de la qualité de la vie,
                                    14 boulevard du Général Leclerc, F-92000 Neuilly/Seine

Chardonnet, E.J.                    Electricité de France, Direction de l'équipement,
                                    3 rue de Messine, F-75 Paris 8e

Coulon, R.                          Département de protection, CEN de Fontenay-aux-Roses,
                                    B.P. n° 6, F-92260 Fontenay-aux-Roses

Doury, A.                           Département de sûreté nucléaire, CEN de Saclay,
                                    B.P. n° 2, F-91190 Gif-sur-Yvette

Fagnani, F.                         CEN de Fontenay-aux-Roses,
                                    B.P. n° 6, F-92260 Fontenay-aux-Roses

Farges, L.F.                        Commissariat à l'énergie atomique,
                                    23-26 rue de la Fédération, F-75015 Paris

Finkelstein, A.                     Commissariat à l'énergie atomique,
                                    29-33 rue de la Fédération,
                                    B.P. n° 510, F-75752 Paris Cedex 15

Garnier, Arlette                    Département de protection, CEN de Fontenay-aux-Roses,
                                    B.P. n° 6, F-92260 Fontenay-aux-Roses

Gauvenet, A.J.                      Commissariat à l'énergie atomique,
                                    29-33 rue de la Fédération, F-75015 Paris

Gérard, R.                          Département de sûreté nucléaire, CEN de Saclay,
                                    B.P. n° 2, F-91190 Gif-sur-Yvette

Gibrat, R.P.

Compagnie Solmer,
105 rue du Ranelagh, F-75016 Paris

Greppo, J.-F.

Electricité de France, Service de la production thermique,
3 rue de Messine, F-75008 Paris

Jammet, H. *

Département de protection, CEN de Fontenay-aux-Roses,
B.P. n°6, F-92260 Fontenay-aux-Roses

Janin, J.-F.

Ministère de la qualité de la vie,
14 boulevard du Général Leclerc, F-92000 Neuilly/Seine

Lavie, J.M.

Etablissement SPS,
B.P. n°17, F-91 Montlhéry

Le Ber, J.

Commissariat à l'énergie atomique,
Centrale nucléaire des Monts d'Arrée,
B.P. n°3, La Feuillée, F-29218 Huelgoat

Lévêque, P.C.

Laboratoire de radiogéologie,
Avenue des Facultés, F-33405 Talence

Martin, D.

Commissariat à l'énergie atomique, Centre de Marcoule,
B.P. n°106, F-30200 Bagnols-sur-Cèze

Merle, D.

Société Bertin,
B.P. n°3, F-78390 Plaisir

Monet, R.F.

CEN de Saclay,
B.P. n°2, F-91190 Gif-sur-Yvette

Morlat, G.

Electricité de France, Direction des études et recherches,
1 avenue du Général de Gaulle, F-92 Clamart

Pomarola, J.

CEN de Fontenay-aux-Roses,
B.P. n°6, F-92260 Fontenay-aux-Roses

Ponsy, J.

Electricité de France, Région équipement Clamart,
2 avenue Général de Gaulle, F-92 Clamart

Schaeffer, R.

Electricité de France, Production-transport,
3 rue de Messine, F-75008 Paris

Vagnieux, J.J.

Electricité de France — NERSA,
177 rue Garibaldi, F-69 Lyon


# GERMAN DEMOCRATIC REPUBLIC

Menzel, S.

VVB Kraftwerke,
Vetschauerstr. 1a, DDR-75 Cottbus

Rabold, H.

Staatliches Amt für Atomsicherheit und Strahlenschutz,
Waldowallee 117, DDR-1157 Berlin-Karlshorst

---

* See also under International Commission on Radiological Protection (ICRP).

## GERMANY, FEDERAL REPUBLIC OF

Beising, R.
Energieversorgung Schwaben AG,
Goethestr. 12, D-7 Stuttgart

Blume, H.
AGF-Programmleitung Angewandte Systemanalyse in der DFVLR,
Linder Höhe, D-505 Porz-Wahn

Bohn, T.
Kernforschungsanlage Jülich,
D-517 Jülich

Bojowald, J.J.
Technischer Überwachungsverein Stuttgart eV,
Bebelstrasse 48, D-7000 Stuttgart 1

Bonnenberg, H.
Bonnenberg und Drescher Ingenieurgesellschaft mbH,
Postfach 72, D-517 Jülich 1

Bork, M.
Institut für Reaktorsicherheit der TÜV,
Glockengasse 2, D-5 Cologne 1

Brust, E.P.
Rheinisch-Westfälischer Technischer Überwachungsverein eV,
Steubenstr. 53, Essen

Fendler, H.G.
Technischer Überwachungsverein Baden eV,
Richard-Wagner-Str. 2, D-68 Mannheim 1

Fischer, P.M.
Institut für Angewandte Systemtechnik und Reaktorphysik,
Gesellschaft für Kernforschung mbH,
D-7500 Karlsruhe 1

Gosslar, K.D.
Hochtemperatur-Reaktorbau GmbH,
Gottlieb-Daimler-Str. 8, D-6800 Mannheim 1

Gutschmidt, W.D.
Institut für Reaktorsicherheit,
Glockengasse 2, D-5 Cologne 1

Handge, P.
Institut für Reaktorsicherheit,
Glockengasse 2, D-5 Cologne 1

Haubelt, R.
Federal Health Office, Department of Radiation Hygiene,
Ingolstädter Landstrasse 1, D-8042 Neuherberg/Munich

Heinz, W.
Gesellschaft zur Wiederaufarbeitung von Kernbrennstoffen mbH,
D-7501 Leopoldshafen

Hennecke, J.
KEWA Kernbrennstoff-Wiederaufarbeitung GmbH,
D-7501 Leopoldshafen

Herrmann, G.
Gesellschaft zur Wiederaufarbeitung von Kernbrennstoffen mbH,
D-7501 Leopoldshafen

Huenlich, W.H.F.
Vereinigung Deutscher Elektrizitätswerke,
Stresemannallee 23, D-6000 Frankfurt/Main

Jaek, W.H.
Programmgruppe Systemforschung und technologische
  Entwicklung,
Kernforschungsanlage Jülich GmbH,
Postfach 365, D-517 Jülich

Jüssen, P. — Rheinische Braunkohlenwerke AG, Konrad Adenauer Ufer 55, D-5000 Cologne

Kohler, H.A.G. — Kraftwerk Union AG, Berlinerstr. 295-299, D-605 Offenbach

Köster, R.H. — Gesellschaft für Kernforschung, Postfach 5620, D-75 Karlsruhe

Kroeger, W. — Kernforschungsanlage Jülich GmbH, Postfach 365, D-517 Jülich

Langer, M. — Bundesanstalt für Bodenforschung, Stilleweg 2, 3 Hanover 23

Laser, M. — Kernforschungsanlage Jülich GmbH, Postfach 365, D-517 Jülich

Mareske, A. — Berliner Kraft- und Licht (Bewag) AG, Technische Planung Kernenergie, Stauffenbergstrasse 26, D-1000 Berlin 30

Maurer, K.G. — Nuklearbrennstoff GmbH, Alfredstr. 110, D-43 Essen 1

Merton, A. — Institut für Reaktorsicherheit, Glockengasse 2, D-5 Cologne 1

Oberbacher, B. — Battelle Institute eV, Am Römerhof 35, D-6 Frankfurt/Main

Österreicher, G. — Kraftanlagen AG, Im Breitspiel 7, Heidelberg

Pohl, A.M. — Universität Dortmund, August-Schmitt-Str., D-46 Dortmund-Eidlinghofen

Riedel, H.J. — Kernforschungsanlage Jülich GmbH, Postfach 365, D-517 Jülich

Schneider, M. — Institut für Reaktorsicherheit, Glockengasse 2, D-5 Cologne 1

Schnurer, H. — Bundesministerium des Innern, Postfach, D-53 Bonn

Thomas, P. — Abteilung Strahlenschutz und Sicherheit, Kernforschungszentrum Karlsruhe, D-7501 Karlsruhe

Zakrocki, H. — Fa. Fichtner, Grazerstr. 22, D-7 Stuttgart 30


## GREECE

Chrysochoides, N. — Nuclear Research Centre Democritos, Athens

Hadjiantoniou, A.                     Nuclear Research Centre Democritos,
                                      Athens

## HUNGARY

Fekete, L.                            Ministry of Building and Urban Development,
                                      Kossuth Lajos tér 4, H-1374 Budapest

Futó, I.                              Authority for Energetics and Technical Safety,
                                      Attila u. 99, H-1012 Budapest

Józsa, M.                             Hungarian Atomic Energy Commission,
                                      Kossuth Lajos tér 4, H-1374 Budapest

Pattantyús, E.                        Secretariat, National Council for Environmental Protection,
                                      Budapest

Raszl, K.                             Hungarian Atomic Energy Commission,
                                      Kossuth Lajos tér 4, H-1374 Budapest

Sztanyik, L.B.                        National Research Institute for Radiobiology and Radiohygiene,
                                      Pentz K.u. 5, H-1354 Budapest XXII

Török, I.                             Dept. of International Relations,
                                      Ministry of Building and Urban Development,
                                      Budapest

## INDIA

Venkatesh, L.                         Bhabha Atomic Research Centre,
                                      Trombay, Bombay

## INDONESIA

Soeroto, R.                           Gama Research Centre, National Atomic Energy Agency,
                                      Jl. Sékip Unit III, Jogjakarta

## IRAN

Arabian, G.                           Atomic Energy Organization of Iran,
                                      Zahed Avenue, Teheran

## IRELAND

Delaney, C.F.G.                       Physics Department, Trinity College,
                                      University of Dublin,
                                      Dublin 2

## ISRAEL

Dagan, D.                             Israel Electric Corporation,
                                      Power House, Haifa

Nelken, M.                        Israel Electric Corporation,
                                  Power House, Haifa

Striem, H. L.                     Israel Atomic Energy Commission,
                                  P. O. Box 17120, Tel-Aviv


# ITALY

Bramati, L.                       Ente Nazionale per l'Energia Elettrica,
                                  Direzione delle Costruzioni,
                                  Via G. B. Martini 3, Rome

Cagnetti, P.                      Laboratorio per lo Studio della Radioattività Ambientale,
                                  CNEN-CSN Casaccia,
                                  C. P. 2400, S. Maria di Galeria, I-00100 Rome

Campos Venuti, Gloria             Istituto Superiore di Sanità,
                                  Viale Regina Elena 299, Rome

Clementel, S.                     AGIP Nucleare,
                                  Corso di Porta Romana 68, Milan

Delitala, E.                      AGIP Nucleare,
                                  Corso di Porta Romana 68, Milan

Dellarciprete, C.                 Ente Nazionale per l'Energia Elettrica,
                                  Via G. B. Martini 3, Rome

Gadda, F.                         Centro Informazione Studi Esperienze,
                                  Via Redecesio 12, Segrate, Milan

Jaccarino, E.                     Comitato Nazionale per l'Energia Nucleare,
                                  Viale Regina Margherita 125, I-00198 Rome

Morelli, V.                       Ente Nazionale per l'Energia Elettrica,
                                  Via G. B. Martini 3, Rome

Tabet, E.                         Istituto Superiore di Sanità,
                                  Via Regina Elena 299, Rome

Zaffiro, C.                       Comitato Nazionale per l'Energia Nucleare,
                                  Viale Regina Margherita 125, I-00198  Rome


# JAPAN

Akimoto, Y.                       Mitsubishi Metal Corporation,
                                  Ohtemachi 1, Chiyoda-ku, Tokyo

Aoki, S.                          Research Laboratory of Nuclear Reactors,
                                  Tokyo Institute of Technology,
                                  Ookayama, Meguro-ku, Tokyo

Kitamura, T.                      Nuclear Power Department, The Kansai Electric Power Co. Inc.,
                                  5-3-chome, Nakanoshima, Kita-ku, Osaka

Saito, O.                         Central Research Institute of the Electric Power Industry,
                                  Otemachi 1-7, Tokyo

Sawaguchi, Y.

The Tokyo Electric Power Co.,
3-1 Uchisaiwai-chu, Chiyoda-ku, Tokyo

Shibayama, T.

Power Reactor and Nuclear Fuel Development Corporation,
1-9-13 Akasaka, Minato-ku, Tokyo

Yoshizawa, S.

Reactor Regulation Division, Atomic Energy Bureau,
Science and Technology Agency,
2-2-1 Kasumigaseki, Chiyoda-ku, Tokyo

## MALAYSIA

Ramanath, A.

National Electricity Board,
P.O. Box 1003, Kuala Lumpur

## NETHERLANDS

Baas, J.

Ministry of Public Health and Environmental Hygiene,
Dokter Reijersstraat 10, Leidschendam, The Hague

Daatselaar, C.J. van

Ministry of Social Affairs,
Balen van Andelplein 2, Voorburg

Lablans, W.N.

Royal Netherlands Meteorological Institute,
Utrechtseweg 279, De Bilt

Pelser, J.

Reactor Centrum Nederland,
Westerduinweg 3, Petten

Rasmussen, C.E.

Interuniversitair Reactor Instituut,
Mekelweg 15, Delft

## NEW ZEALAND

Wylde, A.F.

New Zealand Electricity,
Private Bag, Wellington

## NORWAY

Aamlid, H.

State Institute of Radiation Hygiene,
Montebello, Oslo 3

Garder, Karen

Institutt for Atomenergi,
P.O. Box 40, 2007 Kjeller

Løken, P.C.

Norwegian Water Resources and Electricity Board,
Middelthungsgt. 29, Oslo 3

Michelsen, H.M.

Norwegian Nuclear Installation Inspectorate,
Pottemakerveien 4, Oslo 5

Mölsaeter, M.

Norwegian Nuclear Installation Inspectorate,
Pottemakerveien 4, Oslo 5

Swang, O.

Ministry of Industry,
Oslo Dep., Oslo 1

Tveten, U.

Institutt for Atomenergi,
P.O. Box 40, 2007 Kjeller

## PAKISTAN

Nasim, M.

Karachi Nuclear Power Plant,
P.O. Box 1114, Islamabad

## PHILIPPINES

Bartolome, Z.M.

Philippines Atomic Energy Commission,
Diliman, Quezon City

## PORTUGAL

Fernandes Forte, A.

Companhia Portuguesa de Electricidade,
Lisbon

## ROMANIA

Stoian, D.

Ministère pour l'énergie électrique,
33 Bul. Magheru, Bucharest

## SINGAPORE

Tay, A.O.

Ministry of Science and Technology,
Government Offices Building,
Kay Siang Road, Singapore 10

## SPAIN

Acha Aracama, A. de

Junta de Energía Nuclear,
Avenida Complutense 22, Madrid 3

Bernal, A.

INTECSA,
c/o Condesa de Venadito 1, Madrid

Delgado Hernandez, E.

Burns and Roe de España S.A.,
Pedro Teixeira 8, 7°, Madrid 20

Jiménez Moreno, J.M.

Junta de Energía Nuclear,
Avenida Complutense 22, Madrid 3

Lopez Rodriguez, M.

Burns and Roe de España S.A.,
Pedro Teixeira 8, 7°, Madrid 20

Melches Serrano, C.

Empresa Nacional del Uranio S.A.,
Santiago Rusinol 12, Madrid 3

Pérez del Moral, Consuelo

Junta de Energía Nuclear,
Avenida Complutense 22, Madrid 3

Sanz Valdés, J.                          Fuerzas Eléctricas de Cataluña,
                                         Plaza de Cataluña 2, Barcelona 2

Ximenes de Embun Ramonell, J.            INTECSA,
                                         Beethoven 15, Barcelona


## SWEDEN

Bergström, S.O.W.                        AB Atomenergi,
                                         Studsvik, Fack, S-61101 Nyköping

Gyllander, Christina                     AB Atomenergi,
                                         Studsvik, Fack, S-61101 Nyköping

Jansson, K.B.                            Vattenbyggnars-Byrån,
                                         Box 2203, Gothenburg

Moding, A.P.V.                           Industridepartementet, AKA-utredningen,
                                         Malmö

Nytén, T.                                Swedish State Power Board,
                                         S-16287 Vällingby


## SWITZERLAND

Aus der Au, W.R.                         Electrowatt Engineering Services,
                                         Postfach 8022, CH-8000 Zurich

Babocsay, L.                             Abteilung für die Sicherheit der Kernanlagen,
                                         Eidg. Amt für Energiewirtschaft,
                                         CH-5303 Würenlingen

Fuchs, H.                                Motor-Columbus Construction Engineers Inc.,
                                         Parkstrasse 27, CH-5401 Baden

Gryksa, H.P.                             Nuclear Assurance Corporation,
                                         Mainaustr. 8, CH-8008 Zurich

Müller, R.                               Federal Office of Energy,
                                         Postfach, CH-3001 Berne


## TURKEY

Akçora, G.                               Nuclear Energy Division, Turkish Electricity Authority,
                                         Necati Bey Cad. 26, Ankara

Erturan, A.Y.                            Turkish Atomic Energy Commission,
                                         Bestekar Sokak 29, Kavaklidere, Ankara


## UNION OF SOVIET SOCIALIST REPUBLICS

Dergachev, N.P.                          The State Committee for Utilization of Atomic Energy,
                                         Moscow

## UNITED KINGDOM

Bennett, W.E.

BNFL,
Risley, Warrington, Cheshire

Bryant, Pamela M.

National Radiological Protection Board,
Harwell, Didcot, Oxon. OX11 0RQ

Edwards, J.

Royal Naval College,
Greenwich, London SE 10

Gausden, R.

Nuclear Installations Inspectorate, Department of Energy,
Thames House South, Millbank, London SW1

Haire, T.P.

Nuclear Health and Safety Department,
Department of Energy,
Courtenay House, 18 Warwick Lane, London EC1

Munday, G.

Department of Chemical Engineering and Chemical Technology,
Imperial College,
London SW7 2AZ

Ostle, D.

Institute of Geological Sciences,
154 Clerkenwell Road, London WC1

Shaw, J.

Queen Mary College, University of London,
London E1 4NS

Terry, J.B.

Nuclear Installations Inspectorate, Department of Energy,
Thames House South, Millbank, London SW1

Usher, E.F.F.W.

Central Electricity Generating Board,
Sudbury House, Newgate Street, London EC1

## UNITED STATES OF AMERICA

Burnham, J.B.

Battelle Pacific Northwest Laboratories,
Box 999, Richland, Wash. 99352

Cluff, L.S.

Woodward-Clyde Consultants,
2730 Adeline Street, Oakland, Calif. 94607

Cooper, D.C.

Battelle Columbus Laboratories,
505 King Avenue, Columbus, Ohio 43201

DiNunno, J.J.

NUS Corporation,
4 Research Place, Rockville, Md. 20805

Ellwood, R.B.

Dames and Moore,
123 Mortlake High Street, London SW14 8SN, England

Grimes, B.K.

Directorate of Licensing,
United States Atomic Energy Commission,
Washington, D.C. 20545

Gustafson, P.F.

Argonne National Laboratory,
Argonne, Ill. 60439

Hull, A.P.                                 Health Physics and Safety Division,
                                           Brookhaven National Laboratory,
                                           Upton, Long Island, N.Y. 11973

Jefferson, R.M.                            Sandia Laboratories,
                                           Albuquerque, N.M. 87115

Karpenko, V.N.                             Lawrence Livermore Laboratory, University of California,
                                           P.O. Box 808, Livermore, Calif. 94550

Kissenpfennig, J.F.                        E. D'Appolonia, Consulting Engineers,
                                           10 Duff Road, Pittsburgh, Pa. 15235

Nair, K.                                   Woodward-Clyde Consultants,
                                           2730 Adeline Street, Oakland, Calif. 94607

Newell, J.F.                               John F. Newell Consulting Services,
                                           P.O. Box 26885, Tucson, Ariz. 85726

Rizzo, P.C.                                E. D'Appolonia, Consulting Engineers,
                                           10 Duff Road, Pittsburgh, Pa. 15235

Topper, L.                                 Office of Energy Research and Development Policy,
                                           National Science Foundation,
                                           1800 G Street N.W., Washington, D.C. 20550


## YUGOSLAVIA

Alujevič, A.                               Institute Jozef Stefan,
                                           Jamova 39, YU-61000 Ljubljana


## ORGANIZATIONS

### CEC (Commission of the European Communities)

Luykx, F.                                  Direction protection sanitaire,
                                           Commission des Communautés Européennes,
                                           23 avenue Monterey, Luxembourg

Van Caeneghem, J.                          Direction affaires industrielles et technologiques,
                                           Commission des Communautés Européennes,
                                           200 rue de la Loi, B-1040 Brussels, Belgium

Van Hoeck, F.                              Commission des Communautés Européennes,
                                           200 rue de la Loi, B-1040 Brussels, Belgium

Vinck, W.                                  Commission des Communautés Européennes,
                                           200 rue de la Loi, B-1040 Brussels, Belgium


### Economic Commission for Europe

Lopez-Polo, C.                             Energy Division, Economic Commission for Europe,
                                           Palais des Nations, CH-1211 Geneva 10

<u>European Parliament</u>

Juul, E.

General Direction for Research and Documents, Centre Européen,
European Parliament,
Kirschberg, Luxembourg

Walz, Hanna

Committee for Energy, Research and Technology,
European Parliament,
Magdeburgerstr. 61, D-6400 Fulda,
Federal Republic of Germany

<u>IAEA (International Atomic Energy Agency)</u>

Chernilin, Y.F.

Department of Technical Operations, IAEA,
Kärntnerring 11, A-1010 Vienna

Iansiti, E.

Division of Nuclear Safety and Environmental Protection, IAEA,
Kärntnerring 11, A-1010 Vienna

Nishiwaki, Y.

Division of Nuclear Safety and Environmental Protection, IAEA,
Kärntnerring 11, A-1010 Vienna

Otway, H.J.

Division of Nuclear Safety and Environmental Protection, IAEA,
Kärntnerring 11, A-1010 Vienna

Philip, G.

Division of Nuclear Safety and Environmental Protection, IAEA,
Kärntnerring 11, A-1010 Vienna

Polliart, A.-J.

Division of Nuclear Power and Reactors, IAEA,
Kärntnerring 11, A-1010 Vienna

Rosen, M.

Division of Nuclear Safety and Environmental Protection, IAEA,
Kärntnerring 11, A-1010 Vienna

Servant, J.

Division of Nuclear Safety and Environmental Protection, IAEA,
Kärntnerring 11, A-1010 Vienna

Woite, G.

Division of Nuclear Power and Reactors, IAEA,
Kärntnerring 11, A-1010 Vienna

<u>ICRP (International Commission on Radiological Protection)</u>

Jammet, H.*

Département de protection, CEN de Fontenay-aux-Roses,
B.P. n°6, F-92260 Fontenay-aux-Roses, France

<u>IIASA (International Institute for Applied Systems Analysis)</u>

Avenhaus, R.

International Institute for Applied Systems Analysis,
Schloss Laxenburg, A-2361 Laxenburg, Austria

Baecher, G.B.

International Institute for Applied Systems Analysis,
Schloss Laxenburg, A-2361 Laxenburg, Austria

---

* See also under France.

| | |
|---|---|
| Gros, J.G. | International Institute for Applied Systems Analysis, Schloss Laxenburg, A-2361 Laxenburg, Austria |
| Jewell, W.S. | International Institute for Applied Systems Analysis, Schloss Laxenburg, A-2361 Laxenburg, Austria |
| Velimirović, Helga | International Institute for Applied Systems Analysis, Schloss Laxenburg, A-2361 Laxenburg, Austria |

OEC D/NEA (OECD Nuclear Energy Agency)

| | |
|---|---|
| Crawford, R.E. | OECD Nuclear Energy Agency, 38 boulevard Suchet, Paris 16e, France |

UNSCEAR (United Nations Scientific Committee on the Effects of Atomic Radiation)

| | |
|---|---|
| Beninson, D. | UNSCEAR, P.O. Box 475, A-1011 Vienna, Austria |

WEC (World Energy Conference)

| | |
|---|---|
| Bauer, L. | Technische Hochschule, Gusshausstr. 25, A-1040 Vienna, Austria |

WHO (World Health Organization)

| | |
|---|---|
| Sentici, M. | WHO Liaison Office at the IAEA, Kärntnerring 11, A-1010 Vienna |

WMO (World Meteorological Organization)

| | |
|---|---|
| Steinhauser, F. | Zentralanstalt für Meteorologie und Geodynamik, Hohe Warte 38, A-1190 Vienna |

# AUTHOR INDEX

### (including participants in discussions)

Numerals underlined refer to the first page of the paper by the author concerned.
Further numerals denote comments and questions in the discussions.

*The following conversion table is provided for the convenience of readers and to encourage the use of SI units.*

# FACTORS FOR CONVERTING UNITS TO SI SYSTEM EQUIVALENTS *

SI base units are the metre (m), kilogram (kg), second (s), ampere (A), kelvin (K), candela (cd) and mole (mol).
[For further information, see International Standards ISO 1000 (1973), and ISO 31/0 (1974) and its several parts]

| Multiply | | by | to obtain |
|---|---|---|---|
| **Mass** | | | |
| pound mass (avoirdupois) | 1 lbm | $= 4.536 \times 10^{-1}$ | kg |
| ounce mass (avoirdupois) | 1 ozm | $= 2.835 \times 10^{1}$ | g |
| ton (long) (= 2240 lbm) | 1 ton | $= 1.016 \times 10^{3}$ | kg |
| ton (short) (= 2000 lbm) | 1 short ton | $= 9.072 \times 10^{2}$ | kg |
| tonne (= metric ton) | 1 t | $= 1.00 \times 10^{3}$ | kg |
| **Length** | | | |
| statute mile | 1 mile | $= 1.609 \times 10^{0}$ | km |
| yard | 1 yd | $= 9.144 \times 10^{-1}$ | m |
| foot | 1 ft | $= 3.048 \times 10^{-1}$ | m |
| inch | 1 in | $= 2.54 \times 10^{-2}$ | m |
| mil (= $10^{-3}$ in) | 1 mil | $= 2.54 \times 10^{-2}$ | mm |
| **Area** | | | |
| hectare | 1 ha | $= 1.00 \times 10^{4}$ | $m^2$ |
| (statute mile)$^2$ | 1 mile$^2$ | $= 2.590 \times 10^{0}$ | $km^2$ |
| acre | 1 acre | $= 4.047 \times 10^{3}$ | $m^2$ |
| yard$^2$ | 1 yd$^2$ | $= 8.361 \times 10^{-1}$ | $m^2$ |
| foot$^2$ | 1 ft$^2$ | $= 9.290 \times 10^{-2}$ | $m^2$ |
| inch$^2$ | 1 in$^2$ | $= 6.452 \times 10^{2}$ | $mm^2$ |
| **Volume** | | | |
| yard$^3$ | 1 yd$^3$ | $= 7.646 \times 10^{-1}$ | $m^3$ |
| foot$^3$ | 1 ft$^3$ | $= 2.832 \times 10^{-2}$ | $m^3$ |
| inch$^3$ | 1 in$^3$ | $= 1.639 \times 10^{4}$ | $mm^3$ |
| gallon (Brit. or Imp.) | 1 gal (Brit) | $= 4.546 \times 10^{-3}$ | $m^3$ |
| gallon (US liquid) | 1 gal (US) | $= 3.785 \times 10^{-3}$ | $m^3$ |
| litre | 1 l | $= 1.00 \times 10^{-3}$ | $m^3$ |
| **Force** | | | |
| dyne | 1 dyn | $= 1.00 \times 10^{-5}$ | N |
| kilogram force | 1 kgf | $= 9.807 \times 10^{0}$ | N |
| poundal | 1 pdl | $= 1.383 \times 10^{-1}$ | N |
| pound force (avoirdupois) | 1 lbf | $= 4.448 \times 10^{0}$ | N |
| ounce force (avoirdupois) | 1 ozf | $= 2.780 \times 10^{-1}$ | N |
| **Power** | | | |
| British thermal unit/second | 1 Btu/s | $= 1.054 \times 10^{3}$ | W |
| calorie/second | 1 cal/s | $= 4.184 \times 10^{0}$ | W |
| foot-pound force/second | 1 ft·lbf/s | $= 1.356 \times 10^{0}$ | W |
| horsepower (electric) | 1 hp | $= 7.46 \times 10^{2}$ | W |
| horsepower (metric) (= ps) | 1 ps | $= 7.355 \times 10^{2}$ | W |
| horsepower (550 ft·lbf/s) | 1 hp | $= 7.457 \times 10^{2}$ | W |

* Factors are given exactly or to a maximum of 4 significant figures

| *Multiply* | | *by* | *to obtain* |
|---|---|---|---|

### Density

| pound mass/inch$^3$ | 1 lbm/in$^3$ | = 2.768 $\times$ 10$^4$ | kg/m$^3$ |
| pound mass/foot$^3$ | 1 lbm/ft$^3$ | = 1.602 $\times$ 10$^1$ | kg/m$^3$ |

### Energy

| British thermal unit | 1 Btu | = 1.054 $\times$ 10$^3$ | J |
| calorie | 1 cal | = 4.184 $\times$ 10$^0$ | J |
| electron-volt | 1 eV | $\simeq$ 1.602 $\times$ 10$^{-19}$ | J |
| erg | 1 erg | = 1.00 $\times$ 10$^{-7}$ | J |
| foot-pound force | 1 ft·lbf | = 1.356 $\times$ 10$^0$ | J |
| kilowatt-hour | 1 kW·h | = 3.60 $\times$ 10$^6$ | J |

### Pressure

| newtons/metre$^2$ | 1 N/m$^2$ | = 1.00 | Pa |
| atmosphere[a] | 1 atm | = 1.013 $\times$ 10$^5$ | Pa |
| bar | 1 bar | = 1.00 $\times$ 10$^5$ | Pa |
| centimetres of mercury (0°C) | 1 cmHg | = 1.333 $\times$ 10$^3$ | Pa |
| dyne/centimetre$^2$ | 1 dyn/cm$^2$ | = 1.00 $\times$ 10$^{-1}$ | Pa |
| feet of water (4°C) | 1 ftH$_2$O | = 2.989 $\times$ 10$^3$ | Pa |
| inches of mercury (0°C) | 1 inHg | = 3.386 $\times$ 10$^3$ | Pa |
| inches of water (4°C) | 1 inH$_2$O | = 2.491 $\times$ 10$^2$ | Pa |
| kilogram force/centimetre$^2$ | 1 kgf/cm$^2$ | = 9.807 $\times$ 10$^4$ | Pa |
| pound force/foot$^2$ | 1 lbf/ft$^2$ | = 4.788 $\times$ 10$^1$ | Pa |
| pound force/inch$^2$ (= psi)[b] | 1 lbf/in$^2$ | = 6.895 $\times$ 10$^3$ | Pa |
| torr (0°C) (= mmHg) | 1 torr | = 1.333 $\times$ 10$^2$ | Pa |

### Velocity, acceleration

| inch/second | 1 in/s | = 2.54 $\times$ 10$^1$ | mm/s |
| foot/second (= fps) | 1 ft/s | = 3.048 $\times$ 10$^{-1}$ | m/s |
| foot/minute | 1 ft/min | = 5.08 $\times$ 10$^{-3}$ | m/s |
| mile/hour (= mph) | 1 mile/h | = 4.470 $\times$ 10$^{-1}$ | m/s |
|  |  | = 1.609 $\times$ 10$^0$ | km/h |
| knot | 1 knot | = 1.852 $\times$ 10$^0$ | km/h |
| free fall, standard (= g) |  | = 9.807 $\times$ 10$^0$ | m/s$^2$ |
| foot/second$^2$ | 1 ft/s$^2$ | = 3.048 $\times$ 10$^{-1}$ | m/s$^2$ |

### Temperature, thermal conductivity, energy/area·time

| Fahrenheit, degrees −32 | °F −32 | $\frac{5}{9}$ | °C |
| Rankine | °R |  | K |
| 1 Btu·in/ft$^2$·s·°F |  | = 5.189 $\times$ 10$^2$ | W/m·K |
| 1 Btu/ft·s·°F |  | = 6.226 $\times$ 10$^1$ | W/m·K |
| 1 cal/cm·s·°C |  | = 4.184 $\times$ 10$^2$ | W/m·K |
| 1 Btu/ft$^2$·s |  | = 1.135 $\times$ 10$^4$ | W/m$^2$ |
| 1 cal/cm$^2$·min |  | = 6.973 $\times$ 10$^2$ | W/m$^2$ |

### Miscellaneous

| foot$^3$/second | 1 ft$^3$/s | = 2.832 $\times$ 10$^{-2}$ | m$^3$/s |
| foot$^3$/minute | 1 ft$^3$/min | = 4.719 $\times$ 10$^{-4}$ | m$^3$/s |
| rad | rad | = 1.00 $\times$ 10$^{-2}$ | J/kg |
| roentgen | R | = 2.580 $\times$ 10$^{-4}$ | C/kg |
| curie | Ci | = 3.70 $\times$ 10$^{10}$ | disintegration/s |

[a] atm abs: atmospheres absolute; atm (g): atmospheres gauge.

[b] lbf/in$^2$ (g) (= psig): gauge pressure; lbf/in$^2$ abs (= psia): absolute pressure.

# HOW TO ORDER IAEA PUBLICATIONS

Exclusive sales agents for IAEA publications, to whom all orders and inquiries should be addressed, have been appointed in the following countries:

| | |
|---|---|
| UNITED KINGDOM | Her Majesty's Stationery Office, P.O. Box 569, London SE 1 9NH |
| UNITED STATES OF AMERICA | UNIPUB, Inc., P.O. Box 433, Murray Hill Station, New York, N.Y. 10016 |

In the following countries IAEA publications may be purchased from the sales agents or booksellers listed or through your major local booksellers. Payment can be made in local currency or with UNESCO coupons.

| | |
|---|---|
| ARGENTINA | Comisión Nacional de Energía Atómica, Avenida del Libertador 8250, Buenos Aires |
| AUSTRALIA | Hunter Publications, 58 A Gipps Street, Collingwood, Victoria 3066 |
| BELGIUM | Service du Courrier de l'UNESCO, 112, Rue du Trône, B-1050 Brussels |
| CANADA | Information Canada, 171 Slater Street, Ottawa, Ont. K 1 A OS 9 |
| C.S.S.R. | S.N.T.L., Spálená 51, CS-11000 Prague |
| | Alfa, Publishers, Hurbanovo námestie 6, CS-80000 Bratislava |
| FRANCE | Office International de Documentation et Librairie, 48, rue Gay-Lussac, F-75005 Paris |
| HUNGARY | Kultura, Hungarian Trading Company for Books and Newspapers, P.O. Box 149, H-1011 Budapest 62 |
| INDIA | Oxford Book and Stationery Comp., 17, Park Street, Calcutta 16 |
| ISRAEL | Heiliger and Co., 3, Nathan Strauss Str., Jerusalem |
| ITALY | Libreria Scientifica, Dott. de Biasio Lucio "aeiou", Via Meravigli 16, I-20123 Milan |
| JAPAN | Maruzen Company, Ltd., P.O.Box 5050, 100-31 Tokyo International |
| NETHERLANDS | Martinus Nijhoff N.V., Lange Voorhout 9-11, P.O. Box 269, The Hague |
| PAKISTAN | Mirza Book Agency, 65, The Mall, P.O.Box 729, Lahore-3 |
| POLAND | Ars Polona, Centrala Handlu Zagranicznego, Krakowskie Przedmiescie 7, Warsaw |
| ROMANIA | Cartimex, 3-5 13 Decembrie Street, P.O.Box 134-135, Bucarest |
| SOUTH AFRICA | Van Schaik's Bookstore, P.O.Box 724, Pretoria |
| | Universitas Books (Pty) Ltd., P.O.Box 1557, Pretoria |
| SPAIN | Nautrónica, S.A., Pérez Ayuso 16, Madrid-2 |
| SWEDEN | C.E. Fritzes Kungl. Hovbokhandel, Fredsgatan 2, S-10307 Stockholm |
| U.S.S.R. | Mezhdunarodnaya Kniga, Smolenskaya-Sennaya 32-34, Moscow G-200 |
| YUGOSLAVIA | Jugoslovenska Knjiga, Terazije 27, YU-11000 Belgrade |

75-12079

Orders from countries where sales agents have not yet been appointed and requests for information should be addressed directly to:

Publishing Section,
International Atomic Energy Agency,
Kärntner Ring 11, P.O.Box 590, A-1011 Vienna, Austria